Operational Spacetime

Fundamental Theories of Physics

An International Book Series on The Fundamental Theories of Physics: Their Clarification, Development and Application

Volume 163

For other titles published in this series, go to http://www.springer.com/series/6001

Heinrich Saller

Operational Spacetime

Interactions and Particles

Springer

Heinrich Saller
MPI für Physik
Werner-Heisenberg-Institut
Föhringer Ring 6
80805 München
Germany
hns@mppmu.mpg.de

ISBN 978-1-4614-2493-2 e-ISBN 978-1-4419-0898-8
DOI 10.1007/978-1-4419-0898-8
Springer New York Dordrecht Heidelberg London

Softcover reprint of the hardcover 1st edition 2010

Printed on acid-free paper

Springer is part of Springer Science+Business Media (www.springer.com)

Contents

0 Introduction and Orientation **1**

1 Einstein's Gravity **17**
1.1 Geometrization of Gravity 17
1.2 Schwarzschild–Kruskal Spacetime 20
1.3 Friedmann and de Sitter Universes 24

2 Riemannian Manifolds **29**
2.1 Differentiable Manifolds 29
2.1.1 External Derivative 32
2.2 Riemannian Operation Groups 33
2.2.1 Metric-Induced Isomorphisms 33
2.2.2 Tangent Euclidean and Poincaré Groups 34
2.2.3 Global and Local Invariance Groups 36
2.2.4 Riemannian Connection 39
2.3 Affine Connections 40
2.3.1 Torsion, Curvature, and Ricci Tensor 41
2.3.2 Cartan's Stuctural Equations 43
2.4 Lie Groups as Manifolds 44
2.4.1 Lie Group Operations 44
2.4.2 Lie Algebra Operations 45
2.4.3 The Poincaré Group of a Lie Group 46
2.4.4 Lie–Jacobi Isomorphisms for Lie Groups 46
2.4.5 Examples 47
2.4.6 Adjoint and Killing Connection of Lie Groups 49
2.5 Riemannian Manifolds 51
2.5.1 Lorentz Covariant Derivatives 51
2.5.2 Laplace–Beltrami Operator 52
2.5.3 Riemannian Curvature 53
2.5.4 Einstein Tensor and Conserved Quantities 54
2.6 Tangent and Operational Metrics 55
2.6.1 Invariants 56

2.7 Maximally Symmetric Manifolds 57
2.7.1 Spheres and Hyperboloids 58
2.7.2 Constant-Curvature Manifolds 59
2.8 Rotation-Symmetric Manifolds 60
2.9 Basic Riemannian Manifolds 61
2.9.1 Manifolds with Dimension 1 62
2.9.2 Manifolds with Dimension 2 62
2.9.3 Manifolds with Dimension 3 64
2.9.4 Rotation-Invariant Four-Dimensional Spacetimes . . . 67
2.9.5 Robertson–Walker Metrics 70
2.10 Covariantly Constant-Curvature Manifolds 73
2.10.1 Orthogonal Symmetric Lie Algebras 73
2.10.2 Real Simple Lie Algebras 74
2.10.3 Globally Symmetric Riemannian Manifolds 76
2.10.4 Curvature of Globally Symmetric Riemannian Manifolds 77
2.10.5 Examples 79

3 Mass Points **81**
3.1 Nonrelativistic Classical Interactions 82
3.2 The Symmetries of the Kepler Dynamics 84
3.3 Electrodynamics for Charged Mass Points 86
3.4 Einstein Gravity for Mass Points 87
3.5 Geodesics of Static Spacetimes 88
3.6 Gravity for Charged Mass Points 91

4 Quantum Mechanics **93**
4.1 Nonrelativistic Wave Mechanics 94
4.2 Harmonic Oscillator 96
4.2.1 Position Representation 97
4.2.2 Color SU(3) for 3-Position 97
4.2.3 Harmonic Fermi Oscillator 99
4.2.4 Bose and Fermi Oscillators 99
4.3 Kepler Dynamics 100
4.3.1 Position Representation 101
4.3.2 Orthogonal Lenz–Runge Symmetry 102
4.4 Particles and Ghosts 105
4.4.1 Definite Metric, Fock Space, and Particles 106
4.4.2 Indefinite Metric and Ghosts 108

5 Quantum Fields of Flat Spacetime **111**
5.1 Electrodynamics of Fields 113
5.2 Gravity of Fields 115
5.3 Gravity and Electrodynamics 117
5.4 Linearized Einstein Equations 118

5.5 Free Particles for Flat Spacetime 119
5.6 Massive Particles with Spin 1 and Spin 2 123
5.7 Massless Polarized Photons and Gravitons 126
5.8 Quantum Gauge Fields 130
5.8.1 Fadeev–Popov Ghosts in Quantum Mechanics 131
5.8.2 Fadeev–Popov Ghosts for Quantum Gauge Fields . . . 133
5.8.3 Particle Analysis of Massless Vector Fields 134
5.9 Hilbert Representations of the Poincaré Group 136
5.10 Normalizations and Coupling Constants 138
5.11 Renormalization of Gauge Fields 140
5.11.1 Perturbative Corrections of Normalizations 141
5.11.2 Lie Algebra Renormalization by Vacuum Polarization 144

6 External and Internal Operations 147
6.1 Fiber Bundles 148
6.1.1 Fibers and Base 148
6.1.2 Structural and Gauge Groups 149
6.2 Nonrelativistic and Relativistic Bundles 150
6.3 Connections of Vector Space Bundles 152
6.4 Pure Gauges, Distinguished Frames, and Composite Gauge Fields 154
6.5 Chargelike Internal Connections 156
6.5.1 Currents as Lie Algebra Densities 156
6.5.2 Normalizations of Gauge Fields 158
6.5.3 Gauge Interactions in the Standard Model 158
6.6 Ground-State Degeneracy 161
6.6.1 Electroweak Symmetry Reduction 161
6.6.2 Transmutation from Hyperisospin to Electromagnetic Symmetry 165
6.7 Lie Group Coset Bundles 167
6.8 Electroweak Spacetime 168

7 Relativities and Homogeneous Spaces 173
7.1 Parametrization of Relativity Manifolds 176
7.1.1 Goldstone Manifold for Weak Coordinates 176
7.1.2 Orientation Manifolds of Metrical Tensors 177
7.1.3 Orientation Manifold of Metrical Hyperboloids 177
7.1.4 2-Sphere for Perpendicular Relativity 178
7.1.5 Mass-Hyperboloid for Rotation Relativity 178
7.1.6 Spacetime Future Cone for Unitary Relativity 179
7.2 Relativity Transitions by Transmutators 180
7.2.1 From Interactions to Particles 180
7.2.2 Pauli Transmutator 181
7.2.3 Weyl Transmutators 182
7.2.4 Higgs Transmutators 183

7.2.5 Real Tetrads (Vierbeins) 184
7.2.6 Complex Dyads (Zweibeins) 184
7.3 Linear Representations of Relativities 185
7.3.1 Rectangular Transmutators 186
7.3.2 Representations of Perpendicular Relativity 187
7.3.3 Representations of Electromagnetic Relativity 188
7.3.4 Representations of Rotation Relativity 190
7.3.5 Representations of Unitary Relativity 191
7.3.6 Representations of Lorentz Group Relativity 191
7.4 Relativity Representations by Induction 194
7.4.1 Induced Representations 194
7.4.2 Transmutators as Induced Representations 195
7.5 Hilbert Spaces of Compact Relativities 196
7.6 Flat Spaces as Orthogonal Relativities 198
7.6.1 Particle Analysis of Special Relativity 198

8 Representation Coeffficients 201
8.1 Finite-dimensional Representations 202
8.1.1 Metrics of Representation Spaces 202
8.1.2 Metrics of Lie Algebras 204
8.2 Group Algebras and Representation Spaces 205
8.2.1 Finite Groups 205
8.2.2 Algebras and Vector Spaces for Locally Compact Groups 206
8.3 Schur Product of Group Functions 209
8.3.1 Duality for Group Function Spaces 209
8.3.2 Hilbert Metrics of Cyclic Representation Spaces . . . 210
8.3.3 Induced Positive-Type Measures 212
8.4 Harmonic Analysis of Representations 213
8.5 Schur Orthogonality for Compact Groups 215
8.6 Translation Representations 217
8.6.1 Fourier Transformation 217
8.6.2 Cyclic Translation Representations 219
8.6.3 Spherical and Hyperbolic Positive-Type Functions . . 220
8.6.4 Breit–Wigner Functions 222
8.6.5 Gaussian Functions 223
8.7 Harmonic Representations of Orthogonal Groups 225
8.8 Hilbert Metrics for Flat Manifolds 226
8.8.1 Euclidean Position for Nonrelativistic Scattering . . . 227
8.8.2 Minkowski Spacetime for Relativistic Particles 229
8.9 Parabolic Subgroups 231
8.9.1 Discrete and Continuous Invariants 233
8.10 Eigenfunctions of Homogeneous Spaces (Spherical Functions) 233
8.10.1 Simple Examples 235
8.10.2 Eigenfunctions of Compact Groups 236

8.10.3 Eigenfunctions of Euclidean Groups 236
8.10.4 Eigenfunctions of Noncompact Groups 236
8.11 Hilbert Metrics for Hyperboloids and Spheres 238
8.11.1 Hyperbolic Position in the Hydrogen Atom 238
8.11.2 Representations of Hyperboloids and Spheres 240
8.12 Spherical, Hyperbolic, Feynman, and Causal Distributions . . 244
8.13 Residual Distributions 246
8.13.1 Macdonald, Bessel, and Neumann Functions 246
8.13.2 Some Special Cases 249
8.14 Hypergeometric Functions 252

9 Convolutions and Product Representations 255
9.1 Composite Nambu–Goldstone Bosons 256
9.1.1 Chirality 256
9.1.2 Chiral Degeneracy 258
9.1.3 Massless Chiral Boson 259
9.2 Convolutions for Abelian Groups 261
9.2.1 Convolutions with Linear Invariants 262
9.2.2 Convolutions with Self-Dual Invariants 262
9.3 Convolutions for Position Representations 263
9.3.1 Convolutions for Euclidean Spaces 263
9.3.2 Convolutions for Odd-Dimensional Hyperboloids . . . 264
9.3.3 Convolutions for Odd-Dimensional Spheres 265
9.4 Residual Normalization 266
9.5 Convolution of Feynman Measures 267
9.6 Convolutions for Even-Dimensional Spacetimes 269
9.7 Feynman Propagators 271
9.7.1 Convolutions of Feynman Propagators 272
9.7.2 Convolutions for Free Particles 273
9.7.3 Off-Shell Convolution Contributions 274

10 Interactions and Kernels 277
10.1 Invariant Differential Operators 279
10.2 Kernels 281
10.2.1 Kernels for Bilinear Forms 281
10.2.2 Group Kernels 282
10.3 Green's Kernels 283
10.3.1 Linear Kernels for Spacetime 284
10.3.2 Laplace–Beltrami Kernels 285
10.4 Tangent Kernels and Interactions 287
10.4.1 Position Interactions 287
10.5 Duality and Normalization 288
10.5.1 Schur Products of Feynman Measures 289
10.5.2 Representation Normalizations for Spheres
and Hyperboloids 290

10.6 Kernel Resolvents and Eigenvalues 291
10.6.1 Spacetime Normalization in the Nambu–Jona-Lasinio Model 293

11 Electroweak Spacetime **295**
11.1 Operational Spacetime 295
11.2 Representations of Electroweak Spacetime 297
11.2.1 Harmonic Analysis of the Causal Cartan Plane 299
11.2.2 Harmonic Analysis of Even-Dimensional Causal Spacetimes 301
11.3 Time and Position Subrepresentations 304
11.3.1 Projection to Subgroup Representations 305
11.3.2 Projection to and Embedding of Particles and Interactions 306
11.4 Hilbert Spaces for Causal Spacetimes 307
11.4.1 Hardy Spaces for the Future and Past 308
11.4.2 Hardy Spaces for Spacetime Cones 308
11.5 Spacetime Interactions (Kernels) 310
11.6 Normalization of Electroweak Spacetime 311
11.6.1 Central Correlation of Hypercharge and Isospin 311
11.6.2 The Mass Ratio of Spacetime 312
11.6.3 Internal Multiplicities 314

12 Masses and Coupling Constants **317**
12.1 Harmonic Coefficients of Spacetime 319
12.2 Translation Invariants as Particle Masses 321
12.3 Correlation of Spin and Masses 323
12.4 Massless Interactions 324
12.4.1 Massless Vector Modes 325
12.4.2 Massless Tensor Modes 326
12.5 Spacetime Masses and Normalizations 327

Bibliography **333**

Index **337**

Chapter 0

Introduction and Orientation

Einstein's theory of general relativity identifies spacetime curvature with a gravitational interaction. The theory is formulated primarily in a geometrical language, while the group-theoretical concepts remain more in the background. In an operational spacetime approach, the hierarchy will be reversed.

A reversal of the priorities also for a mathematical treatment of manifolds is announced by S. Helgason in the preface to his book *Groups and Geometric Analysis*:

> The role of group theory in elementary classical analysis is a rather subdued one, the motion group of $\mathbb{R}^3$ enters rather implicitly in standard vector analysis,... In contrast our point of view here is to place a natural transformation group of a given space in the foreground. We use this group as a guide for the principal concepts.

I could not agree more with such a program, here for the operational treatment of the spacetime manifold. In the following, the basic structures of physics will be defined and considered, rather restrictively, via the representations of group and Lie algebra operations, which come in two forms; *external*, acting on spacetimelike degrees of freedom, and *internal*, acting on chargelike ones. The degrees of freedom are given by the dimensions of the representation spaces. The representations are characterized by invariants which will be collected into the concepts of interactions and objects (particles).

A mathematical parallel is given by the *Erlangen Program* (1872) of Felix Klein, with an operational characterization of geometries. A Lie group acting on a manifold, e.g., an affine group acting on a vector space, constitutes a Klein space. The related geometry is characterized, especially, by its invariants, e.g., by dimensions for any affine geometry, and, in addition, by volumes for special geometry, and, in addition again, by distances and angles for a Euclidean geometry. If the vector space is real n-dimensonal, $\mathbb{R}^n$,

H. Saller, *Operational Spacetime: Interactions and Particles*,
Fundamental Theories of Physics 163, DOI 10.1007/978-1-4419-0898-8_1,

the characterizing operation groups are, respectively, the general linear, the special linear, and the special orthogonal groups, denoted by $\mathbf{GL}(n, \mathbb{R}) \supset \mathbf{SL}(n, \mathbb{R}) \supseteq \mathbf{SO}(n)$.

Also, the controversy between analytic and algebraic methods to formalize and investigate operator groups and Lie algebras, familiar from the Schrödinger and Heisenberg–Pauli approach in the foundations of quantum mechanics in the 1920s, is, apparently, not foreign to the mathematicians. A. W. Knapp writes in the preface of his book *Representation Theory of Semisimple Lie Groups*:

> Beginning with Cartan and Weyl and lasting even beyond 1960, there was a continual argument among experts whether the subject should be approached through analysis or through algebra. Some today still take one side or the other. It is clear from history, though, that it is best to use both analysis and algebra; insight comes from each.

Apparently, for compact operations, as for electromagnetic phase transformations, rotations, isospin, and color, the algebraic methods suffice. However, for noncompact, especially nonabelian, operations as given in the Lorentz or Poincaré group with their continuous quantum numbers for boosts and translations, the purely algebraic procedures are sometimes very cumbersome and difficult to apply and the analytical tools prove extremely useful. The difficulty of staying with algebraic methods only is illustrated by the ingenious, but rather complicated, algebraic solution of the nonrelativistic quantum hydrogen atom by Pauli compared with the analytic differential equation approach by Schrödinger.

In the following, differential equations, e.g., equations of motion, are not the basic starting points. I do not think that there exists something like one or a set of "basic equations." However, differential equations remain important; they will be used to characterize representation properties and to solve eigenvalue problems. Also, Lagrangians and the action principle for the derivation of differential equations will not be used as basic tools. The constitution of kinetic terms for free objects (particles) and their separation from interaction terms are, in general, possible only after understanding the origin of those free objects, e.g., of the atoms as the bound states in the nonrelativistic Coulomb potential. Interactions are a primary concept, free objects a secondary one — in parallel to a curved manifold and its flat tangent spaces.

One of the main conceptual difficulties of general relativity is to get rid of the evolutionarily engrained subconscious "absolutization" of spacetime and its coordinates as described in 1949 by Einstein himself: "Why were another seven years required for the construction of the general theory of relativity? The main reason lies in the fact that it is not so easy to free oneself from the idea that coordinates must have an immediate metrical meaning." More in 1950: "According to general relativity, the concept of space detached from any physical content does not exist. The physical reality is represented by a field." And in 1954: "Space as opposed to 'what fills space' which is dependent on the coordinates has no separate existence... There is no such thing as an empty space, i.e., a space without field." Or Born (1962): "It is not space that is there and that impresses its forms on things, but the things and their physical laws that determine space."

A "separate existence" of the coordinates for a spacetime "background" may suggest a "quantization of spacetime," e.g., by nontrivial commutators $[\mathbf{x}^a, \mathbf{x}^b] \neq 0$ of spacetime operators (noncommutative spacetime geometry) as a mathematical generalization of the Born–Heisenberg position-momentum commutators $[\mathbf{x}, \mathbf{p}] \neq 0$ in nonrelativistic quantum mechanics, which are distinguished as the historically first operational formulation. Such a "spacetime quantization" may also lead to discussions of a discrete or grainy or smooth spacetime — all concepts without an experimentally verifiable meaning, at least in my understanding of spacetime and quantum.

What is the physical meaning behind the coordinates; what do they parametrize? In the following, operations are basic: The spacetime manifold will be looked at as a reservoir for the parametrization of operation Lie groups or subgroup classes of Lie groups. For example, the position coordinate does not describe the actual position of a particle; it parametrizes, for flat manifolds, the position translation behavior. Or, the curvature and the Ricci tensor are classical concepts, visualizable with a "rubber spacetime." In a quantum framework, the geometrical curvature has, at most, a heuristic value, which is interpreted via an operational metric, i.e., in terms of bilinear forms of Lorentz transformations and of tangent translations.

A spacetime dynamics is a representation of the time and space defining operations. Spacetime is experienced by its operations, e.g., translations and rotations, implemented by interactions, constituting and acting on particles.

Newton's extremely successful approach to a formalization of physics contained three important assumptions idolizing insights that seemed obvious, but that were seen later to be prejudices and had to be modified.

The first assumption concerns time and space: They were taken to be absolute, unchangeable — two God-given boxes (*sensorium dei*) or stages for the actors of a physical dynamics. Time and space exist on their own. This was in contrast to Leibniz's characterization of space and time as relational concepts. Newton illustrated absolute space or the absoluteness of an acceleration, or, in today's words, the equivalence class of inertial systems, by the properties of a rotating water-filled vessel.

The second assumption was for the actors: They were idealized by mass points; i.e., the actors had no extension. Extended bodies were described by integrals over mass points. This contrasted with Descartes' view, where material bodies were characterized by their space extension. Newton's material points were endowed with only one property — they had mass. Connected with that, it is no wonder that the mass-related gravity structure played the most important role in Newton's mechanics: Gravity was the basic physical law of nature, especially since electromagnetic experiences were restricted to the strange behavior of amber and magnetic stones. The main task in the solution of a dynamics was to give the time development of mass points in space, the script for the actors' movement on the stage.

The third tacit and "obvious" assumption concerns our language, i.e., how we can talk about the physical nature and experiments. The relevant epistemology can be formulated clearly only in hindsight and with the knowledge of its change in a quantum theory: In classical physics, the modality structure in the formulation of the experimental results uses an observer-independent absolute ontology and the classical logic, in the simplest case with yes–no values and in an extended form with probabilities, e.g., in thermostatistics. With the Lorentz group as the main structure in Einstein's special relativity, space and time came closer to each other. However, they remain cleanly separated, no longer as linear spaces, but with the metrical concepts timelike (bicone) and spacelike. Absolute space and the ether were put to rest by the Michelson–Morley experiment. Their absolute nature, now together as linear Minkowski spacetime, was not questioned. The inhomogeneous Galilei group as characterizing the structure of the time and space boxes was replaced by the Poincaré group for the spacetime box.

With special relativity, the mass point idealization with time-parametrized orbits in position gave way to Minkowski spacetime-parametrized fields as proposed by Faraday and encoded in Maxwell's electrodynamics. To use mass points with an eigentime-parametrized motion was still possible, but somewhat artificially restrictive — they became strangers in a field theory. With the spacetime-parametrized fields, valued in spaces not necessarily related to time and position, e.g., with units like Coulomb, came properties in addition to mass, starting with the electric charge.

Einstein's general relativity got rid of the absoluteness of flat Minkowski spacetime.

Quantum theory changed the third epistemological assumption of classical physics: Physics became a theory of operators. The modality structure of our statements about experiments is characterized by an observer-relative ontology, dependent on the experimental setup ("quantum relativity"), and probability *amplitudes*, formalized by scalar products in complex Hilbert spaces. The essential quantum stucture is the Hilbert space representation of operations, infinite-dimensional for nonabelian noncompact Lie groups. It is remarkable that the set-oriented structure of measures and probabilities can be erected on a Hilbert space, i.e., on a linear space with a definite scalar product. This allows the quantum characteristic concepts of probability amplitudes and linear superpositions.

It may well be possible that there remain other unconscious absolutizations and idolizations in the formulation of our theories.

In the development of physics, the actual experimental precision was parallel with and allowed useful step-by-step approximations of better theories, sometimes surprisingly beautiful on each stage. This can be illustrated by the dynamics of the Kepler potential, where the apparently cyclic planetary orbits (Copernicus) were improved by Kepler's ellipses with small eccentricities. They are approximations to the general relativistic rosettes (Einstein) and were completely reinterpreted by quantum theory (Heisenberg, Pauli, Schrödinger) after the ad hoc discretized semiclassical Bohr–Sommerfeld approximation. There is a related remark for the supplementary companionship of experiment and theory by Maxwell (1864):

> For the sake of persons of different types of mind, scientific truth should be presented in different forms and should be regarded as equally scientific whether it appears in the robust form and vivid colouring of a physical illustration or in the tenuity and paleness of a symbolic expression.

Both in quantum theory with, e.g., Hilbert spaces and C^*-algebras for the formulation of probability amplitudes and probabilities, and in general relativity with the geometrization of the gravitational interactions, e.g., by identifying the Einstein curvature tensor with the energy-momentum tensor, every physicist experiences speechlessly the wonder of an unbelievably deep and simple, not trivial, mathematical formalization of physics. He or she ponders the millennia-old question: In which sense are those structures "really there" or "only" imposed by ourselves and, therefore, reflect our methods for the understanding of nature, which, however, are, ultimately, also a part of nature? That groups and Lie algebras in complex representations are such a strong tool to describe the basic structure of physics is, at least for me, a deep wonder.

An operational formulation of quantum gravity is still missing. Such a formulation is proposed to start from operation Lie groups. Given a basic operation group or Lie algebra, all physical structures can be interpreted in terms of corresponding realizations or representations. All basic physical properties are related eigenvalues or invariants. Operation group structures have to be studied; the familiar action and the Lagrangian-based differential equations of classical theories are only one, sometimes, but not always, appropriate formulation for the action of the corresponding Lie algebras.

In an operational spacetime approach, interactions and matter are representations of the operations that constitute spacetime. There is no interaction and matter without spacetime — that is easy to comprehend; there is no spacetime without interaction and matter — that seems to be more difficult to grasp.

In classical general relativity, the spacetime representations, used, e.g., in the metrical coefficients or the curvature, are, in general, not unitary; there is no Hilbert space structure. Quantum gravity will be proposed to rely on Hilbert representations of spacetime operations. Hilbert representations of a Lie group are decomposable into cyclic ones that are determined by characteristic cyclic states. The possible basic cyclic ground states in quantum theory, as exemplified by the weakly degenerate ground state of the standard model of particles and interactions, may have their classical analogue in the cosmological models of general relativity as exemplified by the Friedmann universes. In classical gravity, those models, especially for noncompact universes with nonabelian operation groups, are described by nonunitary metrical tensors. In a quantum description, their Hilbert representations have to be considered.

In both theories — general relativity and quantum theory — metrical structures play an important role: Einstein's gravity is the dynamics of the spacetime metric with causal signature $(1,3)$, whereas quantum theory works with Hilbert spaces acted on by spacetime representations and a Hilbert metric (Hilbert space scalar product). In the interaction-free case, i.e., for flat spacetime, the invariant metric of a Lorentz group representation space is also used for the scalar product of the Hilbert space. For example, the metric[1] $\mathbf{1}_3 \cong \delta^{ab}$ for the three definite "spacelike" degrees in the metric $\begin{pmatrix} 1 & 0 \\ 0 & -\mathbf{1}_3 \end{pmatrix}$ of the Minkowski representation is used for the three spin degrees of freedom of a massive vector field, e.g., for the weak bosons. Matter, representing spacetime operations, comes with a metric, possibly involving a nontrivial spacetime function of positive type. This may lead to an alternative interpretation for the connection of matter and metric as used in Einstein's gravity.

There are basic numbers in basic physics, e.g., the two polarizations of the photon, the three spin directions of the weak bosons, or the charges ± 1 and 0 of the pions in relation to the electron charge. In addition to these integers or rationals, ultimately related to winding numbers in the representations of compact operations (Cartan tori), there seem[2] to be basic numbers from a continuous spectrum like the mass ratios of elementary particles or the strengths, i.e., coupling constants (normalizations) of basic interactions. These numbers characterize noncompact group operations (Cartan planes) with their, in general, complicated Hilbert representations, infinite-dimensional, if faithful.

A framework with both gravity and electromagnetic interactions has to face the huge difference in their strengths, illustrated for mass points with masses m and charges $Q = ze$ by the ratio

$$\frac{\text{EM}}{\text{GR}} = -\frac{Q_1 Q_2}{4\pi\epsilon_0 G m_1 m_2} = -\alpha_S m_P^2 \frac{z_1 z_2}{m_1 m_2},$$

with the square of the electron charge e in Sommerfeld's fine structure constant $\alpha_S = \frac{e^2}{4\pi\epsilon_0 \hbar c} \sim \frac{1}{137}$, with integer charge numbers $z \in \mathbb{Z}$, and with Newton's constant G in the Planck mass $m_P^2 = \frac{\hbar c}{G}$. Its huge ratio with usual elementary particle masses, e.g., for the proton $\frac{m_P^2}{m_p^2} \sim (3.6 \times 10^{19})^2$, seems difficult to obtain in a "natural way," e.g., in polynomial equations $P(x) = 0$ for $x = \frac{m_P^2}{m_p^2}$. Even the logarithm of the ratio, e.g., $\log \frac{m_P^2}{m_p^2} \sim 88$, has to face two decimal orders of magnitude. The most important spacetime operations come

---

[1] $\mathbf{1}_n$ denotes the $(n \times n)$-unit matrix.

[2] With experimental errors and the rationals dense in the reals, $\overline{\mathbb{Q}} = \mathbb{R}$, one can never be sure.

from the orthochronous, i.e., causality-compatible Lorentz group $\mathbf{SO}_0(1,3)$ and its twofold cover group $\mathbf{SL}(2,\mathbb{C})$, i.e., its complex realization. With a maximal abelian subgroup $\mathbf{SO}(2)\times\mathbf{SO}_0(1,1)$ for axial rotations and Lorentz dilations, it has rank 2. Therefore, two invariants characterize its representations. The historically first real four-dimensional Minkowski representation $\mathcal{D}_{[\frac{1}{2}|\frac{1}{2}]}$ of the Lorentz Lie algebra with angular momenta $\vec{\mathcal{L}}$ and boosts $\vec{\mathcal{B}}$ acting on the spacetime translations and its products constitute a very restricted class. These representations have only one nontrivial independent invariant:

$$\mathcal{D}_{[\frac{1}{2}|\frac{1}{2}]}(\vec{\varphi}\vec{\mathcal{L}}+\vec{\psi}\vec{\mathcal{B}}) = \begin{pmatrix} 0 & \psi_1 & \psi_2 & \psi_3 \\ \psi_1 & 0 & \varphi_3 & -\varphi_2 \\ \psi_2 & -\varphi_3 & 0 & \varphi_1 \\ \psi_3 & \varphi_2 & -\varphi_1 & 0 \end{pmatrix} \Rightarrow \begin{cases} \mathcal{D}_{[\frac{1}{2}|\frac{1}{2}]}(\vec{\mathcal{L}}^2-\vec{\mathcal{B}}^2) = -3\mathbf{1}_4, \\ \mathcal{D}_{[\frac{1}{2}|\frac{1}{2}]}(\vec{\mathcal{L}}\vec{\mathcal{B}}) = 0. \end{cases}$$

The later-used complex representations for half-integer spin, faithful for $\mathbf{SL}(2,\mathbb{C})$, e.g., a complex two-dimensional Weyl representation $\mathcal{D}_{[\frac{1}{2}|0]}$ with Pauli matrices $\vec{\sigma}$, have, at least, a nontrivial invariant also for the noncompact classes of the rotation group, parametrizable by a future hyperboloid,[3] $\mathbf{SO}_0(1,3)/\mathbf{SO}(3)\cong\mathcal{Y}^3$:

$$\mathcal{D}_{[\frac{1}{2}|0]}(\vec{\varphi}\vec{\mathcal{L}}+\vec{\psi}\vec{\mathcal{B}}) = \vec{\varphi}i\frac{\vec{\sigma}}{2}+\vec{\psi}\frac{\vec{\sigma}}{2} \Rightarrow \begin{cases} \mathcal{D}_{[\frac{1}{2}|0]}(\vec{\mathcal{L}}^2-\vec{\mathcal{B}}^2) = -\frac{3}{2}\mathbf{1}_2, \\ \mathcal{D}_{[\frac{1}{2}|0]}(\vec{\mathcal{L}}\vec{\mathcal{B}}) = i\frac{3}{4}\mathbf{1}_2. \end{cases}$$

The nontrivial Hilbert representations, i.e., with definite unitary boosts, are, in contrast to Minkowski and Weyl representations, necessarily infinite-dimensional. The spin and $\mathbf{SO}(2)$-related compact invariant $\vec{\mathcal{L}}^2-\vec{\mathcal{B}}^2$ remains rational as for the rotations. The hyperboloid and $\mathbf{SO}_0(1,1)$-related noncompact invariant $\vec{\mathcal{L}}\vec{\mathcal{B}}$ are taken from a continuous complex spectrum.

With an understanding of the quantum structure for spacetime operations, there has to come a deeper understanding of the phenomenon of masses. Einstein's equations identify, up to a constant, the spacetime curvature $\mathcal{R}$ and the energy-momentum tensor $\mathbf{T}$, which, for flat spacetime, gives the space densities of the spacetime translations $\mathcal{P}$, represented with the mass m^2 as invariant. The one-dimensional eigentime translation group $\tau\in\mathbb{R}$ is, as Lie group, isomorphic to the dilation group $e^{\psi}\in\mathbf{D}(1)$, which, in the self-dual form $\mathbf{SO}_0(1,1)\cong\mathbf{D}(1)$ with contractions and extensions ("Procrustes transformations") is a maximal abelian noncompact subgroup of the Lorentz group (Lorentz dilations) $e^{\psi\sigma_3} = \begin{pmatrix} e^{\psi} & 0 \\ 0 & e^{-\psi}\end{pmatrix} \cong \begin{pmatrix}\cosh\psi & \sinh\psi \\ \sinh\psi & \cosh\psi\end{pmatrix} = e^{\psi\sigma_1} \in \mathbf{SO}_0(1,1)\subset\mathbf{SO}_0(1,3)$. In a Lorentz group-compatible theory of spacetime representations, the two-dimensional group $e^{\psi_0\mathbf{1}_2+\psi\sigma_3}\in\mathbf{D}(1)\times\mathbf{SO}_0(1,1)$ (Cartan spacetime), represented with two continuous invariants, plays an important role.

It is useful to have a multiplicative and an additive notation for the abstract real one-dimensional simply connected Lie group $\mathbf{D}(1)=\exp\mathbb{R}\cong$

[3]The symbols Ω^s, with Ω looking faintly like a circle, and $\mathcal{Y}^s$, with $\mathcal{Y}$ looking somewhat like a hyperbola, are used for s-dimensional unit spheres and unit hyperboloids, respectively, elsewhere often denoted by the symbols $S^s=\Omega^s$ and $H^s=\mathcal{Y}^s$.

$\mathbb{R} = \log \mathbf{D}(1)$. This noncompact group will be called the dilation or causal or translation group. Its classes with respect to the integers can be parametrized by the points of a circle $\mathbb{R}/\mathbb{Z} \cong \Omega^1 \cong \mathbf{SO}(2) \cong \mathbf{U}(1) = \exp i\mathbb{R}$. This compact real one-dimensional group will be called the axial rotation or phase or electromagnetic group.

An operation group determines its action spaces: A group G action decomposes a space S into disjoint orbits, $G \bullet x$ for $x \in S$. Each orbit is isomorphic to subgroup classes G/H, where the isotropy group $H \subseteq G$ is isomorphic to all fixgroups of $y \in G \bullet x$. Therefore, the study of subgroup classes G/H (also called homogeneous or symmetric or coset spaces) and, for complex linear quantum theory, of their associated complex vector spaces like the closure $\overline{\mathbb{C}^{(G/H)}}$ of its finite linear combinations (cyclic Hilbert spaces) is of paramount importance. For example, flat Euclidean 3-position and flat Minkowskian 4-spacetime are operationally described, as symmetric spaces, by the orthogonal subgroup classes[4] $\mathbf{SO}(3) \vec{\times} \mathbb{R}^3/\mathbf{SO}(3) \cong \mathbb{R}^3$ and $\mathbf{SO}_0(1,3) \vec{\times} \mathbb{R}^4/\mathbf{SO}_0(1,3) \cong \mathbb{R}^4$, respectively, not[5] by the manifold isomorphic abelian groups $\mathbb{R}^3$ and $\mathbb{R}^4$.

With the action of a group on its irreducible spaces G/H, the choice of a representative $g_r \in gH$ in a coset $gH \in G/H$ leaves the freedom of the local fixgroup H. Coset structures are closely related to gauge structures, where the operations in G/H come in subgroup H-representations.

Representations of Lie operations involve group functions and distributions as elaborated, especially, by Vilenkin. One has to realize that not only the basic exponentials like e^{iEt} and $e^{-|mx|}$, but all relevant, sometimes very complicated, special functions like Bessel functions or Neumann and Macdonald functions for the on-shell contribution $\int d^4q\; \delta(q^2 - m^2)e^{iqx}$ of a Feynman propagator, used for basic physical structures, i.e., for quantum theory and general relativity, are representation coefficients (matrix elements) of Lie groups. With the knowledge of the representational relevance of special functions, which are used quite often in this book, it is easier to get along with their complicated expressions.

The connection between Hilbert representations of spacetime and particles is manifest for interaction-free structures. There, the time and position translations are formalized by the abelian Lie operations in the groups $\mathbb{R}$ and $\mathbb{R}^3$ with energies and momenta, respectively, as eigenvalues of the Hilbert space representations $\mathbb{R} \oplus \mathbb{R}^3 \ni (x_0, \vec{x}) \longmapsto e^{iq_0x_0 - i\vec{q}\vec{x}} \in \mathbf{U}(1)$.

"Free particle" structures, in general defined by Hilbert space representations of translations, start with the quantum harmonic oscillator, which implements the unitary representations of the time translations $\mathbb{R} \ni t \longmapsto e^{iEt} \in \mathbf{U}(1)$, where the equidistanced energy spectrum is given by the invariants of the product representations $\mathbb{R} \ni t \longmapsto (e^{iEt})^k \in \mathbf{U}(1)$, $k = 0, 1, \ldots$, with the eigenvectors $|kE\rangle$.

[4]This $\mathbb{R}^3$ is meant in Helgason's remark on the first page of this chapter.

[5]Such a structural distinction for one underlying set may be illustrated by the plane $\mathbb{R}^2$, considered either as real vector space, where linearity is constitutive, or as topological space, where open sets take this role.

More complicated are the representations of the semidirect product Euclidean group $\mathbf{SO}(3)\ \vec{\times}\ \mathbb{R}^3$ for flat position $\mathbf{SO}(3)\ \vec{\times}\ \mathbb{R}^3/\mathbf{SO}(3) \cong \mathbb{R}^3$, which are used for free scattering states. Nonrelativistic free scattering states have "eigenvectors"[6] $|P^2, |h|; \vec{\omega}, h\rangle$, where the momentum $\vec{q} = P\vec{\omega}$ contains the invariant $P^2 = \vec{q}^{\,2}$ for the position translations $\mathbb{R}^3$ and, as eigenvalues, the directions $\vec{\omega} \in \Omega^2 \cong \mathbf{SO}(3)/\mathbf{SO}(2)$ on the unit 2-sphere. For the axial rotation fixgroup $\mathbf{SO}(2)$ around the momentum, the helicity $|h|$ is the invariant and $h = \pm|h|$ the eigenvalues for left- or right-handed polarization.

For special relativity, the action of the Galilei group $\mathbf{SO}(3)\ \vec{\times}\ \mathbb{R}^3$ with rotations and velocity transformations and its inhomogeneous extension by time and position translations $[\mathbf{SO}(3)\ \vec{\times}\ \mathbb{R}^3]\ \vec{\times}\ [\mathbb{R} \oplus \mathbb{R}^3]$ are expanded to the semidirect Poincaré group $\mathbf{SO}_0(1,3)\ \vec{\times}\ \mathbb{R}^4$ with the homogeneous simple orthochronous Lorentz group $\mathbf{SO}_0(1,3)$ and $\mathbb{R}^4$ as translations for relativistic Minkowski spacetime $\mathbf{SO}_0(1,3)\ \vec{\times}\ \mathbb{R}^4/\mathbf{SO}_0(1,3) \cong \mathbb{R}^4$. According to Wigner, a free elementary particle[7] is described by a complex infinite-dimensional vector space acted on by an irreducible unitary representation of the Poincaré cover group $\mathbf{SL}(2,\mathbb{C})\ \vec{\times}\ \mathbb{R}^4$. The "eigenvectors" $|m^2, J; \vec{q}, J_3\rangle$ contain the eigenvalues momenta $\vec{q}$ and the invariant mass m for spacetime translations $\mathbb{R}^4$, and spin or helicity J with projection J_3, which characterize rotations $\mathbf{SU}(2)$ in a rest system for massive particles and axial rotations $\mathbf{SO}(2)$ around the momentum direction for massless ones.

"Flat" and "curved" are in correspondence with "free" and "interacting" and, also, with time and space, as given in the following table, which should not be taken too precisely, but only to indicate a parallelism of concepts:

	Dynamics	Manifold	Group	Diff. eq.
Time	free	flat	abelian	linear
Space	interacting	curved	nonabelian	nonlinear

Flat manifolds come with operation groups, especially translations, whose representations characterize interaction-free particles. Flat manifolds cannot "explain" the existence and properties of interactions and bound-state structures, which represent operations for curved manifolds, especially with nonabelian action groups. The concept of a bound state with constituents is not very useful if the bound-state energy is of the order of magnitude of the mass of the "constituents." Each group determines its Hilbert spaces. The Hilbert space for free particles, characterized by the Fock ground state, is inappropriate for nonabelian groups. If the particles reveal the spectrum of nonabelian spacetime operations, it is very doubtful that they can be understood via bound states of free particles. Free particles are irreducible Hilbert representations of the Poincaré group — the actually arising invariants, masses, and spins (polarizations) cannot be explained by the Poincaré group.

---

[6] In the eigenvector notation $|I; w\rangle$, the representation-characterizing invariants I stand before the semicolon, the eigenvalues w after it, e.g., the spin $\mathbf{SU}(2)$-eigenvectors $|J; J_3\rangle$.

[7] A necessary extension of Wigner's definition to unstable particles with a width is not considered in the following.

A related quotation from the last public talk of Heisenberg (1975, my translation):

> It is unavoidable that we use a language originating from classical philosophy. We ask: What does the proton consist of? Is the quantum of light elementary or composite?, etc. However, all these questions are wrongly asked since the words "divide" and "consist of" have lost almost all their meaning. Therefore, it would be our task to adjust our language, our thinking, i.e., our scientific philosophy, to this new situation that has been created by the experiments. Unfortunately, that is very difficult. Therefore, there creep into particle physics, always again, wrong questions and wrong conceptions, ...
>
> We have to come to terms with the fact that the experimental experiences for very small and for very large distances no longer provide us with an *anschauliches Bild* and we have to learn to live there without *Anschauung* (something like "without familiar everyday pictures"). In this case we realize that the antinomy of the infinitely small for the elementary particles is resolved in a very subtle way — in a way that neither Immanuel Kant nor the Greek philosophers could think of — the words "to divide" lose their sense.
>
> If one wants to compare the insights of today's particle physics with any earlier philosophy, it could only be the philosophy of Plato, since the particles of today's physics are representations of symmetry groups — that's what quantum theory teaches us — and, therefore, the particles resemble the symmetric Platonic solids.

It is the main prejudice of this book that there is a level for our understanding of physics where it no longer makes sense to assume parts of particles, i.e., particles inside particles, where, however, it makes sense to talk about interactions or, better, about operations and symmetries. Particles implement only a subclass of physically relevant operations, their operations are not complete for curved spaces with nonabelian operations for interactions.

Time as a real one-dimensional, necessarily abelian Lie group is flat; abelian operations cannot implement interactions.

Riemannian manifolds with nonabelian Lie groups as used, e.g., for real three-dimensional position can be curved, e.g., in the maximally symmetric form of a compact 3-sphere $\Omega^3 \cong \mathbf{SO}(4)/\mathbf{SO}(3)$, as used in Einstein's static universe, or of a noncompact 3-hyberboloid $\mathcal{Y}^3 \cong \mathbf{SO}_0(1,3)/\mathbf{SO}(3)$ with nonabelian action (motion) groups $\mathbf{SO}(4)$ and $\mathbf{SO}_0(1,3)$, respectively. Both of these curved positions parametrize classes of $\mathbf{SO}(3)$-subgroups. The hyperbolic position $\mathcal{Y}^3$ is represented in the nonrelativistic hydrogen atom with the Coulomb potential in its Hamiltonian. The bound-state wave functions $\psi^{2J}_{Lm}(\vec{x}) \sim r^L \mathrm{Y}^L_m(\frac{\vec{x}}{r})\, \mathrm{L}^N_{1+2L}(\frac{2r}{n})\, e^{-\frac{r}{n}}$, with the principal quantum number an $\mathbf{SU}(2)$-multiplet multiplicity $n = 1 + 2J = 1 + L + N$, are representation coefficients of position where the harmonic $\mathbf{SO}(3)$-polynomials $\vec{x} \longmapsto (\vec{x})^L_m = r^L \mathrm{Y}^L_m(\frac{\vec{x}}{r})$ with the spherical harmonics Y representing the maximal compact rotations $\mathbf{SO}(3) \subset \mathbf{SO}_0(1,3)$ and the remaining exponential $e^{-\frac{r}{n}}$ with the Laguerre polynomials L the hyperbolic operations $\mathbf{SO}_0(1,1) \subset \mathbf{SO}_0(1,3)$ in a maximal abelian noncompact subgroup. In contrast to the positive momentum invariants $\vec{q}^2 = P^2$ (real momenta) for scattering, used in the spherical Bessel functions $j_0(Pr) = \frac{\sin Pr}{Pr} = \int \frac{d^3q}{2\pi|P|}\delta(\vec{q}^2 - P^2)e^{i\vec{q}\vec{x}}$ as representation coefficients of the flat position $\mathbf{SO}(3) \vec{\times} \mathbb{R}^3/\mathbf{SO}(3)$, the bound structures for the hyperbolic position $\mathbf{SO}_0(1,3)/\mathbf{SO}(3)$ have negative invariants $\vec{q}^2 = -Q^2$ (imaginary momenta) as seen at the momentum function dipole singularity in the exponential wave functions $\mathcal{Y}^3 \cong \mathbb{R}^3 \ni \vec{x} \longmapsto e^{-|Q|r} = \int \frac{d^3q}{\pi^2} \frac{|Q|}{(\vec{q}^2+Q^2)^2} e^{i\vec{q}\vec{x}}$. These coefficients represent hyperbolic position with eigenvalues $i\vec{q}$ on a compact 3-sphere Ω^3 with measure $\int d^3q \frac{2}{(\vec{q}^2+Q^2)^2} = \frac{2\pi^2}{|Q|}$ for radius $\frac{1}{|Q|}$. The invariant energies $-2E = Q^2 \in \{\frac{1}{(1+2J)^2} \,\big|\, J = 0, \frac{1}{2}, \dots\}$ determine the quantized curvature of the position hyperboloid $\mathcal{Y}^3$.

The belief in the universality and completeness of flat Minkowski spacetime with its Poincaré group, of the particle fields, and of the Fock state is reflected by the remark "Each interaction is mediated by a particle." Such a strong statement, illustrated by Yukawa's formulation of nuclear forces with the Compton length of "exchanged" pions as range $\frac{e^{-\frac{m_\pi c}{\hbar}r}}{r}$ is not true and too narrow. Also, a weakened formulation with the concept of "off-shell" or "virtual particles" for interactions is slippery. Parallel to the distinction between time(like) and space(like), the distinction between particles and interactions does not vanish in a relativistic quantum field theory. Spacetime fields can have non-particle degrees of freedom. For example, the four-component electromagnetic field contains both the Coulomb and the two photon degrees of freedom, which, in a particle analysis with the axial rotations $\mathbf{SO}(2)$ around the momenta, come as a rotation scalar and a dublet with polarization ± 1 (left- and right-handed), i.e., the Coulomb interaction is not mediated by a particle. A similar situation occurs for the 10-component gravitational field in the flat spacetime approach with a ± 2-polarization dublet for gravitons (particles) and eight nonparticle degrees of freedom. including an $\mathbf{SO}(2)$-scalar for the Newton interaction.

Masses cannot arise only as translation invariants for flat spacetime to characterize free particles. For example, the masses of quarks, if confined,[8] cannot be used as invariants for spacetime translations. What is the operational meaning of the different kinds of quark masses? What is the meaning of a Feynman propagator for quarks? Does it make sense to ask if quarks are stable, e.g., the up quark, or unstable, e.g., the top quark, and if they are unstable, what is their width? Perhaps, the quarks for the parametrization of the strong interactions implement only homogeneous ("local") operations, but no translations. They are introduced with color, hypercharge, isospin, and Lorentz group properties, but, if confined, without translation properties; i.e.,

[8] So far, there are arguments for color confinement; its rigorous mathematical proof as a consequence of an unbroken $\mathbf{SU}(3)$-gauge interaction is still missing.

they are not unitary Poincaré group representations, no particles according to Wigner's particle definition.

Also, the curvature of manifolds, which can be measured in squared mass units, must not correspond to a particle. In general, masses from a continuous spectrum can occur as invariants of any noncompact operation group. For nonflat space, they may characterize-curvature related interactions and may be measurable in coupling constants.

In an operational framework, particles are described by representation coefficients, i.e., functions on a group, whereas interactions are, in general, distributions, describing the tangent Lie group structure; i.e., they are related to its Lie algebra and Lie algebra forms. This is familiar from the Lie algebra structure of the gauge interactions. From a representation point of view, it is understandable that interactions differ from bound-state wave functions in the order of the singularity, e.g., the Yukawa interaction for hyperbolic position $\mathcal{Y}^3$ with a simple momentum pole $\frac{e^{-|Q|r}}{r} = \int \frac{d^3q}{2\pi^2} \frac{1}{\vec{q}^2+Q^2} e^{i\vec{q}\vec{x}}$ from the ground-state wave function of the nonrelativistic hydrogen atom, with a momentum dipole $e^{-|Q|r} = \int \frac{d^3q}{\pi^2} \frac{|Q|}{(\vec{q}^2+Q^2)^2} e^{i\vec{q}\vec{x}}$.

The "divergences" of interacting spacetime quantum field theories with the undefined local products of spacetime distributions are caused by the expansion of interactions, related to nonabelian operations of curved space, with representation coefficients of abelian operations, especially spacetime translations for free particles. Renormalizable canonical quantum field theories with kinetic terms for the free particles and gauge interactions, expanded with the Fock ground state for free particles, like quantum electrodynamics and the standard model of electroweak interactions, together with the regularization-by-renormalization procedure, seem to be a viable method to describe the scattering of free particles. Such a flat spacetime framework, however, is inappropriate for understanding its ingredients — the particle spectrum and the interactions themselves. For this purpose, one has to start from fundamental operations and symmetries, not from fundamental particles or constituents. Heisenberg (1973): "The fundamental symmetries define the underlying law which determines the spectrum of elementary particles. An analogue: The scattering wave functions for flat Euclidean space $\mathbf{SO}(3)\ \vec{\times}\ \mathbb{R}^3/\mathbf{SO}(3) \cong \mathbb{R}^3$, e.g., the spherical Bessel functions $\frac{\sin Pr}{Pr}$ for the rotation trivial case, are inappropriate for understanding the bound-state wave functions of the nonrelativistic hydrogen atom, e.g., for the ground-state wave function $e^{-|Q|r}$, which are representation coefficients of the curved hyperbolic position $\mathbf{SO}_0(1,3)/\mathbf{SO}(3) \cong \mathcal{Y}^3$.

For general relativistic spacetime, one may expect a generalization from special relativity with the Lorentz group to the invariance under an even larger motion group $G \supset \mathbf{SO}_0(1,3)$, e.g., under all automorphisms $\mathbf{GL}(4,\mathbb{R})$ of the tangent spaces, i.e., under the structural group of the local frames. However, such a generalization seems to be inappropriate. As exemplified by Friedmann universes, there are many different spacetimes with a tangent Poincaré group. Tangent space groups with translations are rather "passive";

they are for free objects. They do not describe the operations responsible for the constitution of particles and their interactions. General relativistic "nonlinear" spacetime should not be considered as an embedding manifold for its tangent spaces, e.g., as an expansion of the Poincaré group to the (anti) de Sitter group $\mathbf{SO}_0(2,3)$ or $\mathbf{SO}_0(1,4)$. The operations that constitute nonlinear spacetime, not its flat spacetime tangent operations, are "responsible" for and determine particles and interactions.

The periodic table of the elements is nonrelativistically explained by the eigenstates of a Hamiltonian with the Coulomb potential. Its operational background is the Lorentz group $\mathbf{SO}_0(1,3)$, represented by scattering states, and its compact partner $\mathbf{SO}(4)$, represented by bound states, as the correlated product of two $\mathbf{SU}(2)$-groups with the rotations $\mathbf{SO}(3)$ and the additional Lenz–Runge classes on a 3-sphere $\Omega^3 \cong \mathbf{SO}(4)/\mathbf{SO}(3)$, classically visible in the perihelion conservation. The quantum structure of the Kepler–Coulomb potential $\frac{1}{r}$ is physically relevant, exactly solvable, and aesthetically appealing. A nonrelativistic nonflat position is implemented by Hilbert space representations of the 3-hyperboloid $\mathcal{Y}^3$. The periodic table reflects representation invariants of the curved position $\mathcal{Y}^3$. Analously, the particle table will be proposed to display invariants of representations of the operational groups for relativistic four-dimensional spacetime $\mathcal{D}^4$, embedding hyperbolic position. The invariants of the $\mathcal{D}^4$-representations are used as masses, spins, charges, and coupling constants for their normalizations.

There are two types of long-range (massless) interactions that come with either "unbroken" or "broken" symmetries: gauge interactions, like electrodynamics for unbroken $\mathbf{U}(1)$-symmetry, and Nambu–Goldstone fields for degeneracy operations of a ground state, like for broken chiral $\mathbf{U}(1)$-symmetry. Usually, degeneracy operations G/H for a ground state with a symmetry group $H \subseteq G$ are characterized, quantitively, by a mass; i.e., the massless Nambu-Goldstone fields in H-multiplets effect also a rearrangement of the broken dilation $\mathbf{D}(1)$ properties. These structures will be proposed for an understanding of the long-range electromagnetic and gravitational interactions.

The chiral model of Nambu and Jona-Lasionio is an interacting model for relativistic eigenstates. In addition to the nonrelativistic Kepler interaction, its structures will play an important role as an illustration for understanding the particle table, especially the chiral degeneracy of its ground state as an example of long-range interactions.

Physical operations act both on spacetimelike degrees of freedom, then called external operations, e.g., Lorentz transformations and translations, and on chargelike ones, then called internal operations, e.g., hypercharge or isospin. An understanding of space, time, interaction, and matter has to come with an "integrative symbiosis" of external and internal operations. A unification was first tried by Weyl; he implemented the electromagnetic operations in the form of a locally acting dilation group $\mathbf{D}(1)$. Although experimentally wrong concerning the specific group, the gauge principle proved extremely fruitful. Together with London, Weyl later replaced the noncompact dilations by the compact $\mathbf{U}(1)$-phase transformations. As of today, all internal operations, hyperisospin $\mathbf{U}(2)$, and color $\mathbf{SU}(3)$, as exemplified by the standard model of electroweak and strong interactions, are from compact groups, implemented as gauge operations that accompany each spacetime translation.

The main new and original proposal of this book is an operational spacetime, called electroweak spacetime $\mathcal{D}^4$, parametrizing the classes of the internal hypercharge-isospin group $\mathbf{U}(2)$ in the general complex linear group $\mathbf{GL}(2,\mathbb{C})$, i.e., the Lorentz cover group $\mathbf{SL}(2,\mathbb{C})$, extended by the causal (dilation) and phase group $\mathbf{D}(1) \times \mathbf{U}(1)$. Its representations and invariants for the bi-regular action group $\mathbf{GL}(2,\mathbb{C}) \times \mathbf{U}(2)$ will be investigated with the aim of connecting them, qualitatively and numerically, with the properties of interactions and particles as arising in the representations of its tangent Minkowski spaces. This is tentatively realized in the last chapter of this book. Electroweak and gravitational interactions are distinguished representations of spacetime $\mathcal{D}^4$ that rearrange dilation degrees of freedom. The position curvature-related representations will be connected with the quarks, which, in the standard model, have been introduced to parametrize the strong interactions and, if confined, are entities that, with Wigner's definition, are not particles, i.e., not translation eigenvectors. The relation between position curvature and strong interaction may not be so unnatural if Einstein's interpretation of gravity by spacetime curvature is remembered. Even if the proposal of interactions and particles describing the harmonic analysis of electroweak spacetime $\mathcal{D}^4$ is premature, too simple, or even plainly wrong, I hope it can illustrate how a unified operational approach for the concepts of spacetime, interaction, and matter may be concretized.

This book does not discuss astrophysical and cosmological problems; it is intended as an operational analysis of spacetime with its consequences for interactions and particles, not as an additional book on the classical differential geometric treatment of relativity and gravitation. For this method, there are many excellent books, where everybody has his or her favorites. In addition to the classic of H. Weyl, *Raum, Zeit, Materie* (1923), with the conceptionally ground-breaking introduction of the gauge principle, using the physically wrong dilation group, I favor, as an introduction, the book of R. Sexl and H. Urbantke, *Gravitation und Kosmologie* (1981), with a good and competent mixture of physical and mathematical concepts.

This book addresses graduate students and scientists with an interest in the structure of those basic physical theories that have some experimental justification. Apparently, the richness of mathematically appealing forms is inexhaustible; their beauty may be a trap for physicists who should be primarily interested in their physical relevance. The scientific level of the book is not undergraduate; it is not written as an introduction. It assumes knowledge of and familiarity with conventional relativity, quantum mechanics and quantum field theory. In a sense, the book is very conservative; there are no flashy new and titillating revolutionary ideas, only concepts that have been used already in the theories mentioned above. However, I try to delve into them more deeply and more radically. For the mathematical tools, used in the following, the two books of S. Helgason, *Differential Geometry, Lie Groups and Symmetric Spaces* (1978) and *Groups and Geometric Analysis* (1984), are recommended. For representation theory, I learned much from the books of G.B. Folland, *A Course in Abstract Harmonic Analysis* (1995), and A. Knapp, *Representation Theory of Semisimple Groups* (1986); for distribution theory, from the book of F. Treves, *Topological Vector Spaces, Distributions and Kernels* (1967). Since the mathematics is also not undergraduate and by no means trivial, at least for me, I hope that I didn't make too many mistakes — in addition to the usual reckless treatment of mathematics by physicists.

The following text contains parts considered in a first run in my books *Operational Quantum Theory I — Nonrelativistic Structures* (2006) and *Operational Quantum Theory II — Relativistic Structures* (2006), where some of the spacetime structures worked with in this book have were introduced. It also uses the standard mathematical concepts and notation given in those books in more detail.

The first six chapters give, in an operational language, a short journey through the conventional theories, while the last six chapters describe how one should proceed to a more basic understanding of the mutual conditioning of spacetime and interactions and matter.

Chapter 1

Einstein's Gravity

In this chapter, Einstein's gravity with the identification of the curvature of a Riemannian manifold as a geometrical concept and interaction as a physical one is shortly exemplified by Schwarzschild–Kruskal spacetime and by some properties of Friedmann spacetimes with Robertson–Walker metrics. As a warm-up, the group-oriented operational language will be used in those important and basic examples for classical macroscopic gravity.

1.1 Geometrization of Gravity

Einstein's gravity is a dynamical theory of the spacetime metric and curvature as determined by the energy and momentum of light and matter.

The geometrical curvature concept has its *anschaulichen* origin in the local change of (derivative with respect to) an area; therefore, it has the dimension of an inverse area: An orange skin breaks up if pressed flat on a table, in contrast to a paper cyclinder, which has no area change. The paper cyclinder is flat.[1] A nontrivial curvature does not exist in one dimension. For example, a circle Ω^1 or a hyperbola $\mathcal{Y}^1$ is flat, equally time, parametrized as a real one-dimensional manifold. Proper curvature needs a two-dimensional area, e.g., a submanifold of a real three-dimensional space. The curvature is called positive or spherical, $k = 1$, for the compact 2-sphere Ω^2 (orange skin) and negative or hyperbolic, $k = -1$, for the noncompact 2-hyperboloid $\mathcal{Y}^2$ as given by a mass hyperboloid in the future cone of $(1 + 2)$-dimensional Minkowski spacetime, which crumples (any part of it) if pressed flat on a table. For $n = 4$ spacetime dimensions, there are $\binom{n}{2} = 6$ linear independent areas for three time-spacelike curvatures and three space-spacelike ones, related to the six real dimensions of the Lorentz group with three boosts and three rotations.

If an area is measured by a metric, the curvature, i.e., the area change, contains its 2nd-order derivatives. The curvature $\mathcal{R}$ (for Riemann and, also,

[1] Its "one-dimensional curvature" is sometimes called *external* in contrast to the "two-dimensional internal" curvature of the orange skin described above.

H. Saller, *Operational Spacetime: Interactions and Particles*, Fundamental Theories of Physics 163, DOI 10.1007/978-1-4419-0898-8_2,

Ricci) involves 1st-order derivatives $\partial^j = \frac{\partial}{\partial x_j}$ of the Riemannian connection Γ, which is defined (see Chapter 2) with 1st-order derivatives of the signature $(1,3)$-metric $\mathbf{g} = ds^2 = \mathbf{g}^{li}(x)dx_l \otimes dx_i$;

$$\begin{aligned} \tfrac{1}{2}\mathbf{g}_{kp}(\partial^i \mathbf{g}^{jp} + \partial^j \mathbf{g}^{ip} - \partial^p \mathbf{g}^{ij}) &= \Gamma_k^{ij}, \\ \partial^i \Gamma_k^{jl} - \partial^j \Gamma_k^{il} - \Gamma_p^{il}\Gamma_k^{jp} + \Gamma_p^{jl}\Gamma_k^{ip} &= \mathcal{R}_k^{lij}. \end{aligned}$$

The Einstein tensor $\check{\mathcal{R}}_\bullet = \mathcal{R}_\bullet - \frac{\mathbf{g}}{2}\mathcal{R}_\bullet^\bullet$ is the combination of the Ricci tensor $\mathcal{R}_\bullet$ with the metric multiplied curvature scalar $\mathcal{R}_\bullet^\bullet = \mathbf{g}_{li}\mathcal{R}_\bullet^{li} = \mathbf{g}_{li}\mathcal{R}_j^{lij}$. In Einstein's gravity, the 2nd-order derivatives of the metric with the dimension of an area density $[\mathcal{R}] = \frac{1}{\mathrm{m}^2}$, multiplied with $\frac{1}{\kappa} = \frac{c^2}{8\pi G} \sim 5.3 \times 10^{27} \frac{\mathrm{kg}}{\mathrm{m}}$, involving Newton's constant G and the maximal action speed $c \sim 3 \times 10^8 \frac{\mathrm{m}}{\mathrm{s}}$, are given by the energy-momentum tensor $\mathbf{T}$ with trace $\mathbf{T}^\bullet = \mathbf{g}_{li}\mathbf{T}^{li}$ and the dimension of a mass density $[\mathbf{T}] = \frac{\mathrm{kg}}{\mathrm{m}^3}$,

$$\begin{aligned} \check{\mathcal{R}}_\bullet + \Lambda\mathbf{g} &= -\kappa\mathbf{T}, \quad \text{with } \check{\mathcal{R}}_\bullet = \mathcal{R}_\bullet - \tfrac{\mathbf{g}}{2}\mathcal{R}_\bullet^\bullet, \\ \text{or } \mathcal{R}_\bullet - \Lambda\mathbf{g} &= -\kappa\check{\mathbf{T}}, \quad \text{with } \check{\mathbf{T}} = \mathbf{T} - \tfrac{\mathbf{g}}{2}\mathbf{T}^\bullet, \end{aligned} \qquad \mathcal{R}_\bullet^\bullet = \kappa\mathbf{T}^\bullet + 4\Lambda.$$

The spacetime properties, as given in $\mathcal{R}$, determine the matter properties, as given in $\mathbf{T}$ — or vice versa. Also, a term with the cosmological constant $[\Lambda] = \frac{1}{\mathrm{m}^2}$ may be included for a ground state energy-momentum tensor, where Λ is interpretable as the cosmological "background" curvature (below).

The Einstein and the metrical tensor are covariantly constant, $\nabla\check{\mathcal{R}}_\bullet = 0$, $\nabla\mathbf{g} = 0$. Only for flat spacetime is the energy-momentum tensor the position density of the translations $\mathcal{P}^a = c\int d^3x\ \mathbf{T}^{0a}(x)$ with the energy-momenta as eigenvalues and the mass as invariant. There, particles can be defined as acted on by representations of the Poincaré group $\mathbf{SO}_0(1,3)\ \vec{\times}\ \mathbb{R}^4$.

In hindsight, Einstein's epochal and ingenious identification of the geometrical spacetime curvature with the gravitational interaction via the energy-momentum tensor, can also be motivated by the insight that already in classical point mechanics the Lagrangian involves, via the kinetic momentum term $\vec{p}^2 = \frac{d\vec{x}^2}{dt^2}$, i.e., the free mass point Lagrangian, the flat position metric $d\vec{x}^2$ (see Chapter 3) which has to be embedded into a spacetime metric with the tangent property $dx_0^2 - d\vec{x}^2$ for local flatness (special relativity). A stationary action involves the extremalization of the length of the mass point orbits, which is generalized to the geodesics on Riemannian spacetimes with their extremal length property.

Gravity is characterizable by an action W with the normalization (gravity coupling constant) given by the Planck area $\ell^2 = \frac{\kappa\hbar}{c} \sim (0.8 \times 10^{-34}\ \mathrm{m})^2$:

$$W^{\mathrm{grav}} = W_{\mathrm{grav}} + W_{\mathrm{matter}}^{\mathrm{grav}}, \quad W_{\mathrm{grav}} = \hbar\int \sqrt{|\mathbf{g}|}\ d^4x\ \tfrac{1}{2\ell^2}\mathcal{R}_\bullet^\bullet, \quad |\mathbf{g}| = -\det\mathbf{g}.$$

Here $\hbar \sim 1.05 \times 10^{-34}\ \frac{\mathrm{kg-m}^2}{\mathrm{s}}$ is used as dimensional action unit only. The action can be written with 2nd-order derivatives (Einstein–Hilbert) and 1st-order ones (Palatini).

The real four-dimensional spacetime manifold $\mathbb{M}^{(1,3)}$ is assumed with a causal structure, compatible with the orthochronous Lorentz group $\mathbf{SO}_0(1,3)$. It is visible by diagonalization[2] of the metrical hyperboloid to a local inertial system with an orthonormal standard form in the tangent spaces $\mathbf{T}_x(\mathbb{M}^{(1,3)}) \cong \mathbb{R}^4$. The existence of a special relativistic ("flat") local Lorentz structure formalizes the principle of equivalence, i.e., the existence of a local spacetime coordinate system without gravity interactions (Einstein's freely falling elevator). The local diagonalization of the metrical tensor uses a tetrad (4-bein) $\mathbf{e}$ with representatives of the 10-dimensional classes[3] $\mathbf{e}(x) \Subset \mathbf{GL}(4,\mathbb{R})/\mathbf{O}(1,3)$ of the six-dimensional Lorentz group[4] in the 16-dimensional general linear group;

$$\mathbf{g}^{li}(x) = \mathbf{e}^l_a(x)\eta^{ab}\mathbf{e}^i_b(x),\ \eta = \begin{pmatrix} 1 & 0 \\ 0 & -\mathbf{1}_3 \end{pmatrix},\ \mathbf{e}(x) = (\mathbf{e}^l_a(x))_{l,a=0,1,2,3},$$
$$\mathbf{GL}(4,\mathbb{R}) \ni \mathbf{e}(x) \cong \mathbf{e}(x)\Lambda(x), \text{with } \Lambda(x) \in \mathbf{O}(1,3).$$

Gravity is operationally implemented by the tetrad field. The tetrad operations involve, in classes of the local Lorentz transformations, the maximal compact group $\mathbf{SO}(4) \subset \mathbf{GL}(4,\mathbb{R})$ with two rotation groups $\mathbf{SO}(4) \sim \mathbf{SO}(3) \times \mathbf{SO}(3)$. The six angles of $\mathbf{SO}(4)$ describe the four-dimensional orientation of the metrical hyperboloid. $\mathbf{GL}(4,\mathbb{R})$ has real rank 4: In diagonal form, the 10 local operations of a tetrad representative are reduced to operations from a maximal noncompact abelian group $\mathbf{D}(1)^4 \subset \mathbf{GL}(4,\mathbb{R})$, i.e., to the *four gravity characteristic dilations*:

$$\mathbf{GL}(4,\mathbb{R}) \ni \mathbf{e}(x) = e^{\lambda_0(x)}O_4(x) \circ D_4(x),\ O_4(x) \in \mathbf{SO}(4),\ e^{\lambda_0(x)} \in \mathbf{D}(1),$$
$$D_4(x) = \begin{pmatrix} e^{-3\lambda} & 0 \\ 0 & e^{\lambda}D_3 \end{pmatrix}(x) \in \mathbf{SO}_0(1,1)^3,\ D_3(x) = \begin{pmatrix} e^{-2\lambda_2} & 0 & 0 \\ 0 & e^{\lambda_2-\lambda_3} & 0 \\ 0 & 0 & e^{\lambda_2+\lambda_3} \end{pmatrix}(x).$$

The four real continuous parameters determine the four lengths of the metrical spacetime hyperboloid, i.e., the four units for the time and position translations in each tangent space $\mathbf{T}_x(\mathbb{M})$. They can be arranged in one overall dilation,

$$e^{\lambda_0(x)} \in \mathbf{D}(1), \text{ with } e^{8\lambda_0(x)} = \det \mathbf{g}(x) \circ \eta = -\det \mathbf{g}(x) = (\det \mathbf{e}(x))^2,$$

one relative time-position normalization $\begin{pmatrix} e^{-3\lambda} & 0 \\ 0 & e^{\lambda}\mathbf{1}_3 \end{pmatrix}(x) \in \mathbf{SO}_0(1,\mathbf{1}_3)$, and two relative normalizations of the three position axes $D_3(x) \in \mathbf{SO}_0(1,\mathbf{1}_2) \times \mathbf{SO}_0(1,1)$ in the metrical position ellipsoid.

For a spacetime with $\mathbf{SO}(3)$-space rotation invariance, there remain three dilations, e.g., $\mathbf{g} \cong \begin{pmatrix} e^{2(\lambda+\lambda_3)} & 0 & 0 \\ 0 & -e^{2(\lambda-\lambda_3)} & 0 \\ 0 & 0 & -e^{2\lambda_2}\mathbf{1}_2 \end{pmatrix}$. For a Friedmann universe

---

[2] A real matrix is orthogonally diagonalizable, $O \circ M \circ O^T = \operatorname{diag} M$, $O \in \mathbf{SO}(n)$, if and only if it is symmetric $M = M^T$.

[3] The doubled symbol $g \Subset G/H$ denotes a representative of a coset (class of a subgroup $H \subseteq G$) $g \in gH \in G/H$.

[4] The indices $a, b, c, \ldots$ from the beginning of the alphabet are "Lorentz group-active"; the indices $i, j, k, \ldots$ from the middle of the alphabet are "$\mathbf{GL}(4,\mathbb{R})$-active."

with a maximally symmetric space (ahead), there remain two dilations, e.g., $\mathbf{g} \cong \begin{pmatrix} e^{2\lambda} & 0 \\ 0 & -e^{2\lambda_2}\mathbf{1}_3 \end{pmatrix}$, and only one dilation, $\lambda_2 = \lambda$, for maximally symmetric spacetimes (de Sitter universes).

1.2 Schwarzschild–Kruskal Spacetime

There exist four-dimensional spacetimes, which have a geometry with a nontrivial curvature and a trivial Ricci tensor. The homogeneous Einstein equations for a rotation-invariant metric of a rotation-invariant mass distribution like a mass point,

$$\mathbf{T}^{ab}(x) = \delta_0^a \delta_0^b m\delta(\vec{x}) \Rightarrow \mathbf{T}^{ab}(x) - \frac{\eta^{ab}}{2}\mathbf{T}^{\bullet}(x) = \frac{\delta^{ab}}{2} m\delta(\vec{x}),$$

and their solutions embed the nonrelativistic position equation and, for the mass point, the Newton potential $-\vec{\partial}^2 \frac{1}{r} = 4\pi\delta(\vec{x})$. It is given outside the Schwarzschild radius $2\ell_m = \frac{m\kappa}{4\pi} = 2\frac{mG}{c^2}$ of the mass inside by the *Schwarzschild metric* — in three different parametrizations, called *geodesic polar* $(t, \rho, \vec{\omega})$, *Cartesian* $(t, r, \vec{\omega})$, and *Eulerian* cooordinates $(t, \psi, \vec{\omega})$:

$$\mathcal{R}_{\bullet} = 0 : \quad \begin{cases} \mathbf{g} &= (1 - \frac{2\ell_m}{\rho})c^2 dt^2 - \left(\frac{d\rho^2}{1 - \frac{2\ell_m}{\rho}} + \rho^2 d\omega^2 \right) \\ &= \left(\frac{1 - \frac{\ell_m}{2r}}{1 + \frac{\ell_m}{2r}} \right)^2 c^2 dt^2 - \left(1 + \frac{\ell_m}{2r}\right)^4 d\vec{x}^2 \\ &= e^{-2\psi} c^2 dt^2 - \frac{e^{2\psi}}{\sinh^4 \psi} \ell_m^2 (d\psi^2 + \sinh^2 \psi d\omega^2), \\ d\omega^2 &= d\theta^2 + \sin^2\theta d\varphi^2, \quad d\vec{x}^2 = dr^2 + r^2 d\omega^2, \\ \text{for } \rho &= r\left(1 + \frac{\ell_m}{2r}\right)^2 = \ell_m \frac{e^{\psi}}{\sinh\psi} > 2\ell_m. \end{cases}$$

It is remarkable that the time-independent nonrelativistic Newton potential survives, in geodesic polar coordinates, in the general relativistic framework in the form of a matrix element of a self-dual dilation:

$$\rho > 2\ell_m : \quad \begin{aligned} e^{\lambda_3(\rho)\sigma^3} &= \begin{pmatrix} e^{\lambda_3} & 0 \\ 0 & e^{-\lambda_3} \end{pmatrix}(\rho) \in \mathbf{SO}_0(1,1), \\ e^{2\lambda_3(\rho)} &= \mathbf{g}_{tt}(\rho) = 1 - \frac{2\ell_m}{\rho} = V(\rho). \end{aligned}$$

In general, the curvature tensor $\mathcal{R}^{dabc} = \eta^{de}\mathcal{R}_e^{abc} = \mathcal{R}^{AB} = \mathcal{R}^{BA}$ with $\binom{4}{2} = 6$ antisymmetric double indices $A = da$, $B = bc$, i.e., $\mathcal{R}^{dabc} = -\mathcal{R}^{adbc}$, can be written as a symmetric (6×6)-matrix with $\binom{6+1}{2} - 1 = 20$ independent entries where one Biancchi identity condition has to be taken into account. It is an orthogonally diagonalizable bilinear form for the six linear independent spacetime "areas" or the six generators of the tangent Lorentz group $\mathbf{SO}_0(1,3)$, with the three "timelike" and three "spacelike curvatures" of two-dimensional submanifolds related to bilinear forms for $\mathbf{SO}_0(1,1)$ and $\mathbf{SO}(2)$, respectively. In contrast to the invariant Killing metric[5] of the tangent space

[5] $\wedge$ denotes total antisymmetrization, $\vee$ total symmetrization.

Lorentz Lie algebra $\eta \wedge \eta = -\begin{pmatrix} 1_3 & 0 \\ 0 & -1_3 \end{pmatrix}$, the curvature tensor is called an *operational metric* since it is related to the Killing metric of the Lie algebra of the spacetime-characterizing motion group (see Chapter 3).

In the example of the Schwarzschild metric, the bilinear curvature form[6] of the Lorentz Lie algebra is invariant under axial rotations $\mathbf{SO}(2)$:

$$\mathcal{R}^{dabc}(\rho) \cong \left(\begin{array}{c|cc||c|cc} \mathcal{R}^{0101} & & & & & \\ \hline & \mathcal{R}^{0202} & & & & \\ & & \mathcal{R}^{0303} & & & \\ \hline\hline & & & \mathcal{R}^{2323} & & \\ \hline & & & & \mathcal{R}^{1212} & \\ & & & & & \mathcal{R}^{1313} \end{array}\right)(\rho)$$

$$= \frac{\ell_m}{\rho^3}\left(\begin{array}{c|c||c|c} 2 & 0 & 0 & 0 \\ \hline 0 & -1_2 & 0 & 0 \\ \hline\hline 0 & 0 & -2 & 0 \\ \hline 0 & 0 & 0 & 1_2 \end{array}\right) = \frac{\ell_m}{\rho^3}\begin{pmatrix} 1 & 0 \\ 0 & -1 \end{pmatrix} \otimes \begin{pmatrix} 2 & 0 \\ 0 & -1_2 \end{pmatrix}.$$

Schwarzschild spacetime embeds as position manifold a rotation paraboloid $\mathbb{P}^3 = \mathbb{P}^1 \times \Omega^2$, where $\Omega^2 \cong \mathbf{SO}(3)/\mathbf{SO}(2)$ is the 2-sphere. The global symmetry group (isometry or motion group) is $\mathbb{R} \times \mathbf{SO}(3)$ with the axial rotations $\mathbf{SO}(2)$ as the local invariance group. All matrix elements in metric and curvature are representation coefficients of the global symmetry group, invariant under the local group.

In the curvature matrix, the six-dimensional adjoint Lorentz group representation is decomposed into two three-dimensional rotation group representations, each of which is decomposed into a one- and two-dimensional $\mathbf{SO}(2)$-representation;

$$\begin{array}{rcccl} D(\mathbf{SO}_0(1,3)) & = & \bigoplus_\iota D_\iota(\mathbf{SO}(3)) & = & \bigoplus_\kappa D_\kappa(\mathbf{SO}(2)), \\ 6 = & & 3 \oplus 3 & = & 1 \oplus 2 \oplus 1 \oplus 2. \end{array}$$

The metrical components $\mathbf{g}^{jk}$ can be connected with the Newton potential as the nonrelativistic approximation $\mathbf{g}^{00}(x) = 1 - \frac{2\ell_m}{\rho}$ and the connection coefficients Γ^{ij}_k with forces (see Chapter 3).

Manifolds have charts and coordinates which, in general, are useful only locally. The Schwarzschild coordinates $(t, \rho, \vec{\omega})$, where $\vec{\omega} \in \Omega^2$ are the coordinates of the unit 2-sphere, have a coordinate singularity[7] at the event horizon $\rho = 2\ell_m$, connected with the dilation $\mathbf{g}^{00}(x) = e^{2\lambda_3(\rho)} = 1 - \frac{2\ell_m}{\rho}$ for $\rho \geq 2\ell_m$. For a maximal extension of the coordinates for Schwarzschild spacetime, the geodesics of photons are used (null-like coordinates with $\mathbf{g} = 0$):

$$(1 - \tfrac{2\ell_m}{\rho})c^2 dt^2 - \frac{d\rho^2}{1 - \frac{2\ell_m}{\rho}} = 0, \quad \frac{d\rho}{dt} = \pm c(1 - \tfrac{2\ell_m}{\rho})$$
$$\Rightarrow \begin{cases} ct = +\rho + 2\ell_m \log|\frac{\rho}{2\ell_m} - 1| + \text{const.} & \text{(outgoing)}, \\ ct = -\rho - 2\ell_m \log|\frac{\rho}{2\ell_m} - 1| + \text{const.} & \text{(ingoing)}. \end{cases}$$

[6]The explicit derivation of the curvatures from the metrical tensors is given in Chapter 2.

[7]A simple example for a coordinate singularity is the radial parametrization of the 2-sphere $d\omega^2 = d\theta^2 + \sin^2\theta d\varphi^2 = \frac{d\rho^2}{1-\rho^2} + \rho^2 d\varphi^2$ with the equatorial singularity at $\rho^2 = \sin^2\theta = 1$.

Via the constants one transforms to either advanced or retarded *gton–Finkelstein coordinates*, related to each other by time reversal $t_+ \leftrightarrow -t_-$,

$$\begin{array}{ll}\text{advanced} & (t_+,\rho,\vec{\omega})\\ \text{retarded} & (t_-,\rho,\vec{\omega})\end{array} \left\{\begin{array}{rl} ct_\pm &= ct \pm 2\ell_m \log|\frac{\rho}{2\ell_m}-1|,\\ \mathbf{g} &= (1-\frac{2\ell_m}{\rho})c^2dt_\pm^2 \mp \frac{4\ell_m}{\rho}cdt_\pm d\rho - (1+\frac{2\ell_m}{\rho})d\rho^2\\ & - \rho^2 d\omega^2.\end{array}\right.$$

Taking both, Schwarzschild coordinates can be replaced by lightlike ones. In the metric, the "radial coordinate" ρ is defined implicitly:

$$\left.\begin{array}{rl} (ct,\rho) \longrightarrow (\xi_+,\xi_-): & \\ \xi_\pm &= ct \pm (\rho + 2\ell_m \log|\frac{\rho}{2\ell_m}-1|),\\ \frac{\xi_+ + \xi_-}{2} &= ct,\\ \frac{\xi_+ - \xi_-}{2} &= \rho + 2\ell_m \log|\frac{\rho}{2\ell_m}-1| \end{array}\right\} \Rightarrow \mathbf{g} = (1-\frac{2\ell_m}{\rho})d\xi_+ d\xi_- - \rho^2 d\omega^2.$$

A maximal extension involves two charts. The coordinate singularity at the event horizon $\rho = 2\ell_m$ is removed by exponentiation of the lightlike coordinates,

$$(\xi_+,\xi_-) \longmapsto e^{\pm\frac{\xi_\pm}{4\ell_m}} = v \pm u = e^{\frac{\rho \pm ct}{4\ell_m}} \left\{\begin{array}{ll} \sqrt{\frac{\rho}{2\ell_m}-1}, & \rho \geq 2\ell_m,\\ (\pm 1)\sqrt{1-\frac{\rho}{2\ell_m}}, & \rho \leq 2\ell_m,\end{array}\right.$$

leading to the *Kruskal coordinates* $(u,v,\vec{\omega})$ for Schwarzschild–Kruskal spacetime — now extended inside the horizon $\rho \leq 2\ell_m$;

$$\begin{array}{rl} (u,v) &= e^{\frac{\rho}{4\ell_m}} \left\{\begin{array}{ll} \sqrt{\frac{\rho}{2\ell_m}-1}(\sinh\frac{ct}{4\ell_m},\cosh\frac{ct}{4\ell_m}), & \rho \geq 2\ell_m,\\ \sqrt{1-\frac{\rho}{2\ell_m}}(\cosh\frac{ct}{4\ell_m},\sinh\frac{ct}{4\ell_m}), & \rho \leq 2\ell_m,\end{array}\right.\\ u^2 - v^2 &= (1-\frac{\rho}{2\ell_m})e^{\frac{\rho}{2\ell_m}}, \quad \frac{2uv}{u^2+v^2} = \tanh\frac{ct}{2\ell_m},\\ \mathbf{g} &= \frac{e^{-\frac{\rho}{2\ell_m}}}{\frac{\rho}{2\ell_m}}(du^2 - dv^2) - \rho^2 d\omega^2 = (1-\frac{2\ell_m}{\rho})c^2dt^2 - \frac{d\rho^2}{1-\frac{2\ell_m}{\rho}} - \rho^2 d\omega^2.\end{array}$$

The radial coordinate ρ is a function of $u^2 - v^2$. The metric involves a hyperbolic Macdonald function $k_0(X) = \frac{e^{-X}}{X}$ with $X = \frac{\rho}{2\ell_m}$ and the Schwarzschild length $\ell_m = m\frac{G}{c^2}$, proportional to the mass m, as familiar from the Yukawa potential, there with the Compton length, $X = \frac{r}{L_m}$, $L_m = \frac{1}{m}\frac{\hbar}{c}$, inversely proportional to the mass m.

With respect to a singularity in the dilation parametrization, there is an analogy of Kruskal and Schwarzschild coordinates $(u,v) \leftrightarrow (ct,\rho)$, on the one side, with, respectively, translation $\mathbb{R}\times\mathbb{R}_+$ and hyperbolic-dilation $\mathbf{SO}_0(1,1)\times\mathbf{D}(1)$-orbit coordinates $(ct_0,r_0) \leftrightarrow (\lambda_3,\xi_0)$ for flat spacetime, on the other side:

$$\begin{array}{rl} (ct_0,r_0) &= \left\{\begin{array}{ll} \sqrt{-2\xi_0}\,(\sinh\lambda_3,\cosh\lambda_3), & \xi_0 \leq 0 \text{ (spacelike)},\\ \sqrt{2\xi_0}\,(\cosh\lambda_3,\sinh\lambda_3), & \xi_0 \geq 0 \text{ (timelike)},\end{array}\right.\\ c^2t_0^2 - r_0^2 &= 2\xi_0, \quad \frac{2ct_0r_0}{c^2t_0^2+r_0^2} = \tanh 2\lambda_3,\\ \eta &= c^2dt_0^2 - dr_0^2 - r_0^2 d\omega^2 = 2\xi_0 d\lambda_3^2 - \frac{d\xi_0^2}{2\xi_0} - r_0^2 d\omega^2.\end{array}$$

r_0 is a function of λ_3 and ξ_0. The Schwarzschild analogue coordinates (λ_3, ξ_0) cannot be used on the lightcone $\xi_0 = 0$. An exponential dilation coordinate is possible either for timelike translations $e^{2\lambda_0} = 2\xi_0$ or for spacelike ones $e^{2\lambda_0} = -2\xi_0$.

The Kruskal coordinates $(u, v, \vec{\omega})$ allow a causal interpretation similar to flat Minkowski spacetime $(ct_0, r_0, \vec{\omega})$: The region with $u^2 - v^2 < 0$ is outside the horizon (like Minkowskian spacelike), the region with $u^2 = v^2$ (like lightlike) is the horizon, and the region with $u^2 - v^2 > 0$ (like timelike) is inside the horizon. With this spacelike-timelike analogy, it is understandable that inside the horizon $\rho \leq 2\ell_m$, the Schwarzschild–Kruskal metric is not static; i.e., it has a time dependence, in contrast to the static character outside $\rho \geq 2\ell_m$.

The Schwarzschild metric is a special case of the *Kerr metric* which are solutions of the homogeneous Einstein equations with only global $\mathbf{SO}(2)$-axial symmetry. In addition to the Schwarzschild length unit $\ell_m = m\frac{G}{c^2}$, they involve a second length unit $a = \frac{L}{mc}$, related to the angular momentum L. In this case, the tetrad contains compact operations from $\mathbf{SO}(4)$ in addition to noncompact dilations.

Other extensions of the Schwarzschild metric lead to the *Reissner metric* with a length $\ell_z^2 = z^2\alpha_S\frac{\hbar G}{c^3} = 8\pi z^2\alpha_S\ell^2$ proportional to an electromagnetic charge number $z \in \mathbb{Z}$ in a Coulomb potential $\frac{\alpha_S}{\rho}$ with Sommerfeld's fine structure constant $\alpha_S \sim \frac{1}{137}$ (see Chapter 3). All those metrics are special cases of *Kerr–Newman metrics* with a trivial curvature scalar that reflects the tracelessness $\mathbf{T}^\bullet(\mathbf{F}) = 0$ of the electromagnetic energy-momentum tensor $\frac{c}{\hbar}\mathbf{T}^{il}(\mathbf{F}) = \mathbf{g}_{jk}\mathbf{F}^{ij}\mathbf{F}^{kl} - \frac{1}{4}\mathbf{g}^{il}\mathbf{F}^{jk}\mathbf{F}_{kj}$:

$$\begin{array}{l} \mathbf{g} = (1 - \frac{2\ell_m L}{R^2})(cdt - a\sin^2\theta d\varphi)^2 - \frac{1}{1 - \frac{2\ell_m L}{R^2}}d\rho^2 \\ \quad - R^2 d\theta^2 - \frac{\sin^2\theta}{R^2}[(\rho^2 + a^2)d\varphi - acdt]^2 \end{array},$$

$$\text{with} \begin{cases} R^2 = \rho^2 + a^2\cos^2\theta, \\ 2\ell_m L = 2\ell_m\rho + a^2\sin^2\theta - \ell_z^2, \end{cases}$$

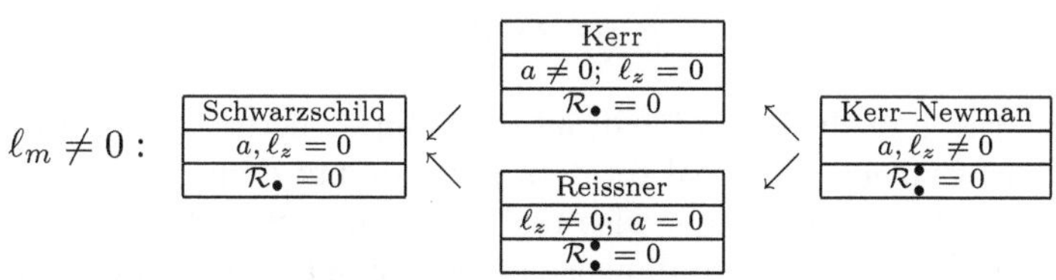

Globally $\mathbf{SO}(3)$-invariant Reissner spacetime with a trivial curvature scalar involves the dilation $e^{2\lambda_3(\rho)} = 1 - \frac{2\ell_m}{\rho} + \frac{\ell_z^2}{\rho^2}$ for $\frac{2\ell_m}{\rho} - \frac{\ell_z^2}{\rho^2} \leq 1$ with Newton's potential and the Coulomb potential related electromagnetic contribution. It has a nontrivial $\mathbf{SO}(2)$-invariant Ricci tensor with the $\mathbf{SO}(2)$-decomposition $4 = 1 \oplus 1 \oplus 2$:

$$\mathbf{g} = \left(1 - \frac{2\ell_m}{\rho} + \frac{\ell_z^2}{\rho^2}\right) c^2 dt^2 - \frac{d\rho^2}{1-\frac{2\ell_m}{\rho}+\frac{\ell_z^2}{\rho^2}} - \rho^2 d\omega^2,$$

$$\mathcal{R}^{dabc}(\rho) \cong \frac{1}{\rho^2}\left(\begin{array}{c|c|c|c} \frac{2\ell_m}{\rho} - \frac{3\ell_z^2}{\rho^2} & 0 & 0 & 0 \\ \hline 0 & -(\frac{\ell_m}{\rho} - \frac{\ell_z^2}{\rho^2})\mathbf{1}_2 & 0 & 0 \\ \hline 0 & 0 & -\frac{2\ell_m}{\rho} + \frac{\ell_z^2}{\rho^2} & 0 \\ \hline 0 & 0 & 0 & (\frac{\ell_m}{\rho} - \frac{\ell_z^2}{\rho^2})\mathbf{1}_2 \end{array}\right),$$

$$\mathcal{R}^{ab}_{\bullet}(\rho) \cong \frac{\ell_z^2}{\rho^4}\left(\begin{array}{c|c|c} -1 & 0 & 0 \\ \hline 0 & 1 & 0 \\ \hline 0 & 0 & -\mathbf{1}_2 \end{array}\right).$$

In general, the Ricci tensor (or the Einstein tensor) is a symmetric bilinear form of the tangent translations. In contrast to the "absolute" flat spacetime Lorentz metric $\eta^{ab} = \begin{pmatrix} 1 & 0 \\ 0 & -\mathbf{1}_3 \end{pmatrix}$, $\mathcal{R}^{ab}_{\bullet} = \mathcal{R}^{abc}_{c}$ is called an operational metric of spacetime, possibly degenerate. It is trivial for chargeless mass points, $\ell_z = 0$, i.e., for Schwarzschild and Kerr spacetime. The field equations identify, up to a normalization, the operational metric with the energy-momentum tensor. The curvature scalar $\mathcal{R}^{\bullet}_{\bullet} = \mathrm{tr}\,\mathcal{R} \circ (\eta \wedge \eta)^{-1}$ is the normalization of the operational Lie algebra metric.

1.3 Friedmann and de Sitter Universes

Robertson–Walker metrical tensors are used for cosmological models. The related *Friedmann universes* embed the three *maximally symmetric operational positions*, given by spherical, flat Euclidean and hyperbolic manifolds, which parametrize rotation group classes, with the time dependence given by the position expansion factor ("radius") $t \longmapsto R(t)$,

$$\begin{aligned} \mathbf{g} &= dt^2 - R^2(t) d\sigma_k^2, \\ d\sigma_k^2 &= \frac{d\rho^2}{1-k\rho^2} + \rho^2 d\omega^2 = \frac{d\vec{x}^2}{(1+k\frac{\vec{x}^2}{4})^2} = \begin{pmatrix} d\theta^2 \\ dr^2 \\ d\psi^2 \end{pmatrix} + \begin{pmatrix} \sin^2\theta \\ r^2 \\ \sinh^2\psi \end{pmatrix} d\omega^2, \\ \text{with} & \begin{cases} \Omega^3 \cong \mathbf{SO}(4)/\mathbf{SO}(3), & k = 1, \\ \mathbb{R}^3 \cong \mathbf{SO}(3) \vec{\times} \mathbb{R}^3/\mathbf{SO}(3), & k = 0, \\ \mathcal{Y}^3 \cong \mathbf{SO}_0(1,3)/\mathbf{SO}(3), & k = -1. \end{cases} \end{aligned}$$

With the introduction of half-integer spin structures after the Stern–Gerlach experiments, the real rotation and Lorentz groups can be replaced by their twofold cover groups, also real Lie groups, with their defining representations on complex two-dimensional vector spaces with Pauli and Weyl spinors, respectively. The maximally symmetric positions can be formulated as classes of the spin group $\mathbf{SU}(2)$, where the isotropy group for Ω^3 is the diagonal $\mathbf{SU}(2)$:

$$\begin{aligned} &\Omega^3 \cong [\mathbf{SU}(2) \times \mathbf{SU}(2)]/\mathbf{SU}(2), && \mathbf{SU}(2)/\{\pm\mathbf{1}_2\} \cong \mathbf{SO}(3), \\ &\mathbb{R}^3 \cong \mathbf{SU}(2) \vec{\times} \mathbb{R}^3/\mathbf{SU}(2), && \\ &\mathcal{Y}^3 \cong \mathbf{SL}(2,\mathbb{C})/\mathbf{SU}(2), && \mathbf{SL}(2,\mathbb{C})/\{\pm\mathbf{1}_2\} \cong \mathbf{SO}_0(1,3). \end{aligned}$$

The Lorentz group $\mathbf{SO}_0(1,3)$ is familiar as a special relativistic spacetime group, i.e., for the spacetime tangent translations $\mathbb{R}^{1+3}$. As the motion group of hyperbolic position $\mathcal{Y}^3$, it does not involve time.

The sphere and the hyperboloid constitute a compact–noncompact pair (see Chapter 2) as seen in the imaginary–real transition, which relates their global Lie algebras,

$$\log \mathbf{SO}(4) \cong (i\mathbb{R})^3 \ \oplus\ (i\mathbb{R})^3 \text{ and } \log \mathbf{SO}_0(1,3) \cong (i\mathbb{R})^3 \ \oplus\ \mathbb{R}^3.$$

The Euclidean position arises by "flattening," i.e., an Inönü-Wigner contraction, from the curved ones, where, for the hyperbolic position, the boosts are contracted to the position translations. This is mathematically isomorphic to the original Inönü–Wigner expansion of the Galilei velocity transformation into the Lorentz group boosts $\mathbf{SO}(3) \vec{\times} \mathbb{R}^3 \overset{0\longleftarrow\frac{1}{c}}{\longleftarrow} \mathbf{SO}_0(1,3)$ with the speed of light as the contraction–expansion parameter.

Friedmann universes have two characteristic lengths $\ell_{t,s}$, or masses, as intrinsic units for dimensionless coordinates, involving one characteristic time $t = \frac{t}{H_0}$, e.g., the Hubble time $\frac{1}{H_0} = \frac{1}{R}\frac{dR}{dt}|_{t=t_0} = \frac{d\log R}{dt}|_{t=t_0}$, $R(t) \sim R(t_0)e^{H_0(t-t_0)}$, with $H_0 \sim 4.2 \times 10^{17}$ s $\sim 1.4 \times 10^{10}$ yr and $\ell_t = H_0 c \sim 1.2 \times 10^{26}$ m, and a characteristic length for position, $\vec{x} = \frac{\vec{x}}{\ell_s}$, and position curvature $\frac{k}{\ell_s^2}$. They are invariants for the action of the motion groups of the spacetime manifold, which contains the direct group product of $\mathbb{R}$ (time operations) and, respectively, rotations $\mathbf{SO}(4)$, Euclidean operations $\mathbf{SO}(3) \vec{\times} \mathbb{R}^3$, and Lorentz group operations $\mathbf{SO}_0(1,3)$. The two lengths determine one universe-characterizing number $\frac{\ell_t^2}{\ell_s^2}$ (se Chapter 11).

As noncompact manifolds, not as symmetric spaces, hyperbolic and flat positions are isomorphic, $\mathcal{Y}^3 \cong \mathbb{R}^3$. The negative curvature $-k = \epsilon^2$ is the invariant $\partial^2 a = \epsilon^2 a$ for representation coefficients a of abelian position subgroups, of compact $\mathbf{SO}(2) = \Omega^1$ with $(k,a) = (1, \sin\theta)$, $\epsilon = \pm i$, of noncompact $\mathbb{R}$ with $(k,a) = (0,r)$, and of noncompact $\mathbf{SO}_0(1,1) = \mathcal{Y}^1$ with $(k,a) = (-1, \sinh\psi)$, $\epsilon = \pm 1$.

Cosmological models in general relativity can be seen as the analogue to ground states with cyclic vectors in quantum theory (see Chapter 8). They are characterized by a global motion group with a local invariance subgroup $H \subseteq G$; e.g., $\mathbf{SO}(3) \subset \mathbf{SO}_0(1,3)$ for a hyperbolic Friedmann universe.

The local $\mathbf{SO}(3)$-invariance of Friedmann universes shows up, via $\mathbf{1}_3$, in the curvature with time dependence $\frac{\ddot{R}}{R}$ for timelike areas and $\frac{\dot{R}^2+k}{R^2}$ for spacelike ones, and in the Ricci tensor as symmetric bilinear forms (operational metrics) of, respectively, tangent Lorentz Lie algebra and translations, invariant under the local group, a Lorentz group subgroup,

$$\mathcal{R}^{dabc} \cong -\frac{1}{R^2}\left(\begin{array}{c|c} \ddot{R}R\mathbf{1}_3 & 0 \\ \hline 0 & -(\dot{R}^2 + k)\mathbf{1}_3 \end{array}\right),$$
$$\mathcal{R}^{ab}_{\bullet} \cong \frac{1}{R^2}\left(\begin{array}{c|c} 3\ddot{R}R & 0 \\ \hline 0 & -(\ddot{R}R + 2\dot{R}^2 + 2k)\mathbf{1}_3 \end{array}\right), \quad \frac{1}{2}\mathcal{R}^{\bullet}_{\bullet} = 3\frac{\ddot{R}R+\dot{R}^2+k}{R^2}.$$

With the energy-momentum tensor of an ideal cosmological fluid, characterized by a density and pressure, both only time-dependent $t \longmapsto (\rho(t), \mathbf{p}(t))$, and a velocity field $u^i u^l \mathbf{g}_{il} = 1$ defining two projectors for time and position,

$$\begin{aligned} \mathbf{T}^{li} &= (\rho + \mathbf{p}) u^l u^i - \mathbf{p}\mathbf{g}^{li} = \rho P_0^{li} - \mathbf{p} P_1^{li}, \\ \mathbf{T}^{ab} &\cong \left(\begin{array}{c|c} \rho & 0 \\ \hline 0 & \mathbf{p}\mathbf{1}_3 \end{array}\right), \end{aligned} \quad \text{with} \left\{ \begin{aligned} P_0^{li} &= u^l u^i = \mathbf{g}^{li} - P_1^{li}, \\ \mathbf{T}^{\bullet} &= \rho - 3\mathbf{p}, \end{aligned} \right.$$

the time dependence of the position "radius" $t \longmapsto R(t)$ of a Robertson–Walker metric is determined by the Einstein–Friedmann equations

$$\begin{aligned} -\mathcal{R}^{ab}_{\bullet} + \frac{\eta^{ab}}{2}\mathcal{R}^{\bullet}_{\bullet} &= \kappa \mathbf{T}^{ab} + \Lambda \eta^{ab} \\ \Rightarrow \left(\begin{array}{c|c} 3\frac{\dot{R}^2 + k}{R^2} & 0 \\ \hline 0 & -\frac{2\ddot{R}R + \dot{R}^2 + k}{R^2}\mathbf{1}_3 \end{array}\right) &= \left(\begin{array}{c|c} \kappa\rho + \Lambda & 0 \\ \hline 0 & (\kappa\mathbf{p} - \Lambda)\mathbf{1}_3 \end{array}\right) \\ \Rightarrow \left\{ \begin{aligned} \frac{\ddot{R}}{R} &= -\kappa\frac{\rho + 3\mathbf{p}}{6} + \frac{\Lambda}{3}, \\ \frac{\dot{R}^2 + k}{R^2} &= \kappa\frac{\rho}{3} + \frac{\Lambda}{3}. \end{aligned} \right. \end{aligned}$$

Solar system experiments give the bounds $|\Lambda| < \frac{10^{-46}}{\text{m}^2}$ for the constant cosmological "background" curvature, $\ell_s = \frac{1}{\sqrt{|\Lambda|}} > 10^{23}$ m for the length, and $\frac{|\Lambda|}{\kappa} < 5.3 \times 10^{-19}\ \frac{\text{kg}}{\text{m}^3}$ (proton mass $m_p \sim 1.67 \times 10^{-27}$ kg) for the mass density.

For a static solution, i.e., for a trivial representation of the time translations and trivial pressure,

$$\mathbf{g} = dt^2 - R^2 d\sigma_k^2, \quad R^2 = \text{const.}, \ \mathbf{p} = 0 \Rightarrow \frac{k}{R^2} = \frac{\kappa\rho}{2} = \Lambda,$$

the ground-state density coincides, up to a factor, with the cosmological constant $\rho = \frac{2\Lambda}{\kappa}$. A static Friedmann universe is either Minkowskian, i.e., it has trivial density and cosmological constant,

$$\mathbb{R}^4: \quad k = 0 \Rightarrow \rho = 0, \ \Lambda = 0,$$

or its position is a 3-sphere with volume $\frac{2\pi^2}{\sqrt{\Lambda^3}}$. This is the *Einstein universe* with global $\mathbb{R} \times \mathbf{SO}(4)$ and local $\mathbf{SO}(3)$-invariance, i.e., the four-dimensional cylinder, $\mathbb{R} \times \Omega^s$ for $s = 3$,

$$k = 1 \Rightarrow \frac{\kappa\rho}{2} = \Lambda = \frac{1}{R^2} > 0, \quad \mathbf{T}^{ab} = \frac{2\Lambda}{\kappa}\delta_0^a\delta_0^b,$$

$$\mathbb{R} \times \Omega^3: \quad \mathbf{g} = dt^2 - \frac{1}{\Lambda}\left(\frac{d\rho^2}{1-\rho^2} + \rho^2 d\omega_2^2\right) = dt^2 - \frac{1}{\Lambda}d\omega_3^2,$$

$$\mathcal{R}^{dabc} \cong -\Lambda\left(\begin{array}{c|c} \mathbf{0}_3 & 0 \\ \hline 0 & -\mathbf{1}_3 \end{array}\right), \quad \mathcal{R}^{ab}_{\bullet} \cong 2\Lambda\left(\begin{array}{c|c} 0 & 0 \\ \hline 0 & -\mathbf{1}_3 \end{array}\right), \quad \frac{1}{2}\mathcal{R}^{\bullet}_{\bullet} = 3\Lambda.$$

The mass of Einstein's universe is $\frac{4\pi^2}{\kappa\sqrt{\Lambda}} > 2 \times 10^{52}$ kg. Its curvature embeds the Killing form of the local invariance Lie algebra $\log \mathbf{SO}(3)$; the Ricci form embeds the Euclidean metric of the position translations. As a manifold, the

3-sphere is isomorphic to the spin group $\Omega^3 \cong \mathbf{SU}(2)$, not as symmetric space, where Ω^3 has the action group $\mathbf{SO}(4)$. The cosmological term, which in its scientific history oscillated between life and death in its presumed physical relevance, was first used by Einstein in his *Eselei* (folly), which proposed, for the first time, with the 3-sphere Ω^3 a finite compact position without boundary.

Einstein's universe is unstable. For a nonflat stable Friedmann universe, the representation of time translations has to be nontrivial.

A curved space can be embedded into a curvature-free spacetime: This is possible for a Friedmann universe with hyperbolic space where the "radius" has to have a linear time dependence $R(t) = R_0 \pm t$:

$$\begin{aligned} \mathcal{R} = 0 \iff \quad & \dot{R}^2 = -k = \epsilon^2 \Rightarrow R(t) = R_0 \pm \epsilon t, \\ & k = -1: \quad \rho = -\mathbf{p} = -\tfrac{\Lambda}{\kappa}. \end{aligned}$$

A curvature-free Friedmann universe with spherical space is not possible since it would need a complex formulation $R(t) = R_0 \pm it$.

The Killing form $\eta \wedge \eta$ of the Lorentz Lie algebra as curvature and the Lorentz metric η of the tangent translations as Ricci tensor, both up to a factor,

$$\mathcal{R}^{dabc} \cong -\tfrac{\ddot{R}}{R}\begin{pmatrix} \mathbf{1}_3 & 0 \\ 0 & -\mathbf{1}_3 \end{pmatrix}, \quad \mathcal{R}_\bullet^{ab} \cong 3\tfrac{\ddot{R}}{R}\begin{pmatrix} 1 & 0 \\ 0 & -\mathbf{1}_3 \end{pmatrix}, \quad \tfrac{1}{2}\mathcal{R}_\bullet^\bullet = 6\tfrac{\ddot{R}}{R},$$

arise for the hyperbolic, trigonometric, and linear time dependence of the scale factor with integration constants m_Λ and c_0:

$$\begin{aligned} \ddot{R}R &= \dot{R}^2 + k \Rightarrow \\ R(t) &= \begin{cases} \frac{\sinh(m_\Lambda t + c_0)}{m_\Lambda}, \frac{\sin(m_\Lambda t + c_0)}{m_\Lambda}, \pm t + c_0, & k = -1, \\ \frac{\cosh(m_\Lambda t + c_0)}{m_\Lambda}, & k = 1, \\ c_0 e^{m_\Lambda t}, & k = 0, \end{cases} \quad \frac{\ddot{R}}{R} = \begin{cases} m_\Lambda^2, -m_\Lambda^2, 0, \\ m_\Lambda^2, \\ m_\Lambda^2. \end{cases} \end{aligned}$$

The constant $\frac{\ddot{R}}{R}$ is the invariant for the time representation.

These operational metrics characterize the three *maximally symmetric universes* (de Sitter, Minkowski, anti-de Sitter) with a 10-dimensional global symmetry and a local six-dimensional Lorentz group invariance. They are the solutions of the Einstein–Friedmann equations for a constant energy-momentum tensor and negative pressure,

$$\mathbf{p} = -\rho \leq 0 \Rightarrow \tfrac{\ddot{R}}{R} = \tfrac{\dot{R}^2 + k}{R^2} = \tfrac{\kappa\rho + \Lambda}{3} = I m_\Lambda^2 = \text{const.}, \quad I \in \{1, -1\},$$

e.g., without matter $(\rho, \mathbf{p}) = 0$ and a possibly nontrivial ground-state energy-momentum with $m_\Lambda^2 = \frac{|\Lambda|}{3}$.

The metric of a *de Sitter universe* with invariant $\frac{\ddot{R}}{R} > 0$ for a hyperbolic time representation $\mathbb{R} \longrightarrow \mathbf{SO}_0(1,1)$ by the position "radius" arises in three parametrizations for $k = -1, 1, 0$, reflecting the possible position submanifolds $\mathcal{Y}^3$, Ω^3, and $\mathbb{R}^3$, related to the $\mathbf{SO}_0(1,4)$-subgroups $\mathbf{SO}_0(1,3)$, $\mathbf{SO}(4)$, and $\mathbf{SO}(3) \vec{\times} \mathbb{R}^3$, respectively, that of an *anti-de Sitter universe* with invariant $\frac{\ddot{R}}{R} < 0$ for a spherical time representation $\mathbb{R} \longrightarrow \mathbf{SO}(2)$ by the position "radius" in one parametrization for $k = -1$ with position $\mathcal{Y}^3$, and that of a *flat Minkowski spacetime* with trivial invariant $\frac{\ddot{R}}{R} = 0$ for a linear time representation $\mathbb{R} \longrightarrow \mathbb{R}$ by the position "radius" in one parametrization for $k = -1$. The corresponding global groups are the de Sitter group $\mathbf{SO}_0(1,4)$, the anti-de Sitter group $\mathbf{SO}_0(2,3)$, and the Poincaré group $\mathbf{SO}_0(1,3) \vec{\times} \mathbb{R}^4$. All these universes have a Lorentz group as the local invariance group, i.e., they are constituted by Lorentz group classes. The "radius" grows or shrinks exponentially, stays constant, or oscillates, respectively, with time:

$$\begin{array}{rrll} \text{de Sitter}, & Im_\Lambda^2 > 0: & R(t) \sim e^{\pm m_\Lambda t}, & \mathbf{SO}_0(1,4)/\mathbf{SO}_0(1,3) \cong \mathcal{Y}^{(1,3)}, \\ \text{Minkowski}, & m_\Lambda^2 = 0: & R(t) \sim 1, & \mathbf{SO}_0(1,3) \vec{\times} \mathbb{R}^4/\mathbf{SO}_0(1,3) \cong \mathbb{R}^4, \\ \text{anti-de Sitter}, & Im_\Lambda^2 < 0: & R(t) \sim e^{\pm i m_\Lambda t}, & \mathbf{SO}_0(2,3)/\mathbf{SO}_0(1,3) \cong \mathcal{Y}^{(3,1)}. \end{array}$$

For $s \geq 1$, both the de Sitter and anti-de Sitter universes are a one-shell hyperboloid, spacelike $\mathbf{SO}_0(1,1+s)/\mathbf{SO}_0(1,s) \cong \mathcal{Y}^{(1,s)}$, and timelike $\mathbf{SO}_0(2,s)/\mathbf{SO}_0(1,s) \cong \mathcal{Y}^{(s,1)}$. As homogeneous spaces, they are isomorphic for $s = 1$ and different for $s \geq 2$. For $s = 0$, the de Sitter and anti-de Sitter universes are hyperbolic time $\mathbf{SO}_0(1,1) = \mathcal{Y}^1$ and cyclic time $\mathbf{SO}(2) = \Omega^1$, respectively.

The three maximally symmetric universes are related to each other by Inönü–Wigner contraction with the time representation invariant m_Λ^2 as the contraction parameter:

$$m_\Lambda^2 \to 0: \quad \left\{ \begin{array}{ccccc} \frac{\sinh m_\Lambda t}{m_\Lambda} & \to & t & \leftarrow & \frac{\sin m_\Lambda t}{m_\Lambda}, \\ \mathbf{SO}_0(1,4) & \to & \mathbf{SO}_0(1,3) \vec{\times} \mathbb{R}^4 & \leftarrow & \mathbf{SO}_0(2,3). \end{array} \right.$$

In the (anti-)de Sitter spacetime, the translations of Minkowski spacetime are expanded into a part of a simple group.

$m_\Lambda^2 = \frac{|\Lambda + \kappa\rho|}{3}$ as intrinsic unit gives the normalization of the Ricci and the curvature tensor, which, for maximal symmetry, coincide, up to a factor, with the metrical tensor and its antisymmetric product, i.e., in orthogonal coordinates with the Lorentz metric of Minkowski spacetime and the Killing form of the Lorentz Lie algebra $\log \mathbf{SO}_0(1,3)$. The maximally symmetric de Sitter and anti-de Sitter spacetimes are conformal (dilation-equivalent) to flat Minkowski spacetime:

$$\begin{aligned} \mathcal{R} &= -Im_\Lambda^2 \mathbf{g} \wedge \mathbf{g} \cong -Im_\Lambda^2 \begin{pmatrix} \mathbf{1}_3 & 0 \\ 0 & -\mathbf{1}_3 \end{pmatrix}, \quad I \in \{1,-1\}, \\ \mathcal{R}_\bullet &= \quad 3Im_\Lambda^2 \mathbf{g} \quad \cong 3Im_\Lambda^2 \begin{pmatrix} 1 & 0 \\ 0 & -\mathbf{1}_3 \end{pmatrix}, \quad \tfrac{1}{2}\mathcal{R}_\bullet^\bullet = 6Im_\Lambda^2, \\ \mathbf{g} &= \frac{dx^2}{(m_\Lambda^2 x^2 - I)^2}, \text{ with } x^2 = x_0^2 - \vec{x}^2. \end{aligned}$$

Chapter 2

Riemannian Manifolds

In this chapter, general structures of differential manifolds are given, especially the operational aspects of Riemannian manifolds. Manifolds as such are rather amorphous structures. Physically important and manifold-characterizing are the operation groups, they are parametrizing. A Riemannian manifold has a global and a local invariance group in addition to its tangent Poincaré or Euclidean group; e.g., for the sphere Ω^2 and the hyperboloid $\mathcal{Y}^2$ given by the rotations $\mathbf{SO}(3)$ and Lorentz transformations $\mathbf{SO}_0(1,2)$, respectively, as the global groups (motion groups), and, for both, the axial rotations $\mathbf{SO}(2)$ as the local group and the Euclidean group $\mathbf{SO}(2) \vec{\times} \mathbb{R}^2$ as the tangent group. The Einstein cosmos $\mathbb{R} \times \Omega^3$ has $\mathbb{R} \times \mathbf{SO}(4)$ as the global group, $\mathbf{SO}(3)$ as the local group, and $\mathbf{SO}(1,3) \vec{\times} \mathbb{R}^4$ as the tangent Poincaré group.

After a discussion of Riemannian manifolds with maximal symmetries and constant curvature, i.e., spheres, flat spaces and timelike hyperboloids, the relationship between coset spaces of real simple Lie groups and manifolds with a covariantly constant curvature as classified by Cartan is presented.

2.1 Differentiable Manifolds

General relativistic spacetime is assumed to be a differentiable manifold. Real manifolds $\mathbb{M} \in \underline{\mathbf{dif}}_{\mathbb{R}}$, assumed in the following as finite-dimensional, connected, and "as smooth as necessary," "look locally" like open subsets of $\mathbb{R}^n$ — as topological spaces. They describe structures used for Lie groups $\underline{\mathbf{lgrp}}_{\mathbb{R}} \subset \underline{\mathbf{dif}}_{\mathbb{R}}$ and their tangent Lie algebras $\underline{\mathbf{lgrp}}_{\mathbb{R}} \longrightarrow \underline{\mathbf{lag}}_{\mathbb{R}}$ (ahead).

The most important manifold structure is the unital commutative *ring with the manifold functions* $\{f : \mathbb{M} \longrightarrow \mathbb{R}\}$,

$$\mathcal{C}(\mathbb{M}) = \bigcup_{x \in \mathbb{M}} (x, \mathcal{C}_x(\mathbb{M})), \quad \mathcal{C}_x(\mathbb{M}) \cong \mathbb{R},$$

H. Saller, *Operational Spacetime: Interactions and Particles*,
Fundamental Theories of Physics 163, DOI 10.1007/978-1-4419-0898-8_3,

involving the manifold parametrizations by charts. There are many vector spaces with manifold functions and distributions, which will be used in later chapters. The functions $\mathcal{C}(\mathbb{M})$ will be assumed "as smooth as necesssary."

With respect to manifold structures, the important difference between $\mathbb{R}$-linear with the real field $\mathbb{R}$ and $\mathcal{C}(\mathbb{M})$-linear with the ring $\mathcal{C}(\mathbb{M})$ has to be taken into account. $\mathbb{R}$-linear structures that are not $\mathcal{C}(\mathbb{M})$-linear are called *gauge dependent* (more ahead).

Differentiable manifolds have "infinitesimal structures", i.e., *tangent structures.* The $\mathbb{R}$-linear derivations of the manifold functions from the unital $\mathbb{R}$-algebra $\mathcal{C}(\mathbb{M})$,

$$\mathcal{C}(\mathbb{M}) \in \underline{\mathbf{aag}}_{\mathbb{R}} : \quad v : \mathcal{C}(\mathbb{M}) \longrightarrow \mathcal{C}(\mathbb{M}), \quad v(fg) = v(f)g + fv(g),$$

are the vector fields of the manifold, denoted in short by $\mathbf{T} = \mathbf{T}(\mathbb{M})$ for tangent bundle (see Chapter 6). They constitute an $\mathbb{R}$-Lie algebra via the commutator. $\mathbf{T}$ is a module over the ring of functions $\mathcal{C}(\mathbb{M})$, patched together by the local tangent spaces, which are $\mathbb{R}$-Lie algebras $\underline{\mathbf{lag}}_{\mathbb{R}} \ni \mathbf{T}_x(\mathbb{M}) \cong \mathbb{R}^n$:

$$\begin{aligned} &\mathbf{T} = \mathbf{T}(\mathbb{M}) = \operatorname{der} \mathcal{C}(\mathbb{M}) = \bigcup_{x \in \mathbb{M}} (x, \mathbf{T}_x(\mathbb{M})) \in \underline{\mathbf{mod}}_{\mathcal{C}(\mathbb{M})}, \\ &\mathbf{T} \in \underline{\mathbf{lag}}_{\mathbb{R}} : \quad \mathbf{T} \wedge \mathbf{T} \longrightarrow \mathbf{T}, \quad [e^i, e^j] = \epsilon_k^{ij} e^k. \end{aligned}$$

In general bases, e.g., left-invariant vector fields for nonabelian Lie groups, the Lie bracket must not be trivial, as in a locally always possible "translation" basis $[\partial^j, \partial^k] = 0$. The Lie algebra structure constants ϵ_k^{ij} in anholonomic bases are called *anholonomy coefficients.*

The manifold morphisms $\mathbf{dif}_{\mathbb{R}}(\mathbb{M}_1, \mathbb{M}_2) = \{\varphi : \mathbb{M}_1 \longrightarrow \mathbb{M}_2\}$ are constituted by the differentiable mappings. In general, the diffeomorphism group of a manifold $\overset{\text{o}}{\mathbf{dif}}_{\mathbb{R}}(\mathbb{M}, \mathbb{M})$, involving the reparametrizations, is too big to form a Lie group in any reasonable topology. Manifold morphisms induce ring morphisms and linear mappings of the tangent spaces:

$$\begin{array}{ccccccc} \mathbb{M}_1 & & \mathcal{C}(\mathbb{M}_1) & & \mathbf{T}(\mathbb{M}_1) & & \mathbf{T}_x(\mathbb{M}_1) \\ \downarrow \varphi & \longmapsto & \uparrow \circ\varphi & \longmapsto & \downarrow \varphi_* & \text{, with} & \downarrow \\ \mathbb{M}_2 & & \mathcal{C}(\mathbb{M}_2) & & \mathbf{T}(\mathbb{M}_2) & & \mathbf{T}_{\varphi(x)}(\mathbb{M}_2) \end{array}, \quad \begin{aligned} \partial^j|_x &\longmapsto \varphi_{*a}^j(x) \partial^a|_{\varphi(x)}, \\ \varphi_{*a}^j(x) &= \frac{\partial \varphi_a(x)}{\partial x_j}. \end{aligned}$$

Diffeomorphisms give automorphisms $\mathbf{T}_x(\mathbb{M}) \overset{\varphi_*}{\leftrightarrows} \mathbf{T}_{\varphi(x)}(\mathbb{M})$ of the tangent spaces, with $\varphi_*(x) \in \mathbf{GL}(n, \mathbb{R})$.

The tangent spaces of a direct product manifold are isomorphic to the direct sum of the individual tangent spaces $\mathbf{T}_{(x_1,x_2)}(\mathbb{M}_1 \times \mathbb{M}_2) \cong \mathbf{T}_{x_1}(\mathbb{M}_1) \oplus \mathbf{T}_{x_2}(\mathbb{M}_2)$.

Each point has a neighborhood $U \ni x$ where the vector fields $\mathbf{T}(U)$ constitute a free $\mathcal{C}(U)$-module of dimension $n = \dim_{\mathbb{R}} \mathbb{M} = \dim_{\mathcal{C}(U)} \mathbf{T}(U)$. There exist *local dual bases*, (*n-beins, moving frames, repères mobiles*), for vector fields $v = v_j e^j \in \mathbf{T}$ and its $\mathcal{C}(\mathbb{M})$-linear forms $\omega = \omega^j \check{e}_j \in \mathbf{T}^T$ with a local decomposition of the identity

$$\mathrm{id}_{\mathbf{T}(U)} = e^j \otimes \check{e}_j, \quad \langle \check{e}_i, e^j \rangle = \delta_i^j.$$

Two frames with dual bases $\mathrm{id}_{\mathbf{T}(U)} = e^j \otimes \check{e}_j = e^a \otimes \check{e}_a$ with $j, a = 1, \ldots, n$, are related to each other by automorphisms,

$$\begin{aligned}
e^a(x) &= e_j^a(x) e^j(x), & e_j^a(x) &= \langle \check{e}_j(x), e^a(x) \rangle, \\
\check{e}_a(x) &= e_a^j(x) \check{e}_j(x), & e_a^j(x) &= \langle \check{e}_a(x), e^j(x) \rangle, \\
e_j^a(x) e_a^i(x) &= \delta_j^i, & e_j^a(x) e_b^j(x) &= \delta_b^a,
\end{aligned}$$
$$e(x) = \begin{pmatrix} e_1^1(x) & \ldots & e_1^n(x) \\ & \ldots & \\ e_n^1(x) & \ldots & e_n^n(x) \end{pmatrix} \in \mathbf{GL}(n, \mathbb{R}).$$

An example is the different *holonomic bases* for different parametrizations:

$$\mathrm{id}_{\mathbf{T}(U)} = \frac{\partial}{\partial x_j} \otimes dx_j = \frac{\partial}{\partial \overline{x}_a} \otimes d\overline{x}_a, \quad \begin{cases} \langle dx_i, \partial^j \rangle = \delta_i^j, \\ \langle d\overline{x}_a, \overline{\partial}^b \rangle = \delta_a^b, \\ \langle d\overline{x}_a, \frac{\partial}{\partial x_j} \rangle = \frac{\partial \overline{x}_a(x)}{\partial x_j} = e_a^j(x). \end{cases}$$

A *parallelizable* manifold even has a global frame, and the tangent bundle is a free module, $\mathbf{T}(\mathbb{M}) \cong \mathcal{C}(\mathbb{M})^n$, not only locally. Vector spaces are parallelizable manifolds. Parallelizable spheres are exactly[1] Ω^1, Ω^3, and Ω^7.

Co- and contravariant tensor fields constitute the $\mathcal{C}(\mathbb{M})$-linear tensor algebra $\bigotimes(\mathbf{T} \oplus \mathbf{T}^T)$. The highest Grassmann power $\bigwedge^n \mathbf{T}^T \cong \mathcal{C}(\mathbb{M})$ contains the volume elements, invariant for the action of $\mathbf{SL}(n, \mathbb{R}) \vec{\times} \mathbb{R}^n$ and related to each other by a dilation factor, e.g., in holonomic bases,

$$\bigwedge^n \mathbf{T}^T \ni \check{e}_1 \wedge \cdots \wedge \check{e}_n = |e| \, d^n x, \text{ with } \check{e}_a = e_a^j dx_j, \quad |e| = \det e_a^j.$$

More generally, a volume element leads to isomorphisms between tangent and cotangent products of equal dimension; for $k = 0, \ldots, n$,

$$\epsilon : \bigwedge^k \mathbf{T}^T \longrightarrow \bigwedge^{n-k} \mathbf{T}, \quad \check{e}_{l_1} \wedge \cdots \wedge \check{e}_{l_k} \longmapsto \epsilon_{l_1 \ldots l_k l_{k+1} \ldots l_n} e^{l_{k+1}} \wedge \cdots \wedge e^{l_n}.$$

The action of the vector fields on the ring of scalars and on themselves is uniquely extended by Leibniz's rule as $\mathbb{R}$-linear *Lie derivations* of all co- and contravariant tensor fields, denoted by $\mathcal{L}_v a = \mathrm{ad}\, v(a) = [v, a]$:

$$v \in \mathbf{T} : \quad \bigotimes(\mathbf{T} \oplus \mathbf{T}^T) \longrightarrow \bigotimes(\mathbf{T} \oplus \mathbf{T}^T),$$
$$\text{with} \quad \begin{cases} [v, f] = v(f), \ f \in \mathcal{C}(\mathbb{M}), \\ [v, \langle \omega, w \rangle] = \langle [v, \omega], w \rangle + \langle \omega, [v, w] \rangle, \ \omega \in \mathbf{T}^T, \\ [v, a_1 \otimes a_2] = [v, a_1] \otimes a_2 + a_1 \otimes [v, a_2], \\ \text{e.g., } [v, fw] = v(f) w + f[v, w]. \end{cases}$$

[1]These three spheres $\mathbf{SO}(1+s)/\mathbf{SO}(s) \cong \Omega^s \subset \mathbb{R}^{1+s}$ are associated with the three division $\mathbb{R}$-algebras for $1+s = 2^n$, $n = 1, 2, 3$, which exist in addition to $\mathbb{R}$ with the "sphere" $\Omega^0 = \{\pm 1\}$: the complex numbers $\mathbb{C} \cong \mathbb{R}^2 \supset \Omega^1$, Hamilton's nonabelian quaternions $\mathbb{H} \cong \mathbb{R}^4 \supset \Omega^3$, and Cayley's nonassociative and nonabelian octonions $\mathbb{O} \cong \mathbb{R}^8 \supset \Omega^7$.

Lie derivation is compatible with tensor grading, deg $\bigotimes^k \mathbf{T} \bigotimes^l \mathbf{T}^T = k - l$. One has the explicit expressions in a local chart:

$$\begin{aligned} v_j\partial^j \in \mathbf{T}: \quad & [v_j\partial^j, \partial^k] = -(\partial^k v_j)\partial^j, \\ & [v_j\partial^j, \langle dx_l, \partial^k\rangle] = 0 = \langle [v_j\partial^j, dx_l], \partial^k\rangle + \langle dx_l, -(\partial^k v_j)\partial^j\rangle \\ \Rightarrow \quad & \langle [v_j\partial^j, dx_l], \partial^k\rangle = \partial^k v_l, \quad [v_j\partial^j, dx_l] = (\partial^j v_l)dx_j, \\ \text{e.g.,} \quad & [\partial^j, \partial^k] = 0, \quad [\partial^j, dx_l] = 0. \end{aligned}$$

A tensor field a is invariant under Lie derivation for $[v, a] = 0$. This leads to the *invariance Lie algebra of a tensor field*, a Lie subalgebra of $\mathbf{T}$:

$$L_a = \{v \in \mathbf{T} \mid [v, a] = 0\} \in \underline{\mathbf{lag}}_{\mathbb{R}}.$$

By the definitions above, a local dual frame is invariant under all Lie derivations $[v, e^j \otimes \check{e}_j] = 0$.

For a Lie subalgebra $L \subseteq \mathbf{T}$, the unital *algebra with its invariants* is a subalgebra of the full tensor algebra:

$$\mathrm{INV}_L \bigotimes(\mathbf{T} \oplus \mathbf{T}^T) = \{a \in \bigotimes(\mathbf{T} \oplus \mathbf{T}^T) \mid [L, a] = \{0\}\} \in \underline{\mathbf{aag}}_{\mathbb{R}}.$$

2.1.1 External Derivative

The *external derivative* from functions to 1-forms,

$$\begin{aligned} d: \mathcal{C}(\mathbb{M}) \longrightarrow \mathbf{T}^T(\mathbb{M}), \quad f \longmapsto df \text{ with } \langle df, v\rangle &= v(f), \quad v \in \mathbf{T}(\mathbb{M}), \\ \text{holonomic bases: } df &= (\partial^j f)dx_j, \end{aligned}$$

acts via nilquadratic extension by Leibniz's rule on the Grassmann algebra of the contravariant tensor fields:

$$d: \bigwedge \mathbf{T}^T \longrightarrow \bigwedge \mathbf{T}^T; \quad \left\{ \begin{aligned} \langle df, v\rangle &= v(f), \quad f \in \mathcal{C}(\mathbb{M}), \\ \omega^k \in \bigwedge^k \mathbf{T}^T &\Rightarrow d\omega^k \in \bigwedge^{1+k} \mathbf{T}^T, \\ d(\omega^k \wedge \omega^l) &= d\omega^k \wedge \omega^l + (-1)^k \omega^k \wedge d\omega^l, \\ d^2 &= 0. \end{aligned} \right.$$

The external derivative of a 1-form ω has as action on two vector fields (v, w):

$$\begin{aligned} \omega \in \mathbf{T}^T: \quad 2d\omega(v, w) &= v(\omega(w)) - w(\omega(v)) - \omega([v, w]) \\ &= 2v_i w_j(\partial^i \omega^j - \partial^j \omega^i), \\ \text{for } \omega &= \omega^k dx_k, \; d\omega = \partial^l \omega^k dx_l \wedge dx_k, \; v = v_i\partial^i, \; w = w_j\partial^j. \end{aligned}$$

This can be expressed in local dual bases with the anholonomy coefficients by the *Maurer–Cartan formula*:

$$\begin{aligned} 2d\check{e}_k(e^i, e^j) &= -\check{e}_k([e^i, e^j]) = -\epsilon_k^{ij}, \\ d\check{e}_k &= -\tfrac{1}{2}\epsilon_k^{ij} \check{e}_i \wedge \check{e}_j. \end{aligned}$$

The general Maurer–Cartan expression for p-forms $\omega \in \bigwedge^{p}\mathbf{T}^T$ reads

$$(1+p)d\omega(v^1,\dots,v^{1+p}) = \sum_{a=1}^{1+p}(-1)^{1+a}v^a\Big(\omega(v^1,\dots,\hat{v}^a,\dots,v^{1+p})\Big) \\ +\sum_{a<b}(-1)^{a+b}\omega([v^a,v^b],v^1,\dots,\hat{v}^a,\dots,\hat{v}^b,\dots,v^{1+p}).$$

2.2 Riemannian Operation Groups

In addition to real parametrizability and differentiability, manifolds can come with more specific operations, e.g., related to metrical structures.

A manifold is endowed with a *Riemannian structure* $(\mathbb{M},\mathbf{g}) \in \underline{\mathbf{rdif}}_{\mathbb{R}}$ by a tensor field $\mathbf{g}$, which yields symmetric nondegenerate bilinear forms as *metrics of the tangent spaces*:

$$\mathbf{g}(x): \mathbf{T}_x \vee \mathbf{T}_x \longrightarrow \mathbb{R},\quad \mathbf{g}(e^l(x),e^i(x)) = \mathbf{g}^{li}(x), \\ \mathbf{g} = \mathbf{g}^{li}\check{e}_l \otimes \check{e}_i \in \mathbf{T}^T \vee \mathbf{T}^T.$$

The metric of a direct product of two Riemannian manifolds $(\mathbb{M}_i,\mathbf{g}_i)$ is a linear combination of the factor metrics with any nontrivial scalar factors $\mathbb{R} \ni \alpha_i \neq 0$:

$$\mathbb{M}_1 \times \mathbb{M}_2 = \mathbb{M}:\quad \mathbf{g}(x_1,x_2) = \alpha_2\mathbf{g}_1(x_1) + \alpha_1\mathbf{g}_2(x_2).$$

If a Riemannian manifold is embedded in a bigger one, there exists a projection $\mathbb{M} \longrightarrow \mathbb{M}_1$ also for the metric $\mathbf{g}(x) \longmapsto \mathbf{g}_1(x_1)$.

By multiplication with the appropriate power of the metric determinant for the dilations $\mathbf{D}(1) \cong \mathbf{GL}(n,\mathbb{R})/\mathbf{SL}(n,\mathbb{R})$, the $\mathbf{SL}(n,\mathbb{R}) \vec{\times} \mathbb{R}^n$-invariant measures d^nx become the up-to-a-constant unique $\mathbf{GL}(n,\mathbb{R}) \vec{\times} \mathbb{R}^n$-invariant measure $\sqrt{|\det\mathbf{g}(x)|}d^nx$.

2.2.1 Metric-Induced Isomorphisms

A nondegenerate metric d of a finite-dimensional vector space involves an isomorphy with the dual vector space $V \overset{d}{\cong} V^T$. Therefore, tangent and cotangent spaces of a Riemannian manifold are $\mathbf{g}$-isomorphic:

$$\mathbf{T} \ni e^j \longmapsto \mathbf{g}^{jk}\check{e}_l \in \mathbf{T}^T.$$

The cotangent (inverse) metric is $\mathbf{g}^{-1} = \mathbf{g}_{li}e^l \otimes e^i \in \mathbf{T} \vee \mathbf{T}$. In composition with the dual isomorphy, the volume element-induced isomorphies give isomorphisms of tangent space powers with equal dimensions $\binom{n}{k} = \binom{n}{n-k}$:

$$k = 0,\dots,n:\ \mathbf{g}\circ\epsilon: \bigwedge^{k}\mathbf{T}^T \longrightarrow \bigwedge^{n-k}\mathbf{T}^T,\ \text{e.g., } \mathbf{T}^T \cong \bigwedge^{n-1}\mathbf{T}^T.$$

The external powers of $(dx_i)_{i=1}^n$ can be used for the integration of tensor fields over the manifold and submanifolds.

The *metrical reflection* is defined for symmetric tensor fields with even grade, starting with the scalars $\mathcal{C}(\mathbb{M}) \ni f \longmapsto -f$, and the grade-2 symmetric tensor fields $\mathbf{S}^{ij} = \mathbf{S}^{ji}$, which use the metric multiplied trace with a dimension-dependent coefficient $\frac{2}{n}$:

$$\mathbf{T}^T \wedge \mathbf{T}^T \ni \mathbf{S} \leftrightarrow \check{\mathbf{S}} \in \mathbf{T}^T \wedge \mathbf{T}^T, \quad \begin{cases} \check{\mathbf{S}} = \mathbf{S} - \frac{2}{n}\mathbf{g}\mathbf{S}^\bullet, \ \ \mathbf{S} = \check{\mathbf{S}} - \frac{2}{n}\mathbf{g}\check{\mathbf{S}}^\bullet, \\ \mathbf{S}^\bullet = \mathbf{g}_{ij}\mathbf{S}^{ij} = -\check{\mathbf{S}}^\bullet = -\mathbf{g}_{ij}\check{\mathbf{S}}^{ij}, \\ \check{\check{\mathbf{S}}} = \mathbf{S}. \end{cases}$$

The reflection-symmetric tensors are traceless, the reflection-antisymmetric tensors are the metric with the normalized trace, while the reflected metric is its negative:

$$\mathbf{S} = \check{\mathbf{S}} \iff \mathbf{S}^\bullet = 0, \ \ \mathbf{S} = -\check{\mathbf{S}} \iff \mathbf{S} - \frac{\mathbf{g}}{n}\mathbf{S}^\bullet = 0, \\ \check{\mathbf{g}} = -\mathbf{g}.$$

Only in four spacetime dimensions, where $\frac{2}{n} = \frac{1}{2}$, does the reflected Ricci tensor $\check{\mathcal{R}}_\bullet$ coincide with the *Einstein tensor* $\mathcal{R}_\bullet - \frac{\mathbf{g}}{2}\mathcal{R}_\bullet^\bullet$, which is defined with the factor $\frac{1}{2}$ in any dimension. In causal spacetime, $\mathbb{M}^{(1,3)}$, a reflected "time-like" tensor in a Minkowski basis is proportional to the definite Euclidean form δ^{ab}, not to the indefinite Lorentz form η^{ab}:

$$\mathbb{M}^{(1,3)}: \ \ \mathbf{S}^{ab} = \delta_0^a\delta_0^b\mathbf{M} \leftrightarrow \check{\mathbf{S}}^{ab} = \mathbf{S}^{ab} - \frac{\eta^{ab}}{2}\mathbf{S}^\bullet = \frac{\delta^{ab}}{2}\mathbf{M}, \ \ \mathbf{S}^\bullet = \mathbf{M} = -\check{\mathbf{S}}^\bullet.$$

2.2.2 Tangent Euclidean and Poincaré Groups

The metrical structure of a Riemannian manifold defines *three operation groups*: its characterizing global group and local invariance group and the not-so-specific tangent Poincaré or Euclidean group.

Two metrics of a Riemannian manifold define isomorphisms of the tangent spaces:

$$\mathbf{g}' \circ \mathbf{g}^{-1}: \mathbf{T} \longrightarrow \mathbf{T}, \ e^j \longmapsto \mathbf{g}^{jk}\mathbf{g}'_{kl}e^l, \\ \mathbf{g}'(x) \circ \mathbf{g}^{-1}(x) \in \mathbf{GL}(n, \mathbb{R}).$$

With a fixed subgroup $H \subseteq \mathbf{GL}(n,\mathbb{R})$, they are *$H$-equivalent* if $\mathbf{g}'(x) \circ \mathbf{g}^{-1}(x) \in H$ for all $x \in \mathbb{M}$. This equivalence relation collects the metrics into disjoint classes $\mathbf{GL}(n,\mathbb{R})/H$, the *$n$-bein manifold modulo H*. The local transformation from one metric to the other gives representatives of the equivalence classes, defined up to local H-transformations,

$$\mathbf{g} = e^T \circ \mathbf{g}' \circ e, \ \ e(x) \Subset \mathbf{GL}(n,\mathbb{R})/H.$$

A special case is given by the center $\mathbf{GL}(\mathbb{R})\mathbf{1}_n \subseteq \mathbf{GL}(n,\mathbb{R})$ as direct factor for the volume invariance group $\mathbf{GL}(n,\mathbb{R}) \cong \mathbf{GL}(1,\mathbb{R}) \times \mathbf{SL}(n,\mathbb{R})$. Two metrics are *conformal* to each other (dilation-equivalent) if the isomorphisms above are the identity up to a scalar factor, a dilation from the unit connection subgroup $\mathbf{D}(1) \subset \mathbf{GL}(1,\mathbb{R})$:

$$\mathbf{g} \overset{\mathbf{D}(1)}{\sim} \mathbf{g}' \iff \mathbf{g}' \circ \mathbf{g}^{-1} = e^{2\gamma}\, \mathrm{id}_{\mathbf{T}(U)}, \quad \mathbf{g}_{jk}(x)\mathbf{g}'^{kl}(x) = e^{2\gamma(x)}\delta^l_j.$$

The local orthonormalization of a Riemannian metric displays a characteristic *signature*[2] (t,s) and allows the definition of a local n-bein field (boldface $\mathbf{e}$) with respect to a fixed constant metric η:

$$\mathbf{g} = ds^2 = \eta^{ab}\check{\mathbf{e}}_a \otimes \check{\mathbf{e}}_b, \quad \mathbf{g}^{jk}(x) = \mathbf{e}^j_a(x)\eta^{ab}\mathbf{e}^k_b(x), \quad \eta = \left(\begin{array}{c|c} \mathbf{1}_t & 0 \\ \hline 0 & -\mathbf{1}_s \end{array}\right).$$

This defines an orthogonal Lorentz group $\mathbf{O}(t,s)$ with the unit connection subgroup $\mathbf{SO}_0(t,s)$. It is a rotation group for definite signature $ts = 0$ and a proper Lorentz group for causal signature $(t,s) = (1,s)$. In contrast to the Lorentz group, the overall dilations $\mathbf{D}(\mathbf{1}_n)$ cannot be characterized as the invariance group of a metric.

The symmetric tensor $\mathbf{g}$ has $\binom{1+n}{2}$ real parameters, which are taken over completely by n-beins from the equally dimensional manifold $\mathbf{GL}(n,\mathbb{R})/\mathbf{SO}_0$ (t,s) as parameters for the local orthonormalization by n dilations and $\binom{n}{2}$ rotations, i.e., by a representative, defined up to local Lorentz transformations:

$$\begin{aligned} &\mathbf{g} = \mathbf{e}^T \circ \eta \circ \mathbf{e}, \quad \mathbf{e}(x) \in \mathbf{D}(1) \times \mathbf{SO}_0(1,1)^{n-1} \times \mathbf{SO}(n) \in \mathbf{GL}(n,\mathbb{R})/\mathbf{SO}_0(t,s), \\ &\binom{1+n}{2} = n^2 - \binom{n}{2} = n + \binom{n}{2}. \end{aligned}$$

The Lie algebra $\log \mathbf{SO}_0(t,s)$ is simple for $n \geq 3$ with the exception of the two semisimple Lie algebras $\log \mathbf{SO}(4) = \log \mathbf{SO}(3) \oplus \log \mathbf{SO}(3)$ and $\log \mathbf{SO}_0(2,2) = \log \mathbf{SO}_0(1,2) \oplus \log \mathbf{SO}_0(1,2)$. There are characteristically different Riemannian manifolds $\mathbb{M}^{(t,s)}$ with isomorphic tangent groups, e.g., spheres Ω^s and hyperboloids $\mathcal{Y}^s$, both with Euclidean tangent groups $\mathbf{SO}(s) \vec{\times} \mathbb{R}^s$.

Together with the tangent translations $\mathbb{R}^n$, one obtains the $\binom{1+n}{2}$ parametric *tangent Poincaré group with homogeneous Lorentz group*, sometimes called pseudo-Euclidean and pseudo-Riemannian for $ts \neq 0$,

$$\mathbf{SO}_0(t,s) \vec{\times} \mathbb{R}^n, \; t + s = n \geq 1.$$

It is a proper Euclidean group with metrical ellipsoids for $ts = 0$, e.g., for the spheres Ω^s and the hyperboloids $\mathcal{Y}^t$, and a proper Poincaré group with metrical "timelike" hyperboloids for $(t,s) = (1,s)$.

Each Lie group with a semisimple Lie algebra has "its" Lorentz and Poincaré group (ahead).

[2]Obviously, the natural numbers (t,s) should remind us of time and space. In general, it will be difficult to confuse the dimension $t \in \mathbb{N}$ with a coordinate $t \in \mathbb{R}$.

The Lie algebra of the tangent Poincaré group,

$$\log \mathbf{SO}_0(t,s) \ \vec{\oplus} \ \mathbb{R}^n \cong \mathbf{T}_x \wedge \mathbf{T}_x \ \vec{\oplus} \ \mathbf{T}_x \cong \mathbb{R}^{\binom{n}{2}} \ \vec{\oplus} \ \mathbb{R}^n \cong \mathbb{R}^{\binom{1+n}{2}},$$

has the brackets in an orthonormal basis $\{\mathcal{L}^{ab}, \ \mathcal{P}^a \mid a,b = 1,\dots,n\}$:

$$\begin{aligned} [\mathcal{L}^{ab}, \mathcal{L}^{cd}] &= \eta^{ac}\mathcal{L}^{bd} - \eta^{bc}\mathcal{L}^{ad} - \eta^{ad}\mathcal{L}^{bc} + \eta^{bd}\mathcal{L}^{ac}, \\ [\mathcal{P}^a, \mathcal{P}^b] &= 0, \\ [\mathcal{L}^{ab}, \mathcal{P}^c] &= \eta^{ac}\mathcal{P}^b - \eta^{bc}\mathcal{P}^a. \end{aligned}$$

The defining n-dimensional Minkowski representation of the Lorentz (orthogonal) group with $A = 1,\dots,t$ and $\alpha = t+1,\dots,t+s = n$,

$$\begin{aligned} \log \mathbf{SO}_0(t,s) \ni \mathcal{L} = \lambda_{ab}\mathcal{L}^{ab} \longmapsto \ & \mathcal{D}_n(\mathcal{L}) \cong \left(\begin{array}{c|c} \lambda_{AB} = -\lambda_{BA} & \lambda_{A\alpha} = \lambda_{\alpha A} \\ \hline \lambda_{\beta B} & \lambda_{\beta\alpha} = -\lambda_{\alpha\beta} \end{array}\right) \\ & \in \left(\begin{array}{c|c} \mathbb{R}^t \otimes \mathbb{R}^t & \mathbb{R}^t \otimes \mathbb{R}^s \\ \hline \mathbb{R}^s \otimes \mathbb{R}^t & \mathbb{R}^s \otimes \mathbb{R}^s \end{array}\right) = \mathbf{AL}(\mathbb{R}^n), \\ \text{e.g., for } \log \mathbf{SO}_0(1,3): \ & \mathcal{D}_4(\mathcal{L}) \cong \left(\begin{array}{c|c} 0 & \lambda_{0\alpha} = \lambda_{\alpha 0} \\ \hline \lambda_{\beta 0} & \lambda_{\beta\alpha} = -\lambda_{\alpha\beta} \end{array}\right), \quad \alpha,\beta = 1,2,3, \end{aligned}$$

is definite unitary only for the compact rotation groups $\mathbf{SO}(n)$. The related $(1+n)$-dimensional representation of the Poincaré group with the noncompact translations

$$\log \mathbf{SO}_0(t,s) \ \vec{\oplus} \ \mathbb{R}^n \longmapsto \left(\begin{array}{c|c} \mathcal{D}_n(\mathcal{L}) & \mathcal{P} \\ \hline 0 & 0 \end{array}\right) \in \mathbf{AL}(\mathbb{R}^{1+n})$$

is never definite unitary.

The faithful definite-unitary, i.e., Hilbert representations of the Euclidean and Poincaré groups, especially of $\mathbf{SO}(3) \ \vec{\times} \ \mathbb{R}^3$ for nonrelativistic scattering in 3-position and of $\mathbf{SO}_0(1,3) \ \vec{\times} \ \mathbb{R}^4$ for particles in Minkowski spacetime, massive and massless, are infinite-dimensional. The tangent Poincaré group contains the operations for free particles.

2.2.3 Global and Local Invariance Groups

The morphisms for Riemannian manifolds $\varphi : \mathbb{M}_1 \longrightarrow \mathbb{M}_2$ are compatible with the metrics $\mathbf{g}_2(\varphi.x, \varphi.y) = \mathbf{g}_1(x,y)$. The metric-associated *global invariance group* $G_{\mathbf{g}}$ of $(\mathbb{M}, \mathbf{g}) \in \underline{\mathbf{rdif}}_{\mathbb{R}}$ consists of the metric-compatible diffeomorphisms with $\mathbf{g} \circ (\varphi \times \varphi) = \mathbf{g}$ (isometries, Riemannian automorphisms).

The infinitesimal formulation: A metrical tensor field defines its *Killing vector fields* by the Lie derivations from its invariance Lie algebra. This leads to a covariant functor from Riemannian manifolds to real Lie algebras:

$$\underline{\mathbf{rdif}}_{\mathbb{R}} \ni (\mathbb{M}, \mathbf{g}) \longmapsto L_{\mathbf{g}} = \{v \in \mathbf{T} \mid [v, \mathbf{g}] = 0\} \in \underline{\mathbf{lag}}_{\mathbb{R}}.$$

With the $\binom{1+n}{2}$ metric components, one obtains a system of $\binom{1+n}{2}$ differential equations for the n functions $\{v_p\}_{p=1}^n$:

$$\begin{aligned} [v_p\partial^p, \mathbf{g}^{ij} dx_i \otimes dx_j] \quad &= \quad [v_p(\partial^p\mathbf{g}^{ij}) + \mathbf{g}^{pj}(\partial^i v_p) + \mathbf{g}^{pi}(\partial^j v_p)] dx_i \otimes dx_j, \\ v_p\partial^p \in L_{\mathbf{g}} \quad &\iff \quad v_p(\partial^p\mathbf{g}^{ij}) + \mathbf{g}^{pj}(\partial^i v_p) + \mathbf{g}^{pi}(\partial^j v_p) = 0 \\ & \qquad \text{for all } i,j \in \{1,\dots,n\}. \end{aligned}$$

Each one-dimensional manifold has one Killing field and $\mathbb{R}$ as invariance Lie algebra:

$$\mathbf{g} = b^2(\tau)d\tau^2 : \quad v\partial_\tau b^2 + 2b^2\partial_\tau v = 2b\partial_\tau(vb) = 0 \Rightarrow v\partial_\tau = \tfrac{1}{b}\partial_\tau.$$

For two dimensions, there are three equations for the Killing fields $v = v_0\partial_\tau + v_1\partial_\rho$:

$$\mathbf{g} = b^2 d\tau^2 \pm a^2 d\rho^2 : \quad \begin{cases} (v_0\partial_\tau + v_1\partial_\rho)b^2 + 2b^2\partial_\tau v_0 &= 0, \\ (v_0\partial_\tau + v_1\partial_\rho)a^2 + 2a^2\partial_\rho v_1 &= 0, \\ b^2\partial_\rho v_0 \pm a^2\partial_\tau v_1 &= 0. \end{cases}$$

For the metrical tensor $\eta = \eta_{ab}dx^a \otimes dx^b \cong \left(\begin{array}{c|c} \mathbf{1}_t & 0 \\ \hline 0 & -\mathbf{1}_s \end{array}\right)$ of the vector space $\mathbb{R}^n$ as Riemannian manifold, the Killing fields constitute a Poincaré Lie algebra:

$$\begin{aligned} (\mathbb{R}^n, \eta) \longmapsto L_\eta &= \log \mathbf{SO}_0(t,s) \ \vec{\oplus} \ \mathbb{R}^n, \\ v_a\partial^a &\in L_\eta \iff \partial^a v^b + \partial^b v^a = 0 \Rightarrow v^a(x) = (\lambda^{ab} - \lambda^{ba})x_b + \xi^a, \\ v_a\partial^a &= \lambda_{ab}(\eta^{ac}x_c\partial^b - \eta^{bc}x_c\partial^a) + \xi_a\partial^a \cong \lambda_{ab}\mathcal{L}^{ab} + \xi_a\mathcal{P}^a. \end{aligned}$$

The Poincaré Lie algebra is represented by Lie derivations,

$$\mathcal{L}^{ab} \longmapsto \eta^{ac}x_c\partial^b - \eta^{bc}x_c\partial^a, \ \mathcal{P}^a \longmapsto \partial^a.$$

For a metric without Killing vectors, the manifold has no symmetry, $L_{\mathbf{g}} = \{0\}$. If a metric does not depend, in a parametrization, on a coordinate, $\frac{\partial}{\partial x_k}\mathbf{g}^{ij}(x) = 0$ for all $i,j \in \{1,\ldots,n\}$, the corresponding derivative $v = \partial^k$ characterizes a one-dimensional invariance Lie algebra. More generally, for each Killing field $v \in L_{\mathbf{g}}$, there exists a parametrization with $v = \partial^k$. Such a one-dimensional Lie algebra $\mathbb{R}$ leads by integration locally to the noncompact covering Lie group $e^{x_k} \in \mathbf{D}(1) = e^{\mathbb{R}} \cong \mathbf{SO}_0(1,1) \ni e^{\sigma_3 x_k}$ or to its compact quotient group $e^{i\sigma_2 x_k} \in \mathbf{SO}(2) \cong \mathbb{R}/\mathbb{Z}$.

Linear independent Killing vectors $\{v^K\}_{K=1}^d$ of $(\mathbb{M}, \mathbf{g})$ define a d-dimensional invariance Lie algebra $L_{\mathbf{g}}$ of the metric with cover group:

$$\underline{\mathbf{rdif}}_{\mathbb{R}} \ni (\mathbb{M}, \mathbf{g}) \longmapsto L_{\mathbf{g}} \longmapsto \exp L_{\mathbf{g}} \in \underline{\mathbf{lgrp}}_{\mathbb{R}}.$$

Under appropriate smoothness and connectivity conditions, at least locally in a chart, a locally isomorphic d-dimensional quotient Lie group $G_{\mathbf{g}}$ (classes of $\exp L_{\mathbf{g}}$) is the *global symmetry or motion or isometry group* of the Riemannian manifold:

$$\exp L_{\mathbf{g}} \sim G_{\mathbf{g}} = \{\varphi \in \overset{\circ}{\mathbf{dif}}_{\mathbb{R}}(\mathbb{M}, \mathbb{M}) \mid \mathbf{g}(\varphi.x, \varphi.y) = \mathbf{g}(x,y)\}.$$

Its maximal dimension is $\binom{1+n}{2}$: The *maximal global symmetry group* is either an orthogonal group, e.g., for spheres and hyperboloids (ahead), or a Poincaré group for a flat vector space,

$$\begin{aligned} \dim_{\mathbb{R}} G_{\mathbf{g}} &\le \tbinom{1+n}{2} = \dim_{\mathbb{R}} G^{\max}(t,s), \\ G^{\max}(t,s) &\in \{\mathbf{SO}_0(t, 1+s), \ \mathbf{SO}_0(1+t, s), \ \mathbf{SO}_0(t,s) \ \vec{\times} \ \mathbb{R}^n\}. \end{aligned}$$

The actual motion group $G_{\mathbf{g}}$ must not be a subgroup of $G^{\max}(t,s)$. It is a subgroup of an $\binom{1+n}{2}$-dimensional product of maximal motion groups — possibly also from smaller-dimensional manifolds.

$G_{\mathbf{g}}$ acts on the manifold $\mathbb{M}$, which, therefore, is the disjoint union of $G_{\mathbf{g}}$-orbits with fixgroups (isotropy subgroup) $H_{\mathbf{g}}^{\iota} \subseteq G_{\mathbf{g}}$ for orbit representatives,

$$\mathbb{M} \cong \biguplus_{\text{representatives } x^{\iota}} G_{\mathbf{g}} \bullet x^{\iota} \cong \biguplus_{\iota} G_{\mathbf{g}}/H_{\mathbf{g}}^{\iota}.$$

The fixgroups of the points of one orbit, $g_{1,2} \bullet x \in G_{\mathbf{g}} \bullet x$, are conjugate to each other, i.e., related by inner $G_{\mathbf{g}}$-automorphisms with $g_1 g_2^{-1}$. There may be disjoint orbits with isomorphic fixgroup.

The *local invariance group* of a Riemannian manifold is defined by the fixgroup of all manifold points, i.e., by the intersection of appropriate representative fixgroups for all orbit types (examples ahaed):

$$H_{\mathbf{g}} = \bigcap_{\kappa} H_{\mathbf{g}}^{\kappa} \subseteq G_{\mathbf{g}}.$$

For a maximal global symmetry group, the local invariance group is maximal and isomorphic to the tangent Lorentz group. In general, the local invariance group is a Lorentz subgroup:

$$H_{\mathbf{g}} \subseteq H^{\max}(t,s) = \mathbf{SO}_0(t,s).$$

The *metrical coefficients are representation matrix elements* of the global symmetry group $G_{\mathbf{g}}$, invariant under the action of the local group $H_{\mathbf{g}}$:

$$\biguplus_{\iota} G_{\mathbf{g}}/H_{\mathbf{g}}^{\iota} \cong \mathbb{M} \ni x \longmapsto \mathbf{g}^{jk}(x), \quad G_{\mathbf{g}}/H_{\mathbf{g}}^{\iota} \subseteq G_{\mathbf{g}}/H_{\mathbf{g}}.$$

Some examples: A manifold with trivial global symmetry group $\{1\}$ is the union of its points $\mathbb{M} = \biguplus_{x \in \mathbb{M}} \{x\}$. A one-dimensional manifold is isomorphic to $\mathbb{R} \cong \mathbf{SO}_0(1,1) \cong \mathcal{Y}^1$ or to $\mathbf{SO}(2) \cong \Omega^1$. A torus is a product of axial rotation groups $\mathbb{T}^n = \mathbf{SO}(2) \times \cdots \times \mathbf{SO}(2)$.

Spheres and hyperboloids are one orbit. The metrical coefficients are spherical harmonics $\mathrm{Y} : \mathbf{SO}(1+s)/\mathbf{SO}(s) \cong \Omega^s \longrightarrow \mathbb{R}$ and their hyperbolic counterparts $\mathbf{SO}_0(1,s)/\mathbf{SO}(s) \cong \mathcal{Y}^s \longrightarrow \mathbb{R}$. For example, the Ω^2-metric involves the spherical harmonics $\sqrt{4\pi}\mathrm{Y}_0^0(\varphi,\theta) = 1$ and $\sqrt{\frac{4\pi}{5}}\mathrm{Y}_0^2(\varphi,\theta) = 1 - \frac{3}{2}\sin^2\theta$. Its three Killing fields are given by the angular momenta — in polar coordinates,

$$\Omega^2: \quad d\omega_2^2 = d\theta^2 + \sin^2\theta d\varphi^2, \quad \vec{\mathcal{L}} \cong \begin{pmatrix} \cos\varphi\cot\theta\frac{\partial}{\partial\varphi} + \sin\varphi\frac{\partial}{\partial\theta} \\ \sin\varphi\cot\theta\frac{\partial}{\partial\varphi} - \cos\varphi\frac{\partial}{\partial\theta} \\ \frac{\partial}{\partial\varphi} \end{pmatrix}, \quad [\mathcal{L}^a, \mathcal{L}^b] = -\epsilon^{abc}\mathcal{L}^c.$$

The local invariance group $\mathbf{SO}(2) \subset \mathbf{SO}(3)$ is parametrizable by a third coordinate in (α,φ,θ). This coordinate does not show up — neither in the metrical coefficients, $\frac{\partial \mathbf{g}^{jk}}{\partial\alpha} = 0$, nor in the basis as $d\alpha^2$.

In contrast to the metrical coefficients for the compact spheres, which are definite-unitary matrix elements, i.e., of Hilbert representations, those of the noncompact hyperboloids are not Hilbert representation coefficients; e.g.,

$$\mathcal{Y}^2: \ d\mathrm{y}_2^2 = d\psi^2 + \sinh^2 \psi d\varphi^2.$$

Minkowski spacetime $\mathbb{R}^{1+s}$, $s \geq 2$, has metrical coefficients $dx^2 = dx_0^2 - d\vec{x}^2$ with a trivial representation of the Poincaré group. The action of the Lorentz group $\mathbf{SO}_0(1,s)$, $s \geq 2$, on the translations has four fixgroup types for the "many" orbits: $\mathbf{SO}_0(1,s)$ for the trivial translation, $\mathbf{SO}(s)$ for timelike translations, $\mathbf{SO}_0(1,s-1)$ for spacelike translations, and $\mathbf{SO}(s-1) \,\vec{\times}\, \mathbb{R}^{s-1}$ for lightlike translations. The common fixgroup for all translations is $\mathbf{SO}(s-1)$, i.e., the axial rotations $\mathbf{SO}(2)$ for four-dimensional spacetime. Vector spaces $\mathbb{R}^n$ are one orbit with respect to their own action as translations. Vector spaces $\mathbb{R}^{t+s}$ as Riemannian manifolds have the Poincaré group as both the tangent and motion groups; the local invariance group is the Lorentz group $\mathbf{SO}_0(t,s) \,\vec{\times}\, \mathbb{R}^n / \mathbf{SO}_0(t,s) \cong \mathbb{R}^n$.

Summarizing: The three Riemannian operation groups for a Riemannian manifold $(\mathbb{M}, \mathbf{g})$ are

$\mathbb{M}^{(t,s)}$:

Tangent Poincaré group
$\mathbf{SO}_0(t,s) \,\vec{\times}\, \mathbb{R}^n$

Global symmetry (motion) group $G_{\mathbf{g}}$
$G^{\max}(t,s) \in \{\mathbf{SO}_0(t, 1+s),\ \mathbf{SO}_0(1+t, s),\ \mathbf{SO}_0(t,s) \,\vec{\times}\, \mathbb{R}^n\}$

Local invariance group
$H_{\mathbf{g}} \subseteq G_{\mathbf{g}} \cap \mathbf{SO}_0(t,s)$

In a physical context, the global group $G_{\mathbf{g}}$ and its generators give the symmetries and conserved quantities of a dynamics, like energy for time translations and static spacetimes or angular momenta for rotation-symmetric spacetimes (see Chapter 3), whereas the local subgroup $H_{\mathbf{g}}$ will be related to gauge transformations, e.g., to electromagnetic and isospin transformations (see Chapter 6).

A structural group $\mathbf{SO}_0(t,s)$ for the tangent frames and signature (t,s) is more specific than all tangent space automorphisms $\mathbf{GL}(n,\mathbb{R})$ for dimension n. However, the tangent group does not differentiate between manifolds of equal signature $(\mathbb{M}^{(t,s)}, \mathbf{g})$, but between different global or local invariance groups.

2.2.4 Riemannian Connection

The conditions for the Killing fields of a Riemannian manifold,

$$(\mathbb{M}, \mathbf{g}): \ v_p \partial^p \in L_{\mathbf{g}} \iff \mathbf{g}^{pj}(\partial^i v_p) + \mathbf{g}^{pi}(\partial^j v_p) + v_p(\partial^p \mathbf{g}^{ij}) = 0,$$

can be written in the simple derivative form of "flat" vector spaces,

$$(\mathbb{R}^n, \eta): \ v_a \partial^a \in L_\eta \iff \partial^a v^b + \partial^b v^a = 0, \ \ v^a = \eta^{ab} v_b,$$

by introducing the *Riemannian connection*,

$$(\mathbb{M}, \mathbf{g}) : \quad v_p \partial^p \in L_{\mathbf{g}} \iff \begin{array}{l} \nabla^i v^j + \nabla^j v^i = 0, \quad v^i = \mathbf{g}^{ij} v_j, \\ \text{with } \nabla^i v^j = \partial^i v^j - \Gamma^{ij}_k v^k. \end{array}$$

It takes into account a possible "nonflatness" by including local changes via derivatives of the metrical tensor via the connection coefficients,

$$\Gamma^{ij}_k = \tfrac{1}{2} \mathbf{g}_{kp} (\partial^i \mathbf{g}^{pj} + \partial^j \mathbf{g}^{pi} - \partial^p \mathbf{g}^{ij}).$$

The Killing field condition for the metric motivates the introduction of a second derivative on a Riemannian manifold in addition to and constructed with the Lie derivative.

2.3 Affine Connections

A Riemann derivative is a special case of an affine connection of a differentiable manifold, in general: "On top" of the derivations of the scalar functions $\mathcal{C}(\mathbb{M})$ and in addition to the Lie derivations, a manifold may be equipped with an *affine connection* ∇ which defines derivations of vector fields $\mathbf{T} = \operatorname{der} \mathcal{C}(\mathbb{M})$, i.e., derivations of derivations,

$$\begin{array}{l} \nabla : \quad \mathbf{T} \longrightarrow \mathbf{T} \otimes \mathbf{T}^T, \quad v \longmapsto \nabla_v, \\ \nabla_v : \quad \mathbf{T} \longrightarrow \mathbf{T}, \quad \begin{cases} w \longmapsto \nabla_v w, \\ \nabla_v (fw) = v(f) w + f \nabla_v w, \\ \nabla_{e^i} e^j = \nabla^i e^j = \Gamma^{ij}_k e^k, \\ \nabla^i (v_k e^k) = [e^i(v_k) + \Gamma^{ij}_k v_j] e^k, \end{cases} \\ \nabla = \Gamma^{ij}_k \check{e}_i \otimes e^k \otimes \check{e}_j \in \mathbf{T}^T \otimes \mathbf{T} \otimes \mathbf{T}^T \in \underline{\mathbf{vec}}_{\mathbb{R}}. \end{array}$$

The $\mathbb{R}$-linear tangent bundle endomorphism ∇_v is called the *covariant derivation in the direction of* v. In contrast to the Lie derivation ad, where, in general, $\operatorname{ad} fv \neq f \operatorname{ad} v$, an affine connection ∇ is $\mathcal{C}(\mathbb{M})$-linear:

$$f, g \in \mathcal{C}(\mathbb{M}) : \quad \nabla_{fv+gw} = f \nabla_v + g \nabla_w .$$

In general, the Lie derivation $\operatorname{ad} v$ is not a covariant derivation.

A one-dimensional manifold has the connections $f(x) \frac{d}{dx}$ with functions $f \in \mathcal{C}(\mathbb{M})$.

A covariant derivative is grading compatibly extendable as $\mathbb{R}$-linear derivation to the full tensor algebra by Leibniz's rule,

$$\begin{array}{l} \nabla_v : \quad \bigotimes(\mathbf{T} \oplus \mathbf{T}^T) \longrightarrow \bigotimes(\mathbf{T} \oplus \mathbf{T}^T), \\ \text{with} \quad \begin{cases} \nabla_v f = v(f), \\ \nabla_v (a \otimes b) = \nabla_v a \otimes b + a \otimes \nabla_v b, \\ \nabla_v \langle \omega, w \rangle = \langle \nabla_v \omega, w \rangle + \langle \omega, \nabla_v w \rangle, \\ \nabla_{e^i} \check{e}_k = \nabla^i \check{e}_k = -\Gamma^{ij}_k \check{e}_j, \\ \nabla^i (\omega^k \check{e}_k) = [e^i(\omega^j) - \Gamma^{ij}_k \omega^k] \check{e}_j. \end{cases} \end{array}$$

A tensor a whose covariant derivatives are trivial in all directions is called *covariantly constant*:

$$\nabla a = 0, \quad \nabla_v a = 0 \text{ for all } v \in \mathbf{T}.$$

A local dual frame is covariantly constant $\nabla e^j \otimes \check{e}_j = 0$.

2.3.1 Torsion, Curvature, and Ricci Tensor

With an affine connection, the vector fields have two real Lie algebra structures via the commutators:

$$\begin{array}{rl} v \in \mathbf{T} \in \underline{\mathbf{lag}}_{\mathbb{R}}, & \text{with } [v,w], \\ \nabla_v \in \mathbf{T} \otimes \mathbf{T}^T \in \underline{\mathbf{lag}}_{\mathbb{R}}, & \text{with } [\nabla_v, \nabla_w]. \end{array}$$

The mapping $v \longmapsto \nabla_v$ for the tangent fields is analogous to the adjoint representation of a Lie algebra $L \ni l \longmapsto \operatorname{ad} l \in L \otimes L^T$ in its endomorphisms.

∇_v and v are morphisms for $\mathbf{T}$ as an $\mathbb{R}$-vector space, not for $\mathbf{T}$ as a $\mathcal{C}(\mathbb{M})$-module; in general, $\nabla_v fw \neq f\nabla_v w$ and $[v, fw] \neq f[v,w]$. That is the origin of their local transformation behavior. The two $\mathbb{R}$-Lie algebra structures are combined in the construction of torsion and curvature, which are even compatible with the $\mathcal{C}(\mathbb{M})$-module property.

The *torsion* (*tensor*) describes the $\mathcal{C}(\mathbb{M})$-linear difference of covariant and Lie derivations:

$$\mathcal{T} : \mathbf{T} \wedge \mathbf{T} \longrightarrow \mathbf{T}, \quad \begin{cases} \mathcal{T}(v \wedge w) &= \nabla_v w - \nabla_w v - [v,w] \\ &= -\mathcal{T}(w \wedge v), \\ \mathcal{T}(e^i \wedge e^j) &= \mathcal{T}_k^{ij} e^k, \end{cases}$$
$$\mathcal{T} = \tfrac{1}{2}\mathcal{T}_k^{ij} e^k \otimes \check{e}_i \wedge \check{e}_j \in \mathbf{T} \otimes \mathbf{T}^T \wedge \mathbf{T}^T \in \underline{\mathbf{mod}}_{\mathcal{C}(\mathbb{M})},$$
$$\mathcal{T}_k^{ij} = -\mathcal{T}_k^{ji} = \Gamma_k^{ij} - \Gamma_k^{ji} - \epsilon_k^{ij}.$$

In general, $\nabla : \mathbf{T} \longrightarrow \mathbf{T} \otimes \mathbf{T}^T$ is not a Lie algebra morphism; i.e., the diagram with the commutator of vector fields and the commutator of covariant derivatives in the direction of vector fields

$$\begin{array}{ccc} & \nabla \otimes \nabla & \\ \mathbf{T} \wedge \mathbf{T} & \longrightarrow & (\mathbf{T} \otimes \mathbf{T}^T) \wedge (\mathbf{T} \otimes \mathbf{T}^T) \\ [\ ,\ ] \downarrow & & \downarrow [\ ,\ ] \\ \mathbf{T} & \longrightarrow & \mathbf{T} \otimes \mathbf{T}^T \\ & \nabla & \end{array}$$

is not commutative. The *curvature* (*tensor*) denotes the difference of both commutators

$$\mathcal{R} : \mathbf{T} \wedge \mathbf{T} \longrightarrow \mathbf{T} \otimes \mathbf{T}^T, \quad \mathcal{R}(v \wedge w) = [\nabla_v, \nabla_w] - \nabla_{[v,w]} = -\mathcal{R}(w \wedge v),$$

$$\mathcal{R} = \tfrac{1}{2}\mathcal{R}_k^{lij} e^k \otimes \check{e}_l \otimes \check{e}_i \wedge \check{e}_j \in \mathbf{T} \otimes \mathbf{T}^T \otimes \mathbf{T}^T \wedge \mathbf{T}^T \in \underline{\mathbf{mod}}_{\mathcal{C}(\mathbb{M})},$$
$$\begin{aligned} \mathcal{R}_k^{lij} &= -\mathcal{R}_k^{lji} = e^i(\Gamma_k^{jl}) - e^j(\Gamma_k^{il}) - \epsilon_p^{ij}\Gamma_k^{pl} - \Gamma_p^{il}\Gamma_k^{jp} + \Gamma_p^{jl}\Gamma_k^{ip} \\ &= \partial^i\Gamma_k^{jl} - \partial^j\Gamma_k^{il} - \Gamma_p^{il}\Gamma_k^{jp} + \Gamma_p^{jl}\Gamma_k^{ip} \text{ (holonomic bases).} \end{aligned}$$

It defines endomorphisms of the tangent spaces:

$$\mathcal{R}(v \wedge w): \mathbf{T} \longrightarrow \mathbf{T}, \quad \begin{cases} \mathcal{R}(e^i \wedge e^j) & = \mathcal{R}_k^{lij} e^k \otimes \check{e}_l, \\ \mathcal{R}(e^i \wedge e^j).e^l & = \mathcal{R}_k^{lij} e^k. \end{cases}$$

There are maximally $\binom{n}{2}$ linear independent coefficients $\{\mathcal{R}(e^i \wedge e^j)_x\}_{i,j}$.

In general, connection, torsion, and curvature are defined without a metrical structure of the manifold. The *Ricci tensor*, a contraction (partial trace) of the curvature tensor, is a connection-induced bilinear form of the tangent spaces

$$\begin{aligned} \mathcal{R}_\bullet : \mathbf{T} \otimes \mathbf{T} &\longrightarrow \mathcal{C}(\mathbb{M}), \quad \mathcal{R}_\bullet(e^l, e^i) = \mathcal{R}_\bullet^{li} = \mathcal{R}_j^{lij}, \\ \mathcal{R}_\bullet &= \mathcal{R}_\bullet^{li} \check{e}_l \otimes \check{e}_i \in \mathbf{T}^T \otimes \mathbf{T}^T \in \underline{\mathbf{mod}}_{\mathcal{C}(\mathbb{M})}, \\ \mathcal{R}_\bullet^{li} &= e^i(\Gamma_j^{jl}) - e^j(\Gamma_j^{il}) - \epsilon_p^{ij} \Gamma_j^{pl} - \Gamma_p^{il} \Gamma_j^{jp} + \Gamma_p^{jl} \Gamma_j^{ip}. \end{aligned}$$

The Ricci form is not necessarily symmetric. It may be degenerate.

A one-dimensional manifold with connection $\nabla_f = f(x)\frac{d}{dx}$ has trivial torsion, curvature, and Ricci tensor.

For a torsionfree connection, the Lie bracket $[v, w]$ coincides with the antisymmetric combination of covariant derivatives. The structure constants of the Lie derivatives (anholonomy coefficients) coincide with the antisymmetric part of the affine connection coefficients:

$$\mathcal{T} = 0: \quad \begin{cases} \nabla_v w - \nabla_w v & = [v, w], \\ \mathcal{T}_k^{ij} & = 0, \\ \Gamma_k^{ij} - \Gamma_k^{ji} & = \epsilon_k^{ij}. \end{cases}$$

The two Lie brackets, for vector fields $[v, w]$ and covariant derivatives $[\nabla_v, \nabla_w]$, locally fulfill a Jacobi identity. For a torsionfree connection, the curvature fulfills the *first Bianchi identity*, which is equivalent to the Jacobi identity for the Lie bracket of the vector fields:

$$\mathcal{T} = 0 \Rightarrow \begin{cases} \mathcal{R}(v \wedge w).u + \mathcal{R}(w \wedge u).v + \mathcal{R}(u \wedge v).w = \\ [[v, w], u] + [[w, u], v] + [[u, v], w] = 0, \\ \mathcal{R}_k^{lij} + \mathcal{R}_k^{ijl} + \mathcal{R}_k^{jli} = 0. \end{cases}$$

The *second Bianchi identity* for a torsionfree connection is equivalent to the Jacobi identity for the Lie bracket of the covariant derivatives:

$$\mathcal{T} = 0 \Rightarrow \begin{cases} \nabla_z[\mathcal{R}(v \wedge w)u] + \nabla_v[\mathcal{R}(w \wedge z)u] + \nabla_w[\mathcal{R}(z \wedge v)u] = \\ [\nabla_z, [\nabla_v, \nabla_w]]u + [\nabla_v, [\nabla_w, \nabla_z]]u + [\nabla_w, [\nabla_z, \nabla_v]]u = 0, \\ \nabla^p \mathcal{R}_k^{lij} + \nabla^i \mathcal{R}_k^{ljp} + \nabla^j \mathcal{R}_k^{lpi} = 0. \end{cases}$$

The Ricci tensor arises in the contracted second Bianchi identity:

$$\mathcal{R}_k^{kij} = 0, \quad \nabla^p \mathcal{R}_\bullet^{li} - \nabla^i \mathcal{R}_\bullet^{lp} + \nabla^j \mathcal{R}_j^{lpi} = 0.$$

2.3.2 Cartan's Stuctural Equations

In general, an affine connection is $\mathcal{C}(\mathbb{M})$-linear "in one index" only, $\nabla_{fv} = f\nabla_v$. With the n-elements of a cotangent basis, an affine connection is determined by n^2 *connection 1-forms*:

$$\Gamma^j_k = \Gamma^{ij}_k \check{e}_i \in \mathbf{T}^T \in \underline{\mathbf{mod}}_{\mathcal{C}(\mathbb{M})}.$$

In general, the n^2 connection 1-forms Γ^j_k are not coefficients for $\mathcal{C}(\mathbb{M})$-tensors: Their *parametrization dependence* leads to an inhomogeneous contribution in addition to the "normal" homogeneous $\mathcal{C}(\mathbb{M})$-transformation in the transition to another dual basis via $e^a \otimes \check{e}_a = e^i \otimes \check{e}_i$. This inhomogeneous contribution is the origin of *gauge transformations*, here for the full linear group as the *structural group for the frames* (see Chapter 6):

$$\mathbf{GL}(n,\mathbb{R}) \ni e(x) \cong e^a_k(x) : \quad \begin{cases} \nabla_{e^i} e^a &= \Gamma^{ia}_b e^b = [e^i(e^a_k) + \Gamma^{ij}_k e^a_j]e^k, \\ \Gamma^{ia}_b &= e^a_j \Gamma^{ij}_k e^k_b + e^i(e^a_k)e^k_b. \end{cases}$$

In physical language, the inhomogenous derivative transformation behavior to the n^2 connection forms Γ^j_k with respect of the n^2-dimensional structural group $\mathbf{GL}(n,\mathbb{R})$ is called a *gauge transformation*,

$$e \in \mathbf{GL}(n,\mathbb{R}) : \quad \Gamma^i \longmapsto e \circ \Gamma^i \circ e^{-1} + (\partial^i e) \circ e^{-1} \text{ (holonomic bases).}$$

The maximal automorphism group $\mathbf{GL}(n,\mathbb{R})$ of the tangent spaces is not very characteristic of a manifold and, probably, physically irrelevant. It contains the tangent structures of all reparametrizations.

Cartan's structural equations relate the external derivative of a cotangent basis and its connection 1-forms to the 2-forms for torsion and curvature with the characteristic sum of two contributions:

$$\begin{aligned} d\check{e}_k + \Gamma^i_k \wedge \check{e}_i &= \tfrac{1}{2}\mathcal{T}_k &&= \tfrac{1}{2}\mathcal{T}^{ij}_k \check{e}_i \wedge \check{e}_j \in \mathbf{T}^T \wedge \mathbf{T}^T, \\ d\Gamma^l_k + \Gamma^p_k \wedge \Gamma^l_p &= \tfrac{1}{2}\mathcal{R}^l_k &&= \tfrac{1}{2}\mathcal{R}^{lij}_k \check{e}_i \wedge \check{e}_j \in \mathbf{T}^T \wedge \mathbf{T}^T. \end{aligned}$$

The torsion equations refine the Maurer-Cartan equations $d\check{e}_k = -\frac{1}{2}\epsilon^{ij}_k \check{e}_i \wedge \check{e}_j$. In contrast to the n^2 fields Γ^j_k with inhomogeneous transformation behavior, the n 2-forms $\mathcal{T}_k$ and the n^2 2-forms $\mathcal{R}^l_k$ are $\mathcal{C}(\mathbb{M})$ tensors. They have a homogeneous $\mathbf{GL}(n,\mathbb{R})$-transformation behavior without n-bein derivatives.

The covariant derivative on the covariant tensor fields can be written with the external derivative in the form of an *absolute derivative*,

$$D : \bigotimes \mathbf{T} \longrightarrow \mathbf{T}^T \otimes \bigotimes \mathbf{T}, \quad \begin{cases} Da(v) &= \nabla_v a, \quad v \in \mathbf{T}, \\ Df &= df, \\ D(a \otimes b) &= \nabla a \otimes b + a \otimes \nabla b, \\ D(fw) &= df \otimes w + fDw, \end{cases}$$

e.g., the absolute derivative of a vector field

$$D : \mathbf{T} \longrightarrow \mathbf{T}^T \otimes \mathbf{T}, \quad Dv = D(v_j e^j) = dv_j \otimes e^j + v_j \Gamma^{ij}_k \check{e}_i \otimes e^k \\ = [e^i(v_k) + \Gamma^{ij}_k v_j]\check{e}_i \otimes e^k.$$

A parameter-dependent vector field $\mathbb{R} \ni \tau \longmapsto v(\tau) \in \mathbf{T}$ has the absolute derivative

$$\begin{aligned} Dv &= d_\tau \otimes d_\tau(v_j \partial^j) = dx_i \otimes \nabla^i(v_j \partial^j) \\ &= (\partial^i v_k + \Gamma^{ij}_k v_j) dx_i \otimes \partial^k = (d_\tau v_k + \Gamma^{ij}_k v_j d_\tau x_i) d\tau \otimes \partial^k. \end{aligned}$$

A one-parameter-dependent *geodesic* $\mathbb{R} \ni \tau \longmapsto x_i(\tau)$ fulfills

$$d_\tau(v_j \partial^j) = 0, \text{ with } v_j = d_\tau x_j \Rightarrow d_\tau v_k + \Gamma^{ij}_k v_j v_i = 0.$$

2.4 Lie Groups as Manifolds

Lie groups G are special manifolds. A Lie group action on itself restricts the relevant manifold structures to group action-compatible ones, representable at the group unit $1 \in G$.

2.4.1 Lie Group Operations

A group acts on itself by left and right translations (multiplications)

$$(g_1, g_2) \in G \times G: \quad L_{g_1} \circ R_{g_2} : G \longrightarrow G, \quad k \longmapsto g_1 k g_2^{-1},$$

which are diffeomorphisms for Lie groups. The local invariance group in the global symmetry (motion) group $G \times G$ is the diagonal group that arises as the fixgroup of the neutral element,

$$\begin{aligned} k \in G: \quad (G \times G)_k &= \{(g_1, g_2) \in G \times G \mid g_1 k g_2^{-1} = k\} \\ &= \{(kg, gk) \mid g \in G\} \\ \cong (G \times G)_1 &= \{(g, g) \mid g \in G\} = \operatorname{diag}[G \times G] \cong G. \end{aligned}$$

Therefore, the group is isomorphic to the orbit of the doubled group as global symmetry group with the diagonal group as the local invariance group:

$$\mathbb{G} = [G \times G] / \operatorname{diag}[G \times G].$$

diag $[G \times G]$ is not a normal subgroup; i.e., the classes of the diagonal group (denoted with "doubled letters" like $\mathbb{G}$) are *isomorphic to the group G as manifold, in general not as Lie group,*

$$G \times G \ni (g_1, g_2) \longmapsto g_1 g_2^{-1} \in G \cong \mathbb{G} = [G \times G] / \operatorname{diag}[G \times G].$$

An example is the manifold isomorphy of the spin group $\mathbf{SU}(2)$ and the 3-sphere:

$$[\mathbf{SU}(2) \times \mathbf{SU}(2)] / \mathbf{SU}(2) \cong \mathbf{SO}(4)/\mathbf{SO}(3) = \mathbb{SU}(2) \cong \Omega^3 \cong \mathbf{SU}(2).$$

Ω^3 characterizes the directions of the Lenz–Runge perihelion vector in the nonrelativistic Kepler potential $\frac{1}{r}$ (see Chapters 3 and 4).

The Lie group functions $\mathcal{C}(G)$ are acted on by the left and right regular group representations, $\mathcal{C}(G) \ni f \longmapsto {}_{g_1}f_{g_2}$ with ${}_{g_1}f_{g_2}(k) = f(g_1^{-1}kg_2)$. The functions are both-sided regular representation matrix elements (representation coefficients) of $G \times G$ and can be harmonically analyzed with respect to the irreducible representations.

2.4.2 Lie Algebra Operations

The left-invariant vector fields of a Lie group define its Lie algebra:

$$\underline{\mathbf{lgrp}}_{\mathbb{R}} \ni G \longmapsto \log G = L \in \underline{\mathbf{lag}}_{\mathbb{R}}.$$

The Lie algebra can be realized by derivations of group functions at the group unit $l = \partial|_{g=1} \in \mathbf{T}_1(G) = L$.

A Lie algebra acts on itself in the *adjoint representation*

$$L \times L \longrightarrow L, \quad \operatorname{ad} l(m) = [l, m],$$

written in dual Lie algebra bases with the structure constants $\epsilon^{ab}_c = (\epsilon^a)^b_c$ as matrices,

$$\begin{aligned} \text{dual bases of } L, L^T: &\quad (l^a, \check{l}_a)^n_{a=1}, \quad \langle \check{l}_a, l^b\rangle = \delta^b_a, \\ \text{holonomic bases at } 1 \in G: &\quad \left(\frac{\partial}{\partial \alpha_a} = \partial^a, d\alpha_a\right)^n_{a=1}, \\ \text{adjoint action:} &\quad \operatorname{ad} l^a = \epsilon^{ab}_c l^c \otimes \check{l}_b. \end{aligned}$$

The adjoint action can be extended, as usual, to the tensor algebra by dual product invariance and Leibniz's rule

$$[l, \ ]: \bigotimes(L \oplus L^T) \longrightarrow \bigotimes(L \oplus L^T),$$
$$\text{with} \begin{cases} [l, \alpha] &= 0, \quad \alpha \in \mathbb{R}, \\ [l, \omega] &= -(\operatorname{ad} l)^T(\omega), \ \omega \in L^T, \\ [l^a, \check{l}_c] &= -\epsilon^{ab}_c \check{l}_b, \\ [l, a_1 \otimes a_2] &= [l, a_1] \otimes a_2 + a_1 \otimes [l, a_2]. \end{cases}$$

The identification of the Lie bracket with the tensor product commutator in the tensor algebra of the Lie algebra defines its *enveloping algebra*:

$$\begin{aligned} \mathbf{E}(L) &= \bigotimes L \text{ modulo } \{[l, m] - (l \otimes m - m \otimes l) \text{ for all } l, m \in L\}, \\ \text{in } \mathbf{E}(L): \ [l, m] &= l \otimes m - m \otimes l. \end{aligned}$$

Its center contains the Lie algebra invariants,

$$\operatorname{centr} \mathbf{E}(L) = \{I \in \mathbf{E}(L) \mid [L, I] = \{0\}\}.$$

A Lie group acts adjointly on its Lie algebra, illustrated in a representation with $\log G \subseteq \mathbf{AL}(\mathbb{R}^n)$ and $G \subseteq \mathbf{GL}(n, \mathbb{R})$:

$$G \times \log G \longrightarrow \log G, \quad (g, l) \longmapsto \operatorname{Ad} g(l) = g \circ l \circ g^{-1}.$$

2.4.3 The Poincaré Group of a Lie Group

The Lie algebra intrinsic bilinear form, invariant under Lie algebra action (inner derivations), is the symmetric Killing form,

$$\begin{aligned}
\kappa : L \vee L \longrightarrow \mathbb{R}, \quad & \kappa(l,m) = \operatorname{tr} \operatorname{ad} l \circ \operatorname{ad} m = \kappa(m,l), \\
& \kappa([k,l],m) + \kappa(l,[k,m]) = 0, \\
\text{in dual bases} \quad & \kappa = \kappa^{ab} \check{l}_a \otimes \check{l}_b = \kappa^{ab} d\alpha_a \otimes d\alpha_b \in L^T \vee L^T, \\
& \kappa^{ab} = \kappa^{ba}, \quad \epsilon^{ab}_c \kappa^{cd} + \epsilon^{bd}_c \kappa^{ac} = 0.
\end{aligned}$$

The Killing form is trivial for abelian Lie algebras and nondegenerate precisely for semisimple Lie algebras; there the adjoint representation is injective $L \cong \operatorname{ad} L$.

Linear forms L^T of a semisimple Lie algebra inherit, with the Killing isomorphism, the inverse Killing form as bilinear form,

$$\begin{aligned}
L &\longrightarrow L^T, \quad l \longmapsto l_\kappa = \kappa(l, \ ), \\
\kappa^{-1} : L^T \vee L^T &\longrightarrow \mathbb{R}, \quad \kappa^{-1}(l_\kappa, m_\kappa) = \kappa(l,m).
\end{aligned}$$

The diagonalized Killing form displays the dimensions of noncompact, nulllike (including abelian), and compact degrees of freedom $\kappa \cong \begin{pmatrix} \mathbf{1}_t & 0 & 0 \\ 0 & \mathbf{0}_l & 0 \\ 0 & 0 & -\mathbf{1}_s \end{pmatrix}$.

The adjoint representation of a semisimple Lie algebra is a subalgebra of the Lie algebra of an *orthogonal group*,

$$\text{semisimple } \mathbb{R}^n \cong L \cong \operatorname{ad} L \subseteq \log \mathbf{SO}_0(t,s) \cong \mathbb{R}^{\binom{n}{2}}.$$

With the exception of the smallest simple Lie algebras $\log \mathbf{SO}(3), \log \mathbf{SO}(1,2) \cong \mathbb{R}^3$, the invariance Lie algebra of the L-Killing form is strictly larger $\binom{n}{2} \geq n$ than the Lie algebra L, e.g., $\log \mathbf{SU}(n) \subset \log \mathbf{SO}(n^2-1)$ for $n > 2$, and $\log \mathbf{SO}_0(1,3) \subset \log \mathbf{SO}_0(3,3)$ for the Lorentz group. The invariance group of the Killing form ("signature group") for a semisimple Lie algebra defines *the Lorentz and Poincaré group of the Lie group G*:

$$\log G \cong \mathbb{R}^n, \quad \operatorname{sign} \kappa = (t,s) : \quad \mathbf{SO}_0(t,s) \vec{\times} \mathbb{R}^n, \quad \log \mathbf{SO}_0(t,s) \supseteq \log G.$$

2.4.4 Lie–Jacobi Isomorphisms for Lie Groups

The isomorphisms $g_* \in \mathbf{GL}(n,\mathbb{R})$, induced by the left multiplications (also left translations), transport all structures of the tangent space at the unit to any group element $g \in G$, $\mathbf{T}_1 \cong \mathbf{T}_g$,

$$\begin{array}{ccc}
G & & \mathbf{T}(G) \\
\downarrow g & \longmapsto & \downarrow g_*, \\
G & & \mathbf{T}(G)
\end{array}$$

$$\begin{aligned}
G \ni 1 &\longmapsto g \cdot 1 = g \text{ (left multiplication)}, \\
\mathbb{R}^n \cong \log G = L = \mathbf{T}_1(G) \ni l^a &\longmapsto l(g)^j = g_{*a}^{\,j}\, l^a \in \mathbf{T}_g(G) \cong L,
\end{aligned}$$

e.g., the Lie algebra structures at the group unit with a basis $\{l^a\}_a^n$ and brackets $[l^a, l^b] = \text{ad}\, l^a(l^b) = \epsilon_c^{ab} l^c$ to isomorphic Lie algebras with transformed structure constants,

$$[l(g)^i, l(g)^j] = \epsilon(g)_k^{ij}\ l(g)^k, \quad \epsilon(g)_k^{ij} = g_{*}{}_a^i g_{*}{}_b^j \epsilon_c^{ab} g_{*}{}_k^c, \quad g_{*}{}_a^j|_{g=1} = \delta_j^a.$$

The *Lie–Jacobi isomorphisms* $g_* = e(g) \in \mathbf{GL}(n, \mathbb{R})/\,\text{Ad}\, G$ represent classes (not all classes) of the adjoint group $\text{Ad}\, G \cong G/\,\text{centr}\, G$ in the general group. They are n-beins for the Lie group with a Lie algebra parametrization at least in the neighborhood of the group unit,

$$\begin{aligned}
\text{Ad}\, G \ni \text{Ad}\, g &= e^{\,\text{ad}\, l} = \sum_{k\geq 0} \frac{(\text{ad}\, l)^k}{k!}, \\
g_* &= e_*^{\,\text{ad}\, l} = \sum_{k\geq 0} \frac{(\text{ad}\, l)^k}{(1+k)!} = \mathbf{1}_n + \tfrac{1}{2}\,\text{ad}\, l + \ldots \\
&\cong \text{“}\, \frac{\partial\, \text{Ad}\, g}{\partial\, \text{ad}\, l}\,\text{”} = \text{“}\, \frac{e^{\,\text{ad}\, l} - \mathbf{1}_n}{\text{ad}\, l}\,\text{”} \text{ (symbolic notation).}
\end{aligned}$$

The adjoint group is trivial for abelian G. The adoint Lie algebra $\text{ad}\, L \cong L/\,\text{ad}\, L$ is trivial for abelian L, and isomorphic to the Lie algebra in the semisimple case $L \cong \text{ad}\, L$.

The Haar measure of the adjoint group can be written, at least in the neighborhood of the group unit, with an overall dilation as the volume factor:

$$\text{for } \text{Ad}\, G: \quad dg = |\det g_*| d^n l.$$

The Lie–Jacobi transportation of the Killing form defines, for semisimple Lie algebras, a Riemannian structure of the Lie group, $(G, \kappa) \in \underline{\mathbf{rdif}}_{\mathbb{R}}$:

$$\mathbf{T}_g \times \mathbf{T}_g \longrightarrow \mathbb{R}, \quad \kappa(g)^{jk} = \text{tr}\ \text{ad}\, l^j(g) \circ \text{ad}\, l^k(g) = g_{*}{}_a^j \kappa^{ab} g_{*}{}_b^k.$$

2.4.5 Examples

The compact space rotation group $\mathbf{SO}(3)$ has the angular momentum Lie algebra, block-diagonalizable in the real and diagonalizable in the complex to the rotation eigenvalues, i.e., third spin components $\{1, 0, -1\}$, in a Euclidean parametrization:

$$\begin{aligned}
&\mathbb{R}^3 \cong \log \mathbf{SO}(3) \ni \mathcal{O} = \text{ad}\, \mathcal{O} \\
&\Rightarrow \begin{cases}
\mathcal{O} = \begin{pmatrix} 0 & \alpha_3 & -\alpha_2 \\ -\alpha_3 & 0 & \alpha_1 \\ \alpha_2 & -\alpha_1 & 0 \end{pmatrix} = \alpha R \circ \begin{pmatrix} 0 & 0 & -1 \\ 0 & 0 & 0 \\ 1 & 0 & 0 \end{pmatrix} \circ R^T = i\alpha u \circ \Delta \circ u^\star, \\
\Delta^k = \begin{pmatrix} 1 & 0 & 0 \\ 0 & 0 & 0 \\ 0 & 0 & (-1)^k \end{pmatrix}, \quad R \in \mathbf{SO}(3),\ u \in \mathbf{SU}(3),\ \alpha^2 = \vec{\alpha}^2 = -\tfrac{1}{2}\,\text{tr}\, \mathcal{O}^2.
\end{cases}
\end{aligned}$$

It leads to the group parametrization:

$$\begin{aligned}
&O = \text{Ad}\, O = e^{\mathcal{O}} = \mathbf{1}_3 + \tfrac{\sin\alpha}{\alpha}\mathcal{O} + \tfrac{1-\cos\alpha}{\alpha^2}\mathcal{O}^2 \in \mathbf{SO}(3), \\
&\text{with} \begin{cases} \mathcal{O}^2 = \mathcal{O} \circ \mathcal{O} = -\alpha^2 R \circ \Delta^2 \circ R^T = -\alpha^2 u \circ \Delta^2 \circ u^\star, \\ \mathcal{O}^3 = \mathcal{O} \circ \mathcal{O} \circ \mathcal{O} = -\alpha^2 \mathcal{O}. \end{cases}
\end{aligned}$$

The determinant of the Lie–Jacobi isomorphisms O_* of $\mathbf{SO}(3)$ as representative of $\mathbf{GL}(3,\mathbb{R})/\mathbf{SO}(3)$,

$$\begin{aligned}
O_* &= e_*^{\mathcal{O}} = \text{“}\frac{\partial O}{\partial \mathcal{O}}\text{”} = \mathbf{1}_3 - \frac{\sin\alpha-\alpha}{\alpha^3}\mathcal{O}^2 + \frac{1-\cos\alpha}{\alpha^2}\mathcal{O} = \mathbf{1}_3 + \tfrac{1}{2}\mathcal{O} + \dots \\
&= u \circ \begin{pmatrix} \frac{e^{i\alpha}-1}{i\alpha} & 0 & 0 \\ 0 & 1 & 0 \\ 0 & 0 & \frac{e^{-i\alpha}-1}{-i\alpha} \end{pmatrix} \circ u^\star = R \circ \begin{pmatrix} \frac{\sin\alpha}{\alpha} & 0 & \frac{1-\cos\alpha}{\alpha} \\ 0 & 1 & 0 \\ -\frac{1-\cos\alpha}{\alpha} & 0 & \frac{\sin\alpha}{\alpha} \end{pmatrix} \circ R^T, \\
\det O_* &= \left(\frac{\sin\frac{\alpha}{2}}{\frac{\alpha}{2}}\right)^2, \quad d^3O = \left(\frac{\sin\frac{\alpha}{2}}{\frac{\alpha}{2}}\right)^2 d\alpha_1 d\alpha_2 d\alpha_3,
\end{aligned}$$

involves the spherical Bessel function, which arises in the representation of three-dimensional quantum scattering structures (see Chapter 8):

$$j_0(r) = \frac{\sin\frac{\alpha}{2}}{\frac{\alpha}{2}} = \int \frac{d^3q}{2\pi}\, \delta(\vec{q}^{\,2}-1)e^{i\vec{q}\vec{x}}, \quad r = |\vec{x}| = \tfrac{\alpha}{2}.$$

The Killing form for the angular momenta,

$$\kappa(\mathcal{O},\mathcal{O}) = \operatorname{tr}\mathcal{O}^2 = -2\alpha_a\delta^{ab}\alpha_b, \quad \kappa(\mathcal{O}^a,\mathcal{O}^b) = \kappa^{ab} = -2\delta^{ab},$$

gives the Killing metric for the rotation group. It is the product of Lie–Jacobi isomorphisms with their transposed ones and contains the spherical Bessel function as metrical coefficient:

$$\begin{aligned}
\mathbf{g}(\vec{\alpha}) = \tfrac{1}{2}O_* \circ \kappa \circ O_*^T &= -\mathbf{1}_3 + \frac{4\sin^2\frac{\alpha}{2}-\alpha^2}{\alpha^4}\mathcal{O} \circ \mathcal{O}^T \\
&= -\left(\tfrac{\vec{\alpha}d\vec{\alpha}}{\alpha}\right)^2 - \left(\frac{\sin\frac{\alpha}{2}}{\frac{\alpha}{2}}\right)^2 \left[d\vec{\alpha}^2 - \left(\tfrac{\vec{\alpha}d\vec{\alpha}}{\alpha}\right)^2\right].
\end{aligned}$$

The volume element and metrical tensor in Euler angles $(\frac{\alpha}{2},\theta,\varphi)$ show the manifold isomorphy $\mathbf{SU}(2) \cong \Omega^3$:

$$\begin{pmatrix}\alpha_1\\ \alpha_2\\ \alpha_3\end{pmatrix} = \alpha\begin{pmatrix}\cos\varphi\sin\theta\\ \sin\varphi\sin\theta\\ \cos\theta\end{pmatrix}, \quad \begin{cases} \frac{1}{8}d^3O &= \sin^2\frac{\alpha}{2}d^2\omega\; d\frac{\alpha}{2}, \\ \text{with } d^2\omega &= \sin\theta\; d\varphi\; d\theta, \\ \mathbf{g}(\vec{\alpha}) &= -d\alpha^2 - 4\sin^2\frac{\alpha}{2}d\omega^2, \\ \text{with } d\omega^2 &= d\theta^2 + \sin^2\theta d\varphi^2. \end{cases}$$

The noncompact partner for the 3-rotations is the Lorentz group $\mathbf{SO}_0(1,2)$ in one and two dimensions with real and imaginary angles for the two Cartan subgroup types $\mathbf{SO}_0(1,1)$ and $\mathbf{SO}(2)$, respectively, diagonalizable for $\det\mathcal{Q} \neq 0$ in the complex to the eigenvalues, either boost components or polarization components, in a Euclidean parametrization:

$$\begin{aligned}
&\mathbb{R}^3 \cong \log\mathbf{SO}_0(1,2) \ni \mathcal{Q} = \operatorname{ad}\mathcal{Q} \\
&\Rightarrow \begin{cases} \mathcal{Q} &= \begin{pmatrix} 0 & \beta_1 & \beta_2 \\ \beta_1 & 0 & \alpha_3 \\ \beta_2 & -\alpha_3 & 0 \end{pmatrix} = \gamma u_\gamma \circ \Delta_\gamma \circ u_\gamma^\star, \quad u_\gamma \in \mathbf{SU}(3), \\ \gamma^2 &= \beta_1^2 + \beta_2^2 - \alpha_3^2 = \frac{1}{2}\operatorname{tr}\mathcal{Q}^2, \\ \Delta_\gamma &= \begin{cases} \operatorname{diag}(1,0,-1) & \text{(boosts)}, \\ \operatorname{diag}(0,1,-1) & \text{(rotations)} \end{cases} \text{for } \gamma = \begin{cases} \sqrt{\gamma^2}, & \gamma^2 > 0, \\ i\sqrt{-\gamma^2}, & \gamma^2 < 0. \end{cases} \end{cases}
\end{aligned}$$

The group and the Lie–Jacobi isomorphisms are parametrized by the exponent:

$$\begin{aligned}
Q &= \mathrm{Ad}\,Q = e^{\mathcal{Q}} = \mathbf{1}_3 + \tfrac{\sinh\gamma}{\gamma}\mathcal{Q} - \tfrac{1-\cosh\gamma}{\gamma^2}\mathcal{Q}^2 \in \mathbf{SO}_0(1,2),\\
Q_* &= e^{\mathcal{Q}}_* = \text{“}\, \tfrac{\partial Q}{\partial \mathcal{Q}} \,\text{”} = \mathbf{1}_3 + \tfrac{\sinh\gamma-\gamma}{\gamma^3}\mathcal{Q}^2 - \tfrac{1-\cosh\gamma}{\gamma^2}\mathcal{Q} = \mathbf{1}_3 + \tfrac{1}{2}\mathcal{Q} + \ldots,\\
\det Q_* &= \left(\tfrac{\sinh\frac{\gamma}{2}}{\frac{\gamma}{2}}\right)^2, \quad d^3Q = \left(\tfrac{\sinh\frac{\gamma}{2}}{\frac{\gamma}{2}}\right)^2 d\beta_1 d\beta_2 d\alpha_3.
\end{aligned}$$

The determinant contains for $\gamma \in \mathbb{R}$ the hyperbolic Macdonald function with the Yukawa potential and for $\gamma \in i\mathbb{R}$ with $\sinh i\alpha = i\sin\alpha$ the spherical Bessel function:

$$\begin{aligned}
k_0(\xi) &= \tfrac{e^{-\xi}}{\xi}, \quad \tfrac{k_0(-\xi)-k_0(\xi)}{2} = \tfrac{\sinh\xi}{\xi},\\
k_0(r) &= \int \tfrac{d^3q}{2\pi^2}\tfrac{1}{\vec{q}^2+1}e^{i\vec{q}\vec{x}}, \quad r = |\vec{x}|.
\end{aligned}$$

The Lie algebra Killing form gives the Killing metric for the Lorentz group $\mathbf{SO}_0(1,2)$:

$$\begin{aligned}
\kappa(\mathcal{Q}^a, \mathcal{Q}^b) = \kappa^{ab} &\cong 2\begin{pmatrix} 1 & 0 & 0\\ 0 & 1 & 0\\ 0 & 0 & -1\end{pmatrix}, \quad \gamma\delta = \gamma_1\delta_1 + \gamma_2\delta_2 - \gamma_3\delta_3,\\
\mathbf{g}(\gamma) = \tfrac{1}{2}Q_* \circ \kappa \circ Q_*^T &= \tfrac{(\gamma d\gamma)^2}{\gamma^2} + \left(\tfrac{\sinh\frac{\gamma}{2}}{\frac{\gamma}{2}}\right)^2 \left[d\gamma^2 - \tfrac{(\gamma d\gamma)^2}{\gamma^2}\right]\\
&= d\gamma^2 + 4\sinh^2\tfrac{\gamma}{2} d\omega^2.
\end{aligned}$$

2.4.6 Adjoint and Killing Connection of Lie Groups

For Lie groups $G \in \underline{\mathbf{lgrp}}_{\mathbb{R}}$, the geometrical concepts of connection, torsion, and curvature can be given by properties at the group unit $\mathbf{T}_1(G) = L \in \underline{\mathbf{lag}}_{\mathbb{R}}$.

A Lie group has many connections: There is a bijection between bilinear Lie algebra mappings $\gamma : L \times L \longrightarrow L$ and left-invariant connections $\nabla_l m = \gamma(l,m)$. The Lie bracket gives inner Lie algebra derivations. For a semisimple Lie algebra, all derivations are inner. Distinguished Lie bracket-related connections are defined, up to a real factor, by an inner derivation:

$$\text{at } 1 \in G: \ \nabla_l = \alpha\,\mathrm{ad}\,l, \ \alpha \in \mathbb{R} \Rightarrow \begin{cases} \mathcal{T}(l \wedge m) &= (2\alpha - 1)[l,m],\\ \mathcal{R}(l \wedge m) &= \alpha^2[\mathrm{ad}\,l, \mathrm{ad}\,m] - \alpha\,\mathrm{ad}\,[l,m]\\ &= \alpha(\alpha-1)\,\mathrm{ad}\,[l,m]. \end{cases}$$

Precisely for abelian groups with $L = \mathbb{R}^n$ (“translations”), Lie bracket-induced connections are trivial. There are two “extreme” cases for Lie bracket-related connections, either with trivial curvature or with trivial torsion,

$$\begin{aligned}
\alpha = 1 &\Rightarrow \mathcal{R} = 0,\\
\alpha = \tfrac{1}{2} &\Rightarrow \mathcal{T} = 0.
\end{aligned}$$

A direct identification of a covariant derivative with the adjoint action $\alpha = 1$ defines the *adjoint connection.* It identifies the covariant and Lie derivation. The Lie bracket is the torsion. The mapping $l \longmapsto \nabla_l$ is a Lie algebra morphism; i.e., the curvature is trivial,

$$\text{at } 1 \in G: \ \nabla_l = \operatorname{ad} l \Rightarrow \begin{cases} \mathcal{T}(l \wedge m) = [l, m], \\ \mathcal{R}(l \wedge m) = 0. \end{cases}$$

Some structures of the adjoint connection arise in teleparallel theories.

The roles of torsion and curvature are "reversed" for the *Killing connection* of a Lie group. A connected Lie group has a unique connection, invariant under left and right translations and group inversion. It is given by inner derivations with trivial torsion:

$$\text{at } 1 \in G: \ \nabla_l = \tfrac{1}{2} \operatorname{ad} l \Rightarrow \mathcal{T}(l \wedge m) = 0.$$

A nontrivial connection with *one half* of the adjoint representation is not a Lie algebra morphism:

$$\begin{aligned} \text{at } 1 \in G: \ \nabla: \ & L \longrightarrow L \otimes L^T, \ \ l \longmapsto \nabla_l = \tfrac{1}{2} \operatorname{ad} l, \\ \nabla_l: \ & L \longrightarrow L, \ \ \begin{cases} \nabla_l\, m = \tfrac{1}{2} \operatorname{ad} l(m) = \tfrac{1}{2}[l, m], \\ \nabla_{l^a} l^b = \tfrac{1}{2} \epsilon_c^{ab} l^c, \ \ \Gamma_c^{ab} = -\Gamma_c^{ba} = \tfrac{1}{2} \epsilon_c^{ab}, \end{cases} \\ \nabla = \tfrac{1}{2} \operatorname{ad} & = \tfrac{1}{2} \epsilon_c^{ab} l^c \otimes \check{l}_b \wedge \check{l}_a \in L \otimes L^T \wedge L^T. \end{aligned}$$

For the Killing connection, the manifold is constituted by the classes of the left–right operation group with the diagonal group $\mathbb{G} = G \times G / \operatorname{diag}(G \times G) \cong G$. The tangent structures at the unit are characterized by the vector space with the corresponding classes in the doubled Lie algebra:

$$L \oplus L \ni (l_1, l_2) \longmapsto l_1 - l_2 \in L \cong \mathbb{L} = L \oplus L / \operatorname{diag}(L \oplus L).$$

In general, the classes are isomorphic $\mathbb{L} \cong L$ as vector space, not as Lie algebra.

The *Killing curvature* at the group unit is, up to a factor $-\frac{1}{4}$, the commutator in the adjoint representation:

$$\begin{aligned} &\text{at } 1 \in G: \\ &\mathcal{R}: \ L \wedge L \longrightarrow L \otimes L^T, \\ &\mathcal{R}(l \wedge m): \ L \longrightarrow L, \end{aligned} \quad \begin{cases} [\nabla_l, \nabla_m] = \tfrac{1}{2} \nabla_{[l,m]} = \tfrac{1}{4} \operatorname{ad}[l, m], \\ \mathcal{R}(l \wedge m) = -\tfrac{1}{4} \operatorname{ad}[l, m] = -\tfrac{1}{4}[\operatorname{ad} l, \operatorname{ad} m], \\ \mathcal{R}(l \wedge m).n = -\tfrac{1}{4}[[l, m], n], \\ \mathcal{R}(l^a \wedge l^b) = -\tfrac{1}{4} \epsilon_c^{ab} \operatorname{ad} l^c = -\tfrac{1}{4} \epsilon_c^{ab} \epsilon_e^{cd}\, l^e \otimes \check{l}_d, \\ \mathcal{R}_e^{dab} = -\tfrac{1}{4} \epsilon_c^{ab} \epsilon_e^{cd} = -\tfrac{1}{4}(\epsilon_e^{ac} \epsilon_c^{bd} - \epsilon_e^{bc} \epsilon_c^{ad}). \end{cases}$$

The name of the connection is motivated by the fact that the Ricci tensor (bilinear form) at the group unit coincides up to a constant $\frac{1}{4}$ with the Killing form:

$$\begin{aligned} &\text{at } 1 \in G: \\ &\mathcal{R}_\bullet: L \vee L \longrightarrow \mathbb{R}, \end{aligned} \quad \begin{cases} \mathcal{R}_\bullet(l, m) = \tfrac{1}{4} \operatorname{tr} \operatorname{ad} l \circ \operatorname{ad} m = \tfrac{1}{4} \kappa(l, m), \\ \mathcal{R}_\bullet(l^d, l^a) = \mathcal{R}_\bullet^{da} = \mathcal{R}_b^{dab} = \tfrac{1}{4} \epsilon_c^{ab} \epsilon_b^{dc} = \tfrac{1}{4} \kappa^{da}. \end{cases}$$

In contrast to the adjoint connection $\nabla = \text{ad}$ (at $1 \in G$), the Killing connection $\nabla = \frac{1}{2}\text{ad}$ (at $1 \in G$) yields, for nondenerate Killing form, i.e., for a semisimple Lie algebra, a Riemannian connection for $\mathbb{G} \cong G$:

$$\begin{array}{l}\text{at } 1 \in G,\\ \text{semisimple } L \cong \mathbb{R}^n:\end{array} \quad \mathcal{R}_{\bullet} = \frac{1}{4}\kappa, \quad \mathcal{R}_{\bullet}^{\bullet} = \frac{n}{4}.$$

An example is the rotation group:

$$\begin{array}{l}\text{at } \mathbf{1}_3 \in \mathbf{SO}(3),\\ \log \mathbf{SO}(3) \cong \mathbb{R}^3:\end{array} \quad \left\{\begin{array}{ll}\epsilon_c^{ab} &= -\epsilon^{abc}, \; a,b,c \in \{1,2,3\}, \; \epsilon^{123} = 1,\\ \mathcal{R}_e^{dab} &= -\frac{1}{4}\epsilon^{abc}\epsilon^{cde} = -\frac{1}{4}(\delta^{ad}\delta^{be} - \delta^{bd}\delta^{ae}),\\ \mathcal{R}_{\bullet}^{ad} &= \frac{1}{4}\kappa(\mathcal{O}^a, \mathcal{O}^d) = \frac{1}{4}\epsilon_c^{ab}\epsilon_b^{dc} = -\frac{1}{2}\delta^{ad}.\end{array}\right.$$

2.5 Riemannian Manifolds

2.5.1 Lorentz Covariant Derivatives

The *Riemannian connection* of a Riemannian manifold $(\mathbb{M}^{(t,s)}, \mathbf{g}) \in \underline{\mathbf{rdif}}_{\mathbb{R}}$, motivated above by the Killing fields of its metric, is uniquely characterized by its torsion freedom and a covariantly constant (derivation trivial) metric. It is given in a holonomic basis $e^j = \partial^j$ by

$$\begin{array}{rl}\mathcal{T} &= 0,\\ \nabla \mathbf{g} &= 0,\end{array} \quad \left\{\begin{array}{ll}\Gamma_k^{ij} &= \Gamma_k^{ji}\\ (jli) &= \partial^j \mathbf{g}^{li} - \Gamma_p^{jl}\mathbf{g}^{pi} - \Gamma_p^{ji}\mathbf{g}^{lp} = 0,\\ (jli) &-(lij) - (ijl) = 0,\\ \Gamma_k^{ij} &= \frac{1}{2}\mathbf{g}_{kp}(\partial^i \mathbf{g}^{pj} + \partial^j \mathbf{g}^{pi} - \partial^p \mathbf{g}^{ij}).\end{array}\right.$$

Both the identity and the metric (dual isomorphism) are covariantly constant $\nabla e^i \otimes \check{e}_i = 0$, $\nabla \mathbf{g}^{il}\check{e}_l \otimes \check{e}_i = 0$, $\nabla \mathbf{g}_{il} e^i \otimes e^l = 0$. With the invariant dual isomorphisms, one has $\mathbf{g}_{li}\nabla a^i = \nabla a_l$.

A (pseudo-)Riemannian manifold with torsion cannot be embedded into a (pseudo-)Euclidean space.

The transformation of the n^2 connection 1-forms from a general, e.g., a holonomic basis $\partial^j \otimes dx_j = \mathbf{e}^a \otimes \check{\mathbf{e}}_a$, to orthonormal bases with invariance Lorentz group involves derivatives of the n-bein (parametrization dependence),

$$\begin{array}{ll}\mathbf{e}(x) &\in \mathbf{GL}(n,\mathbb{R})/\mathbf{SO}_0(t,s):\\ \mathbf{T}^T &\ni \Gamma_k^j = \Gamma_k^{ij}dx_i = \Gamma_k^{cj}\check{\mathbf{e}}_c \longmapsto \boldsymbol{\Gamma}_b^a = \boldsymbol{\Gamma}_b^{ia}dx_i = \boldsymbol{\Gamma}_b^{ca}\check{\mathbf{e}}_c \in \mathbf{T}^T,\\ \boldsymbol{\Gamma}_b^{ia} &= \mathbf{e}_j^a\Gamma_k^{ij}\mathbf{e}_b^k + (\partial^i \mathbf{e}_k^a)\mathbf{e}_b^k \Rightarrow \partial^i \mathbf{e}_k^a + \mathbf{e}_j^a\Gamma_k^{ij} - \boldsymbol{\Gamma}_b^{ia}\mathbf{e}_k^b = 0,\\ \Gamma_k^{ij} &= \mathbf{e}_a^j\boldsymbol{\Gamma}_b^{ia}\mathbf{e}_k^b + (\partial^i \mathbf{e}_b^j)\mathbf{e}_k^b \Rightarrow \partial^i \mathbf{e}_b^j - \Gamma_k^{ij}\mathbf{e}_b^k + \mathbf{e}_a^j\boldsymbol{\Gamma}_b^{ia} = 0.\end{array}$$

Since the n-bein are equivalence classes with the representatives defined up to local $\mathbf{SO}_0(t,s)$-transformations, the Lorentz degrees of freedom show up in an inhomogeneous $\mathbf{SO}_0(t,s)$-transformation behavior,

$$\Lambda(x) \in \mathbf{SO}_0(t,s), \quad \begin{cases} \mathbf{e}_k^c \longmapsto & \Lambda_a^c \mathbf{e}_k^a, \quad \mathbf{e}_b^j \longmapsto \mathbf{e}_d^j \Lambda^{-1}{}_b^d, \\ \mathbf{\Gamma}_b^{ia} \longmapsto & \Lambda_c^a \mathbf{\Gamma}_d^{ic} \Lambda^{-1}{}_b^d + (\partial^i \Lambda_c^a) \Lambda^{-1}{}_b^c, \\ \mathbf{\Gamma}^i \longmapsto & \Lambda \mathbf{\Gamma}^i \Lambda^{-1} + (\partial^i \Lambda) \Lambda^{-1}. \end{cases}$$

Einstein's relativity, as formulated by Weyl for spinor fields, defined in representation spaces for the orthogonal covering group $\overline{\mathbf{SO}_0(1,3)}$, uses the Lorentz group as the structural group. The *Fock–Iwanenkov coefficients* $\mathbf{\Gamma}_{ab}$ for $\mathbf{e} \in \mathbf{GL}(n,\mathbb{R})/\mathbf{SO}_0(t,s)$ are given by the tangent space forms $\mathbf{\Gamma}_b^c$ with Lorentz group gauge transformation behavior. The $\binom{n}{2}$ linearly independent elements, written with $\mathbf{e}^c = \partial^c = \mathbf{e}_j^c \partial^j$,

$$\begin{aligned} \mathbf{\Gamma}_b^{ic} \eta_{ca} &= \mathbf{\Gamma}_{ab}^i = -\mathbf{\Gamma}_{ba}^i \\ &= \mathbf{e}_e^i (\eta_{ac}\eta_{bd} - \eta_{bc}\eta_{ad})(\eta^{df} \mathbf{e}_j^e \partial^c \mathbf{e}_f^j + \eta^{ef} \mathbf{e}_j^d \partial^c \mathbf{e}_f^j - \eta^{df} \mathbf{e}_j^c \partial^e \mathbf{e}_f^j), \end{aligned}$$

can be used for any Lie algebra representation of the tangent Lorentz group $\mathbf{SO}_0(t,s)$ on a vector space V,

$$\begin{aligned} & \log \mathbf{SO}_0(t,s) \ni \mathcal{L}^{ab} \longmapsto \mathcal{D}_V(\mathcal{L}^{ab}) \subseteq \mathbf{AL}(V), \\ \text{e.g., Minkowski representation:} \quad & \mathbf{\Gamma}_{ab}^i \mathcal{D}_n(\mathcal{L}^{ab}) \in \mathbf{T}_x \wedge \mathbf{T}_x \subseteq \mathbf{AL}(\mathbb{R}^n). \end{aligned}$$

The Lorentz covariant derivative of the tensor algebra $\bigotimes(\mathbf{T} \oplus \mathbf{T}^T)$ with the tangent fields and its forms, induced by the affine connection, can be defined, via the Fock–Iwanenkov coeficients, also for the spinor representations, one of $\overline{\mathbf{SO}(t,s)}$ for odd $t+s$, e.g., for the Pauli spinors for $\overline{\mathbf{SO}(3)} \cong \mathbf{SU}(2)$, and two for even $t+s$, e.g., for the Weyl spinor representations of $\overline{\mathbf{SO}_0(1,3)} \cong \mathbf{SL}(2,\mathbb{C})$, for their sums and product representations, i.e., for all fields acted on by a finite-dimensional representation of the Lorentz cover group $\overline{\mathbf{SO}_0(t,s)}$ and its Lie algebra.

The tetrad is more appropriate than the metrical tensor for the formulation of the gravitational interaction (general relativity). It implements the 10 degrees of freedom in the form of operations and allows the inclusion of spinor structures. The Lorentz group as subgroup of the maximal structural group $\mathbf{GL}(4,\mathbb{R})$ is physically relevant for the spinor structures.

2.5.2 Laplace–Beltrami Operator

With the isomorphy of tangent and cotangent fields $\mathbf{T}^T \overset{\mathbf{g}}{\cong} \mathbf{T}$, defined by the metric, the external derivative of a function defines a vector field as its *gradient*:

$$\begin{aligned} \text{grad} &= \mathbf{g}^{-1} \circ d : \mathcal{C}(\mathbb{M}) \xrightarrow{d} \mathbf{T}^T(\mathbb{M}) \xrightarrow{\mathbf{g}^{-1}} \mathbf{T}(\mathbb{M}), \\ f \longmapsto \text{grad } f &= \mathbf{g}_{kj}(\partial^j f)\partial^k \text{ (in holonomic bases)}, \\ \text{from } \langle df, v\rangle &= v(f) = \mathbf{g}(\text{grad } f, v) \text{ for } v \in \mathbf{T}(\mathbb{M}). \end{aligned}$$

The trace of the covariant derivative,

$$\nabla^k v = \nabla^k v_i \partial^i = (\partial^k v_j + \Gamma_j^{ki} v_i)\partial^j, \quad \partial^k v_k + \Gamma_k^{ki} v_i = (\partial^k + \Gamma_j^{jk}) v_k,$$

leads to the *divergence* from vector fields to scalar fields,

$$\begin{aligned}\text{div }: \mathbf{T}(\mathbb{M}) \longrightarrow \mathcal{C}(\mathbb{M}),\ \ v &= v_j\partial^j \longmapsto \text{div } v = (\partial^k + \Gamma_j^{jk})v_k,\\ f\in\mathcal{C}(\mathbb{M}): \ \text{div } fv &= v(f) + f\text{div } v.\end{aligned}$$

It can be written as a metric-normalized derivative,

$$\begin{aligned}|\mathbf{g}| &= -\det \mathbf{g}^{ij} = |\mathbf{e}|^2, \ \ |\mathbf{e}| = \det \mathbf{e}_a^i = \sqrt{|\mathbf{g}|},\\ \mathbf{g}_{ij}d\mathbf{g}^{ij} &= -\mathbf{g}^{ij}d\mathbf{g}_{ij} = 2\mathbf{e}_i^a d\mathbf{e}_a^i = -2\mathbf{e}_a^i d\mathbf{e}_i^a = 2\Gamma_j^{jk}dx_k\\ &= \tfrac{1}{|\mathbf{g}|}d|\mathbf{g}| = d\log|\mathbf{g}| = 2d\log|\mathbf{e}|,\\ \Rightarrow \text{div } v &= (\partial^k + \Gamma_j^{jk})v_k = \tfrac{1}{\sqrt{|\mathbf{g}|}}\partial^k\left(\sqrt{|\mathbf{g}|}v_k\right).\end{aligned}$$

The manifold-integrated divergence of any vector field is trivial:

$$\int_{\mathbb{M}}\sqrt{|\mathbf{g}|}d^nx \ \text{div } v = 0.$$

A *conformal vector field* has the metric as eigentensor. The eigenvalue is, up to a dimension-related factor, the divergence of the vector field:

$$v\in\mathbf{T}: \ \ [v,\mathbf{g}] = \rho(v)\mathbf{g} \Rightarrow \rho(v) = \tfrac{2}{n}\text{div } v.$$

Therefore, a Killing vector field for a global invariance has trivial divergence.

The *Laplace–Beltrami operator* for functions on a Riemannian manifold is the divergence of the gradient. It contains the Lorentz covariant derivative of a vector field and is invariant under the global group $G_{\mathbf{g}}$:

$$\begin{aligned}\partial_{\mathbf{g}}^2 &: \mathcal{C}(\mathbb{M}) \xrightarrow{\text{grad}} \mathbf{T}(\mathbb{M}) \xrightarrow{\text{div}} \mathcal{C}(\mathbb{M}),\\ \partial_{\mathbf{g}}^2 &= \text{div grad } = \tfrac{1}{\sqrt{|\mathbf{g}|}}\partial^k\sqrt{|\mathbf{g}|}\ \ \mathbf{g}_{kj}\partial^j\\ &= \mathbf{g}_{kj}(\partial^k\partial^j - \Gamma_i^{kj}\partial^i) = (\partial_a - \mathbf{\Gamma}_{ab}^b)\partial^a.\end{aligned}$$

It is integration-symmetric:

$$\int_{\mathbb{M}}\sqrt{|\mathbf{g}|}d^nx\ f_1(x)(\partial_{\mathbf{g}}^2 f_2)(x) = \int_{\mathbb{M}}\sqrt{|\mathbf{g}|}d^nx\ (\partial_{\mathbf{g}}^2 f_1)(x)f_2(x).$$

The Laplace–Beltrami operator for a Lie group with semisimple Lie algebra is closely related to its Killing form (see Chapter 10).

2.5.3 Riemannian Curvature

The Riemannian connection is given by the 1-forms in the Cartan equation for trivial torsion, which allows the computation of the curvature 2-forms — in orthogonal bases,

$$\begin{aligned}\mathbf{g} &= \eta^{ab}\check{\mathbf{e}}_a\otimes\check{\mathbf{e}}_b,\\ d\check{\mathbf{e}}_c + \mathbf{\Gamma}_c^b\wedge\check{\mathbf{e}}_b &= 0, \qquad \eta^{ac}\mathbf{\Gamma}_c^b = \mathbf{\Gamma}^{ab} = -\mathbf{\Gamma}^{ba},\\ d\mathbf{\Gamma}_c^d + \mathbf{\Gamma}_c^a\wedge\mathbf{\Gamma}_a^d &= \tfrac{1}{2}\mathcal{R}_c^{dab}\check{\mathbf{e}}_a\wedge\check{\mathbf{e}}_b,\\ d\mathbf{\Gamma}^{cd} - \eta_{ab}\mathbf{\Gamma}^{ca}\wedge\mathbf{\Gamma}^{db} &= \tfrac{1}{2}\mathcal{R}^{cdab}\check{\mathbf{e}}_a\wedge\check{\mathbf{e}}_b = \mathbf{R}^{cd} = -\mathbf{R}^{dc}.\end{aligned}$$

The tangent space endomorphisms, defined by the *Riemannian curvature*, leave the metric invariant. This leads to an additional antisymmetry of the curvature tensor:

$$\mathcal{R}(v \wedge w) : \mathbf{T} \longrightarrow \mathbf{T} : \quad \begin{cases} \mathbf{g}(\mathcal{R}(v \wedge w).u, z) + \mathbf{g}(u, \mathcal{R}(v \wedge w).z) = 0, \\ \mathbf{g}^{kp}\mathcal{R}_p^{lij} = \mathcal{R}^{klij} = -\mathcal{R}^{lkij}. \end{cases}$$

Thus, the curvature defines a symmetric bilinear form of the antisymmetric tensors $\mathbf{T}_x \wedge \mathbf{T}_x \cong \mathbb{R}^{\binom{n}{2}}$:

$$\mathcal{R} : (\mathbf{T} \wedge \mathbf{T}) \vee (\mathbf{T} \wedge \mathbf{T}) \longrightarrow \mathbb{R}, \quad \mathcal{R}(e^k \wedge e^l, e^i \wedge e^j) = \mathcal{R}^{klij} = \mathcal{R}^{ijkl}.$$

For $\dim_{\mathbb{R}} \mathbb{M} = n = 1, 2, 3, 4, \ldots$, the maximal number of independent component takes into account $\binom{n}{4}$ first Bianchi identities, caused by the Lie structure of the derivations and relevant for dimension $n \geq 4$:

$$\binom{\binom{n}{2}+1}{2} - \binom{n}{4} = \frac{n^2(n^2-1)}{12} = 0, 1, 6, 20, \ldots.$$

The *Riemannian Ricci tensor* is a symmetric bilinear form of the tangent translations:

$$\mathcal{R}_\bullet : \mathbf{T} \vee \mathbf{T} \longrightarrow \mathbb{R}, \quad \mathcal{R}_\bullet(e^i, e^l) = \mathcal{R}_j^{lij} = \mathcal{R}^{klij}\mathbf{g}_{kj} = \mathcal{R}_\bullet^{li} = \mathcal{R}_\bullet^{il}.$$

The trace of the composition of inverse metric and Ricci tensor defines the *curvature scalar*:

$$\mathcal{R}_\bullet \circ \mathbf{g}^{-1} : \mathbf{T} \longmapsto \mathbf{T}, \quad \mathcal{R}_\bullet^\bullet : \mathbb{M} \longrightarrow \mathbb{R}, \quad \mathcal{R}_\bullet^\bullet = \mathbf{g}_{li}\mathcal{R}_\bullet^{li} \in \mathcal{C}(\mathbb{M}).$$

There exist *stationary coordinates* at each point of a Riemannian manifold $\mathbb{M} \ni P^0 \cong x^0$, where the Taylor expansion of the metrical tensor gives the orthonormal standard metric, trivial connection coefficients, and the curvature as second-order coefficients (area change),

$$\begin{aligned} \mathbf{g}^{il}(x) &= \eta^{il} + \tfrac{1}{3}\mathcal{R}^{iklp}(x^0)(x_k - x_k^0)(x_p - x_p^0) + \ldots, \\ \mathbf{g}^{il}(x^0) &= \eta^{il}, \quad \Gamma_k^{il}(x^0) = 0. \end{aligned}$$

2.5.4 Einstein Tensor and Conserved Quantities

The second Bianchi identity, as a twice-contracted Jacobi identity, leads, for any dimension, to a covariantly divergence-free Einstein tensor,

$$\mathbf{g}_{pl}\nabla^p \left(\mathcal{R}_\bullet^{li} - \frac{\mathbf{g}^{li}}{2}\mathcal{R}_\bullet^\bullet \right) = 0.$$

With the exception of the metrical tensor $\nabla^p \mathbf{g}^{li} = 0$, the Einstein tensor is, in four dimensions, the only covariantly conserved second-order tensor, that can be built from $\mathbf{g}$ and its first- and second-order derivatives.

For nonflat Riemannian manifolds in general, there are no conserved currents as familiar from flat spacetime theories. The Killing fields $(v^K)_{K=1}^d$ of the global invariance Lie algebra of a Riemannian manifold allow the construction of *conserved quantities* $(\mathcal{V}^K)_{K=1}^d$ in the Killing field directions of the Einstein tensor if it vanishes sufficiently asymptotically, e.g., for "asymptotically flat" spacetimes $\mathbb{M}^{(1,s)}$. The integration over an $(n-1)$-dimensional submanifold, e.g., an s-dimensional position submanifold, uses the volume elements, given with the isomorphies $\mathbf{g} \circ \epsilon$,

$$v^K = v_i^K \partial^i \in L_{\mathbf{g}}: \quad \mathcal{V}^K = \int_{\mathbb{M}^{n-1}} (d^{n-1}x)_l \; v_i^K \left(\mathcal{R}_\bullet^{li} - \tfrac{\mathbf{g}^{li}}{2}\mathcal{R}_\bullet^\bullet\right),$$

$$(d^{n-1}x)_l \in \bigwedge^{n-1} \mathbf{T}^T, \text{ e.g. for } n{=}4: \quad dx_l \leftrightarrow (d^3x)_l {=} \mathbf{g}_{lr}\epsilon^{rijk} dx_i \wedge dx_j \wedge dx_k.$$

For Minkowski spacetime, one obtains the 10 conserved quantities of the Poincaré group: four momenta, three angular momenta, and three boosts. In Einstein's gravity, the integration goes over the energy-momentum tensor $\mathcal{R}_\bullet^{li} - \frac{\mathbf{g}^{li}}{2}\mathcal{R}_\bullet^\bullet = -\kappa \mathbf{T}^{li} - \Lambda \mathbf{g}^{li}$, e.g., $\int_{\mathbb{R}^3} d^3x \; \mathbf{T}^{0i}(x)$.

2.6 Tangent and Operational Metrics

The *tangent metrics* of a Riemannian manifold are the metric of the translations and the Killing form of the orthogonal tangent group for $n \geq 3$. They are symmetric, nondegenerate, and invariant under the full Lie algebra $\log \mathbf{SO}_0(t,s) \cong \mathbb{R}^{\binom{n}{2}}$ with dimension $\binom{n}{2} = ts + \binom{t}{2} + \binom{s}{2}$, the number of linear independent areas,

$$\mathbb{R}^n \times \mathbb{R}^n \longrightarrow \mathbb{R}, \quad \begin{cases} \eta(\mathcal{P}^a, \mathcal{P}^b) = \eta^{ab}, \\ \eta \cong \begin{pmatrix} \mathbf{1}_t & 0 \\ 0 & -\mathbf{1}_s \end{pmatrix}, \end{cases}$$

$$\log \mathbf{SO}_0(t,s) \times \log \mathbf{SO}_0(t,s) \longrightarrow \mathbb{R}, \quad \begin{cases} \kappa(\mathcal{L}^{da}, \mathcal{L}^{bc}) = -(\eta^{db}\eta^{ac} - \eta^{dc}\eta^{ab}), \\ \kappa = -\eta \wedge \eta \cong \begin{pmatrix} \mathbf{1}_{ts} & 0 \\ 0 & -\mathbf{1}_{\binom{t}{2}+\binom{s}{2}} \end{pmatrix}. \end{cases}$$

The curvature and Ricci tensors give additional, possibly degenerate, symmetric bilinear forms of the tangent Lorentz Lie algebra and the translations, respectively, called the *operational metrics* of the manifold:

$$\begin{aligned} \log \mathbf{SO}_0(t,s) \times \log \mathbf{SO}_0(t,s) \longrightarrow \mathbb{R}, \quad \mathcal{R}(\mathcal{L}^{da}, \mathcal{L}^{bc}) &= \mathcal{R}^{dabc} \\ &= \mathbf{e}_k^d \mathbf{e}_l^a \mathcal{R}^{klij} \mathbf{e}_i^b \mathbf{e}_j^c, \\ \mathbb{R}^n \times \mathbb{R}^n \longrightarrow \mathbb{R}, \quad \mathcal{R}_\bullet(\mathcal{P}^a, \mathcal{P}^b) &= \mathcal{R}_\bullet^{ab} \\ &= \mathbf{e}_i^b \mathbf{e}_l^a \mathcal{R}_\bullet^{li} = \eta_{dc}\mathcal{R}^{dabc}. \end{aligned}$$

In contrast to the signatures of the tangent metrics (Killing form and Lorentz metric) on an analytic and connected manifold, the signatures of the operational metrics (curvature and Ricci tensor) may change from point to point, e.g., spherical, flat, and hyperbolic (saddlelike) areas on a deflated rubber ball.

In an orthonormal tangent basis, half the curvature scalar is given by the sum of the $\binom{n}{2}$ diagonal elements of the curvature, with signs $\eta(A) = \pm 1$:

$$\eta^{ab} = \eta(a)\delta^{ab}, \quad \eta(a) = \pm 1 \Rightarrow \begin{cases} \mathcal{R}_{\bullet}^{ab} = \sum_{d=1}^{n} \eta(d)\mathcal{R}^{dabd}, \\ \frac{1}{2}\mathcal{R}_{\bullet}^{\bullet} = \sum_{d<a} \eta(d)\eta(a)\mathcal{R}^{daad} = \sum_{A=1}^{\binom{n}{2}} \eta(A)\mathcal{R}^{AA}, \\ \text{with } A = da, \quad \eta(A) = -\eta(d)\eta(a). \end{cases}$$

At any manifold point, the global symmetry group with the local invariance group is represented by linear transformations $G_{\mathbf{g}} \longrightarrow \mathbf{GL}(n, \mathbb{R})$, $\mathbf{T}_x \cong \mathbb{R}^n$. An n-bein with a left action by the global symmetry group and a right action by the tangent Lorentz group transmutes from $\mathbf{GL}(n, \mathbb{R})$ to $\mathbf{SO}_0(t, s)$. The components of the metric, curvature, and Ricci tensor are representation coefficients of the global symmetry group, invariant under the local invariance group $H_{\mathbf{g}} \subseteq H^{\max}(t, s) \cong \mathbf{SO}_0(t, s)$:

$$\biguplus_{\iota} G_{\mathbf{g}}/H_{\mathbf{g}}^{\iota} \cong \mathbb{M} \ni x \longmapsto \mathbf{g}(x), \ \mathcal{R}(x), \ \mathcal{R}_{\bullet}(x), \ \mathcal{R}_{\bullet}^{\bullet}(x),$$

$$\mathcal{H} \in \log H_{\mathbf{g}} : \begin{cases} \mathcal{R}([\mathcal{H}, \mathcal{L}_1], \mathcal{L}_2) + \mathcal{R}(\mathcal{L}_1, [\mathcal{H}, \mathcal{L}_2]) = 0, \\ \qquad \mathcal{L}_{1,2} \in \log \mathbf{SO}_0(t, s), \\ \mathcal{R}_{\bullet}([\mathcal{H}, \mathcal{P}_1], \mathcal{P}_2) + \mathcal{R}_{\bullet}(\mathcal{P}_1, [\mathcal{H}, \mathcal{P}_2]) = 0, \\ \qquad \mathcal{P}_{1,2} \in \mathbb{R}^n. \end{cases}$$

2.6.1 Invariants

In general, a bilinear vector space form is invariant under a linear group H for

$$\gamma : V \times V \longrightarrow \mathbb{R}, \quad \gamma(h.v, h.w) = \gamma(v, w), \quad h \in H \subseteq \mathbf{GL}(V),$$
$$h \circ \gamma \circ h^T = \gamma.$$

With another H-invariant bilinear nondenerate form $\delta : V \times V \longrightarrow \mathbb{R}$, the coefficients I_k of the characteristic polynomial, especially the determinant I_0 and the trace I_1, for the (γ, δ)-composed V-endomorphisms are H-invariants:

$$\gamma \circ \delta^{-1} : V \longrightarrow V, \quad h \circ (\gamma \circ \delta^{-1}) \circ h^{-1} = \gamma \circ \delta^{-1},$$
$$\det(\gamma \circ \delta^{-1} - \lambda \mathbf{1}_V) = \sum_{k=0}^{\dim_{\mathbb{R}} V} I_k \lambda^k.$$

Therefore, invariants of the local invariance group $H_{\mathbf{g}}$ of a Riemannian manifold can be obtained from the characteristic polynomials with the curvature tensor, which defines endomorphisms of the Lorentz Lie algebra, and the Ricci tensor, which defines endomorphisms of the tangent translations:

$$\begin{array}{rl} \log \mathbf{SO}_0(t,s) & \longrightarrow \log \mathbf{SO}_0(t,s), \\ \mathbf{e}^b \wedge \mathbf{e}^c & \longmapsto \mathcal{R}^{bc}_{da}\mathbf{e}^d \wedge \mathbf{e}^a, \end{array} \quad \left\{ \begin{array}{l} \det [\mathcal{R} \circ (\mathbf{g} \wedge \mathbf{g})^{-1} - \lambda \mathbf{1}_{\binom{n}{2}}], \\ \det [\mathcal{R}^{efbc}(\eta \wedge \eta)_{efda} - \lambda (\mathbf{1}_n \wedge \mathbf{1}_n)^{bc}_{da}], \end{array} \right.$$

$$\begin{array}{rl} \mathbb{R}^n & \longrightarrow \mathbb{R}^n, \\ \mathbf{e}^b & \longmapsto \mathcal{R}_{\bullet}{}^b_a \mathbf{e}^a, \end{array} \quad \left\{ \begin{array}{l} \det (\mathcal{R}_{\bullet} \circ \mathbf{g}^{-1} - \lambda \mathbf{1}_n), \\ \det (\mathcal{R}_{\bullet}^{fb}\eta_{fa} - \lambda \delta^b_a). \end{array} \right.$$

The composed mappings in the characteristic polynomials of degree $\binom{n}{2}$ and n can be taken in an orthonormal basis with $\det \mathbf{e} \circ f \circ \mathbf{e}^{-1} = \det f$ and $\operatorname{tr} \mathbf{e} \circ f \circ \mathbf{e}^{-1} = \operatorname{tr} f$. Functions, especially $\mathbb{R}$-polynomials of invariants, are invariants too. The curvature scalar $\mathcal{R}^{\bullet}_{\bullet} = \operatorname{tr} \mathcal{R}_{\bullet} \circ \mathbf{g}^{-1}$ is an invariant. The maximal $\mathbf{SO}_0(t,s) \supseteq H_{\mathbf{g}}$ has $R = 1, 2, \ldots$ generating invariants for $t + s = n = 2R$ and $n = 2R + 1$.

2.7 Maximally Symmetric Manifolds

A Riemannian manifold $(\mathbb{M}^{(t,s)}, \mathbf{g})$ where the Ricci tensor coincides, up to normalization, with the metric is called an *Einstein manifold*; i.e.,

$$\lambda \in \mathbb{R}: \ \mathcal{R}_{\bullet} = \tfrac{\lambda}{\ell^2}\mathbf{g}, \ \tfrac{1}{2}\mathcal{R}^{\bullet}_{\bullet} = \tfrac{n}{2}\tfrac{\lambda}{\ell^2}.$$

The normalization factor with a length unit ℓ can be absorbed in the definition of $\mathbf{g}$.

A Riemannian manifold has *maximal global symmetry* or, equivalently, *constant curvature,* if the curvature is, up to a constant, the antisymmetric square of the metric $\mathbf{g} \wedge \mathbf{g}$. Then the manifold is Einsteinian with $\lambda = -k(n-1)$, even conformally flat $\mathbf{g} = e^{\gamma}\eta = e^{\gamma}\begin{pmatrix} \mathbf{1}_t & 0 \\ 0 & -\mathbf{1}_s \end{pmatrix}$ with a scalar function e^{γ} (more ahead). For $n \geq 3$, the curvature is, up to a scalar function, the Killing form[3] of the local invariance Lie algebra $\log \mathbf{SO}_0(t,s)$. The invariant scalar $\mathcal{R}^{\bullet}_{\bullet}$ involves a constant $k \in \{\pm 1, 0\}$:

$$\begin{array}{rl} \mathcal{R} & = \frac{k}{\ell^2}\mathbf{g} \wedge \mathbf{g}, \\ \mathcal{R}^{klij} & = \frac{k}{\ell^2}(\mathbf{g}^{ki}\mathbf{g}^{lj} - \mathbf{g}^{kj}\mathbf{g}^{li}) \end{array} \Rightarrow \left\{ \begin{array}{rl} \mathcal{R}_{\bullet} & = -(n-1)\frac{k}{\ell^2}\mathbf{g}, \\ \mathcal{R}_{\bullet}^{li} & = -(n-1)\frac{k}{\ell^2}\mathbf{g}^{li}, \\ \frac{1}{2}\mathcal{R}^{\bullet}_{\bullet} & = -\binom{n}{2}\frac{k}{\ell^2}. \end{array} \right.$$

For an Einstein manifold, the metrically reflected Ricci tensor is the negative Ricci tensor. The Einstein tensor is trivial exactly for $n = 1, 2$, i.e., for

[3]With Schur's lemma, the Killing form of a simple complex Lie algebra is, as an invariant bilinear form, unique up to a scalar factor.

the trivial $\mathbf{SO}(1) = \{1\}$ and the abelian orthogonal tangent Lie algebras, compact $\mathbf{SO}(2)$, and noncompact $\mathbf{SO}_0(1,1)$:

$$\mathcal{R}_\bullet = \frac{\lambda}{\ell^2}\mathbf{g} \Rightarrow \begin{cases} \mathcal{R}_\bullet - \frac{2}{n}\mathbf{g}\mathcal{R}_\bullet^\bullet = \check{\mathcal{R}}_\bullet = -\mathcal{R}_\bullet, \\ \mathcal{R}_\bullet - \frac{\mathbf{g}}{2}\mathcal{R}_\bullet^\bullet = -\frac{n-2}{2}\frac{\lambda}{\ell^2}\mathbf{g}. \end{cases}$$

2.7.1 Spheres and Hyperboloids

A simply connected complete Riemannian manifold with definite metric and constant curvature is isometric either to a *sphere*, if compact, or, if noncompact, to a *Euclidean space* or to a one-shell "timelike" *hyperboloid*,

$$s = 1, 2, \ldots : \quad \begin{cases} \Omega^s \cong \mathbf{SO}(1+s)/\mathbf{SO}(s), & k = 1, \\ \mathbb{R}^s \cong \mathbf{SO}(s) \vec{\times} \mathbb{R}^s/\mathbf{SO}(s), & k = 0, \\ \mathcal{Y}^s \cong \mathbf{SO}_0(1,s)/\mathbf{SO}(s), & k = -1. \end{cases}$$

The structures for $s = 1$ with $\mathbf{SO}(1) = \{1\}$ are abelian. As manifolds, not as symmetric spaces, hyperboloids and flat spaces are isomorphic, $\mathcal{Y}^s \cong \mathbb{R}^s$. One can define for $s = 0$ the discrete groups $\Omega^0 = \{\pm 1\}$, $\mathbb{R}^0 = \{0\}$, and $\mathcal{Y}^0 = \{1\}$.

The invariant metric and measure for the Euclidean space $\mathbb{R}^s$ are $d\vec{x}_s^{\,2}$ and $d^s x$. Unit spheres and hyperboloids can be parametrized by polar coordinates from the *Eulerian parametrization* of their global symmetry groups with nonabelian degrees of freedom for $s \geq 2$:

$$s \geq 1: \quad \Omega^s \ni \vec{\omega}_s = \begin{pmatrix} \cos\theta \\ \sin\theta \; \vec{\omega}_{s-1} \end{pmatrix} \in \mathbb{R}^{1+s}, \quad \vec{\omega}_0 = 1,$$
$$\mathcal{Y}^s \ni \mathrm{y}_s = \begin{pmatrix} \cosh\psi \\ \sinh\psi \; \vec{\omega}_{s-1} \end{pmatrix} \in \mathbb{R}^{1+s}.$$

Therefrom, one obtains metric and measure,

$$d\omega_s^2 = \begin{cases} d\theta^2, & s = 1, \\ d\theta^2 + \sin^2\theta \; d\varphi^2, & s = 2, \\ d\theta^2 + \sin^2\theta \; d\omega_{s-1}^2, \end{cases} \quad d\mathrm{y}_s^2 = \begin{cases} d\psi^2, & s = 1, \\ d\psi^2 + \sinh^2\psi \; d\varphi^2, & s = 2, \\ d\psi^2 + \sinh^2\psi \; d\omega_{s-1}^2, \end{cases}$$

$$d^s\omega = \sin^{s-1}\theta \; d\theta \; d^{s-1}\omega, \qquad d^s\mathrm{y} = \sinh^{s-1}\psi \; d\psi \; d^{s-1}\omega.$$

The measures involve $\sin\theta$ and $\sinh\psi$ as "radii" of spheres. *Geodesic polar coordinates* are obtained by a transformation of the "leading angle" to a length coordinate:

$$\theta \in [-\tfrac{\pi}{2}, \tfrac{\pi}{2}], \qquad \sin\theta = \rho \in [-1, 1]: \quad \begin{cases} d\omega_s^2 = \frac{d\rho^2}{1-\rho^2} + \rho^2 \, d\omega_{s-1}^2, \\ d^s\omega = \rho^{s-1}\frac{d\rho}{\sqrt{1-\rho^2}} \, d^{s-1}\omega, \end{cases}$$

$$\psi \in]-\infty, \infty[, \quad \sinh\psi = \rho \in]-\infty, \infty[: \quad \begin{cases} d\mathrm{y}_s^2 = \frac{d\rho^2}{1+\rho^2} + \rho^2 \, d\omega_{s-1}^2, \\ d^s\mathrm{y} = \rho^{s-1}\frac{d\rho}{\sqrt{1+\rho^2}} \, d^{s-1}\omega. \end{cases}$$

The compact spheres have a coordinate singularity at $\rho^2 = 1$.

Spheres and hyperboloids are conformally Euclidean, as seen explicitly in a *Cartesian parametrization,*

$$k = (1, 0, -1) \cong (\Omega^s, \mathbb{R}^s, \mathcal{Y}^s): \begin{cases} \frac{d\rho^2}{1-k\rho^2} + \rho^2 d\omega_{s-1}^2 &= \frac{d\vec{x}_s^{\,2}}{(1+k\frac{\vec{x}^2}{4})^2}, \\ \text{where } d\vec{x}_s^{\,2} &= dr^2 + r^2 d\omega_{s-1}^2, \\ \text{with } \rho &= \frac{r}{1+k\frac{r^2}{4}},\ r^2 = \vec{x}^2, \\ \text{and } r &= \left(2\tan\frac{\theta}{2}, r, 2\tanh\frac{\psi}{2}\right), \\ (d^s\omega, d^s x, d^s \mathrm{y}) &= \frac{d^s x}{(1+k\frac{\vec{x}^2}{4})^s}, \\ d^s x &= r^{s-1} dr\ d^{s-1}\omega. \end{cases}$$

The disk $r < 2$ for the 2-hyperboloid $\mathcal{Y}^2$ (Poincaré's model) has the non-Euclidean geometry $\frac{d\vec{x}_2^{\,2}}{(1-\frac{r^2}{4})^2}$ where the conformal factor for the length expansion goes to infinity at the boundary $r \to 2$. For half of the 2-sphere Ω^2, projected on the disk $r \le 2$, the non-Euclidean geometry $\frac{d\vec{x}_2^{\,2}}{(1+\frac{r^2}{4})^2}$ arises with the conformal factor for the length contraction. The finite distances on the nonflat geodesics are given by the "leading angle":

$$d(0, r) = \int_0^r \frac{dR}{1+k\frac{R^2}{4}} = \begin{cases} 4\arctan\frac{r}{2} = 2\theta, & k = +1, \\ 2\log\frac{2+r}{2-r} = 4\operatorname{artanh}\frac{r}{2} = 2\psi, & k = -1. \end{cases}$$

The one-shell spacelike hyperboloids have as pseudometric and invariant measure:

$$\begin{aligned} s \ge 1: \quad \mathbf{SO}_0(1, 1+s)/\mathbf{SO}_0(1, s) \cong \mathcal{Y}^{(1,s)} \ni \mathrm{y}_{(\vec{1},s)} &= \begin{pmatrix} \sinh\psi \\ \cosh\psi\ \vec{\omega}_s \end{pmatrix} \in \mathbb{R}^{2+s}, \\ d\mathrm{y}_{(1,s)}^2 &= d\psi^2 - \cosh^2\psi\ d\omega_s^2, \\ d^{(1,s)}\mathrm{y} &= \cosh^s\psi\ d\psi\ d^s\omega. \end{aligned}$$

2.7.2 Constant-Curvature Manifolds

More generally, precisely for a constant curvature, the Riemannian manifold is a general hyperboloid (sphere for positive signature), or a flat vector space,

$$\begin{aligned} \mathcal{Y}^{(t,s)} &\cong \mathbf{SO}_0(t, 1+s)/\mathbf{SO}_0(t, s), & k &= 1, \\ \mathbb{R}^n &\cong \mathbf{SO}_0(t, s) \,\vec{\times}\, \mathbb{R}^n/\mathbf{SO}_0(t, s), & k &= 0, \\ \mathcal{Y}^{(s,t)} &\cong \mathbf{SO}_0(1+t, s)/\mathbf{SO}_0(t, s), & k &= -1. \end{aligned}$$

The numbers (t, s) and (s, t) denote the (noncompact, compact) dimensions in the hyperboloids, e.g., noncompact $t(1 + s) - ts = t$ for $\mathcal{Y}^{(t,s)}$. Examples are the (anti-)de Sitter universes $\mathcal{Y}^{(1,3)}$ and $\mathcal{Y}^{(3,1)}$. The flat spaces $\mathbb{R}^{(t,s)} = \mathbf{SO}_0(t, s) \,\vec{\times}\, \mathbb{R}^n/\mathbf{SO}_0(t, s) \cong \mathbb{R}^n$ have to be distinguished, as symmetric spaces, from the abelian groups $\mathbb{R}^n$ for $n \ge 2$.

The tangent Poincaré groups, isomorphic for these three manifolds, arise as Inönü–Wigner contractions of the global symmetry groups,

$$\begin{array}{ll} & \mathbf{SO}_0(1+t,s) \to \quad \mathbf{SO}_0(t,s)\ \vec{\times}\ \mathbb{R}^n \quad \leftarrow \mathbf{SO}_0(t,1+s),\\ \left.\begin{array}{rl}(t,s) &= (1,0),\\ (\theta,\psi) &= \frac{x}{c},\\ c &\to \infty\end{array}\right\} & \begin{pmatrix}\cos\theta & c\sin\theta\\ -\frac{1}{c}\sin\theta & \cos\theta\end{pmatrix} \to \begin{pmatrix}1 & x\\ 0 & 1\end{pmatrix} \leftarrow \begin{pmatrix}\cosh\psi & c\sinh\psi\\ \frac{1}{c}\sinh\psi & \cosh\psi\end{pmatrix}.\end{array}$$

In the complex, the two hyperboloids are related to each other by the compact–noncompact transition $i\theta \leftrightarrow \psi$. All three maximal global symmetry groups are subgroups of $\mathbf{SO}_0(1+t,1+s)$, characterizable as fixgroups of nontrivial vectors $x \in \mathbb{R}^{2+t+s}$ with negative, trivial, and positive square x^2 as familiar from the $\mathbf{SO}_0(1,3)$-subgroups $\mathbf{SO}_0(1,2)$, $\mathbf{SO}(2)\ \vec{\times}\ \mathbb{R}^2$, and $\mathbf{SO}(3)$ as fixgroups for spacelike, nontrivial lightlike, and timelike $x \in \mathbb{R}^4$, respectively.

The metrical tensor is invariant under the maximal local group $\mathbf{SO}_0(t,s)$: i.e., it is conformal to a flat bilinear form. The overall Lorentz invariant dilation factor for the metrical tensor depends on the invariant x^2:

$$\begin{aligned} \mathbf{g}^{li}(x) &= \mathbf{e}(x)^2\eta^{li}, \quad \mathbf{e}(x) = e^{\gamma(x)}\mathbf{1}_n &&= \tfrac{\mathbf{1}_n}{1+k\frac{x^2}{4}} \in \mathbf{D}(1) \text{ with } x^2 = x^r\eta_{rm}x^m,\\ & &\gamma(x) &= -\log(1+k\tfrac{x^2}{4}) = k\tfrac{x^2}{4} + \ldots,\\ \mathcal{R} &= k\mathbf{g}\wedge\mathbf{g} = \quad ke^{2\gamma}\ \eta\wedge\eta &&\cong \mathcal{R}^{klij}(x) = ke^{2\gamma(x)}(\eta^{ki}\eta^{lj} - \eta^{ki}\eta^{lj}).\end{aligned}$$

Is is possible to use points with $x = 0$.

From the cases with definite metric $\mathcal{Y}^{(0,s)} = \Omega^s$ and $\mathcal{Y}^{(s,0)} = \mathcal{Y}^s$, one generalizes to hyperboloids $\mathcal{Y}^{(t,s)}$ and $\mathcal{Y}^{(s,t)}$ with $\vec{\omega_0} = 1$, $d\omega_0^2 = 0$, and $d^0\omega = 0$ (sloppy ω-notation):

$$\begin{aligned} t \geq 1,\ \mathbf{SO}_0(t,1+s)/\mathbf{SO}_0(t,s) &\cong \mathcal{Y}^{(t,s)} \ni \vec{\mathrm{y}_{(t,s)}} = \begin{pmatrix}\sinh\psi\ \vec{\omega_{t-1}}\\ \cosh\psi\ \vec{\omega_s}\end{pmatrix} \in \mathbb{R}^{1+t+s},\\ d\mathrm{y}^2_{(t,s)} &= d\psi^2 + \sinh^2\psi d\omega^2_{t-1} - \cosh^2\psi\ d\omega_s^2,\\ d^{(t,s)}\mathrm{y} &= \sinh^{t-1}\psi\ d^{t-1}\omega\ \cosh^s\psi\ d^s\omega\ d\psi,\\ s \geq 1,\ \mathbf{SO}_0(1+t,s)/\mathbf{SO}_0(t,s) &\cong \mathcal{Y}^{(s,t)} \ni \vec{\mathrm{y}_{(s,t)}} = \begin{pmatrix}\cosh\psi\ \vec{\omega_t}\\ \sinh\psi\ \vec{\omega_{s-1}}\end{pmatrix} \in \mathbb{R}^{1+t+s},\\ d\mathrm{y}^2_{(s,t)} &= d\psi^2 - \cosh^2\psi d\omega_t^2 + \sinh^2\psi\ d\omega^2_{s-1},\\ d^{(s,t)}\mathrm{y} &= \cosh^t\psi\ d^t\omega\ \sinh^{s-1}\psi\ d\psi d^{s-1}\omega.\end{aligned}$$

2.8 Rotation-Symmetric Manifolds

The Schwarzschild metric characterizes a spacetime manifold that embeds a parabolic position manifold $\mathbb{P}^3$. Starting from the one-dimensional parabola,

$$\begin{aligned} \mathbb{P}^1:\ \zeta^2 &= 4\ell_0(\rho - 2\ell),\ \ \ell_0\ell \neq 0,\ \ \mathrm{p} = \begin{pmatrix}\zeta\\ \rho\end{pmatrix} = \begin{pmatrix}\pm\sqrt{4\ell_0(\rho-2\ell)}\\ \rho\end{pmatrix} \in \mathbb{R}^2,\\ d\mathrm{p}^2 &= d\rho^2 + d\zeta^2 = \left(1 + \tfrac{\ell_0}{\rho - 2\ell}\right)d\rho^2 = \tfrac{d\rho^2}{1-\frac{2\ell}{\rho}} \text{ for } \ell_0 = 2\ell,\end{aligned}$$

paraboloids arise in Euclidean $\mathbb{R}^{1+s}$ by axial rotations. They have as geodesic polar and Cartesian parametrization:

$$\begin{aligned}\mathbb{P}^s: \ \mathtt{p}_s &= \begin{pmatrix} \pm\sqrt{4\ell_0(\rho-2\ell)} \\ \rho\, \omega_{\vec{s}-1}\end{pmatrix} \in \mathbb{R}^{1+s}, \quad \omega_{\vec{s}-1}\in\Omega^{s-1}, \quad \tfrac{\rho^2}{4\ell^2}\geq 1,\\ d\mathtt{p}_s^2 &= \left(1+\tfrac{\ell_0}{\rho-2\ell}\right)d\rho^2+\rho^2 d\omega_{s-1}^2\\ &= \tfrac{d\rho^2}{1-\frac{2\ell}{\rho}}+\rho^2 d\omega_{s-1}^2 = \left(1+\tfrac{\ell}{2r}\right)^4 d\vec{x}_s^2 \text{ for } \ell_0=2\ell \text{ and } \rho=r\left(1+\tfrac{\ell}{2r}\right)^2.\end{aligned}$$

The rotation axis of the not simply connected paraboloid is inside, but not part of, the paraboloid (visualize with $s=2$). The parameter space of $\mathbb{P}^s$ is twice the Euclidean space $\mathbb{R}^s$ (projection of $\mathbb{P}^s$ on $\mathbb{R}^s$) up to a ball around the origin $\{\vec{x}\in\mathbb{R}^s \mid \vec{x}^2 < 4\ell^2\}$. The rotation paraboloid $\mathbb{P}^s$ has $\mathbf{SO}(s)$ as the global symmetry (motion) group with $\mathbf{SO}(s)$ as the local invariance group,

$$\mathbb{P}^s \cong 2 \biguplus_{\rho\geq 2\ell} \{\rho\}\times\Omega^{s-1}.$$

As manifold, $\mathbb{P}^s$ is isomorphic to $\mathbb{R}\times\Omega^{s-1}$ — however, not as homogeneous space, since its global symmetry group is not $\mathbb{R}\times\mathbf{SO}(s)$.

In general, a line $\mathbb{M}^1$, rotated in Euclidean $\mathbb{R}^{1+s}$ with $\mathbf{O}(s)$ for $s=1,2,\dots$,

$$\mathbb{M}^1: \ \zeta = f(\rho), \quad \mathtt{l}_s = \begin{pmatrix} f(\rho)\\ \rho\,\omega_{\vec{s}-1}\end{pmatrix} \in \mathbb{R}^{1+s}, \quad \omega_{\vec{s}-1}\in\Omega^{s-1},$$

is an s-dimensional manifold with the metric:

$$\mathbb{M}^s \supset \Omega^{s-1}: \ d\mathtt{m}_s^2 = a^2(\rho)d\rho^2+\rho^2 d\omega_{s-1}^2 \text{ with } a^2(\rho)=1+f'(\rho)^2,$$

e.g., cylinders $\mathbb{R}\times\Omega^{s-1}$ with $f=1=a$ or a 2-torus with the two radii $c_0\leq c_1$,

$$\mathbb{T}^2\cong\Omega^1\times\Omega^1: \ d\mathtt{t}_2^2 = \tfrac{c_0^2}{c_0^2-(\rho-c_1)^2}d\rho^2+\rho^2 d\varphi^2.$$

In general, the transformations for $\mathbf{SO}(s-1)$-invariant metrics between geodesic polar and Cartesian parametrizations read with integration constant $c_0\in\mathbb{R}$,

$$a^2(\rho)d\rho^2+\rho^2 d\omega_{s-1}^2 = \tfrac{\xi^2(r)}{r^2}(dr^2+r^2 d\omega_{s-1}^2), \quad \begin{cases} a(\xi)\frac{d\xi}{dr} &= \frac{\xi}{r},\\ \int\frac{d\xi}{\xi}a(\xi) &= \log c_0 r.\end{cases}$$

2.9 Basic Riemannian Manifolds

A Riemannian metric, symmetric and nondegenerate, determines "its" connection. The related curvature, Ricci tensor, and curvature scalar involve up to second-order derivatives of the metric

$$\mathbf{g} \stackrel{\partial^2}{\longmapsto} \mathcal{R},\ \mathcal{R}_\bullet,\ \mathcal{R}^\bullet_\bullet.$$

This transition will be considered in all details for the manifolds with dimensions 1 and 2 and for rotation-invariant manifolds of dimensions 3 and 4 in an orthonormal basis (n-bein), $\mathbf{g} = ds^2 = \eta^{ab}\check{\mathbf{e}}_a \otimes \check{\mathbf{e}}_b,\ \check{\mathbf{e}}_a(x) = \mathbf{e}_a^j(x)dx_j$. The curvature and Ricci tensors with representation coefficients of the global symmetry group $G_\mathbf{g}$ yield symmetric bilinear forms of the orthogonal Lie algebra $\mathbf{T}_x \wedge \mathbf{T}_x \cong \log \mathbf{SO}_0(t,s)$ and of the translations $\mathbf{T}_x \cong \mathbb{R}^n$, respectively, invariant under the action of the local invariance group of the manifold $H_\mathbf{g} \subseteq G_\mathbf{g} \cap \mathbf{SO}_0(t,s)$. The curvature scalar, if nontrivial, defines an $H_\mathbf{g}$-invariant dilation.

The solution of Einstein's equations requires the opposite transition (integration) from curvature and Ricci tensor (energy-momentum tensor) to Riemannian metric.

2.9.1 Manifolds with Dimension 1

A "sufficiently smooth" dilation factor for a one-dimensional Riemannian manifold with abelian tangent group $\mathbb{R}$ can be absorbed by a reparametrization:

$$\mathbb{M}^1: \quad \mathbf{g} = b^2(\tau)d\tau^2 = dt^2 \text{ with } t = \int_{C_0}^{\tau} ds\ b(s), \quad \partial_t = \tfrac{1}{b}\partial_\tau.$$

The manifold is isomorphic either to the circle $\Omega^1 \cong \mathbf{SO}(2) \cong \mathbb{R}/\mathbb{Z}$ or to the one-branch hyperbola and real one-dimensional line $\mathcal{Y}^1 \cong \mathbf{SO}_0(1,1) \cong \mathbb{R}$. The curvature is trivial.

2.9.2 Manifolds with Dimension 2

Two-dimensional Riemannian manifolds, visualizable in ordinary 3-position space, are the origin of the curvature concept with the change of an area and the nontrivial real groups $\mathbf{SO}(2)$ for axial rotations and $\mathbf{SO}_0(1,1)$ for Lorentz dilations. The Riemannian operation groups for nonflat manifolds with $(0,2)$- or $(1,1)$-signature are basic for the curvature concept in $n \geq 2$ dimensions:

Manifold	Tangent Poincaré group	Global symmetry group $G_\mathbf{g}$	Local invariance group
$\mathbb{M}^{(0,2)}$	$\mathbf{SO}(2) \vec{\times} \mathbb{R}^2$	$G^{\max}(0,2) \in \{\mathbf{SO}(3), \mathbf{SO}_0(1,2)\}$	$H_\mathbf{g} \subseteq G_\mathbf{g} \cap \mathbf{SO}(2)$
$\mathbb{M}^{(1,1)}$	$\mathbf{SO}_0(1,1) \vec{\times} \mathbb{R}^2$	$G^{\max}(1,1) \cong \mathbf{SO}_0(1,2)$	$H_\mathbf{g} \subseteq G_\mathbf{g} \cap \mathbf{SO}_0(1,1)$

The transition from a Riemannian metric to its Ricci tensor reproduces the tangent space metric up to a factor, given by the one curvature component. The derivatives are denoted by $\dot{a} = \partial_\tau a$, $a' = \partial_\rho a$, etc.:

$$\begin{aligned}
\mathbf{g} &= b^2(\tau,\rho)d\tau^2 \pm a^2(\tau,\rho)d\rho^2, \quad \begin{pmatrix}\check{\mathbf{e}}_1\\ \check{\mathbf{e}}_2\end{pmatrix} = \begin{pmatrix} b\ d\tau\\ a\ d\rho\end{pmatrix},\\
\begin{pmatrix}d\check{\mathbf{e}}_1\\ d\check{\mathbf{e}}_2\end{pmatrix} &= \begin{pmatrix} b'\ d\rho\wedge d\tau\\ \dot{a}\ d\tau\wedge d\rho\end{pmatrix}, \quad \mathbf{\Gamma}^{12} = \tfrac{\dot{a}}{b}\ d\rho \mp \tfrac{b'}{a} d\tau = \tfrac{1}{ba}(\dot{a}\check{\mathbf{e}}_2 \mp b'\check{\mathbf{e}}_1),\\
\mathbf{R}^{12} &= d\mathbf{\Gamma}^{12} = (\partial_\tau \tfrac{\dot{a}}{b} \pm \partial_\rho \tfrac{b'}{a})d\tau\wedge d\rho = \kappa_\pm \check{\mathbf{e}}_1\wedge\check{\mathbf{e}}_2,\\
\mathcal{R}^{1212} &= \kappa_\pm, \quad \mathcal{R}_\bullet^{ab} \cong \pm\kappa_\pm \begin{pmatrix} 1 & 0\\ 0 & \pm 1\end{pmatrix} = \begin{cases} \kappa_+ \mathbf{1}_2, \\ -\kappa_- \eta_2, \end{cases} \quad \tfrac{1}{2}\mathcal{R}_\bullet^\bullet = \pm\kappa_\pm.
\end{aligned}$$

There is only one linear independent area for two dimensions. Therefore, curvature, Ricci tensor, and curvature scalar contain the same relevant coefficient:

$$\kappa_\pm(\tau,\rho) = \tfrac{1}{ba}(\partial_\tau \tfrac{\dot{a}}{b} \pm \partial_\rho \tfrac{b'}{a}).$$

The Einstein tensor of two-dimensional gravity with abelian Lorentz and local invariance group is trivial:

$$\mathcal{R}_\bullet - \tfrac{\mathbf{g}}{2}\mathcal{R}_\bullet^\bullet = 0.$$

The following three examples are characteristic for the curvature concept: the two-dimensional sphere (Riemannian) and the two one-shell hyperboloids, timelike (Riemannian and simply connected) and spacelike (pseudo-Riemannian and doubly connected). They have the smallest real simple Lie groups as maximal operation groups and the abelian Cartan subgroups $\mathbf{SO}(2)$ and $\mathbf{SO}_0(1,1)$ as local invariance groups:

$$\begin{aligned}
\Omega^2 &\cong \mathbf{SO}(3)/\mathbf{SO}(2): & d\omega_2^2 &= d\theta^2 + \sin^2\theta d\varphi^2,\\
\mathcal{Y}^2 &\cong \mathbf{SO}_0(1,2)/\mathbf{SO}(2): & d\mathrm{y}_2^2 &= d\psi^2 + \sinh^2\psi d\varphi^2,\\
\mathcal{Y}^{(1,1)} &\cong \mathbf{SO}_0(1,2)/\mathbf{SO}_0(1,1): & d\mathrm{y}_{(1,1)}^2 &= d\psi^2 - \cosh^2\psi d\varphi^2.
\end{aligned}$$

The curvature $\kappa_\pm = \frac{\ddot{a}}{a} = -k$ is the representation-characterizing invariant $-k = \lambda^2$, $\lambda = (\pm i, \pm 1)$, of the compact and noncompact local abelian group $\mathbf{SO}(2)$ and $\mathbf{SO}_0(1,1)$ with the second-order derivative,

$$(\partial^2 + k)a = 0: \quad (a,k) = \begin{cases} (\sin\theta, 1), \\ (\sinh\psi, -1), \quad (\cosh\psi, -1). \end{cases}$$

This gives the metric proportional curvature and Ricci tensors:

$$\begin{aligned}
\Omega^2: &\quad \mathcal{R}^{1212} = -1, \quad \mathcal{R}_\bullet^{ab} = -\mathbf{1}_2,\\
\mathcal{Y}^2: &\quad \mathcal{R}^{1212} = +1, \quad \mathcal{R}_\bullet^{ab} = +\mathbf{1}_2,\\
\mathcal{Y}^{(1,1)}: &\quad \mathcal{R}^{1212} = +1, \quad \mathcal{R}_\bullet^{ab} = -\eta_2.
\end{aligned}$$

With $\partial_\tau \in L_{\mathbf{g}}$ as the global symmetry group, e.g., for a static manifold, one obtains

$$\mathbf{g} = b^2(\rho)d\tau^2 \pm a^2(\rho)d\rho^2, \quad \kappa_\pm(\rho) = \pm\tfrac{1}{ba}\partial_\rho \tfrac{b'}{a}.$$

Reciprocal metrical coefficients in a self-dual dilation group $\mathbf{SO}_0(1,1) \ni e^{\lambda_3\sigma_3}$ with $e^{\lambda_3} = a$,

$$\mathbf{g} = b^2(\rho)d\tau^2 \pm \tfrac{d\rho^2}{b^2(\rho)}, \; ab = 1 \Rightarrow \kappa_\pm(\rho) = \pm\tfrac{1}{2}\partial_\rho^2 b^2,$$

are used in the Schwarzschild geometry, which, for (1, 1)-spacetime, has a nontrivial Ricci tensor:

$$\mathbb{M}^{(1,1)} \cong \mathbb{R} \times \mathbb{P}^1 : \quad \begin{cases} \mathbf{g} &= (1 - \frac{2}{\rho})d\tau^2 - \frac{d\rho^2}{1-\frac{2}{\rho}}, \\ \mathcal{R}^{1212} &= \kappa_- = \frac{2}{\rho^3}, \quad \mathcal{R}^{ab}_{\bullet} = -\frac{2}{\rho^3}\eta_2. \end{cases}$$

Two-dimensional rotation manifolds have $\mathbf{SO}(2)$ as the global symmetry group, e.g., paraboloid and torus,

$$\begin{aligned} \mathbb{M}^2 \supset \Omega^1 : &\begin{cases} \mathbf{g} &= a^2(\rho)d\rho^2 + \rho^2 d\varphi^2, \\ \mathcal{R}^{1212} &= \kappa_+ = \frac{\partial}{\partial\rho^2}\frac{1}{a^2} = \frac{1}{2\rho}\partial_\rho \frac{1}{a^2} = -\frac{a'}{\rho a^3}, \quad \mathcal{R}^{ab}_{\bullet} = \kappa_+ \mathbf{1}_2, \end{cases} \\ \mathbb{P}^2 : &\ \frac{1}{a^2} = 1 - \frac{2}{\rho}, \qquad \kappa_+ = \frac{1}{\rho^3}, \\ \mathbb{T}^2 = \Omega^1 \times \Omega^1 : &\ \frac{1}{a^2} = 1 - (\frac{\rho - c_0}{c_1})^2, \ \kappa_+ = \frac{1}{c_1^2}\frac{c_0 - \rho}{\rho}. \end{aligned}$$

A constant curvature arises for

$$b^2(\rho) = \frac{1}{a^2(\rho)} = c_2\rho^2 + c_1\rho + c_0 \Rightarrow \mathcal{R}^{1212} = \kappa_\pm = \pm c_2.$$

2.9.3 Manifolds with Dimension 3

Three-dimensional nonflat manifolds have nonabelian Lorentz groups and the following Riemannian operation groups for (0, 3)- or (1, 2)-signature:

Manifold
$\mathbb{M}^{(0,3)}$
$\mathbb{M}^{(1,2)}$

Tangent Poincaré group
$\mathbf{SO}(3) \vec{\times} \mathbb{R}^3$
$\mathbf{SO}_0(1,2) \vec{\times} \mathbb{R}^3$

Global symmetry group $G_{\mathbf{g}}$
$G^{\max}(0,3) \in \{\mathbf{SO}(4),\ \mathbf{SO}_0(1,3)\}$
$G^{\max}(1,2) \in \{\mathbf{SO}_0(1,3),\ \mathbf{SO}_0(2,2)\}$

Killing form
$\log \mathbf{SO}(3) : \ \kappa \cong -\mathbf{1}_3$
$\log \mathbf{SO}_0(1,2) : \ \kappa \cong \begin{pmatrix} \mathbf{1}_2 & 0 \\ 0 & -1 \end{pmatrix} = \begin{pmatrix} 1 & 0 \\ 0 & \eta_2 \end{pmatrix}$

Local invariance group
$H_{\mathbf{g}} \subseteq G_{\mathbf{g}} \cap \mathbf{SO}(3)$
$H_{\mathbf{g}} \subseteq G_{\mathbf{g}} \cap \mathbf{SO}_0(1,2)$

The local invariance group can be nonabelian and the Einstein tensor nontrivial.

A (0, 3)-manifold with a 2-sphere factor $d\omega_2^2 = d\theta^2 + \sin^2\theta d\varphi^2$ has the rotations in the global symmetry group and the axial rotations in the local invariance group:

$$\begin{aligned} &\mathbb{M}^{(0,3)} \supset \Omega^2 : \quad G_{\mathbf{g}} \supseteq \mathbf{SO}(3), \ H_{\mathbf{g}} \supseteq \mathbf{SO}(2), \\ &\mathbf{g} = a^2(\rho)d\rho^2 + \rho^2 d\omega_2^2, \quad \begin{pmatrix} \check{\mathbf{e}}_1 \\ \check{\mathbf{e}}_2 \\ \check{\mathbf{e}}_3 \end{pmatrix} = \begin{pmatrix} a\ d\rho \\ \rho\ d\theta \\ \rho \sin\theta\ d\varphi \end{pmatrix}, \\ &\begin{pmatrix} d\check{\mathbf{e}}_1 \\ d\check{\mathbf{e}}_2 \\ d\check{\mathbf{e}}_3 \end{pmatrix} = \begin{pmatrix} 0 \\ d\rho \wedge d\theta, \\ [\sin\theta\ d\rho + \rho\ \cos\theta\ d\theta] \wedge d\varphi \end{pmatrix} \Rightarrow \begin{cases} \mathbf{\Gamma}^{12} &= \frac{1}{a}d\theta &= \frac{1}{\rho a}\ \check{\mathbf{e}}_2, \\ \mathbf{\Gamma}^{13} &= \frac{1}{a}\sin\theta\ d\varphi &= \frac{1}{\rho a}\ \check{\mathbf{e}}_3, \\ \mathbf{\Gamma}^{23} &= \cos\theta\ d\varphi &= \frac{\cot\theta}{\rho}\ \check{\mathbf{e}}_3 \end{cases} \end{aligned}$$

It arises by rotations (above) of $\zeta = f(\rho)$ with $f'(\rho) = \sqrt{a^2(\rho) - 1}$. With the 2-forms by exterior derivative,

$$\begin{aligned}
d\mathbf{\Gamma}^{12} &= -\tfrac{a'}{a^2}\, d\rho \wedge d\theta &&= -\tfrac{a'}{\rho a^3}\, \check{\mathbf{e}}_1 \wedge \check{\mathbf{e}}_2,\\
d\mathbf{\Gamma}^{13} &= [-\tfrac{a'}{a^2} \sin\theta\, d\rho + \tfrac{1}{a} \cos\theta\, d\theta] \wedge d\varphi &&= [-\tfrac{a'}{\rho a^3}\, \check{\mathbf{e}}_1 + \tfrac{1}{\rho^2 a} \cot\theta\, \check{\mathbf{e}}_2] \wedge \check{\mathbf{e}}_3,\\
d\mathbf{\Gamma}^{23} &= -\sin\theta\, d\theta \wedge d\varphi &&= -\tfrac{1}{\rho^2}\, \check{\mathbf{e}}_2 \wedge \check{\mathbf{e}}_3,
\end{aligned}$$

one obtains the curvature 2-forms that display the local $\mathbf{SO}(2)$-invariance $2 \leftrightarrow 3$:

$$\begin{aligned}
\mathbf{R}^{12} &= d\mathbf{\Gamma}^{12} + \mathbf{\Gamma}^{13} \wedge \mathbf{\Gamma}^{23} = d\mathbf{\Gamma}^{12} &&= -\tfrac{a'}{\rho a^3}\, \check{\mathbf{e}}_1 \wedge \check{\mathbf{e}}_2 = \tfrac{\partial}{\partial \rho^2} \tfrac{1}{a^2}\, \check{\mathbf{e}}_1 \wedge \check{\mathbf{e}}_2,\\
\mathbf{R}^{13} &= d\mathbf{\Gamma}^{13} + \mathbf{\Gamma}^{12} \wedge \mathbf{\Gamma}^{32} = d\mathbf{\Gamma}^{13} - \mathbf{\Gamma}^{12} \wedge \mathbf{\Gamma}^{23} &&= -\tfrac{a'}{\rho a^3}\, \check{\mathbf{e}}_1 \wedge \check{\mathbf{e}}_3 = \tfrac{\partial}{\partial \rho^2} \tfrac{1}{a^2}\, \check{\mathbf{e}}_1 \wedge \check{\mathbf{e}}_3,\\
\mathbf{R}^{23} &= d\mathbf{\Gamma}^{23} + \mathbf{\Gamma}^{21} \wedge \mathbf{\Gamma}^{31} &&= \tfrac{1}{\rho^2}(\tfrac{1}{a^2} - 1)\, \check{\mathbf{e}}_2 \wedge \check{\mathbf{e}}_3.
\end{aligned}$$

The (3×3) curvature matrix as a bilinear form of the tangent Lie algebra $\log \mathbf{SO}(3)$ decomposes as $3 = 1 \oplus 2$ into $\mathbf{SO}(2)$-invariant spaces:

$$\mathcal{R}^{dabc} \cong \left(\begin{array}{cc|c} \mathcal{R}^{1212} & & \\ & \mathcal{R}^{1313} & \\ \hline & & \mathcal{R}^{2323} \end{array}\right) = \left(\begin{array}{c|c} \frac{1}{2\rho}\partial_\rho \frac{1}{a^2} \mathbf{1}_2 & 0 \\ \hline 0 & \frac{1}{\rho^2}(\frac{1}{a^2} - 1) \end{array}\right).$$

The two sums of the two curvature elements determine the Ricci tensor for the $\mathbf{SO}(2)$-invariant bilinear form of the translations $\mathbb{R}^3$:

$$\mathcal{R}^{ab}_\bullet \cong \left(\begin{array}{c|cc} \mathcal{R}^{11}_\bullet & & \\ \hline & \mathcal{R}^{22}_\bullet & \\ & & \mathcal{R}^{33}_\bullet \end{array}\right) = \left(\begin{array}{c|c} \frac{1}{\rho}\partial_\rho \frac{1}{a^2} & 0 \\ \hline 0 & [\frac{1}{2\rho}\partial_\rho \frac{1}{a^2} + \frac{1}{\rho^2}(\frac{1}{a^2} - 1)]\mathbf{1}_2 \end{array}\right),$$

$$\tfrac{1}{2}\mathcal{R}^{\bullet}_{\bullet} = \tfrac{1}{\rho}\partial_\rho \tfrac{1}{a^2} + \tfrac{1}{\rho^2}(\tfrac{1}{a^2} - 1), \quad \mathcal{R}^{ab}_\bullet - \tfrac{1}{2}\delta^{ab}\mathcal{R}^{\bullet}_{\bullet} \cong \left(\begin{array}{c|c} -\frac{1}{\rho^2}(\frac{1}{a^2} - 1) & 0 \\ \hline 0 & -\frac{1}{2\rho}\partial_\rho \frac{1}{a^2} \mathbf{1}_2 \end{array}\right).$$

A trivial curvature scalar characterizes the rotation 3-paraboloid:

$$\tfrac{1}{\rho}\partial_\rho \tfrac{1}{a^2} + \tfrac{1}{\rho^2}(\tfrac{1}{a^2} - 1) = 0 \iff \tfrac{1}{a^2} = 1 - \tfrac{2c_1}{\rho},$$

$$\mathbb{P}^3: \left\{\begin{array}{l} d\mathrm{p}_3^2 = \frac{d\rho^2}{1 - \frac{2}{\rho}} + \rho^2 d\omega_2^2, \quad \frac{1}{a^2} = 1 - \frac{2}{\rho}\\ \quad \Rightarrow \ -2\frac{a'}{a^3} = \frac{2}{\rho^2} = -\frac{1}{\rho}(\frac{1}{a^2} - 1),\\ \mathcal{R}^{dabc} \cong \frac{1}{\rho^3}\left(\begin{array}{c|c} -2 & 0 \\ \hline 0 & \mathbf{1}_2 \end{array}\right), \ \mathcal{R}^{ab}_\bullet \cong \frac{1}{\rho^3}\left(\begin{array}{c|c} 2 & 0 \\ \hline 0 & -\mathbf{1}_2 \end{array}\right), \ \mathcal{R}^{\bullet}_{\bullet} = 0. \end{array}\right.$$

The 3-sphere, Euclidean 3-space, and 3-hyperboloid have a maximal global symmetry group and all the local invariance group $\mathbf{SO}(3)$:

$$\left.\begin{array}{l} \Omega^3 \cong \mathbf{SO}(4)/\mathbf{SO}(3),\\ \mathbb{R}^3 \cong \mathbf{SO}(3) \vec{\times} \mathbb{R}^3/\mathbf{SO}(3),\\ \mathcal{Y}^3 \cong \mathbf{SO}_0(1,3)/\mathbf{SO}(3), \end{array}\right\} \begin{array}{l} \frac{1}{a^2} = 1 - k\rho^2 \Rightarrow \frac{\partial}{\partial \rho^2}\frac{1}{a^2} = -k = \frac{1}{\rho^2}(\frac{1}{a^2} - 1),\\ \mathcal{R}^{dabc} \cong -k\mathbf{1}_3, \ \mathcal{R}^{ab}_\bullet \cong -2k\mathbf{1}_3,\\ \frac{1}{2}\mathcal{R}^{\bullet}_{\bullet} = -3k, \ \mathcal{R}^{ab}_\bullet - \frac{1}{2}\delta^{ab}\mathcal{R}^{\bullet}_{\bullet} \cong k\mathbf{1}_3. \end{array}$$

A spherical parametrization of a manifold, conformal to a 3-sphere, gives the curvature and the Ricci tensor. Now $R' = \partial_\alpha R$,

$$\begin{aligned}
\mathbf{g} &= R^2(\alpha)(d\alpha^2 + \sin^2 \alpha d\omega_2^2),\\
\mathcal{R}_\bullet^{ab} &\cong -\frac{1}{R^2}\left(\begin{array}{c|c} 2 - 2(\cot\alpha + \partial_\alpha)\frac{R'}{R} & 0 \\ \hline 0 & (2 - 3\cot\alpha\frac{R'}{R} - \frac{R''}{R})\mathbf{1}_2 \end{array}\right).
\end{aligned}$$

A hyperbolic parametrization of an $\mathbf{SO}(2)$-invariant $(0,3)$-manifold, conformal to the 3-hyperboloid, is obtained by noncompact–compact transition with $\psi \leftrightarrow i\alpha$ and $R' \leftrightarrow -iR'$:

$$\begin{aligned}
\mathbf{g} &= R^2(\psi)(d\psi^2 + \sinh^2 \psi d\omega_2^2),\\
\mathcal{R}^{dabc} &\cong \frac{1}{R^2}\left(\begin{array}{c|c} 1 + (2\coth\psi + \frac{R'}{R})\frac{R'}{R} & 0 \\ \hline 0 & [1 + (\coth\psi + \partial_\psi)\frac{R'}{R}]\mathbf{1}_2 \end{array}\right),\\
\mathcal{R}_\bullet^{ab} &\cong \frac{1}{R^2}\left(\begin{array}{c|c} 2 + 2(\coth\psi + \partial_\psi)\frac{R'}{R} & 0 \\ \hline 0 & (2 + 3\coth\psi\frac{R'}{R} + \frac{R''}{R})\mathbf{1}_2 \end{array}\right).
\end{aligned}$$

For three dimensions, a diagonal metric can lead to a nondiagonal Ricci tensor, as exemplified with a causal $(1,2)$-manifold $\mathbb{M}^{(1,2)} \supset \Omega^1$ with global symmetry group $\mathbf{SO}(2) \subseteq G_\mathbf{g}$ [details ahead as special case of the (1,3)-case]:

$$\begin{aligned}
\mathbf{g} &= b^2(\tau,\rho)d\tau^2 - [a^2(\tau,\rho)d\rho^2 + \rho^2 d\varphi^2], \quad \kappa_- = \frac{1}{ba}(\partial_\tau \frac{\dot a}{b} - \partial_\rho \frac{b'}{a}),\\
\mathcal{R}^{dabc} &\cong \left(\begin{array}{c|cc} \mathcal{R}^{0101} & & \\ \hline & \mathcal{R}^{0202} & \mathcal{R}^{0212} \\ & \mathcal{R}^{0212} & \mathcal{R}^{1212} \end{array}\right) = \frac{1}{\rho a^2}\left(\begin{array}{c|cc} \rho a^2 \kappa_- & 0 & 0 \\ \hline 0 & -\frac{b'}{b} & -\frac{\dot a}{b} \\ 0 & -\frac{\dot a}{b} & -\frac{a'}{a} \end{array}\right),\\
\mathcal{R}_\bullet^{ab} &\cong \left(\begin{array}{cc|c} \mathcal{R}_\bullet^{00} & \mathcal{R}_\bullet^{01} & \\ \mathcal{R}_\bullet^{10} & \mathcal{R}_\bullet^{11} & \\ \hline & & \mathcal{R}_\bullet^{22} \end{array}\right) = -\kappa_-\left(\begin{array}{cc|c} 1 & 0 & 0 \\ 0 & -1 & 0 \\ \hline 0 & 0 & 0 \end{array}\right) + \frac{1}{\rho a^2}\left(\begin{array}{cc|c} 2\frac{b'}{b} & 2\frac{\dot a}{b} & 0 \\ 2\frac{\dot a}{b} & 2\frac{a'}{a} & 0 \\ \hline 0 & 0 & -\frac{b'}{b} + \frac{a'}{a} \end{array}\right).
\end{aligned}$$

Static spacetime with the motion group $\mathbf{SO}_0(1,1)$ has reciprocal metrical time-independent coefficients,

$$\mathbf{g} = b^2(\rho)d\tau^2 - [\frac{d\rho^2}{b^2(\rho)} + \rho^2 d\varphi^2], \quad ab = 1, \quad \dot b = 0, \quad \kappa_- = -\frac{1}{2}\partial_\rho^2 b^2,$$

$$\begin{aligned}
\mathcal{R}^{dabc} &\cong \left(\begin{array}{c|cc} \mathcal{R}^{0101} & & \\ \hline & \mathcal{R}^{0202} & \\ & & \mathcal{R}^{1212} \end{array}\right) = \left(\begin{array}{c|c} -\frac{1}{2}\partial_\rho^2 b^2 & 0 \\ \hline 0 & -\frac{1}{\rho}\partial_\rho b^2 \eta_2 \end{array}\right),\\
\mathcal{R}_\bullet^{ab} &\cong \left(\begin{array}{cc|c} \mathcal{R}_\bullet^{00} & & \\ & \mathcal{R}_\bullet^{11} & \\ \hline & & \mathcal{R}_\bullet^{22} \end{array}\right) = \left(\begin{array}{c|c} (\frac{1}{2}\partial_\rho^2 b^2 + \frac{1}{\rho}\partial_\rho b^2)\eta_2 & 0 \\ \hline 0 & -\frac{2}{\rho}\partial_\rho b^2 \end{array}\right),\\
\frac{1}{2}\mathcal{R}_\bullet^\bullet &= \frac{1}{2}\partial_\rho^2 b^2 + \frac{2}{\rho}\partial_\rho b^2.
\end{aligned}$$

$(1,2)$-Schwarzschild spacetime has a nontrivial Ricci tensor:

$$\mathbb{M}^{(1,2)} \cong \mathbb{R} \times \mathbb{P}^2 : \quad \left\{\begin{aligned}
\mathbf{g} &= (1 - \frac{2}{\rho})d\tau^2 - [\frac{d\rho^2}{1 - \frac{2}{\rho}} + \rho^2 d\varphi^2],\\
\mathcal{R}^{dabc} &\cong \frac{2}{\rho^3}\left(\begin{array}{c|c} \eta_2 & 0 \\ \hline 0 & 1 \end{array}\right),\\
\mathcal{R}_\bullet^{ab} &\cong \frac{4}{\rho^3}\left(\begin{array}{c|c} \mathbf{0}_2 & 0 \\ \hline 0 & -1 \end{array}\right), \quad \frac{1}{2}\mathcal{R}_\bullet^\bullet = \frac{2}{\rho^3}.
\end{aligned}\right.$$

2.9.4 Rotation-Invariant Four-Dimensional Spacetimes

A rotation-invariant nonflat causal $(1,3)$-spacetime has as Riemannian operation groups:

$$\mathbb{M}^{(1,3)} \supset \Omega^2 : \quad \begin{array}{|c|}\hline \text{Tangent Poincaré group} \\ \hline \mathbf{SO}_0(1,3) \vec{\times} \mathbb{R}^4 \\ \hline \end{array} \quad \begin{array}{|c|}\hline \text{Global symmetry group } G_{\mathbf{g}} \supseteq \mathbf{SO}(3) \\ \hline G^{\max}(1,3) \in \{\mathbf{SO}_0(1,4), \mathbf{SO}_0(2,3)\} \\ \hline \end{array}$$

$$\begin{array}{|c|}\hline \text{Killing form of } \log \mathbf{SO}_0(1,3) \\ \hline \kappa \cong \begin{pmatrix} \mathbf{1}_3 & 0 \\ 0 & -\mathbf{1}_3 \end{pmatrix} = \eta_2 \otimes \mathbf{1}_3 \\ \hline \end{array} \quad \begin{array}{|c|}\hline \text{Local invariance group} \\ \hline \mathbf{SO}(2) \subseteq H_{\mathbf{g}} \subseteq G_{\mathbf{g}} \cap \mathbf{SO}_0(1,3) \\ \hline \end{array}$$

A metric can be given with an orthonormal basis of the cotangent spaces,

$$\mathbf{g} = b^2(\tau,\rho)d\tau^2 - [a^2(\tau,\rho)d\rho^2 + \rho^2 d\omega_2^2] \Rightarrow \begin{pmatrix} \check{\mathbf{e}}_0 \\ \check{\mathbf{e}}_1 \\ \check{\mathbf{e}}_2 \\ \check{\mathbf{e}}_3 \end{pmatrix} = \begin{pmatrix} b\ d\tau \\ a\ d\rho \\ \rho\ d\theta \\ \rho \sin\theta\ d\varphi \end{pmatrix}.$$

The nontrivial connection 1-forms are with $\dot{b} = \partial_\tau b$, $b' = \partial_\rho b$, etc.:

$$\begin{pmatrix} d\check{\mathbf{e}}_0 \\ d\check{\mathbf{e}}_1 \\ d\check{\mathbf{e}}_2 \\ d\check{\mathbf{e}}_3 \end{pmatrix} = \begin{pmatrix} b'\ d\rho \wedge d\tau \\ \dot{a}\ d\tau \wedge d\rho \\ d\rho \wedge d\theta, \\ [\sin\theta\ d\rho + \rho\ \cos\theta\ d\theta] \wedge d\varphi \end{pmatrix} \Rightarrow \begin{cases} \mathbf{\Gamma}^{01} &= \frac{\dot{a}}{b}\ d\rho + \frac{b'}{a}\ d\tau &= \frac{\dot{a}}{ba}\ \check{\mathbf{e}}_1 + \frac{b'}{ba}\ \check{\mathbf{e}}_0, \\ \mathbf{\Gamma}^{12} &= \frac{1}{a} d\theta &= \frac{1}{\rho a}\ \check{\mathbf{e}}_2, \\ \mathbf{\Gamma}^{13} &= \frac{1}{a} \sin\theta\ d\varphi &= \frac{1}{\rho a}\ \check{\mathbf{e}}_3, \\ \mathbf{\Gamma}^{23} &= \cos\theta\ d\varphi &= \frac{\cot\theta}{\rho}\ \check{\mathbf{e}}_3; \end{cases}$$

with the 2-forms by exterior derivative:

$$\begin{aligned} d\mathbf{\Gamma}^{01} &= (\partial_\tau \tfrac{\dot{a}}{b} - \partial_\rho \tfrac{b'}{a})\ d\tau \wedge d\rho &&= \kappa_-\ \check{\mathbf{e}}_0 \wedge \check{\mathbf{e}}_1, \\ d\mathbf{\Gamma}^{12} &= [\partial_\tau \tfrac{1}{a}\ d\tau + \partial_\rho \tfrac{1}{a}\ d\rho] \wedge d\theta &&= \tfrac{1}{\rho}[\tfrac{1}{b}\partial_\tau \tfrac{1}{a}\ \check{\mathbf{e}}_0 + \tfrac{1}{a}\partial_\rho \tfrac{1}{a}\ \check{\mathbf{e}}_1] \wedge \check{\mathbf{e}}_2, \\ d\mathbf{\Gamma}^{13} &= [\partial_\tau \tfrac{1}{a} \sin\theta\ d\tau + \partial_\rho \tfrac{1}{a} \sin\theta\ d\rho + \tfrac{1}{a} \cos\theta\ d\theta] \wedge d\varphi &&= \tfrac{1}{\rho}[\tfrac{1}{b}\partial_\tau \tfrac{1}{a}\ \check{\mathbf{e}}_0 + \tfrac{1}{a}\partial_\rho \tfrac{1}{a}\ \check{\mathbf{e}}_1 \\ &&& \quad + \tfrac{1}{\rho a} \cot\theta\ \check{\mathbf{e}}_2] \wedge \check{\mathbf{e}}_3, \\ d\mathbf{\Gamma}^{23} &= -\sin\theta\ d\theta \wedge d\varphi &&= -\tfrac{1}{\rho^2}\ \check{\mathbf{e}}_2 \wedge \check{\mathbf{e}}_3; \end{aligned}$$

with the $(1,1)$-curvature

$$\kappa_- = \tfrac{1}{ba}(\partial_\tau \tfrac{\dot{a}}{b} - \partial_\rho \tfrac{b'}{a}).$$

Thus, one obtains the curvature 2-forms [there is the local $\mathbf{SO}(2)$-invariance $2 \leftrightarrow 3$]:

$$\begin{aligned} \mathbf{R}^{01} &= d\mathbf{\Gamma}^{01} + \mathbf{\Gamma}^{02} \wedge \mathbf{\Gamma}^{12} + \mathbf{\Gamma}^{03} \wedge \mathbf{\Gamma}^{13} = d\mathbf{\Gamma}^{01} &&= \kappa_-\ \check{\mathbf{e}}_0 \wedge \check{\mathbf{e}}_1, \\ \mathbf{R}^{02} &= d\mathbf{\Gamma}^{02} + \mathbf{\Gamma}^{01} \wedge \mathbf{\Gamma}^{21} + \mathbf{\Gamma}^{03} \wedge \mathbf{\Gamma}^{23} = -\mathbf{\Gamma}^{01} \wedge \mathbf{\Gamma}^{12} &&= -[\tfrac{b'}{ba}\ \check{\mathbf{e}}_0 + \tfrac{\dot{a}}{ba}\ \check{\mathbf{e}}_1] \wedge \tfrac{1}{\rho a}\ \check{\mathbf{e}}_2, \\ \mathbf{R}^{12} &= d\mathbf{\Gamma}^{12} - \mathbf{\Gamma}^{10} \wedge \mathbf{\Gamma}^{20} + \mathbf{\Gamma}^{13} \wedge \mathbf{\Gamma}^{23} = d\mathbf{\Gamma}^{12} &&= [\tfrac{1}{\rho b}\partial_\tau \tfrac{1}{a}\ \check{\mathbf{e}}_0 + \tfrac{1}{\rho a}\partial_\rho \tfrac{1}{a}\ \check{\mathbf{e}}_1] \wedge \check{\mathbf{e}}_2, \\ \mathbf{R}^{23} &= d\mathbf{\Gamma}^{23} - \mathbf{\Gamma}^{20} \wedge \mathbf{\Gamma}^{30} + \mathbf{\Gamma}^{21} \wedge \mathbf{\Gamma}^{31} = d\mathbf{\Gamma}^{23} + \mathbf{\Gamma}^{12} \wedge \mathbf{\Gamma}^{13} &&= (\tfrac{1}{a^2} - 1)\tfrac{1}{\rho^2}\ \check{\mathbf{e}}_2 \wedge \check{\mathbf{e}}_3, \end{aligned}$$

with the (6×6)-curvature matrix as a bilinear form of the Lorentz Lie algebra $\log \mathbf{SO}_0(1,3)$. It decomposes into $\mathbf{SO}(2)$-invariant subspaces of dimensions 1 and 2:

$$\mathcal{R}^{dabc} \cong \left(\begin{array}{c|c||c|c} \mathcal{R}^{0101} & & & \\ \hline & \begin{matrix}\mathcal{R}^{0202} & \\ & \mathcal{R}^{0303}\end{matrix} & & \begin{matrix}\mathcal{R}^{0212} & \\ & \mathcal{R}^{0313}\end{matrix} \\ \hline\hline & & \mathcal{R}^{2323} & \\ \hline & \begin{matrix}\mathcal{R}^{0212} & \\ & \mathcal{R}^{0313}\end{matrix} & & \begin{matrix}\mathcal{R}^{1212} & \\ & \mathcal{R}^{1313}\end{matrix} \end{array}\right)$$

$$= \frac{1}{\rho a^2}\left(\begin{array}{c|c||c|c} \rho a^2 \kappa_- & 0 & 0 & 0 \\ \hline 0 & -\frac{b'}{b}\mathbf{1}_2 & 0 & -\frac{\dot a}{b}\mathbf{1}_2 \\ \hline\hline 0 & 0 & \frac{1-a^2}{\rho} & 0 \\ \hline 0 & -\frac{\dot a}{b}\mathbf{1}_2 & 0 & -\frac{a'}{a}\mathbf{1}_2 \end{array}\right).$$

The bilinear Ricci tensor gives a (4×4)-matrix with subtraces of the curvature as an $\mathbf{SO}(2)$-invariant bilinear form of the spacetime translations $\mathbb{R}^4$:

$$\begin{aligned}
\mathcal{R}_\bullet^{00} &= -\mathcal{R}^{0101} - 2\mathcal{R}^{0202} = -\kappa_- + \tfrac{2b'}{\rho a^2 b},\\
\mathcal{R}_\bullet^{11} &= \mathcal{R}^{0101} - 2\mathcal{R}^{1212} = \kappa_- + \tfrac{2a'}{\rho a^3},\\
\mathcal{R}_\bullet^{01} &= -2\mathcal{R}^{0212} = \tfrac{2\dot a}{\rho a^2 b},\\
\mathcal{R}_\bullet^{22} &= \mathcal{R}_\bullet^{33} = \mathcal{R}^{0202} - \mathcal{R}^{1212} - \mathcal{R}^{2323} = \tfrac{1}{\rho a^2}\left(\tfrac{a'}{a} - \tfrac{b'}{b}\right) - \tfrac{1-a^2}{\rho^2 a^2},
\end{aligned}$$

$$\mathcal{R}_\bullet^{ab} \cong \left(\begin{array}{c|c|c} \mathcal{R}_\bullet^{00} & \mathcal{R}_\bullet^{01} & \\ \hline \mathcal{R}_\bullet^{10} & \mathcal{R}_\bullet^{11} & \\ \hline & & \begin{matrix}\mathcal{R}_\bullet^{22} & \\ & \mathcal{R}_\bullet^{33}\end{matrix} \end{array}\right)$$

$$= -\kappa_- \left(\begin{array}{c|c|c} 1 & 0 & 0 \\ \hline 0 & -1 & 0 \\ \hline 0 & 0 & \mathbf{0}_2 \end{array}\right) + \frac{1}{\rho a^2}\left(\begin{array}{c|c|c} 2\frac{b'}{b} & 2\frac{\dot a}{b} & 0 \\ \hline 2\frac{\dot a}{b} & 2\frac{a'}{a} & 0 \\ \hline 0 & 0 & \left(-\frac{b'}{b} + \frac{a'}{a} - \frac{1-a^2}{\rho}\right)\mathbf{1}_2 \end{array}\right).$$

A trivial Ricci tensor characterizes Schwarzschild $(1,3)$-spacetime:

$$\mathcal{R}_\bullet = 0: \quad \left\{\begin{array}{rl} \mathcal{R}^{0212} &= 0, \\ 2\mathcal{R}^{0202} &= -\mathcal{R}^{0101} = -2\mathcal{R}^{1212} = \mathcal{R}^{2323} \end{array}\right.$$

$$\Rightarrow \quad \begin{array}{c} \dot a = 0, \ \partial_\rho \frac{b'}{a} = -\frac{2}{\rho}\frac{b'}{a}, \\ -\frac{b'}{b} = \frac{a'}{a} = \frac{1-a^2}{2\rho} \end{array} \quad \Rightarrow \left\{\begin{array}{rl} ba &= \text{constant}, \\ \frac{1}{a^2} = b^2 &= 1 - \frac{2c_1}{\rho}. \end{array}\right.$$

It has position paraboloids $\mathbb{P}^3$. The global symmetry group is $\mathbb{R}\times\mathbf{SO}(3)$, with $\mathbf{SO}(2)$ as the local invariance group:

$$\mathbb{M}^{(1,3)} \cong \mathbb{R}\times\mathbb{P}^3: \left\{\begin{array}{l} \mathbf{g} = (1-\frac{2}{\rho})d\tau^2 - \left(\frac{d\rho^2}{1-\frac{2}{\rho}} + \rho^2 d\omega_2^2\right), \\ \mathcal{R}^{dabc} \cong \frac{1}{\rho^3}\left(\begin{array}{c|c||c|c} 2 & 0 & 0 & 0 \\ \hline 0 & -\mathbf{1}_2 & 0 & 0 \\ \hline\hline 0 & 0 & -2 & 0 \\ \hline 0 & 0 & 0 & \mathbf{1}_2 \end{array}\right) = \frac{1}{\rho^3}\,\eta_2 \otimes \left(\begin{array}{c|c} 2 & 0 \\ \hline 0 & -\mathbf{1}_2 \end{array}\right). \end{array}\right.$$

For only two position dimensions with abelian rotations $\mathbf{SO}(2)$ and $(1,2)$-signature, a trivial Ricci tensor requires a flat spacetime — there is no analogue for Schwarzschild $(1,3)$-spacetime,

$$(1,2)\text{-spacetime: } \mathcal{R}_\bullet = 0 \Rightarrow \left\{\begin{array}{l} \dot a = 0, \ \frac{b'}{b} = \frac{a'}{a} = 0, \\ \mathbf{g} = b^2(\tau)d\tau^2 - (d\rho^2 + \rho^2 d\varphi^2). \end{array}\right.$$

The curvature and Ricci tensor are diagonal for time independent position dilation $\dot{a} = 0$, they lead to *static spacetimes*, and the time dependence in b is ineffective,

$$\begin{aligned}
\mathbf{g} &= b^2(\tau,\rho)d\tau^2 - [a^2(\rho)d\rho^2 + \rho^2 d\omega_2^2],\\
\mathcal{R}^{dabc} &\cong \frac{1}{\rho a^2}\left(\begin{array}{c|c||c|c}
-\rho\,\frac{a}{b}\partial_\rho\frac{b'}{a} & 0 & 0 & 0\\ \hline
0 & -\frac{b'}{b}\mathbf{1}_2 & 0 & 0\\ \hline\hline
0 & 0 & \frac{1-a^2}{\rho} & 0\\ \hline
0 & 0 & 0 & -\frac{a'}{a}\mathbf{1}_2
\end{array}\right),\\
\mathcal{R}^{ab}_{\bullet} &\cong \frac{1}{\rho a^2}\left(\begin{array}{c|c|c}
\rho\,\frac{a}{b}\partial_\rho\frac{b'}{a} + 2\frac{b'}{b} & 0 & 0\\ \hline
0 & -\rho\,\frac{a}{b}\partial_\rho\frac{b'}{a} + 2\frac{a'}{a} & 0\\ \hline
0 & 0 & -(\frac{b'}{b} - \frac{a'}{a} + \frac{1-a^2}{\rho})\mathbf{1}_2
\end{array}\right),\\
\tfrac{1}{2}\mathcal{R}^{\bullet}_{\bullet} &= \frac{1}{ab}\partial_\rho\frac{b'}{a} + \frac{2}{\rho a^2}(\frac{b'}{b} - \frac{a'}{a}) + \frac{1-a^2}{\rho^2 a^2};
\end{aligned}$$

e.g., Einstein's static universe with $b = 1$ and $\frac{1}{a^2} = 1 - \rho^2$,

$$\mathbb{M}^{(1,3)} \cong \mathbb{R} \times \Omega^3 : \quad \mathbf{g} = d\tau^2 - d\omega_3^2.$$

A position dependence of time only leads to

$$\mathbf{g} = b^2(\rho)d\tau^2 - (d\rho^2 + \rho^2 d\omega_2^2), \quad \left\{\begin{aligned}
\mathcal{R}^{dabc} &\cong \left(\begin{array}{c|c||c|c}
-\frac{b''}{b} & 0 & 0 & 0\\ \hline
0 & -\frac{1}{\rho}\,\frac{b'}{b}\mathbf{1}_2 & 0 & 0\\ \hline\hline
0 & 0 & 0 & 0\\ \hline
0 & 0 & 0 & 0
\end{array}\right),\\
\mathcal{R}^{ab}_{\bullet} &\cong \left(\begin{array}{c|c|c}
\frac{b''}{b} + \frac{2}{\rho}\,\frac{b'}{b} & 0 & 0\\ \hline
0 & -\frac{b''}{b} & 0\\ \hline
0 & 0 & -\frac{1}{\rho}\,\frac{b'}{b}\mathbf{1}_2
\end{array}\right),\\
\tfrac{1}{2}\mathcal{R}^{\bullet}_{\bullet} &= \frac{b''}{b} + \frac{2}{\rho}\,\frac{b'}{b}.
\end{aligned}\right.$$

Static manifolds with time-independent dilations $\mathbf{SO}_0(1,1)$ like Schwarzschild spacetime have an $(\eta_2 \oplus \mathbf{1}_2)$-Ricci form. They are given in the form of endomorphisms of the Lorentz Lie algebra and the tangent translations:

$$\begin{aligned}
&\mathbf{g} = b^2(\rho)d\tau^2 - [\frac{d\rho^2}{b^2(\rho)} + \rho^2 d\omega_2^2],\\
&\left\{\begin{aligned}
\mathcal{R}^{dabc} &\cong \left(\begin{array}{c|c||c|c}
-\frac{1}{2}\partial_\rho^2 b^2 & 0 & 0 & 0\\ \hline
0 & -\frac{1}{2\rho}\partial_\rho b^2\mathbf{1}_2 & 0 & 0\\ \hline\hline
0 & 0 & \frac{b^2-1}{\rho^2} & 0\\ \hline
0 & 0 & 0 & \frac{1}{2\rho}\partial_\rho b^2\mathbf{1}_2
\end{array}\right),\\
\mathcal{R}^{ab}_{\bullet} &\cong \left(\begin{array}{c|c}
(\frac{1}{2}\partial_\rho^2 b^2 + \frac{1}{\rho}\partial_\rho b^2)\eta_2 & 0\\ \hline
0 & -(\frac{1}{\rho}\partial_\rho b^2 + \frac{b^2-1}{\rho^2})\mathbf{1}_2
\end{array}\right),\\
\tfrac{1}{2}\mathcal{R}^{\bullet}_{\bullet} &= \tfrac{1}{2}\partial_\rho^2 b^2 + \tfrac{2}{\rho}\partial_\rho b^2 + \frac{b^2-1}{\rho^2}.
\end{aligned}\right.
\end{aligned}$$

A trivial curvature scalar,

$$\tfrac{1}{2}\partial_\rho^2 b^2 + \tfrac{2}{\rho}\partial_\rho b^2 + \frac{b^2-1}{\rho^2} = 0 \iff b^2(\rho) = 1 - \frac{2c_1}{\rho} + \frac{c_2}{\rho^2},$$

characterizes Reissner spacetimes with Schwarzschild length $c_1 = \ell_m$ and charge area $c_2 = \ell_z^2$:

$$
\begin{aligned}
\mathbf{g} &= \Big(1 - \tfrac{2c_1}{\rho} + \tfrac{c_2}{\rho^2}\Big) d\tau^2 - \Big(\tfrac{d\rho^2}{1 - \frac{2c_1}{\rho} + \frac{c_2}{\rho^2}} + \rho^2 d\omega_2^2\Big),\\
\mathcal{R}^{dabc} &\cong \tfrac{1}{\rho^3} \left(\begin{array}{c|c||c|c}
2c_1 - \frac{3c_2}{\rho} & 0 & 0 & 0 \\ \hline
0 & -(c_1 - \frac{c_2}{\rho})\mathbf{1}_2 & 0 & 0 \\ \hline\hline
0 & 0 & -2c_1 + \frac{c_2}{\rho} & 0 \\ \hline
0 & 0 & 0 & (c_1 - \frac{c_2}{\rho})\mathbf{1}_2
\end{array}\right),\\
\mathcal{R}_{\bullet}^{ab} &\cong \tfrac{c_2}{\rho^4} \left(\begin{array}{c||c|c}
1 & 0 & 0 \\ \hline\hline
0 & -1 & 0 \\ \hline
0 & 0 & \mathbf{1}_2
\end{array}\right), \quad \mathcal{R}_{\bullet}^{\bullet} = 0.
\end{aligned}
$$

A curvature tensor with a factor $\eta_2 \otimes$ requires the metric for a Schwarzschild spacetime with a cosmological constant $\frac{k}{\ell^2} = \frac{\Lambda}{3}$,

$$
\begin{aligned}
\tfrac{1}{2}\partial_\rho^2 b^2 &= \tfrac{b^2 - 1}{\rho^2} \iff b^2(\rho) = 1 + k\rho^2 - \tfrac{2c_1}{\rho},\\
\mathbf{g} &= \Big(1 + k\rho^2 - \tfrac{2c_1}{\rho}\Big) d\tau^2 - \Big(\tfrac{d\rho^2}{1 + k\rho^2 - \frac{2c_1}{\rho}} + \rho^2 d\omega_2^2\Big),\\
\mathcal{R}^{dabc} &\cong \eta_2 \otimes \left(\begin{array}{c|c}
k - \frac{2c_1}{\rho^3} & 0 \\ \hline
0 & (k + \frac{c_1}{\rho^3})\mathbf{1}_2
\end{array}\right),\\
\mathcal{R}_{\bullet}^{ab} &\cong 3k\eta_4, \quad \tfrac{1}{2}\mathcal{R}_{\bullet}^{\bullet} = 6k.
\end{aligned}
$$

Local $\mathbf{SO}(3)$-invariance implies maximal local $\mathbf{SO}_0(1,3)$-invariance for maximal symmetric (anti-)de Sitter and flat spacetimes:

$$
\mathbb{M}^{(1,3)} \cong (\mathcal{Y}^{(1,3)}, \mathbb{R}^4, \mathcal{Y}^{(3,1)}) : \left\{
\begin{aligned}
\mathbf{g} &= b^2(\rho) d\tau^2 - [\tfrac{d\rho^2}{b^2(\rho)} + \rho^2 d\omega_2^2],\\
\tfrac{b^2 - 1}{\rho^2} &= \tfrac{1}{2\rho}\partial_\rho b^2 \Rightarrow b^2(\rho) = 1 + k\rho^2,\\
\mathcal{R}^{dabc} &\cong k\mathbf{1}_{(3,3)} \text{ for } k = (1, 0, -1).
\end{aligned}\right.
$$

Nondiagonal elements arise for only time-dependent dilations. The time dilation factor can be absorbed in a reparametrization, $b = 1$:

$$
\begin{aligned}
\mathbf{g} &= d\tau^2 - [a^2(\tau) d\rho^2 + \rho^2 d\omega_2^2],\\
\mathcal{R}_{\bullet}^{ab} &\cong -\tfrac{\ddot{a}}{a} \left(\begin{array}{c|c|c}
1 & 0 & 0 \\ \hline
0 & -1 & 0 \\ \hline
0 & 0 & \mathbf{0}_2
\end{array}\right) + \tfrac{1}{\rho a^2} \left(\begin{array}{c|c|c}
0 & 2\dot{a} & 0 \\ \hline
2\dot{a} & 0 & 0 \\ \hline
0 & 0 & -\frac{1 - a^2}{\rho}\mathbf{1}_2
\end{array}\right).
\end{aligned}
$$

2.9.5 Robertson–Walker Metrics

Examples for $\mathbf{SO}(2)$-invariant $(1,3)$-spacetimes with hyperbolic or spherical coordinates,

$$
\begin{aligned}
\mathbf{g} &= T^2(\tau, \psi) d\tau^2 - R^2(\tau, \psi)(d\psi^2 + \sinh^2 \psi \, d\omega_2^2)\\
\Rightarrow \begin{pmatrix} \check{\mathbf{e}}_0 \\ \check{\mathbf{e}}_1 \\ \check{\mathbf{e}}_2 \\ \check{\mathbf{e}}_3 \end{pmatrix} &= \begin{pmatrix} T \; d\tau \\ R \; d\psi \\ R \sinh \psi \; d\theta \\ R \sinh \psi \sin \theta \; d\varphi \end{pmatrix},
\end{aligned}
$$

are the $\mathbf{SO}(3)$-invariant Friedmann universes with position manifolds Ω^3, $\mathbb{R}^3$, and $\mathcal{Y}^3$, the global symmetry groups $\mathbf{SO}(4)$, $\mathbf{SO}(3) \vec{\times} \mathbb{R}^3$, and $\mathbf{SO}_0(1,3)$, respectively, and the local invariance group $\mathbf{SO}(3)$. They have a nontrivial time representation coefficient, $t \longmapsto R(t)$:

$$\begin{aligned} \mathbf{g} &= dt^2 - R^2(t)d\sigma_k^2 = T^2(\tau)(d\tau^2 - d\sigma_k^2), \\ \text{with } \ \frac{dt}{d\tau} &= R(t) = T(\tau), \\ d\sigma_k^2 &= (d\omega_3^2, d\mathbf{y}_3^2, d\vec{x}_3^{\,2}) = \frac{d\rho^2}{1-k\rho^2} + \rho^2 d\omega_2^2, \quad k = (1,0,-1). \end{aligned}$$

The nontrivial connection 1-forms of $(1,3)$-spacetimes with local $\mathbf{SO}(2)$-invariance are with $\dot{T} = \partial_\tau T$, $T' = \partial_\psi T$, etc.:

$$\begin{pmatrix} d\check{\mathbf{e}}_0 \\ d\check{\mathbf{e}}_1 \\ d\check{\mathbf{e}}_2 \\ d\check{\mathbf{e}}_3 \end{pmatrix} = \begin{pmatrix} T'\; d\psi \wedge d\tau \\ \dot{R}\; d\tau \wedge d\psi \\ [\dot{R}\sinh\psi\; d\tau + (R'\sinh\psi + R\cosh\psi)\; d\psi] \wedge d\theta, \\ [\dot{R}\; \sinh\psi \sin\theta\; d\tau + (R'\sinh\psi + R\cosh\psi)\; \sin\theta\; d\psi \\ + R\sinh\psi\; \cos\theta\; d\theta] \wedge d\varphi \end{pmatrix}$$

$$\Rightarrow \begin{cases} \mathbf{\Gamma}^{01} = \frac{\dot{R}}{T}\; d\psi + \frac{T'}{R}\; d\tau & = \frac{\dot{R}}{TR}\; \check{\mathbf{e}}_1 + \frac{T'}{TR}\; \check{\mathbf{e}}_0, \\ \mathbf{\Gamma}^{02} = \frac{\dot{R}}{T}\sinh\psi\; d\theta & = \frac{\dot{R}}{TR}\; \check{\mathbf{e}}_2, \\ \mathbf{\Gamma}^{03} = \frac{\dot{R}}{T}\sinh\psi\; \sin\theta\; d\varphi & = \frac{\dot{R}}{TR}\; \check{\mathbf{e}}_3, \\ \mathbf{\Gamma}^{12} = (\frac{R'}{R}\sinh\psi + \cosh\psi)\; d\theta & = \frac{1}{R}(\frac{R'}{R} + \coth\psi)\; \check{\mathbf{e}}_2, \\ \mathbf{\Gamma}^{13} = (\frac{R'}{R}\sinh\psi + \cosh\psi)\; \sin\theta\; d\varphi & = \frac{1}{R}(\frac{R'}{R} + \coth\psi)\; \check{\mathbf{e}}_3, \\ \mathbf{\Gamma}^{23} = \cos\theta\; d\varphi & = \frac{\cot\theta}{R\sinh\psi}\; \check{\mathbf{e}}_3, \end{cases}$$

with the 2-forms by exterior derivative:

$$d\mathbf{\Gamma}^{01} = (\partial_\tau \frac{\dot{R}}{T} - \partial_\psi \frac{T'}{R})\; d\tau \wedge d\psi = \frac{1}{TR}(\partial_\tau \frac{\dot{R}}{T} - \partial_\psi \frac{T'}{R})\; \check{\mathbf{e}}_0 \wedge \check{\mathbf{e}}_1,$$

$$\begin{aligned} d\mathbf{\Gamma}^{02} &= [\sinh\psi \partial_\tau \frac{\dot{R}}{T}\; d\tau + \partial_\psi(\frac{\dot{R}}{T}\sinh\psi)\; d\psi] \wedge d\theta \\ &= \frac{1}{R}[\frac{1}{T}\partial_\tau \frac{\dot{R}}{T}\; \check{\mathbf{e}}_0 + \frac{1}{R\sinh\psi}\partial_\psi(\frac{\dot{R}}{T}\sinh\psi)\; \check{\mathbf{e}}_1] \wedge \check{\mathbf{e}}_2, \\ d\mathbf{\Gamma}^{03} &= [\sinh\psi \partial_\tau \frac{\dot{R}}{T}\sin\theta\; d\tau + \partial_\psi(\frac{\dot{R}}{T}\sinh\psi)\sin\theta\; d\psi + \frac{\dot{R}}{T}\sinh\psi\cos\theta\; d\theta] \wedge d\varphi \\ &= \frac{1}{R}[\frac{1}{T}\partial_\tau \frac{\dot{R}}{T}\; \check{\mathbf{e}}_0 + \frac{1}{R\sinh\psi}\partial_\psi(\frac{\dot{R}}{T}\sinh\psi)\; \check{\mathbf{e}}_1 + \frac{\dot{R}}{TR\sinh\psi}\cot\theta\; \check{\mathbf{e}}_2] \wedge \check{\mathbf{e}}_3, \end{aligned}$$

$$\begin{aligned} d\mathbf{\Gamma}^{12} &= [\sinh\psi \partial_\tau \frac{R'}{R}\; d\tau + \partial_\psi(\frac{R'}{R}\sinh\psi + \cosh\psi)\; d\psi] \wedge d\theta \\ &= \frac{1}{R}[\frac{1}{T}\partial_\tau \frac{R'}{R}\; \check{\mathbf{e}}_0 + \frac{1}{R\sinh\psi}\partial_\psi(\frac{R'}{R}\sinh\psi + \cosh\psi)\; \check{\mathbf{e}}_1] \wedge \check{\mathbf{e}}_2, \\ d\mathbf{\Gamma}^{13} &= [\sinh\psi \partial_\tau \frac{R'}{R}\sin\theta\; d\tau + \partial_\psi(\frac{R'}{R}\sinh\psi + \cosh\psi)\sin\theta\; d\psi \\ &\quad + (\frac{R'}{R}\sinh\psi + \cosh\psi)\cos\theta\; d\theta] \wedge d\varphi \\ &= \frac{1}{R}[\frac{1}{T}\partial_\tau \frac{R'}{R}\; \check{\mathbf{e}}_0 + \frac{1}{R\sinh\psi}\partial_\psi(\frac{R'}{R}\sinh\psi + \cosh\psi)\; \check{\mathbf{e}}_1 \\ &\quad + \frac{1}{R\sinh\psi}(\frac{R'}{R} + \coth\psi)\cot\theta\; \check{\mathbf{e}}_2] \wedge \check{\mathbf{e}}_3, \end{aligned}$$

$$d\mathbf{\Gamma}^{23} = -\sin\theta\; d\theta \wedge d\varphi = -\frac{1}{R^2\sinh^2\psi}\; \check{\mathbf{e}}_2 \wedge \check{\mathbf{e}}_3.$$

Thus, one obtains the nontrivial curvature 2-forms [there is the local $\mathbf{SO}(2)$-invariance $2 \leftrightarrow 3$]:

$$\begin{aligned}
\mathbf{R}^{01} &= d\mathbf{\Gamma}^{01} + \mathbf{\Gamma}^{02} \wedge \mathbf{\Gamma}^{12} + \mathbf{\Gamma}^{03} \wedge \mathbf{\Gamma}^{13} = d\mathbf{\Gamma}^{01} \\
&= \tfrac{1}{TR}(\partial_\tau \tfrac{\dot{R}}{T} - \partial_\psi \tfrac{T'}{R})\ \check{\mathbf{e}}_0 \wedge \check{\mathbf{e}}_1, \\
\mathbf{R}^{02} &= d\mathbf{\Gamma}^{02} + \mathbf{\Gamma}^{01} \wedge \mathbf{\Gamma}^{21} + \mathbf{\Gamma}^{03} \wedge \mathbf{\Gamma}^{23} = d\mathbf{\Gamma}^{02} - \mathbf{\Gamma}^{01} \wedge \mathbf{\Gamma}^{12} \\
&= -[\tfrac{\dot{R}}{TR}\ \check{\mathbf{e}}_1 + \tfrac{T'}{TR}\ \check{\mathbf{e}}_0] \wedge \tfrac{1}{R}(\tfrac{R'}{R} + \coth\psi)\ \check{\mathbf{e}}_2 + [\tfrac{1}{TR}\partial_\tau \tfrac{\dot{R}}{T}\ \check{\mathbf{e}}_0 + \tfrac{1}{R^2 \sinh\psi}\partial_\psi \\
&\quad \times (\tfrac{\dot{R}}{T}\sinh\psi)\ \check{\mathbf{e}}_1] \wedge \check{\mathbf{e}}_2, \\
\mathbf{R}^{12} &= d\mathbf{\Gamma}^{12} - \mathbf{\Gamma}^{10} \wedge \mathbf{\Gamma}^{20} + \mathbf{\Gamma}^{13} \wedge \mathbf{\Gamma}^{23} = d\mathbf{\Gamma}^{12} - \mathbf{\Gamma}^{01} \wedge \mathbf{\Gamma}^{02} \\
&= -[\tfrac{\dot{R}}{TR}\ \check{\mathbf{e}}_1 + \tfrac{T'}{TR}\ \check{\mathbf{e}}_0] \wedge \tfrac{\dot{R}}{TR}\ \check{\mathbf{e}}_2 + [\tfrac{1}{TR}\partial_\tau \tfrac{R'}{R}\ \check{\mathbf{e}}_0 + \tfrac{1}{R^2 \sinh\psi}\partial_\psi \\
&\quad \times (\tfrac{R'}{R}\sinh\psi + \cosh\psi)\ \check{\mathbf{e}}_1] \wedge \check{\mathbf{e}}_2, \\
\mathbf{R}^{23} &= d\mathbf{\Gamma}^{23} - \mathbf{\Gamma}^{20} \wedge \mathbf{\Gamma}^{30} + \mathbf{\Gamma}^{21} \wedge \mathbf{\Gamma}^{31} = d\mathbf{\Gamma}^{23} - \mathbf{\Gamma}^{02} \wedge \mathbf{\Gamma}^{03} + \mathbf{\Gamma}^{12} \wedge \mathbf{\Gamma}^{13} \\
&= -\tfrac{\dot{R}}{TR}\ \check{\mathbf{e}}_2 \wedge \tfrac{\dot{R}}{TR}\ \check{\mathbf{e}}_3 + \tfrac{1}{R}(\tfrac{R'}{R} + \coth\psi)\ \check{\mathbf{e}}_2 \wedge \tfrac{1}{R}(\tfrac{R'}{R} + \coth\psi)\ \check{\mathbf{e}}_3 \\
&\quad - \tfrac{1}{R^2 \sinh^2\psi}\ \check{\mathbf{e}}_2 \wedge \check{\mathbf{e}}_3,
\end{aligned}$$

with the following nontrivial curvature components:

$$\begin{aligned}
\mathbf{g} &= T^2(\tau,\psi)d\tau^2 - R^2(\tau,\psi)(d\psi^2 + \sinh^2\psi d\omega_2^2), \\
\mathcal{R}^{0101} &= \tfrac{1}{TR}[\partial_\tau \tfrac{\dot{R}}{T} - \partial_\psi \tfrac{T'}{R}], \\
\mathcal{R}^{0202} = \mathcal{R}^{0303} &= \tfrac{1}{TR}[\partial_\tau \tfrac{\dot{R}}{T} - (\coth\psi + \tfrac{R'}{R})\tfrac{T'}{R}], \\
\mathcal{R}^{0212} = \mathcal{R}^{0313} &= \tfrac{1}{TR}[\partial_\tau \tfrac{R'}{R} - \tfrac{T'\dot{R}}{TR}], \\
\mathcal{R}^{1212} = \mathcal{R}^{1313} &= \tfrac{1}{R^2}[1 - \tfrac{\dot{R}^2}{T^2} + (\coth\psi + \partial_\psi)\tfrac{R'}{R}], \\
\mathcal{R}^{2323} &= \tfrac{1}{R^2}[1 - \tfrac{\dot{R}^2}{T^2} + (2\coth\psi + \tfrac{R'}{R})\tfrac{R'}{R}].
\end{aligned}$$

The Ricci tensor has the following nontrivial components:

$$\begin{aligned}
\mathcal{R}_\bullet^{00} &= -\tfrac{1}{TR}[3\partial_\tau \tfrac{\dot{R}}{T} - \partial_\psi \tfrac{T'}{R}] + \tfrac{2}{R^2}(\coth\psi + \tfrac{R'}{R})\tfrac{T'}{T}, \\
\mathcal{R}_\bullet^{11} &= +\tfrac{1}{TR}[\partial_\tau \tfrac{\dot{R}}{T} - \partial_\psi \tfrac{T'}{R}] - \tfrac{1}{R^2}[2 - 2\tfrac{\dot{R}^2}{T^2} + 2(\coth\psi + \partial_\psi)\tfrac{R'}{R}], \\
\mathcal{R}_\bullet^{01} &= -\tfrac{1}{TR}[2\partial_\tau \tfrac{R'}{R} - 2\tfrac{T'\dot{R}}{TR}], \\
\mathcal{R}_\bullet^{22} = \mathcal{R}_\bullet^{33} &= +\tfrac{1}{TR}\partial_\tau \tfrac{\dot{R}}{T} - \tfrac{1}{R^2}[2 - 2\tfrac{\dot{R}^2}{T^2} + \coth\psi(\tfrac{T'}{T} + 3\tfrac{R'}{R}) + \tfrac{R''}{R} + \tfrac{T'R'}{TR}].
\end{aligned}$$

For the three Robertson–Walker metrics, one obtains — in addition to hyperbolic position $k = -1$ — the metric for spherical $k = 1$ and flat position $k = 0$:

$$\mathbf{g} = dt^2 - R^2(t)d\sigma_k^2, \qquad \begin{cases} \mathcal{R}^{dabc} \cong \frac{1}{R^2}\left(\begin{array}{c|c} \ddot{R}R\mathbf{1}_3 & 0 \\ \hline 0 & -(\dot{R}^2 + k)\mathbf{1}_3 \end{array}\right), \\ \mathcal{R}_\bullet^{ab} \cong \frac{1}{R^2}\left(\begin{array}{c|c} -3\ddot{R}R & 0 \\ \hline 0 & (\ddot{R}R + 2\dot{R}^2 + 2k)\mathbf{1}_3 \end{array}\right), \\ \frac{1}{2}\mathcal{R}_\bullet^\bullet = -3\frac{\ddot{R}R + \dot{R}^2 + k}{R^2}, \end{cases}$$

$$\mathbf{g} = T^2(\tau)(d\tau^2 - d\sigma_k^2), \qquad \begin{cases} \mathcal{R}^{dabc} \cong \frac{1}{T^2}\left(\begin{array}{c|c} [\frac{\ddot{T}}{T} - (\frac{\dot{T}}{T})^2]\mathbf{1}_3 & 0 \\ \hline 0 & -[(\frac{\dot{T}}{T})^2 + k]\mathbf{1}_3 \end{array}\right), \\ \mathcal{R}_\bullet^{ab} \cong \frac{1}{T^2}\left(\begin{array}{c|c} -3[\frac{\ddot{T}}{T} - (\frac{\dot{T}}{T})^2] & 0 \\ \hline 0 & [\frac{\ddot{T}}{T} + (\frac{\dot{T}}{T})^2 + 2k]\mathbf{1}_3 \end{array}\right), \\ \frac{1}{2}\mathcal{R}_\bullet^\bullet = -\frac{3}{T^2}(\frac{\ddot{T}}{T} + 2k). \end{cases}$$

The de Sitter, flat, and anti-de Sitters manifold with maximal symmetry arise for

$$\mathbb{M}^{(1,3)} \cong (\mathcal{Y}^{(1,3)}, \mathbb{R}^4, \mathcal{Y}^{(3,1)}) : \left\{ \begin{array}{rl} \mathbf{g} &= dt^2 - R^2(t) d\mathrm{y}_3^2 \\ &= T^2(\tau)(d\tau^2 - d\mathrm{y}_3^2), \\ R(t) &= (\sinh t, t, \sin t) \\ = T(\tau) &= \left(\frac{e^\tau}{\sqrt{1-e^{2\tau}}}, e^\tau, \frac{e^\tau}{\sqrt{1+e^{2\tau}}}\right). \end{array} \right.$$

2.10 Covariantly Constant-Curvature Manifolds

A torsionless manifold with *covariantly constant curvature* is called *affine locally symmetric*,

$$\mathcal{T} = 0, \quad \nabla \mathcal{R} = 0.$$

Special cases of manifolds with covariantly constant curvature are globally symmetric Riemannian manifolds with a *definite metric*, completely classified by Cartan. They are isomorphic to symmetric spaces G/K with classes of a compact local invariance subgroup $K \subseteq G$ in a simply connected real Lie group as the global symmetry (motion) group and inherit adjoint structures of G. Their algebraic origin lies in orthogonal symmetric Lie algebras.

2.10.1 Orthogonal Symmetric Lie Algebras

An *orthogonal symmetric Lie algebra* $(L, \mathtt{R})$ is a real Lie algebra with a non-trivial involutive automorphism where the fixed elements $C = \{c \in L \mid \mathtt{R}(c) = c\}$ constitute a compact Lie algebra. It is called *effective* if the distinguished compact subalgebra does not contain central elements $\operatorname{centr} L \cap C = \{0\}$.

An orthogonal symmetric Lie algebra has a Killing form orthogonal direct decomposition into involution eigenspaces, the compact Lie subalgebra C, and the R-antisymmetric vector subspace V, compatible with the Lie bracket,

$$L = L_+^{\mathtt{R}} \perp L_-^{\mathtt{R}} = C \perp V, \quad \left\{ \begin{array}{l} [C,C] \subseteq C, \quad [C,V] \subseteq V, \quad [V,V] \subseteq C, \\ \kappa(C,V) = \{0\}. \end{array} \right.$$

There are three types of effective orthogonal symmetric Lie algebras, denoted with subindex $(c, nc, 0)$,

$$\begin{array}{rl} L_c \text{ semisimple and compact} & \Rightarrow (L_c, \mathtt{R}_c) \text{ of } \textit{compact type}, \\ \left.\begin{array}{r} L_{nc} \text{ semisimple and noncompact,} \\ L_{nc} = C \perp V \text{ is a Cartan decomposition} \end{array}\right\} & \Rightarrow (L_{nc}, \mathtt{R}_{nc}) \text{ of } \textit{noncompact type}, \\ [V,V] = \{0\}, \ V \text{ abelian ideal} & \Rightarrow (L_0, \mathtt{R}_0) \text{ of } \textit{Euclidean type}. \end{array}$$

Therefore, all semisimple real Lie algebras are effective orthogonal symmetric. Every effective orthogonal symmetric Lie algebra has a Killing form orthogonal decomposition into ideals of compact, noncompact, and Euclidean types:

$$(L, \mathtt{R}) = (L_c, \mathtt{R}_c) \perp (L_{nc}, \mathtt{R}_{nc}) \perp (L_0, \mathtt{R}_0).$$

In the spherical, hyperbolic, and flat example the antisymmetric subspaces contain the classes of Lie algebras for orthogonal groups $\mathcal{O} = -\mathcal{O}^T \in \log \mathbf{SO}(s)$. They are the $\mathbb{R}^s$-isomorphic tangent spaces of sphere, hyperboloid, and Euclidean space:

$$\mathbb{R}^s \ni \left(\begin{array}{c|c} 0 & \vec{\theta} \\ \hline -\vec{\theta}^T & \mathcal{O} \end{array}\right) \in \log \mathbf{SO}(1+s) \quad = \log \mathbf{SO}(s) \perp \log \Omega^s,$$

$$\mathbb{R}^s \ni \left(\begin{array}{c|c} 0 & \vec{\beta} \\ \hline \vec{\beta}^T & \mathcal{O} \end{array}\right) \in \log \mathbf{SO}_0(1,s) \quad = \log \mathbf{SO}(s) \perp \log \mathcal{Y}^s,$$

$$\mathbb{R}^s \ni \left(\begin{array}{c|c} 0 & \vec{x} \\ \hline 0 & \mathcal{O} \end{array}\right) \in \log[\mathbf{SO}(s) \vec{\times} \mathbb{R}^s] = \log \mathbf{SO}(s) \;\vec{\oplus}\; \mathbb{R}^s.$$

The polar decompositions of the complex linear groups give rise to orthogonal symmetric, not effective, Lie algebras:

$$\log[\mathbf{GL}(n, \mathbb{C})] = \log \mathbf{U}(n) \perp \mathbb{R}(n), \;\; \mathbb{R}(n) \cong \mathbb{R}^{n^2}.$$

For an orthogonal symmetric Lie algebra $(L, \mathtt{R})$, the complexification $(L^*, \mathtt{R}^*)$ with "imaginary" iV is, as a real Lie algebra, orthogonal symmetric:

$$\Big(L = C \perp V, \mathtt{R}\Big) \leftrightarrow \Big(L^* = C \perp iV, \mathtt{R}^*(\mathcal{C} + i\mathcal{V}) = \mathcal{C} - i\mathcal{V}\Big).$$

There is a *compact–noncompact duality*: If $(L, \mathtt{R})$ is of the compact type, its dual $(L^*, \mathtt{R}^*)$ is of the noncompact type, and vice versa.

An orthogonal symmetric Lie algebra (L, C) is *irreducible* if L is semisimple and if C does not contain a nontrivial L-ideal. Then $[V, V] = C$, and $\mathrm{ad}\,_L C$ acts irreducibly on V.

2.10.2 Real Simple Lie Algebras

The transition from a complex to real simple Lie algebra is characterized by a Lie algebra involution (conjugation).

According to Cartan, there are two types of real simple Lie algebras. The first type arises from a complex simple Lie algebra L_r with rank r by using the three possible kinds of involutions $\mathtt{R}$ (anticonjugation, symplectic, and orthogonal) of its compact form:

$$\underline{\mathbf{lag}}_{\mathbb{R}}(\text{compact}) \ni L_r^{\mathtt{c}} \longmapsto L_r^{\mathtt{R}} \in \underline{\mathbf{lag}}_{\mathbb{R}}(\text{noncompact, simple}).$$

L	Complex L_r $\dim_{\mathbb{C}} L = d$	Compact L_r^c	Anti $L_r^{\mathbb{R}}$ in $\log \mathbf{SL}(n, \mathbb{R})$	Symplectic $L_r^{\mathbb{H}}$ in $\log \mathbf{SL}(n, \mathbb{H})$	Orthogonal $L_r^{p,q}$
A	$r \geq 1$ $\log \mathbf{SL}(1+r, \mathbb{C})$ $r(2+r)$	$\log \mathbf{SU}(1+r)$ $r(2+r)$	$\log \mathbf{SL}(1+r, \mathbb{R})$ $\supset \log \mathbf{SO}(1+r)$ $\binom{1+r}{2}$	for $1+r$ even $\log \mathbf{SU}^\star(1+r)$ $\supset \log \mathbf{SpU}(1+r)$ $\binom{2+r}{2}$	$p+q = 1+r$ $\log \mathbf{SU}(p,q)$ $\supset \log \mathbf{SU}(p) \times \mathbf{U}(1)$ $\times \mathbf{SU}(q)$ p^2+q^2-1
C	$r \geq 3$ $\log \mathbf{Sp}(\mathbb{C}^{2r})$ $\binom{1+2r}{2}$	$\log \mathbf{SpU}(2r)$ $\binom{1+2r}{2}$	$\log \mathbf{Sp}(2r)$ $\supset \log \mathbf{U}(r)$ r^2	$= \log \mathbf{Sp}(2r)$	$p+q=r$ $\log \mathbf{SpU}(2p, 2q)$ $\supset \log \mathbf{SpU}(2p)$ $\times \mathbf{SpU}(2q)$ $\binom{2p+1}{2} + \binom{2q+1}{2}$
B	$r \geq 2$ $\log \mathbf{SO}(1+2r, \mathbb{C})$ $\binom{1+2r}{2}$	$\log \mathbf{SO}(1+2r)$ $\binom{1+2r}{2}$	$= \log \mathbf{SO}(1+2r)$	–	$p+q = 1+2r$ $\log \mathbf{SO}(p,q)$ $\supset \log \mathbf{SO}(p) \times \mathbf{SO}(q)$ $\binom{p}{2} + \binom{q}{2}$
D	$r \geq 4$ $\log \mathbf{SO}(2r, \mathbb{C})$ $\binom{2r}{2}$	$\log \mathbf{SO}(2r)$ $\binom{2r}{2}$	$= \log \mathbf{SO}(2r)$	$\log \mathbf{SO}^\star(2r)$ $\supset \log \mathbf{U}(r)$ r^2	$p+q=2r$ $\log \mathbf{SO}(p,q)$ $\supset \log \mathbf{SO}(p) \times \mathbf{SO}(q)$ $\binom{p}{2} + \binom{q}{2}$

Simple complex and real Lie algebras
of real form type I: L_r^c, $L_r^{\mathbb{R}}$, $L_r^{\mathbb{H}}$, $L_r^{p,q}$
($\supset$ maximal compact Lie subalgebra with dimension)
(without the exceptional Lie algebras)

For small rank $r \leq 4$, there are Lie algebra isomorphies, for the complex algebras given by

$$A_1 \cong B_1 \cong C_1, \quad B_2 \cong C_2, \quad D_2 \cong A_1 \oplus A_1, \quad D_3 \cong A_3.$$

One has the inclusions

$$A_r \subset D_{1+r} \subset B_{1+r} \cap C_{1+r}.$$

Via their adjoint representations and the orthogonal invariance group of the Killing form, all semisimple Lie algebras can be considered orthogonal subalgebras:

$$\mathbb{C}^d \cong L \cong \operatorname{ad} L \subseteq \log \mathbf{SO}(d, \mathbb{C}) \cong \mathbb{C}^{\binom{d}{2}}.$$

In addition to the real forms of simple complex Lie algebras, there are the simple real Lie algebras, which are the *canonically complexified* (with the doubled reals $\mathbb{C}_{\mathbb{R}} = \mathbb{R} \oplus i\mathbb{R}$) compact Lie algebras L_r^c. The basic compact Lie algebras are maximal in the doubling $L_{(r,r)} = L_r^c \oplus iL_r^c$ with doubled dimension $2d_c$ and doubled rank $2r$.

Lie algebra		Lie group	Dimension $d = 2d_c$
$A_{(r,r)}$	$\cong$	$\log \mathbf{SL}(1+r, \mathbb{C}_{\mathbb{R}})$	$2r(2+r)$
$C_{(r,r)}$	$\cong$	$\log \mathbf{Sp}(2r, \mathbb{C}_{\mathbb{R}})$	$2\binom{1+2r}{2}$
$B_{(r,r)}$	$\cong$	$\log \mathbf{SO}(1+2r, \mathbb{C}_{\mathbb{R}})$	$2\binom{1+2r}{2}$
$D_{(r,r)}$	$\cong$	$\log \mathbf{SO}(2r, \mathbb{C}_{\mathbb{R}})$	$2\binom{2r}{2}$

Simple real Lie algebras
of canonical complexification type II: $L_{(r,r)} = L_r^c \oplus iL_r^c$
(without the exceptional Lie algebras)

2.10.3 Globally Symmetric Riemannian Manifolds

Corresponding to the classification of the real simple Lie algebras, there are two types, I and II, of irreducible orthogonal symmetric Lie algebras $(L, \mathtt{R})$. Both types come in pairs (c, nc) connected with each other by the compact–noncompact duality.

The compact L_c are

type $\mathrm{I_c}$: L_c simple, R any involutive L_c-automorphism,
type $\mathrm{II_c}$: $L_c = L_1 \oplus L_2$ with simple ideals $L_1 \overset{\mathtt{R}}{\leftrightarrow} L_2$.

The two isomorphic compact ideals $L_1 \cong L_2$ give a compact subalgebra with the reflection-symmetric elements $C = L_+ = [L_\pm, L_\pm]$ and a vector subspace with reflection-antisymmetric ones, $V = L_- = [L_+, L_-]$. There is the vector space isomorphy $V \cong C$.

The duality-related noncompact L_{nc} are

type $\mathrm{I_{nc}}$: L_{nc} simple with simple complexification,
R keeps fixed a maximal compactly embedded subalgebra,
type $\mathrm{II_{nc}}$: $L_{nc} = L_1 \oplus iL_2$ with $\mathbb{C} \bullet [L_1 \oplus iL_2] = L \in \underline{\mathbf{lag}}_{\mathbb{C}}$ simple,
R is an L-conjugation
with respect to a maximal compactly embedded subalgebra.

A pair (G, K) of a connected Lie group with a compact connected subgroup is *associated with an orthogonal symmetric Lie algebra* $L = C \perp V$ for $(L, C) = (\log G, \log K)$. For each G-invariant positive definite metric $\mathbf{g}$ (which exists), $(G/K, \mathbf{g})$ is a locally symmetric Riemannian manifold, i.e., torsionfree with covariantly constant curvature. The tangent space at the unit coset $1K \in G/K$ is the antisymmetric vector subspace $\mathbf{T}_{1K}(G/K) = V$. The universal covering group defines the universal covering manifold $\exp L/K$ for all (L, C)-associated locally symmetric Riemannian manifolds; it is a *globally symmetric Riemannian manifold.* The Cartan classification above of the real simple Lie algebras and the irreducible orthogonal symmetric Lie algebras gives all *globally symmetric Riemannian manifolds.*

Subtype	Noncompact $\mathrm{I_{nc}}$	Compact $\mathrm{I_c}$	Dimension n	Real rank r
A I	$\mathbf{SL}(d, \mathbb{R})/\mathbf{SO}(d)$	$\mathbf{SU}(d)/\mathbf{SO}(d)$	$\frac{(d+2)(d-1)}{2}$	$d-1$
A II	$\mathbf{SU}^*(2d)/\mathbf{SpU}(2d)$	$\mathbf{SU}(2d)/\mathbf{SpU}(2d)$	$(2d+1)(d-1)$	$d-1$
A III	$\mathbf{SU}(p,q)/\mathbf{SU}(p)$ $\times \mathbf{U}(1) \times \mathbf{SU}(q)$	$\mathbf{SU}(p+q)/\mathbf{SU}(p)$ $\times \mathbf{U}(1) \times \mathbf{SU}(q)$	$2pq$	$\min(p,q)$
C I	$\mathbf{Sp}(2d)/\mathbf{U}(d)$	$\mathbf{SpU}(2d)/\mathbf{U}(d)$	$2\binom{d+1}{2}$	d
C II	$\mathbf{Sp}(2p, 2q)/\mathbf{SpU}(2p)$ $\times \mathbf{SpU}(2q)$	$\mathbf{SpU}(2p+2q)/\mathbf{SpU}(2p)$ $\times \mathbf{SpU}(2q)$	$4pq$	$\min(p,q)$
BD I	$\mathbf{SO}_0(p,q)/\mathbf{SO}(p)$ $\times \mathbf{SO}(q)$	$\mathbf{SO}(p+q)/\mathbf{SO}(p)$ $\times \mathbf{SO}(q)$	pq	$\min(p,q)$
D III	$\mathbf{SO}^*(2d)/\mathbf{U}(d)$	$\mathbf{SO}(2d)/\mathbf{U}(d)$	$2\binom{d}{2}$	$[\frac{d}{2}]$

Riemannian globally symmetric manifolds of type I
(without the exceptional manifolds)

Examples for $BD\ I$ are the timelike one-shell hyperboloids $\mathcal{Y}^s$ and their compact partners the spheres Ω^s with the same compact subgroup $\mathbf{SO}(s)$ for the classes.

Starting from a Cartan factorization of a noncompact group with simple Lie algebra into maximal compact and maximal abelian noncompact subgroups whose dimension is the real rank

$$G = KAK, \ \ \dim_{\mathbb{R}}(G, K, A) = (d_G, d_K, r),$$

the Riemannian manifold is given by the maximal compact classes

$$\text{type I}_{\text{nc}}\text{: } \mathbb{M} \cong G/K, \quad \dim_{\mathbb{R}} \mathbb{M} = n = d_G - d_K.$$

For the compact manifold, the abelian group A has to be replaced by its compact partner; e.g., $\mathbf{SO}_0(1,1) \ni e^{\beta\sigma_3}$ by $e^{i\alpha\sigma_3} \in \mathbf{SO}(2)$.

The type II involves a doubling: The compact manifolds contain the diagonal group classes of the doubling of the simple, compact Lie groups $K = \exp L$, e.g., $\mathbb{SU}(n) = \mathbf{SU}(n) \times \mathbf{SU}(n)/\mathbf{SU}(n)$. They are isomorphic to the group $\mathbb{K} \cong K$.

Noncompact II_{nc}	Compact II_{c}	Dimension	Real rank
$K \times K^*/K$	$[K \times K]/\operatorname{diag}[K \times K]$	d_K	r_K

Riemannian globally symmetric manifolds of type II

The II_{nc} noncompact partner manifolds are G/K with a connected Lie group G and maximal compact subgroup K, where the complexification $\log G = \log K + i \log K$ is simple; e.g., $\mathbf{SL}(n, \mathbb{C})/\mathbf{SU}(n)$.

2.10.4 Curvature of Globally Symmetric Riemannian Manifolds

The adjoint Lie algebra structures can be taken over by the involution antisymmetric subspace V of an orthogonal symmetric Lie algebra,

$$(\log G, \log K) = (L, C), \quad L = C \perp V, \quad \text{bases: } \begin{cases} (l^a)_{a=1}^{n=\dim_{\mathbb{R}} L}, \\ (\mathcal{C}^\alpha)_{\alpha=1}^{s=\dim_{\mathbb{R}} C}, \ (\mathcal{V}^A)_{A=1}^{\dim_{\mathbb{R}} V}. \end{cases}$$

The doubled adjoint V-action defines endomorphisms for V and C:

$$\begin{aligned} [[V, V], V] &\subseteq [C, V] &&\subseteq V, \\ [[V, V], C] &\subseteq [C, C] &&\subseteq C. \end{aligned}$$

The antisymmetric subspace V is an example for a *triple Lie subspace*, defined in general by a double adjoint stability $[[V, V], V] \subseteq V$.

The exponential mapping of V parametrizes the manifold $G/K \cong \exp V$. The n-bein for the group action $1K \longmapsto gK$ relate to each other the tangent spaces. They involve the even contribution of $e_*^{\operatorname{ad} l} \cong$ "$\frac{e^{\operatorname{ad} l}-\mathbf{1}_n}{\operatorname{ad} l}$" with the adjoint square:

$$\begin{aligned}
gK \cong e^{\mathcal{V}} &= \sum_{k\geq 0} \frac{\mathcal{V}^k}{k!},\\
V = \mathbf{T}_{1K}(G/K) \ni \mathcal{V}^A &\longmapsto \mathcal{V}^j(gK) = (gK_*)^j_A \mathcal{V}^A \in \mathbf{T}_{gK}(G/K) \cong V,\\
gK_* \cong e_*^{\operatorname{ad}\mathcal{V}} &= \sum_{k\geq 0} \frac{((\operatorname{ad}\mathcal{V})^2|_V)^k}{(1+2k)!} \cong \text{“}\frac{\sinh \operatorname{ad}\mathcal{V}}{\operatorname{ad}\mathcal{V}}|_V\text{”} \text{ (symbolic notation)}\\
&= \mathbf{1}_{n-s} + \tfrac{1}{6}(\operatorname{ad}\mathcal{V})^2|_V + \cdots \in \mathbf{GL}(n-s,\mathbb{R}).
\end{aligned}$$

The $\operatorname{Ad} G$-invariant measure on the coset space $\operatorname{Ad} G/K$ can be written in orthogonal coordinates:

$$\text{for } \operatorname{Ad} G/K: \quad dgK = |\det e_*^{\operatorname{ad}\mathcal{V}}| d^{n-s}\rho.$$

Since, in general, $[V,V] \not\subseteq V$, the Killing connection cannot be taken over for G/K. Via the triple Lie property, the V-restricted adjoint square defines the *Killing curvature* at the neutral class $1K \in G/K$ of the globally symmetric Riemannian manifold:

$$\mathcal{R}: V \wedge V \longrightarrow V \otimes V^T, \begin{cases} \mathcal{R}(\mathcal{V} \wedge \mathcal{V}') &= -\operatorname{ad}|_V[\mathcal{V},\mathcal{V}'],\\ \mathcal{R}(\mathcal{V}^A \wedge \mathcal{V}^B) &= -\epsilon_\alpha^{AB} \operatorname{ad}|_V l^\alpha = -\epsilon_\alpha^{AB}\epsilon_E^{\alpha D}\mathcal{V}^E \otimes \check{\mathcal{V}}_D,\end{cases}$$

$$\mathcal{R}(\mathcal{V}\wedge\mathcal{V}'): V \longrightarrow V, \begin{cases} \mathcal{R}(\mathcal{V}\wedge\mathcal{V}')(\mathcal{V}'') &= -[[\mathcal{V},\mathcal{V}'],\mathcal{V}''],\\ \mathcal{R}(\mathcal{V}^A \wedge \mathcal{V}^B).\mathcal{V}^D &= -\epsilon_\alpha^{AB}\epsilon_E^{\alpha D}\mathcal{V}^E,\\ \mathcal{R}_E^{DAB} &= -\epsilon_\alpha^{AB}\epsilon_E^{\alpha D}.\end{cases}$$

$$\text{Killing Ricci tensor: } \mathcal{R}_\bullet^{DA} = \mathcal{R}_B^{DAB} = \epsilon_\alpha^{AB}\epsilon_B^{D\alpha}.$$

For V as a representation space of the compact Lie algebra C, there exist strictly positive definite symmetric C-invariant bilinear forms:

$$\gamma: V \vee V \longrightarrow \mathbb{R}, \quad \begin{cases} \gamma(\mathcal{V},\mathcal{V}) > 0, \quad \gamma(\mathcal{V}^A,\mathcal{V}^D) = \gamma^{AD},\\ \gamma([\mathcal{C},\mathcal{V}],\mathcal{V}') + \gamma(\mathcal{V},[\mathcal{C},\mathcal{V}']) = 0 \text{ for } \mathcal{C} \in C.\end{cases}$$

Any such form can be used for a *G-invariant Riemannian metric*

$$G/K \ni gK \cong e^{\mathcal{V}} \longmapsto \mathbf{g}(e^{\mathcal{V}})^{jk} = (e_*^{\operatorname{ad}\mathcal{V}})^j_A \gamma^{AD} (e_*^{\operatorname{ad}\mathcal{V}})^k_D,$$

and all G-invariant metrics on G/K arise from a C-invariant positive definite bilinear form on V. All G-invariant metrics $\mathbf{g}$ on G/K lead to the same Riemannian connection and to the Killing curvature tensor on the associated symmetric space G/K.

For a semisimple $L = C \perp V$ the L-invariant symmetric Killing form, restricted to V, adds up two C-invariant contributions, related to the two subspaces and reflected to each other. They are not necessarily symmetric. The Killing form of L, restricted to V, is the symmetrized Killing Ricci tensor

$$\text{semisimple } G/K,\ L = C \perp V, \quad \begin{cases} V \vee V \longrightarrow \mathbb{R}, \quad \kappa(\mathcal{V}^A, \mathcal{V}^D), \\ \kappa^{AD} = \epsilon_b^{Aa}\epsilon_a^{Db} = \epsilon_\alpha^{AB}\epsilon_B^{D\alpha} + \epsilon_B^{A\alpha}\epsilon_\alpha^{DB} \\ \quad = \mathcal{R}_\bullet^{DA} + \mathcal{R}_\bullet^{AD}, \\ \mathcal{R}_\bullet^\bullet = \kappa_{AD}\mathcal{R}_\bullet^{AD} = \frac{n-s}{2} \\ \frac{\mathcal{R}_\bullet^{DA}+\mathcal{R}_\bullet^{AD}}{2} - \frac{1}{2}\kappa^{AD}\mathcal{R}_\bullet^\bullet = -\frac{n-s-1}{2}\kappa^{AD}. \end{cases}$$

With the appropriate renormalization $\gamma = \mp\kappa$ for compact and noncompact orthogonal Lie algebras, one obtains a positive definite Riemannian structure.

2.10.5 Examples

The simplest type $\mathrm{I_c}$ example is the 2-sphere $\Omega^2 \cong \mathbf{SO}(3)/\mathbf{SO}(2)$:

$$(\log \mathbf{SO}(3), \log \mathbf{SO}(2)) : \begin{cases} \text{bases:} \quad \left((\mathcal{O}^a \cong -\epsilon^{abc})_{a=1}^3, \ \mathcal{O}^3\right), \\ \qquad (\mathcal{T}^A = \mathcal{O}^A)_{A=1,2}, \\ [\mathcal{O}^3, \mathcal{O}^3] = 0, \quad [\mathcal{O}^3, \mathcal{T}^{1,2}] = \mp\mathcal{T}^{2,1}, \\ [\mathcal{T}^1, \mathcal{T}^2] = \mathcal{O}^3, \end{cases}$$

$$\begin{aligned} \mathcal{T} &= \operatorname{ad}\mathcal{T} = \left(\begin{array}{cc|c} 0 & 0 & -\theta_2 \\ 0 & 0 & \theta_1 \\ \hline \theta_2 & -\theta_1 & 0 \end{array}\right), \quad \mathbb{R}^2 \cong \log\Omega^2 \ni \begin{pmatrix} -\theta_2 \\ \theta_2 \end{pmatrix}, \\ e^{\mathcal{T}} &= \mathbf{1}_3 + \tfrac{\sin\theta}{\theta}\mathcal{T} + \tfrac{1-\cos\theta}{\theta^2}\mathcal{T}^2, \ \theta^2 = \theta_1^2 + \theta_2^2, \ \Omega^2 \ni \tfrac{\sin\theta}{\theta}\begin{pmatrix} -\theta_2 \\ \theta_2 \end{pmatrix} = \sin\theta\begin{pmatrix} \cos\varphi \\ \sin\varphi \end{pmatrix}, \\ \mathcal{T}^2 &= \left(\begin{array}{cc|c} -\theta_2^2 & \theta_1\theta_2 & 0 \\ \theta_1\theta_2 & -\theta_1^2 & 0 \\ \hline 0 & 0 & -\theta^2 \end{array}\right) = \left(\begin{array}{c|c} \mathcal{T}^2|_{\mathbb{R}^2} & 0 \\ \hline 0 & -\theta^2 \end{array}\right) = -\theta^2 R \circ \left(\begin{array}{cc|c} 1 & 0 & 0 \\ 0 & 0 & 0 \\ \hline 0 & 0 & 1 \end{array}\right) \circ R^T. \end{aligned}$$

The 2-bein for Ω^2 is orthogonally diagonalizable with its determinant for the dilation:

$$\begin{aligned} e_*^{\mathcal{T}} &= \mathbf{1}_2 - \tfrac{\sin\theta - \theta}{\theta^3}\mathcal{T}^2|_{\mathbb{R}^2} = R \circ \begin{pmatrix} \frac{\sin\theta}{\theta} & 0 \\ 0 & 1 \end{pmatrix} \circ R^T, \\ e_*^{\mathcal{T}} &\in \mathbf{GL}(2,\mathbb{R}), \quad R \in \mathbf{SO}(2), \\ \det e_*^{\mathcal{T}} &= \tfrac{\sin\theta}{\theta}, \ d^2\omega = \tfrac{\sin\theta}{\theta} d\theta_1 d\theta_2 = d\cos\theta \, d\varphi. \end{aligned}$$

The 2-sphere has the Ricci tensor

$$\Omega^2 : \ \mathcal{R}_\bullet^{AD} = \tfrac{1}{2}\kappa(\mathcal{T}^A, \mathcal{T}^D) = \frac{\epsilon^{AB3}\epsilon^{D3B} + \epsilon^{A3B}\epsilon^{DB3}}{2} = -\delta^{AD}.$$

As example for types $\mathrm{I_{nc}}$ and $\mathrm{II_{nc}}$, the 3-hyperboloid $\mathcal{Y}^3 \cong \mathbf{SO}_0(1,3)/\mathbf{SO}(3) \cong \mathbf{SL}(2,\mathbb{C})/\mathbf{SU}(2)$ with the action of the orthochronous Lorentz group and the classes of the rotation groups is associated with the Lie algebra in a Cartesian parametrization. The rotations are unitarily diagonalizable in the complex displaying the spin eigenvalues, whereas the boosts are orthogonally diagonalizable in the real and display the boost eigenvalues:

$$\begin{aligned}
[\mathcal{O}^a, \mathcal{O}^b] &= -\epsilon^{abc}\mathcal{O}^c, \quad [\mathcal{O}^a, \mathcal{B}^b] = -\epsilon^{abc}\mathcal{B}^c, \quad [\mathcal{B}^a, \mathcal{B}^b] = \epsilon^{abc}\mathcal{O}^c,\\
\mathcal{O} &= \operatorname{ad}\mathcal{O} = \left(\begin{array}{c|c} -\vec{\alpha} & 0 \\ \hline 0 & -\vec{\alpha}\end{array}\right) = i\alpha\, u \circ \left(\begin{array}{c|c} \Delta & 0 \\ \hline 0 & \Delta\end{array}\right) \circ u^*,\\
\vec{\alpha} &\cong \alpha_a \epsilon^{abc}, \quad \alpha^2 = \alpha_a^2, \quad \Delta = \begin{pmatrix} 1 & 0 & 0\\ 0 & 0 & 0\\ 0 & 0 & -1\end{pmatrix}, \quad u \in \mathbf{SU}(6),\\
\mathcal{B} &= \operatorname{ad}\mathcal{B} = \left(\begin{array}{c|c} 0 & \vec{\beta} \\ \hline \vec{\beta}^T & 0\end{array}\right) = \beta\, r \circ \left(\begin{array}{c|c} \Delta & 0 \\ \hline 0 & -\Delta\end{array}\right) \circ r^*,\\
\vec{\beta} &\cong \beta_a \epsilon^{abc}, \quad \beta^2 = \beta_a^2, \quad r \in \mathbf{SO}(6),\\
\text{Killing form:}\quad & \kappa(\mathcal{B}^a, \mathcal{B}^b) = 4\delta^{ab} = -\kappa(\mathcal{O}^a, \mathcal{O}^b).
\end{aligned}$$

The boost exponents give a $\mathcal{Y}^3$-parametrization:

$$\begin{aligned}
\mathcal{B} &= \left(\begin{array}{c|c} 0 & \vec{\beta} \\ \hline \vec{\beta}^T & 0\end{array}\right), \quad \mathbb{R}^3 \cong \log \mathcal{Y}^3 \ni \vec{\beta} = \begin{pmatrix} 0 & \beta_3 & -\beta_2\\ -\beta_3 & 0 & \beta_1\\ \beta_2 & -\beta_1 & 0\end{pmatrix},\\
e^{\mathcal{B}} &= \mathbf{1}_6 + \tfrac{\sinh\beta}{\beta}\mathcal{B} + \tfrac{1-\cosh\beta}{\beta^2}\mathcal{B}^2, \quad \mathcal{Y}^3 \ni \tfrac{\vec{\beta}}{\beta}\sinh\beta \cong \sinh\beta \begin{pmatrix} \cos\varphi\sin\theta\\ \sin\varphi\sin\theta\\ \cos\theta\end{pmatrix}.
\end{aligned}$$

The Lie–Jacobi isomorphism leads to the 3-bein $e_*^{\mathcal{B}}$ for $\mathcal{Y}^3$:

$$\begin{aligned}
e_*^{\mathcal{B}} &= \mathbf{1}_3 + \tfrac{\sinh\beta - \beta}{\beta^3}\mathcal{B}^2|_{\mathbb{R}^3} = R \circ \begin{pmatrix} \frac{\sinh\beta}{\beta} & 0 & 0\\ 0 & 1 & 0\\ 0 & 0 & \frac{\sinh\beta}{\beta}\end{pmatrix} \circ R^T,\\
\mathcal{B}^2|_{\mathbb{R}^3} &= \beta^2 R \circ \Delta^2 \circ R^T, \quad e_*^{\mathcal{B}} \in \mathbf{GL}(2, \mathbb{R}), \quad R \in \mathbf{SO}(3),\\
\det e_*^{\mathcal{B}} &= \tfrac{\sinh^2\beta}{\beta^2}, \quad d^3y = \tfrac{\sinh^2\beta}{\beta^2} d\beta_1 d\beta_2 d\beta_3 = \sinh^2\beta\, d\beta\, d^2\omega.
\end{aligned}$$

The structures for the 3-sphere $\Omega^3 \cong \mathbf{SO}(4)/\mathbf{SO}(3) = \mathbb{SU}(2)$ (type $\mathrm{II_c}$ example) are obtained with the noncompact–compact duality:

$$\vec{\mathcal{B}} \leftrightarrow i\vec{\mathcal{T}}, \ \vec{\beta} \leftrightarrow i\vec{\alpha}, \ \sinh\beta \leftrightarrow i\sin\alpha.$$

The starting point is a direct product Lie algebra $\log \mathbf{SO}(3) \oplus \log \mathbf{SO}(3)$ with the diagonal Lie algebra $\mathcal{O} \in \log \mathbf{SO}(3)$ and the isomorphic vector subspace $\mathcal{T} \in V \cong \mathbb{R}^3$:

$$\begin{aligned}
[\mathcal{O}^a, \mathcal{O}^b] &= -\epsilon^{abc}\mathcal{O}^c, \quad [\mathcal{O}^a, \mathcal{T}^b] = -\epsilon^{abc}\mathcal{T}^c, \quad [\mathcal{T}^a, \mathcal{T}^b] = -\epsilon^{abc}\mathcal{O}^c,\\
\mathcal{L}_{1,2} &= \tfrac{\mathcal{O} \pm \mathcal{T}}{2}, \quad [\mathcal{L}_{1,2}^a, \mathcal{L}_{1,2}^b] = -\epsilon^{abc}\mathcal{L}_{1,2}^c, \quad [\mathcal{L}_1, \mathcal{L}_2] = 0,\\
\mathcal{T} &= \operatorname{ad}\mathcal{T} = \left(\begin{array}{c|c} 0 & i\vec{\alpha} \\ \hline i\vec{\alpha}^T & 0\end{array}\right), \quad \mathcal{L}_{1,2} = \left(\begin{array}{c|c} \vec{\gamma_{1,2}} & \pm i\vec{\gamma_{1,2}} \\ \hline \pm i\vec{\gamma_{1,2}}^T & \vec{\gamma_{1,2}}\end{array}\right),\\
\text{Killing form:}\quad & \kappa(\mathcal{T}^a, \mathcal{T}^b) = -4\delta^{ab} = \kappa(\mathcal{O}^a, \mathcal{O}^b).
\end{aligned}$$

Chapter 3

Mass Points

A dynamics is a representation of spacetime operations. A nonrelativistic mass point dynamics is a representation of the time translations. The equations of motion, of second order in time for position as basic observable and of first order for the position-momentum pair, express the Lie algebra action $\frac{d}{dt}$ of the time operations. The classical mass point orbits as solutions are realizations of the (eigen)time translation group $t \in \mathbb{R} \cong \mathbf{D}(1) \ni e^t$ in position, faithful noncompact with image $\mathbb{R} \cong \mathbf{SO}_0(1,1) = \mathcal{Y}^1$, e.g., for free mass points and hyperbolic orbits of never-returning comets, and unfaithful compact with image $\mathbb{R}/\mathbb{Z} \cong \mathbf{SO}(2) = \Omega^1$ for periodic orbits, e.g., for elliptic orbits of planets. Newton's idealization of mechanics working with mass points was successful even after the introduction of the electromagnetic fields by Faraday and Maxwell, and of the metrical tensor field in Einstein's gravity. The time orbits of mass points in position are derivable by an extremalization of an action, leading to the Euler–Lagrange equations of motion. For a general relativistic mass point dynamics, this extremalization merges into the property of geodesics to have an extremal length; the Lagrangian is essentially the spacetime metric.

The gravitational interactions in the general relativistic formulation were first tested by the geodesics of mass points and light rays in a Schwarzschild geometry, in the classical tests of the perihelion shift of the planets and the light ray bending at the sun's boundary. The metrical tensor arises as the $\mathbf{GL}(\mathbb{R}^4)$-orbit of the flat Minkowski metric for free mass points (particles) by linear operations $\mathbf{g}^{li}(x) = V^{li}_{ab}(x)\eta^{ab}$ with the tetrad product $V^{li}_{ab}(x) = \mathbf{e}^l_a(x)\mathbf{e}^i_b(x)$, where $\mathbf{e}(x) \in \mathbf{GL}(4,\mathbb{R})/\mathbf{O}(1,3)$ contains compact $\mathbf{SO}(4)$-rotations with six parameters and the four noncompact dilations $\mathbf{D}(1)^4$ (see Chapter 1). Therefore, the geodesics as free mass point eigentime orbits in curved spacetime can be interpreted as orbits in flat position under the influence of a tetrad implemented interaction V: Curvature becomes the interaction for flat space; the metrical coefficients of a static spacetime

H. Saller, *Operational Spacetime: Interactions and Particles*,
Fundamental Theories of Physics 163, DOI 10.1007/978-1-4419-0898-8_4,

describe interactions that can be formulated in terms of a nonrelativistic potential. A nonrelativistic precursor for such a reinterpretation is the centrifugal potential.

The motion group with the global invariances of spacetime is generated by the conserved quantities of a dynamics, e.g., the motion group $\mathbb{R}\times\mathbf{SO}(3)$ with the time translations generated by the Hamiltonian for the conserved energy and the rotations in the position motion group generated by the conserved angular momenta.

In contrast to quantum physics, measurements in classical physics are complete and absolute. All observables, like positions, momenta, energy, angular momenta, etc., have their values in the real or complex numbers $\mathbb{K} = \mathbb{R},\mathbb{C}$, which are directly the possible results of their measurement. In a quantum language, the classical observables are simultaneously diagonalizable; they commute with each other. Vectorial observables, $V \cong \mathbb{K}^n$, are measurable with all their coefficients; the values in different bases can be computed from each other by basis transformations.

3.1 Nonrelativistic Classical Interactions

Newton space–time is operationally characterized by classes of the inhomogeneous Galilei group, isomorphic as flat manifold, not as homogeneous space, to the time and position translations,

$$[\mathbf{SO}(3)\ \vec{\times}\ \mathbb{R}^3]\ \vec{\times}\ [\mathbb{R}\ \oplus\ \mathbb{R}^3]\big/\mathbf{SO}(3)\ \vec{\times}\ \mathbb{R}^3 \cong \mathbb{R}\ \oplus\ \mathbb{R}^3.$$

The (doubled) semidirect group multiplication law looks rather complicated — more suggestive in the four- and five-dimensional faithful representations:

$$\begin{aligned}
\text{Galilei group:}\quad & \mathbf{SO}(3)\ \vec{\times}\ \mathbb{R}^3 \ni (O,\vec{v}) \longmapsto \begin{pmatrix} O & \vec{v} \\ 0 & 1\end{pmatrix} \in \mathbf{GL}(4,\mathbb{R}),\\
& (O_1,\vec{v_1})\circ(O_2,\vec{v_2}) = (O_1\circ O_2, \vec{v_1}+O_1.\vec{v_2}),\\
\text{inhomogeneous:}\quad & [\mathbf{SO}(3)\ \vec{\times}\ \mathbb{R}^3]\ \vec{\times}\ [\mathbb{R}\ \oplus\ \mathbb{R}^3] \longrightarrow \mathbf{GL}(5,\mathbb{R}),\\
& \qquad\qquad (O,\vec{v};t,\vec{x}) \longmapsto \begin{pmatrix} O & \vec{v} & \vec{x} \\ 0 & 1 & t \\ 0 & 0 & 1\end{pmatrix},\\
& (O_1,\vec{v_1};t_1,\vec{x_1})\circ(O_2,\vec{v_2};t_2,\vec{x_2})\\
& \qquad = (O_1\circ O_2, \vec{v_1}+O_1.\vec{v_2}; t_1+t_2, \vec{x_1}+\vec{v_1}t_2+O_1.\vec{x_2}).
\end{aligned}$$

A time translation $\mathbb{R}$ and rotation $\mathbf{SO}(3)$-invariant dynamics of a nonrelativistic mass point in a potential V are characterized by an action W with time derivatives $\dot{\xi} = \frac{d\xi}{dt} = d_t\xi$, etc.,

$$\begin{aligned}
W &= \int dt\ (\vec{p}\dot{\vec{x}} - H) \cong \int dt[m\tfrac{\dot{\vec{x}}^2}{2} - V(r)],\\
H &= \tfrac{\vec{p}^2}{2m} + V,
\end{aligned}
\qquad
\left\{\begin{aligned}
\dot{\vec{x}} &= \tfrac{\vec{p}}{m},\\
\dot{\vec{p}} &= -\vec{\partial}V = -\tfrac{\vec{x}}{r}\tfrac{\partial V}{\partial r},\\
\ddot{\vec{x}} &= -\tfrac{\vec{x}}{r}\tfrac{\partial V}{\partial mr}.
\end{aligned}\right.$$

Parallel with the metrical tensor $d\vec{x}^2 = dr^2 + r^2(d\theta^2 + \sin^2\theta d\varphi^2)$ of flat position, the kinetic energy $\vec{p}^{\,2} = \frac{d\vec{x}^2}{dt^2}$ with the momenta $\vec{p}$ can be decomposed with polar coordinates $\mathbb{R}^3 = \mathbb{R}_+ \times \Omega^2$ into the contributions with radial momentum p_r and with angular momenta $\vec{\mathcal{L}}$:

$$\vec{x} = r\begin{pmatrix}\sin\theta\cos\varphi\\ \sin\theta\sin\varphi\\ \cos\theta\end{pmatrix}, \quad \vec{\mathcal{L}} = \vec{x}\times\vec{p} = mr^2\begin{pmatrix}-\dot\theta\sin\varphi - \dot\varphi\sin\theta\cos\theta\cos\varphi\\ \dot\theta\cos\varphi - \dot\varphi\sin\theta\cos\theta\sin\varphi\\ \dot\varphi\sin^2\theta\end{pmatrix},$$
$$\vec{p}^2 = p_r^2 + \frac{\vec{\mathcal{L}}^2}{r^2}, \qquad p_r^2 = m^2\dot r^2, \quad \vec{\mathcal{L}}^2 = m^2r^4(\dot\theta^2 + \dot\varphi^2\sin^2\theta).$$

The equations of motion are derived by a stationary action, i.e., as Euler–Lagrange equations $\frac{\partial L}{\partial\xi} - \frac{d}{dt}\frac{\partial L}{\partial\dot\xi} = 0$:

$$W = \int dt\,[\frac{m\dot r^2}{2} + \frac{mr^2}{2}(\dot\theta^2 + \dot\varphi^2\sin^2\theta) - V]$$
$$\Rightarrow \begin{cases}\frac{d}{dt}r^2\dot\varphi\sin^2\theta = 0,\\ \frac{d}{dt}r^2\dot\theta = r^2\dot\varphi^2\sin\theta\cos\theta,\\ \ddot r = \frac{\partial}{\partial mr}[\frac{mr^2}{2}(\dot\theta^2 + \dot\varphi^2\sin^2\theta) - V].\end{cases}$$

Rotation invariance is equivalent to angular momentum conservation. The Poisson–Lie brackets, $[F,G]_P = \frac{\partial F}{\partial\vec{p}}\frac{\partial G}{\partial\vec{x}} - \frac{\partial G}{\partial\vec{p}}\frac{\partial F}{\partial\vec{x}}$, normalized as $[\vec{p},\vec{x}]_P = \mathbf{1}_3$, vanish for angular momenta and Hamiltonian:

$$\vec{\mathcal{L}} \in \log\mathbf{SO}(3): \quad d_t\vec{\mathcal{L}} = [H,\vec{\mathcal{L}}]_P = 0, \quad [\mathcal{L}^a,\mathcal{L}^b]_P = -\epsilon^{abc}\mathcal{L}^c.$$

Therefore, in classical physics, all orbits with conserved angular momentum $d_t\vec{\mathcal{L}} = 0$ are planar since $\vec{\mathcal{L}}\vec{x} = 0$, $\vec{\mathcal{L}}\vec{p} = 0$ (first Kepler law). For the mass point motion, one can choose an "equatorial" plane:

$$d_t\vec{\mathcal{L}} = 0, \quad \vec{\mathcal{L}}^2 = L^2 = \text{constant},$$
$$(\theta,\dot\theta) = (\tfrac{\pi}{2},0) \Rightarrow L = mr^2\dot\varphi.$$

The time translations $\mathbb{R}$, generated by the Hamiltonian H, have the energy as their invariant:

$$H = \frac{m\dot r^2}{2} + \frac{L^2}{2mr^2} + V = E.$$

The remaining one-dimensional radial position translations are characterized by a radial action with an additional repulsive centrifugal potential $V_L(r) = \frac{L^2}{2mr^2}$,

$$m\ddot r = -\frac{\partial}{\partial r}(V + \frac{L^2}{2mr^2}), \quad W_r = \int dt(\frac{m\dot r^2}{2} - \frac{L^2}{2mr^2} - V).$$

The centrifugal potential originates from the rotation degrees of freedom. It arises also in a free theory with $V = 0$.

The planar orbits have the time parametrization $t \longmapsto r(t)$,

$$(\frac{dr}{dt})^2 = \frac{2(E-V)}{m} - \frac{L^2}{m^2r^2} \Rightarrow t - t_0 = \int_{r(t_0)}^{r(t)}\frac{mrdr}{\sqrt{2m(E-V)r^2 - L^2}},$$

and the polar representation $\varphi \longmapsto r(\varphi)$,

$$\dot r = \frac{dr}{d\varphi}\dot\varphi = \frac{L}{mr^2}\frac{dr}{d\varphi} \Rightarrow (\frac{dr}{d\varphi})^2 = \frac{r^4}{L^2}[2m(E-V) - \frac{L^2}{r^2}],$$
$$\varphi - \varphi_0 = \int_{r(\varphi_0)}^{r(\varphi)}\frac{Ldr}{r\sqrt{2m(E-V)r^2 - L^2}}.$$

3.2 The Symmetries of the Kepler Dynamics

Mass point mechanics, by itself, has no principle to distinguish "basic" potentials. In hindsight, the basically most important nonrelativistic interactions for mass points are described by the rotation-invariant Kepler potential. It is used as a Newton potential with coupling constant $G = \frac{\kappa c^2}{8\pi}$ for a gravistatic interaction, only attractive, and as a Coulomb potential for an electrostatic interaction, attractive or repulsive,

$$V(\vec{x_1} - \vec{x_2}) = \frac{\gamma_0}{|\vec{x_1} - \vec{x_2}|}, \quad \gamma_0 = \begin{cases} -m_1 m_2 G & \text{with masses } m_{1,2}, \\ Q_1 Q_2 \frac{1}{4\pi\epsilon_0} & \text{with charges } Q_{1,2}. \end{cases}$$

The dimensionless interaction constants $\frac{\gamma_0}{\hbar c}$, with $\hbar$ as dimensional unit only, allow, for the Coulomb potential, a factorization in the normalization of the electromagnetic interaction with Sommerfeld's fine structure constant α_S and integer charge numbers with respect to the electron charge e:

$$\frac{\gamma_0}{\hbar c} = \begin{cases} -m_1 m_2 \frac{1}{m_P^2} & \text{with Planck mass } m_P^2 = \frac{\hbar c}{G} = \frac{8\pi\hbar}{c\kappa} \sim (6.1 \times 10^{-8} \text{ kg})^2, \\ z_1 z_2 \frac{g^2}{4\pi} & \text{with } g^2 = \frac{e^2}{\hbar c \epsilon_0} = 4\pi\alpha_S \sim \frac{1}{10.9}, \quad z = \frac{Q}{e} \in \mathbb{Z}. \end{cases}$$

In contrast to the integer charge numbers $z \in \mathbb{Z}$, related to the compactness of the electromagnetic operation group $\mathbf{U}(1)$ (see Chapter 5), there arise continuous numbers $\frac{m}{m_P}$ for the noncompact operations of gravity with the masses $m^2 \in \mathbb{R}_+$ as invariants of the spacetime translation group $\mathbb{R}^4$. The Planck mass with $m_P^2 \ell^2 = 8\pi \frac{\hbar^2}{c^2}$ defines the limit for the relative magnitude of Schwarzschild and Compton lengths:

$$\ell_m = \frac{\kappa m}{8\pi}, \quad L_m = \frac{\hbar}{cm}, \text{ with } \begin{cases} \ell_m \geq L_m & \iff m^2 \geq m_P^2, \\ \ell_m \leq L_m & \iff m^2 \leq m_P^2. \end{cases}$$

The Kepler potential determines a dynamics of two classical mass points with position-momentum pairs $(\vec{x_i}, \vec{p_i})_{i=1,2}$ in Euclidean position, compatible with the inhomogeneous Galilei group $[\mathbf{SO}(3) \vec{\times} \mathbb{R}^3] \vec{\times} \mathbb{R}^4$ for nonrelativistic theories. A Lagrangian encodes the time development by the Hamiltonian H:

$$L(1,2) = \vec{p_1} d_t \vec{x_1} + \vec{p_2} d_t \vec{x_2} - H(1,2), \quad H(1,2) = \frac{\vec{p_1}^2}{2m_1} + \frac{\vec{p_2}^2}{2m_2} + V(\vec{x_1} - \vec{x_2}).$$

It is the sum of the free center of mass motion $(\vec{x_c}, \vec{p_c})$ for the translations of flat position and the reduced dynamics $(\vec{x}, \vec{p})$ for the position representing Kepler interaction:

$$\begin{aligned} L(1,2) &= \vec{p_c} d_t \vec{x_c} - \frac{\vec{p_c}^2}{2M} + L(\vec{x}, \vec{p}), \\ M &= m_1 + m_2, \end{aligned} \quad \begin{cases} L(\vec{x}, \vec{p}) = \vec{p} d_t \vec{x} - H(\vec{x}, \vec{p}), \\ H(\vec{x}, \vec{p}) = \frac{\vec{p}^2}{2m} + V(\vec{x}), \\ Mm = m_1 m_2. \end{cases}$$

Only for gravity is the center of mass transformation compatible with the transition to mass sum and reduced mass:

$$V(r) = -m_1 m_2 \frac{c^2\kappa}{8\pi r} = -mM\frac{c^2\kappa}{8\pi r} = -mc^2\frac{\ell_M}{r}.$$

There is no analogue factorization into "sum charge" and "reduced charge" for the integer charge numbers in the product $z_1 z_2$.

The Kepler potential is distinguished by being the inverse of the invariant Laplace operator (Laplace kernel) for flat position $\mathbb{R}^3 \cong \mathbf{SO}(3) \vec{\times} \mathbb{R}^3/\mathbf{SO}(3)$ with respect to the point-supported Dirac position distribution (see Chapter 10):

$$-\vec{\partial}^2\frac{1}{r} = 4\pi\delta(\vec{x}) = 4\pi\delta(\vec{\omega})\frac{1}{r^2}\delta(r).$$

A classical dynamics with the Kepler potential $md_t^2\vec{x} = \gamma_0\frac{\vec{x}}{r^3}$ has one intrinsic unit $\frac{\gamma_0}{m} = (-c^2\ell_M, \frac{z_1z_2}{m}\alpha_S c\hbar)$ with the dimension $\frac{\mathrm{m}^3}{\mathrm{s}^2}$. For bound orbits, it gives the relation of the orbit "radius" to the periodic orbit time as illustrated by Kepler's third law, where the reduced mass m drops out.

The equations of motion for the reduced system of the Kepler dynamics give as action of the time translations

$$\begin{aligned} H &= \frac{\vec{p}^2}{2m} + \frac{|\gamma_0|\delta}{r}, \\ \delta &= \pm 1 \text{ (repulsion, attraction)}, \end{aligned} \quad \left\{ \begin{aligned} d_t\vec{x} &= [H, \vec{x}]_P = \frac{\vec{p}}{m}, \\ d_t\vec{p} &= [H, \vec{p}]_P = -\vec{\partial}V = \gamma_0\frac{\vec{x}}{r^3}, \end{aligned} \right.$$

with the first-order position derivatives of the potential (Kepler force $\gamma_0\frac{\vec{x}}{r^3}$).

The Kepler Hamiltonian is distinguished by real six-dimensional global position groups $G \supset \mathbf{SO}(3)$ with the rotations as subgroup. It has an invariant perihelion vector $\vec{\mathcal{P}}$, the *Lenz–Runge vector* in the orbit plane,

$$\vec{\mathcal{P}} = \frac{1}{m|\gamma_0|}\vec{p} \times \vec{\mathcal{L}} + \delta\frac{\vec{x}}{r}, \quad \vec{\mathcal{L}}\vec{\mathcal{P}} = 0, \quad [\vec{\mathcal{P}}, H]_P = 0.$$

The energy E as time translation invariant for the Hamiltonian is a function of the angular momentum and perihelion invariant, which both describe position properties,

$$\vec{\mathcal{P}}^2 = 1 + \frac{2}{m\gamma_0^2}H\vec{\mathcal{L}}^2 \Rightarrow E = \frac{m\gamma_0^2}{2}\frac{P^2-1}{L^2}, \text{ with } \vec{\mathcal{P}}^2 = P^2,\ \vec{\mathcal{L}}^2 = L^2.$$

The time orbits in position space are conic sections, given by polar equations with one focus as origin (second Kepler law):

$$\begin{aligned} \vec{\mathcal{P}}\vec{x} = Pr\cos\varphi &= \frac{1}{m|\gamma_0|}(\vec{p} \times \vec{\mathcal{L}})\vec{x} + \delta r = \frac{1}{m|\gamma_0|}L^2 + \delta r \\ \Rightarrow r(\varphi) &= \frac{1}{m|\gamma_0|}\frac{L^2}{P\cos\varphi - \delta}. \end{aligned}$$

$\vec{x}$ directs to the peri- and aphelion for $\varphi = 0$ and $\varphi = \pi$, respectively.

As will become more clear in quantum theory, the invariance group of the Kepler dynamics is a consequence of the operational structure of position. Its classical representation by the mass point orbits depends on the energy sign.

There arise the three maximal global invariance groups of three-dimensional Riemannian manifolds, i.e., $\mathbf{SO}(4)$ (rotations), $\mathbf{SO}_0(1,3)$ (Lorentz transformations), and $\mathbf{SO}(3) \vec{\times} \mathbb{R}^3$ (Galilei group) as the motion groups of, respectively, sphere Ω^3, hyperboloid $\mathcal{Y}^3$, and flat $\mathbb{R}^3$, used as position manifolds in Friedmann universes, all with $\mathbf{SO}(3)$ as the local invariance group. The Lie algebra for a time-translation eigenvalue $E \in \mathbf{spec}\, H$,

$$\begin{aligned} [\mathcal{L}^a, \mathcal{L}^b]_P &= -\epsilon^{abc}\mathcal{L}^c, \quad [\mathcal{L}^a, \mathcal{P}^b]_P = -\epsilon^{abc}\mathcal{P}^c, \\ [\mathcal{P}^a, \mathcal{P}^b]_P &= \tfrac{2H}{m\gamma_0^2}\epsilon^{abc}\mathcal{L}^c \cong \tfrac{2E}{m\gamma_0^2}\epsilon^{abc}\mathcal{L}^c, \end{aligned}$$

is semisimple for negative energy $\epsilon(E) = \frac{E}{|E|} = -1$ and cyclic (elliptic) "bound" orbits, characterized by the subgroup $\mathbf{SO}(2) \subset \mathbf{SO}(4)$ and illustrated by planets, simple for positive energy $\epsilon(E) = 1$ and hyperbolic "scattering" orbits, characterized by the subgroup $\mathbf{SO}_0(1,1) \subset \mathbf{SO}_0(1,3)$ and illustrated by comets, and semidirect for trivial energy $E = 0$ and parabolic "scattering" orbits, characterized by the subgroup $\mathbb{R} \subset \mathbf{SO}(3) \vec{\times} \mathbb{R}^3$:

$$\begin{aligned} E \neq 0: \quad \vec{\mathcal{B}} = \sqrt{\tfrac{m\gamma_0^2}{2|E|}}\vec{\mathcal{P}}, \quad [\mathcal{L}^a, \mathcal{B}^b]_P &= -\epsilon^{abc}\mathcal{B}^c, \quad [\mathcal{B}^a, \mathcal{B}^b]_P = \epsilon(E)\epsilon^{abc}\mathcal{L}^c, \\ E = 0: \qquad\qquad\qquad [\mathcal{L}^a, \mathcal{P}^b]_P &= -\epsilon^{abc}\mathcal{P}^c, \quad [\mathcal{P}^a, \mathcal{P}^b]_P = 0. \end{aligned}$$

The contractions of the noncompact–compact pair for trivial energy lead to the Galilei group,

$$\mathbf{SO}_0(1,3) \stackrel{E\to 0}{\longrightarrow} \mathbf{SO}(3) \vec{\times} \mathbb{R}^3 \stackrel{0\leftarrow E}{\longleftarrow} \mathbf{SO}(4),$$

which is isomorphic to the Euclidean group with the contracted boosts $\vec{\mathcal{P}}$ as flat position momenta.

3.3 Electrodynamics for Charged Mass Points

In special relativistic electrodynamics, compatible with the Poincaré group,

$$\begin{aligned} \mathbf{SO}_0(1,3) \vec{\times} \mathbb{R}^4 \ni (\Lambda, x) &\longmapsto \begin{pmatrix} \Lambda & x \\ 0 & 1 \end{pmatrix} \in \mathbf{GL}(5, \mathbb{R}), \\ (\Lambda_1, x_1) \circ (\Lambda_2, x_2) &= (\Lambda_1 \circ \Lambda_2, x_1 + \Lambda_1.x_2), \end{aligned}$$

and its Lorentz group classes $\mathbf{SO}_0(1,3) \vec{\times} \mathbb{R}^4/\mathbf{SO}_0(1,3) \cong \mathbb{R}^4$ as flat Minkowski spacetime, the time-independent Coulomb potential is embedded into the spacetime-dependent vector potential $\mathbf{A}$. The orbit of a charged mass point $\tau \longmapsto X_a(\tau) = (ct, \vec{X}(t))$ is parametrized by the eigentime $[\tau] = \mathrm{s}$; i.e., the orbit X as a spacetime "field" depends only on $\tau = \epsilon(x_0)\vartheta(x^2)\sqrt{x^2}$. Electrodynamics is characterizable by an action as the sum of the free actions and the interaction:

$$\begin{aligned} W_{\text{matter}}^{\text{elmag}} &= W_{\text{elmag}} + W_{\text{matter}} + W_{\text{int}}, \\ W_{\text{elmag}} &= \hbar \int d^4x \ \tfrac{1}{g^2}(\mathbf{F}_{ab}\tfrac{\partial^a \mathbf{A}^b - \partial^b \mathbf{A}^a}{2} + \tfrac{\mathbf{F}_{ab}\mathbf{F}^{ab}}{4}), \\ W_{\text{matter}} &= \int d\tau \ [P^a d_\tau X_a - \tfrac{P_a P^a}{2m}] \cong \int d\tau \ \tfrac{m}{2} d_\tau X_a d_\tau X^a \\ W_{\text{int}} &= -\hbar z \int d\tau \ \mathbf{A}^a(X) d_\tau X_a. \end{aligned}$$

The vector potential and field strengths, acted on by a four-dimensional Minkowski and a six-dimensional adjoint representation of the Lorentz group, have the dimensions of a length and an area density, $[\mathbf{A}] = \frac{1}{\text{m}}, [\mathbf{F}] = \frac{1}{\text{m}^2}$. The Lorentz invariant interaction is the line integral along the spacetime coordinates of the mass point:

$$z\int d\tau (d_\tau X_a)\mathbf{A}^a(X) = z\int dX_a \mathbf{A}^a(X) = \int d^4x\ \mathbf{J}_a(x)\mathbf{A}^a(x).$$

The current for a mass point with the dimension of a volume density $[\mathbf{J}] = \frac{1}{\text{m}^3}$,

$$\mathbf{J}_a(x) = z\int d\tau\ d_\tau X_a \delta(x-X) = z\int dX_a \delta(x-X) = z\frac{P_a}{P_0}\delta(\vec{x}-\vec{X}(t)),$$

is proportional to its energy-momentum $P_a = md_\tau X_a = \frac{mc}{\sqrt{1-\frac{\vec{V}^2}{c^2}}}(1, \frac{\vec{V}}{c})$ and the Lorentz invariant inverse energy-multiplied Dirac position distribution $\frac{1}{P_0}\delta(\vec{x}-\vec{X})$.

The equations of motion display the action of spacetime translations with the Lorentz force $\frac{1}{m}\mathbf{F}^{ba}(X)P_b = \frac{\vec{\mathbf{E}}(X)+\vec{V}\times\vec{\mathbf{B}}(X)}{\sqrt{1-\frac{\vec{V}^2}{c^2}}}$ effecting the mass point orbit,

$$\begin{aligned}
\partial^a\mathbf{A}^b - \partial^b\mathbf{A}^a &= \mathbf{F}^{ba}, & -\tfrac{1}{g^2}\partial^b\mathbf{F}_{ba} &= \mathbf{J}_a,\\
d_\tau X_a &= \tfrac{P_a}{m}, & d_\tau P^a &= \tfrac{z}{m}\hbar\mathbf{F}^{ba}(X)P_b,\\
\Rightarrow [\delta_a^b d_\tau - \tfrac{z}{m}\hbar\mathbf{F}^{ba}(X)]d_\tau X_b &= 0.
\end{aligned}$$

The system involves the mass point equations with the Coulomb force as an electrostatic approximation with $\mathbf{A}^a(x) = (\mathbf{A}^0(\vec{x}), 0)$ for a current $\mathbf{J}_0(x) = z_1\delta(\vec{x})$:

$$\begin{aligned}
-\tfrac{1}{g^2}\vec{\partial}^2\mathbf{A}^0(\vec{x}) &= z_1\delta(\vec{x}),\ \ \mathbf{A}^0(\vec{x}) = z_1\tfrac{g^2}{4\pi r},\\
d_t\vec{P} &= -z_2\hbar c\vec{\partial}\mathbf{A}^0(\vec{X}) = z_1z_2\tfrac{\hbar c g^2}{4\pi}\tfrac{\vec{X}}{R^3}.
\end{aligned}$$

3.4 Einstein Gravity for Mass Points

In general relativity with a Riemannian manifold $(\mathbb{M}^{(1,3)}, \mathbf{g})$, gravity for a mass point on the orbit $\tau \longmapsto X_k(\tau)$ is characterizable by the action

$$\begin{aligned}
W &= \int \sqrt{|\mathbf{g}|}d^4x\ \tfrac{c}{2\kappa}\mathcal{R}^{\bullet}_{\bullet} + W^{\text{grav}}_{\text{matter}},\\
W^{\text{grav}}_{\text{matter}} &= \int d\tau\ \tfrac{m}{2}\mathbf{g}^{li}(X)d_\tau X_l d_\tau X_i \cong \int d\tau\ (P^l d_\tau X_l - \tfrac{1}{2m}\mathbf{g}_{li}P^lP^i).
\end{aligned}$$

Its variation leads to the geodesic orbits (shortest paths):

$$\begin{aligned}
d_\tau X_k &= \tfrac{1}{m}\mathbf{g}_{ki}P^i,\ \ d_\tau \mathbf{g}_{ki}P^i = -\tfrac{1}{m}\Gamma_k^{ij}(X)\mathbf{g}_{ir}\mathbf{g}_{jl}P^rP^l,\\
&\Rightarrow [\delta_k^i d_\tau + \Gamma_k^{ij}(X)d_\tau X_j]d_\tau X_i = 0.
\end{aligned}$$

The energy-momentum tensor of the mass point is proportional to the energy-momentum square,

$$\begin{aligned}\sqrt{|\mathbf{g}|}\mathbf{T}^{li}(x) &= m\int cd\tau\ \frac{P^lP^i}{m^2c^2}\ \delta(x-X) = \frac{P^lP^i}{cP^0}\ \delta(\vec{x}-\vec{X}),\\ -\kappa\mathbf{T}^{li} &= \mathcal{R}^{li}_{\bullet} - \tfrac{1}{2}\mathbf{g}^{li}\mathcal{R}^{\bullet}_{\bullet}.\end{aligned}$$

In contrast to the charge dependence of the electromagnetic interaction via $\frac{z}{m_{\text{inert}}}$, the gravitative interaction is independent of the mass of the mass point via $\frac{m_{\text{grav}}}{m_{\text{inert}}} = 1$. However, there remains a difference between mass points with $m^2 > 0$ and light rays with $m = 0$.

Flat spacetime gives the free Newtonian mass point in a special relativistic framework:

$$\mathbf{g} = \eta:\ \ d_\tau X_k = \tfrac{1}{m}P_k,\ \ d_\tau P_k = 0.$$

For the nonrelativistic Newton interaction of mass points, there is the gravistatic approximation with $\mathbf{g}^{ki}(x) \cong \left(\begin{array}{c|c}\mathbf{g}^{00}(\vec{x}) & 0\\ \hline 0 & -\mathbf{1}_3\end{array}\right)$ for an energy-momentum tensor $\mathbf{T}^{00}(x) = m_1\delta(\vec{x})$:

$$\begin{aligned}-\tfrac{2}{\kappa}\partial^a\Gamma^{00}_{\ \ a}(\vec{x}) = \tfrac{1}{\kappa}\vec{\partial}^2\mathbf{g}^{00}(\vec{x}) &= m_1\delta(\vec{x}),\ \ \mathbf{g}^{00}(\vec{x}) = 1 - m_1\tfrac{\kappa}{4\pi r} = 1 - \tfrac{2\ell_1}{r},\\ d_t\vec{P} &= m_2\tfrac{c^2}{2}\vec{\partial}\mathbf{g}^{00}(\vec{X}) = m_1m_2\tfrac{c^2\kappa}{8\pi}\tfrac{\vec{X}}{R^3}.\end{aligned}$$

3.5 Geodesics of Static Spacetimes

Spacetime $\mathbb{M}^{(1,3)}$ with the time translations and the rotations in the motion group $\mathbb{R}\times\mathbf{SO}(3) \subseteq G_{\mathbf{g}}$, e.g., Schwarzschild or Reissner spacetime outside the Schwarzschild radius, has the metric

$$\mathbf{g} = c^2d\tau^2 = e^{2\lambda_3(\rho)}c^2dt^2 - e^{2\lambda(\rho)}d\rho^2 - \rho^2(d\theta^2 + \sin^2\theta\ d\varphi^2).$$

The geodesics are parametrizable by eigentime τ for $\dot{\xi} = \frac{d\xi}{cd\tau} = \frac{1}{c}d_\tau\xi$ etc. in the action

$$W^{\text{grav}}_{\text{matter}} = \int d\tau\ \tfrac{m}{2}\mathbf{g}^{li}(X)\tfrac{dX_l}{d\tau}\tfrac{dX_i}{d\tau} = c^2\int d\tau\ \tfrac{m}{2}[e^{2\lambda_3}\dot{t}^2 - e^{2\lambda}\dot{\rho}^2 - \rho^2(\dot{\theta}^2 + \sin^2\theta\ \dot{\varphi}^2)].$$

The metrical coefficients are radial-dependent normalizations, i.e., dilations $\rho \longmapsto (e^{\lambda_3(\rho)},\ e^{\lambda(\rho)},\ \rho,\ \rho\sin\theta) \in \mathbf{D}(1)^4$ of the "energy-momenta" $(\dot{t},\dot{\rho},\dot{\theta},\dot{\varphi})$. The geodesics as shortest paths arise from a stationary action, i.e., from the Euler–Lagrange equations for the time coordinate t and for the position coordinates (ρ,θ,φ) with the *metrical tensor as Lagrangian* $L(\tau) = \frac{\mathbf{g}(\tau)}{c^2d\tau^2}$:

$$\tfrac{d}{cd\tau}e^{2\lambda_3}\dot{t} = 0 \text{ and } \left\{\begin{aligned}\tfrac{d}{cd\tau}\rho^2\dot{\varphi}\ \sin^2\theta &= 0,\\ \tfrac{d}{cd\tau}\rho^2\dot{\theta} &= \rho^2\dot{\varphi}^2\sin\theta\cos\theta,\\ \tfrac{d}{cd\tau}e^{2\lambda}\dot{\rho} &= \tfrac{\partial}{\partial\rho}[-e^{2\lambda_3}\dot{t}^2 + e^{2\lambda}\dot{\rho}^2 + \rho^2(\dot{\theta}^2 + \sin^2\theta\ \dot{\varphi}^2)].\end{aligned}\right.$$

The Killing fields for the global invariance group $G_{\mathbf{g}}$ of spacetime give rise to a planar motion with invariant eigenvalues $\mathcal{E}$ for time translations $\mathbb{R}$ and $\mathcal{L}$ for rotations $\mathbf{SO}(3)$:

$$\mathcal{E} = e^{2\lambda_3}\dot{t},\quad (\theta,\dot{\theta}) = (\tfrac{\pi}{2},0),\quad \mathcal{L} = \rho^2\dot{\varphi}.$$

The eigentime parametrization of the metrical coefficients gives the $\mathbb{R}\times\mathbf{SO}(3)$-invariant normalization of the Lagrangian (GR for "general relativistic"):

$$\begin{aligned}\text{GR: } \tfrac{\mathbf{g}}{c^2 d\tau^2} &= e^{2\lambda_3}\dot{t}^2 - e^{2\lambda}\dot{\rho}^2 - \rho^2(\dot{\theta}^2 + \sin^2\theta\,\dot{\varphi}^2)\\ &= \mathcal{E}^2 e^{-2\lambda_3} - e^{2\lambda}\dot{\rho}^2 - \tfrac{\mathcal{L}^2}{\rho^2} = \vartheta(m^2) = \begin{cases}1 & \text{for mass points } m^2 > 0,\\ 0 & \text{for light } m^2 = 0.\end{cases}\end{aligned}$$

It is the radially renormalized extension of the mass as the translation invariant of Minkowski spacetime translations and generalizes the nonrelativistic expression for the time translations and the position translations with the Euclidean group (NR for "nonrelativistic," SR for "special relativistic")

$$\begin{array}{lrl}\text{SR:} & \mathbf{SO}_0(1,3)\,\vec{\times}\,\mathbb{R}^4: & p_0^2 - \vec{p}^2 = \frac{E^2}{c^2} - p_r^2 - \frac{\vec{\mathcal{L}}^2}{r^2} = m^2,\\ \text{NR:} & \mathbb{R}\times[\mathbf{SO}(3)\,\vec{\times}\,\mathbb{R}^3]: & E - \frac{\vec{p}^2}{2m} = E - \frac{p_r^2}{2m} - \frac{\vec{\mathcal{L}}^2}{2mr^2} = V(r).\end{array}$$

The *general relativistic invariant for the eigentime translations* can be compared with the corresponding relation in nonrelativistic mass point mechanics with time translations:

$$\begin{array}{lccccccc}\text{GR with } d\tau: & \mathcal{E}^2 e^{-2\lambda_3} & - & e^{2\lambda}\dot{\rho}^2 & - & \frac{\mathcal{L}^2}{\rho^2} & = & \vartheta(m^2),\\ \text{NR with } dt: & E & - & m\frac{\dot{r}^2}{2} & - & \frac{L^2}{2mr^2} & = & V.\end{array}$$

The second-order radial equations of motions arise by time and eigentime derivation, respectively,

$$\begin{aligned}\text{NR:}\quad \ddot{r} &= -\tfrac{1}{2}\tfrac{\partial}{\partial r}\left(\tfrac{L^2}{m^2r^2} + \tfrac{2V}{m}\right),\\ \text{GR:}\quad \ddot{\rho} &= -\tfrac{1}{2}\tfrac{\partial}{\partial\rho}\left[\tfrac{\mathcal{L}^2}{\rho^2} + \vartheta(m^2) - \mathcal{E}^2 e^{-2\lambda_3}\right]e^{-2\lambda}.\end{aligned}$$

They can be derived as Euler–Lagrange equations of a radial action with an *effective potential*:

$$\begin{array}{lll}\text{NR:} & W_t = \int dt\, m\,[\frac{\dot{r}^2}{2} - \frac{V_{\text{eff}}(r)}{m}], & 2V_{\text{eff}} = \frac{L^2}{mr^2} + 2(V - E),\\ \text{GR:} & \frac{W_\tau}{c^2} = \int d\tau\, m\,[\frac{\dot{\rho}^2}{2} - \frac{v_{\text{eff}}(\rho)}{c^2}], & 2\frac{v_{\text{eff}}}{c^2} = [\frac{\mathcal{L}^2}{\rho^2} + \vartheta(m^2) - \mathcal{E}^2 e^{-2\lambda_3}]e^{-2\lambda}.\end{array}$$

The effective potential involves the additional centrifugal potential for a rotation-invariant nonrelativistic dynamics, connected with $\mathbf{SO}(3)$ and the conserved angular momentum, and, in the general relativistic case, a "genuine" potential for static spacetimes, connected with the invariance under time translations $\mathbb{R}$:

$$\begin{aligned}\text{NR:}\quad \ddot{r} &= -\tfrac{\partial}{\partial r}\tfrac{V_{\text{eff}}(r)}{m}, \quad \text{with } \tfrac{\dot{r}^2}{2} + \tfrac{V_{\text{eff}}(r)}{m} = 0,\\ \text{GR:}\quad \ddot{\rho} &= -\tfrac{\partial}{\partial\rho}\tfrac{v_{\text{eff}}(\rho)}{c^2}, \quad \text{with } \tfrac{\dot{\rho}^2}{2} + \tfrac{v_{\text{eff}}(\rho)}{c^2} = 0.\end{aligned}$$

The effective potential is given by tetrad action on the flat Minkowski spacetime tensor (free theory), here by dilations $e^{\lambda(\rho)}$, $e^{\lambda_3(\rho)}$:

$$\begin{array}{lll}\text{NR:} & V=0\Rightarrow 2V_{\text{eff}} &= \frac{L^2}{mr^2}-2E,\\ \text{GR:} & \mathbf{g}=\eta\ \Rightarrow 2\frac{v_{\text{eff}}}{c^2} &= \frac{\mathcal{L}^2}{\rho^2}-\vartheta(m^2)-\mathcal{E}^2.\end{array}$$

The remaining integration gives the planar orbits in position $t,\varphi \longmapsto r(t), r(\varphi)$ and $\tau,\varphi \longmapsto \rho(\tau),\rho(\varphi)$:

$$\begin{array}{llll}\text{NR:} & dt = \frac{mdr}{\sqrt{-2mV_{\text{eff}}(r)}}, & d\varphi = \frac{Ldr}{r^2\sqrt{-2mV_{\text{eff}}(r)}},\\ \text{GR:} & d\tau = \frac{d\rho}{\sqrt{-2v_{\text{eff}}(\rho)}}, & d\varphi = \frac{c\mathcal{L}d\rho}{\rho^2\sqrt{-2v_{\text{eff}}(\rho)}}.\end{array}$$

In the opposite direction, a given potential can be translated into a metrical spacetime tensor via $V_{\text{eff}} \sim v_{\text{eff}}$.

Thus, a static spacetime metric describes a gravity interaction whose geodesics can be expressed as caused by a potential of a nonrelativistic dynamics, e.g., for reciprocal metrical coefficients of time and position, i.e., with a self-dual dilation $\rho \longmapsto e^{\sigma_3\lambda_3(\rho)} \in \mathbf{SO}_0(1,1)$:

$$\begin{aligned}\mathbf{g} &= e^{2\lambda_3(\rho)}c^2dt^2 - e^{-2\lambda_3(\rho)}d\rho^2 - \rho^2 d\omega^2\\ \Rightarrow 2\frac{v_{\text{eff}}(\rho)}{c^2} &= [\vartheta(m^2)+\frac{\mathcal{L}^2}{\rho^2}]e^{2\lambda_3}-\mathcal{E}^2.\end{aligned}$$

The classical example is Schwarzschild spacetime, i.e., outside the event horizon $\rho \geq 2\ell_\odot$, with the angular momentum-dependent corrections $\frac{\mathcal{L}^2\ell_\odot}{\rho^3}$ of the Newton potential $\frac{1}{mc^2}[V_{\text{eff}}(r)-\frac{L^2}{2mr^2}] = -\frac{\ell_\odot}{r}$:

$$\begin{aligned}\mathbf{g} &= (1-\frac{2\ell_\odot}{\rho})c^2dt^2 - \frac{d\rho^2}{1-\frac{2\ell_\odot}{\rho}} - \rho^2 d\omega^2,\\ \vartheta(m^2) &= \frac{\mathcal{E}^2-\dot\rho^2}{1-\frac{2\ell_\odot}{\rho}} - \frac{\mathcal{L}^2}{\rho^2}\\ \Rightarrow 2\frac{v_{\text{eff}}(\rho)}{c^2} &= [\vartheta(m^2)+\frac{\mathcal{L}^2}{\rho^2}](1-\frac{2\ell_\odot}{\rho})-\mathcal{E}^2.\end{aligned}$$

For Schwarzschild spacetime without the Lenz–Runge invariance, the elliptic orbits of mass points for a Kepler dynamics with $\mathbf{SO}(4)$-invariance are replaced by the geodesic rosette orbits with an only rotation $\mathbf{SO}(3)$-invariant dynamics and position as a rotation paraboloid:

$$m^2>0:\quad 2\frac{v_{\text{eff}}(\rho)}{c^2} = (1+\frac{\mathcal{L}^2}{\rho^2})(1-\frac{2\ell_\odot}{\rho})-\mathcal{E}^2.$$

The perihelion shift $\Delta\varphi_\bullet \sim \frac{\ell_\odot}{R_\bullet}$ is of the order of magnitude of the Schwarzschild length of the central mass, e.g., the sun with $\ell_\odot \sim 1.5\times 10^3$ m (general relativistic correction), divided by the radius of the planet's orbit (Newton gravity), e.g., $\Delta\varphi_\bullet \sim 2.6\times 10^{-8}$ for mercury with $R_\bullet \sim 5.8\times 10^{10}$ m.

Also, light with trivial eigentime invariant $\vartheta(m^2)=0$ has a nontrivial effective potential for nontrivial angular momenta,

$$m=0:\quad 2\frac{v_{\text{eff}}(\rho)}{c^2} = \frac{\mathcal{L}^2}{\rho^2}(1-\frac{2\ell_\odot}{\rho})-\mathcal{E}^2.$$

There is no nonrelativistic contribution from Newton's gravity. The bending angle $\Delta\varphi_\odot \sim \frac{\ell_\odot}{r_\odot}$ is of the order of magnitude of the Schwarzschild length of the central mass divided by the distance of the light ray to its center, e.g., $\Delta\varphi_\odot \sim 2.1 \times 10^{-6}$ for the Schwarzschild length of the sun divided by the sun radius $r_\odot \sim 7 \times 10^8$ m in the case of a ray at the sun's boundary.

Other examples are static spacetimes with spherical, flat, or hyperbolic position:

$$\begin{aligned}
\mathbf{g} &= e^{2\lambda_3(\rho)} c^2 dt^2 - \tfrac{d\rho^2}{1-k\rho^2} - \rho^2 d\omega^2,\\
\vartheta(m^2) &= \mathcal{E}^2 e^{-2\lambda_3} - \tfrac{\dot\rho^2}{1-k\rho^2} - \tfrac{\mathcal{L}^2}{\rho^2}\\
\Rightarrow 2\tfrac{v_{\text{eff}}(\rho)}{c^2} &= \tfrac{\mathcal{L}^2}{\rho^2} + [\vartheta(m^2) - \mathcal{E}^2 e^{2\lambda_3}](1-k\rho^2) - k\mathcal{L}^2.
\end{aligned}$$

The static universe with flat position uses the inverse metrical coefficient $-\frac{1}{\mathbf{g}^{00}(r)}$, i.e., the eigentime derivation of time as effective potential,

$$\begin{aligned}
\mathbf{g} &= e^{2\lambda_3(\rho)} c^2 dt^2 - d\rho^2 - \rho^2 d\omega^2,\\
\vartheta(m^2) &= \mathcal{E}^2 e^{-2\lambda_3} - \dot\rho^2 - \tfrac{\mathcal{L}^2}{\rho^2}\\
\Rightarrow 2\tfrac{v_{\text{eff}}(\rho)}{c^2} &= \tfrac{\mathcal{L}^2}{\rho^2} + \vartheta(m^2) - 2\mathcal{E}^2 e^{-2\lambda_3}, \quad \mathcal{E}^2 e^{-2\lambda_3} = \dot t = \tfrac{dt}{cd\tau},\\
\ddot{\vec\xi} &= \tfrac{\mathcal{E}^2}{2}\tfrac{\vec\xi}{\rho}\tfrac{\partial}{\partial\rho} e^{-2\lambda_3} \text{ for } \vec\xi^2 = \rho^2.
\end{aligned}$$

The nonflat static Friedmann universes, especially for a spherical position $k = 1$ (Einstein's universe),

$$\begin{aligned}
\mathbb{R} \times (\Omega^3, \mathcal{Y}^3):\ \mathbf{g} &= c^2 dt^2 - \tfrac{d\rho^2}{1-k\rho^2} - \rho^2 d\omega^2\\
\Rightarrow 2\tfrac{v_{\text{eff}}(\rho)}{c^2} &= \tfrac{\mathcal{L}^2}{\rho^2} + [\vartheta(m^2) - \mathcal{E}^2](1-k\rho^2) - k\mathcal{L}^2,
\end{aligned}$$

lead to a harmonic oscillator potential for $k[\mathcal{E}^2 - \vartheta(m^2)] > 0$. In this case, the geodesics are oscillator orbits with one condition for the integration constants:

$$\begin{aligned}
k[\mathcal{E}^2 - \vartheta(m^2)] &= \omega^2 \Rightarrow 2\tfrac{v_{\text{eff}}(\rho)}{c^2} = \tfrac{\mathcal{L}^2}{\rho^2} + \omega^2\rho^2 - k(\omega^2 + \mathcal{L}^2),\\
\vec\xi^{\,2} &= \rho^2: \ \ddot{\vec\xi} + \omega^2\vec\xi = 0 \text{ with } \dot{\vec\xi}^2 + \omega^2\vec\xi^2 = k(\omega^2 + \mathcal{L}^2),\\
\vec\xi &= \vec{c_+} e^{i\omega t} + \vec{c_-} e^{-i\omega t} \Rightarrow \dot{\vec\xi}^2 + \omega^2\vec\xi^2 = 4\omega^2 \vec{c_+}\vec{c_-}, \ \ \vec{c_+}\vec{c_-} = k\tfrac{\omega^2 + \mathcal{L}^2}{4\omega^2}.
\end{aligned}$$

3.6 Gravity for Charged Mass Points

For the Reissner metric,

$$\mathbf{g} = e^{2\lambda_3(\rho)} c^2 dt^2 - e^{-2\lambda_3(\rho)} d\rho^2 - \rho^2 d\omega^2,$$

the Lie algebra of the dilations involves two contributions:

$$e^{2\lambda_3(\rho)} = 1 - \tfrac{2\ell_m}{\rho} + \tfrac{\ell_z^2}{\rho^2}.$$

The gravitative Newton potential $-\frac{2\ell_m}{\rho}$ with the Schwarzschild length

$$\ell_m = m\tfrac{\kappa}{8\pi} = m\tfrac{G}{c^2} = \tfrac{\ell^2}{8\pi L_m}, \text{ with } L_m = \tfrac{\hbar}{mc} \text{ Compton length}$$

is a solution of the homogeneous Einstein equations $\mathcal{R}_{\bullet a}^{\ b} = 0$, e.g., outside a mass point, with the position derivation

$$-\vec{\partial}^2 \frac{\ell_m}{\rho} = 4\pi\ell_m\delta(\vec{x}).$$

The electromagnetic contribution $\frac{\ell_z^2}{\rho^2}$ arises from the Coulomb potential of a charged mass point with a radial field strength:

$$\begin{aligned} \mathbf{A}^0 &= z\frac{\alpha_S}{\rho}, \quad \mathbf{F}^{\alpha 0} \cong -\vec{\partial}\frac{z\alpha_S}{\rho} = z\alpha_S\frac{\vec{x}}{\rho^3}, \\ -\frac{1}{\kappa}\mathcal{R}_\bullet^{ab} = \mathbf{T}^{ab}(\mathbf{F}) &= \frac{1}{g^2}(\eta_{cd}\mathbf{F}^{ac}\mathbf{F}^{bd} - \frac{\eta^{ab}}{4}\mathbf{F}^{cd}\mathbf{F}_{cd}) \\ &\cong z^2\frac{\alpha_S}{8\pi\rho^4}\left(\begin{array}{c|c|c} 1 & 0 & 0 \\ \hline 0 & -1 & 0 \\ \hline 0 & 0 & \mathbf{1}_2 \end{array}\right). \end{aligned}$$

It is normalized by an area, where the Planck area is multiplied by the electromagnetic normalization $\alpha_S = \frac{g^2}{4\pi} \sim \frac{1}{137}$ and the squared charge number $z = \frac{Q}{e}$ of a mass point:

$$\ell_z^2 = z^2\frac{\alpha_S\kappa\hbar}{8\pi c} = z^2\frac{\alpha_S\hbar G}{c^3} = z^2\frac{\alpha_S}{8\pi}\ell^2.$$

The position dependence $\frac{1}{\rho^2}$ originates from the second-order integrated traceless energy-momentum tensor of the electromagnetic field with

$$\vec{\partial}^2\frac{\ell_z^2}{2\rho^2} = \frac{\ell_z^2}{\rho^4} = (\vec{\partial}\frac{\ell_z}{\rho})^2.$$

The validity of the parametrization is dependent on the ratio of electromagnetic and gravitative area involved:

$$\begin{aligned} 0 = 1 - \frac{2\ell_m}{\rho} + \frac{\ell_z^2}{\rho^2} &\Rightarrow \frac{\rho}{\ell_m} = 1 \pm \sqrt{1 - \frac{\ell_z^2}{\ell_m^2}}, \\ \text{with } \frac{\ell_z}{\ell_m} &= \frac{zL_m}{\ell}\sqrt{8\pi\alpha_S} \sim zL_m\frac{5.1\times 10^{33}}{\mathrm{m}}. \end{aligned}$$

For electromagnetism “stronger” than gravity as in $\frac{\ell_z}{\ell_m} \geq 1$, e.g., for charged elementary particles, there is a “naked” singularity, i.e., without event horizon, and a static universe,

$$\text{proton: } L_p = \frac{\hbar}{m_p c} \sim 6\times 10^{-16}\ \mathrm{m},\ z_p^2 = 1 \Rightarrow (\frac{\ell_z}{\ell_m})_p \sim 3\times 10^{18}.$$

The effective potential for a charged mass point is a fourth-order polynomial in $\frac{1}{\rho}$:

$$m^2 > 0: \quad 2\frac{v_{\mathrm{eff}}(\rho)}{c^2} = \left(1 + \frac{\mathcal{L}^2}{\rho^2}\right)\left(1 - \frac{2\ell_m}{\rho} + \frac{\ell_z^2}{\rho^2}\right) - \mathcal{E}^2.$$

Chapter 4

Quantum Mechanics

It is not really a good strategy to follow the historical development of classical theory to quantum theory by "quantizing" a classical theory. Quantum theory is not only the change of a government; it is even the change of the constitution.

For example, a classical Lagrangian with the separation in a free kinetic and an interaction term is not a good starting point to solve a quantal bound-state problem. The ontology and interpretation of a quantum description are completely different from those of a classical one. A classical–quantum relationship and distinction may be found in the different representation structures of time and position operations. In classical mechanics, the time orbits are valued in position, $t \longmapsto \vec{x}(t)$, mass points have a position, and the concept of a "point particle" makes sense. In quantum mechanics, the time orbits are valued in a Hilbert space with probability amplitudes. Now, in quantum mechanics, there are also orbits of position ("of," not "in"): The Schrödinger wave functions ("information catalogues") for bound-state vectors are matrix elements of infinite-dimensional Hilbert representations of noncompact position operations $\vec{x} \longmapsto \psi(\vec{x})$. The concept of a mass point is very restricted. For example, it does not make sense to call an electron, e.g., "in" a hydrogen atom, a point particle, as there are no orbits in position.

Nonrelativistic quantum mechanics is a theory of representations of time and position operations acting on a Hilbert space. The position-momentum operators $(\vec{\mathbf{x}}, \vec{\mathbf{p}})$ come with the duality-induced nontrivial commutator $[\vec{\mathbf{x}}, \vec{\mathbf{p}}] = i\hbar\mathbf{1}_3$, which defines the Heisenberg Lie algebra. The solution of a dynamics involves the determination of the energy eigenvalues of a Hamiltonian, generating time translations in irreducible Hilbert representations,

$$\begin{aligned} &\mathbf{H} = \frac{\vec{\mathbf{p}}^2}{2m} + V(\vec{\mathbf{x}}), \\ &E \in \mathbf{spec}\,\mathbf{H} \cap \mathbb{R}, \quad \mathbb{R} \ni t \longmapsto e^{iEt} \in \mathbf{U}(1). \end{aligned}$$

In nonrelativistic mechanics, time translations are implemented as $\frac{d}{dt} = i \operatorname{ad} \mathbf{H}$; i.e., by the adjoint action of the Hamiltonian, built by position

H. Saller, *Operational Spacetime: Interactions and Particles*,
Fundamental Theories of Physics 163, DOI 10.1007/978-1-4419-0898-8_5,

operations, with the corresponding relations of eigenvalues and representation invariants for time and position operations.

The vectors, acted on by time translations, especially the bound-state eigenvectors $|E\rangle$, lie in a Hilbert space that is related to the quantum algebra generated by position and momentum $(\vec{\mathbf{x}},\vec{\mathbf{p}})$ with appropriate topological features.

The familiar Hilbert space with the square-integrable function classes $L^2(\mathbb{R}^s)$ of position translations $\vec{x}\in\mathbb{R}^s$, $s=1,2,\ldots$, as eigenvalues of $\vec{\mathbf{x}}$ and the momentum representation $\vec{\mathbf{p}}\longmapsto -i\hbar\vec{\partial}$ is defined by the faithful Hilbert representations of the noncompact real $(1+2s)$-dimensional Heisenberg group $\mathbf{H}(s)\cong e^{\mathbb{R}^s}\vec{\times}\mathbb{R}^{1+s}$, a semidirect product affine subgroup. Its nilpotent Lie algebra $\log\mathbf{H}(s)\cong\mathbb{R}^{1+2s}$ is characterized by s nontrivial brackets $[\vec{\mathbf{x}},\vec{\mathbf{p}}]=\mathbf{1}_s\mathbf{I}$, for illustration in a nonunitary faithful real $(2+s)$-dimensional representation:

$$\begin{aligned}\mathbf{H}(s)&\longrightarrow\mathbf{GL}(\mathbb{R}^{2+s}),\\ \exp(\vec{\alpha}\vec{\mathbf{x}}+\vec{\beta}\vec{\mathbf{p}}+\gamma\mathbf{I})&\longmapsto\exp\left(\begin{array}{cc|c}0&\vec{\alpha}&\gamma\\0&0&\vec{\beta}\\\hline 0&0&0\end{array}\right)=\left(\begin{array}{cc|c}1&\vec{\alpha}&\gamma+\vec{\alpha}\vec{\beta}\\0&1&\vec{\beta}\\\hline 0&0&1\end{array}\right),\\ e^{\vec{\alpha_1}\vec{\mathbf{x}}+\vec{\beta_1}\vec{\mathbf{p}}+\gamma_1\mathbf{I}}\circ e^{\vec{\alpha_2}\vec{\mathbf{x}}+\vec{\beta_2}\vec{\mathbf{p}}+\gamma_2\mathbf{I}}&=e^{(\vec{\alpha_1}+\vec{\alpha_2})\vec{\mathbf{x}}+(\vec{\beta_1}+\vec{\beta_2})\vec{\mathbf{p}}+(\gamma_1+\gamma_2-\vec{\beta_1}\vec{\alpha_2})\mathbf{I}},\\ \mathbf{H}(s)\cong e^{\mathbb{R}^s}\vec{\times}\mathbb{R}^{1+s}&\ni e^{\vec{\alpha}\vec{\mathbf{x}}+\vec{\beta}\vec{\mathbf{p}}+\gamma\mathbf{I}}=e^{\vec{\alpha}\vec{\mathbf{x}}}\circ e^{\vec{\beta}\vec{\mathbf{p}}+\gamma\mathbf{I}}.\end{aligned}$$

Here, the position-momentum commutator is represented by a nilquadratic matrix $\mathbf{I}\longmapsto\begin{pmatrix}0&0&1\\0&0&0\\0&0&0\end{pmatrix}$. In the faithful complex infinite dimensional Hilbert representations, the invariant basic element $\mathbf{I}\longmapsto i\hbar\mathbf{1}$ takes one nontrivial spectral value $i\hbar\in\mathbf{spec}\,\mathbf{I}=i\mathbb{R}$ as action unit (Planck's constant). Then the time translation eigenvectors are given by position-dependent Schrödinger functions:

$$|E\rangle\cong\psi_E\in L^2(\mathbb{R}^s):\quad\psi_E(t)=e^{iEt}\psi_E,\quad\mathbf{H}|E\rangle\cong(-\tfrac{\hbar^2}{2m}\vec{\partial}^2+V)\psi_E=E\psi_E.$$

The Schrödinger functions lead to position measures $d^sx\ |\psi(\vec{x})|^2$ for probabilities.

A dynamics for nonrelativistic quantum mechanics comes with three intrinsic units — in addition to the universal action unit $\hbar$ and the "individual" mass m, there is a "normalization unit" of the specific potential. With those units, only dimensionless variables will be used in the following.

4.1 Nonrelativistic Wave Mechanics

For a rotation-invariant dynamics with time translation generator (Hamiltonian) $\mathbf{H}=\frac{\vec{\mathbf{p}}^2}{2}+V(\mathbf{r})$, $[\mathbf{H},\vec{\mathbf{L}}]=0$, the position translations are decomposed $\mathbb{R}^3=\mathbb{R}_+\times\Omega^2$ into radial translations and rotations. In a derivative representation the radial and angular momentum squares act on differentiable functions as follows (see Chapter 10):

$$\vec{\mathbf{p}}^2 = \mathbf{p}_r^2 + \frac{\vec{\mathbf{L}}^2}{r^2}, \quad [\vec{\mathbf{p}}^2, \vec{\mathbf{L}}] = 0, \quad \begin{cases} i\mathbf{p}_r \cong \frac{1}{r} d_r r = d_r + \frac{1}{r}, \\ \mathbf{p}_r^2 \cong -d_r^2 - \frac{2}{r} d_r, \\ \vec{\mathbf{L}}^2 \cong -\frac{1}{\sin^2\theta}[(\sin\theta \frac{\partial}{\partial\theta})^2 + (\frac{\partial}{\partial\varphi})^2]. \end{cases}$$

The rotation representations for Ω^2 are harmonically analyzable by the spherical harmonics as Hilbert basis of the 2-sphere functions

$$\{\mathrm{Y}_m^L \mid L = 0, 1, \ldots; |m| \le L\} \text{ basis of } L^2(\Omega^2) \overset{\text{dense}}{\cong} \bigoplus_{L=0}^{\infty} \mathbb{C}^{1+2L},$$

$$\text{with} \begin{cases} \int d^2\omega \, \overline{\mathrm{Y}_{m'}^{L'}(\varphi,\theta)} \mathrm{Y}_m^L(\varphi,\theta) &= \delta^{LL'} \delta_{mm'}, \\ \sum_{L=0}^{\infty} \sum_{m=-L}^{L} \overline{\mathrm{Y}_m^L(\varphi,\theta)} \mathrm{Y}_m^L(\varphi',\theta') &= \delta(\vec{\omega} - \vec{\omega}') \\ &= \delta(\varphi - \varphi') \frac{1}{\sin\theta} \delta(\theta - \theta'), \\ \vec{\mathbf{L}}^2 \mathrm{Y}_m^L(\varphi,\theta) &= L(1+L) \mathrm{Y}_m^L(\varphi,\theta), \\ -i\frac{\partial}{\partial\varphi} \mathrm{Y}_m^L(\varphi,\theta) &= m \mathrm{Y}_m^L(\varphi,\theta). \end{cases}$$

In contrast to the spherical harmonics, appropriate for asymptotically free scattering waves, with the $r \to 0$ ambiguity in $\frac{\vec{x}}{r}$, the *harmonic* $\mathbf{SO}(3)$-*polynomials* as product with the corresponding radial power are defined also for $\vec{x} \to 0$. They are eigenfunctions for a trivial translation invariant and, therefore, appropriate for bound waves,

$$(\vec{x})_m^L = r^L \mathrm{Y}_m^L(\varphi,\theta) : \quad \left. \begin{aligned} \vec{\mathbf{L}}^2 \mathrm{Y}_m^L(\varphi,\theta) &= L(1+L) \mathrm{Y}_m^L(\varphi,\theta), \\ \mathbf{p}_r^2 r^L &= -L(1+L) r^{L-2} \end{aligned} \right\} \Rightarrow \vec{\partial}^2 (\vec{x})_m^L = 0.$$

The $\vec{x}$-homogeneous harmonic polynomials span the irreducible $\mathbf{SO}(3)$-representation spaces $\mathbb{C}^{1+2L}$. Position polynomials as a sum of homogeneous polynomials can be decomposed into harmonic polynomials and r^2-parametrized invariant coefficients (see Chapter 8).

Representations of the radial translations $\mathbb{R}_+$ in bound waves come after the separation of the harmonic polynomials, which are acted on by the irreducible $\mathbf{SO}(3)$-representations. This leads to the Schrödinger equations for the representation coefficients of radial translations $\mathbb{R}_+$:

$$\psi(\vec{x}) = \sum_{L=0}^{\infty} \sum_{m=-L}^{L} (\vec{x})_m^L d_L(r) \Rightarrow [d_r^2 + \frac{2(1+L)}{r} d_r - 2V(r) + 2E] d_L(r) = 0.$$

The representations can involve an $\mathbb{R}_+$-reparametrization with a monotonic function ξ:

$$\mathbb{R}_+ \ni r \longmapsto \xi(r) \in \mathbb{R}_+.$$

The $\mathbb{R}_+$-representation coefficient is a product with an exponential,[1]

$$\mathbb{R}_+ \ni \frac{\xi}{2} \longmapsto d_L(r) = F_L(\xi) \, e^{-\frac{\xi}{2}},$$

[1]The normalization $\frac{\xi}{2}$ is chosen with respect to the Laguerre polynomials below.

with the Schrödinger equation for the remaining function:

$$\Big[(d_r\xi)^2 d_\xi^2 + \Big(d_r^2\xi - (d_r\xi)^2 + \tfrac{2(1+L)}{r} d_r\xi\Big) d_\xi \\ + \tfrac{1}{4}(d_r\xi)^2 - \tfrac{1}{2} d_r^2\xi + 2(E-V) - \tfrac{1+L}{r} d_r\xi\Big] F_L(\xi) = 0.$$

Quantum numbers of bound states are determined with the condition of square integrability $L^2(\mathbb{R}^3)$ of the position wave functions.

For both the harmonic oscillator and the Kepler potential (below), there arises a Laplace differential equation,

$$[\xi \tfrac{d^2}{d\xi^2} + (1+\lambda-\xi)\tfrac{d}{d\xi} + N]\mathrm{L}_N^\lambda(\xi) = 0, \quad N = 0,1,\ldots,$$

which is solved by the Laguerre polynomials of degree N (radial quantum number, knot number) and real order $\lambda \neq -1,-2,\ldots$. The λ-dependence is used for the angular momentum L-dependence. The Rodrigues formula for the Laguerre polynomials contains derivatives of the function $\xi \longmapsto \xi^\lambda e^{-\xi}$:

$$\begin{aligned} \mathrm{L}_N^\lambda(\xi) &= (\xi^{-\lambda} e^\xi \tfrac{d}{d\xi} \xi^\lambda e^{-\xi})^N \tfrac{\xi^N}{N!} = \tfrac{1}{N!} \xi^{-\lambda} e^\xi \tfrac{d^N}{d\xi^N} \xi^{\lambda+N} e^{-\xi} \\ &= \sum_{k=0}^{N} \tfrac{\Gamma(1+\lambda+N)}{\Gamma(1+N-k)\Gamma(1+\lambda+k)} \tfrac{(-\xi)^k}{k!}. \end{aligned}$$

The bound waves involve the representation coefficients (see Chapter 8)

$$\mathbb{R}_+ \ni \tfrac{\xi}{2} \longmapsto \mathrm{L}_N^\lambda(\xi)\, e^{-\frac{\xi}{2}},$$

which give a basis for a Hilbert space for each $\lambda \notin -\mathbb{N}$,

$$L^2(\mathbb{R}_+, \mathbb{R}) \text{ has basis } \{\xi \longmapsto \xi^{\frac{\lambda}{2}} \mathrm{L}_N^\lambda(\xi) e^{-\frac{\xi}{2}} \mid N = 0,1,\ldots\}$$

$$\text{with } \begin{cases} \int_0^\infty \xi^\lambda e^{-\xi} d\xi\ \mathrm{L}_N^\lambda(\xi)\ \mathrm{L}_{N'}^\lambda(\xi) = \tfrac{\Gamma(1+\lambda+N)}{N!} \delta_{NN'}, \\ \sum_{N=0}^\infty \tfrac{N!}{\Gamma(1+\lambda+N)} \mathrm{L}_N^\lambda(\xi)\ \mathrm{L}_N^\lambda(\xi') = \xi^{-\lambda} e^\xi \delta(\xi - \xi'). \end{cases}$$

4.2 Harmonic Oscillator

The three- dimensional isotropic harmonic oscillator Hamiltonian,

$$\mathbf{H} = \tfrac{\vec{\mathbf{p}}^2 + \vec{\mathbf{x}}^2}{2},$$

has only bound waves, no scattering solutions. The normalization of the potential $V(r) = \frac{k_0 r^2}{2} = \frac{r^2}{2}$ yields an intrinsic frequency unit $[\frac{k_0}{m}] = \frac{1}{\mathrm{s}^2}$.

4.2.1 Position Representation

An irreducible exponential with squared radial dependence as the radial representation coefficient $\xi(r) = r^2$, $d_L(r) = F_L(r^2)e^{-r^2}$ determines the harmonic oscillator potential, normalized[2] with a momentum unit $|Q|$:

$$\begin{aligned}&[d_r^2 + \tfrac{2(1+L)}{r}d_r - 2V(r) + 2E]F_L(r^2) = 0\\ &\text{for } F_0(r^2) = e^{-\frac{Q^2 r^2}{2}} \Rightarrow V(r) = \tfrac{r^2}{2}Q^4,\quad E = \tfrac{3}{2}Q^2.\end{aligned}$$

The momentum unit can be chosen to be L-independent; e.g., $Q = 1$. The general square-integrable solutions are products of a Laguerre polynomial of degree N with a quantum number independent exponential:

$$[d_r^2 + \tfrac{2(1+L)}{r}d_r - r^2 + 2E]F_L(r^2) = 0 \Rightarrow \begin{cases} F_L(r^2) = \mathrm{L}_N^{\frac{1+2L}{2}}(r^2)e^{-\frac{r^2}{2}},\\ E_{LN} = \frac{3}{2} + k = \frac{3}{2} + L + 2N,\\ \qquad L, N = 0, 1, \ldots,\\ \psi^k_{Lm}(\vec{x}) \sim (\vec{x})^L_m \, \mathrm{L}_N^{\frac{1+2L}{2}}(r^2)\, e^{-\frac{r^2}{2}}.\end{cases}$$

The harmonic oscillator solutions for each angular momentum $L = 0, 1, 2, \ldots$ constitute a Hilbert space basis:

$$\begin{aligned}&L^2(\mathbb{R}_+, \mathbb{R})\text{-basis}: \quad \{\xi \longmapsto \xi^{\frac{1+2L}{4}}\mathrm{L}_N^{\frac{1+2L}{2}}(\xi)e^{-\frac{\xi}{2}} \mid N = 0, 1, \ldots\},\\ &\int_0^\infty dr\, \mathrm{L}_N^{\frac{1+2L}{2}}(r^2)\, e^{-r^2} r^{2+2L} \mathrm{L}_{N'}^{\frac{1+2L}{2}}(r^2) = \tfrac{1}{2}\int_0^\infty \xi^{\frac{1+2L}{2}} e^{-\xi} d\xi\, \mathrm{L}_N^{\frac{1+2L}{2}}(\xi)\, \mathrm{L}_{N'}^{\frac{1+2L}{2}}(\xi)\\ &\qquad = \frac{\Gamma(N+\frac{3}{2}+L)}{2N!}\delta_{NN'}.\end{aligned}$$

4.2.2 Color SU(3) for 3-Position

The harmonic oscillator Hamiltonian with creation and annihilation operators $(\mathrm{u}^a, \mathrm{u}^\star_a)$, $a = 1, 2, 3$,

$$\mathbf{H} = \frac{\vec{\mathbf{p}}^2 + \vec{\mathbf{x}}^2}{2} = \frac{\{\mathrm{u}^a, \mathrm{u}^\star_a\}}{2},\quad \mathrm{u}^a = \frac{\mathbf{x}^a - i\mathbf{p}^a}{\sqrt{2}},\quad [\mathrm{u}^\star_b, \mathrm{u}^a] = \delta^a_b,$$

generates time orbits of the Hilbert vectors with k quanta. The creation operator polynomials of degree k, acting on the *Fock ground-state vector* $|0\rangle$, give the Schrödinger wave functions as position representation with degree-k polynomials:

$$\begin{aligned}|k; n_1, n_2, n_3\rangle &= \frac{(\mathrm{u}^1)^{n_1}(\mathrm{u}^2)^{n_2}(\mathrm{u}^3)^{n_3}}{\sqrt{k!}}|0\rangle\\ &\cong \{\vec{x} \longmapsto \mathrm{H}^k(\vec{x})e^{-\frac{r^2}{2}}\} \in L^2(\mathbb{R}^3),\quad k = n_1 + n_2 + n_3,\\ \mathrm{u}^\star_a|0\rangle &= 0,\\ \mathrm{u}^a(t) &= e^{it}\mathrm{u}^a,\quad |k; n_1, n_2, n_3\rangle(t) = e^{ikt}|k; n_1, n_2, n_3\rangle.\end{aligned}$$

[2]Usually, the additive term $V_0 = \frac{3}{2}Q^2$ is introduced as ground-state energy.

The complex embedding of the **3** positions and momenta $\mathbb{R}^3 \oplus \mathbb{R}^3 \longrightarrow \mathbb{C}^3$ leads to a color $\mathbf{SU}(3)$-invariance with Gell–Mann matrices $\{\lambda^A\}_{A=1}^8$:

$$\chi_A\lambda^A = \begin{pmatrix} \chi_3 + \frac{\chi_8}{\sqrt{3}} & \chi_1 - i\chi_2 & \chi_4 - i\chi_5 \\ \chi_1 + i\chi_2 & -\chi_3 + \frac{\chi_8}{\sqrt{3}} & \chi_6 - i\chi_7 \\ \chi_4 + i\chi_5 & \chi_6 + i\chi_7 & -2\frac{\chi_8}{\sqrt{3}} \end{pmatrix}, \quad \mathbf{C} = \tfrac{i}{2}\mathrm{u}^a\lambda_a^b\mathrm{u}_b^\star, \quad [\mathbf{C}, \mathbf{H}] = 0.$$

The irreducible representations are characterized by two integers as generating invariants for the $\mathbf{SU}(3)$-operations with rank 2:

$$\begin{aligned} \mathbf{irrep}\,\mathbf{SU}(3) &= \{[2C_1, 2C_2] \,\big|\, 2C_{1,2} \in \mathbb{N}\}, \\ \dim_{\mathbb{C}}[2C_1, 2C_2] &= (1 + 2C_1)(1 + 2C_2)(1 + C_1 + C_2). \end{aligned}$$

The harmonic oscillator representations $[k, 0]$ (singlet, triplet, sextet, etc.) with one trivial $\mathbf{SU}(3)$-invariant $C_2 = 0$ are the totally symmetric products of $\mathbf{SU}(3)$-triplets,

$$\begin{aligned} [k, 0] &= \bigvee^{k}[1, 0] \in \bigvee^{k}\mathbb{C}^3 \cong \mathbb{C}^{\binom{2+k}{2}}, \\ \dim_{\mathbb{C}}[k, 0] &= \tbinom{2+k}{2} = 1, 3, 6, \ldots, \quad k = 0, 1, 2, \ldots. \end{aligned}$$

The embedded rotation group, generated by the transposition antisymmetric Lie subalgebra $\log \mathbf{SO}(3) = \{\varphi_a\mathbf{L}^a \,\big|\, \varphi_a \in \mathbb{R}\}$,

$$\mathbf{SO}(3) \hookrightarrow \mathbf{SU}(3) \text{ with } \begin{cases} \chi_A i\frac{\lambda^A - (\lambda^A)^T}{2} = \begin{pmatrix} 0 & \chi_2 & \chi_5 \\ -\chi_2 & 0 & \chi_7 \\ -\chi_5 & -\chi_7 & 0 \end{pmatrix} = \begin{pmatrix} 0 & \varphi_3 & -\varphi_2 \\ -\varphi_3 & 0 & \varphi_1 \\ \varphi_2 & -\varphi_1 & 0 \end{pmatrix}, \\ \mathbf{L}^a = \epsilon^{abc}\mathrm{u}^b\mathrm{u}_c^\star, \quad (\mathbf{L}^1, \mathbf{L}^2, \mathbf{L}^3) = (\mathbf{C}^7, -\mathbf{C}^5, \mathbf{C}^2), \end{cases}$$

comes with the real five-dimensional orientation manifold, given by the rotation group orbits $\mathbb{A}(3) = \mathbf{SU}(3)/\mathbf{SO}(3)$ in the color group, which describes the $\binom{4}{2} - 1$ relative phases of the three angular momenta directions in complex quantum structures. $\mathbb{A}(s) = \mathbf{SU}(s)/\mathbf{SO}(s)$ are the globally symmetric compact Riemannian manifolds of subtype $A\ I$ (see Chapter 2).

The energy (suitably normalized) as principal quantum number $E = k = L + 2N$ (polynomial degree) is the power of the product representations $\bigvee^{k}[1, 0]$ of the time translations $\mathbb{R}$. It is the sum of the angular momentum quantum number L for $\mathbf{SO}(3)$ and the radial quantum number (knot number) N for the rotation group classes in the real five-dimensional space $\mathbf{SU}(3)/\mathbf{SO}(3)$. One has with the angular momentum degeneracy $1 + 2L$ the energy degeneracy given by the dimension $\binom{2+k}{2}$ of the $\mathbf{SU}(3)$-representation $[k, 0]$:

$$[k, 0] \overset{\mathbf{SO}(3)}{\cong} \begin{cases} \displaystyle\bigoplus_{L=0,2,\ldots,k} [L], & k = 0, 2, \ldots, \\ \displaystyle\bigoplus_{L=1,3,\ldots,k} [L], & k = 1, 3, \ldots, \end{cases} \quad \text{with } \tbinom{2+k}{2} = \begin{cases} \sum_{\cdots}(1 + 2L), \\ \sum_{\cdots}(1 + 2L), \end{cases}$$

$$E_{LN} - \tfrac{3}{2} = k = L + 2N \Rightarrow (L, N) = \begin{cases} (k, 0), (k-2, 1), \ldots, (0, \frac{k}{2}), \\ (k, 0), (k-2, 1), \ldots, (1, \frac{k-1}{2}). \end{cases}$$

4.2.3 Harmonic Fermi Oscillator

The free fields of Minkowski spacetime (see Chapter 5) are harmonically analyzed with momentum-dependent creation and annihilation operators $(\mathrm{u}(\vec{q}), \mathrm{u}^\star(\vec{q}))$ as eigenvector distributions for the translations:

$$\vec{x} \in \mathbb{R}^3: \quad (\mathrm{u}(\vec{q}), \mathrm{u}^\star(\vec{q})) \longmapsto (e^{i\vec{q}\vec{x}}\mathrm{u}(\vec{q}), e^{-i\vec{q}\vec{x}}\mathrm{u}^\star(\vec{q})).$$

The harmonic *Bose oscillators* of the former section are used for Bose particle fields and, in addition, harmonic *Fermi oscillators* for Fermi particle fields. They have the quantum mechanical Hamiltonian for the time translation representation:

$$\mathbf{H} = \frac{[\mathrm{u}^a, \mathrm{u}^\star_a]}{2}, \quad \{\mathrm{u}^\star_b, \mathrm{u}^a\} = \delta^a_b, \quad \{\mathrm{u}^b, \mathrm{u}^a\} = 0, \quad a, b = 1, 2, 3.$$

There is no position representation for Fermi oscillators. In three dimensions, they also have $\mathbf{SU}(3)$-symmetry with the Lie algebra representation,

$$\mathbf{C} = \tfrac{i}{2}\mathrm{u}^a \lambda^b_a \mathrm{u}^\star_b, \quad [\mathbf{C}, \mathbf{H}] = 0.$$

The eigenvectors are given by the totally antisymmetrized products, where, because of the anticommutators in the quantization, there arise only singlets, a triplet, and an antitriplet (Pauli exclusion principle for fermions):

$$\begin{aligned} [1,0] \wedge [1,0] &= [0,1], \quad [1,0] \wedge [1,0] \wedge [1,0] = [0,0], \\ \dim_{\mathbb{C}}[0,0] &= 1, \quad \dim_{\mathbb{C}}[1,0] = \dim_{\mathbb{C}}[0,1] = 3. \end{aligned}$$

For the time translation eigenvectors in the Hilbert space, the creation operators are applied to the Fock ground state with $\mathrm{u}^\star_a|0\rangle = 0$:

$$\left.\begin{array}{rrl} \text{singlet:} & & |0\rangle, \\ \text{triplet:} & \mathrm{u}^a|0\rangle = & |1; a\rangle, \\ \text{antitriplet:} & \frac{\epsilon_{abc}\mathrm{u}^b\mathrm{u}^c}{2}|0\rangle = & |2; a\rangle, \\ \text{singlet:} & \mathrm{u}^1\mathrm{u}^2\mathrm{u}^3|0\rangle = & |3\rangle, \end{array}\right\} \quad \begin{array}{r} \text{with } |k;\ .\ \rangle \overset{\mathbb{R}}{\longmapsto} e^{ikt}|k;\ .\ \rangle, \\ k = 0, 1, 2, 3. \end{array}$$

The singlet and triplet representation dimensions are also valid for the rotation subgroup $\mathbf{SO}(3) \subset \mathbf{SU}(3)$ with $L = 0, 1$.

4.2.4 Bose and Fermi Oscillators

In general for isotropic harmonic Bose and Fermi oscillators: The rotation group embedding $\mathbf{SO}(s) \hookrightarrow \mathbf{SU}(s)$ in $s = 2, 3, \dots$ positions, $\mathbb{R}^s \subset \mathbb{C}^s \cong V^1$, leads to an energy degeneracy, i.e., the s-position comes with a quantum-implemented $\mathbf{SU}(s)$-symmetry. There arise the totally (anti)symmetric products of the defining representation, for Fermi the two trivial ($k = 0, s$) and the $(s-1)$ fundamental $\mathbf{SU}(s)$-representations,

$$\begin{aligned} \text{Bose}: &\ \bigvee^k [1, 0, \dots, 0] \text{ with } \dim_{\mathbb{C}} \bigvee^k [1, 0, \dots, 0] = \tbinom{s+k-1}{k}, \quad k = 0, 1, \dots, \\ \text{Fermi}: &\ \bigwedge^k [1, 0, \dots, 0] \text{ with } \dim_{\mathbb{C}} \bigwedge^k [1, 0, \dots, 0] = \tbinom{s}{k}, \quad k = 0, 1, \dots, s. \end{aligned}$$

The invariant of the time translation representation in $\mathbf{U}(\mathbf{1}_s) \subseteq \mathbf{U}(s)$ is related to a representation invariant of $\mathbf{SU}(s)$ since both groups have the center of $\mathbf{SU}(s)$ as nontrivial intersection, i.e., the cyclotomic group $\mathbb{I}(s)$,

$$\mathbf{U}(\mathbf{1}_s) \cap \mathbf{SU}(s) \cong \mathbb{I}(s) = \{z \in \mathbb{C} \mid z^s = 1\},$$

in the $\mathbf{SU}(3)$-Bose example: $E = k$, $\dim_{\mathbb{C}}[k, 0] = \binom{2+E}{2}$. A similar central correlation shows up in the hypercharge-color relation of the quarks in the standard model of elementary particles (see Chapter 6).

The creation operators, acting on the Fock state vector, give orthogonal bases for the (anti)symmetric powers

$$\begin{aligned}(\mathrm{u}^1)^{n_1} \cdots (\mathrm{u}^s)^{n_s} |0\rangle &= |k; n_1, \ldots, n_s\rangle \in V^k, \\ \sum_{i=1}^{s} n_i &= k, \text{ for Fermi } n_i \in \{0, 1\}.\end{aligned}$$

The direct sum of those finite-dimensional Hilbert spaces is the finite dimensional Hilbert space for the Fermi oscillator,

$$\text{Fermi: } V^k = \bigwedge^k V^1 \cong \mathbb{C}^{\binom{s}{k}}, \quad \bigoplus_{k=0}^{s} V^k = \bigwedge V^1 \cong \mathbb{C}^{2^s}.$$

The direct sum for the Bose oscillator gives the complex polynomials in the position coordinates. Its completion with the Fock ground state $|0\rangle \cong e^{-\frac{r^2}{2}}$ in the scalar product is isomorphic to $L^2(\mathbb{R}^s)$:

$$\begin{aligned}\text{Bose: } V^k = \bigvee^k V^1 \cong \mathbb{C}^{\binom{s+k-1}{k}}, \quad \bigoplus_{k=0}^{\infty} V^k = \bigvee V^1 &\cong \mathbb{C}[x^1, \ldots, x^s] \cong \mathbb{C}^{\aleph_0}, \\ \overline{\mathbb{C}[x^1, \ldots, x^s]} &= L^2(\mathbb{R}^s).\end{aligned}$$

4.3 Kepler Dynamics

The Kepler dynamics with time translation generator,

$$\mathbf{H} = \frac{\vec{\mathbf{p}}^2}{2} + \frac{\delta}{\mathbf{r}}, \quad \delta = \pm 1 \text{ repulsive, attractive},$$

has both bound waves and scattering solutions. The normalization of the potential $V(r) = \frac{|\gamma_0|\delta}{r} = \frac{\delta}{r}$ gives an intrinsic velocity unit $[\frac{\gamma_0}{\hbar}] = \frac{\mathrm{m}}{\mathrm{s}}$ with $\frac{\gamma_0}{\hbar} = c\alpha_S z_1 z_2$, $\alpha_S \sim \frac{1}{137}$, for a Coulomb potential and charge numbers $z_{1,2} \in \mathbb{Z}$. The binding energies contain a factor $\alpha_S^2 \sim 6 \times 10^{-5}$ that reduces the electron mass-energy $m_e c^2 \sim 0.5$ MeV to the Rydberg energy $\frac{1}{2} M_R c^2 \sim 14$ eV, $M_R = m_e \alpha_S^2$. The maximally symmetric noncompact position as represented by the wave functions of the nonrelativistic hydrogen atom is not the flat

Euclidean position $\mathbb{R}^3$, but the hyperbolic position $\mathcal{Y}^3 \cong \mathbf{SO}_0(1,3)/\mathbf{SO}(3)$, i.e., the rotation group classes in the Lorentz group. They are isomorphic as manifolds, not as homogeneous spaces,

$$\begin{pmatrix} \cosh\psi \\ \sinh\psi\ \vec{\omega_2} \end{pmatrix} \in \mathcal{Y}^3 \subset \mathbb{R}^4, \quad \vec{x} = r\vec{\omega_2} \in \mathbb{R}^3, \quad r = |\sinh\psi|.$$

This will be discussed in more detail in Chapters 8 and 10.

4.3.1 Position Representation

In the position representation, the Kepler Hamiltonian is the sum of the Laplacian and the Laplacian kernel:

$$\mathbf{H} \cong -\tfrac{1}{2}\vec{\partial}^2 + \tfrac{\delta}{r}, \quad \left\{ \begin{array}{rl} \vec{\mathbf{p}}^2 \cong -\vec{\partial}^2, \ \frac{1}{r} &= \int \frac{d^3q}{2\pi^2}\frac{1}{\vec{q}^2}e^{i\vec{q}\vec{x}}, \\ \text{with } \vec{\partial}^2\ \frac{1}{r} &= -4\pi\delta(\vec{x}). \end{array} \right.$$

For an irreducible noncompact radial representation $r \longmapsto e^{-r}$ as bound solution of 3-position, an attractive Kepler potential is necessary to compensate the contribution $(\frac{1}{r}d_r r)^2 - d_r^2 = \frac{2}{r}d_r$, related to the $\mathbf{SO}(3)$-rotations,

$$\begin{array}{l} [d_r^2 + \frac{2(1+L)}{r}d_r - 2V(r) + 2E]d_L(r) = 0, \\ \text{for } d_0(r) = e^{-|Q|r} \Rightarrow V(r) = -\frac{|Q|}{r}, \ \ 2E = -Q^2. \end{array}$$

After separating, in general, for a rotation-invariant interaction $V(r)$, an exponential with a complex "momentum" iq, there remain the radial equations:

$$\begin{array}{rl} d_L(r) = & F_L(\xi)e^{-\frac{\xi}{2}}, \ \ \frac{\xi}{2} = iqr, \ \ q \in \mathbb{C}, \\ \Rightarrow & [\xi d_\xi^2 + (2+2L-\xi)d_\xi - (1+L-\frac{\xi V(r)}{2q^2}) + \frac{\xi}{4}(1-\frac{2E}{q^2})]F_L(\xi) = 0. \end{array}$$

For the Kepler potential $V(r) = \frac{\delta}{r}$ and a purely "kinetic" energy, given by half of the squared "momentum,"

$$E = \tfrac{q^2}{2} \Rightarrow [\xi d_\xi^2 + (2+2L-\xi)d_\xi - (1+L-i\tfrac{\delta}{q})]F_L(\xi) = 0,$$

there arises a Laplacian differential equation, solved, for $r = 0$ regularity, by confluent hypergeometric functions (see Chapter 8), which, for negative integer α, are Laguerre polynomials:

$$\begin{array}{rl} [\xi d_\xi^2 + (\gamma-\xi)d_\xi - \alpha]\ f(\xi) = 0; & \xi, \alpha, \gamma \in \mathbb{C}, \\ f(\xi) \sim {}_1F_1(\alpha;\gamma;\xi) &= \sum_{k=0}^{\infty} \frac{\Gamma(\alpha+k)}{\Gamma(\alpha)}\frac{\Gamma(\gamma)}{\Gamma(\gamma+k)}\frac{\xi^k}{k!} = {}_1F_1(\gamma-\alpha;\gamma,-\xi)e^\xi, \\ {}_1F_1(-N;1+\lambda;\xi) &= N!\frac{\Gamma(1+\lambda)}{\Gamma(1+\lambda+N)}\mathrm{L}_N^\lambda(\xi) \text{ for } N \in \mathbb{N}. \end{array}$$

One obtains for the Kepler potential

$$V(r) = \tfrac{\delta}{r} \Rightarrow d_L(r) = {}_1F_1(1+L-i\tfrac{\delta}{q}; 2+2L; 2iqr)e^{-iqr}.$$

For the attractive interaction $V(r) = -\frac{1}{r}$ and *states with negative energy* (*binding energy*) and square-integrable wave functions, the radial *imaginary "momentum"* and, therefore, the energy is "quantized" (quantum numbers in the original sense) as seen in the Laguerre polynomials,

$$\begin{array}{l} \delta = -1,\ q = -i|Q|, \\ E = -\frac{Q^2}{2} < 0, \\ \frac{\xi}{2} = |Q|r, \end{array} \Rightarrow \left\{ \begin{array}{rl} d_L(r) & = \mathrm{L}_N^{1+2L}(2|Q|r)e^{-|Q|r}, \\ \frac{1}{Q_n} & = n = 1 + 2J = 1 + L + N, \\ & L, N = 0, 1, \ldots, \\ \psi_{Lm}^{2J}(\vec{x}) & \sim (\frac{\vec{x}}{n})_m^L\ \mathrm{L}_N^{1+2L}(\frac{2r}{n})\ e^{-\frac{r}{n}}. \end{array} \right.$$

In contrast to the bound-state exponential $e^{-|Q|r}$, the *scattering solutions* in the Kepler potential $V(r) = \frac{\delta}{r}$, with positive energy and real momentum from a continuous spectrum, are obtained after the separation of a unitary exponential e^{iPr}. The radial equations have the real ($r = 0$)-regular solutions:

$$\begin{array}{l} q = P, \\ E = \frac{P^2}{2} > 0, \\ \frac{\xi}{2} = iPr \end{array} \Rightarrow \left\{ \begin{array}{rl} d_L(r) & = {}_1F_1(1 + L - i\frac{\delta}{P}; 2 + 2L; +2iPr)e^{-iPr} \\ & = {}_1F_1(1 + L + i\frac{\delta}{P}; 2 + 2L; -2iPr)e^{+iPr}, \\ & P \in \mathbb{R},\ \ L = 0, 1, \ldots, \\ \psi_{L,m}^P(\vec{x}) & \sim (P\vec{x})_m^L\ {}_1F_1(1 + L + i\frac{\delta}{P}; 2 + 2L; -2iPr)e^{iPr}. \end{array} \right.$$

The large distance behavior [44] does not lead to spherical Bessel functions for free particles, representing Euclidean position $\mathbb{R}^3$ (see Chapter 8). For the hyperbolic position $\mathcal{Y}^3$-representations, the asymptotic Bessel functions are modified by a radial-dependent logarithm and a constant phase α_L:

$$\begin{aligned} r \to \infty: \quad & (P\vec{x})_m^L {}_1F_1(1 + L + i\tfrac{\delta}{P}; 2 + 2L; -2iPr)e^{iPr} \\ & \sim \mathrm{Y}_m^L(\varphi, \theta)\frac{\sin(Pr - \frac{L\pi}{2} - \frac{\delta}{P}\log 2Pr + \alpha_L)}{Pr}. \end{aligned}$$

4.3.2 Orthogonal Lenz–Runge Symmetry

In an algebraic quantum treatment,[3] the Kepler Hamiltonian with angular momentum and Lenz–Runge vector,

$$\mathbf{H} = \frac{\vec{\mathbf{p}}^2}{2} + \frac{\delta}{\mathbf{r}},\ \ \vec{\mathbf{L}} = \vec{\mathbf{x}} \times \vec{\mathbf{p}},\ \ \vec{\mathbf{P}} = \frac{\vec{\mathbf{p}}\times\vec{\mathbf{L}} - \vec{\mathbf{L}}\times\vec{\mathbf{p}}}{2} + \delta\frac{\vec{\mathbf{x}}}{\mathbf{r}},$$

build the same three Lie algebra structures as in the classical case (see Chapter 3):

$$[\mathbf{H}, \vec{\mathbf{L}}] = 0,\ \ [\mathbf{H}, \vec{\mathbf{P}}] = 0,\ \ \left\{ \begin{array}{rl} [i\mathbf{L}^a, i\mathbf{L}^b] & = -\epsilon^{abc} i\mathbf{L}^c, \\ [i\mathbf{L}^a, i\mathbf{P}^b] & = -\epsilon^{abc} i\mathbf{P}^c, \\ [i\mathbf{P}^a, i\mathbf{P}^b] & = 2\mathbf{H}\epsilon^{abc} i\mathbf{L}^c, \end{array} \right.$$

[3]There is another algebraic treatment of the quantum hydrogen atom by Pauli — working, in some analogy to the harmonic oscillator, with creation and annihilation operators.

with the Lorentz group for scattering, the orthogonal group for bound states, and the Galilei or Euclidean group as contraction:

$$\begin{array}{rcccccc} & \mathbf{SO}_0(1,3) & \to & \mathbf{SO}(3)\,\vec{\times}\,\mathbb{R}^3 & \leftarrow & \mathbf{SO}(4), \\ E \in \mathbf{spec}\, H: & E>0 & & E=0 & & E<0. \end{array}$$

The additional i-factors in the Lie brackets are related to the different dual normalization in the Poisson bracket $[p,x]_P = 1$ and the quantum commutator $i[\mathbf{p},\mathbf{x}] = 1$.

Again, the squares of angular momentum and Lenz–Runge vector as invariants for position operations determine the invariant Hamiltonian for the time translations,

$$\vec{\mathbf{P}}^2 = 1 + 2\mathbf{H}(\vec{\mathbf{L}}^2 + 1) \Rightarrow -\tfrac{1}{2\mathbf{H}} = 1 + \vec{\mathbf{L}}^2 - \tfrac{\vec{\mathbf{P}}^2}{2\mathbf{H}},$$

with an additional constant 1 compared to the classical case.

For attraction, $\delta = -1$, and negative energies, one has representations of the doubled compact Lie algebra $\log \mathbf{SO}(3) \cong A_1^c \cong \mathbb{R}^3$:

$$\mathbf{spec}\,\mathbf{H} \ni E<0: \begin{cases} \vec{\mathbf{B}} = \frac{\vec{\mathbf{P}}}{\sqrt{-2\mathbf{H}}}, \quad \vec{\mathbf{J}}_\pm = \frac{\vec{\mathbf{L}}\pm\vec{\mathbf{B}}}{2}, \\ [i\mathbf{J}^a_\pm, i\mathbf{J}^b_\pm] = -\epsilon^{abc} i\mathbf{J}^c_\pm, \quad [\mathbf{J}^a_+, \mathbf{J}^b_-] = 0, \\ \text{invariants: } \vec{\mathbf{L}}^2 + \vec{\mathbf{B}}^2 = 2(\vec{\mathbf{J}}_+^{\,2} + \vec{\mathbf{J}}_-^{\,2}), \quad \vec{\mathbf{L}}\vec{\mathbf{B}} = \vec{\mathbf{J}}_+^{\,2} - \vec{\mathbf{J}}_-^{\,2}, \end{cases}$$
$$\log[\mathbf{SO}(3) \times \mathbf{SO}(3)] \cong A_1^c \oplus A_1^c.$$

The quantum algebra for the space with the defining four-dimensional representation of $A_1^c \oplus A_1^c$ is generated by two pairs of Pauli spinors (creation and annihilation operators) with Bose statistics,

$$\text{nontrivial: } [\mathrm{u}^\star_A, \mathrm{u}^B] = \delta^B_A, \quad [\mathrm{a}^\star_A, \mathrm{a}^B] = \delta^B_A, \quad A,B = 1,2.$$

The double-"spin" Lie algebra is implemented by the six basic vectors

$$\begin{array}{rll} \log[\mathbf{SU}(2)\times\mathbf{SU}(2)]: & i\vec{\mathbf{J}}_+ = i\mathrm{u}\frac{\vec{\sigma}}{2}\mathrm{u}^\star, & i\vec{\mathbf{J}}_- = i\mathrm{a}\frac{\vec{\sigma}}{2}\mathrm{a}^\star, \\ \log\mathbf{SO}(4): & \vec{\mathbf{L}} = \vec{\mathbf{J}}_+ + \vec{\mathbf{J}}_-, & \vec{\mathbf{B}} = \vec{\mathbf{J}}_+ - \vec{\mathbf{J}}_-. \end{array}$$

The angular momenta are a basis of the diagonal Lie algebra.

The Hilbert space uses four creation operators:

$$\begin{array}{l} \text{basis } \{(\mathrm{u}^1)^{n_1}(\mathrm{u}^2)^{n_2}(\mathrm{a}^1)^{m_1}(\mathrm{a}^2)^{m_2}|0\rangle \,\big|\, n_{1,2}, m_{1,2} = 0,1,\ldots\}, \\ \text{with } \mathrm{u}^\star_A|0\rangle = 0 = \mathrm{a}^\star_A|0\rangle, \quad \langle 0|\mathrm{u}^\star_A\mathrm{u}^B|0\rangle = \delta^B_A = \langle 0|\mathrm{a}^\star_A\mathrm{a}^B|0\rangle. \end{array}$$

The eigenvalue of a time-conserved operator $\mathbf{Q}$, i.e., $[\mathbf{H},\mathbf{Q}] = 0$, for a simultaneous eigenvector of the Hamiltonian $\mathbf{H}$ and $\mathbf{Q}$ is denoted in the following as $\langle\mathbf{Q}\rangle = \langle E|\mathbf{Q}|E\rangle$.

The weight diagrams of the irreducible $\mathbf{SU}(2)\times\mathbf{SU}(2)$-representations,

$$\begin{aligned} \mathbf{irrep}\,[\mathbf{SU}(2)\times\mathbf{SU}(2)] &= \mathbf{irrep}\,\mathbf{SU}(2)\times\mathbf{irrep}\,\mathbf{SU}(2) \\ &= \{(J_1,J_2) \,\big|\, J_{1,2} = 0, \tfrac{1}{2}, 1, \ldots\}, \end{aligned}$$

occupy $(1+2J_1)(1+2J_2)$ points of a rectangular grid. The two invariants determine the occurring representations. The triviality of the invariant $\vec{\mathbf{L}}\vec{\mathbf{P}}=0$ (classical orthogonality of angular momentum and Lenz–Runge perihelion vector) "synchronizes" the centers $\mathbb{I}(2)=\{\pm 1\}$ of both $\mathbf{SU}(2)$s (central correlation) and leads to the relevant group $\mathbf{SO}(4)$ with the integer spin sum in the irreducible representations:

$$\begin{aligned}\tfrac{\mathbf{SU}(2)\times\mathbf{SU}(2)}{\mathbb{I}(2)} &\cong \mathbf{SO}(4),\ \text{with } \mathbb{I}(2)=\{(1,1),(-1,-1)\}\subset \mathbf{SU}(2)\times\mathbf{SU}(2),\\ \mathbf{irrep}\,\mathbf{SO}(4) &= \{(J_1,J_2)\mid J_{1,2}=0,\tfrac{1}{2},1,\ldots,\ \text{with } J_1+J_2=0,1,\ldots\}.\end{aligned}$$

The orthogonality condition enforces even the equality of both $\mathbf{SU}(2)$-invariants $J_+=J_-=J$:

$$0=\vec{\mathbf{L}}\vec{\mathbf{B}}=\vec{\mathbf{J}}_+^2-\vec{\mathbf{J}}_-^2\Rightarrow\langle\vec{\mathbf{J}}_+^2\rangle=\langle\vec{\mathbf{J}}_-^2\rangle=J(1+J),\quad J=0,\tfrac{1}{2},1,\tfrac{3}{2},\ldots.$$

Therefore, the energy-degenerated representations are of the type (J,J); the multiplets of both $A_1^{\mathbb{C}}$-representations have equal dimension $1+2J$. The $\mathbf{SU}(2)$-multiplet dimension is the principal quantum number $n=1+2J$. The weight diagrams occupy $(1+2J)^2$ points of a square grid:

$$\mathbf{irrep}_{\,\mathrm{Kep}}\mathbf{SO}(4)=\{(J,J)\mid J=0,\tfrac{1}{2},1,\ldots\}.$$

These *Kepler or harmonic* $\mathbf{SO}(4)$*-representations* (see Chapter 8) are the totally symmetrized products of the defining four-dimensional $\mathbf{SO}(4)$-representation $(\frac{1}{2},\frac{1}{2})$ with the decompositions into irreducible ones with the dimensions $\dim_{\mathbb{C}}(\frac{L}{2},\frac{L}{2})=(1+L)^2$:

$$\bigvee^{2J}(\tfrac{1}{2},\tfrac{1}{2})=\begin{cases}\bigoplus\limits_{L=0,2,\ldots,2J}(\tfrac{L}{2},\tfrac{L}{2}), & 2J=0,2,\ldots,\\ \bigoplus\limits_{L=1,3,\ldots,J}(\tfrac{L}{2},\tfrac{L}{2}), & 2J=1,3,\ldots,\end{cases}\quad \text{with } \binom{3+2J}{3}=\begin{cases}\sum\limits_{\cdots}(1+L)^2,\\ \sum\limits_{\cdots}(1+L)^2.\end{cases}$$

Similar to the one-dimensional harmonic oscillator with the 1-quantum state vector $|1\rangle=\mathrm{u}|0\rangle$ as the defining $\mathbf{U}(1)$-orbit, there is the state vector with the defining fourdimensional $\mathbf{SO}(4)$-representation $(\frac{1}{2},\frac{1}{2})$ (s- and p-shell) for the atomic bound-state vectors. The highest-weight vector in an irreducible representation space comes with highest "spins" $j_\pm=J$:

$$(\mathrm{u}^1\mathrm{a}^1)^{2J}|0\rangle=|J;J\rangle|J;J\rangle.$$

Its extremality involves the triviality for the action of the two raising operators:

$$\begin{aligned}\text{raising: } \mathbf{J}_\pm^+ &= (\mathrm{u}^1\mathrm{u}_2^\star,\mathrm{a}^1\mathrm{a}_2^\star), & (\mathbf{L}^+,\mathbf{P}^+) &= \mathrm{u}^1\mathrm{u}_2^\star\pm\mathrm{a}^1\mathrm{a}_2^\star,\\ \text{lowering: } \mathbf{J}_\pm^- &= (\mathrm{u}^2\mathrm{u}_1^\star,\mathrm{a}^2\mathrm{a}_1^\star), & (\mathbf{L}^-,\mathbf{P}^-) &= \mathrm{u}^2\mathrm{u}_1^\star\pm\mathrm{a}^2\mathrm{a}_1^\star.\end{aligned}$$

By the two lowering operators $(\mathbf{J}_\pm^-)^{J-j_\pm}|J;J\rangle=|J;j_\pm\rangle$ (in the weight diagram: horizontal to the left and vertical downwards), one reaches all eigenvectors of a square grid:

$$\text{basis of } \mathbb{C}^{(1+2J)^2}:\quad \{|J;j_+\rangle|J;j_-\rangle\mid j_\pm=-J,\ldots,J\}.$$

The energy eigenvalues are given with the inverse of the Casimir invariant $1 = -2\mathbf{H}(1 + \vec{\mathbf{L}}^2 + \vec{\mathbf{B}}^2)$, i.e., by the normalization $1 = -2E_n n^2$:

$$\begin{aligned} -\frac{1}{2\langle\mathbf{H}\rangle} &= 1 + 2\langle\vec{\mathbf{J}}_+^2 + \vec{\mathbf{J}}_-^2\rangle = 1 + 4J(1+J), \quad J = 0, \tfrac{1}{2}, 1, \tfrac{3}{2}, \ldots, \\ E_n &= -\frac{1}{2n^2}, \quad \text{multiplicity: } n^2 = (1+2J)^2 = 1, 4, 9, 16, \ldots. \end{aligned}$$

As seen in experiments, there is an additional twofold degeneracy in the atoms. It originates from an additional "internal" spin $\mathbf{SU}(2)$-property of the electron not contained in the nonrelativistic scheme above. It can be added by an ad hoc doubling leading to doubled multiplicities $2n^2 = 2, 8, 18, \ldots$.

The $\mathbf{SO}(4)$-representations are decomposable with respect to the position rotation $\mathbf{SO}(3)$-properties into irreducible representations of dimension $(1 + 2L)$ with integer $L = 0, 1, \ldots$ for angular momentum $\vec{\mathbf{L}} = \vec{\mathbf{J}}_+ + \vec{\mathbf{J}}_-$:

$$(J, J) \overset{\mathbf{SO}(3)}{\cong} \bigoplus_{L=0}^{2J} [L], \quad (1+2J)^2 = \sum_{L=0}^{2J} (1 + 2L).$$

The Lenz–Runge invariance-related difference $2J - L = N$, characterizing the classes $\mathbf{SO}(4)/\mathbf{SO}(3) \cong \Omega^3$, is the radial quantum number or knot number.

The $\mathbf{SO}(4)$ multiplets comprise all wave functions ψ^{2J} with equal sum $L + N = 2J$ for the principal quantum number $n = 1 + 2J$ with angular momentum $(1 + 2L)$-multiplets for $\mathbf{SO}(3)$ and radial quantum numbers N for Lenz–Runge classes, parametrizable by the 3-sphere $\Omega^3 \cong \mathbf{SO}(4)/\mathbf{SO}(3)$.

There is an orthogonal basis transformation from eigenvectors of double-"spin" $\{\mathbf{J}_\pm^3\}$ to eigenvectors of angular momentum $\{\vec{\mathbf{L}}^2, \mathbf{L}^3\}$:

$$\begin{aligned} n = 1 + 2J: \quad |J; j_+\rangle|J; j_-\rangle &\sim |n; L, m\rangle, \quad m = j_+ + j_-, \\ \vec{\mathbf{L}}^2|n; L, m\rangle &= L(1 + L)|n; L, m\rangle, \ L = 0, \ldots, 2J, \\ \mathbf{L}^3|n; L, m\rangle &= m|n; L, m\rangle, \qquad m = -L, \ldots, +L. \end{aligned}$$

All vectors of this basis are obtained from the highest vector $|J; J\rangle|J; J\rangle = |n; 2J, 2J\rangle$ with the angular momentum and Lenz–Runge lowering operators (in the weight diagram: diagonal and skew-diagonal downwards, respectively):

$$n = 1 + 2J: \quad \begin{aligned} |n; L, L\rangle &= (\mathbf{P}^-)^{2J-L}|n; 2J, 2J\rangle, \\ |n; L, m\rangle &= (\mathbf{L}^-)^{L-m}|n; L, L\rangle, \end{aligned} \quad \text{with} \begin{cases} L = 0, \ldots, n-1, \\ m = -L, \ldots, +L. \end{cases}$$

4.4 Particles and Ghosts

Particles are defined with irreducible Hilbert representations of time translations. Nonparticle degrees of freedom use indefinite unitary, nondecomposable time representations, e.g., the two nonparticle degrees in the four-component electromagnetic field (see Chapter 5).

4.4.1 Definite Metric, Fock Space, and Particles

The irreducible Hilbert representation of the time translations with real energy $E \in \mathbb{R}$,

$$\mathbb{R} \ni t \longmapsto e^{iEt} \in \mathbf{U}(1) \subset \mathbf{GL}(V),$$

as used for the harmonic oscillator act on a complex one-dimensional space V with the creation operator u as basis. The dual representation $\mathbb{R} \ni t \longmapsto e^{-iEt}$ acts on the dual space V^T, spanned by the annihilation operator $\mathrm{u}^\star$:

$$V = \mathbb{C}\mathrm{u},\ \mathrm{u}(t) = e^{iEt}\mathrm{u},\ V^T = \mathbb{C}\mathrm{u}^\star,\ \mathrm{u}^\star(t) = e^{-iEt}\mathrm{u}^\star.$$

The product representations with energy eigenvalues $\{zE \bigm| z \in \mathbb{Z}\}$ act on the tensor algebra $\bigotimes \mathbf{V}$ of the self-dual space $\mathbf{V} = V \oplus V^T \cong \mathbb{C}^2$, i.e., on all complex linear combinations of all products of u and $\mathrm{u}^\star$. The *Bose and Fermi quantum algebras* $\mathbf{Q}_\epsilon(\mathbf{V})$ for $\epsilon = \mp 1$, respectively, are the quotient algebras of the tensor algebras with respect to the equality of the dual product of the basic vectors,

$$\text{duality of } V \text{ and } V^T\text{: } \langle \mathrm{u}^\star, \mathrm{u}\rangle = 1,\ \langle \mathrm{u}, \mathrm{u}\rangle = 0,\ \langle \mathrm{u}^\star, \mathrm{u}^\star\rangle = 0,$$

and their corresponding (anti-)commutators, $[a,b]_\epsilon = a \otimes b + \epsilon b \otimes a$. This can be implemented by going over to classes of $\bigotimes \mathbf{V}$ with respect to an appropriate ideal (the elements of the ideal constitute the 0-class):

$$\begin{aligned} S_\epsilon &= \{[\mathrm{u}^\star, \mathrm{u}]_\epsilon - 1,\ [\mathrm{u}, \mathrm{u}]_\epsilon,\ [\mathrm{u}^\star, \mathrm{u}^\star]_\epsilon\},\\ I(S_\epsilon) &= \text{minimal ideal in } \textstyle\bigotimes \mathbf{V}, \text{ generated by } S_\epsilon,\\ \mathbf{Q}_\epsilon(\mathbf{V}) &= \textstyle\bigotimes \mathbf{V}/I(S_\epsilon) \cong \begin{cases} \mathbb{C}^4, & \epsilon = +1 \text{ (Fermi quantum algebra)},\\ \mathbb{C}^{\aleph_0}, & \epsilon = -1 \text{ (Bose quantum algebra)}. \end{cases} \end{aligned}$$

In the finite-dimensional Fermi quantum algebra, the Pauli exclusion principle is formalized by nilquadratic basic vectors:

$$\begin{aligned} \epsilon = +1: &\quad (\mathrm{u})^2 = 0 = (\mathrm{u}^\star)^2,\\ \text{basis of } \mathbf{Q}_+(\mathbb{C}^2): &\quad \{1,\ \mathrm{u},\ \mathrm{u}^\star,\ [\mathrm{u}, \mathrm{u}^\star]\}. \end{aligned}$$

The quantum algebras are associative and unital $1 \in \mathbb{C} = \overset{0}{\bigotimes}\mathbf{V}$. Their $\mathbf{U}(1)$-conjugation $\star$ with $(\alpha a)^\star = \overline{\alpha} a^\star$ for $\alpha \in \mathbb{C}$ and $(ab)^\star = b^\star a^\star$ is induced by the conjugation of the basic vectors $\mathrm{u} \overset{\star}{\leftrightarrow} \mathrm{u}^\star$. The familiar (anti-)commutators hold in the quantum algebras, e.g., for the adjoint action of the Hamiltonian, constructed with the quantum-opposite commutator:

$$\text{in } \mathbf{Q}_\epsilon(\mathbb{C}^2): \quad \begin{cases} [\mathrm{u}^\star, \mathrm{u}]_\epsilon = 1,\ [\mathrm{u}, \mathrm{u}]_\epsilon = 0 = [\mathrm{u}^\star, \mathrm{u}^\star]_\epsilon,\\ \mathbf{H} = E\mathbf{I}_1,\ \mathbf{I}_1 = \frac{1}{2}[\mathrm{u}, \mathrm{u}^\star]_{-\epsilon},\\ [\mathbf{I}_1, (\mathrm{u})^k(\mathrm{u}^\star)^l] = (k-l)\ (\mathrm{u})^k(\mathrm{u}^\star)^l, \quad k,l = 0,1,2,\ldots \end{cases}$$

The Bose quantum algebra can be generated by position and momentum as basic hermitian combinations:

$$\text{for } \mathbf{Q}_-(\mathbb{C}^2): \quad \mathbf{x} = \frac{\mathrm{u}^\star + \mathrm{u}}{\sqrt{2}}, \quad i\mathbf{p} = \frac{\mathrm{u}^\star - \mathrm{u}}{\sqrt{2}} \Rightarrow [i\mathbf{p}, \mathbf{x}] = 1,$$
$$\mathbf{H} = E\frac{\mathbf{p}^2 + \mathbf{x}^2}{2}, \quad [i\mathbf{H}, \mathbf{x}] = E\mathbf{p}, \quad [i\mathbf{H}, \mathbf{p}] = -E\mathbf{x}.$$

The quantum algebras inherit the Hilbert metric of the basic representation space,

$$V \times V \longrightarrow \mathbb{C}, \quad \langle \mathrm{u}|\mathrm{u}\rangle = 1,$$

in the form of the Fock state, a linear, conjugation-compatible form:

$$\mathbf{Q}_\epsilon(\mathbb{C}^2) \longrightarrow \mathbb{C}, \; a \longmapsto \langle a \rangle \text{ with } \begin{cases} \langle a + b \rangle &= \langle a \rangle + \langle b \rangle, \\ \langle \alpha a \rangle &= \alpha \langle a \rangle, \qquad \alpha \in \mathbb{C}, \\ \langle a^\star \rangle &= \overline{\langle a \rangle}, \end{cases}$$

defined by $\langle [\mathbf{I}_1, a] \rangle = 0$, $\langle (\mathrm{u}^\star \mathrm{u})^k \rangle = 1$ for $k = 0, 1, 2, \ldots \Rightarrow \langle (\mathrm{u}^\star)^k (\mathrm{u})^l \rangle = \delta_{kl} k!$.

It makes the quantum algebras pre-Hilbert spaces with a semidefinite product:

$$\begin{aligned} \mathbf{Q}_\epsilon(\mathbb{C}^2) \times \mathbf{Q}_\epsilon(\mathbb{C}^2) \longrightarrow \mathbb{C}, \qquad \langle a|b \rangle &= \langle a^\star b \rangle, \\ \Rightarrow \qquad \langle a|a \rangle &= \langle a^\star a \rangle \geq 0, \\ \text{e.g.,} \qquad \langle \mathrm{u}|\mathrm{u} \rangle &= \langle \mathrm{u}^\star \mathrm{u} \rangle = 1, \\ \langle \mathrm{u}^\star|\mathrm{u}^\star \rangle &= \langle \mathrm{u}\mathrm{u}^\star \rangle = 0. \end{aligned}$$

The classes of the quantum algebras with respect to the annihilator left ideal $\mathbf{Q}_\epsilon(\mathbb{C}^2)\mathrm{u}^\star$ with $\langle \mathbf{Q}_\epsilon(\mathbb{C}^2)\mathrm{u}^\star \rangle = 0$ is a vector space with a definite scalar product. Its Cauchy completion is the familiar Fock space. The ground-state vector $|0\rangle$ is the class of the unit, and the vectors $|k\rangle$ with k quanta arise by the creation operator products:

$$\mathrm{Fock}_\epsilon(\mathbb{C}^2) = \overline{\mathbf{Q}_\epsilon(\mathbb{C}^2)/\mathbf{Q}_\epsilon(\mathbb{C}^2)\mathrm{u}^\star} : \begin{cases} |0\rangle &= 1 + \mathbf{Q}_\epsilon(\mathbb{C}^2)\mathrm{u}^\star, \\ |k\rangle &= \frac{(\mathrm{u})^k}{\sqrt{k!}}|0\rangle, \; \mathrm{u}^\star|0\rangle = 0, \\ \langle k|l \rangle &= \delta_{kl}. \end{cases}$$

The two-dimensional Fermi Fock space has the ground-state vector and the 1-quantum vector as basis. The Bose Fock space has countably infinite Hilbert dimension:

$$\mathrm{Fock}_+(\mathbb{C}^2) = \mathbb{C}|0\rangle \oplus \mathbb{C}|1\rangle, \quad \mathrm{Fock}_-(\mathbb{C}^2) = \overline{\bigoplus_{k=0}^{\infty} \mathbb{C}|k\rangle}.$$

The position representation for the Bose oscillator comes with the Hermitian polynomials H^k:

$$\begin{aligned} \mathrm{u} &\cong \frac{x - d_x}{\sqrt{2}} = -e^{\frac{x^2}{2}} \frac{d_x}{\sqrt{2}} e^{-\frac{x^2}{2}}, \\ \mathrm{u}^\star &\cong \frac{x + d_x}{\sqrt{2}} = e^{-\frac{x^2}{2}} \frac{d_x}{\sqrt{2}} e^{\frac{x^2}{2}}, \end{aligned} \qquad \begin{cases} \mathrm{u}^\star|0\rangle &= 0 \Rightarrow |0\rangle \cong \frac{e^{-\frac{x^2}{2}}}{\sqrt{\sqrt{\pi}}}, \\ |k\rangle &= \frac{(\mathrm{u})^k}{\sqrt{k!}}|0\rangle \cong \frac{\mathrm{H}^k(x)}{\sqrt{2^k k!}} \frac{e^{-\frac{x^2}{2}}}{\sqrt{\sqrt{\pi}}}, \\ \mathrm{H}^k(x) &= e^{x^2}(-d_x)^k e^{-x^2}. \end{cases}$$

All free particles, embedded in relativistic fields (see Chapter 5), use this structure, characterized by the (anti-)commutator and the Fock form:

$$\mathbf{Q}_\epsilon(\mathbb{C}^2) \text{ with } [\mathrm{u}^\star, \mathrm{u}]_\epsilon = 1 \text{ and } \langle \mathrm{u}^\star \mathrm{u}\rangle = 1.$$

The time-dependent Fock state value with $\langle a\rangle = \langle 0|a|0\rangle$ can be formulated as an integral with a Dirac energy measure or as the residue of a simple pole:

$$\begin{aligned} \langle 0|[\mathrm{u}^\star(0), \mathrm{u}(t)]_\epsilon|0\rangle &= \langle 0|[\mathrm{u}^\star(0), \mathrm{u}(t)]_{-\epsilon}|0\rangle = \langle 0|\mathrm{u}^\star(0)\mathrm{u}(t)|0\rangle \\ &= e^{iEt} = \int dq_0\ \delta(q_0 - E)e^{iq_0 t} = \oint \frac{dq_0}{2i\pi}\frac{1}{q_0 - E}e^{iq_0 t}, \\ (d_t - iE)e^{iEt} &= 0. \end{aligned}$$

The time-ordered operator products have a pole structure in the upper and lower complex energy planes:

$$\begin{aligned} \vartheta(\pm t)\langle 0|\mathrm{u}^\star(0)\mathrm{u}(t)|0\rangle &= \vartheta(\pm t)e^{iEt} = \pm\int \frac{dq_0}{2i\pi}\frac{1}{q_0 \mp io - E}e^{iq_0 t}, \\ (d_t - iE)\vartheta(\pm t)e^{iEt} &= d_t\vartheta(\pm t) = \pm\delta(t). \end{aligned}$$

$\vartheta(-t)\langle 0|\mathrm{u}^\star(0)\mathrm{u}(t)|0\rangle$ connects time order with creation-annihilation order — first creation, then annihilation. For free particle fields, those integrals show up as energy-momentum integrals, e.g., in Feynman propagators (see Chapter 5).

4.4.2 Indefinite Metric and Ghosts

Nonparticle degrees of freedom come with *reducible, but nondecomposable* translation representations, the simplest ones being complex two-dimensional. The faithful image of noncompact time $\mathbb{R}$ is in an *indefinite unitary* group. In addition to the energy eigenvalue $E \in \mathbb{R}$, there occurs an arbitrary real nontrivial nilconstant (gauge-fixing constant) $\nu \in \mathbb{R}$, $\nu \neq 0$:

$$\mathbb{R} \ni t \longmapsto e^{iEt}\begin{pmatrix} 1 & i\nu t \\ 0 & 1\end{pmatrix} \in \mathbf{U}(1,1) \subset \mathbf{GL}(W).$$

The vector space W and its dual space W^T with dual representation $t \longmapsto e^{-iEt}\begin{pmatrix} 1 & 0 \\ -i\nu t & 1\end{pmatrix}$ can be spanned by two basic vectors (b, g) and their dual vectors $(\mathrm{g}^\times, \mathrm{b}^\times)$:

$$\begin{aligned} &W = \mathbb{C}\mathrm{b} \oplus \mathbb{C}\mathrm{g} \cong \mathbb{C}^2 \text{ and } W^T = \mathbb{C}\mathrm{g}^\times \oplus \mathbb{C}\mathrm{b}^\times \cong \mathbb{C}^2, \\ &\text{duality (nontrivial): } \langle \mathrm{g}^\times, \mathrm{b}\rangle = 1,\ \langle \mathrm{b}^\times, \mathrm{g}\rangle = 1, \end{aligned}$$

which are eigenvectors (g for "good") and nilvectors (b for "bad," no eigenvector):

$$\begin{aligned} \mathrm{g}(t) &= e^{iEt}\ \mathrm{g}, & \mathrm{b}(t) &= e^{iEt}\ (\mathrm{b} + i\nu t\mathrm{g}), \\ \mathrm{g}^\times(t) &= e^{-iEt}\mathrm{g}^\times, & \mathrm{b}^\times(t) &= e^{-iEt}(\mathrm{b}^\times - i\nu t\mathrm{g}^\times). \end{aligned}$$

The quantum algebras for the complex four-dimensional self-dual basic space $\mathbf{W} = W \oplus W^T \cong \mathbb{C}^4$ with all complex linear combinations of products of the four basic vectors are constructed, as in the irreducible case in the former subsection, with the (anti-)commutators implemented by the dual products:

$$\text{nontrivial for } \mathbf{Q}_\epsilon(\mathbb{C}^4) = \mathbf{Q}_\epsilon(\mathbb{C}^2) \otimes \mathbf{Q}_\epsilon(\mathbb{C}^2)\text{: } [\mathrm{g}^\times, \mathrm{b}]_\epsilon = 1,\ [\mathrm{b}^\times, \mathrm{g}]_\epsilon = 1.$$

Again, the quantum algebras are associative and unital. They inherit the $\mathbf{U}(1,1)$-conjugation $\times$ from the basic vectors $(\mathrm{g}, \mathrm{b}) \overset{\times}{\leftrightarrow} (\mathrm{g}^\times, \mathrm{b}^\times)$. The Fermi quantum algebra is finite-dimensional, $\mathbf{Q}_+(\mathbb{C}^4) \cong \mathbb{C}^{16}$, in contrast to the Bose quantum algebra, $\mathbf{Q}_-(\mathbb{C}^4) \cong \mathbb{C}^{\aleph_0}$.

The Hamiltonian is the sum of a diagonal part and a nilpotent part:

$$\begin{aligned} \mathbf{H}_N &= E\mathbf{I}_2 + \nu\mathbf{N} = E\tfrac{[\mathrm{g},\mathrm{b}^\times]_{-\epsilon}+[\mathrm{b},\mathrm{g}^\times]_{-\epsilon}}{2} + \tfrac{\nu}{2}[\mathrm{g},\mathrm{g}^\times]_{-\epsilon} \cong \begin{pmatrix} E & \nu \\ 0 & E \end{pmatrix}, \\ [\mathbf{H}_N, \mathrm{b}] &= E\mathrm{b} + \nu\mathrm{g}, \quad [\mathbf{H}_N, \mathrm{g}] = E\mathrm{g}. \end{aligned}$$

The action of $\mathbf{N}$ with the admixture of good eigenvectors is the precursor of the gauge transformations in quantum field theories. Its adjoint action $\mathrm{ad}\,\mathbf{N} = [\mathbf{N},\ ]$ is nilquadratic:

$$[\mathbf{N}, \mathbf{H}_N] = 0,\quad [\mathbf{N}, \mathrm{b}] = \nu\mathrm{g},\quad [\mathbf{N}, \mathrm{g}] = 0,\ (\mathrm{ad}\,\mathbf{N})^2(\mathrm{b},\mathrm{g}) = 0.$$

The sesquilinear $\mathbf{U}(1,1)$-form is indefinite:

$$\begin{aligned} W \times W \longrightarrow \mathbb{C},\quad & \begin{pmatrix} \langle \mathrm{b}|\mathrm{b}\rangle & \langle \mathrm{b}|\mathrm{g}\rangle \\ \langle \mathrm{g}|\mathrm{b}\rangle & \langle \mathrm{g}|\mathrm{g}\rangle \end{pmatrix} = \begin{pmatrix} 0 & 1 \\ 1 & 0 \end{pmatrix}, \\ & \begin{pmatrix} \langle \mathrm{b}+\mathrm{g}|\mathrm{b}+\mathrm{g}\rangle & \langle \mathrm{b}+\mathrm{g}|\mathrm{b}-\mathrm{g}\rangle \\ \langle \mathrm{b}-\mathrm{g}|\mathrm{b}+\mathrm{g}\rangle & \langle \mathrm{b}-\mathrm{g}|\mathrm{b}-\mathrm{g}\rangle \end{pmatrix} = 2\begin{pmatrix} 1 & 0 \\ 0 & -1 \end{pmatrix}. \end{aligned}$$

The eigenvector and nilvector constitute a *ghost pair* (*Witt pair*) with a neutral signature metric, as familiar from a lightlike basis $\{\mathbf{e}^\pm = \mathbf{e}^0 \pm \mathbf{e}^3\}$ for the lightcone in Minkowski spacetime with the indefinite $\mathbf{O}(1,1) \subset \mathbf{O}(1,3)$ Lorentz metric.

As in the former subsection, the sesquilinear $\mathbf{U}(1,1)$-form can be extended to the quantum algebras, which, because of the indefiniteness, does not lead to a nontrivial Hilbert space.

In an integral formulation of the time representation matrix elements, the indefinite metric shows up in the derivative of a Dirac distribution and the higher-order pole (dipole):

$$\begin{aligned} \begin{pmatrix} \langle \mathrm{g}^\times(0)\mathrm{b}(t)\rangle & \langle \mathrm{b}^\times(0)\mathrm{b}(t)\rangle \\ \langle \mathrm{g}^\times(0)\mathrm{g}(t)\rangle & \langle \mathrm{b}^\times(0)\mathrm{g}(t)\rangle \end{pmatrix} = \begin{pmatrix} 1 & i\nu t \\ 0 & 1 \end{pmatrix} e^{iEt} &= \int dq_0 \begin{pmatrix} \delta(q_0 - E) & -\nu\delta'(q_0 - E) \\ 0 & \delta(q_0 - E) \end{pmatrix} e^{iq_0 t} \\ &= \oint \frac{dq_0}{2i\pi} \begin{pmatrix} \frac{1}{q_0 - E} & \frac{\nu}{(q_0-E)^2} \\ 0 & \frac{1}{q_0 - E} \end{pmatrix} e^{iq_0 t}. \end{aligned}$$

Particles and ghosts can be combined by using decomposable time representations and the corresponding product quantum algebras, e.g.,

$$\mathbb{R} \ni t \longmapsto \left(\begin{array}{c|cc} e^{i\omega t} & 0 & 0 \\ \hline 0 & e^{iEt} & i\nu t\, e^{iEt} \\ 0 & 0 & e^{iEt} \end{array}\right) \in \mathbf{U}(1) \times \mathbf{U}(1,1) \subset \mathbf{U}(2,1) \subset \mathbf{GL}(\mathbb{C}^3),$$

$$\text{nontrivial for } \mathbf{Q}_\epsilon(\mathbb{C}^6) = \bigotimes^3 \mathbf{Q}_\epsilon(\mathbb{C}^2)\text{: } [\mathrm{u}^\star, \mathrm{u}]_\epsilon = 1,\ [\mathrm{g}^\times, \mathrm{b}]_\epsilon = 1,\ [\mathrm{b}^\times, \mathrm{g}]_\epsilon = 1.$$

The linear combinations of eigenvectors (particles) and ghost pairs have time representation matrix elements that combine simple and higher-order poles, for example, with $\alpha \in \mathbb{C}$:

$$\begin{aligned}\psi = \mathrm{u} + \alpha\,\mathrm{b} - \tfrac{1}{2\alpha}\,\mathrm{g} \Rightarrow \langle \psi^*(0)\psi(t)\rangle &= \oint \tfrac{dq_0}{2i\pi}\left[\tfrac{1}{q_0-\omega} - \tfrac{1}{q_0-E} + \tfrac{\nu|\alpha|^2}{(q_0-E)^2}\right]e^{iq_0 t}\\ &= \oint \tfrac{dq_0}{2i\pi}\,\tfrac{(\omega-E)^2}{(q_0-\omega)(q_0-E)^2}\,e^{iq_0 t}\\ &\text{for } |\alpha|^2 = \tfrac{E-\omega}{\nu} > 0.\end{aligned}$$

Chapter 5

Quantum Fields of Flat Spacetime

Spacetime theories are field theories as conceptionally initiated by Faraday and first formulated by Maxwell. Charged or neutral mass points with orbits $x \longmapsto X(\tau)$ in spacetime, i.e., depending only on the Lorentz invariant eigentime $\tau = \epsilon(x_0)\vartheta(x^2)\sqrt{x^2}$, and the related point-supported Dirac distributions $\delta(\vec{x} - \vec{X}(\tau))$ for position in the electromagnetic current and the energy-momentum tensor are strangers for relativistic electromagnetic and gravity fields, $x \longmapsto \mathbf{A}(x),\ \mathbf{g}(x)$. As underlined by Einstein himself and suggested by quantum theory, such dichotomic theories have to be replaced by pure field theories, i.e., with fields also for matter. Einstein (1936): "What appears certain to me, however, is that, in the foundations of any consistent field theory, the particle concept must not appear in addition to the field concept." Classical gravity is a real theory. The interpretation of quantum theory with probability amplitudes is established by a complex Hilbert space formulation. Einstein probably thought primarily of classical fields. Nevertheless, he also considered complex structures in his attempts to unify gravity and electrodynamics.

Fields can be valued in "new" spaces. They have, in general, external and internal degrees of freedom. Quantum fields can also implement, in addition to global operations like rotations $\mathbf{SO}(3)$ in classical and quantum mechanics, local operation groups, e.g., a gauge symmetry like $\mathbf{U}(1)$-electromagnetism or the electroweak $\mathbf{U}(2)$-interactions in the standard model of particle interactions.

Quantum theory does not use pointlike particles and time orbits in position as basic concepts. In nonrelativistic quantum mechanics, the position is an operator that may — but does not have to — be diagonalized by an appropriate experimental setup as exemplified by the double-slit experiments. The position dependent wave functions yield probability catalogues for the possible results of position measurements. In relativistic quantum field theory,

H. Saller, *Operational Spacetime: Interactions and Particles*,
Fundamental Theories of Physics 163, DOI 10.1007/978-1-4419-0898-8_6,

neither time nor position comes as basic operator. Quantum fields are operators, not spacetime-dependent probability amplitudes (wave functions). The name "second quantization" is misleading, as there is only one "quantization" — in two steps: Quantum mechanics is the quantum structure of time-dependent operators, built by position and momentum, which give rise to time representation coefficients; quantum field theory is the quantum structure of spacetime-dependent operators, built by, e.g., electron-positron or gauge fields, which give rise to spacetime representation coefficients.

The time-dependent position-momentum pairs $t \longmapsto (\mathbf{x}, \mathbf{p})(t)$ of quantum mechanics are replaced, in canonical quantum field theories, by spacetime-dependent field pairs, e.g., by the pair with electromagnetic potential and field strength $x \longmapsto (\mathbf{A}^a, \mathbf{F}^{ba})(x)$, or by a pair with a scalar field and its derivative $x \longmapsto (\boldsymbol{\Phi}, \partial^a \boldsymbol{\Phi})(x)$, or by a Dirac field with its conjugate $x \longmapsto (\boldsymbol{\Psi}, \boldsymbol{\Psi}^*)(x)$. Interactions, in mechanics described by position-dependent potentials $\mathbf{x} \longmapsto V(\mathbf{x})$, e.g., $\frac{\omega}{2}\mathbf{x}^2$ (harmonic oscillator) or $\frac{\delta}{|\mathbf{x}|}$ (Kepler potential), are given, in the field theories used, by field polynomials, $(\mathbf{A}, \boldsymbol{\Phi}, \boldsymbol{\Psi}) \longmapsto P(\mathbf{A}, \boldsymbol{\Phi}, \boldsymbol{\Psi})$, implementing operation invariants, e.g., by $\frac{m^2}{2}\boldsymbol{\Phi}^2$ (mass term) or $\frac{g_0}{8}(\boldsymbol{\Phi}^*\boldsymbol{\Phi} - M^2)^2$ (Higgs potential) or by gauge vertices $\overline{\boldsymbol{\Psi}}\gamma_a \boldsymbol{\Psi} \mathbf{A}^a$.

To characterize quantum field theory as a theory of pointlike particles is wide of the mark. It is also not obvious how to concretize operationally the concept of a "pointwise" interaction as attributed, e.g., to the local product of an electron-positron field with a gauge field $\overline{\boldsymbol{\Psi}}(x)\gamma_a \boldsymbol{\Psi}(x)\mathbf{A}^a(x)$, used as gauge vertex for an electromagnetic interaction. Also a naive interpretation of Feynman diagrams with pointlike particles propagating in spacetime may lead to wrong associations.

The Poincaré group $\mathbf{SO}_0(1,3) \vec{\times} \mathbb{R}^4$ for Minkowski spacetime with the translation subgroup, or, better, its twofold cover $\mathbf{SL}(2, \mathbb{C}) \vec{\times} \mathbb{R}^4$, is implemented by interaction-free elementary particles. All spacetimes with Minkowski tangent spacetimes can be endowed with free particles à la Wigner, i.e., as irreducible unitary Poincaré group representations acting on infinite-dimensional Hilbert spaces. Massive particles with spin 0, e.g., pions as an example for integer spin, and with spin $\frac{1}{2}$, e.g., electron-positrons as an example for half-integer spin, are relativistically embedded, respectively, in a scalar field $\boldsymbol{\Phi}$ and a Dirac spinor field $\boldsymbol{\Psi}$ with dimension of a length density for integer spin, $[\boldsymbol{\Phi}] = \frac{1}{\mathrm{m}}$, and of a square root volume density for halfinteger spin, $[\boldsymbol{\Psi}] = \frac{1}{\sqrt{\mathrm{m}^3}}$.

The embedding of particles in relativistic fields is closely related to the induction of Hilbert representations of the Poincaré group from Hilbert representations of a "little" group and translations, given by the spin rotations and timelike translations for massive particles in a rest system and by the axial rotations around the momentum and lightlike translations for massless particles in a "lightsystem." The Fourier expansion of a free field, i.e., its particle analysis, involves creation and annihilation operators for particles (better: operator distributions) with a definite momentum $\vec{q} \in \mathbb{R}^3$. Those translation eigenoperators do not describe the position of a pointparticle. For free particles, the position coordinate characterizes the translation behavior in the irreducible unitary representations $\mathbb{R}^3 \ni \vec{x} \longmapsto e^{i\vec{q}\vec{x}} \in \mathbf{U}(1)$.

The Wigner definition of gravitons as particles makes sense only for flat spacetime, i.e., for an expansion of Einstein's gravity in absolute Minkowski spacetime. Obviously, the related definition of special relativistic fields for interaction-free gravitons does not solve the problem of the quantization of gravity. In addition and as well known, such a perturbative flat spacetime approach of gravitative interactions leads to a nonrenormalizable framework.

Special relativistic quantum field theory has Planck's action constant $\hbar$ and the speed of light c as two universal intrinsic units. They reduce the "individual" units of a special dynamics to a mass or a length unit with the Compton conversion factor $\frac{\hbar}{c} \sim 3.3 \times 10^{-42}$ kg m.

5.1 Electrodynamics of Fields

Flat spacetime field theory uses the gauge principle for the construction of "basic" interactions. However, there is no principle for the determination of the "basic" gauge groups.

In contrast to electrodynamics with charged mass points, which is usually formulated in the real, electrodynamics with charged matter fields is formulated more appropriately in the complex with a manifest $\mathbf{U}(1)$-origin of the electromagnetic gauge interactions. Since real electrodynamics does not suggest immediately the compact phase group $\mathbf{U}(1)$, it is understandable that Weyl, in his first attempt, gauged the locally isomorphic noncompact dilation group $\mathbf{D}(1) = \exp \mathbb{R} \cong \mathbb{R}$, the simply connected $\mathbb{Z}$-fold covering group of $\mathbf{U}(1) = \exp i\mathbb{R} \cong \mathbb{R}/\mathbb{Z}$.

Maxwell's electrodynamics is characterized by an action, given by the Killing form $\eta \wedge \eta \cong \begin{pmatrix} -\mathbf{1}_3 & 0 \\ 0 & \mathbf{1}_3 \end{pmatrix}$ with signature $(3, 3)$, for the field strengths $\mathbf{F}$ in an adjoint representation of the Lorentz Lie algebra. The Killing form is normalized with the dimensionless Sommerfeld constant, in the second-order derivative Lagrangian:

$$W_{\text{elmag}} = \int d^4x \; \mathbf{L}(\mathbf{A}), \quad \mathbf{L}(\mathbf{A}) = \frac{1}{4g^2} \mathbf{F}_{ab} \mathbf{F}^{ab}, \quad g^2 = 4\pi\alpha_S \sim \frac{4\pi}{137} \sim 0.9 \times 10^{-1}.$$

The field strengths $\mathbf{F}^{ab} = \partial^b \mathbf{A}^a - \partial^a \mathbf{A}^b$ arise from the vector potential $\mathbf{A}$, which is acted on by the Minkowski representation of the Lorentz group. In a first-order derivative Lagrangian, the vector potential and field strengths are used as independent variables:

$$\mathbf{L}(\mathbf{A}, \mathbf{F}) = \mathbf{F}_{ab} \frac{\partial^a \mathbf{A}^b - \partial^b \mathbf{A}^a}{2} + \frac{g^2}{4} \mathbf{F}_{ab} \mathbf{F}^{ab} \Rightarrow \begin{cases} g^2 \mathbf{F}^{ab} &= \partial^b \mathbf{A}^a - \partial^a \mathbf{A}^b, \\ \partial_a \mathbf{F}^{ab} &= 0. \end{cases}$$

With opposite charge numbers $\pm z = \pm \frac{Q}{e} \in \mathbb{Z}$ in dual irreducible representations of the group $\mathbf{U}(1) \ni e^{i\alpha} \longmapsto e^{\pm iz\alpha}$, charged fields come in complex pairs, e.g., $(\mathbf{\Phi},\mathbf{\Phi}^*)$ and $(\mathbf{\Psi},\overline{\mathbf{\Psi}})$, involving particles and antiparticles. The fields have an electromagnetic interaction via a covariant derivative[1] with the vector potential. Each spacetime translation is accompanied by an electromagnetic $\mathbf{U}(1)$-operation. This replaces the Lorentz force for the charged mass points:

$$\begin{aligned}
W^{\text{elmag}}_{\text{matter}}(\mathbf{\Phi}) &= \int d^4x\, [(\partial^a + iz\mathbf{A}^a)\mathbf{\Phi}^*(\partial_a - iz\mathbf{A}_a)\mathbf{\Phi} - m^2\mathbf{\Phi}\mathbf{\Phi}^*],\\
&\quad (\partial_a - iz\mathbf{A}_a)(\partial^a - iz\mathbf{A}^a)\mathbf{\Phi} = -m^2\mathbf{\Phi},\\
W^{\text{elmag}}_{\text{matter}}(\mathbf{\Psi}) &= \int d^4x\, i[\overline{\mathbf{\Psi}}\gamma_a(\partial^a - iz\mathbf{A}^a)\mathbf{\Psi} - im\overline{\mathbf{\Psi}}\mathbf{\Psi}],\\
&\quad \gamma_a(\partial^a - iz\mathbf{A}^a)\mathbf{\Psi} = im\mathbf{\Psi}.
\end{aligned}$$

The electromagnetic current for a charged mass point is given by a Dirac distribution in position:

$$\begin{aligned}
\mathbf{J}_a(x) &= \tfrac{z}{m}\int d\tau P_a\delta(x - X(\tau)) = z\int dX_a\delta(x-X) = z\tfrac{P_a}{P_0}\delta(\vec{x} - \vec{X}(t)),\\
\int d^3x\, \mathbf{J}_0(x) &= z.
\end{aligned}$$

For matter fields, the point-supported Dirac distribution is spread to a spacetime field by a $\mathbf{U}(1)$-invariant product of dual fields with nontrivial $\mathbf{U}(1)$-transformations:

$$\begin{aligned}
\mathbf{J}_a(\mathbf{\Phi}) &= iz[(\partial_a + iz\mathbf{A}_a)\mathbf{\Phi}^*\mathbf{\Phi} - \mathbf{\Phi}^*(\partial_a - iz\mathbf{A}_a)\mathbf{\Phi}],\\
\mathbf{J}_a(\mathbf{\Psi}) &= z\overline{\mathbf{\Psi}}\gamma_a\mathbf{\Psi}.
\end{aligned}$$

The eigentime orbits $x \longmapsto \tau \longmapsto X(\tau)$ of charged mass points have trivial electromagnetic $\mathbf{U}(1)$-properties $X \overset{\mathbf{U}(1)}{\longmapsto} X$. Fields $x \longmapsto (\mathbf{\Phi}(x),\mathbf{\Psi}(x))$ "open up local spaces" for *local transformation groups* with chargelike (internal) degrees of freedom, here $(\mathbf{\Phi},\mathbf{\Psi}) \overset{\mathbf{U}(1)}{\longmapsto} e^{iz\alpha}(\mathbf{\Phi},\mathbf{\Psi})$ (see Chapter 6).

The current involves the density for the $\mathbf{U}(1)$-Lie algebra:

$$i\int d^3x\, \mathbf{J}_0(x) = l^0 \in \log \mathbf{U}(1).$$

It can be considered to be the answer $\mathbf{J}_a = \frac{\partial \mathbf{L}}{\partial \mathbf{A}^a}$ of a classical field Lagrangian $W = \int d^4x\, \mathbf{L}$ to the change of the vector potential (gauge field). The gauge interaction vertex $\mathbf{J}^a\mathbf{A}_a$ implements the representation of the $\mathbf{U}(1)$-Lie algebra acting on the vector spaces with the matter fields (see Chapter 6).

A real formulation exists for integer spin: Dual $\mathbf{U}(1)$-representations can be combined in a representation of the $\mathbf{U}(1)$-isomorphic axial rotation group $\mathbf{SO}(2)$ with a basis of the real 2-dimensional space $\mathbf{\Phi} = \frac{\mathbf{\Phi}_1 + i\mathbf{\Phi}_2}{\sqrt{2}}$:

$$\begin{aligned}
\varphi &= \begin{pmatrix}\mathbf{\Phi}_1\\ \mathbf{\Phi}_2\end{pmatrix} \longmapsto e^{zI\alpha}\varphi,\ \ I = \begin{pmatrix}0 & -1\\ 1 & 0\end{pmatrix},\ \ e^{zI\alpha} \in \mathbf{SO}(2) \cong \mathbf{U}(1),\\
L(\mathbf{\Phi}) &= \tfrac{1}{2}[(\partial^a - zI\mathbf{A}^a)\varphi]^T(\partial_a - zI\mathbf{A}_a)\varphi - m^2\varphi^2,\\
\mathbf{J}_a(\mathbf{\Phi}) &= -z\varphi^T I(\partial_a - zI\mathbf{A}_a)\varphi.
\end{aligned}$$

T denotes transposition; e.g., $\varphi^T = (\mathbf{\Phi}_1,\mathbf{\Phi}_2)$.

[1]To save parentheses, derivatives act only on the immediately following field.

5.2 Gravity of Fields

Einstein's gravity is characterized by an action with the curvature scalar in terms of the metrical tensor, normalized with the Planck area or the squared Planck mass $m_P^2\ell^2 = 8\pi\frac{\hbar^2}{c^2}$, in the second-order derivative Lagrangian,

$$\begin{aligned} W_{\text{grav}} &= \int d^4x \, \mathbf{L}(\mathbf{g}), \\ \mathbf{L}(\mathbf{g}) &= \frac{1}{2\ell^2}\sqrt{|\mathbf{g}|}\ \mathbf{g}_{kj}\mathcal{R}^{kj}_{\bullet}, \end{aligned} \quad \text{with } \ell^2 = \frac{8\pi G\hbar}{c^3} \sim (0.8\times 10^{-34}\ \text{m})^2 \sim \frac{8\pi}{m_P^2}.$$

The Planck mass is given in proton mass m_p units by $\frac{m_P}{m_p} \sim 10^{20}$ or $\log\frac{m_P^2}{m_p^2} \sim 88$.

A first-order derivative Lagrangian with a tetrad $\mathbf{e}(x) \in \mathbf{GL}(4,\mathbb{R})/\mathbf{SO}_0(1,3)$ and the Fock–Iwanenkov connection $\mathbf{\Gamma}$ for a structural group $\mathbf{SO}_0(1,3)$ as independent variables reads

$$\begin{aligned} \mathbf{L}(\mathbf{e},\mathbf{\Gamma}) &= \tfrac{1}{2}|\mathbf{e}|\mathbf{e}^a_k\mathbf{e}^b_i(\partial^{[k}\mathbf{\Gamma}^{i]}_{ab} - \ell^2\mathbf{\Gamma}^{[k}_{ac}\eta^{cd}\mathbf{\Gamma}^{i]}_{db}), \quad |\mathbf{e}| = \det \mathbf{e}^i_c \\ \Rightarrow \ell^2\mathbf{\Gamma}^i_{ab} &= (\eta_{ac}\eta_{bd} - \eta_{bc}\eta_{ad})\mathbf{e}^i_e(\eta^{df}\mathbf{e}^e_j\partial^c\mathbf{e}^j_f + \eta^{ef}\mathbf{e}^d_j\partial^c\mathbf{e}^j_f - \eta^{df}\mathbf{e}^c_j\partial^e\mathbf{e}^j_f). \end{aligned}$$

The gravitative equations of motion use the covariant derivative for the tangent translations of the spacetime manifold, e.g., for the scalar field:

$$\begin{aligned} W^{\text{grav}}_{\text{matter}}(\mathbf{\Phi}) &= \int \sqrt{|\mathbf{g}|}\ d^4x\ \tfrac{1}{2}(\mathbf{g}_{kj}\partial^k\mathbf{\Phi}\partial^j\mathbf{\Phi} - m^2\mathbf{\Phi}^2) \\ &= -\int \sqrt{|\mathbf{g}|}\ d^4x\ \tfrac{1}{2}(\mathbf{\Phi}\partial^2_{\mathbf{g}}\mathbf{\Phi} + m^2\mathbf{\Phi}^2), \\ \partial^2_{\mathbf{g}}\mathbf{\Phi} &= -m^2\mathbf{\Phi}. \end{aligned}$$

In mass point gravitation, the mass drops out in the geodesic equations. The field equations of motion contain the masses.

The field equation involves the invariant Laplace-Beltrami operator (see Chapter 2):

$$\partial^2_{\mathbf{g}} = \frac{1}{\sqrt{|\mathbf{g}|}}\partial^k\sqrt{|\mathbf{g}|}\ \ \mathbf{g}_{kj}\partial^j = (\partial_j - \Gamma^{ki}_j\mathbf{g}_{ki})\partial^j = (\partial_a - \mathbf{\Gamma}^b_{ab})\partial^a.$$

Six of the 16 connection forms act as gauge fields of the tangent Lorentz group $\mathbf{SO}_0(1,3)$:

$$\Lambda \in \mathbf{SO}_0(1,3): \quad \mathbf{\Gamma}^k \longmapsto \quad \Lambda\mathbf{\Gamma}^k\Lambda^{-1} + (\partial^k\Lambda)\Lambda^{-1}.$$

The Lorentz covariant derivative is defined for the two fundamental Weyl spinor representations $[\frac{1}{2}|0]$ and $[0|\frac{1}{2}]$, for their sums and product representations, i.e., for all fields acted on by a finite-dimensional representation $[L|R]$ of Lorentz group and Lie algebra:

$$\begin{aligned} &\log \mathbf{SO}_0(1,3) \ni \mathcal{L}^{ab} \longmapsto \mathcal{D}_{[L|R]}(\mathcal{L}^{ab}) \in \mathbf{AL}(V), \quad V \cong \mathbb{C}^{(1+2R)(1+2L)}, \\ &\log \mathbf{SO}_0(1,3) \ni \mathbf{\Gamma}^k_{ab}\mathcal{L}^{ab} \longmapsto \mathbf{\Gamma}^k_{ab}\mathcal{D}_{[L|R]}(\mathcal{L}^{ab}) = \mathbf{\Gamma}^k_{[L|R]} \in \mathbf{AL}(V), \\ &\text{Lorentz group covariant derivative: } \mathbf{1}_V\partial^k - \mathbf{\Gamma}^k_{[L|R]}. \end{aligned}$$

Integer spin particles are in fields with a Lorentz group representation that can be chosen as real, i.e., $V \cong \mathbb{R}^{(1+2R)(1+2L)}$ for the integer sum $L+R \in \mathbb{N}$.

For the action on the vector field $\partial^i\mathbf{\Phi}$ with the scalar field derivatives, the Lorentz group gauge fields are valued in the Lie algebra of the real four-dimensional Minkowski representation:

$$\mathbf{\Gamma}^k_{[\frac{1}{2}|\frac{1}{2}]} \cong \left(\begin{array}{c|c} 0 & \mathbf{\Gamma}^k_{0\alpha} = \mathbf{\Gamma}^k_{\alpha 0} \\ \hline \mathbf{\Gamma}^k_{\beta 0} & \mathbf{\Gamma}^k_{\alpha\beta} = -\mathbf{\Gamma}^k_{\beta\alpha} \end{array}\right) \in \mathbf{AL}(\mathbb{R}^4), \quad \alpha,\beta = 1,2,3.$$

The energy-momentum tensor with dimension $[\mathbf{T}] = \frac{1}{\mathrm{m}^4}$ for the gravitative interaction

$$-\tfrac{1}{\ell^2}(\mathcal{R}^{li}_{\bullet} - \tfrac{\mathbf{g}^{li}}{2}\mathcal{R}^{\bullet}_{\bullet}) = \mathbf{T}^{li}, \quad -\tfrac{1}{\ell^2}\mathcal{R}^{li}_{\bullet} = \mathbf{T}^{li} - \tfrac{\mathbf{g}^{li}}{2}\mathbf{T}_{\bullet}$$

is given by the derivative of the Lagrangian with respect to the metric, e.g., for a scalar field,

$$\begin{aligned}
\frac{2}{\sqrt{|\mathbf{g}|}}\frac{\partial \mathbf{L}(\mathbf{\Phi})}{\partial \mathbf{g}_{li}} = \mathbf{T}^{li}(\mathbf{\Phi}) &= \partial^l\mathbf{\Phi}\partial^i\mathbf{\Phi} - \frac{\mathbf{g}^{li}}{\sqrt{|\mathbf{g}|}}\mathbf{L}(\mathbf{\Phi}) \\
&= \partial^l\mathbf{\Phi}\partial^i\mathbf{\Phi} - \tfrac{\mathbf{g}^{li}}{2}(\partial_j\mathbf{\Phi}\partial^j\mathbf{\Phi} - m^2\mathbf{\Phi}^2), \\
\mathbf{T}^{\bullet}(\mathbf{\Phi}) &= -\partial_j\mathbf{\Phi}\partial^j\mathbf{\Phi} + 2m^2\mathbf{\Phi}^2, \\
\mathbf{T}^{li} - \tfrac{\mathbf{g}^{li}}{2}\mathbf{T}_{\bullet} &= \partial^l\mathbf{\Phi}\partial^i\mathbf{\Phi} - \tfrac{\mathbf{g}^{li}}{2}m^2\mathbf{\Phi}^2.
\end{aligned}$$

The fields spread the Dirac position distribution in the energy-momentum tensor of a mass point to spacetime. In flat spacetime, it involves the position density of the translation generators,

$$\begin{aligned}
\sqrt{|\mathbf{g}|}\mathbf{T}^{li}(x) &= \tfrac{1}{m}\int d\tau\ P^l P^i\ \delta(x - X(\tau)) = \tfrac{P^l P^i}{P^0}\ \delta(\vec{x} - \vec{X}(t)), \\
\mathbf{g} &= \eta: \quad \int d^3x\ \mathbf{T}^{0a}(x) = P^a.
\end{aligned}$$

For orthonormal bases, the metric derivative can be replaced by the tetrad derivative and $\sqrt{|\mathbf{g}|} = |\mathbf{e}|$:

$$\begin{aligned}
W^{\text{grav}}_{\text{matter}} &= \int |\mathbf{e}| d^4x\ \tfrac{1}{2}(\partial_a\mathbf{\Phi}\partial^a\mathbf{\Phi} - m^2\mathbf{\Phi}^2), \\
\frac{1}{|\mathbf{e}|}\frac{\partial \mathbf{L}(\mathbf{\Phi})}{\partial \mathbf{e}^a_i} &= \mathbf{T}^i_a(\mathbf{\Phi}) = \partial_a\mathbf{\Phi}\partial^i\mathbf{\Phi} - \frac{\mathbf{e}^i_a}{|\mathbf{e}|}\mathbf{L}(\mathbf{\Phi}), \quad \mathbf{e}^a_l\mathbf{T}^i_a(\mathbf{\Phi}) = \mathbf{T}^i_l(\mathbf{\Phi}).
\end{aligned}$$

The energy-momentum tensor for the gravitative interactions contains, compared with the current for the electromagnetic interactions, an additional derivative.

Dirac fields for half-integer spin particles can be defined in orthonormal bases with the Dirac representation $[\frac{1}{2}|0] \oplus [0|\frac{1}{2}]$ of the Lorentz cover group $\mathbf{SL}(2,\mathbb{C})$ and an action $\mathbf{\Psi} \longmapsto e^{-\frac{1}{4}[\gamma^a,\gamma^b]\lambda_{ab}}\mathbf{\Psi}$. They gravitate via the tetrad

$$\begin{aligned}
W^{\text{grav}}_{\text{matter}}(\mathbf{\Psi}) &= \int |\mathbf{e}|\ d^4x\ [i\overline{\mathbf{\Psi}}\gamma_c\mathbf{e}^c_k(\partial^k - \mathbf{\Gamma}^k)\mathbf{\Psi} + m\overline{\mathbf{\Psi}}\mathbf{\Psi}], \\
\gamma_c\mathbf{e}^c_k(\partial^k - \mathbf{\Gamma}^k)\mathbf{\Psi} &= \gamma_c(\partial^c - \mathbf{\Gamma}^c)\mathbf{\Psi} = im\mathbf{\Psi}.
\end{aligned}$$

Their covariant derivative involves the gauge fields $\mathbf{\Gamma}^k$ valued in the Lorentz Lie algebra of the complex four-dimensional Dirac representation:

$$\begin{aligned} \mathbf{\Gamma}^k &= -\mathbf{\Gamma}^k_{ab}\tfrac{1}{4}[\gamma^a,\gamma^b] \in \mathbf{AL}(\mathbb{C}^4), \\ \text{Lorentz Lie algebra: } \mathcal{L}^{ab} &\longmapsto i\int d^3x \;\; \mathbf{J}^{ab}_0(\vec{x}), \\ \mathbf{J}^{ab}_c &= i\overline{\mathbf{\Psi}}\gamma_c\tfrac{[\gamma^a,\gamma^b]}{4}\mathbf{\Psi}. \end{aligned}$$

The Lagrangian change for a tetrad variation reads

$$\tfrac{1}{|\mathbf{e}|}\tfrac{\partial \mathbf{L}(\mathbf{\Psi})}{\partial \mathbf{e}^a_i} = i\overline{\mathbf{\Psi}}\Big(\gamma_a\partial^i - \gamma_c\tfrac{\partial \mathbf{\Gamma}^c}{\partial \mathbf{e}^a_i}\Big)\mathbf{\Psi} - \tfrac{\mathbf{e}^i_a}{|\mathbf{e}|}\mathbf{L}(\mathbf{\Psi}).$$

5.3 Gravity and Electrodynamics

In the action of electromagnetic and gravitative fields,

$$W^{\text{grav,elmag}} = \int \sqrt{|\mathbf{g}|}\; d^4x\; [\tfrac{1}{2\ell^2}\mathbf{g}_{li}\mathcal{R}^{li}_{\bullet} + \tfrac{1}{4g^2}\mathbf{g}_{li}\mathbf{g}_{kj}\mathbf{F}^{kl}\mathbf{F}^{ji}],$$

the Ricci tensor contains second-order derivatives of the metric. The first-order derivative of the vector potential in the field strength $\mathbf{F}^{il} = \partial^l\mathbf{A}^i - \partial^i\mathbf{A}^l$ as an external derivative $\mathbf{F} = d\mathbf{A}$ has no metric-dependent connection contributions. The field strengths have the equation of motion

$$\tfrac{1}{\sqrt{|\mathbf{g}|}}\partial^k\sqrt{|\mathbf{g}|}\;\mathbf{g}_{li}\mathbf{g}_{kj}\mathbf{F}^{ij} = \partial^k\mathbf{F}_{lk} + \Gamma^{kj}_{k}\mathbf{F}_{lj} - \Gamma^{kj}_{l}\mathbf{F}_{jk} = 0.$$

The energy-momentum tensor of the electromagnetic field is traceless:

$$\mathbf{T}^{li}(\mathbf{F}) = \tfrac{1}{g^2}(\mathbf{g}_{kj}\mathbf{F}^{kl}\mathbf{F}^{ji} - \tfrac{\mathbf{g}^{li}}{4}\mathbf{F}^{kj}\mathbf{F}_{kj}),\quad \mathbf{T}^{\bullet}(\mathbf{F}) = 0.$$

The electromagnetic and gravitative interaction of a charged scalar matter field,

$$\begin{aligned} W^{\text{grav,elmag}}_{\text{matter}} = \int \sqrt{|\mathbf{g}|}\; d^4x \quad & [\tfrac{1}{2\ell^2}\mathbf{g}_{li}\mathcal{R}^{li}_{\bullet} + \tfrac{1}{4g^2}\mathbf{g}_{li}\mathbf{g}_{kj}\mathbf{F}^{kl}\mathbf{F}^{ji} \\ & + \mathbf{g}_{kj}(\partial^k + iz\mathbf{A}^k)\mathbf{\Phi}^*(\partial^j - iz\mathbf{A}^j)\mathbf{\Phi} - m^2\mathbf{\Phi}\mathbf{\Phi}^*], \end{aligned}$$

comes with the electromagnetic current and the energy-momentum tensor of the matter field, both invariant under phase $\mathbf{U}(1) \cong \mathbf{SO}(2)$-gauge transformations:

$$\begin{aligned} -\tfrac{1}{g^2}(\partial^b\mathbf{F}_{ba} + \Gamma^{bc}_{b}\mathbf{F}_{ca} - \Gamma^{bc}_{a}\mathbf{F}_{cb}) &= \mathbf{J}_a(\mathbf{\Phi}), \\ -\tfrac{1}{\ell^2}(\mathcal{R}^{ab}_{\bullet} - \tfrac{\eta^{ab}}{2}\mathcal{R}^{\bullet}_{\bullet}) &= \mathbf{T}^{ab}(\mathbf{F}) + \mathbf{T}^{ab}(\mathbf{\Phi}), \end{aligned}$$

$$\begin{aligned} \mathbf{J}_a(\mathbf{\Phi}) &= iz[(\partial_a + iz\mathbf{A}_a)\mathbf{\Phi}^*\mathbf{\Phi} - \mathbf{\Phi}^*(\partial_a - iz\mathbf{A}_a)\mathbf{\Phi}], \\ \mathbf{T}^{ab}(\mathbf{\Phi}) &= (\partial^b + iz\mathbf{A}^b)\mathbf{\Phi}^*(\partial^a - iz\mathbf{A}^a)\mathbf{\Phi} \\ &\quad - \eta^{ab}[(\partial^c + iz\mathbf{A}^c)\mathbf{\Phi}^*(\partial_c - iz\mathbf{A}_c)\mathbf{\Phi} - m^2\mathbf{\Phi}^*\mathbf{\Phi}]. \end{aligned}$$

5.4 Linearized Einstein Equations

For a linear approximation of gravity on the background of absolute Minkowski spacetime with fixed Lorentz metric $(\mathbb{M}^{(1,3)}, \eta)$, the metric is expanded with an order parameter $\lambda \in \mathbb{R}$ by a 10-component symmetric tensor field $\mathbb{R}^4 \ni x \longmapsto \lambda \underline{\mathbf{E}}(x) = \mathbf{E}(x)$ (order parameter λ is included):

$$\mathbf{g}^{il} = \delta^i_c \delta^l_d (\eta^{cd} + 2\lambda \underline{\mathbf{E}}^{cd}) + \ldots = \delta^i_c \delta^l_d (\eta^{cd} + 2\mathbf{E}^{cd}) + \ldots .$$

It is the first-order approximation of the tetrad,

$$\mathbf{g}^{il} = \mathbf{e}^i_c \eta^{cd} \mathbf{e}^j_d, \quad \mathbf{e}^i_c = \delta^i_a (\delta^a_c + \mathbf{E}^a_c) + \ldots .$$

The gravity tensor field will be put side by side with a flat spacetime vector field $x \longmapsto \mathbf{A}(x)$ as used for the electromagnetic gauge field. The analogue to the electromagnetic field strengths with first-order derivatives is the linearized connection coefficients:

$$\begin{aligned} \mathbf{F}^{cb} &= \partial^b \mathbf{A}^c - \partial^c \mathbf{A}^b, \\ \mathbf{\Gamma}^{cdb} = \mathbf{\Gamma}^{cd}_a \eta^{ab} &= -\partial^b \mathbf{E}^{cd} + \partial^c \mathbf{E}^{bd} + \partial^d \mathbf{E}^{cb}. \end{aligned}$$

The derivative of the field strength corresponds to the linearized curvature tensor with only spacetime derivatives and without the characteristic nonlinearity, i.e., without the self-coupling squares $\Gamma^{lj}_p \Gamma^{pi}_k - \Gamma^{li}_p \Gamma^{pj}_k = \mathcal{O}(\lambda^2)$:

$$(\mathbf{A}, \mathbf{F}, \partial\mathbf{F}): \quad \begin{cases} \partial_a \mathbf{F}^{cb} = \partial_a \partial^b \mathbf{A}^c - \partial_a \partial^c \mathbf{A}^b, \\ \partial_a \mathbf{F}^{ca} = \partial^2 \mathbf{A}^c - \partial^c \partial_a \mathbf{A}^a, \end{cases}$$

$$(\mathbf{E}, \mathbf{\Gamma}, \mathcal{R}): \quad \begin{cases} \mathcal{R}^{cdb}_a = \partial^d \mathbf{\Gamma}^{cb}_a - \partial^b \mathbf{\Gamma}^{cd}_a, \\ \mathcal{R}^{cd}_\bullet = \partial^d \mathbf{\Gamma}^{ca}_a - \partial^a \mathbf{\Gamma}^{cd}_a \\ \qquad = \partial^2 \mathbf{E}^{cd} - \partial^c \partial_a \mathbf{E}^{da} - \partial^d \partial_a \mathbf{E}^{ca} + \partial^c \partial^d \mathbf{E}^\bullet, \\ \frac{1}{2} \mathcal{R}^\bullet_\bullet = \partial^2 \mathbf{E}^\bullet - \partial_c \partial_a \mathbf{E}^{ca}, \end{cases}$$

with the trace denoted by $\mathbf{E}^\bullet = \mathbf{E}^a_a$.

The Einstein tensor $\check{\mathcal{R}}_\bullet$ of the curvature, which, in four spacetime dimensions, coincides with the reflected Ricci tensor (see Chapter 2),

$$\mathcal{R}^{ab}_\bullet \leftrightarrow \check{\mathcal{R}}^{ab}_\bullet = \mathcal{R}^{ab}_\bullet - \frac{\eta^{ab}}{2} \mathcal{R}^\bullet_\bullet, \quad \check{\check{\mathcal{R}}}_\bullet = \mathcal{R}_\bullet,$$

involves derivatives only of the *reflected linearized tensor* (*Einstein combination* $\check{\mathbf{E}}$):

$$\begin{aligned} \partial_a \mathbf{F}^{ca} &= \partial^2 \mathbf{A}^c - \partial^c \partial_a \mathbf{A}^a, \\ \mathcal{R}^{cd}_\bullet &= \partial^2 \mathbf{E}^{cd} - (\delta^d_b \partial^c + \delta^c_b \partial^d) \partial_a \check{\mathbf{E}}^{ab}, \\ \check{\mathcal{R}}^{cd}_\bullet &= \partial^2 \check{\mathbf{E}}^{cd} - (\delta^d_b \partial^c + \delta^c_b \partial^d - \eta^{cd} \partial_b) \partial_a \check{\mathbf{E}}^{ab}, \\ \text{with } \check{\mathbf{E}}^{ab} &= \mathbf{E}^{ab} - \frac{\eta^{ab}}{2} \mathbf{E}^\bullet, \quad \check{\check{\mathbf{E}}} = \mathbf{E}. \end{aligned}$$

The field strength derivative in the Maxwell equations is determined by the electromagnetic current, and the Einstein tensor by the energy-momentum tensor:

$$\begin{aligned} \partial_a \mathbf{F}^{ca} &= -g^2 \mathbf{J}^c, \\ \check{\mathcal{R}}^{cd}_{\bullet} &= -\ell^2 \mathbf{T}^{cd}. \end{aligned}$$

In the lineariziation of $\check{\mathcal{R}}_{\bullet} = -\ell^2 \mathbf{T} - \Lambda \mathbf{g}$ no cosmological term arises if the constant is assumed of more than second-order $\Lambda = \mathcal{O}(\lambda^{2+k})$, $k \geq 0$.

The analogue to the Lagrangian of electrodynamics,

$$L_{\text{elmag}} = L(\mathbf{A}) + L_{\text{matter}} = \tfrac{1}{4g^2} \mathbf{F}_{ab} \mathbf{F}^{ab} - \mathbf{A}^c \mathbf{J}_c + L_{\text{matter}},$$

is the Lagrangian of gravity, linearized with $|\mathbf{g}| = -\det \mathbf{g} = 1 - 2\mathbf{E}^{\bullet}$, and a gravitational normalization ℓ^2 of second-order in the flat space expansion:,

$$\begin{aligned} L_{\text{gravity}} &= L(\mathbf{E}) + L_{\text{matter}} + \mathcal{O}(\lambda^2), \\ L(\mathbf{E}) &= \tfrac{1}{2\ell^2} \mathbf{E}^{\bullet} \mathcal{R}^{\bullet}_{\bullet} + \mathbf{E}_{cd} \mathbf{T}^{cd}, \quad \text{with} \left\{ \begin{array}{ll} \mathbf{E} &= \lambda \underline{\mathbf{E}} = \mathcal{O}(\lambda), \\ \ell^2 &= \mathcal{O}(\lambda^2), \ \Lambda = \mathcal{O}(\lambda^{2+k}). \end{array} \right. \end{aligned}$$

The Minkoswki spacetime expansion can be considered as an expansion in Newton's constant $\lambda^2 \sim \ell^2 = \frac{8\pi G \hbar}{c^3}$.

The linearized gravitative interaction is illustrated by a scalar field:

$$\begin{aligned} W^{\text{grav}}_{\text{matter}}(\mathbf{\Phi}) &= \int \sqrt{|\mathbf{g}|}\, d^4x\, \tfrac{1}{2} (\mathbf{g}^{kj} \partial_k \mathbf{\Phi} \partial_j \mathbf{\Phi} - m^2 \mathbf{\Phi}^2) \\ &= \int d^4x\, (1 - \mathbf{E}^{\bullet}) \tfrac{1}{2} [(\eta^{cd} + 2\mathbf{E}^{cd}) \partial_c \mathbf{\Phi} \partial_d \mathbf{\Phi} - m^2 \mathbf{\Phi}^2] + \mathcal{O}(\lambda^2), \end{aligned}$$

$$\begin{aligned} L(\mathbf{\Phi}) &= (1 - \mathbf{E}^{\bullet}) \tfrac{1}{2} (\partial^c \mathbf{\Phi} \partial_c \mathbf{\Phi} - m^2 \mathbf{\Phi}^2) + \mathbf{E}_{cd} \partial^c \mathbf{\Phi} \partial^d \mathbf{\Phi} \\ &= \tfrac{1}{2} (\partial^c \mathbf{\Phi} \partial_c \mathbf{\Phi} - m^2 \mathbf{\Phi}^2) + \mathbf{E}_{cd} \mathbf{T}^{cd}(\mathbf{\Phi}), \\ \text{with} \quad & \mathbf{T}^{cd}(\mathbf{\Phi}) = \partial^c \mathbf{\Phi} \partial^d \mathbf{\Phi} - \tfrac{\eta^{cd}}{2} (\partial_c \mathbf{\Phi} \partial^c \mathbf{\Phi} - m^2 \mathbf{\Phi}^2). \end{aligned}$$

The field equations are compared with those of electrodynamics for a charged scalar field:

$$\begin{aligned} \partial^2 \mathbf{\Phi} + (2\partial_a \mathbf{\Phi} \partial_b + 2\partial_a \partial_b \mathbf{\Phi})(\mathbf{E}^{ab} - \tfrac{\eta^{ab}}{2} \mathbf{E}^{\bullet}) &= -(1 - \mathbf{E}^{\bullet}) m^2 \mathbf{\Phi}, \\ \partial^2 \mathbf{\Phi} - 2iz \partial_a \mathbf{\Phi} \mathbf{A}^a - z^2 \mathbf{\Phi} \mathbf{A}_a \mathbf{A}^a &= -m^2 \mathbf{\Phi}. \end{aligned}$$

5.5 Free Particles for Flat Spacetime

Canonical quantum field theory works with free particle fields. They are used also as asymptotic fields for an expansion of interactions. A free scalar quantum field has the Lorentz transformation behavior

$$\Lambda \in \mathbf{SO}_0(1,3): \quad \mathbf{\Phi} \stackrel{\Lambda}{\longmapsto} \mathbf{\Phi}_\Lambda, \text{with } \mathbf{\Phi}_\Lambda(x) = \mathbf{\Phi}(\Lambda^{-1}.x).$$

The particle mass is the translation invariant in the flat spacetime equation:

$$\mathbf{g} = \eta: \quad \left\{ \begin{array}{ll} \partial_a \partial^a \mathbf{\Phi} &= -m^2 \mathbf{\Phi}, \\ \mathbf{T}^{ab}(\mathbf{\Phi}) &= \partial^a \mathbf{\Phi} \partial^b \mathbf{\Phi} - \tfrac{\eta^{ab}}{2} (\partial_c \mathbf{\Phi} \partial^c \mathbf{\Phi} - m^2 \mathbf{\Phi}^2). \end{array} \right.$$

The Feynman propagator with a normalization factor $\rho(m^2) \neq 0$,

$$\langle 0|\mathbf{\Phi}(y)\mathbf{\Phi}(x)|0\rangle_{\text{Feynman}} = \frac{i}{\pi}\int \frac{d^4q}{(2\pi)^3}\frac{\rho(m^2)}{q^2+io-m^2}e^{iq(x-y)},$$

involves the on-shell contribution (particles) from the Dirac distribution in the decomposition

$$\frac{i}{\pi}\frac{1}{q^2+io-m^2} = \delta(q^2-m^2) + \frac{i}{\pi}\frac{1}{q_{\rm P}^2-m^2}.$$

This contribution with the expectation value $\langle 0|\dots|0\rangle$ of the anticommutator and the *Fock ground-state vector* $|0\rangle$ *for translation eigenvectors* is a *representation matrix element of the Poincaré group* $\mathbf{SO}_0(1,3) \vec{\times} \mathbb{R}^4$ (see Chapter 8), for a scalar field nontrivial for the translations $\mathbb{R}^4$ only:

$$\mathbb{R}^4 \ni x-y \longmapsto \langle 0|\{\mathbf{\Phi}(y),\mathbf{\Phi}(x)\}|0\rangle = \rho(m^2)\int \frac{d^4q}{(2\pi)^3}\delta(q^2-m^2)e^{iq(x-y)}.$$

The off-shell contribution ("virtual particles") with the principal value part,

$$\begin{aligned}-\epsilon(x_0-y_0)\langle 0|[\mathbf{\Phi}(y),\mathbf{\Phi}(x)]|0\rangle &= \frac{i}{\pi}\int \frac{d^4q}{(2\pi)^3}\frac{\rho(m^2)}{q_{\rm P}^2-m^2}e^{iq(x-y)}\\ &= 0 \text{ for } (x-y)^2<0,\end{aligned}$$

is not a coefficient of a Poincaré group representation. It is a distribution, arising by causal ordering from the causally supported, i.e., $x \in \mathbb{R}^4_+ \cup \mathbb{R}^4_-$, on-shell quantization of the free scalar field via the commutator:

$$\begin{aligned}[\mathbf{\Phi}(y),\mathbf{\Phi}(x)] &= \rho(m^2)\int \frac{d^4q}{(2\pi)^3}\epsilon(q_0)\delta(q^2-m^2)e^{iq(x-y)}\\ &= -\epsilon(x_0-y_0)\frac{i}{\pi}\int \frac{d^4q}{(2\pi)^3}\frac{\rho(m^2)}{q_{\rm P}^2-m^2}e^{iq(x-y)},\\ [i\partial^a\mathbf{\Phi}(y),\mathbf{\Phi}(x)]\Big|_{x_0=y_0} &= \rho(m^2)\int \frac{d^4q}{(2\pi)^3}\epsilon(q_0)q^a\delta(q^2-m^2)e^{-i\vec{q}(\vec{x}-\vec{y})}\\ &= \rho(m^2)\delta^a_0\delta(\vec{x}-\vec{y}).\end{aligned}$$

For free fields, the normalization $\rho(m^2) \neq 0$ can be chosen arbitrarily, e.g., $\rho(m^2)=1$ for a position translation normalization with $\int d^3x\ \delta(\vec{x})=1$.

For a particle field, the time projection, effected by position integration, gives as real part a harmonic oscillator cosine, i.e., a time representation coefficient with the mass as invariant. The position projection, effected by time integration, is nontrivial for the off-shell contribution (imaginary part) only. It gives as position representation coefficient a Yukawa potential with the inverse mass as its range:

$$\begin{aligned}\int d^3x \int \frac{d^4q}{i\pi(2\pi)^3}\frac{1}{-q^2-io+m^2}e^{iqx} &= \frac{\cos mx_0-\epsilon(x_0)i\sin|m|x_0}{|m|} = \frac{e^{-i|mx_0|}}{|m|},\\ \int dx_0 \int \frac{d^4q}{i\pi(2\pi)^3}\frac{1}{-q^2-io+m^2}e^{iqx} &= \int dx_0 \int \frac{d^4q}{i\pi(2\pi)^3}\frac{1}{-q_{\rm P}^2+m^2}e^{iqx}\\ &= \int \frac{d^3q}{4i\pi^3}\frac{1}{\vec{q}^2+m^2}e^{-i\vec{q}\vec{x}} = -i\frac{e^{-|m|r}}{2\pi r}.\end{aligned}$$

The position dependence of the Yukawa potential comes from the time projection of a causally supported $x^2 \geq 0$ distribution, not from a spacelike $x^2<0$ dependence.

The on-shell function for particles fulfills a homogeneous equation for the translation invariant of the Poincaré group representation $\partial^2 = -m^2$, whereas the off-shell distribution for "virtual particles" and interactions is a Green's kernel of the corresponding inhomogeneous equation,

$$(\partial^2 + m^2) \int \frac{d^4q}{(2\pi)^3} \begin{pmatrix} \delta(q^2 - m^2) \\ \frac{i}{\pi} \frac{1}{q_{\mathrm{P}}^2 - m^2} \end{pmatrix} e^{iqx} = \begin{pmatrix} 0 \\ -2i\delta(x) \end{pmatrix}.$$

A free-field Feynman propagator has three parts: In addition to the on-shell contribution with the spherical Bessel function $\frac{\sin|\vec{q}|r}{r}$, there is the causally supported off-shell contribution for energies under the mass threshold with the Yukawa potential $\frac{e^{-|Q|r}}{r}$ and for energies over the mass threshold with the spherical Neumann function $\frac{\cos|\vec{q}|r}{r}$, both singular at $r = 0$:

$$\begin{aligned} &\int \frac{d^4q}{(2\pi)^3} \begin{pmatrix} \delta(q^2 - m^2) \\ \frac{i}{\pi} \frac{1}{q_{\mathrm{P}}^2 - m^2} \end{pmatrix} e^{iqx} \\ &= \int \frac{dq_0}{(2\pi)^2} \left[\vartheta(q_0^2 - m^2) \begin{pmatrix} \frac{\sin|\vec{q}|r}{r} \\ -i\frac{\cos|\vec{q}|r}{r} \end{pmatrix} + \vartheta(m^2 - q_0^2) \begin{pmatrix} 0 \\ -i\frac{e^{-|Q|r}}{r} \end{pmatrix}\right] e^{iq_0x_0}, \\ &\quad \text{with } |\vec{q}| = \sqrt{q_0^2 - m^2} \text{ and } |Q| = \sqrt{m^2 - q_0^2}. \end{aligned}$$

The harmonic expansion of a free field, i.e., its *particle analysis*, uses creation and annihilation eigenoperators $(\mathrm{u}(\vec{q}), \mathrm{u}^\star(\vec{q}))$ of the spacetime translations in the direct integral decomposition,

$$\mathbf{\Phi}(x) = {}^{\oplus}\!\!\int \frac{d^3q}{2q_0(2\pi)^3} \, [e^{iqx}\mathrm{u}(\vec{q}) + e^{-iqx}\mathrm{u}^\star(\vec{q})], \text{ with } q_0 = \sqrt{\vec{q}^2 + m^2},$$

with the quantization and the Hilbert space metric involving the "inverse" of the Lorentz invariant momentum measure $2(2\pi)^3\sqrt{\vec{q}^2 + m^2}\delta(\vec{q}) \leftrightarrow \frac{d^3q}{2(2\pi)^3\sqrt{\vec{q}^2+m^2}}$,

$$\begin{aligned} [\mathrm{u}^\star(\vec{p}), \mathrm{u}(\vec{q})] &= \rho(m^2)2q_0\delta(\tfrac{\vec{q}-\vec{p}}{2\pi}), \\ \langle 0|\{\mathrm{u}^\star(\vec{p}), \mathrm{u}(\vec{q})\}|0\rangle = \langle 0|\mathrm{u}^\star(\vec{p})\mathrm{u}(\vec{q})|0\rangle = \langle m^2; \vec{p}|m^2; \vec{q}\rangle &= \rho(m^2)2q_0\delta(\tfrac{\vec{q}-\vec{p}}{2\pi}), \\ \text{distributive Hilbert basis: } \{|m^2; \vec{q}\rangle = \mathrm{u}(\vec{q})|0\rangle \,\big|\, \vec{q} \in \mathbb{R}^3\}. \end{aligned}$$

The field action on the Fock ground state gives cyclic vectors (see Chapter 8),

$$\mathbf{\Phi}(x)|0\rangle = {}^{\oplus}\!\!\int \frac{d^3q}{2q_0(2\pi)^3} \, e^{iqx}|m^2; \vec{q}\rangle, \quad \langle 0|\mathbf{\Phi}(x) = {}^{\oplus}\!\!\int \frac{d^3q}{2q_0(2\pi)^3} \, e^{-iqx}\langle m^2; \vec{q}|.$$

Corresponding structures occur for a Dirac field, $\mathbf{\Psi} = \mathbf{r} \oplus \mathbf{l}$, acted on by a right-left decomposable $\mathbf{SL}(2,\mathbb{C})$-representation $[\frac{1}{2}|0] \oplus [0|\frac{1}{2}]$ (ahead), as defined by a commutative diagram,

$$s \in \mathbf{SL}(2,\mathbb{C}): \quad \begin{array}{ccc} \mathbb{R}^4 & \xrightarrow{\mathbf{\Psi}} & \mathbb{C}^4 \\ {\scriptstyle \Lambda(s)}\downarrow & & \downarrow{\scriptstyle \mathbf{s}} \\ \mathbb{R}^4 & \xrightarrow[\mathbf{\Psi}_s]{} & \mathbb{C}^4 \end{array}, \quad \begin{array}{l} \mathbf{\Psi} \overset{s}{\longmapsto} \mathbf{\Psi}_s, \\ \text{with } \mathbf{\Psi}_s(x) = \mathbf{s}.\mathbf{\Psi}(\Lambda^{-1}(s).x), \end{array}$$

$$\begin{aligned} &\text{with } s \longmapsto \mathbf{s} = \begin{pmatrix} s & 0 \\ 0 & \hat{s} \end{pmatrix} \in \mathbf{SL}(2,\mathbb{C}) \oplus \mathbf{SL}(2,\mathbb{C}), \quad \hat{s} = s^{-1*}, \\ &s \longmapsto \Lambda(s) \in \mathbf{SO}_0(1,3), \\ &\Lambda_a^b(s) = \tfrac{1}{2}\operatorname{tr}\sigma_a s \check{\sigma}^b s^*, \text{ Weyl matrices: } \sigma_a = (\mathbf{1}_2, \vec{\sigma}), \ \check{\sigma}_a = (\mathbf{1}_2, -\vec{\sigma}), \end{aligned}$$

and $\mathbf{SU}(2)$ rotation properties for its spin-$\frac{1}{2}$ particles. The flat spacetime equation with translation invariant mass and the Feynman propagator are

$$\mathbf{g} = \eta: \begin{cases} \gamma_c \partial^c \mathbf{\Psi} &= im\mathbf{\Psi}, \\ \mathbf{T}_a^b(\mathbf{\Psi}) &= i\overline{\mathbf{\Psi}}\gamma_a\partial^b\mathbf{\Psi} - \delta_a^b i(\overline{\mathbf{\Psi}}\gamma_c\partial^c\mathbf{\Psi} - im\overline{\mathbf{\Psi}}\mathbf{\Psi}), \\ \langle 0|\mathbf{\Psi}(y)\overline{\mathbf{\Psi}}(x)|0\rangle_{\text{Feynman}} &= \rho(m^2)\frac{i}{\pi}\int \frac{d^4q}{(2\pi)^3}\frac{\gamma_a q^a + m}{q^2 + io - m^2} e^{iq(x-y)} \\ = \langle 0|[\mathbf{\Psi}(y), \overline{\mathbf{\Psi}}(x)]|0\rangle &- \epsilon(x_0 - y_0)\langle 0|\{\mathbf{\Psi}(y), \overline{\mathbf{\Psi}}(x)\}|0\rangle. \end{cases}$$

The harmonic analysis of the Dirac spinor field with left and right irreducible Weyl spinor contributions involves translation and rotation eigenoperators for particles $(\mathrm{u}^A(\vec{q}), \mathrm{u}_A^\star(\vec{q}))$, e.g., negatively charged, $z = -1$, and antiparticles $(\mathrm{a}_A(\vec{q}), \mathrm{a}^{A\star}(\vec{q}))$, then positively charged, $z = +1$, with spin direction (eigenvalue) "up" and "down," $A = 1, 2$:

$$\mathbf{\Psi}(x) \cong \begin{pmatrix} \mathbf{r}^A \\ \mathbf{l}^{\dot{A}} \end{pmatrix}(x) = \sqrt{m}\,{}^{\oplus}\!\!\int \frac{d^3q}{2q_0(2\pi)^3} \begin{pmatrix} s^A_C(\frac{q}{m})\ [e^{iqx}\mathrm{u}^C(\vec{q}) + e^{-iqx}\mathrm{a}^{\star C}(\vec{q})] \\ \hat{s}^{\dot{A}}_C(\frac{q}{m})\ [e^{iqx}\mathrm{u}^C(\vec{q}) - e^{-iqx}\mathrm{a}^{\star C}(\vec{q})] \end{pmatrix},$$

$$\begin{aligned} \mathbf{\Psi}^*(x) \cong \begin{pmatrix} \mathbf{r}^\star_{\dot{A}} \\ \mathbf{l}^\star_{A} \end{pmatrix}(x) &= \sqrt{m}\,{}^{\oplus}\!\!\int \frac{d^3q}{2q_0(2\pi)^3} \begin{pmatrix} s^{-1}{}^C_{\dot{A}}(\frac{q}{m})\ [e^{-iqx}\mathrm{u}^\star_C(\vec{q}) + e^{iqx}\mathrm{a}_C(\vec{q})] \\ \hat{s}^{-1}{}^C_{A}(\frac{q}{m})\ [e^{-iqx}\mathrm{u}^\star_C(\vec{q}) - e^{iqx}\mathrm{a}_C(\vec{q})] \end{pmatrix}, \\ \text{with } q_0 &= \sqrt{\vec{q}^2 + m^2}. \end{aligned}$$

$s(\frac{q}{m})$ and $\hat{s}(\frac{q}{m})$ are the momentum dependent boost representations (Weyl transmutators; see Chapter 7), which embed the spin group $\mathbf{SU}(2)$ into the Lorentz group $\mathbf{SL}(2,\mathbb{C})$.

The particle operators have the quantization and Hilbert metric:

$$\begin{aligned} \{\mathrm{u}^\star_A(\vec{p}), \mathrm{u}^B(\vec{q})\} &= \rho(m^2)2q_0\delta^B_A\delta(\tfrac{\vec{q}-\vec{p}}{2\pi}), \\ \langle 0|\mathrm{u}^\star_A(\vec{p})\mathrm{u}^B(\vec{q})|0\rangle = \langle m^2, \tfrac{1}{2}; -1, \vec{p}, A|m^2, \tfrac{1}{2}; -1, \vec{q}, B\rangle &= \rho(m^2)2q_0\delta^{AB}\delta(\tfrac{\vec{q}-\vec{p}}{2\pi}), \end{aligned}$$

$$\begin{aligned} \{\mathrm{a}^{B\star}(\vec{p}), \mathrm{a}_A(\vec{q})\} &= \rho(m^2)2q_0\delta^B_A\delta(\tfrac{\vec{q}-\vec{p}}{2\pi}), \\ \langle 0|\mathrm{a}^{B\star}(\vec{p})\mathrm{a}_A(\vec{q})|0\rangle = \langle m^2, \tfrac{1}{2}; +1, \vec{p}, B|m^2, \tfrac{1}{2}; +1, \vec{q}, A\rangle &= \rho(m^2)2q_0\delta_{AB}\delta(\tfrac{\vec{q}-\vec{p}}{2\pi}), \end{aligned}$$

$$\text{distributive Hilbert basis: } \{|m^2, \tfrac{1}{2}; z, \vec{q}, A\rangle \mid z = \pm 1, \ \vec{q} \in \mathbb{R}^3, \ A = 1, 2\}.$$

The Hilbert metric involves the $\mathbf{SU}(2)$-scalar products δ^{AB} and δ_{AB} for the two spin degrees of freedom.

In general, the finite-dimensional irreducible representations $[L|R]$ of the Lorentz cover group $\mathbf{SL}(2,\mathbb{C})$, in parallel to its compact partner $\mathbf{SU}(2)\times\mathbf{SU}(2)$, have complex dimensions $(1+2L)(1+2R)$:

$$\mathbf{irrep}_{\,\mathrm{fin}}\mathbf{SL}(2,\mathbb{C}) = \{[L|R] \bigm| L,R=0,\tfrac{1}{2},1,\dots\}.$$

They are indefinite unitary for $[L|R]\neq[0|0]$. In a rest system for a massive particle, the Lorentz group representations are decomposable into irreducible $(1+2J)$-dimensional $\mathbf{SU}(2)$-representations $[J]$:

$$[L|R] \overset{\mathbf{SU}(2)}{\cong} \bigoplus_{J=|L-R|}^{J=L+R} [J].$$

The subset with the representations, faithful only for the proper Lorentz group $\mathbf{SO}_0(1,3)\cong\mathbf{SL}(2,\mathbb{C})/\mathbb{I}(2)$, has integer spins $\mathbf{SO}(3)\cong\mathbf{SU}(2)/\mathbb{I}(2)$, in analogy to the central correlation for the $\mathbf{SO}(4)$-representations (see Chapter 4):

$$\mathbf{irrep}_{\,\mathrm{fin}}\mathbf{SO}_0(1,3) = \{[L|R] \bigm| L,R=0,\tfrac{1}{2},\dots;\quad L+R=0,1,\dots\}.$$

A four-component vector field is acted on by a Minkowski representation $[\frac{1}{2}|\frac{1}{2}]$. Its totally symmetric products contain the irreducible Minkowski representations, which are the *harmonic* $\mathbf{SO}_0(1,3)$*-representations* (see Chapter 8). They are the noncompact partners of the Kepler (harmonic) $\mathbf{SO}(4)$-representations,

$$\mathbf{irrep}_{\,\mathrm{Mink}}\mathbf{SO}_0(1,3) = \{[J|J] \bigm| J=0,\tfrac{1}{2},\dots\},$$

$$\bigvee^{2J}[\tfrac{1}{2}|\tfrac{1}{2}] = \begin{cases} \displaystyle\bigoplus_{L=0,2,\dots,2J} [\tfrac{L}{2}|\tfrac{L}{2}], & 2J=0,2,\dots,\\ \displaystyle\bigoplus_{L=1,3,\dots,2J} [\tfrac{L}{2}|\tfrac{L}{2}], & 2J=1,3,\dots. \end{cases}$$

5.6 Massive Particles with Spin 1 and Spin 2

The Lorentz-compatible projectors $\mathcal{S}(q)$ and $\mathcal{V}(q)$ from a Lorentz vector $q\in\mathbb{R}^4$ to its spin $J=0$ and spin $J=1$ contributions, respectively, give a time-space decomposition in a rest system of a massive particle $q^2=m^2>0$,

$$\begin{aligned}
&[\tfrac{1}{2}|\tfrac{1}{2}] \overset{\mathbf{SO}(3)}{\cong} [0]\oplus[1],\quad \mathbf{1}_4=\mathcal{S}+\mathcal{V},\quad \mathcal{S}\circ\mathcal{V}=0,\\
&\quad\mathcal{S}\cong\mathcal{S}^c_a=\tfrac{q^cq_a}{q^2} \overset{q_0=m}{=} \left(\begin{array}{c|c}1&0\\\hline 0&0_3\end{array}\right),\quad \mathcal{S}.q=q,\quad \mathcal{S}\circ\mathcal{S}=\mathcal{S},\quad \mathcal{S}^c_c=1,\\
&\quad\mathcal{V}\cong\mathcal{V}^c_a=\delta^c_a-\tfrac{q^cq_a}{q^2} \overset{q_0=m}{=} \left(\begin{array}{c|c}0&0\\\hline 0&\mathbf{1}_3\end{array}\right),\quad \mathcal{V}.q=0,\quad \mathcal{V}\circ\mathcal{V}=\mathcal{V},\quad \mathcal{V}^c_c=3.
\end{aligned}$$

The projectors arise as products of the Minkowski boost representations $[\frac{1}{2}|\frac{1}{2}]$, parametrized by energy-momenta:

$$\Lambda(\frac{q}{m}) \cong \frac{1}{m}\begin{pmatrix} q_0 & q_\beta \\ q_\alpha & \delta_{\alpha\beta}m + \frac{q_\alpha q_\beta}{m+q_0}\end{pmatrix}, \quad \alpha,\beta = 1,2,3,$$
$$\Rightarrow \Lambda^c_\gamma(\frac{q}{m})\delta^{\gamma\alpha}\Lambda^a_\alpha(\frac{q}{m}) = -\eta^{ca} + \frac{q^c q^a}{m^2} = -\mathcal{V}^{ca}, \quad \Lambda^c_0(\frac{q}{m})\Lambda^a_0(\frac{q}{m}) = \frac{q^c q^a}{m^2} = \mathcal{S}^{ca}.$$

For higher-order Lorentz group representations, the projectors to rotation group representations can be combined by $\mathcal{S}$ and $\mathcal{V}$, e.g., the projector $\mathcal{S}^{cd}\mathcal{S}_{cd} = 1$ from $[0|0]$ to spin 0 and the projector $\mathcal{P}_2$ from $[1|1]$ to spin 2, where the spin-0 contribution has to be subtracted from the symmetric $\mathbf{SO}_0(1,3)$-combination $\mathcal{V} \vee \mathcal{V} = [1] \vee [1] = [0] \oplus [2]$:

$$[1|1] \overset{\mathbf{SO}(3)}{\cong} [0] \oplus [1] \oplus [2],$$
$$\mathbf{1}_9 = \mathcal{P}_0 + \mathcal{P}_1 + \mathcal{P}_2, \quad \begin{cases} (\mathcal{P}_0)^{cd}_{ab} &= \frac{1}{3}\mathcal{V}^{cd}\mathcal{V}_{ab}, \\ (\mathcal{P}_2)^{cd}_{ab} &= \frac{\mathcal{V}^c_a\mathcal{V}^d_b + \mathcal{V}^d_a\mathcal{V}^c_b}{2} - \frac{1}{3}\mathcal{V}^{cd}\mathcal{V}_{ab}, \\ (\mathcal{P}_1)^{cd}_{ab} &= \frac{\mathcal{V}^c_a\mathcal{V}^d_b - \mathcal{V}^d_a\mathcal{V}^c_b}{2}, \\ (\mathcal{P}_A)^{cd}_{ab}(\mathcal{P}_B)^{ab}_{ef} &= \delta_{AB}(\mathcal{P}_A)^{cd}_{ef}, \end{cases}$$

$$\begin{pmatrix} \mathcal{P}_0^{cd,ab} \\ \mathcal{P}_2^{cd,ab} \\ \mathcal{P}_1^{cd,ab} \end{pmatrix} = \Lambda^c_\gamma(\frac{q}{m})\Lambda^d_\delta(\frac{q}{m})\begin{pmatrix} \frac{1}{3}\delta^{\gamma\delta}\delta^{\alpha\beta} \\ \frac{\delta^{\gamma\alpha}\delta^{\delta\beta}+\delta^{\gamma\beta}\delta^{\delta\alpha}}{2} - \frac{1}{3}\delta^{\gamma\delta}\delta^{\alpha\beta} \\ \frac{\delta^{\gamma\alpha}\delta^{\delta\beta}-\delta^{\gamma\beta}\delta^{\delta\alpha}}{2}\end{pmatrix}\Lambda^a_\alpha(\frac{q}{m})\Lambda^b_\beta(\frac{q}{m}).$$

For a massive spin-1 and spin-2 particle the Lorentz group covariant equations (with $q \cong i\partial$ in the projectors) for the embedding fields $\mathbf{Z}$ and $\mathbf{Y}$ involve the transversality conditions from $\mathcal{V}.q = 0$:

$$\begin{aligned} \mathcal{V}^c_a\partial^2\ \mathbf{Z}^a &= -m_Z^2\mathbf{Z}^c &\Rightarrow \partial_c\mathbf{Z}^c = 0, \\ (\mathcal{P}_2)^{cd}_{ab}\partial^2\mathbf{Y}^{ab} &= -m_Y^2\mathbf{Y}^{cd} &\Rightarrow \partial_c\mathbf{Y}^{cd} = 0. \end{aligned}$$

The $\mathbf{Z}$-boson mass, neglecting its width, is given in proton mass units by $\frac{m_Z}{m_p} \sim 97$, $\log\frac{m_Z^2}{m_p^2} \sim 9.1$.

In the Feynman propagators for the massive particles with normalizations $\rho(m^2) \neq 0$,

$$\begin{aligned} \langle 0|\mathbf{Z}^c(y)\mathbf{Z}^a(x)|0\rangle_{\text{Feynman}} &= \rho(m_Z^2)\frac{i}{\pi}\int \frac{d^4q}{(2\pi)^3}\frac{-\mathcal{V}^{ca}}{q^2+io-m_Z^2}e^{iq(x-y)}, \\ \langle 0|\mathbf{Y}^{cd}(y)\mathbf{Y}^{ab}(x)|0\rangle_{\text{Feynman}} &= \rho(m_Y^2)\frac{i}{\pi}\int \frac{d^4q}{(2\pi)^3}\frac{(\mathcal{P}_2)^{cd,ab}}{q^2+io-m_Y^2}e^{iq(x-y)}, \end{aligned}$$

the on-shell contribution from the Dirac distribution is a representation matrix element of the Poincaré group for a massive particle with fixgroup $\mathbf{SO}(3)$ for the energy-momenta with $q^2 = m^2 > 0$. To avoid long range interactions via massless fields, the on-shell projectors with the mass are also used for the off-shell contribution, e.g., $\mathcal{V}^c_a = \delta^c_a - \frac{q^c q_a}{m_Z^2}$, not $\delta^c_a - \frac{q^c q_a}{q^2}$.

The projectors in the numerators, $\mathcal{V}^a_c$ for spin-1 particles and $(\mathcal{P}_2)^{ab}_{cd}$ for spin-2 particles, embed, in the respective forms $-\mathcal{V}^{ca}$ and $(\mathcal{P}_2)^{cd,ab}$, the

Hilbert metric for the particle degrees of freedom. The Lorentz invariant indefinite metric is used for the metrical tensor of the spin degrees of freedom (subindex $\pm$). The vector representation has $\text{sign}\,(-\eta) = (3,1)$ and the tensor representation $\text{sign}\,\eta \vee \eta = (7,3)$:

$$\text{metric of } [\tfrac{1}{2}|\tfrac{1}{2}]: \quad \begin{cases} -\eta = \begin{pmatrix} -1 & 0 \\ 0 & \mathbf{1}_3 \end{pmatrix}, \\ [\tfrac{1}{2}|\tfrac{1}{2}] \overset{\mathbf{SO}(3)}{\cong} [0]_+ \oplus [1]_-, \end{cases}$$

$$\text{metric of } [\tfrac{1}{2}|\tfrac{1}{2}] \vee [\tfrac{1}{2}|\tfrac{1}{2}]: \quad \begin{cases} \eta \vee \eta = \begin{pmatrix} 1 & 0 & 0 & 0 \\ 0 & 1 & 0 & 0 \\ 0 & 0 & -\mathbf{1}_3 & 0 \\ 0 & 0 & 0 & \mathbf{1}_5 \end{pmatrix}. \\ [0|0] \oplus [1|1] \overset{\mathbf{SO}(3)}{\cong} 2 \times [0]_+ \oplus [1]_- \oplus [2]_+. \end{cases}$$

They contain the definite metrical subtensors $\mathbf{1}_{1+2J}$, projected to for massive spin-1 particles, $\mathbf{1}_3$, and for massive spin-2 particles, $\mathbf{1}_5$, enclosed by double lines $\begin{pmatrix} & \\ & \times \end{pmatrix}$.

In general, the metrical structure of the Lorentz group $[J|J]$-representation is given by the signature of indefinite orthogonal groups:

$$\mathbf{SO}_0(1,3) \xrightarrow{[J|J]} \begin{cases} \mathbf{SO}_0(t,s), & J = 0, 1, \dots, \\ \mathbf{SO}_0(s,t), & J = \frac{1}{2}, \frac{3}{2}, \dots, \end{cases}$$

$$\text{with } (t,s) = (\tbinom{2+2J}{2}, \tbinom{1+2J}{2}), \quad t+s = (1+2J)^2, \ t-s = 1+2J.$$

The maximal compact subgroup $\mathbf{SO}(\binom{2+2J}{2}) \times \mathbf{SO}(\binom{1+2J}{2})$ contains, for integer J, the rotation $\mathbf{SO}(3)$-representations $\bigoplus_{L=0,2,\dots}^{2J} [L]$ (positive definite) and $\bigoplus_{L=1,3,\dots}^{2J-1} [L]$ (negative definite), and, with opposite association, for half-integer J.

The spin-1 particle field has a harmonic expansion with translation and rotation eigenoperators for energy-momenta q and spin 1 with three directions in the quantization and Hilbert metric:

$$\begin{aligned} \mathbf{Z}^a(x) &= {}^{\oplus}\!\!\int \frac{d^3q}{2q_0(2\pi)^3} \Lambda^a_\alpha(\tfrac{q}{m_Z})[e^{iqx}\mathrm{u}^\alpha(\vec{q}) + e^{-iqx}\mathrm{u}^\star_\alpha(\vec{q})], \\ [\mathrm{u}^\star_\alpha(\vec{p}), \mathrm{u}^\beta(\vec{q})] &= \rho(m_Z^2) 2q_0 \delta^\beta_\alpha \delta(\tfrac{\vec{q}-\vec{p}}{2\pi}), \text{ with } q_0 = \sqrt{m_Z^2 + \vec{q}^2}, \\ \langle 0|\mathrm{u}^\star_\alpha(\vec{p})\mathrm{u}^\beta(\vec{q})|0\rangle &= \langle m_Z^2, 1; \vec{p}, \alpha | m_Z^2, 1; \vec{q}, \beta\rangle = \rho(m_Z^2) 2q_0 \delta^{\alpha\beta} \delta(\tfrac{\vec{q}-\vec{p}}{2\pi}), \\ \text{distributive Hilbert basis: } & \{|m_Z^2, 1; \vec{q}, \alpha\rangle \mid \vec{q} \in \mathbb{R}^3, \ \alpha = 1,2,3\}. \end{aligned}$$

The spin-2 particle field has a similar harmonic expansion for the particles with five spin directions:

$$\begin{aligned} \mathbf{Y}^{ab}(x) = {}& {}^{\oplus}\!\!\int \frac{d^3q}{2q_0(2\pi)^3} \Lambda^a_\alpha(\tfrac{q}{m_Y}) \Lambda^b_\beta(\tfrac{q}{m_Y}) \\ & \times (\tfrac{\delta^{\gamma\alpha}\delta^{\delta\beta} + \delta^{\gamma\beta}\delta^{\delta\alpha}}{2} - \tfrac{1}{3}\delta^{\gamma\delta}\delta^{\alpha\beta})[e^{iqx}\mathrm{u}^{\gamma\delta}(\vec{q}) + e^{-iqx}\mathrm{u}^\star_{\gamma\delta}(\vec{q})], \\ \text{distributive Hilbert basis: } & \{|m_Y^2, 2; \vec{q}, A\rangle \mid \vec{q} \in \mathbb{R}^3, \ A = \pm 2, \pm 1, 0\}. \end{aligned}$$

5.7 Massless Polarized Photons and Gravitons

With Wigner, particles — here, polarized massless photons and gravitons for the electromagnetic and gravitational field, respectively — are defined for flat Minkowski spacetime only.

Massless particles have no rest system. Projectors like $\mathcal{V} \cong \delta^c_a - \frac{q^c q_a}{q^2}$ for spin 1 and $\mathcal{S} \cong \frac{q^c q_a}{q^2}$ for spin 0 are not defined for $q^2 = 0$. With the fixgroup $\mathbf{SO}(2) \vec{\times} \mathbb{R}^2 \subset \mathbf{SO}_0(1,3)$ of the nontrivial energy-momenta with $q^2 = 0$, and the fixgroup in the fixgroup $\mathbf{SO}(2) \subset \mathbf{SO}(2) \vec{\times} \mathbb{R}^2$, massless particles have no spin $\mathbf{SO}(3)$, they have only polarization $\mathbf{SO}(2)$ around the flight direction, e.g., polarization (± 1) for the photons and polarization (± 2) for gravitons. The four-component Lorentz vector for the electromagnetic field decomposes into an $\mathbf{SO}(2)$-dublet (± 1) for the two polarized particle degrees of freedom and two $\mathbf{SO}(2)$-singlets for the Coulomb and gauge degree of freedom. The 10-component symmetric Lorentz tensor for the flat spacetime gravitative field contains an irreducible Lorentz group nonet $[1|1]$ and singlet $[0|0]$ and decomposes with respect to polarization $\mathbf{SO}(2)$ into two particle degrees of freedom (± 2) and eight nonparticle degrees of freedom as follows:

$$\begin{array}{llcl} \mathbf{A}^a: & [\frac{1}{2}|\frac{1}{2}] & \overset{\mathbf{SO}(2)}{\cong} & (\pm 1) \oplus 2 \times (0), \\ \mathbf{E}^{ab}: & [1|1] \oplus [0|0] & \overset{\mathbf{SO}(2)}{\cong} & (\pm 2) \oplus 2 \times (\pm 1) \oplus 4 \times (0). \end{array}$$

The free equation for the electromagnetic vector field contains the ∂^2-multiplied spin-1 projector,

$$\partial_a \mathbf{F}^{ca} = (\delta^c_a \partial^2 - \partial^c \partial_a)\mathbf{A}^a = \mathcal{V}^c_a \partial^2 \mathbf{A}^a = 0,$$

whereas the free equation for the Einstein tensor field does not contain pure projectors, but instead contains the following combination of the derivatives (spin-2 and spin-0 projectors for massive particles):

$$\begin{aligned} \check{\mathcal{R}}^{cd}_{\bullet} &= [\delta^c_a \delta^d_b \partial^2 - \delta^d_b \partial^c \partial_a - \delta^c_b \partial^d \partial_a - \eta^{cd}\eta_{ab}\partial^2 + \eta^{cd}\partial_b\partial_a + \eta_{ab}\partial^c\partial^d]\mathbf{E}^{ab} \\ &= (\tfrac{\mathcal{V}^c_a \mathcal{V}^d_b + \mathcal{V}^d_a \mathcal{V}^c_b}{2} - \mathcal{V}^{cd}\mathcal{V}_{ab})\partial^2 \mathbf{E}^{ab} \\ &= [(\mathcal{P}_2)^{cd}_{ab} - 2(\mathcal{P}_0)^{cd}_{ab}]\partial^2 \mathbf{E}^{ab} = 0. \end{aligned}$$

With the divergenceless antisymmetric field strengths and the divergenceless Einstein tensor,

$$\begin{aligned} \partial_c \partial_a \mathbf{F}^{ca} &= 0, \\ \partial_c \check{\mathcal{R}}^{cd}_{\bullet} &= 0, \end{aligned}$$

the vector and tensor fields for massless particles are determined up to the divergences of a scalar and a vector field, respectively:

$$\mathbf{A}^c \longmapsto \mathbf{A}^c + \partial^c \alpha \Rightarrow \mathbf{F}^{cb} \longmapsto \mathbf{F}^{cb},$$

$$\left.\begin{array}{l} \mathbf{E}^{cd} \longmapsto \mathbf{E}^{cd} + \partial^c \xi^d + \partial^d \xi^c, \\ \mathbf{\Gamma}^{cd}_b \longmapsto \mathbf{\Gamma}^{cd}_b + 2\partial^c \partial^d \xi_b \end{array}\right\} \Rightarrow \check{\mathcal{R}}^{cd}_{\bullet} \longmapsto \check{\mathcal{R}}^{cd}_{\bullet}.$$

The electromagnetic current and the flat spacetime energy-momentum tensor are conserved, $\partial_c \mathbf{J}^c = 0$ and $\partial_c \mathbf{T}^{cd} = 0$.

The freedom in the definition of the polarization (± 1)-embedding Lorentz vector field $\mathbf{A}$ is connected with the $\mathbf{U}(1)$-gauge transformation of the electromagnetic field. The corresponding freedom in the definition of the polarization (± 2)-embedding Lorentz tensor field $\mathbf{E}$ in the tetrad expansion $\mathbf{e} = \mathbf{1} + \mathbf{E}$ can be related to the reparametrizations of the spacetime manifold:

$$x_b \longmapsto \overline{x}_b(x) = x_b + \lambda \xi_b(x), \quad \mathbf{e}^c_b = \partial^c \overline{x}_b = \delta^c_b + \lambda \partial^c \xi_b.$$

These reparametrizations up to order λ are the remainder of the parametrization independence of Einstein's gravity, i.e., of the homogeneous $\mathcal{C}(\mathbb{M})$-module transformations $\mathbf{g}^{il} \longmapsto e^i_j \mathbf{g}^{jk} e^l_k$ with the structural group $e^i_j \in \mathbf{GL}(4, \mathbb{R})$ for the local frames (see Chapter 2).

The gauge transformation analogy between the electromagnetic potential $\mathbf{A}$ and the metrical tensor $\mathbf{g}$, or the tetrad $\mathbf{e}$, is superficial: With respect to the Lorentz group $\mathbf{SO}_0(1,3)$ for orthogonal frames, not the tetrad $\mathbf{e}$, but the connection $\mathbf{\Gamma}$ is the analogue to the $\mathbf{U}(1)$-connection, the electromagnetic gauge field $\mathbf{A}$, both with inhomogeneous transformation behavior (see Chapter 6):

$$\begin{array}{ccccccc}
 & & & & \mathbf{e} & & \\
 & & & & \downarrow \partial & & \\
 & & \mathbf{A} & \leftrightarrow & \mathbf{\Gamma} & & \\
\mathbf{U}(1) & & \partial \downarrow & & \downarrow \partial & & \mathbf{SO}_0(1,3). \\
 & & \mathbf{F} & \leftrightarrow & \mathcal{R}_\bullet & &
\end{array}$$

The matter equations for linearized gravity are invariant up to order λ only with a corresponding coordinate reparametrization, e.g., $\mathbf{\Phi}(x) \longmapsto \mathbf{\Phi}(\overline{x})$. A full reparametrization invariance requires the full nonlinear theory.

The tetrad field can be considered as a Goldstone field for the rearrangement of its 10 operations $\mathbf{e} \in \mathbf{GL}(4, \mathbb{R})/\mathbf{O}(1,3)$, especially for the four metrical dilations (see Chapter 1). The $10 = 4 + \binom{4}{2}$ operations, dilations $\mathbf{D}(1)^4$, and rotations $\mathbf{SO}(4)$, come in four $\mathbf{O}(1,3)$-vector fields $(\mathbf{e}^l_a)_{a=0,1,2,3}$ with respect to the ground-state Lorentz invariance for a flat spacetime metric η (see Chapter 7).

The free-field equations can be reduced to Klein–Gordon equations $\partial^2 \mathbf{\Phi} = 0$ for the translation-invariant mass $m^2 = 0$, where the Lorentz vector properties of the scalar squared spacetime derivative ∂^2 do not combine with the Lorentz properties of the fields. To obtain such a decoupling for electrodynamics, the divergence of the vector field has to vanish. For gravity, *harmonic coordinates* are used, where the divergence of the Einstein tensor vanishes:

$$\begin{aligned}
\text{for } \partial_c \mathbf{A}^c = 0 \quad &\Rightarrow \quad \partial^2 \mathbf{A}^c = -g^2 \mathbf{J}^c, \\
\text{for } \partial_c (\mathbf{E}^{cd} - \tfrac{\eta^{cd}}{2} \mathbf{E}^\bullet) = 0 \quad &\Rightarrow \quad \left\{ \begin{array}{l} \partial^2 (\mathbf{E}^{cd} - \frac{\eta^{dc}}{2} \mathbf{E}^\bullet) = -\ell^2 \mathbf{T}^{cd}, \\ \partial^2 \mathbf{E}^{cd} = -\ell^2 (\mathbf{T}^{cd} - \frac{\eta^{dc}}{2} \mathbf{T}^\bullet). \end{array} \right.
\end{aligned}$$

The Schwarzschild metric for energy-momentum tensor $\mathbf{T}^{cd}(x) = m\delta_0^c\delta_0^d\delta(\vec{x})$ is considered in this framework: A reflected "timelike" tensor is proportional to the Euclidean δ^{cd}, not to the Lorentz metric η^{cd}:

$$\mathbf{T}^{cd} = \delta_0^c\delta_0^d\mathbf{T}^\bullet \Rightarrow \mathbf{T}^{cd} - \frac{\eta^{cd}}{2}\mathbf{T}^\bullet = \frac{\delta^{cd}}{2}\mathbf{T}^\bullet.$$

The gravity field and the linearized metric for a static "timelike" energy-momentum tensor $\partial^0\mathbf{T}^\bullet = 0$ is given as follows:

$$\begin{aligned}\partial^2\mathbf{E}^{cd}(x) &= -\vec{\partial}^2\mathbf{E}^{cd}(x) = -\frac{\ell^2}{2}\mathbf{T}^\bullet(\vec{x})\delta^{cd}\\ \Rightarrow \mathbf{E}^{cd} &= \mathbf{E}^{00}\delta^{cd} \text{ with } \mathbf{E}^{00}(\vec{x}) = -\frac{\ell^2}{2}\int d^3y\frac{\mathbf{T}^\bullet(\vec{y})}{4\pi|\vec{x}-\vec{y}|},\\ \mathbf{g} &= (1+2\mathbf{E}^{00})c^2dt^2 - (1-2\mathbf{E}^{00})d\vec{x}^2.\end{aligned}$$

For a mass point, one obtains the linearized Schwarzschild metric and, analogously, the Kepler potential:

$$\begin{array}{ll}\mathbf{T}^{cd}(x) = m\delta_0^c\delta_0^d\delta(\vec{x}) \Rightarrow \mathbf{E}^{cd}(x) = -m\frac{\ell^2}{8\pi r}\delta^{cd}, & \\ \mathbf{J}^c(x) = z\delta_0^c\delta(\vec{x}) \Rightarrow \mathbf{A}^c(x) = z\frac{g^2}{4\pi r}\delta_0^c, & \end{array}\quad \text{with } -\vec{\partial}^2\frac{1}{4\pi r} = \delta(\vec{x}).$$

The propagators of massless vector and tensor fields in spacetime are, up to gauge terms,

$$\begin{aligned}\langle 0|\mathbf{A}^c(y)\mathbf{A}^a(x)|0\rangle_{\text{Feynman}} &= \rho^A(0)\frac{i}{\pi}\int\frac{d^4q}{(2\pi)^3}\frac{-\eta^{ca}}{q^2+io}e^{iq(x-y)},\\ \langle 0|\mathbf{E}^{cd}(y)\mathbf{E}^{ab}(x)|0\rangle_{\text{Feynman}} &= \rho^E(0)\frac{i}{\pi}\int\frac{d^4q}{(2\pi)^3}\frac{\frac{\eta^{ca}\eta^{db}+\eta^{da}\eta^{cb}-\eta^{cd}\eta^{ab}}{2}}{q^2+io}e^{iq(x-y)}.\end{aligned}$$

They can be obtained from the projectors for massive spin-0, -1, and -2 fields by omitting the gauge-related contributions $\frac{q\otimes q}{m^2}$ in the vector projector $\mathcal{V}_c^a = \delta_c^a - \frac{q_cq^a}{m^2}$. The projector inverse of the kinetic term in the free equation for the Einstein tensor field,

$$\begin{aligned}[(\mathcal{P}_2)_{ab}^{cd} - 2(\mathcal{P}_0)_{ab}^{cd}]\partial^2\mathbf{E}^{ab} &= 0,\\ (\mathcal{P}_2 - 2\mathcal{P}_0)(\mathcal{P}_2 - \tfrac{1}{2}\mathcal{P}_0) &= \mathcal{P}_2 + \mathcal{P}_0,\end{aligned}$$

gives the relevant combination [20]:

$$(\mathcal{P}_2 - \tfrac{1}{2}\mathcal{P}_0)_{cd}^{ab} = \frac{\delta_c^a\delta_d^b + \delta_d^a\delta_c^b - \eta_{cd}\eta^{ab}}{2} + \frac{q\otimes q}{m^2}\text{-terms}.$$

The nonrelativistic Kepler potential $\frac{1}{r}$ can be embedded into massless relativistic fields with any Lorentz group representation $[J|J]$: The Coulomb and Newton potentials are the position projections — via time integration — of the off-shell contributions ("virtual particles") in the vector and tensor field propagators,

$$\int dx_0\int\frac{d^4q}{i\pi(2\pi)^3}\frac{1}{-q^2-io}e^{iqx} = \int dx_0\int\frac{d^4q}{i\pi(2\pi)^3}\frac{1}{-q_{\text{P}}^2}e^{iqx} = \int\frac{d^3q}{4i\pi^3}\frac{1}{\vec{q}^2}e^{-i\vec{q}\vec{x}} = \frac{1}{2i\pi r},$$

which is, up to $-2i$, the inverse of the invariant spacetime Laplacian (see Chapter 10),

$$\partial^2\int\frac{d^4q}{i\pi(2\pi)^3}\frac{1}{-q^2-io}e^{iqx} = -2i\delta(x).$$

There are no projectors from a four-component Lorentz vector field $\mathbf{A}_c$ to the two polarized massless photons, which can be expressed with their energy-momenta, and, similarily, no projectors from a 10-component symmetric Lorentz tensor field $\mathbf{E}_{cd}$ to the two polarized massless gravitons. The boost representations $\Lambda(\frac{q}{m})$ for a transformation to a rest system are not defined for $m = 0$. The embedding of the two particle degrees of freedom with polarization into Lorentz group-compatible fields with four and 10 components, respectively, involves two and eight additional nonparticle components.

For the four-component electromagnetic field, the propagator involves the four-dimensional unit

$$[\tfrac{1}{2}|\tfrac{1}{2}],\quad \mathbf{1}_4 = \mathbf{1}_{[\frac{1}{2}|\frac{1}{2}]} \cong \delta^a_c.$$

The indefinite metric $-\eta^{ca}$ contains a definite Hilbert submetric $\mathbf{1}_2$ (lower right corner in the metrical matrix) for the two polarized particle components (± 1) (subindex $\pm$ gives the metric for $-\eta$):

$$\text{metric of } [\tfrac{1}{2}|\tfrac{1}{2}] : \left\{ \begin{array}{ccc} -\eta & = & \left(\begin{array}{cc|c} -1 & 0 & 0 \\ 0 & 1 & 0 \\ \hline 0 & 0 & \mathbf{1}_2 \end{array} \right), \\ [\tfrac{1}{2}|\tfrac{1}{2}] & \overset{\mathbf{SO}(2)}{\cong} & (0)_- \oplus\ (0)_+ \oplus\ (\pm 1)_+ . \end{array} \right.$$

The other two components constitute a *Witt pair* $\begin{pmatrix} -1 & 0 \\ 0 & 1 \end{pmatrix} \cong \begin{pmatrix} 0 & 1 \\ 1 & 0 \end{pmatrix}$ with neutral signature for the Coulomb and gauge degree of freedom, both nonparticlelike.

For the 10-component gravity field, the 10-dimensional symmetric tensor unit can be decomposed into the two harmonic units (see Chapter 8) for the irreducible Lorentz group representations, $10 = 9 + 1$:

$$\begin{aligned} [\tfrac{1}{2}|\tfrac{1}{2}] \vee [\tfrac{1}{2}|\tfrac{1}{2}] &= [1|1] \ \oplus\ [0|0], \\ \mathbf{1}_4 \vee \mathbf{1}_4 &= \mathbf{1}_{[1|1]} + \mathbf{1}_{[0|0]} \cong [\tfrac{\delta^a_c \delta^b_d + \delta^a_d \delta^b_c}{2} - \tfrac{\eta_{cd}\eta^{ab}}{4}] + \tfrac{\eta_{cd}\eta^{ab}}{4}. \end{aligned}$$

The graviton propagator comes with a projector combination:

$$\frac{\delta^a_c \delta^b_d + \delta^a_d \delta^b_c - \eta_{cd}\eta^{ab}}{2} = (\mathbf{1}_{[1|1]} - \mathbf{1}_{[0|0]})^{ab}_{cd}.$$

$\mathbf{1}_{[1|1]}$ gives the pair of polarized gravitons (± 2) with definite metric $\mathbf{1}_2$ (lower right corner in the metrical matrix). There is a triplet of Witt pairs with neutral signature $(\mathbf{1}_3, -\mathbf{1}_3)$. The remaining degree of freedom with metric 1 is paired with the one degree of freedom in $-\mathbf{1}_{[0|0]}$ with metric -1:

$$\text{metric of } [1|1] : \left\{ \begin{array}{ccl} (\eta \vee \eta)_{[1|1]} & = & \left(\begin{array}{ccc|c} -\mathbf{1}_3 & 0 & 0 & 0 \\ 0 & 1 & 0 & 0 \\ 0 & 0 & \mathbf{1}_3 & 0 \\ \hline 0 & 0 & 0 & \mathbf{1}_2 \end{array} \right), \\ [1|1] & \overset{\mathbf{SO}(2)}{\cong} & (0)_- \oplus\ 2 \times (0)_+ \oplus\ (\pm 1)_- \oplus\ (\pm 1)_+ \oplus\ (\pm 2)_+ , \end{array} \right.$$

$$\text{metric of } [0|0] : \left\{ \begin{array}{ccc} -(\eta \vee \eta)_{[0|0]} & = & -1, \\ [0|0] & \overset{\mathbf{SO}(2)}{\cong} & (0)_- . \end{array} \right.$$

In analogy to the one neutral pair for electrodynamics, the four neutral Witt pairs in gravity $\begin{pmatrix} -\eta & 0 \\ 0 & \eta \end{pmatrix} \cong \begin{pmatrix} 0 & \eta \\ \eta & 0 \end{pmatrix}$ for $\eta = \begin{pmatrix} 1 & 0 \\ 0 & -\mathbf{1}_3 \end{pmatrix}$, with the Newton and gauge degrees of freedom, are nonparticlelike.

The Lorentz group representations for flat spacetime gravity constitute an irreducible real 10-dimensional $\mathbf{SL}_0(\mathbb{R}^4)$-representation (see Chapter 7), which is considered with a Lorentz invariant metric of signature (6,4):

$$\begin{array}{ccccc} \mathbf{irrep}\,\mathbf{SL}_0(\mathbb{R}^4) \ni [2,0,0] & \overset{\mathbf{SO}_0(1,3)}{\cong} & [0|0] & \oplus & [1|1], \\ \begin{pmatrix} -\mathbf{1}_4 & 0 \\ 0 & \mathbf{1}_6 \end{pmatrix} & \cong & (-1) & \oplus & \begin{pmatrix} -\mathbf{1}_3 & 0 \\ 0 & \mathbf{1}_6 \end{pmatrix}. \end{array}$$

For long-range interactions in flat spacetime, only two Lorentz group representations are used — the vector representation $[\frac{1}{2}|\frac{1}{2}]$, e.g., for the electromagnetic field, and the tensor with scalar representation $[1|1] \oplus [0|0]$ for the gravitational field, coming, respectively, with one pair of massless photons of polarization $(\pm 1)_+$ and one Witt pair $(0)_\pm$ and, still to be detected, one pair of massless gravitons $(\pm 2)_+$ and four Witt pairs $(\pm 1)_\pm$, $2 \times (0)_\pm$. So far, no long-range interactions have been found with harmonic Lorentz group representations $[J|J]$ for $J \geq \frac{3}{2}$, coming together with a massless particle pair with polarizations $(\pm 2J)$ and $(1+2J)^2 - 2$ non-particle modes:

$$\begin{array}{l} [J|J] \overset{\mathbf{SO}(3)}{\cong} \bigoplus_{L=0}^{2J} [L] \overset{\mathbf{SO}(2)}{\cong} (2J+1) \times (0) \oplus 2J \times (\pm 1) \oplus \ldots \oplus (\pm 2J), \\ \quad \text{with } [L] \overset{\mathbf{SO}(2)}{\cong} (0) \oplus (\pm 1) \oplus \ldots \oplus (\pm L). \end{array}$$

5.8 Quantum Gauge Fields

The extension of global "phase" transformations $e^{i\alpha} \in \mathbf{U}(1)$ to spacetime-dependent local "gauge" transformations $\{x \longmapsto e^{i\alpha(x)}\} \in \mathbf{U}(1)^{\mathbb{R}^4}$ is not enough for quantum gauge structures. Gauge transformations for quantum fields arise in the form of Becchi–Rouet–Stora (BRS)transformations, the gauge parameters are implemented by quantum fields: the Lorentz scalar Fadeev–Popov fields with Fermi statistics. The origin of BRS transformations for four-component massless vector fields lies in the two $\mathbf{SO}(2)$-scalar non-particle degrees of freedom (Witt pair with neutral signature), which arise in addition to the two polarized particle degrees of freedom in an $\mathbf{SO}(2)$-dublet. BRS invariance goes with the projection to the particle degrees of freedom. BRS nontrivial degrees of freedom are without particle content and can have equations of motion with "usual," i.e., not covariant, derivatives.

For a quantum formulation of a gauge field theory, demonstrated for the simplest abelian case, i.e., for quantum electrodynamics with a massless vector field $\mathbf{A}$, the spacetime dependence of a classical gauge transformation with parameter α as invariance of the Lagrangian

$$\begin{array}{rcl} \mathbf{L}(\mathbf{A}^a, \mathbf{F}^{ab}) & = & \frac{1}{2}\mathbf{F}^{ab}\epsilon_{ab}^{cd}\partial_c \mathbf{A}_d + g^2 \frac{\mathbf{F}^{ab}\mathbf{F}_{ab}}{4}, \\ \mathbf{A}^a & \longmapsto & \mathbf{A}^a + \partial^a \alpha, \quad \mathbf{F}_{ab} \longmapsto \mathbf{F}_{ab}, \end{array}$$

is drastically reduced. A quantum gauge theory has a duality-completing scalar field ("gauge-fixing" field) $\mathbf{S}$. There remains a transformation with a massless "Lie parameter field" $\boldsymbol{\beta}$:

$$\begin{array}{rcl} \mathbf{L}(\mathbf{A}^a, \mathbf{F}^{ab}, \mathbf{S}) & = & \frac{1}{2}\mathbf{F}^{ab}\epsilon_{ab}^{cd}\partial_c \mathbf{A}_d + \mathbf{S}\partial_a \mathbf{A}^a + g^2 \frac{\mathbf{F}^{ab}\mathbf{F}_{ab}}{4} - g^2 \lambda \frac{\mathbf{S}^2}{2}, \\ \mathbf{A}^a & \longmapsto & \mathbf{A}^a + \partial^a \boldsymbol{\beta}, \quad \mathbf{F}_{ab} \longmapsto \mathbf{F}_{ab}, \quad \mathbf{S} \longmapsto \mathbf{S}, \text{ with } \partial^2 \boldsymbol{\beta} = 0. \end{array}$$

5.8.1 Fadeev–Popov Ghosts in Quantum Mechanics

The "gauge-fixing" part of the dynamics with the gauge transformations,

$$\mathbf{L}(\mathbf{A}, \mathbf{S}) = \mathbf{S}\partial_a \mathbf{A}^a - g^2 \lambda \frac{\mathbf{S}^2}{2}, \quad \left\{ \begin{array}{rl} \partial_a \mathbf{A}^a = g^2 \lambda \mathbf{S}, & \partial_a \mathbf{S} = 0, \\ \mathbf{A}^a \longmapsto \mathbf{A}^a + \boldsymbol{\gamma}^a, & \mathbf{S} \longmapsto \mathbf{S}, \\ \partial^a \boldsymbol{\beta} = \boldsymbol{\gamma}^a, & \partial_a \boldsymbol{\gamma}^a = 0, \end{array} \right.$$

is the Lorentz compatible spacetime distribution of the noncompact time development for a free mass point:

$$L(\mathbf{x}, \mathbf{p}) = \mathbf{p} d_t \mathbf{x} - \frac{\mathbf{p}^2}{2}, \quad \left\{ \begin{array}{rclrcl} d_t \mathbf{x} & = & \mathbf{p}, & d_t \mathbf{p} & = & 0, \\ \mathbf{x} & \longmapsto & \mathbf{x} + \boldsymbol{\gamma}, & \mathbf{p} & \longmapsto & \mathbf{p}, \\ d_t \boldsymbol{\beta} & = & \boldsymbol{\gamma}, & d_t \boldsymbol{\gamma} & = & 0. \end{array} \right.$$

The gauge transformation is the relativistic distribution of a position translation transformation for the free mass point position.

A noncompact time development has translation eigenvectors and nilvectors. The subspace built by the eigenvectors has a trivial eigenvalue (nildimension) for the action of the nilpotent part of the Hamiltonian. In the self-dual vector space $V \cong \mathbb{R}^2$, spanned by position and momentum, the Hamiltonian matrix of the free mass point, a linear 2×2 transformation, is nilquadratic:

$$\begin{array}{l} \mathbf{x} \cong x = \begin{pmatrix} 0 \\ 1 \end{pmatrix}, \quad \mathbf{p} \cong p = \begin{pmatrix} 1 \\ 0 \end{pmatrix}, \\ \begin{array}{rl} H_B = \frac{\mathbf{p}^2}{2} \cong h = & \begin{pmatrix} 0 & 1 \\ 0 & 0 \end{pmatrix}, \\ h \circ h = 0, \quad e^{ht} = & \begin{pmatrix} 1 & t \\ 0 & 1 \end{pmatrix} \end{array} \Rightarrow \left\{ \begin{array}{rl} d_t x & = h.x = p, \\ d_t p & = h.p = 0, \\ \begin{pmatrix} x(t) \\ p(t) \end{pmatrix} & = e^{ht} \begin{pmatrix} x \\ p \end{pmatrix} = \begin{pmatrix} x + tp \\ p \end{pmatrix}. \end{array} \right. \end{array}$$

In a Bose quantum algebra, generated by position and momentum, the Hamiltonian is not nilquadratic with respect to the quantum product:

$$\begin{array}{r} [i\mathbf{p}, \mathbf{x}] = 1, \quad H_B = \frac{\mathbf{p}^2}{2}, \\ \text{but } H_B^2 \neq 0 \end{array} \Rightarrow \left\{ \begin{array}{l} d_t \mathbf{x} = [iH_B, \mathbf{x}] = \mathbf{p}, \\ d_t \mathbf{p} = [iH_B, \mathbf{p}] = 0. \end{array} \right.$$

By introducing additional Fermi degrees of freedom as partners for the Bose position-momentum pair, it is possible to construct nontrivial nilquadratic quantum operators. To formulate the distinction between translation eigenvectors and nilvectors (particle- and nonparticle-interpretable), a quantum gauge theory has a *Bose–Fermi twin structure*: The spinless part of the gauge Bose field and its "gauge-fixing" dual partner are accompanied by *Fadeev–Popov fields* as their Fermi counterparts, whose classical limits are the spacetime-dependent Lie parameters of the gauge group.

The Bose–Fermi twin structure is discussed first in a nonrelativistic quantum-mechanical model: A noncompact time development for the additional Fermi degrees of freedom needs two dual pairs:

$$\begin{aligned} \text{Bose: } [i\mathbf{p},\mathbf{x}] &= 1, \quad \text{Fermi: } \{\boldsymbol{\beta},\check{\boldsymbol{\gamma}}\} = 1 = \{\boldsymbol{\gamma},\check{\boldsymbol{\beta}}\},\\ \text{Hamiltonian: } H_{B+F} &= H_B + H_F = \tfrac{\mathbf{p}^2}{2} + i\check{\boldsymbol{\gamma}}\boldsymbol{\gamma}. \end{aligned}$$

The equations of motion are

$$\text{Bose: } \left\{ \begin{aligned} d_t\mathbf{x} &= [iH_{B+F},\mathbf{x}] &&= \mathbf{p},\\ d_t\mathbf{p} &= [iH_{B+F},\mathbf{p}] &&= 0, \end{aligned} \right. \quad \text{Fermi: } \left\{ \begin{aligned} d_t\boldsymbol{\beta} &= [iH_{B+F},\boldsymbol{\beta}] &&= \boldsymbol{\gamma},\\ d_t\boldsymbol{\gamma} &= [iH_{B+F},\boldsymbol{\gamma}] &&= 0,\\ d_t\check{\boldsymbol{\beta}} &= [iH_{B+F},\check{\boldsymbol{\beta}}] &&= -\check{\boldsymbol{\gamma}},\\ d_t\check{\boldsymbol{\gamma}} &= [iH_{B+F},\check{\boldsymbol{\gamma}}] &&= 0. \end{aligned} \right.$$

They can be derived from a classical Lagrangian (first- or second-order time derivatives):

$$\begin{aligned} L(\mathbf{x},\mathbf{p},\boldsymbol{\beta},\boldsymbol{\gamma}) &= \mathbf{p}d_t\mathbf{x} - \tfrac{\mathbf{p}^2}{2} + i\boldsymbol{\gamma}d_t\check{\boldsymbol{\beta}} + i\check{\boldsymbol{\gamma}}d_t\boldsymbol{\beta} - i\check{\boldsymbol{\gamma}}\boldsymbol{\gamma},\\ L(\mathbf{x},\boldsymbol{\beta}) &= \tfrac{1}{2}(d_t\mathbf{x})^2 + i(d_t\boldsymbol{\beta})(d_t\check{\boldsymbol{\beta}}). \end{aligned}$$

The nilquadratic *Becchi–Rouet–Stora charge* N_{BF} implementing the gauge transformation $\mathbf{x} \longmapsto \mathbf{x} + \delta\mathbf{x}$ with $\delta\mathbf{x} = \boldsymbol{\gamma}$ is given by the time development invariant

$$N_{BF} = \boldsymbol{\gamma}\mathbf{p} \Rightarrow N_{BF}^2 = 0, \quad [H_{B+F}, N_{BF}] = 0.$$

Its linear hybrid adjoint action in a hybrid algebra generated by Bose and Fermi vectors,

$$[\![a,b]\!] = \left\{ \begin{aligned} &[a,b] &&\iff a \text{ or } b \text{ is Bose},\\ &\{a,b\} &&\iff a \text{ and } b \text{ are Fermi}, \end{aligned} \right.$$

defines the BRS transformations:

$$\text{Bose: } \left\{ \begin{aligned} \delta\mathbf{x} &= [iN_{BF},\mathbf{x}] &&= \boldsymbol{\gamma},\\ \delta\mathbf{p} &= [iN_{BF},\mathbf{p}] &&= 0, \end{aligned} \right. \quad \text{Fermi: } \left\{ \begin{aligned} \delta\boldsymbol{\beta} &= \{iN_{BF},\boldsymbol{\beta}\} &&= 0,\\ \delta\boldsymbol{\gamma} &= \{iN_{BF},\boldsymbol{\gamma}\} &&= 0,\\ \delta\check{\boldsymbol{\beta}} &= \{iN_{BF},\check{\boldsymbol{\beta}}\} &&= i\mathbf{p},\\ \delta\check{\boldsymbol{\gamma}} &= \{iN_{BF},\check{\boldsymbol{\gamma}}\} &&= 0. \end{aligned} \right.$$

With the *Fadeev–Popov number operator* for the Fermi degrees of freedom,

$$\begin{aligned} P &= i(\check{\boldsymbol{\gamma}}\boldsymbol{\beta} + \check{\boldsymbol{\beta}}\boldsymbol{\gamma}) \Rightarrow \left\{ \begin{aligned} [iP,\boldsymbol{\beta}] &= \boldsymbol{\beta}, & [iP,\check{\boldsymbol{\gamma}}] &= -\check{\boldsymbol{\gamma}},\\ [iP,\check{\boldsymbol{\beta}}] &= -\check{\boldsymbol{\beta}}, & [iP,\boldsymbol{\gamma}] &= \boldsymbol{\gamma}, \end{aligned} \right.\\ [P, H_{B+F}] &= 0, \end{aligned}$$

the space $\mathbf{Q}(H_{B+F})$ spanned by the eigenvectors of the Hamiltonian is defined by trivial eigenvalues both for the BRS charge N_{BF} and the Fadeev–Popov number P:

$$\mathbf{Q}(H_{B+F}) = \{a \mid [\![N_{BF}, a]\!] = 0 \text{ and } [P, a] = 0\}.$$

5.8.2 Fadeev–Popov Ghosts for Quantum Gauge Fields

The Lorentz-compatible distribution of the nonrelativistic model for the electromagnetic quantum gauge field,

$$\begin{aligned}\mathbf{L}(\mathbf{A}, \mathbf{F}, \mathbf{S}, \boldsymbol{\beta}, \boldsymbol{\gamma}) &= \mathbf{F}_{ab}\tfrac{\partial^a \mathbf{A}^b - \partial^b \mathbf{A}^a}{2} + \mathbf{S}\partial_a \mathbf{A}^a + g^2(\tfrac{\mathbf{F}_{ab}\mathbf{F}^{ab}}{4} - \lambda \tfrac{\mathbf{S}^2}{2}) \\ &\quad + i\boldsymbol{\gamma}^a \partial_a \check{\boldsymbol{\beta}} + i\check{\boldsymbol{\gamma}}^a \partial_a \boldsymbol{\beta} - ig^2 \lambda \check{\boldsymbol{\gamma}}^a \boldsymbol{\gamma}_a,\end{aligned}$$

$$\text{Bose: } \begin{cases} \partial^b \mathbf{A}^a - \partial^a \mathbf{A}^b &= g^2 \mathbf{F}^{ab}, \\ \partial_a \mathbf{A}^a &= g^2 \lambda \mathbf{S}, \\ \partial^b \mathbf{F}_{ab} - \partial_a \mathbf{S} &= 0, \end{cases} \quad \text{Fermi: } \begin{cases} \partial^a \boldsymbol{\beta} &= g^2 \lambda \boldsymbol{\gamma}^a, \\ \partial_a \boldsymbol{\gamma}^a &= 0, \\ \partial^a \check{\boldsymbol{\beta}} &= -g^2 \lambda \check{\boldsymbol{\gamma}}^a, \\ \partial_a \check{\boldsymbol{\gamma}}^a &= 0, \end{cases}$$

uses Lorentz scalar *Fadeev–Popov fields* $(\boldsymbol{\beta}, \check{\boldsymbol{\beta}}, \boldsymbol{\gamma}^a, \check{\boldsymbol{\gamma}}^a)$ with Fermi quantization:

$$[i\mathbf{S}, \mathbf{A}^a](\vec{x}) = \{\boldsymbol{\beta}, \check{\boldsymbol{\gamma}}^a\}(\vec{x}) = \{\boldsymbol{\gamma}^a, \check{\boldsymbol{\beta}}\}(\vec{x}) = \delta_0^a \delta(\vec{x}).$$

A second-order derivative Lagrangian reads

$$\mathbf{L}(\mathbf{A}, \boldsymbol{\beta}) = -\tfrac{1}{4g^2}(\partial^a \mathbf{A}^b - \partial^b \mathbf{A}^a)(\partial_a \mathbf{A}_b - \partial_b \mathbf{A}_a) + \tfrac{1}{2g^2\lambda}(\partial_a \mathbf{A}^a)^2 + \tfrac{i}{g^2\lambda}(\partial^a \boldsymbol{\beta})(\partial_a \check{\boldsymbol{\beta}}).$$

The hybrid adjoint action of the nilquadratic linear BRS charge generates the linear BRS transformations

$$\begin{aligned} N_{BF} &= g^2\lambda \int d^3x\ \boldsymbol{\gamma}_0(x)\mathbf{S}(x), \\ N_{BF}^2 &= 0 \end{aligned} \quad \Rightarrow \quad \begin{aligned} &\text{Bose: } \begin{cases} \delta \mathbf{A}^a &= [iN_{BF}, \mathbf{A}^a] &= g^2 \lambda \delta_0^a \boldsymbol{\gamma}_0, \\ \delta \mathbf{S} &= [iN_{BF}, \mathbf{S}] &= 0, \\ \delta \mathbf{F}_{ab} &= [iN_{BF}, \mathbf{F}_{ab}] &= 0, \end{cases} \\ &\text{Fermi: } \begin{cases} \delta \boldsymbol{\beta} &= \{iN_{BF}, \boldsymbol{\beta}\} &= 0, \\ \delta \boldsymbol{\gamma}^a &= \{iN_{BF}, \boldsymbol{\gamma}^a\} &= 0, \\ \delta \check{\boldsymbol{\beta}} &= \{iN_{BF}, \check{\boldsymbol{\beta}}\} &= ig^2\lambda \mathbf{S}, \\ \delta \check{\boldsymbol{\gamma}}^a &= \{iN_{BF}, \check{\boldsymbol{\gamma}}^a\} &= 0. \end{cases} \end{aligned}$$

The subspace with the particle-interpretable degrees of freedom, i.e., without nilvectors, is characterized by a trivial BRS charge and a trivial Faddev–Popov number:

$$P = \int d^3x\ \mathbf{P}_0(x), \quad \mathbf{P}_a = i(\check{\boldsymbol{\gamma}}_a \boldsymbol{\beta} + \check{\boldsymbol{\beta}} \boldsymbol{\gamma}_a).$$

"Gauge-invariant" fields are harmonically analyzable with translation eigenvectors only.

The spinless and "gauge-fixing" Bose degrees of freedom and the Fermi Fadeev–Popov ones display a twin structure. The BRS current $\mathbf{N}_b(x)$ of the Fermi type has its counterpart in the nonderivative part $\mathbf{H}(x)$ of the Lagrangian (Bose type):

$$\mathbf{N}_a = g^2\lambda\boldsymbol{\gamma}_a\mathbf{S}, \quad \mathbf{H}_{B+F} = g^2\lambda(\tfrac{\mathbf{S}^2}{2} + i\check{\boldsymbol{\gamma}}^a\boldsymbol{\gamma}_a).$$

The dynamics H_{B+F} in the mass point model arises by a BRS transformation from an operator K connecting Bose and Fermi degrees of freedom:

$$\begin{aligned} N_{BF} &= \boldsymbol{\gamma}\mathbf{p}, \quad H_{B+F} = \tfrac{\mathbf{p}^2}{2} + i\check{\boldsymbol{\gamma}}\boldsymbol{\gamma}, \\ H_{B+F} &= \{N_{BF}, K\}, \quad K = \tfrac{\check{\boldsymbol{\beta}}\mathbf{p}}{2} + \check{\boldsymbol{\gamma}}\mathbf{x}. \end{aligned}$$

Since $N_{BF}^2 = 0$, the BRS invariance of the Hamiltonian is obvious:

$$[N_{BF}, H_{B+F}] = [N_{BF}, \{N_{BF}, K\}] = [N_{BF}^2, K] = 0.$$

The corresponding relativistic field operators arise from the position distributions:

$$\mathbf{K} = \tfrac{\check{\boldsymbol{\beta}}\mathbf{S}}{2} + \check{\gamma}_a\mathbf{A}^a, \quad (H_{B+F}, N_{BF}, K) = \int d^3x\, (\mathbf{H}, \mathbf{N}_0, \mathbf{K})(x).$$

In parallel to the equations of motion, the BRS transformations for nonabelian groups, e.g., for a massless $\mathbf{SU}(2)$–Yang–Mills triplet vector field, also contain nonlinear contributions.

5.8.3 Particle Analysis of Massless Vector Fields

The harmonic analysis (see Chapter 4) of a massless vector field contains four momentum operators, two with a time representation in $\mathbf{U}(\mathbf{1}_2)$ (first and second components) and two with a time representation in $\mathbf{U}(1,1)$ (zeroth and third components):

$$\mathbf{A}^a(x) = {}^{\oplus}\!\!\int \frac{d^3q}{2|\vec{q}|(2\pi)^3} O_w^a(\tfrac{\vec{q}}{|\vec{q}|}) \begin{pmatrix} e^{iqx}[\mathrm{B}(\vec{q}) + i\nu|\vec{q}|x_0\mathrm{G}(\vec{q})] + (1-\nu)e^{-iqx}\mathrm{G}^\times(\vec{q}) \\ e^{iqx}\mathrm{u}^1(\vec{q}) + e^{-iqx}\mathrm{u}_1^\star(\vec{q}) \\ e^{iqx}\mathrm{u}^2(\vec{q}) + e^{-iqx}\mathrm{u}_2^\star(\vec{q}) \\ (1-\nu)e^{iqx}\mathrm{G}(\vec{q}) + e^{-iqx}[\mathrm{B}^\times(\vec{q}) - i\nu|\vec{q}|x_0\mathrm{G}^\times(\vec{q})] \end{pmatrix},$$

with $q_0 = |\vec{q}|$ and $\lambda = 1 - 2\nu$.

The transition from the Lorentz group $\mathbf{SO}_0(1,3)$ to the axial rotation fixgroup $\mathbf{SO}(2)$ uses the transmutator (see Chapter 7):

$$\begin{aligned} O_w(\tfrac{\vec{q}}{|\vec{q}|}) &= \left(\begin{array}{c|c} 1 & 0 \\ \hline 0 & O(\frac{\vec{q}}{|\vec{q}|}) \end{array}\right) \circ w \in \mathbf{SO}_0(1,3)/\mathbf{SO}(2), \\ O(\tfrac{\vec{q}}{|\vec{q}|}) &= \left(\begin{array}{c|c} \delta^{AB} - \frac{q_A q_B}{|\vec{q}|(|\vec{q}|+q_3)} & \frac{q_A}{|\vec{q}|} \\ -\frac{q_B}{|\vec{q}|} & \frac{q_3}{|\vec{q}|} \end{array}\right) \in \mathbf{SO}(3)/\mathbf{SO}(2), \\ \begin{pmatrix} 0 & 0 & 1 \\ 0 & \mathbf{1}_2 & 0 \\ 1 & 0 & 0 \end{pmatrix} &= w \circ \begin{pmatrix} -1 & 0 & 0 \\ 0 & \mathbf{1}_2 & 0 \\ 0 & 0 & 1 \end{pmatrix} \circ w^T, \text{ with } w = \frac{1}{\sqrt{2}}\begin{pmatrix} 1 & 0 & 1 \\ 0 & \mathbf{1}_2 & 0 \\ -1 & 0 & 1 \end{pmatrix}. \end{aligned}$$

The massless vector fields involve an $\mathbf{SO}(2)$-polarized particle pair (left and right polarized photons) and an $\mathbf{SO}(2)$-trivial pair without particle interpretation:

$$\begin{aligned} \text{for } \mathbf{A}^a:\ \mathbb{R}^4\times\mathbf{SO}(2) &\longrightarrow \mathbf{U}(1,1)\times\mathbf{U}(2)\\ \mathbf{U}(1,1)\times\mathbf{U}(2) &\subset \mathbf{U}(1,3)\supset \mathbf{SO}_0(1,3). \end{aligned}$$

The first and second components of the massless field with nontrivial polarization around the momentum $\vec{q}$ carry two irreducible $\mathbf{U}(1)$ time representations with energy $q_0=|\vec{q}\,|$. They constitute a harmonic $\mathbf{U}(2)$-oscillator:

$$A,B\in\{1,2\}:\ \begin{cases} \mathbb{R}^4 \longrightarrow \mathbf{U}(\mathbf{1}_2)\ni e^{i|\vec{q}\,|x_0-i\vec{q}\vec{x}}\begin{pmatrix}1&0\\0&1\end{pmatrix},\\ \mathrm{u}^A(\vec{q},x_0) = e^{i|\vec{q}\,|x_0}\mathrm{u}^A(\vec{q}\,),\\ \mathbb{R}^4\times\mathbf{SO}(2) \longrightarrow \mathbf{U}(\mathbf{1}_2)\circ\mathbf{SU}(2)=\mathbf{U}(2),\\ [\mathrm{u}^\star_A(\vec{p}\,),\mathrm{u}^B(\vec{q}\,)] = g^2 2|\vec{q}\,|\delta^B_A\delta(\frac{\vec{q}-\vec{p}}{2\pi}). \end{cases}$$

The zeroth and third components with trivial polarization are connected in a reducible, but decomposable faithful time representation in the indefinite unitary group $\mathbf{U}(1,1)$ with energy $q_0=|\vec{q}\,|$ and nilconstant $\nu|\vec{q}\,|$ involving the "gauge fixing" constant $2\nu=1-\lambda$ (see Chapter 4):

$$a,b\in\{0,3\}:\ \begin{cases} \mathbb{R}^4 \longrightarrow \mathbf{U}(1,1)\ni\begin{pmatrix}1&i\nu|\vec{q}\,|x_0\\0&1\end{pmatrix}e^{i|\vec{q}\,|x_0-i\vec{q}\vec{x}},\\ \mathrm{B}(\vec{q},x_0) = e^{i|\vec{q}\,|x_0}[\mathrm{B}(\vec{q}\,)+i\nu|\vec{q}\,|x_0\mathrm{G}(\vec{q}\,)],\\ \mathrm{G}(\vec{q},x_0) = e^{i|\vec{q}\,|x_0}\mathrm{G}(\vec{q}\,),\\ [\mathrm{B}^\times(\vec{p}\,),\mathrm{G}(\vec{q}\,)] = [\mathrm{G}^\times(\vec{p}\,),\mathrm{B}(\vec{q}\,)]=g^2 2|\vec{q}\,|\delta(\frac{\vec{q}-\vec{p}}{2\pi}). \end{cases}$$

A $\mathbf{U}(1)$-time development with particle interpretation has a Fock state:

$$\begin{aligned} \langle 0,1;\vec{p},A|0,1;\vec{q},B\rangle &= \langle 0|\mathrm{u}^\star_A(\vec{p}\,)\mathrm{u}^B(\vec{q}\,)|0\rangle=\langle 0|\{\mathrm{u}^\star_A(\vec{p}\,),\mathrm{u}^B(\vec{q}\,)\}|0\rangle\\ &- g^2 2|\vec{q}\,|\delta^{AB}\delta(\tfrac{\vec{q}-\vec{p}}{2\pi}). \end{aligned}$$

A Fock form also for the $\mathbf{U}(1,1)$-time representations leads to an indefinite metric$\begin{pmatrix}\langle \mathrm{B}|\mathrm{B}\rangle&\langle \mathrm{B}|\mathrm{G}\rangle\\\langle \mathrm{G}|\mathrm{B}\rangle&\langle \mathrm{G}|\mathrm{G}\rangle\end{pmatrix}\sim\begin{pmatrix}0&1\\1&0\end{pmatrix}\cong\begin{pmatrix}-1&0\\0&1\end{pmatrix}$:

$$\begin{aligned} \langle 0|\mathrm{B}^\times(\vec{p}\,)\mathrm{G}(\vec{q}\,)|0\rangle &= \langle 0|\{\mathrm{B}^\times(\vec{p}\,),\mathrm{G}(\vec{q}\,)\}|0\rangle\\ =\langle 0|\mathrm{G}^\times(\vec{p}\,)\mathrm{B}(\vec{q}\,)|0\rangle &= \langle 0|\{\mathrm{G}^\times(\vec{p}\,),\mathrm{B}(\vec{q}\,)\}|0\rangle = g^2 2|\vec{q}\,|\delta(\tfrac{\vec{q}-\vec{p}}{2\pi}). \end{aligned}$$

It gives a Fock value for the anticommutator and the Feynman propagator:

$$\begin{aligned} \langle 0|\{\mathbf{A}^c,\mathbf{A}^a\}(x)|0\rangle &= g^2\int\frac{d^4q}{(2\pi)^3}(-\eta^{ca}-2\nu q^c q^a\frac{\partial}{\partial q^2})\delta(q^2)e^{iqx},\\ \langle 0|\{\mathbf{A}^c,\mathbf{A}^a\}(x)-\epsilon(x_0)[\mathbf{A}^c,\mathbf{A}^a](x)|0\rangle &= g^2\frac{i}{\pi}\int\frac{d^4q}{(2\pi)^3}\left[\frac{-\eta^{ca}}{q^2+io}+2\nu\frac{q^c q^a}{(q^2+io)^2}\right]e^{iqx}. \end{aligned}$$

5.9 Hilbert Representations of the Poincaré Group

As initiated by Wigner, the Hilbert representations of the semidirect product Poincaré group $\mathbf{SO}_0(1,3) \vec{\times} \mathbb{R}^4$, or of its cover $\mathbf{SL}(2,\mathbb{C}) \vec{\times} \mathbb{R}^4$, can be induced from Hilbert representations of direct product subgroups where the fixgroups $\mathbf{SL}(2,\mathbb{C})_q$ of spacetime translations or, equivalently, of energy-momenta, are relevant.

There are four fixgroup types: for the trivial translation, and for time-, space- and lightlike translations, e.g., for $q = \mathbf{1}_2,\ \sigma^3,\ \mathbf{1}_2 + \sigma^3$, respectively,

$$\begin{aligned} q &= q_a\sigma^a = q_0\mathbf{1}_2 + \vec{q}\vec{\sigma} = \begin{pmatrix} q_0+q_3 & q_1-iq_2 \\ q_1+iq_2 & q_0-q_3 \end{pmatrix}, \quad \det q = q^2 = q_0^2 - \vec{q}^{\,2}, \\ \mathbf{SL}(2,\mathbb{C})_q &= \{s \in \mathbf{SL}(2,\mathbb{C}) \mid s \circ q \circ s^* = q\} \\ &\cong \begin{cases} \mathbf{SL}(2,\mathbb{C}) \sim \mathbf{SO}_0(1,3), & q = 0, \\ \mathbf{SU}(2) \sim \mathbf{SO}(3), & q^2 > 0, \\ \mathbf{SU}(1,1) \sim \mathbf{SO}_0(1,2), & q^2 < 0, \\ \mathbf{SO}(2) \vec{\times} \mathbb{R}^2,\ q \neq 0, & q^2 = 0. \end{cases} \end{aligned}$$

The representations, always Hilbert in the following, with nontrivial translation representations are induced from the representations of direct products:

$$q = 0: \ \mathbf{SL}(2,\mathbb{C}), \quad q \neq 0: \begin{cases} q^2 > 0: & \mathbf{SU}(2) \times \mathbb{R}_t, \\ q^2 < 0: & \mathbf{SU}(1,1) \times \mathbb{R}_s, \\ q^2 = 0: & [\mathbf{SO}(2) \vec{\times} \mathbb{R}^2] \times \mathbb{R}_l. \end{cases}$$

The representations of the one-dimensional time-, space-, and lightlike translations $\mathbb{R}_{t,s,l} \cong \mathbb{R} \longrightarrow \mathbf{U}(1)$ can by written with four-dimensional ones, $\mathbb{R}^4$, e.g., $e^{imt} \cong e^{iqx}$ with $q^2 = m^2 > 0$. The representations of the Poincaré group integrate the translation representations over the corresponding three-dimensional energy-momentum orbits:

$$\begin{aligned} &q^2 > 0: && \mathbf{SL}(2,\mathbb{C})/\mathbf{SU}(2) \cong \mathcal{Y}^3 && \text{(energylike hyperboloid)}, \\ &q^2 < 0: && \mathbf{SL}(2,\mathbb{C})/\mathbf{SU}(1,1) \cong \mathcal{Y}^{(1,2)} && \text{(momentumlike hyperboloid)}, \\ &q^2 = 0, && q \neq 0: \ \mathbf{SL}(2,\mathbb{C})/\,[\mathbf{SO}(2) \vec{\times} \mathbb{R}^2] && \text{(pointed future lightcone)}. \end{aligned}$$

The representations of $\mathbf{SU}(2) \times \mathbb{R}_t$ are used for massive particles; the irreducible ones are characterized by the invariants (m^2, J) with mass $m^2 > 0$ for the translations and spin $2J \in \mathbb{N}$ for the rotations.

The representations of $\mathbf{SU}(1,1) \times \mathbb{R}_s$ have a momentumlike translation invariant $x \longmapsto e^{iqx}$, $q^2 < 0$. The representations of the homogeneous group $\mathbf{SO}_0(1,2) \sim \mathbf{SU}(1,1) \sim \mathbf{SL}_0(\mathbb{R}^2)$ (locally isomorphic) were first given by Bargmann and are superficially described in the following: $\mathbf{SL}_0(\mathbb{R}^2)$ has two types of Cartan subgroups (see Chapter 8), compact $\mathbf{SO}(2)$ and noncompact $\mathbf{SO}_0(1,1)$, whose representations characterize, respectively, the *discrete series representations* with integer invariant (winding number) and the *principal series representations* with imaginary continuous invariant $iQ \in i\mathbb{R}_+$. In addition, there is a *supplementary series* with real invariant $0 < I < 1$. Its representations are nonamenable, i.e., with trivial Plancherel measure. So far, the representations of the Poincaré group, induced from $\mathbf{SL}_0(\mathbb{R}^2) \times \mathbb{R}_s$, have not been used to describe experiments.

The representations of $[\mathbf{SO}(2) \vec{\times} \mathbb{R}^2] \times \mathbb{R}_l$ have lightlike translation invariant $x \longmapsto e^{iqx}$, $q^2 = 0$, $q \neq 0$. The representation of the semidirect fixgroup, a subgroup of the Galilei group, isomorphic to the Euclidean group in two dimensions (see Chapter 8), $e^{i\varphi_3\sigma^3 + \psi_1\sigma^1 + \psi_2\sigma^2} \in \mathbf{SO}(2) \vec{\times} \mathbb{R}^2$, with one rotation and two boosts can be induced from subgroup representations with boost fixgroups:

$$\psi = \psi_1\sigma^1 + \psi_2\sigma^2,$$
$$\mathbf{SO}(2)_\psi = \{o \in \mathbf{SO}(2) \mid o \circ \psi \circ o^* = \psi\} = \begin{cases} \mathbf{SO}(2), & \psi = 0, \\ \{1\}, & \psi \neq 0. \end{cases}$$

The representations with nontrivial boosts start from trivial homogeneous group representations with a continuous invariant for the boosts $\mathbb{R}_l \subset \mathbb{R}^2 \times \mathbb{R}_l \subset \mathbf{SL}(2,\mathbb{C}) \vec{\times} \mathbb{R}^4$. So far, they were not necessary in particle physics. The fixgroup $\mathbf{SO}(2)$-representations for trivial boosts are characterized by integer invariant pairs (polarization) $(\pm 2J)$. They are used for massless particles.

Also, the representations of the Lorentz group $\mathbf{SL}(2,\mathbb{C})$ as fixgroup for trivial energy-momenta have not been used for free particles. $\mathbf{SL}(2,\mathbb{C})$ has one type of Cartan subgroup $\mathbf{SO}(2) \times \mathbf{SO}_0(1,1)$, which leads to the characterization of the irreducible representations, as given by Bargman and Gel'fand, by one integer "compact" invariant $2J \in \mathbb{N}$, and by a "noncompact" continuous imaginary one for the *principal series* and, for the *supplementary series* (nonamenable), by a trivial "compact" invariant and a real continuous one $0 < I < 1$.

There are altogether eight types of Hilbert representations of the Poincaré group, two for fixgroup $\mathbf{SL}(2,\mathbb{C})$, one for fixgroup $\mathbf{SU}(2)$, three for fixgroup $\mathbf{SU}(1,1)$, and two for fixgroup $\mathbf{SO}(2) \vec{\times} \mathbb{R}^2$:

Translation invariant	Translation fixgroup	Homogeneous group invariants
$- - -$	$\mathbf{SL}(2,\mathbb{C})$	$\mathbf{SO}(2) \times \mathbf{SO}_0(1,1)$: $\mathbb{N} \times i\mathbb{R}_+$ (principal) $\{0\} \times]0,1[$ (supplementary)
$m^2 > 0$	$\mathbf{SU}(2)$	$\mathbf{SO}(2)$: $\mathbb{N}$
$m^2 < 0$	$\mathbf{SL}_0(\mathbb{R}^2)$	$\mathbf{SO}(2)$: $\mathbb{N}$ (discrete) $\mathbf{SO}_0(1,1)$: $i\mathbb{R}_+$ (principal) $]0,1[$ (supplementary)
$m^2 = 0$	$\mathbf{SO}(2) \vec{\times} \mathbb{R}^2$	$\mathbf{SO}(2)$: $\mathbb{N}$ $\mathbb{R}$: $i\mathbb{R}_+$

Hilbert representations of the Poincaré group
$\mathbf{SL}(2,\mathbb{C}) \vec{\times} \mathbb{R}^4 \sim \mathbf{SO}_0(1,3) \vec{\times} \mathbb{R}^4$

The real three-dimensional fixgroups for nontrivial translations with rank-1 Lie algebras and one generating invariant, related either to compact $\mathbf{SO}(2)$

or to noncompact $\mathbf{SO}_0(1,1) \cong \mathbb{R}$, are subgroups of the real six-dimensional $\mathbf{SL}(2,\mathbb{C})$ with rank-2 Lie algebra and two generating invariants $(2J, iQ) \in \mathbb{N} \times i\mathbb{R}_+$, related to $\mathbf{SO}(2) \times \mathbf{SO}_0(1,1)$. They have corresponding subsets of invariants for their representations.

The Wigner classification of the representations of the Poincaré group with the Minkowski translations as the tangent group of the spacetime manifold gives no hintas to why only two types are used for flat spacetime, those for free particles, with both causal translation invariants $m^2 \geq 0$ and compact fixgroups,

$$\begin{array}{rr} m^2 > 0: & \mathbf{SU}(2) \times \mathbb{R}_t \subset \mathbf{SL}(2,\mathbb{C}) \,\vec{\times}\, \mathbb{R}^4, \\ m^2 = 0: & \mathbf{SO}(2) \times \mathbb{R}_l \subset [\mathbf{SO}(2) \,\vec{\times}\, \mathbb{R}^2] \times \mathbb{R}_l \subset \mathbf{SL}(2,\mathbb{C}) \,\vec{\times}\, \mathbb{R}^4. \end{array}$$

Flat spacetime theory also cannot explain the particle spectrum, i.e., for these two types, which invariants occur, i.e., which masses and spins or polarizations.

5.10 Normalizations and Coupling Constants

Flat spacetime fields with particles are acted on by irreducible Hilbert representations of the Poincaré cover group $\mathbf{SL}(2,\mathbb{C}) \,\vec{\times}\, \mathbb{R}^4$, which are induced by representations of $\mathbf{SU}(2) \times \mathbb{R}^4$ for massive particles and $\mathbf{SO}(2) \times \mathbb{R}^4$ for massless particles.

Coefficients of Hilbert representations for massive particles $m^2 > 0$ have a spectral decomposition into irreducible components for $\mathbf{SU}(2)$ with invariant spin J and, for fixed J, a Lehmann–Källen decomposition for the translations with invariant mass m^2, normalized by $\rho(m^2)$:

$$J = 0, \tfrac{1}{2}, 1, \cdots: \quad \int_0^\infty dm^2 \; \rho(m^2) \int \tfrac{d^4q}{(2\pi)^3} \; \zeta_J(\tfrac{q}{m}) \delta(q^2 - m^2) e^{iqx}.$$

In a rest system, $q = (m,0,0,0)$, the polynomials $\zeta_J(\frac{q}{m})$ are spin-$\mathbf{SU}(2)$ units, used as sesquilinear forms for the Hilbert metric of the spin degrees of freedom:

$$\zeta_J(1,0,0,0) = \mathbf{1}_{1+2J}.$$

The $\mathbf{SU}(2)$-embedding polynomials involve real linear combinations of harmonic Lorentz group polynomials $(q)^L$ with maximal degree $L = 2J$,

$$\zeta_J(\tfrac{q}{m}) = \begin{cases} \displaystyle\sum_{L=0,2,\ldots,2J} \alpha_L^{2J} (\tfrac{q}{m})^L, & J = 0, 1, \ldots, \\ \displaystyle\sum_{L=1,3,\ldots,2J} \alpha_L^{2J} (\tfrac{q}{m})^L, & J = \tfrac{1}{2}, \tfrac{3}{2}, \ldots, \end{cases} \qquad \alpha_{2J}^{2J} = 1,$$

$$\text{harmonic } \mathbf{SO}_0(1,3)\text{-polynomials:} \begin{cases} (q)^0 = 1, \\ (q)^1 = q = \{q_a \mid a = 0,1,2,3\}, \\ (q)^2 = (q \vee q) = \{q_a q_b - \frac{\eta_{ab}}{4} q^2\}, \\ (q)^3 = \{q_a q_b q_c - \frac{\eta_{ab} q_c + \eta_{ac} q_b + \eta_{bc} q_a}{4} q^2\}, \ldots, \end{cases}$$

for example, for spin $\frac{1}{2}$ with Weyl matrices and for spin 1:

$$\begin{aligned}\zeta_{\frac{1}{2}}(\tfrac{q}{m}) &= \sigma^a \tfrac{q_a}{m} \to \sigma^0 = \mathbf{1}_2 \cong \delta_{AB},\\ \zeta_1(\tfrac{q}{m}) &= \tfrac{q_a q_b}{m^2} - \eta_{ab}\tfrac{q^2}{m^2} \cong (\tfrac{q}{m})^2 - \tfrac{3}{4}(\tfrac{q}{m})^0 \to \mathbf{1}_3 \cong \delta_{\alpha\beta}.\end{aligned}$$

The harmonic polynomials are totally symmetric products of the left and right Weyl representations $[\frac{1}{2}|0]$ and $[0|\frac{1}{2}]$ of the boosts (Weyl transmutators; see Chapter 7):

$$\begin{aligned}(\tfrac{q}{m})^L &= \overset{L}{\bigvee} s^*(\tfrac{q}{m}) \otimes \overset{L}{\bigvee} s(\tfrac{q}{m}),\\ s(\tfrac{q}{m}) &= \sqrt{\tfrac{q_0+m}{2m}}\left[\mathbf{1}_2 + \tfrac{\vec{\sigma}\vec{q}}{q_0+m}\right] = \frac{1}{\sqrt{2m(q_0+m)}}\begin{pmatrix} q_0+q_3+m & q_1-iq_2\\ q_1+iq_2 & q_0-q_3+m\end{pmatrix},\\ \hat{s}(\tfrac{q}{m}) &= \sqrt{\tfrac{q_0+m}{2m}}\left[\mathbf{1}_2 - \tfrac{\vec{\sigma}\vec{q}}{q_0+m}\right],\quad s(1,0,0,0) = \mathbf{1}_2,\quad \hat{s}(1,0,0,0) = \mathbf{1}_2.\end{aligned}$$

They are acted on by the $\mathbf{SO}_0(1,3)$-representations $[\frac{L}{2}|\frac{L}{2}]$ (see Chapter 8).

Feynman propagators, as used for the expansion of interactions with particle fields, also include the imaginary principal value off-shell part:

$$J = 0, \tfrac{1}{2}, 1, \cdots: \quad \int_0^\infty dm^2\ \rho(m^2)\tfrac{i}{\pi}\int \tfrac{d^4q}{(2\pi)^3}\ \tfrac{\zeta_J(\frac{q}{m})}{q^2+io-m^2}e^{iqx}.$$

Representations of stable particles with different masses are Schur-orthogonal (see Chapter 8), as illustrated for scalar particles with the divergent volume of the mass hyperboloid,

$$\begin{aligned}|m^2\}(x) &= \int \tfrac{d^4q}{(2\pi)^3}\ \delta(q^2-m^2)e^{iqx},\quad |\mathcal{Y}^3| = \int d^4q\ \delta(q^2-1) = \infty,\\ \{m_1^2|m_2^2\} &= \int d^4x\ |m_1^2\}(x)\ |m_2^2\}(x) = \tfrac{|\mathcal{Y}^3|}{(2\pi)^2}\delta(\tfrac{m_1^2-m_2^2}{m_1^2}).\end{aligned}$$

Relativistic $\mathbf{SL}(2,\mathbb{C})$-fields for massless particles with $\mathbf{SO}(2)$-polarization pairs $(\pm L_3)$ contain, with the exception of scalar fields $L_3 = 0$ and spinor fields $L_3 = \frac{1}{2}$, nonparticle degrees of freedom. The Feynman propagators are linear combinations with harmonic Lorentz group polynomials and $\beta_{2J}^{2J} = 1$:

$$\begin{aligned}|L_3| = J = 0, 1, \cdots: &\quad \rho^J(0)\tfrac{i}{\pi}\int \tfrac{d^4q}{(2\pi)^3}\ \tfrac{1}{q^2+io}\ \sum_{L=0,2,\dots,2J} \beta_L^{2J}\frac{(q)^L}{(q^2+io)^{\frac{L}{2}}}e^{iqx},\\ |L_3| = J = \tfrac{1}{2}, \tfrac{3}{2}, \cdots: &\quad \rho^J(0)\tfrac{i}{\pi}\int \tfrac{d^4q}{(2\pi)^3}\ \tfrac{1}{q^2+io}\ \sum_{L=1,3,\dots,2J} \beta_L^{2J}\frac{(q)^L}{(q^2+io)^{\frac{L-1}{2}}}e^{iqx}.\end{aligned}$$

Only the q^2-independent contributions give a Dirac distribution $\delta(q^2)$ (on-shell), e.g., for $J = \frac{1}{2}$ and $J = 1$,

$$\tfrac{i}{\pi}\tfrac{(q)^1}{q^2+io} \cong q_a\delta(q^2) + \dots,\quad \tfrac{i}{\pi}\tfrac{(q)^2}{(q^2+io)^2} \cong -\tfrac{\eta_{ca}}{4}\delta(q^2) + \dots.$$

In the Feynman propagators, there are also "gauge-dependent" contributions, e.g., for the vector field with $\nu \in \mathbb{R}$,

$$\rho^1(0)\tfrac{i}{\pi}\int \tfrac{d^4q}{(2\pi)^3}\ \frac{-\eta_{ca}+2\nu\frac{q_c q_a}{q^2+io}}{q^2+io}e^{iqx}.$$

The normalizations of the Poincaré group representations are the residues of the Fourier-transformed Feynman propagators at the particles masses

$q^2 = m^2$ in the complex q^2-plane, e.g., for a massive Dirac spinor (with left and right components) and a vector field or a massless Weyl spinor, a vector, and a tensor field:

$$\int \frac{d^4q}{2\pi} e^{-iqx} \begin{cases} \langle 0|\mathbf{\Psi}(0)\overline{\mathbf{\Psi}}(x)|0\rangle_{\text{Feynman}} &= \frac{i}{\pi}\frac{\rho(m^2)}{q^2+io-m^2} & (\gamma q + m), \\ \langle 0|\mathbf{Z}_c(0)\mathbf{Z}_a(x)|0\rangle_{\text{Feynman}} &= \frac{i}{\pi}\frac{\rho(m_Z^2)}{q^2+io-m_Z^2} & (\frac{q_c q_a}{m_Z^2} - \eta_{ca}), \\ \langle 0|\mathbf{l}(0)\mathbf{l}^*(x)|0\rangle_{\text{Feynman}} &= \frac{i}{\pi}\frac{\rho^l(0)}{q^2+io} & \sigma q, \\ \langle 0|\mathbf{A}_c(0)\mathbf{A}_a(x)|0\rangle_{\text{Feynman}} &= \frac{i}{\pi}\frac{\rho^A(0)}{q^2+io} & (-\eta_{ca} + \ldots), \\ \langle 0|\mathbf{E}_{cd}(0)\mathbf{E}_{ab}(x)|0\rangle_{\text{Feynman}} &= \frac{i}{\pi}\frac{\rho^E(0)}{q^2+io} & (\frac{\eta_{ca}\eta_{db}+\eta_{da}\eta_{cb}-\eta_{cd}\eta_{ab}}{2} + \ldots). \end{cases}$$

For free fields (flat spacetime) with the on-shell contributions $\rho(m^2)\delta(q^2-m^2)$ for Poincaré group representation coefficients, the normalizations are free. For interactions, as mediated by the causally supported off-shell contributions $\frac{\rho(m^2)}{q_{\mathrm{P}}^2-m^2}$, the normalizations give the coupling constants.

For a massive field, an expansion for small energy-momenta $q = 0$, e.g.,

$$\langle 0|\mathbf{Z}_c(0)\mathbf{Z}_a(x)|0\rangle_{\text{Feynman}} = 2i\frac{\rho(m_Z^2)}{m_Z^2}\eta_{ca}\delta(x) + \ldots,$$

gives the local approximation of a field-mediated interaction as used in the pointlike description of weak interactions with coupling constant $\frac{\rho(m_Z^2)}{m_Z^2}$ involving the inverse particle mass.

The normalization of the electromagnetic interaction is given by $4\pi\alpha_S$ with the dimensionless Sommerfeld constant. The Newton constant for the gravitational interaction has the dimension of an area $\ell^2 = \frac{8\pi G\hbar}{c^3}$ or the inverse Planck mass squared:

$$\begin{aligned} \langle \mathbf{A}\ \mathbf{A}\rangle &\sim \frac{\rho^A(0)}{q^2} = \frac{g^2}{q^2} = \frac{4\pi\alpha_S}{q^2}, \\ \langle \mathbf{E}\ \mathbf{E}\rangle &\sim \frac{\rho^E(0)}{q^2} = \frac{\ell^2}{q^2} = \frac{8\pi}{m_P^2 q^2}. \end{aligned}$$

With a mass unit M, dimensionless quantities (underlined) can be used, leading to dimensionless Feynman propagators with dimensionless normalizations and coupling constants:

$$\begin{aligned} x &= \frac{\underline{x}}{M}, \\ q &= M\underline{q},\ \frac{d^4q}{q^2} = M^2\frac{d^4\underline{q}}{\underline{q}^2}, \\ \mathbf{A} &= M\underline{\mathbf{A}},\ \mathbf{E} = \underline{\mathbf{E}} \end{aligned} \Rightarrow \begin{cases} \langle \underline{\mathbf{A}}\ \underline{\mathbf{A}}\rangle \sim \frac{\underline{\rho}^A(0)}{\underline{q}^2}, & \underline{\rho}^A(0) = g^2, \\ \langle \underline{\mathbf{E}}\ \underline{\mathbf{E}}\rangle \sim \frac{\underline{\rho}^E(0)}{\underline{q}^2}, & \underline{\rho}^E(0) = \frac{8\pi M^2}{m_P^2}. \end{cases}$$

5.11 Renormalization of Gauge Fields

Quantum field theories for flat spacetime start from an action as the sum of the free-field part and the interaction. Flat spacetime operations cannot describe interactions. This shows up in a perturbative expansion of an interaction with flat spacetime quantum fields. By giving up to calculate, e.g., the mass ratios and the coupling constants, the "divergencies" can be tamed in "renormalizable" theories by the familiar regularization-renormalization procedure, e.g., for quantum electrodynamics. This is no longer possible for "nonrenormalizable" quantum gravity in the flat spacetime approach.

5.11.1 Perturbative Corrections of Normalizations

In a perturbative expansion of an interacting theory, the coupling constants are renormalized as exemplified by the electromagnetic interaction $\mathbf{J}_a\mathbf{A}^a$: The field equation for the gauge field,

$$\tfrac{1}{g^2}(-\partial^2\mathbf{A}_a + \partial_a\partial^b\mathbf{A}_b) = \mathbf{J}_a,$$

contains the current as position density of the $\mathbf{U}(1)$-generator. The normalization of the current-carrying fields, e.g., a massive Dirac field with charge number z (electron-positron with $z = \pm 1$),

$$\begin{aligned}
\langle 0|\mathbf{\Psi}(0)\overline{\mathbf{\Psi}}(x)|0\rangle_{\text{Feynman}} &= \rho(m^2)\tfrac{i}{\pi}\int \tfrac{d^4q}{(2\pi)^3}\tfrac{\gamma_a q^a + m}{q^2+io-m^2}e^{iqx},\\
\{\mathbf{\Psi}(0),\overline{\mathbf{\Psi}}(x)\}\big|_{x_0=0} &= \rho(m^2)\gamma_0\delta(\vec{x}),\\
(i\gamma_a\partial^a + m)\langle 0|\mathbf{\Psi}(0)\overline{\mathbf{\Psi}}(x)|0\rangle_{\text{Feynman}} &= -2i\rho(m^2)\delta(x),
\end{aligned}$$

has to be compatible with the $\mathbf{U}(1)$-generator properties; i.e., the current has to involve the inverse normalization factor, e.g., a position translation Dirac normalization $\int d^3x\ \delta(\vec{x})$ with $\rho(m^2) = 1$:

$$Q = i\int d^3x\ \mathbf{J}_0(x) \in \log\mathbf{U}(1),\quad \begin{cases} \mathbf{\Psi} \longmapsto e^{iz\alpha}\mathbf{\Psi},\\ [Q,\mathbf{\Psi}(x)] = iz\mathbf{\Psi}(x),\\ \mathbf{J}_a = \frac{z}{\rho(m^2)}\overline{\mathbf{\Psi}}\gamma_a\mathbf{\Psi}. \end{cases}$$

Green's function of the free gauge field equation, up to gauge-dependent terms,

$$-\tfrac{1}{g^2}\partial^2\kappa^0(x) = \delta(x) \Rightarrow \kappa^0(x) = \int \tfrac{d^4q}{(2\pi)^4}\ \tfrac{g^2}{q^2+io}e^{iqx},\quad \tilde{\kappa}^0(q) = \tfrac{g^2}{q^2+io},$$

can be made to coincide up to a factor $2i$ with the gauge field Feynman propagator:

$$\langle 0|\mathbf{A}_a(0)\mathbf{A}_b(x)|0\rangle_0 = \tfrac{i}{\pi}\int \tfrac{d^4q}{(2\pi)^3}\tfrac{g^2(-\eta_{ab})}{q^2+io}e^{iqx} = -2i\eta_{ab}\kappa^0(x) = 2i\kappa^0_{ab}(x).$$

In a perturbative approach, the Feynman progagator of the free electromagnetic field (subindex 0) is modified in the first order (subindex 1),

$$\begin{aligned}
\langle 0|\mathbf{A}_a(x_1)\mathbf{A}_b(x_2)|0\rangle_1 &= \langle 0|\mathbf{A}_a(x_1)\mathbf{A}_b(x_2)|0\rangle_0\\
&\quad + \int d^4y_1 d^4y_2 \kappa^0_{ac}(x_1-y_1)\langle 0|\mathbf{J}^c(y_1)\mathbf{J}^d(y_2)|0\rangle\kappa^0_{db}(y_2-x_2),\\
\langle\mathbf{A}_a\mathbf{A}_b\rangle_1(q) &= \langle\mathbf{A}_a\mathbf{A}_b\rangle_1(q) + \tilde{\kappa}^0_{ac}(q)\Pi^{cd}(q,m^2)\tilde{\kappa}^0_{db}(q)\\
&= \left[\delta^d_a + \tfrac{g^2(-\eta_{ac})}{q^2+io}\Pi^{cd}(q,m^2)\right]\tfrac{i}{\pi}\tfrac{g^2(-\eta_{db})}{q^2+io},
\end{aligned}$$

by the vacuum polarization as the ground-state value of the bilinear bilocal product of currents:

$$\langle 0|\mathbf{J}^c(0)\mathbf{J}^d(x)|0\rangle = \frac{i}{\pi}\int \frac{d^4q}{(2\pi)^3}\,\Pi^{cd}(q,m^2)e^{iqx},\quad \langle \mathbf{A}_a\mathbf{A}_b\rangle_0(q) = \frac{i}{\pi}\frac{g^2(-\eta_{ab})}{q^2+io}.$$

The vacuum polarization involves the convolution of the energy-momentum propagators (see Chapter 9), the normalization factors $\rho^2(m^2)$ drop out:

$$\Pi^{cd}(q,m^2) = -2iz^2\int \frac{d^4p}{(2\pi)^4}\,\mathrm{tr}\,\gamma^c\frac{\gamma(q-p)+m}{(q-p)^2+io-m^2}\gamma^d\frac{\gamma p+m}{p^2+io-m^2}.$$

Its quadratic "divergence," formally illustrated by

$$\begin{aligned}\Pi^c_c(0,m^2) &= z^2\int \frac{d^4p}{i\pi^4}\frac{p^2-2m^2}{(p^2+io-m^2)^2} = z^2\int \frac{d^4p}{i\pi^4}\int_\infty^{m^2} d\kappa^2\,\frac{2\kappa^2}{(-p^2-io+\kappa^2)^3}\\ &= \frac{z^2}{\pi^2}\int_\infty^{m^2} d\kappa^2\int \frac{d^4p}{i\pi^2}\frac{2\kappa^2}{(-p^2-io+\kappa^2)^3} = \frac{z^2}{\pi^2}\int_\infty^{m^2} d\kappa^2,\end{aligned}$$

shows the failure to expand a spacetime interaction in terms of free fields with a Fock ground-state vector $|0\rangle$.

The Pauli–Villars regularization, denoted with a ground-state vector $|\mathcal{M}\rangle$ for a field theory with interactions, keeps the naive gauge invariance and current conservation $q_c\Pi^{cd}(q) = 0$ by regularizing the vacuum polarization "as a whole," i.e., pairwise, not by a modification of the individual fermion propagators. The regularization trivializes the "divergent" moments of a dilation (mass) expansion:

$$\int \frac{d^4x}{2i}\langle\mathcal{M}|\mathbf{J}^c(0)\mathbf{J}^d(x)|\mathcal{M}\rangle e^{-iqx} = \Pi^{cd}(q) = \sum_{i=0}^{N}\rho_i\Pi^{cd}(q,m_i^2),$$

$$\text{with}\ \sum_{i=0}^{N}\rho_i(m_i^2)^k = 0,\quad k=0,1,\ldots,\ \text{where}\ \rho_0 = 1,\ m_0^2 = m^2,$$

$$\Pi^{cd}(q) = \frac{z^2}{\pi^2}(\eta^{cd}q^2 - q^cq^d)\sum_{i=0}^{N}\rho_i\int d\zeta\,\zeta(1-\zeta)\log[-\zeta(1-\zeta)q^2+io+m_i^2].$$

The gauge-invariant part of the renormalization of the free gauge field propagator can be regularized by one regulator, $k=0$:

$$\begin{aligned}\Pi^{cd}(q) &= \Pi^{cd}(q,m^2) - \Pi^{cd}(q,M^2),\\ \langle\mathbf{A}_a\mathbf{A}_b\rangle_1(q) &= \left(1+\frac{g^2z^2}{\pi^2}\int_0^1 d\zeta\,\zeta(1-\zeta)\log\frac{-\zeta(1-\zeta)q^2-io+m^2}{-\zeta(1-\zeta)q^2-io+M^2}\right)\frac{i}{\pi}\frac{g^2(-\eta_{ab})}{q^2+io}.\end{aligned}$$

It is even possible to restrict the regularization to lightlike energy-momentum $q^2 = 0$:

$$\langle\mathbf{A}_a\mathbf{A}_b\rangle_1(q) = \left(1+\frac{g^2z^2}{\pi^2}\int_0^1 d\zeta\,\zeta(1-\zeta)\log\frac{-\zeta(1-\zeta)q^2-io+m^2}{M^2}\right)\frac{i}{\pi}\frac{g^2(-\eta_{ab})}{q^2+io}.$$

The vaccum polarization involves the characteristic convolution for four-dimensional spacetime,

$$\begin{aligned}
\tfrac{g^2(-\eta_{ac})}{q^2}\Pi^{cd}(q) &= \tfrac{g^2}{6\pi^2}\Pi(q^2)\delta_a^d,\\
\tfrac{1}{z^2}\Pi(q^2+io) &= \int_0^1 d\zeta\, 6\zeta(1-\zeta)\log\tfrac{-\zeta(1-\zeta)q^2-io+m^2}{M^2}\\
&= \log\tfrac{m^2}{M^2} + q^2\int_0^1 d\zeta\, \tfrac{\zeta^2(3-2\zeta)(1-2\zeta)}{-\zeta(1-\zeta)q^2-io+m^2}\\
&= \log\tfrac{m^2}{M^2} - \tfrac{5}{3} - \tfrac{4m^2}{q^2} - \tfrac{(q^2+2m^2)(q^2-4m^2)}{2q^2}\int_0^1 \tfrac{d\zeta}{-\zeta(1-\zeta)q^2-io+m^2},
\end{aligned}$$

with three parts: momentumlike, energylike over and under the two particle threshold $q^2 = 4m^2$ (see Chapter 9),

$$\begin{aligned}
\tfrac{1}{z^2}\Pi(q^2+io) &= \log\tfrac{m^2}{M^2} - \tfrac{5}{3} - \tfrac{4m^2}{q^2}\\
&\quad -\tfrac{q^2+2m^2}{q^2}\sqrt{\left|\tfrac{q^2-4m^2}{q^2}\right|}\begin{bmatrix} \vartheta(-q^2)\log\left|\tfrac{2m^2-q^2+\sqrt{q^2(q^2-4m^2)}}{2m^2}\right| \\ +\,\vartheta(q^2)\vartheta(4m^2-q^2)\arctan\tfrac{\sqrt{q^2(4m^2-q^2)}}{q^2-2m^2} \\ +\,i\pi\,\vartheta(q^2-4m^2)\end{bmatrix}\\
&= \log\tfrac{m^2}{M^2} - \tfrac{q^2}{5m^2} + \ldots \text{ for } |q^2| \ll m^2.
\end{aligned}$$

The three contributions in the gauge field Feynman propagator, on-shell, off-shell momentumlike, and off-shell energylike,

$$\tfrac{i}{\pi}\tfrac{g^2(-\eta_{ab})}{q^2+io} = g^2(-\eta_{ab})\delta(q^2) + [\vartheta(-q^2)+\vartheta(q^2)]\tfrac{i}{\pi}\tfrac{g^2(-\eta_{ab})}{q^2},$$

are modified differently. The Poincaré group representation coefficient (on-shell contribution, real) is renormalized by

$$g^2 \longmapsto g_R^2 = (1+\tfrac{g^2}{6\pi^2}\Pi(0))g^2,\quad \Pi(0) = z^2\log\tfrac{m^2}{M^2}.$$

The imaginary contribution over the threshold for the creation of oppositely charged particle pairs is multiplied with the off-shell $\frac{i}{\pi}\frac{g^2(-\eta_{ab})}{q_{\rm P}^2}$ and gives an additional real two particle on-shell contribution, all proportional to $-\eta_{ab}$:

$$g^2\delta(q^2) \longmapsto (1+\tfrac{g^2z^2}{6\pi^2}\log\tfrac{m^2}{M^2})g^2\delta(q^2) + \vartheta(q^2-4m^2)\tfrac{g^2z^2}{6\pi^2}\sqrt{1-\tfrac{4m^2}{q^2}}(1+\tfrac{2m^2}{q^2})\tfrac{g^2}{q^2}.$$

The off-shell contribution (imaginary) with the principal value and the Coulomb interaction $\int dx_0 \int \frac{d^4q}{i\pi(2\pi)^3}\frac{g^2}{-q^2-io}e^{iqx} = \frac{g^2}{2i\pi r}$ has an energy-momentum–dependent modification, for the momentumlike $\vartheta(-q^2)$ contributions:

$$\begin{aligned}
\vartheta(-q^2)\tfrac{g^2}{q^2} \longmapsto \vartheta(-q^2)\Big(1+\tfrac{g^2z^2}{6\pi^2}\Big[&\log\tfrac{m^2}{M^2} - \tfrac{5}{3} - \tfrac{4m^2}{q^2}\\
&-\tfrac{q^2+2m^2}{q^2}\sqrt{\tfrac{q^2-4m^2}{q^2}}\log\tfrac{2m^2-q^2+\sqrt{q^2(q^2-4m^2)}}{2m^2}\Big]\Big)\tfrac{g^2}{q^2},
\end{aligned}$$

and for the energylike $\vartheta(q^2)$-contributions:

$$\begin{aligned}
\vartheta(q^2)\tfrac{g^2}{q^2} \longmapsto \vartheta(q^2)\Big(1+\tfrac{g^2z^2}{6\pi^2}\Big[&\log\tfrac{m^2}{M^2} - \tfrac{5}{3} - \tfrac{4m^2}{q^2}\\
&-\vartheta(4m^2-q^2)\tfrac{q^2+2m^2}{q^2}\sqrt{\tfrac{4m^2-q^2}{q^2}}\arctan\tfrac{\sqrt{q^2(4m^2-q^2)}}{q^2-2m^2}\Big]\Big)\tfrac{g^2}{q^2}.
\end{aligned}$$

The first-order perturbative correction can be considered as the first term of a geometrical series expansion for the gauge field propagator:

$$\langle \mathbf{A}_a \mathbf{A}_b \rangle(q) = \frac{i}{\pi} \frac{g^2(-\eta_{ab})}{(q^2+io)[1-\frac{g^2}{6\pi^2}\Pi(q^2+io)]} = g_R^2(-\eta_{ab})\delta(q^2) + \dots.$$

The renormalized coupling constant is the residue at $q^2 = 0$, i.e., the inverse derivative there of the denominator function,

$$\frac{6\pi^2}{g_R^2} = \frac{\partial}{\partial q^2} q^2 [\frac{6\pi^2}{g^2} - \Pi(q^2)]\Big|_{q^2=0} = \frac{6\pi^2}{g^2} - z^2 \log \frac{m^2}{M^2}.$$

5.11.2 Lie Algebra Renormalization by Vacuum Polarization

Lie algebra L representations on finite-dimensional vector spaces V with their nondecomposable parts on V_ι,

$$L \ni l^A \longmapsto \mathcal{D}(l^A) = \bigoplus_{\iota=1}^{d} \mathcal{D}_\iota(l^A) \cong \mathcal{D}^{A\beta}_{\ \ \alpha}, \quad V = \bigoplus_{\iota=1}^{d} V_\iota,$$

are implemented in field theories by charges as position integrals over the timelike component of currents,

$$L \ni l^A \longmapsto iQ^A = i \int d^3x \ \mathbf{J}_0^A(x), \quad [iQ^A, iQ^B] = \epsilon_C^{AB} iQ^C,$$

e.g., the currents of massive Dirac fields, normalized by $\{\mathbf{\Psi}(0), \overline{\mathbf{\Psi}}(x)\}\big|_{x_0=0} = \gamma_0\delta(\vec{x})$, for electromagnetic $\mathbf{U}(1)$ or isospin $\mathbf{SU}(2)$ with Pauli matrices (see Chapter 6):

$$i\mathbf{J}_a^A = \overline{\mathbf{\Psi}}_\beta \mathcal{D}^{A\beta}_{\ \ \alpha} \gamma_a \mathbf{\Psi}^\alpha, \text{ e.g.,} \begin{cases} \mathbf{J}_a &= \sum_{\iota=1}^{d} z_\iota \overline{\mathbf{\Psi}}_\iota \gamma_a \mathbf{\Psi}_\iota, \quad z_\iota \in \mathbb{Z}, \quad \text{for } \mathbf{U}(1), \\ \mathbf{J}_a^A &= \overline{\mathbf{\Psi}}_\beta \frac{\tau^{A\beta}_{\ \ \alpha}}{2} \gamma_a \mathbf{\Psi}^\alpha \qquad\qquad \text{for } \mathbf{SU}(2). \end{cases}$$

The vacuum polarization for current-gauge field coupling $\mathbf{J}_a^B \mathbf{A}_B^a$ contains the ground-state expectation value of the current square:

$$\begin{aligned} \Pi_{ab}^{AB}(q) &= \int \frac{d^4x}{2i} \langle \mathcal{M} | \mathbf{J}_a^A(0) \mathbf{J}_b^B(x) | \mathcal{M} \rangle e^{-iqx}, \\ \frac{g^2(-\eta^{ac})\delta_{AC}}{q^2} \Pi_{cd}^{CD}(q) &= \frac{g^2}{6\pi^2} \Pi(q^2) \delta_d^a \delta_A^D. \end{aligned}$$

It involves invariant symmetric bilinear forms of the Lie algebra represen-tatations, which are orthogonally diagonalizable,

$$L \times L \longrightarrow \mathbb{R}, \quad \kappa_V(l^A, l^B) = \sum_{\iota=1}^{d} \kappa_\iota^{AB} \cong \delta^{AB} \sum_{\iota=1}^{d} \kappa_\iota,$$

e.g., $-\sum_{\iota=1}^{d} z_\iota^2$ for electromagnetic $\mathbf{U}(1)$ and $\mathrm{tr}\,\frac{i\tau^A}{2} \circ \frac{i\tau^B}{2} = -\frac{\delta^{AB}}{2}$ for isospin $\mathbf{SU}(2)$.

The perturbative corrections by the corresponding vacuum polarizations are illustrated by Dirac field currents with one regulator:

$$\langle \mathbf{A}_A^a \mathbf{A}_B^b \rangle(q) = \frac{i}{\pi} \frac{g^2(-\eta^{ab})\delta_{AB}}{(q^2+io)\left[1-\frac{g^2}{6\pi^2}\Pi(q^2+io)\right]}, \quad \Pi(q^2) = -\sum_{\iota=1}^{d} \kappa_\iota \Pi^\iota(q^2),$$

$$\frac{6\pi^2}{g_R^2} = \frac{6\pi^2}{g^2} - \Pi(0), \quad \Pi(0) = -\sum \kappa_\iota \Pi^\iota(0) = \begin{cases} \sum z_\iota^2 \log \frac{m_\iota^2}{M^2}, & \mathbf{U}(1), \\ \frac{1}{2} \log \frac{m^2}{M^2}, & \mathbf{SU}(2). \end{cases}$$

Chapter 6

External and Internal Operations

Basic physical theories involve both *external or spacetimelike* and *internal or chargelike* degrees of freedom in complex representation vector spaces that are acted on, respectively, by operations from the Poincaré group, i.e., by Lorentz transformations and spacetime translations, and by electroweak and strong operations from the hypercharge, isospin, and color groups. The external–internal dichotomy goes with a noncompact–compact distinction of the relevant groups.

The properties of all basic interactions and particles are determined and characterized by invariants and eigenvalues for these operation groups. Although the product of external and internal operations in the acting group is direct, $G_{\text{ext}} \times G_{\text{int}}$, the internal "chargelike" operations are coupled to the external "spacetimelike" ones: Any spacetime translation is accompanied by a chargelike operation. This is implemented by the gauge fields and the corresponding covariant derivatives for the particle fields in the standard model of electroweak and strong interactions. All interactions can be formulated, in a classical geometrical language, by connections of bundles, by a Riemannian connection for the tangent spaces of "horizontal" spacetime as the base, yielding the external interactions, i.e., gravity, and by connections of "vertical" complex vector spaces as fibers, yielding the internal interactions, e.g., the electroweak and strong ones.

This chapter reviews the internal gauge symmetries and their mathematical formalization in the standard model and in the framework of fiber bundles with connections.

H. Saller, *Operational Spacetime: Interactions and Particles*, Fundamental Theories of Physics 163, DOI 10.1007/978-1-4419-0898-8_7,

6.1 Fiber Bundles

Chargelike operations can be described with bundles. Bundles generalize the "infinitesimal" (tangent space) structures $\mathbf{T}(\mathbb{M})$ with the derivations of the functions of a manifold (see Chapter 2). The local structure of a manifold $\mathbb{M}$ can be enriched by an additional differential manifold F. The fiber F may be a topological vector space, an operational Lie group, or an operational Lie algebra. One specimen of the fiber F is "planted" at each point of the base manifold $\mathbb{M}$.

6.1.1 Fibers and Base

A *fiber bundle* $F(\mathbb{M}) \in \underline{\mathbf{dif}}_{\mathbb{K}}(\underline{\mathbf{top}})$ with a topological space $\mathbb{M} \in \underline{\mathbf{top}}$ as *base* and a manifold $F \in \underline{\mathbf{dif}}_{\mathbb{K}}$ as *typical fiber* is characterized by a surjective continuous *projection* from bundle to base whose local inverses are the *local fibers*, all isomorphic to the typical fiber:

$$\begin{aligned} \pi : F(\mathbb{M}) &\longrightarrow \mathbb{M}, \quad \xi \longmapsto \pi(\xi), \\ \mathbb{M} \ni x &\longmapsto \pi^{-1}(x) = F_x \subset F(\mathbb{M}), \quad F_x \cong F. \end{aligned}$$

The continuity and "sufficient smoothness" of the mappings used will not be discussed. The additional conditions for local triviality and the structural group of a bundle are given in the next subsection. In the cases considered, the base is also a manifold, $\mathbb{M} \in \underline{\mathbf{dif}}_{\mathbb{R}}$.

With the equivalence relation to belong to the same fiber, the *base characterizes equivalence classes*:

$$\xi, \xi' \in F(\mathbb{M}) : \quad \xi \pi \xi' \iff \pi(\xi) = \pi(\xi') \Rightarrow F(\mathbb{M})/\pi \cong \mathbb{M}.$$

A bundle is the union of the local fibers with the base, the hedgehog with prickles and skin,

$$F(\mathbb{M}) = \bigcup_{x \in \mathbb{M}} (x, F_x) = \pi^{-1}[\mathbb{M}].$$

The *trivial F-bundle for* $\mathbb{M}$ is the set product $F \times \mathbb{M}$.

Properties related to the base $\mathbb{M}$ are called *horizontal*, later used as "external" and "spacetimelike," whereas properties related to the fiber F are called *vertical*, later used as "internal" and "chargelike."

The category from which the fiber comes gives the name for the bundle, e.g., a topological *vector space bundle* $\underline{\mathbf{tvec}}_{\mathbb{K}}(\underline{\mathbf{top}})$ with a vector space fiber $F \in \underline{\mathbf{tvec}}_{\mathbb{K}}$ or a Lie group bundle $\underline{\mathbf{lgrp}}_{\mathbb{K}}(\underline{\mathbf{top}})$. Examples of vector space bundles are the tangent and cotangent bundle $\mathbf{T}(\mathbb{M})$, $\mathbf{T}^T(\mathbb{M})$ of a manifold with the (co)tangent spaces as the local fibers $\mathbf{T}_x(\mathbb{M}) \cong \mathbb{R}^n \cong \mathbf{T}^T_x(\mathbb{M})$. The frame bundle $\mathbf{AL}(\mathbb{R}^n)(\mathbb{M})$ with the linear tangent space mappings $\mathbf{AL}(\mathbb{R}^n)$ is both an algebra and a Lie algebra bundle.

The *subcategories* $\underline{\mathbf{dif}}_{\mathbb{K}}(\mathbb{M})$ *are important, i.e., different typical fibers for one fixed base space* $\mathbb{M}$, e.g., for spacetime $\mathbb{M}^{(t,s)}$.

The *sections* ψ *of a bundle* are projection-compatible mappings of the base into the bundle, i.e., the mappings $F^{\mathbb{M}}$ from base to fiber,

$$\begin{aligned}\psi : \mathbb{M} &\longrightarrow F(\mathbb{M}), \quad \pi \circ \psi = \mathrm{id}_{\mathbb{M}}, \quad \mathbb{M} \ni x \longmapsto \psi(x) \in F,\\ \psi &= \{(x, \psi(x)) \mid x \in \mathbb{M}\} \in \bigcup_{x \in \mathbb{M}} (x, F_x).\end{aligned}$$

Examples are vector and tensor fields.

The *fiber bundle morphisms* (f, φ) have to be compatible with the projections. The restrictions to the local fibers F_x have to yield morphisms in the fiber category:

$$\begin{array}{ccc} F(\mathbb{M}) & \stackrel{f}{\longrightarrow} & F'(\mathbb{M}') \\ {\scriptstyle\pi}\downarrow & & \downarrow{\scriptstyle\pi'} \\ \mathbb{M} & \underset{\varphi}{\longrightarrow} & \mathbb{M}' \end{array}, \quad f|_{F_x} : F_x \longrightarrow F'_{\varphi(x)}.$$

Operations with the fiber category $\underline{\mathbf{dif}}_{\mathbb{K}}$, i.e., functors acting on $\underline{\mathbf{dif}}_{\mathbb{K}}$, can be transferred to the corresponding bundles $\underline{\mathbf{dif}}_{\mathbb{K}}(\mathbb{M})$. In this way, there arise *associated bundles*, e.g., dual bundles with vector space duality $(V(\mathbb{M}), V^T(\mathbb{M}))$ in $\underline{\mathbf{tvec}}_{\mathbb{K}}(\mathbb{M})$ (^{T} is the dual functor for vector spaces), direct sum bundles $V_1 \oplus V_2(\mathbb{M})$, and tensor product bundles $V_1 \otimes V_2(\mathbb{M})$, and, correspondingly, the transition from a vector space bundle to its tensor algebra bundle $\bigotimes V(\mathbb{M}) \in \underline{\mathbf{aag}}_{\mathbb{K}}(\mathbb{M})$ (direct sum $\oplus$, tensor product $\otimes$, and tensor algebra $\bigotimes$ are functors for vector spaces), or from a Lie group bundle to its Lie algebra bundle $\underline{\mathbf{lgrp}}_{\mathbb{R}}(\mathbb{M}) \ni G(\mathbb{M}) \longmapsto \log G(\mathbb{M}) \in \underline{\mathbf{lag}}_{\mathbb{R}}(\mathbb{M})$ via the functor $\log : \underline{\mathbf{lgrp}}_{\mathbb{R}} \longrightarrow \underline{\mathbf{lag}}_{\mathbb{R}}$.

6.1.2 Structural and Gauge Groups

Now we have the additional condition for a fiber bundle with respect to its operational structure — its local triviality up to its structural group: A manifold is required to be locally like $\mathbb{R}^n$; its tangent bundle involves free modules, $\mathbf{T}(U) \in \underline{\mathbf{mod}}_{C(U)}$ for neighborhoods $U \ni x$. Similarly, a bundle is required to be *locally trivial* $F \times U$; i.e., a bundle $F(\mathbb{M})$ comes with a covering of the base $\mathbb{M}$ and a fiber isomorphism for each set from the covering, called a *trivialization* $(U_\iota, \chi_\iota)_{\iota \in I}$, leading to local trivializations,

$$x \in U_\iota : \quad F_x \stackrel{\chi_\iota}{\cong} F, \quad \pi^{-1}[U_\iota] \cong F \times U_\iota.$$

If a base point is in the intersection of two covering sets, the two local trivializations define a diffeomorphism of the typical fiber:

$$x \in U_\iota \cap U_\kappa, \quad g_\iota^\kappa(x) = \chi_\iota \circ \chi_\kappa^{-1} \in \overset{\circ}{\mathbf{dif}}_{\mathbb{K}}(F, F).$$

The group of all these F-isomorphisms from all local trivializations is required to be equal for all manifold points. It is called the *structural group of the bundle*, or also its gauge group (ahead):

$$\text{for all } x \in \mathbb{M}: \quad \{g_\iota^\kappa(x) \mid x \in U_\iota \cap U_\kappa\} = G \subseteq \overset{\circ}{\mathbf{dif}}_{\mathbb{K}}(F, F).$$

It has to act effectively on the fiber, i.e., in a faithful realization.

The definition of a fiber bundle has to specify the structural group. The fiber with the structural group is a *Klein space* $G \bullet F$, i.e., a differential manifold $F \in \underline{\mathbf{dif}}_G$ with effective and "smooth" Lie group G action:

$$G \bullet F(\mathbb{M}) \in \underline{\mathbf{dif}}_G(\mathbb{M}), \text{ especially } F \in \underline{\mathbf{tvec}}_{\mathbb{K}}, \underline{\mathbf{lgrp}}_{\mathbb{R}}, \underline{\mathbf{lag}}_{\mathbb{R}}.$$

The bundle morphisms have to include a compatible morphism ρ for the structural groups G and G':

$$\begin{aligned} (f, \varphi, \rho): \quad & G \bullet F(\mathbb{M}) \longrightarrow G' \bullet F'(\mathbb{M}'), \\ & \text{with } \rho: G \longrightarrow G', \quad \rho(g_\iota^\kappa) = g_{\iota'}^{\kappa'}. \end{aligned}$$

The structural group is closely related to the local trivialization, given in the definition of the bundle. For example, the trivial bundle $F \times \mathbb{M}$ with trivialization $(\mathbb{M}, \mathrm{id}_F \times \mathrm{id}_{\mathbb{M}})$ has as structural group only the fiber identity $\{\mathrm{id}_F\}$; for the trivialization $(\mathbb{M}, \overset{\circ}{\mathbf{dif}}_{\mathbb{K}}(F, F) \times \mathrm{id}_{\mathbb{M}})$, one obtains the full group $\overset{\circ}{\mathbf{dif}}_{\mathbb{K}}(F, F)$.

It is also possible to impose the structural group via fiber properties, e.g., maximality, a fiber metric, etc., with a correspondingly chosen trivialization. For example, the maximal structural group of a tangent bundle $\mathbf{T}(\mathbb{M})$ is the full linear group $\mathbf{GL}(n, \mathbb{R})$ of the typical fiber. For a Riemannian manifold, the structural group can be restricted via the tangent space metric and orthonormal bases to the tangent Lorentz group $\mathbf{SO}_0(t, s)$. In any case, the structural group is realized faithfully in the diffeomorphism group of the fiber, $G \longrightarrow \overset{\circ}{\mathbf{dif}}_{\mathbb{R}}(F, F)$.

6.2 Nonrelativistic and Relativistic Bundles

Relativistic fields are mappings of a spacetime manifold into a real or complex vector space, i.e., sections of a bundle $V(\mathbb{M}^{(1,3)})$,

$$\mathbb{M}^{(1,3)} \ni x \longmapsto \mathbf{\Phi}(x) \in V.$$

Special relativistic fields (see Chapter 5) map the spacetime translations, i.e., the flat manifold $\mathbb{R}^4 \cong \mathbf{SO}_0(1,3) \vec{\times} \mathbb{R}^4 / \mathbf{SO}_0(1,3)$, into a vector space with the action (a representation) of the Lorentz group. The fields carry the Lorentz group transformation behavior, which is induced by the Lorentz group action on both spaces, as given in the commutative diagram

$$\begin{array}{ccc} \mathbb{R}^4 & \xrightarrow{\Lambda(s)} & \mathbb{R}^4 \\ \Phi \downarrow & & \downarrow \Phi_s, \\ V & \xrightarrow[D(s)]{} & V \end{array} \qquad \begin{array}{rl} \mathbf{SL}(2,\mathbb{C}) \ni s & \longmapsto \Lambda(s) \in \mathbf{SO}_0(1,3), \\ \mathbf{SL}(2,\mathbb{C}) \ni s & \longmapsto D(s) \in \mathbf{SL}(V), \\ V^{\mathbb{R}^4} \ni \Phi & \stackrel{s}{\longmapsto} \Phi_s \in V^{\mathbb{R}^4}, \\ \Phi_s(x) & = D(s).\Phi(\Lambda^{-1}(s).x). \end{array}$$

For example, a Dirac spinor field is transformed by the complex four-dimensional Dirac representation,

$$\begin{aligned} \mathbb{R}^4 \ni x &\longmapsto \Psi(x) \in V \cong \mathbb{C}^4, \\ V^{\mathbb{R}^4} \ni \Psi &\stackrel{s}{\longmapsto} \Psi_s \in V^{\mathbb{R}^4}, \quad \Psi_s(x) = (s \oplus s^{-1\star})\Psi(\Lambda^{-1}(s).x). \end{aligned}$$

In addition to the external Lorentz transformations, the vector space V with the field values can be acted on by an internal operation group $\mathbf{U}$, which acts trivially on spacetime:

$$\mathbf{U} \ni u \longmapsto R(u) \in \mathbf{GL}(V), \quad V^{\mathbb{R}^4} \ni \Phi \stackrel{u}{\longmapsto} R(u).\Phi \in V^{\mathbb{R}^4}.$$

Summarizing: Special relativistic fields use trivial vector space bundles with the Minkowski translations $\mathbb{R}^4$ as base and, as fiber, a representation space V for the Lorentz group and for the chargelike operations $\mathbf{U}$ as structural group, e.g., for the hyperisospin group $\mathbf{U}(2) \longrightarrow \mathbf{U}(V)$,

$$[\mathbf{SL}(2,\mathbb{C}) \times \mathbf{U}] \bullet V(\mathbb{R}^4), \text{ with } V(\mathbb{R}^4) \cong V \times \mathbb{R}^4, \quad V \cong \mathbb{C}^N.$$

If nonrelativistic mechanics is described by the same concepts, positions and momenta are mappings of the real one-dimensional time manifold $\mathbb{M}^1 \cong \mathbb{R}$ into a value space:

$$\mathbb{M}^1 \ni t \longmapsto (\mathbf{x},\mathbf{p})(t) \in V.$$

In contrast to the Lorentz group $\mathbf{SO}_0(1,3)$ for spacetime translations $\mathbb{R}^4$, the external homogeneous group on time translations $\mathbb{R}$ is trivial, $\mathbf{SO}(1) = \{1\}$.

The vector space with the position-momentum values can be acted on by an internal operation group $\mathbf{O}$, which acts trivially on time,

$$\mathbf{O} \ni O \longmapsto R(O) \in \mathbf{GL}(V), \quad V^{\mathbb{R}} \ni (\mathbf{x},\mathbf{p}) \stackrel{O}{\longmapsto} (R(O).\mathbf{x}, R(O).\mathbf{p}) \in V^{\mathbb{R}},$$

e.g., the rotation group $\mathbf{O}(3)$ in three-dimensional mechanics, $a, b = 1, 2, 3$,

$$\begin{aligned} \mathbb{R} \ni t &\longmapsto (\mathbf{x}^a, \mathbf{p}^a)(t) \in V \cong \mathbb{R}^3, \\ \mathbf{O}(3) \ni O, \quad V^{\mathbb{R}} \ni (\mathbf{x}^a, \mathbf{p}^a) &\stackrel{O}{\longmapsto} (O^a_b \mathbf{x}^b, O^a_b \mathbf{p}^b) \in V^{\mathbb{R}}. \end{aligned}$$

In such a parallel interpretation of fields on spacetime $x \longmapsto \Phi(x)$ and "fields on time" $t \longmapsto \mathbf{x}(t)$, the rotations in mechanics can be viewed as an internal ("vertical") operation group and the angular momenta viewed as an internal quantum number with the vector space bundles

$$\mathbf{SO}(3) \bullet V(\mathbb{R}), \text{ with } \mathbf{SO}(3) \longrightarrow \mathbf{SO}(V), \quad V(\mathbb{R}) \cong V \times \mathbb{R}, \quad V \cong \mathbb{R}^N.$$

in comparison with the internal quantum numbers for hypercharge and isospin and the special relativistic bundles $\mathbf{U}(2) \bullet V(\mathbb{R}^4)$. In contrast to the internal rotations in mechanics on time, the rotations $\mathbf{SO}(3) \subset \mathbf{SO}_0(1,3)$ in spacetime relativity are an external ("horizontal") group.

6.3 Connections of Vector Space Bundles

A connection of a vector space bundle connects the Lie algebra operations as given by the manifold vector fields, i.e., derivations of manifold functions, with the structural Lie algebra acting on the vector space fibers.

A vector space bundle $G \bullet V(\mathbb{M}) \in \underline{\mathbf{vec}}_{\mathbb{K}}(\mathbb{M})$ with a base manifold $\mathbb{M} \in \underline{\mathbf{dif}}_{\mathbb{R}}$ and a finite-dimensional G-space $V \cong \mathbb{K}^N$ as typical fiber is a module $V(\mathbb{M}) \in \underline{\mathbf{mod}}_{\mathcal{C}(\mathbb{M})}$ over the ring with the manifold functions. One has local free modules $V(U) \cong \mathcal{C}(U)^N$, $U \subseteq \mathbb{M}$, with dual fiber bases (V-frames) $\mathrm{id}_{V(U)} = E^\alpha \otimes \check{E}_\alpha$, $\alpha = 1, \ldots, N$, as familiar from the tangent bundle $\mathbf{T}(\mathbb{M}) \in \underline{\mathbf{mod}}_{\mathcal{C}(\mathbb{M})}$ with local dual tangent bases $\mathrm{id}_{\mathbf{T}(U)} = e^i \otimes \check{e}_i$, $i = 1, \ldots, n$ and maximal stuctural group $e(x) \in \mathbf{GL}(n, \mathbb{R})$ (see Chapter 2). Holonomic bases $\partial^i \otimes dx_i$ are a specialty of the tangent bundle.

To connect, on the manifold, the structural group of the vector space fibers to each other, one defines, in addition to and by generalizing an affine connection for a tangent bundle $\nabla : \mathbf{T}(\mathbb{M}) \longrightarrow \mathbf{T} \otimes \mathbf{T}^T(\mathbb{M})$, a *connection of the vector space bundle*: It is given by a mapping of the manifold vector fields (derivations) into $\mathbb{R}$-linear mappings of the local vector spaces,

$$\begin{aligned} \mathfrak{D} : \mathbf{T}(\mathbb{M}) &\longrightarrow V \otimes V^T(\mathbb{M}), \quad v \longmapsto \mathfrak{D}_v, \\ f, g \in \mathcal{C}(\mathbb{M}) : \ \mathfrak{D}_{fv+gw} &= f\mathfrak{D}_v + g\mathfrak{D}_w, \end{aligned}$$

which defines actions on sections, in holonomic bases,

$$\mathfrak{D}_v : \quad V(\mathbb{M}) \longrightarrow V(\mathbb{M}), \qquad \begin{cases} \psi \longmapsto \mathfrak{D}_v \psi, \\ \mathfrak{D}_v(f\psi) = v(f)\psi + f\mathfrak{D}_v\psi, \\ \mathfrak{D}_{e^i} E^\alpha = \mathfrak{D}^i E^\alpha = \mathbf{A}^{i\alpha}_{\ \beta} E^\beta, \\ \mathfrak{D}^i(\psi_\alpha E^\alpha) = (\partial^i \psi_\beta + \mathbf{A}^{i\alpha}_{\ \beta}\psi_\alpha) E^\beta. \end{cases}$$

The connection is determined by N^2 1-forms:

$$\begin{aligned} \mathbf{A}^\alpha_\beta &= \mathbf{A}^{i\alpha}_{\ \beta} \check{e}_i \in \mathbf{T}^T(\mathbb{M}), \\ \mathfrak{D}^i &= \mathbf{A}^i = \mathbf{A}^{i\alpha}_{\ \beta} E^\beta \otimes \check{E}_\alpha \in V \otimes V^T(\mathbb{M}), \\ \mathfrak{D} &= \mathbf{A} \ = \mathbf{A}^{i\alpha}_{\ \beta} E^\beta \otimes \check{E}_\alpha \otimes \check{e}_i \in V \otimes V^T \otimes \mathbf{T}^T(\mathbb{M}). \end{aligned}$$

More: The endomorphisms $\mathfrak{D}_v$ have to be compatible with the structural Lie group $G \subseteq \mathbf{GL}(V)$ of the typical fiber vector space; i.e., they have to be valued in the *structural Lie algebra* $L = \log G \subseteq V \otimes V^T$. Therefore, $\mathfrak{D}$ and $\mathbf{A}$ are called the *G-covariant derivation* and *G-gauge field on the base* $\mathbb{M}$. The coefficients of an L-basis define $\mathbb{R}$-linear local mappings from the Lie algebra $\mathbf{T}_x(\mathbb{M})$ to the structural Lie algebra $L_x(\mathbb{M})$ — in general, no Lie algebra morphisms —

$$\begin{array}{c}\mathfrak{D} : \mathbf{T}(\mathbb{M}) \longrightarrow L(\mathbb{M}), \quad L \subseteq V \otimes V^T, \\ \mathbb{R}^n \cong \mathbf{T}_x(\mathbb{M}) \xrightarrow{\mathbf{A}(x)} L_x(\mathbb{M}) \cong \mathbb{R}^d,\end{array}$$

$$\begin{array}{l}L\text{-basis: } \{l^a \in V \otimes V^T)\}_{a=1}^d, \\ L\text{-frame: } \mathrm{id}_L = l^a \otimes \check{l}_a,\end{array} \quad \left\{\begin{array}{rl}\mathbf{A} &= \mathbf{A}_a^i l^a \otimes \check{e}_i \in L \otimes \mathbf{T}^T(\mathbb{M}), \\ \mathbf{A}_a^i &= \mathrm{tr}\, \check{l}_a \circ \mathbf{A}^i, \\ \mathbf{A}^{i\alpha}_{\ \beta} &= \mathbf{A}_a^i l^{a\alpha}_{\ \beta}.\end{array}\right.$$

A connection is grading-compatibly extendable, by Leibniz's rule, to the full tensor algebra,

$$\begin{array}{l}\mathfrak{D}_v : \quad \bigotimes L(\mathbb{M}) \longrightarrow \bigotimes L(\mathbb{M}), \quad L \in V \otimes V^T, \\ \text{with} \quad \left\{\begin{array}{rl}\mathfrak{D}_v f &= \nabla_v f = v(f), \ f \in \mathcal{C}(\mathbb{M}), \\ \mathfrak{D}_v(a \otimes b) &= \mathfrak{D}_v a \otimes b + a \otimes \mathfrak{D}_v b, \\ \mathfrak{D}_v \langle \check{\psi}, \psi \rangle &= \langle \mathfrak{D}_v \check{\psi}, \psi \rangle + \langle \check{\psi}, \mathfrak{D}_v \psi \rangle, \ \check{\psi} \in V^T(\mathbb{M}), \\ \mathfrak{D}^i(\check{\psi}^\beta \check{E}_\beta) &= (\partial^i \check{\psi}^\alpha - \mathbf{A}^{i\alpha}_{\ \beta} \check{\psi}^\beta) \check{E}_\alpha.\end{array}\right.\end{array}$$

The local V-frames are covariant constant, $\mathfrak{D}_v \, \mathrm{id}_{V(U)} = \mathfrak{D}_v(E^\beta \otimes \check{E}_\beta) = 0$.

The connection can be used as a G-covariant derivative for any representation $\mathcal{L}$ of the structural Lie algebra on a vector space W:

$$\begin{array}{c}V \otimes V^T \supseteq L \ni l \longmapsto \mathcal{L}(l) \in \mathcal{L} \subseteq W \otimes W^T, \\ \mathfrak{D}^W : \mathbf{T}(\mathbb{M}) \longrightarrow \mathcal{L}(\mathbb{M}).\end{array}$$

The fiber endomorphisms $(\mathbf{A}^i)_{i=1}^n$ are not $\mathcal{C}(\mathbb{M})$-tensors. Their *gauge transformation*, effected by G-changing the local V-frames $E^\alpha \otimes \check{E}_\alpha = E^\mu \otimes \check{E}_\mu$, involves derivatives of the N-bein from the structural group:

$$E \in G(\mathbb{M}) : \left\{\begin{array}{rl}E^\mu(x) &= E^\mu_\alpha(x) E^\alpha(x), \\ \mathbf{A}^{i\mu}_{\ \nu} &= E^\mu_\alpha \mathbf{A}^{i\alpha}_{\ \beta} E^{-1\beta}_{\ \ \nu} + (\partial^i E^\mu_\alpha) E^{-1\alpha}_{\ \ \nu}, \\ \mathbf{A}^i &\longmapsto E \circ \mathbf{A}^i \circ E^{-1} + (\partial^i E) \circ E^{-1}, \\ \partial^i E^\mu_\beta &- \mathbf{A}^{i\mu}_{\ \nu} E^\nu_\beta + E^\mu_\alpha \mathbf{A}^{i\alpha}_{\ \beta} = 0.\end{array}\right.$$

With the exception of torsion, first Bianchi identity, and Ricci tensor, which are specialties of the tangent bundle $\mathbf{T}(\mathbb{M})$, the tangent bundle analogous structures can be defined on vector space bundles in general: The *field strengths* (*curvature*) of a G-connection on $\mathbb{M}$ involve the difference of two tangent field commutators. They define endomorphisms of the vector space fibers, which can be spanned by an L-basis, i.e., they are Lie algebra operations:

$$\begin{aligned}
\mathbf{F}: \mathbf{T}\wedge\mathbf{T}(\mathbb{M}) &\longrightarrow L(\mathbb{M}) \subseteq V\otimes V^T(\mathbb{M}),\\
\mathbf{F}(v\wedge w) &= [\mathfrak{D}_v, \mathfrak{D}_w] - \mathfrak{D}_{[v,w]} = -\mathbf{F}(w\wedge v),\\
\mathbf{F}(e^i\wedge e^j) &= \mathbf{F}^{ij} = e^i(\mathbf{A}^j) - e^j(\mathbf{A}^i) - \epsilon_k^{ij}\mathbf{A}^k + [\mathbf{A}^i, \mathbf{A}^j]\\
&= l^a\mathbf{F}_a^{ij} = \mathbf{F}_\beta^{\alpha ij} E^\beta \otimes \check{E}_\alpha,\\
\mathbf{F}_\beta^{\alpha ij} &= e^i(\mathbf{A}^{j\alpha}_{\ \beta}) - e^j(\mathbf{A}^{i\alpha}_{\ \beta}) - \epsilon_k^{ij}\mathbf{A}^{k\alpha}_{\ \beta} + \mathbf{A}^{i\gamma}_{\ \beta}\mathbf{A}^{j\alpha}_{\ \gamma} - \mathbf{A}^{j\gamma}_{\ \beta}\mathbf{A}^{i\alpha}_{\ \gamma}\\
&= \partial^i\mathbf{A}^{j\alpha}_{\ \beta} - \partial^j\mathbf{A}^{i\alpha}_{\ \beta} + \mathbf{A}^{i\gamma}_{\ \beta}\mathbf{A}^{j\alpha}_{\ \gamma} - \mathbf{A}^{j\gamma}_{\ \beta}\mathbf{A}^{i\alpha}_{\ \gamma} \text{ (holonomic bases)},\\
\mathbf{F} &= \tfrac{1}{2}\mathbf{F}_\beta^{\alpha ij} E^\beta\otimes\check{E}_\alpha\otimes\check{e}_i\wedge\check{e}_j \in L\otimes\mathbf{T}^T\wedge\mathbf{T}^T(\mathbb{M}).
\end{aligned}$$

$[\mathbf{A}^i, \mathbf{A}^j]$ is the L-bracket. The field strengths (curvature) are 2-forms

$$\mathbf{F}_\beta^\alpha = \tfrac{1}{2}\mathbf{F}_\beta^{\alpha ij}\check{e}_i\wedge\check{e}_j = d\mathbf{A}_\beta^\alpha + \mathbf{A}_\beta^\gamma\wedge\mathbf{A}_\gamma^\alpha \in \mathbf{T}^T\wedge\mathbf{T}^T(\mathbb{M}).$$

The field strengths (curvature) map the Lorentz Lie algebra, acting on the tangent spaces $\mathbf{F}_\beta^{\alpha ab} : \log \mathbf{SO}_0(t,s) \longrightarrow \log G$, into the structural Lie algebra, acting on the fiber (no Lie algebra morphism). The geometrical meaning of "curvature" (local area change) applies only for the tangent bundle, where the curvature is a bilinear form (operational metric) of the tangent Lorentz Lie algebra (see Chapter 1).

In contrast to the gauge fields (connection) $\mathbf{A}: \mathbf{T}(\mathbb{M}) \longrightarrow L(\mathbb{M})$, the field strengths (curvature) $\mathbf{F}: \mathbf{T}\wedge\mathbf{T}(\mathbb{M}) \longrightarrow L(\mathbb{M})$ are $\mathcal{C}(\mathbb{M})$-linear; i.e., they have "homogeneous" local transformation behavior with respect to the structural group, given by the adjoint action of the group on its Lie algebra,

$$G(\mathbb{M})\times L(\mathbb{M}) \longrightarrow L(\mathbb{M}), \quad \mathbf{F}^{ij} \longmapsto E\circ\mathbf{F}^{ij}\circ E^{-1}.$$

Via the Jacobi identities for the brackets with the covariant derivatives $\mathfrak{D}_v$, the curvature fulfills the (second) *Bianchi identity*:

$$\begin{aligned}
&\mathfrak{D}_z[\mathbf{F}(v\wedge w)\psi] + \mathfrak{D}_v[\mathbf{F}(w\wedge z)\psi] + \mathfrak{D}_w[\mathbf{F}(z\wedge v)\psi]\\
&= [\mathfrak{D}_z, [\mathfrak{D}_v, \mathfrak{D}_w]]\psi + [\mathfrak{D}_v, [\mathfrak{D}_w, \mathfrak{D}_z]]\psi + [\mathfrak{D}_w, [\mathfrak{D}_z, \mathfrak{D}_v]]\psi = 0,\\
&\mathfrak{D}^k\mathbf{F}_\beta^{\alpha ij} + \mathfrak{D}^i\mathbf{F}_\beta^{\alpha jk} + \mathfrak{D}^j\mathbf{F}_\beta^{\alpha ki} = 0.
\end{aligned}$$

6.4 Pure Gauges, Distinguished Frames, and Composite Gauge Fields

If there exists for a connection of a vector space bundle,

$$\mathbf{T}(\mathbb{M}) \ni v \longmapsto \mathfrak{D}_v \in \log G(\mathbb{M}) \subseteq V\otimes V^T(\mathbb{M}), \quad V\cong\mathbb{K}^N,$$

a local V-frame $E^\alpha\otimes\check{E}_\alpha$ with trivial gauge fields for the structural group G, the connection has the form of a *pure gauge*. Pure gauge fields have trivial field strengths (curvature):

$$\begin{aligned}
\mathbf{A}^{i\mu}_{\ \nu} &= (\partial^i E^\mu_\alpha)E^{-1\alpha}_{\ \ \nu}, \quad \mathbf{A}^i = l^i(E) = (\partial^i E)\circ E^{-1}\\
\Rightarrow \mathbf{F}^{ij} &= \partial^i\mathbf{A}^j - \partial^j\mathbf{A}^i + [\mathbf{A}^i, \mathbf{A}^j] = 0.
\end{aligned}$$

It may occur, e.g., in the case of a degenerate ground state, that there exist distinguished V-frames, which consist of transmutators from G to a subgroup H (see Chapter 7):

$$\mathbf{GL}(N,\mathbb{K}) \cong \mathbf{GL}(V) \ni E(x) \cong E^{\mu}_{\alpha}(x) \in G/H.$$

Transmutators transform from G-active vectors (here, indices $\mu = 1,\ldots,N$) to only H-active ones (here, indices $\alpha = 1,\ldots,N$).

Reducing gauge fields to the "little" group $H \subseteq G$: If there exist, in such a case, basic gauge fields for the "large" group G with the affine $G \vec{\times} \log G$ transformation behavior,

$$\begin{aligned}
&\mathbb{M} \longrightarrow \log G, \quad x \longmapsto \mathbf{A}^i(x) = \mathbf{A}^i_a(x) l^a \cong \mathbf{A}^i_a(x) l^{a\mu}{}_{\nu},\\
&\mathbb{M} \longmapsto G, \qquad\ \ x \longmapsto g(x),\\
&\mathbb{M} \longrightarrow \log G, \quad x \longmapsto l^i(g(x)), \quad l^i(g) = (\partial^i g) \circ g^{-1},\\
&\qquad\qquad\qquad\quad \mathbf{A}^i \longmapsto g \circ \mathbf{A}^i \circ g^{-1} + l^i(g),
\end{aligned}$$

they can be "frozen" up to gauge fields for the "little" group H: They are stripped of the G/H degrees of freedom, provided by the distinguished frames,

$$\begin{aligned}
\mathbb{M} \longrightarrow \log H, \quad x \longmapsto \underline{\mathbf{A}}^i(x), \quad \underline{\mathbf{A}}^i &= E^{-1} \circ \mathbf{A}^i \circ E - l^i(E),\\
\underline{\mathbf{A}}^{i\alpha}{}_{\beta} &= E^{-1\alpha}{}_{\mu} \mathbf{A}^{i\mu}{}_{\nu} E^{\nu}_{\beta} - E^{-1\alpha}{}_{\mu} \partial^i E^{\mu}_{\beta}.
\end{aligned}$$

The *basic G-gauge fields are "muted" up to the H-gauge degrees of freedom:*

$$\begin{aligned}
&\mathbb{M} \longrightarrow H, \qquad\ \ x \longmapsto h(x) = E^{-1}(x) \circ g(x),\\
&\mathbb{M} \longrightarrow \log H, \quad x \longmapsto l^i(h(x)),\\
&\qquad\qquad\qquad\quad \underline{\mathbf{A}}^i \longmapsto h \circ \underline{\mathbf{A}}^i \circ h^{-1} + l^i(h).
\end{aligned}$$

An example is the reduction of the four gauge fields for hyperisospin $\mathbf{U}(2)$ in the electroweak standard model to an electromagnetic $\mathbf{U}(1)$-gauge with massless gauge field (ahead).

Constructing gauge fields for the "little" group $H \subseteq G$: There may exist distinguished frames (transmutators) for a coset space $E(x) \in G/H$ *without* basic gauge fields for the group G. An example is a Riemannian manifold with the distinguished frames $\mathbf{e}(x) \in \mathbf{GL}(n,\mathbb{R})/\mathbf{SO}_0(t,s)$ yielding transmutators from the full linear group to the tangent Lorentz group (see Chapter 2). The Riemannian connection is given by composite Lorentz group gauge fields and can be expressed by the n-bein $\Gamma \sim \mathbf{e}^{-1}\partial\mathbf{e}$ (Fock–Iwanenkov coefficients):

$$\begin{aligned}
\mathbf{\Gamma}^{ic}_{b}\eta_{ca} &= \mathbf{\Gamma}^{i}_{ab} = -\mathbf{\Gamma}^{i}_{ba}\\
&= \mathbf{e}^i_e (\check{l}_{ab})_{cd} (\eta^{df} \mathbf{e}^e_k \partial^c + \eta^{ef} \mathbf{e}^d_k \partial^c - \eta^{df} \mathbf{e}^c_k \partial^e) \mathbf{e}^k_f,\\
(\check{l}_{ab})_{cd} &= \eta_{ac}\eta_{bd} - \eta_{bc}\eta_{ad} \in [\log \mathbf{SO}_0(t,s)]^T.
\end{aligned}$$

It has nontrivial curvature; i.e., it is not a pure gauge.

In general: For vertical operations with cosets G/H, the distinguished frames are determined up to local H-transformation and allow the construction of *composite gauge fields* for a nontrivial fixgroup H by $\log H$-projection of the pure $\log G$-gauge:

$$
\begin{aligned}
\log H\text{-basis: } & \{l^A \in V \otimes V^T \cong \mathbf{AL}(\mathbb{K}^N) \bigm| A = 1, \ldots, \dim_{\mathbb{R}} H\},\\
\text{dual basis: } & \check{l}_A \in (\log H)^T, \quad \log H\text{-frame: } \mathcal{P}_{\log H} = l^A \otimes \check{l}_A,\\
\mathbf{B}^i &= \mathcal{P}_{\log H}(E^{-1} \circ \partial^i E), \quad \mathbf{B}^i(x) = \mathbf{B}^i_A(x) l^A,\\
\mathbf{B}^i_A &= \operatorname{tr} \check{l}_A \circ E^{-1} \circ \partial^i E = \check{l}_{A\alpha}^{\ \beta} E^{-1\alpha}_{\ \ \mu} \partial^i E^{\mu}_{\beta}.
\end{aligned}
$$

6.5 Chargelike Internal Connections

Gauge theories for flat spacetime (Lorentz indices $j, k, l, \ldots$) connect with each other spacetime translations and internal Lie algebra operations.

6.5.1 Currents as Lie Algebra Densities

A Lie algebra with dual bases and Lie bracket,

$$
\begin{aligned}
\underline{\mathbf{lag}}_{\mathbb{R}} \ni L \cong \mathbb{R}^d : \quad & \langle \check{l}_a, l^b \rangle = \delta^b_a, \quad [l^a, l^b] = \epsilon^{ab}_c l^c,\\
& \text{internal indices } a = 1, \ldots, d,
\end{aligned}
$$

is represented in the endomorphisms of a vector space and its dual,

$$
\begin{array}{ll}
W, W^T \cong \mathbb{C}^n, & \text{dual bases } \langle \check{e}_\beta, e^\gamma \rangle = \delta^\gamma_\beta = \epsilon \langle e^\gamma, \check{e}_\beta \rangle,\\
& \alpha = 1, \ldots, n, \ \epsilon = \pm 1 \text{ for Fermi or Bose},\\
\mathcal{D} : L \longrightarrow \mathbf{AL}(W), & l^a \longmapsto \mathcal{D}(l^a) = \mathcal{D}^{a\beta}_{\ \ \gamma} e^\gamma \otimes \check{e}_\beta,\\
L \times W \longrightarrow W, & l^a \bullet e^\beta = \mathcal{D}^{a\beta}_{\ \ \gamma} e^\gamma,\\
\check{\mathcal{D}} : L \longrightarrow \mathbf{AL}(W^T), & l^a \longmapsto -\mathcal{D}(l^a)^T = -\mathcal{D}^{a\beta}_{\ \ \gamma} \epsilon \check{e}_\beta \otimes e^\gamma,\\
L \times W^T \longrightarrow W^T, & l^a \bullet \check{e}_\gamma = -\mathcal{D}^{a\beta}_{\ \ \gamma} \check{e}_\beta,
\end{array}
$$

e.g., the Lie algebras of $\mathbf{U}(1)$ and $\mathbf{SU}(n)$,

$$
\begin{array}{ll}
\log \mathbf{U}(1) \longrightarrow \mathbf{AL}(\mathbb{C}), & \mathcal{D}(l^0) \cong iz, \quad z \in \mathbb{Z},\\
\log \mathbf{SU}(n) \longrightarrow \mathbf{AL}(\mathbb{C}^n), & \mathcal{D}(l^a) \cong \frac{i}{2} \tau^a(n)^\beta_\gamma,
\end{array}
$$

$$
\text{Pauli matrices: } \{\tau^a(n) \bigm| a = 1, \ldots, n^2 - 1\}, \text{ with } \left\{ \begin{aligned} \operatorname{tr} \tau(n) &= 0,\\ \operatorname{tr} \tau^a(n) \circ \tau^b(n) &= 2\delta^{ab}. \end{aligned} \right.
$$

For flat spacetime fields, the Lie algebra representation is given by the charges:

$$
l^a \longmapsto iQ^a = i \int d^3x \ \mathbf{J}^a_0(x), \quad [iQ^a, iQ^b] = \epsilon^{ab}_c iQ^c.
$$

They are position integrals over the currents, which are defined with the quantization opposite (anti)commutators (normalized with $\rho(m^2) = 1$, see Chapter 5), e.g.,

Field	Quantization	Current
Scalar (Hermitian)	$[i\mathbf{\Phi}_{k\beta}, \mathbf{\Phi}^\gamma](\vec{x}) = \delta^\gamma_\beta \delta^0_k \delta(\vec{x})$	$i\mathbf{J}^a_k = \mathcal{D}^{a\beta}_{\ \ \gamma} \frac{\{\mathbf{\Phi}^\gamma, i\mathbf{\Phi}_{k\beta}\}}{2}$
Scalar (complex)	$[i\mathbf{\Phi}^\star_{k\beta}, \mathbf{\Phi}^\gamma](\vec{x}) = [i\mathbf{\Phi}^\gamma_k, \mathbf{\Phi}^\star_\beta](\vec{x}) = \delta^\gamma_\beta \delta^0_k \delta(\vec{x})$	$i\mathbf{J}^a_k = \mathcal{D}^{a\beta}_{\ \ \gamma} \frac{\{\mathbf{\Phi}^\gamma, i\mathbf{\Phi}^\star_{k\beta}\} + \{-i\mathbf{\Phi}^\gamma_k, \mathbf{\Phi}^\star_\beta\}}{2}$
Vector (Hermitian)	$[i\mathbf{G}^\gamma_{kj}, \mathbf{A}^l_\beta](\vec{x}) = \delta^\gamma_\beta \delta^0_k \delta^a_j \delta^l_b \delta^b_a \delta(\vec{x})$	$i\mathbf{J}^a_k = \mathcal{D}^{a\beta}_{\ \ \gamma} \frac{\{\mathbf{A}^j_\beta, i\mathbf{G}^\gamma_{kj}\}}{2}$
Weyl (left)	$\{\mathbf{l}^\star_\beta, \mathbf{l}^\gamma\}(\vec{x}) = \delta^\gamma_\beta \sigma^0 \delta(\vec{x})$	$i\mathbf{J}^a_k = \mathcal{D}^{a\beta}_{\ \ \gamma} \frac{[\mathbf{l}^\gamma \check{\sigma}_k, \mathbf{l}^\star_\beta]}{2}$
Weyl (right)	$\{\mathbf{r}^\star_\beta, \mathbf{r}^\gamma\}(\vec{x}) = \delta^\gamma_\beta \check{\sigma}^0 \delta(\vec{x})$	$i\mathbf{J}^a_k = \mathcal{D}^{a\beta}_{\ \ \gamma} \frac{[\mathbf{r}^\gamma \sigma_k, \mathbf{r}^\star_\beta]}{2}$
Dirac	$\{\overline{\mathbf{\Psi}}_\beta, \mathbf{\Psi}^\gamma\}(\vec{x}) = \delta^\gamma_\beta \gamma^0 \delta(\vec{x})$	$i\mathbf{J}^a_k = \mathcal{D}^{a\beta}_{\ \ \gamma} \frac{[\mathbf{\Psi}^\gamma \gamma_k, \overline{\mathbf{\Psi}}_\beta]}{2}$

The adjoint action on the fields reads

$$[iQ^a, \mathbf{\Phi}^\beta] = \mathcal{D}^{a\beta}_{\ \ \gamma}\mathbf{\Phi}^\gamma, \quad [iQ^a, \mathbf{\Phi}^\star_\gamma] = -\mathcal{D}^{a\beta}_{\ \ \gamma}\mathbf{\Phi}^\star_\beta.$$

The simultaneous external–internal action (covariant derivatives),

$$(\partial\delta^\beta_\gamma - \mathcal{D}^{a\beta}_{\ \ \gamma}\mathbf{A}_a)\mathbf{\Phi}^\gamma, \quad (\partial\delta^\beta_\gamma + \mathcal{D}^{a\beta}_{\ \ \gamma}\mathbf{A}_a)\mathbf{\Phi}^\star_\beta,$$

is implemented by *gauge vertices* (gauge interactions). They are Lorentz-compatible spacetime distributions of the power-3 Lie algebra representation tensor:

$$\begin{array}{rccccc} \mathcal{D} = & \check{l}_a \otimes \mathcal{D}(l^a) & = & \check{l}_a \otimes \mathcal{D}^{a\beta}_{\ \ \gamma} & e^\gamma \otimes \check{e}_\beta, \\ \text{implemented by} & \mathbf{A}^k_a \ \mathbf{J}^a_k & \overset{\text{e.g.}}{=} & \mathbf{A}^k_a \ \ \mathcal{D}^{a\beta}_{\ \ \gamma} & \frac{[\mathbf{\Psi}^\gamma\gamma_k, \overline{\mathbf{\Psi}}_\beta]}{2}. \end{array}$$

Gauge fields go with a Lie algebra and its currents: The number of gauge fields is given by the Lie algebra dimension $L \cong \mathbb{R}^d$: they transform under the adjoint Lie algebra representation (for field strength $\mathbf{F}^b$ and the currents $\mathbf{j}^b, \mathbf{J}^b$) and its dual coadjoint representation (for gauge fields $\mathbf{A}_c$):

$$\begin{array}{lll} \mathrm{ad} : L \longrightarrow \mathbf{AL}(L), & \mathrm{ad}\, l^a = \epsilon^{ab}_c l^c \otimes \check{l}_b, & l^b \longmapsto \epsilon^{ab}_c l^c, \\ \check{\mathrm{ad}} : L \longrightarrow \mathbf{AL}(L^T), & \check{\mathrm{ad}}\, l^a = -\epsilon^{ab}_c \check{l}_b \otimes l^c, & \check{l}_c \longmapsto \epsilon^{ab}_c \check{l}_b. \end{array}$$

The gauge field currents are products of the dual pairs $(\mathbf{A}_a, \mathbf{F}^a)$:

$$\begin{array}{l} \mathbf{j}^a_k - \mathbf{A}^j_b \epsilon^{ab}_c \mathbf{F}^c_{kj}, \\ Q^a = \int d^3x \ (\mathbf{j}^a_0 + \mathbf{J}^a_0) \end{array} \Rightarrow \left\{ \begin{array}{rl} [iQ^a, (\mathbf{F}^b_{kj}, \mathbf{j}^b, \mathbf{J}^b)] &= \epsilon^{ab}_c (\mathbf{F}^c_{kj}, \mathbf{j}^c, \mathbf{J}^c), \\ {}[iQ^a, \mathbf{A}^k_c] &= -\epsilon^{ab}_c \mathbf{A}^k_b. \end{array} \right.$$

With the adjoint Lie algebra representation, the gauge field self-coupling is nontrivial only for a nonabelian Lie algebra.

In the Lagrangian for the gauge field sector,

$$\mathbf{L}(\mathbf{A}, \mathbf{F}) = \mathbf{F}^c_{kj} \frac{\partial^k \mathbf{A}^j_c - \partial^j \mathbf{A}^k_c}{2} + g_{cb} \frac{\mathbf{F}^c_{kj} \mathbf{F}^{kjb}}{4} - \tfrac{1}{2} \mathbf{A}^k_a \mathbf{j}^a_k,$$

the statistical factor $\frac{1}{2}$ in $\frac{1}{2}\mathbf{A}^k_a \mathbf{j}^a_k$ takes into account the tensor power 2 of the gauge field $\mathbf{A} \vee \mathbf{A}$ in the interaction. The current arises by gauge field derivation:

$$\tfrac{1}{2}\mathbf{A}^k_a \mathbf{j}^a_k = \epsilon^{ab}_c \frac{\mathbf{A}^k_a \mathbf{A}^j_b}{2} \mathbf{F}^c_{kj}, \quad \frac{\partial \frac{1}{2}\mathbf{A}^k_a \mathbf{j}^a_k}{\partial \mathbf{A}^k_a} = \epsilon^{ab}_c \mathbf{A}^j_b \mathbf{F}^c_{kj}.$$

In a second-order derivative formulation, derivatives of the gauge fields and cubic gauge field products occur in the current:

$$\mathbf{j}_k^a = \mathbf{A}_b^j \epsilon_c^{ab} (\partial_k \mathbf{A}_j^c - \partial_j \mathbf{A}_k^c + \epsilon^{cde} \mathbf{A}_k^d \mathbf{A}_j^e + \delta_{kj} \partial^l \mathbf{A}_l^c).$$

The Lagrangian for the gauge field sector,

$$\mathbf{L}(\mathbf{A}, \mathbf{F}) = \mathbf{F}_{kj}^c \frac{\partial^k \mathbf{A}_c^j - \partial^j \mathbf{A}_c^k - \epsilon_c^{ab} \mathbf{A}_a^k \mathbf{A}_b^j}{2} + g^2 \frac{\mathbf{F}_{kj}^c \mathbf{F}_c^{kj}}{4},$$

gives the field equations:

$$\partial^k \mathbf{A}_c^j - \partial^j \mathbf{A}_c^k - \epsilon_c^{ab} \mathbf{A}_a^k \mathbf{A}_b^j = g^2 \mathbf{F}_c^{jk} \ , \quad \partial^j \mathbf{F}_{kj}^b + \epsilon_c^{ab} \mathbf{A}_a^j \mathbf{F}_{kj}^c = 0.$$

6.5.2 Normalizations of Gauge Fields

A gauge field is normalized by an invariant nondegenerate symmetric bilinear form of the Lie algebra and its dual,

$$L \times L \longrightarrow \mathbb{R}, \quad \langle l^a | l^b \rangle = \kappa^{ab} = \kappa^{ba},$$

e.g., by the Killing form $\kappa(l^a, l^b) = \epsilon_d^{ac} \epsilon_c^{bd}$ for a semisimple Lie algebra like $\mathbf{SU}(n)$, $n \geq 2$, or by a squared linear form $\langle l | l \rangle = [\kappa(l)]^2$ for an abelian Lie algebra like $\mathbf{U}(1)$. In the following with compact gauge group $\mathbf{U}$, a positive definite diagonal normalization $\kappa^{ab} = \kappa_{\mathbf{U}}^2 \delta^{ab}$ is possible with Lie algebra bases for totally antisymmetric structure constants $\epsilon_d^{ab} \delta^{dc} = -\epsilon^{abc}$. The gauge field coupling constant in the field strength square can be seen as the normalization ratio of the represented internal Lie algebra $L = \log \mathbf{U}$ with $\kappa_{\mathbf{U}}^2$ and the external Lorentz Lie algebra $\log \mathbf{SO}_0(1,3)$ with its Killing form $\eta \wedge \eta$, normalized by $\kappa_{\mathbf{SO}_0(1,3)}^2$:

$$\begin{aligned} \langle \mathbf{F} | \mathbf{F} \rangle &= g^2 \ \delta_{ab} \eta^{kl} \eta^{jm} \mathbf{F}_{kj}^a \mathbf{F}_{lm}^b = g^2 \mathbf{F}_{kj}^a \mathbf{F}_a^{kj}, \\ g^2 &= \frac{\kappa_{\mathbf{SO}_0(1,3)}^2}{\kappa_{\mathbf{U}}^2}. \end{aligned}$$

In a theory with interaction, the gauge field coupling constant is renormalized (see Chapter 5).

6.5.3 Gauge Interactions in the Standard Model

The *standard model* of the electroweak and strong interactions in Minkowski spacetime is a theory of compatibly represented external and internal operations. The electromagnetic $\mathbf{U}(1)$ is embedded into the product of the abelian hypercharge $\mathbf{U}(1)$ and the nonabelian isospin-color group $\mathbf{SU}(2) \times \mathbf{SU}(3)$:

$$\mathbf{U}(1) \hookrightarrow \mathbf{U}(2 \times 3) = \mathbf{U}(\mathbf{1}_6) \circ [\mathbf{SU}(2) \times \mathbf{SU}(3)] = \frac{\mathbf{U}(1) \times \mathbf{SU}(2) \times \mathbf{SU}(3)}{\mathbb{I}(2) \times \mathbb{I}(3)}.$$

The electromagnetic interaction for a Dirac electron field (quantum electrodynamics) becomes a part of the electroweak and strong gauge interactions of quark and lepton Weyl fields. The fields involved are acted on by irreducible representations $[L|R]$ of the Lorentz group $\mathbf{SL}(2,\mathbb{C})$ and irreducible representations of the hypercharge group $\mathbf{U}(1)$ (rational hypercharge number in $[y]$), the isospin group $\mathbf{SU}(2)$ (integer or half-integer isospin in $[T]$), and the color group $\mathbf{SU}(3)$, as given in the following table:

Field	**symbol**	$\mathbf{SL}(2,\mathbb{C})$ $[L\|R]$	$\mathbf{U}(1)$ $[y]$	$\mathbf{SU}(2)$ $[T]$	$\mathbf{SU}(3)$ $[2C_1, 2C_2]$
Left lepton	$\mathbf{l}$	$[\frac{1}{2}\|0]$	$-\frac{1}{2}$	$[\frac{1}{2}]$	$[0,0]$
Right lepton	$\mathbf{e}$	$[0\|\frac{1}{2}]$	-1	$[0]$	$[0,0]$
Left quark	$\mathbf{q}$	$[\frac{1}{2}\|0]$	$\frac{1}{6}$	$[\frac{1}{2}]$	$[1,0]$
Right up quark	$\mathbf{u}$	$[0\|\frac{1}{2}]$	$\frac{2}{3}$	$[0]$	$[1,0]$
Right down quark	$\mathbf{d}$	$[0\|\frac{1}{2}]$	$-\frac{1}{3}$	$[0]$	$[1,0]$
Hypercharge gauge	$\mathbf{A}_0$	$[\frac{1}{2}\|\frac{1}{2}]$	0	$[0]$	$[0,0]$
Isospin gauge	$\vec{\mathbf{A}}$	$[\frac{1}{2}\|\frac{1}{2}]$	0	$[1]$	$[0,0]$
Color gauge	$\mathbf{G}$	$[\frac{1}{2}\|\frac{1}{2}]$	0	$[0]$	$[1,1]$
Higgs	$\mathbf{\Phi}$	$[0\|0]$	$\frac{1}{2}$	$[\frac{1}{2}]$	$[0,0]$

The fields of the minimal standard model

With the exception of the Higgs field, the isospin $\mathbf{SU}(2)$-representation is a subrepresentation of the Lorentz group $\mathbf{SL}(2,\mathbb{C})$-representation. This is a characteristic structure of induced representations that start with the two-sided regular representation of the doubled group (see Chapter 7).

The two factors in the internal group $\mathbf{U}(\mathbf{1}_6)\circ[\mathbf{SU}(2)\times\mathbf{SU}(3)]$ are not direct, but centrally correlated; i.e., the representations of hypercharge $\mathbf{U}(1)$ are related to the representations of the $\mathbf{SU}(2)\times\mathbf{SU}(3)$-center, the cyclotomic group $\mathbb{I}(2)\times\mathbb{I}(3)=\mathbb{I}(6)$ (hexality = two-triality, "star of David"). All $\mathbf{U}(2\times 3)$-representations $[y||T;2C_1,2C_2]$ carried by the standard model fields with the isospin and color multiplicities,

$$d_{\mathbf{SU}(2)}=1+2T,\quad d_{\mathbf{SU}(3)}=(1+2C_1)(1+2C_2)(1+C_1+C_2),$$

can be generated by the dual defining representations of $\mathbf{U}(2\times 3)$,

$$u=[\tfrac{1}{6}||\tfrac{1}{2};1,0],\quad \check{u}=[-\tfrac{1}{6}||\tfrac{1}{2};0,1],$$

as seen in the powers $\bigwedge^{n} u\otimes\bigwedge^{m}\check{u}$ (all fermion fields are taken as left-handed),

Field	$\mathbf{U}(2\times 3)$ $[y\|\|T;2C_1,2C_2]$	(n,m)	$n-m$ $=6y$	$6y$ mod 2	$6y$ mod 3
$\mathbf{l}$	$[-\frac{1}{2}\|\|\frac{1}{2};0,0]$	$(0,3)$	-3	1	0
$\mathbf{e}^\star$	$[1\|\|0;0,0]$	$(6,0)$	6	0	0
$\mathbf{q}$	$[\frac{1}{6}\|\|\frac{1}{2};1,0]$	$(1,0)$	1	1	1
$\mathbf{u}^\star$	$[-\frac{2}{3}\|\|0;0,1]$	$(0,4)$	-4	0	-1
$\mathbf{d}^\star$	$[\frac{1}{3}\|\|0;0,1]$	$(2,0)$	2	0	-1
$\mathbf{A}_0$	$[0\|\|0;0,0]$	$(0,0)$	0	0	0
$\vec{\mathbf{A}}$	$[0\|\|1;0,0]$	$(1,1)$	0	0	0
$\mathbf{G}$	$[0\|\|0;1,1]$	$(1,1)$	0	0	0
$\mathbf{\Phi}$	$[\frac{1}{2}\|\|\frac{1}{2};0,0]$	$(3,0)$	3	1	0

The central correlations of the internal symmetries are expressed by the modulo relations

$$6y \bmod 2 = 2T \bmod 2, \quad 6y \bmod 3 = 2(C_1 - C_2) \bmod 3,$$
$$y \cdot d_{\mathbf{SU}(2)} \cdot d_{\mathbf{SU}(3)} \in \mathbb{Z}.$$

The nongauge fields with the free Lagrangians,

$$\begin{aligned} \text{left fermions:} \quad \mathbf{L}(\mathbf{l}) &= i\mathbf{l}^\star_\alpha \partial\check{\sigma}\mathbf{l}^\alpha + i\mathbf{q}^\star_{\alpha c}\partial\check{\sigma}\mathbf{q}^{\alpha c}, \quad \alpha = 1,2;\ c = 1,2,3, \\ \text{right fermions:} \quad \mathbf{L}(\mathbf{r}) &= i\mathbf{e}^\star \partial\sigma\mathbf{e} + i\mathbf{u}^\star_c \partial\sigma\mathbf{u}^c + i\mathbf{d}^\star_c\partial\sigma\mathbf{d}^c, \\ \text{Higgs:} \quad \mathbf{L}(\mathbf{\Phi}) &= \mathbf{\Phi}^\star_{k\alpha}\partial^k\mathbf{\Phi}^\alpha + \mathbf{\Phi}^\alpha_k\partial^k\mathbf{\Phi}^\star_\alpha - \mathbf{\Phi}^\alpha_k\mathbf{\Phi}^{k\star}_\alpha \cong (\partial^k\mathbf{\Phi}^\star_\alpha)(\partial_k\mathbf{\Phi}^\alpha), \end{aligned}$$

interact with the four gauge fields $\mathbf{A}_0$ and $\vec{\mathbf{A}}$ for the electroweak interactions and the eight gauge fields $\mathbf{G}$ for the strong interactions:

$$\begin{aligned} \mathbf{L}(\mathbf{A}_0) &= \mathbf{F}_{kj}\frac{\partial^k\mathbf{A}_0^j - \partial^j\mathbf{A}_0^k}{2} && + g_1^2\frac{\mathbf{F}_{kj}\mathbf{F}^{kj}}{4}, \\ \mathbf{L}(\vec{\mathbf{A}}) &= \mathbf{F}^c_{kj}\frac{\partial^k\mathbf{A}_c^j - \partial^j\mathbf{A}_c^k - \epsilon_c^{ab}\mathbf{A}_a^k\mathbf{A}_b^j}{2} && + g_2^2\frac{\mathbf{F}^b_{kj}\mathbf{F}_b^{kj}}{4}, \\ \mathbf{L}(\mathbf{G}) &= \mathbf{F}^C_{kj}\frac{\partial^k\mathbf{G}_C^j - \partial^j\mathbf{G}_C^k - \epsilon_C^{AB}\mathbf{G}_A^k\mathbf{G}_B^j}{2} && + g_3^2\frac{\mathbf{F}^B_{kj}\mathbf{F}_B^{kj}}{4}. \end{aligned}$$

The indices differentiate between the different Lie algebras in the case of the field strengths $\mathbf{F}$. The structure constants are taken in a Pauli and Gell–Mann basis:

$$\log \mathbf{SU}(2): \quad \{\tfrac{i}{2}\tau^a{}^\beta_\gamma | a = 1,2,3, \ \ \beta = 1,2\}, \qquad [\tfrac{i}{2}\tau^a, \tfrac{i}{2}\tau^b] = \epsilon_c^{ab}\tfrac{i}{2}\tau^c,$$

$$\log \mathbf{SU}(3): \quad \{\tfrac{i}{2}\lambda^A{}^b_c | A = 1,\dots,8; \ \ b = 1,2,3\}, \quad [\tfrac{i}{2}\lambda^A, \tfrac{i}{2}\lambda^B] = \epsilon_C^{AB}\tfrac{i}{2}\lambda^C.$$

The gauge field coupling constants are the normalization ratios of the internal Lie algebras and the Lorentz Lie algebra:

$$\left(\frac{1}{g_1^2}, \frac{1}{g_2^2}, \frac{1}{g_3^2}\right) = \frac{\left(\kappa^2_{\mathbf{U}(1)}, \kappa^2_{\mathbf{SU}(2)}, \kappa^2_{\mathbf{SU}(3)}\right)}{\kappa^2_{\mathbf{SO}_0(1,3)}}.$$

The gauge interactions of the matter fields,

$$\mathbf{L}(\mathbf{A}_0) + \mathbf{L}(\vec{\mathbf{A}}) + \mathbf{L}(\mathbf{G}) + \mathbf{L}(\mathbf{l}) + \mathbf{L}(\mathbf{r}) + \mathbf{L}(\mathbf{\Phi}) - (\mathbf{A}_0^k\mathbf{J}_k + \mathbf{A}_a^k\mathbf{J}^a_k + \mathbf{G}_A^k\mathbf{J}^A_k),$$

involve the currents for the nongauge fields:

$$\begin{aligned} \log \mathbf{U}(1): \quad \mathbf{J}_k &= -\ \tfrac{1}{2}\mathbf{l}\check{\sigma}_k\mathbf{l}^\star + \tfrac{1}{6}\mathbf{q}\check{\sigma}_k\mathbf{q}^\star - \mathbf{e}\sigma_k\mathbf{e}^\star + \tfrac{2}{3}\mathbf{u}\sigma_k\mathbf{u}^\star - \tfrac{1}{3}\mathbf{d}\sigma_k\mathbf{d}^\star \\ &\quad -\ \tfrac{i}{2}(\mathbf{\Phi}^\star\mathbf{\Phi}_k - \mathbf{\Phi}\mathbf{\Phi}^\star_k), \\ \log \mathbf{SU}(2): \quad \mathbf{J}^a_k &= \mathbf{l}\check{\sigma}_k\tfrac{\tau^a}{2}\mathbf{l}^\star + \mathbf{q}\check{\sigma}_k\tfrac{\tau^a}{2}\mathbf{q}^\star \\ &\quad +\ (i\mathbf{\Phi}\tfrac{\tau^a}{2}\mathbf{\Phi}^\star_k - i\mathbf{\Phi}_k\tfrac{\tau^a}{2}\mathbf{\Phi}^\star), \\ \log \mathbf{SU}(3): \quad \mathbf{J}^A_k &= \mathbf{q}\check{\sigma}_k\tfrac{\lambda^A}{2}\mathbf{q}^\star + \mathbf{u}\sigma_k\tfrac{\lambda^A}{2}\mathbf{u}^\star + \mathbf{d}\sigma_k\tfrac{\lambda^A}{2}\mathbf{d}^\star. \end{aligned}$$

The gauge field strength equations are given with the corresponding currents:

$$\partial^j\mathbf{F}_{kj} + \mathbf{F}_{kj} \times \mathbf{A}^j = \mathbf{J}_k.$$

The spacetime translations for the lepton and quark come with internal gauge field actions

$$\begin{aligned} (\partial + \tfrac{i}{2}\mathbf{A}_0 - i\tfrac{\tau^a}{2}\mathbf{A}_a)\mathbf{l}\check{\sigma} &= 0, & (\partial + i\mathbf{A}_0)\mathbf{e}\sigma &= 0,\\ (\partial - \tfrac{i}{6}\mathbf{A}_0 - i\tfrac{\tau^a}{2}\mathbf{A}_a - i\tfrac{\lambda^A}{2}\mathbf{G}_A)\mathbf{q}\check{\sigma} &= 0, & (\partial - \tfrac{2i}{3}\mathbf{A}_0 - i\tfrac{\lambda^A}{2}\mathbf{G}_A)\mathbf{u}\sigma &= 0,\\ & & (\partial + \tfrac{i}{3}\mathbf{A}_0 - i\tfrac{\lambda^A}{2}\mathbf{G}_A)\mathbf{d}\sigma &= 0, \end{aligned}$$

as well as the Higgs field:

$$(\partial^k - i\tfrac{\mathbf{1}_2\mathbf{A}_0^k+\tau^a\mathbf{A}_a^k}{2})\mathbf{\Phi} = \mathbf{\Phi}^k, \quad (\partial^k - i\tfrac{\mathbf{1}_2\mathbf{A}_0^k+\tau^a\mathbf{A}_a^k}{2})\mathbf{\Phi}_k = 0.$$

For the scalar Higgs field, the second-order Lagrangian reads

$$\mathbf{L}(\mathbf{\Phi}, \mathbf{A}) = [(\partial - i\tfrac{\mathbf{1}_2\mathbf{A}_0+\tau^a\mathbf{A}_a}{2})\mathbf{\Phi}][(\partial + i\tfrac{\mathbf{1}_2\mathbf{A}_0+\tau^a\mathbf{A}_a}{2})\mathbf{\Phi}^\star].$$

6.6 Ground-State Degeneracy

For an interaction with symmetry group G, e.g., hypercharge-isospin $\mathbf{U}(2)$, the symmetry of a ground state may be characterized by a subgroup $H \subseteq G$ as the fixgroup, e.g., by electromagnetic $\mathbf{U}(1)$. Then, with the interaction symmetry G, there exists a ground state degeneracy, effected by the operations from the coset manifold G/H, e.g., from the weak Goldstone manifold $\mathbf{U}(2)/\mathbf{U}(1)$, implemented by long-range ("massless") scalar Nambu–Goldstone fields. All particles, defined with respect to a ground state, are rearranged into fixgroup H-multiplets. The interaction symmetry G is *spontaneously broken*, leaving the particle symmetry H. If operations from the degeneracy manifold are implemented, in addition, by Lorentz vector gauge fields, e.g., by four $\mathbf{U}(2)$-gauge fields, their ground-state–related rearrangement with the corresponding Nambu–Goldstone degrees of freedom leads, in a particle analysis, to massive vector fields, e.g., to three weak spin-1 massive bosons.

6.6.1 Electroweak Symmetry Reduction

In the electroweak standard model, the definition of particles requires the transition from the hyperisospin interaction symmetry $\mathbf{U}(2) = \mathbf{U}(\mathbf{1}_2)\circ\mathbf{SU}(2)$ (no direct product) to an abelian electromagnetic subsymmetry for the ground state and particles. Taking into account the nontrivial central correlation $\mathbf{U}(\mathbf{1}_2) \cap \mathbf{SU}(2) = \{\pm\mathbf{1}_2\}$, a Cartan torus of $\mathbf{U}(2)$ is $\mathbf{U}(1)_+ \times \mathbf{U}(1)_-$ with $e^{\frac{\mathbf{1}_2\pm\tau^3}{2}\alpha} \in \mathbf{U}(1)_\pm$ (projective generators). For a ground-state fixgroup $\mathbf{U}(1)_+$, as anticipated in the hypercharge representation numbers, chosen above, the electromagnetic charge number is the sum of the hypercharge number and the third isospin eigenvalue:

$$z = y + T^3.$$

Asymmetric boundary conditions are implemented by choosing from a ground-state manifold $\mathbf{U}(2)/\mathbf{U}(1)_+$ (Goldstone degrees of freedom) one representative and hence violating (stripping, rearranging) the $\mathbf{U}(2)/\mathbf{U}(1)_+$-related transformations in hyperisospin $\mathbf{U}(2)$. The *symmetry rearrangement* is implemented by an ad hoc scalar field, the Higgs field $\mathbf{\Phi}$ in the defining $\mathbf{U}(2)$-representation, with a Higgs potential:

$$\mathcal{V}(\mathbf{\Phi}) = \tfrac{g_0}{8}(\mathbf{\Phi}^\star\mathbf{\Phi} - M^2)^2, \quad g_0 > 0.$$

The minima of the potential,

$$\mathcal{V}(\mathbf{\Phi}) = \min \Rightarrow \langle \mathbf{\Phi}^\star\mathbf{\Phi}\rangle(x) = M^2,$$

give the breakdown-characterizing mass unit and, taking the appropriate representative, leaves an electromagnetic $\mathbf{U}(1)_+$ (ground-state fixgroup) as the remaining particle symmetry:

$$\langle \mathbf{\Phi}\rangle(x) = \begin{pmatrix} 0 \\ M \end{pmatrix} = \tfrac{\mathbf{1}_2 - \tau^3}{2} M.$$

The $\mathbf{U}(2)$-asymmetric effects in Weinberg's original "model of leptons,"

$$\begin{aligned} \mathbf{L}(\mathbf{\Phi}) &= [(\partial - i\tfrac{\mathbf{A}_0\mathbf{1}_2 + \vec{\mathbf{A}}\vec{\tau}}{2})\mathbf{\Phi}][(\partial + i\tfrac{\mathbf{A}_0\mathbf{1}_2 + \vec{\mathbf{A}}\vec{\tau}}{2})\mathbf{\Phi}^\star] - \mathcal{V}(\mathbf{\Phi}) \\ &\quad - g_e(\mathbf{e}\mathbf{\Phi}\mathbf{l}^\star + \mathbf{l}\mathbf{\Phi}^\star\mathbf{e}^\star), \end{aligned}$$

come in the particle structure of the $\mathbf{U}(2)$-gauge fields via the covariant derivative of the Higgs field and in the lepton particles via a Yukawa interaction. The ground-state value of the Higgs field gives the mass contributions:

$$\mathbf{L}(\mathbf{\Phi})\Big|_{\mathbf{\Phi}=\langle\mathbf{\Phi}\rangle} = M^2 \operatorname{tr} \tfrac{\mathbf{1}_2 - \tau^3}{2}(\tfrac{\mathbf{A}_0\mathbf{1}_2 + \vec{\mathbf{A}}\vec{\tau}}{2})^2 - Mg_e(\mathbf{e}\mathbf{l}_2^\star + \mathbf{l}^2\mathbf{e}^\star).$$

For the lepton fields, the $\mathbf{U}(1)_+$-trivial component (up component in the isospin doublet, neutrino) remains massless. The electron mass m_e for the massive electron Dirac field $\mathbf{\Psi}_e$ with the down component in the isospin doublet as the left-handed part can replace the Yukawa coupling constant g_e; both are theoretically undetermined parameters in the model:

$$\begin{aligned} \text{massless neutrino:} \quad \boldsymbol{\nu}_e &= \mathbf{l}^1, & m_\nu &= 0, \\ \text{massive electron:} \quad \mathbf{\Psi}_e &= (\mathbf{e}_L, \mathbf{e}_R) = (\mathbf{l}^2, \mathbf{e}), & m_e &= Mg_e. \end{aligned}$$

In general, a Yukawa coupling for a left-handed isodoublet $\mathbf{Q} = \begin{pmatrix} \mathbf{U}_L \\ \mathbf{D}_L \end{pmatrix}$ with hypercharge number y and two corresponding right-handed isosinglets $(\mathbf{U}_R, \mathbf{D}_R)$ with adapted hypercharge numbers $y \pm \frac{1}{2}$ generates a mass term after the symmetry reduction:

$$\begin{aligned} \mathbf{L}^{\text{Yuk}}(\mathbf{\Phi})\Big|_{\mathbf{\Phi}=\langle\mathbf{\Phi}\rangle} &= -g_D(\mathbf{D}_R\mathbf{\Phi}\mathbf{Q}^\star + \mathbf{Q}\mathbf{\Phi}^\star\mathbf{D}_R^\star) - g_U(\mathbf{U}_R\mathbf{\Phi}^\star\mathbf{Q}^\star + \mathbf{Q}\mathbf{\Phi}\mathbf{U}_R^\star)\Big|_{\mathbf{\Phi}=\langle\mathbf{\Phi}\rangle} \\ &= -m_D(\mathbf{D}_R\mathbf{D}_L^\star + \mathbf{D}_L\mathbf{D}_R^\star) - m_U(\mathbf{U}_R\mathbf{U}_L^\star + \mathbf{U}_L\mathbf{U}_R^\star) \\ &= -m_D\mathbf{\Psi}_D\overline{\mathbf{\Psi}}_D - m_U\mathbf{\Psi}_U\overline{\mathbf{\Psi}}_U. \end{aligned}$$

Left and right components constitute two massive Dirac fields:

$$\mathbf{\Psi}_U = (\mathbf{U}_L, \mathbf{U}_R),\quad \mathbf{\Psi}_D = (\mathbf{D}_L, \mathbf{D}_R),\quad m_D = Mg_D,\quad m_U = Mg_U.$$

An $\mathbf{SU}(2)$-index notation for the Yukawa couplings looks like the following:

$$\mathbf{D}_R\mathbf{\Phi}\mathbf{Q}^\star = \mathbf{D}_R\mathbf{\Phi}^\alpha\mathbf{Q}^\star_\alpha,\quad \mathbf{U}_R\mathbf{\Phi}^\star\mathbf{Q}^\star = \mathbf{U}_R\mathbf{\Phi}^\star_\beta\epsilon^{\beta\alpha}\mathbf{Q}^\star_\alpha.$$

Pairing the left-handed lepton fields $\mathbf{l} = \binom{\nu_L}{\mathbf{e}_L}$ not only with a right-handed electron $\mathbf{e}_R$ but also with a right-handed neutrino partner ν_R, a nontrivial neutrino mass $m_\nu = Mg_\nu$ is possible. Such a right-handed isosinglet field ν_R comes with trivial hypercharge $y = 0$, i.e., without any internal gauge interaction ("sterile neutrino").

The vector-field–related terms in the Higgs field coupling,

$$\begin{aligned} M^2 \operatorname{tr} \tfrac{\mathbf{1}_2-\tau^3}{2}\tfrac{(\mathbf{A}_0\mathbf{1}_2+\vec{\mathbf{A}}\vec{\tau})^2}{4} &= M^2 \operatorname{tr} \tfrac{\mathbf{1}_2-\tau^3}{2}\tfrac{[(\mathbf{A}_0)^2+(\vec{\mathbf{A}})^2]\mathbf{1}_2+2\mathbf{A}_0\vec{\mathbf{A}}\vec{\tau}}{4} \\ &= M^2\tfrac{(\mathbf{A}_1)^2+(\mathbf{A}_2)^2+(\mathbf{A}_3-\mathbf{A}_0)^2}{4}, \end{aligned}$$

contribute to the free theory of two massive charged vector fields $\mathbf{W} \in \{\mathbf{A}_{1,2}\}$:

$$\mathbf{L}(\mathbf{W}) = \mathbf{F}_{kj}\tfrac{\partial^k\mathbf{W}^j-\partial^j\mathbf{W}^k}{2} + g_2^2\tfrac{\mathbf{F}_{kj}\mathbf{F}^{kj}}{4} + \tfrac{M^2}{2}\tfrac{\mathbf{W}^k\mathbf{W}_k}{2} \Rightarrow m_W^2 = \tfrac{g_2^2}{2}M^2.$$

The two neutral vector fields come with the free theory:

$$\begin{aligned} \mathbf{L}(\mathbf{A}_0, \mathbf{A}_3) &= \mathbf{F}^0_{kj}\tfrac{\partial^k\mathbf{A}_0^j-\partial^j\mathbf{A}_0^k}{2} + \mathbf{F}^3_{kj}\tfrac{\partial^k\mathbf{A}_3^j-\partial^j\mathbf{A}_3^k}{2} \\ &+ g_1^2\tfrac{\mathbf{F}^0_{kj}\mathbf{F}_0^{kj}}{4} + g_2^2\tfrac{\mathbf{F}^3_{kj}\mathbf{F}_3^{kj}}{4} + M^2\tfrac{(\mathbf{A}_3^k-\mathbf{A}_0^k)^2}{4}. \end{aligned}$$

The diagonalization from interaction to particle fields, required by the nondiagonal mass term, is performed by the Weinberg $\mathbf{SO}(2)$-rotation:

$$\begin{aligned} g_1^2\tfrac{\mathbf{F}_0^2}{4} + g_2^2\tfrac{\mathbf{F}_3^2}{4} &= \gamma^2\tfrac{\mathbf{G}^2}{4} + g^2\tfrac{\mathbf{F}^2}{4}, &&\text{with} \quad \begin{pmatrix}\cos\theta & -\sin\theta\\ \sin\theta & \cos\theta\end{pmatrix}\begin{pmatrix} g_2\mathbf{F}^3\\ g_1\mathbf{F}^0\end{pmatrix} = \begin{pmatrix}\gamma\mathbf{G}\\ g\mathbf{F}\end{pmatrix}, \\ \mathbf{F}_0\partial\mathbf{A}_0 + \mathbf{F}_3\partial\mathbf{A}_3 &= \mathbf{G}\partial\mathbf{Z} + \mathbf{F}\partial\mathbf{A}, &&\text{with} \quad \begin{pmatrix}\cos\theta & -\sin\theta\\ \sin\theta & \cos\theta\end{pmatrix}\begin{pmatrix}\frac{1}{g_2}\mathbf{A}_3\\ \frac{1}{g_1}\mathbf{A}_0\end{pmatrix} = \begin{pmatrix}\frac{1}{\gamma}\mathbf{Z}\\ \frac{1}{g}\mathbf{A}\end{pmatrix}. \end{aligned}$$

It involves the Weinberg angle θ and dual normalizations $(\kappa, \frac{1}{\kappa})$ for the coupling constants $\kappa \in \{g_1, g_2, \gamma, g\}$. The combination $\mathbf{Z} = \mathbf{A}_3 - \mathbf{A}_0$ arises as a massive vector field; the massless gauge field $\mathbf{A}$ carries the ground-state fixgroup $\mathbf{U}(1)_+$ transformations; this defines the particle field normalizations $\{g, \gamma\}$ in terms of the gauge field normalizations and the Weinberg angle:

$$\begin{aligned} \mathbf{Z} &= \mathbf{A}_3 - \mathbf{A}_0 &&\Rightarrow \tfrac{\cos\theta}{g_2} = \tfrac{\sin\theta}{g_1} = \tfrac{1}{\gamma}, \\ \mathbf{A} &= \sin^2\theta\ \mathbf{A}_3 + \cos^2\theta\ \mathbf{A}_0 &&\Rightarrow \sin\theta = \tfrac{g}{g_2}. \end{aligned}$$

The Weinberg rotation,

$$\text{electroweak:}\ \left(\tfrac{1}{g_1^2}, \tfrac{1}{g_2^2}\middle|\tfrac{1}{\gamma^2}, \tfrac{1}{g^2}\right) \quad \left\{\begin{aligned} (\mathbf{A}_0, \mathbf{A}_3) &\longmapsto (\mathbf{Z}, \mathbf{A}), \\ \mathbf{U}(\mathbf{1}_2) \circ \mathbf{U}(1)_3 &= \mathbf{U}(1)_+ \times \mathbf{U}(1)_-, \\ e^{i(\gamma_0\mathbf{1}_2+\gamma_3\tau^3)} &= e^{i(\gamma_0+\gamma_3)\frac{\mathbf{1}_2+\tau^3}{2}} \times e^{i(\gamma_0-\gamma_3)\frac{\mathbf{1}_2-\tau^3}{2}}, \end{aligned}\right.$$

defines the *electroweak orthogonal triangle* with $(\frac{1}{g_1}, \frac{1}{g_2})$ the orthogonal sides. The hypotenuse $\frac{1}{g}$ is related to Sommerfeld's fine structure constant $\alpha_S = \frac{g^2}{4\pi} \sim \frac{1}{137}$ for the electromagnetic $\mathbf{U}(1)_+$-gauge field $\mathbf{A}$:

$$g_1 g_2 = \gamma g, \quad \frac{g_2}{g_1} = \cot\theta \quad \begin{cases} \frac{1}{g^2} &= \frac{1}{g_1^2} + \frac{1}{g_2^2}, \\ \gamma^2 &= g_1^2 + g_2^2. \end{cases}$$

Multiplication by the area dilation factor $\gamma g = g_1 g_2$ gives the similar dual triangle (g_2, g_2) with the squared lengths $(g_2^2, g_1^2 | g^2, \gamma^2)$.

The Weinberg rotation diagonalizes the free theory with two neutral vector particle fields:

$$\left.\begin{aligned} \mathbf{L}(\mathbf{A}_0, \mathbf{A}_3) &= \mathbf{F}_{kj} \frac{\partial^k \mathbf{A}^j - \partial^j \mathbf{A}^k}{2} + g^2 \frac{\mathbf{F}_{kj}\mathbf{F}^{kj}}{4} \\ &+ \mathbf{G}_{kj} \frac{\partial^k \mathbf{Z}^j - \partial^j \mathbf{Z}^k}{2} + G^2 \frac{\mathbf{G}_{kj}\mathbf{G}^{kj}}{4} + M^2 \frac{\mathbf{Z}^k \mathbf{Z}_k}{4} \end{aligned}\right\} \Rightarrow \begin{cases} m_A^2 &= 0, \\ m_Z^2 &= \frac{\gamma^2}{2} M^2. \end{cases}$$

The electroweak gauge field interactions are rearranged with the neutral and charged vector particle fields:

$$\left.\begin{aligned} \mathbf{A}_0 &= \mathbf{A} - \sin^2\theta\, \mathbf{Z} \\ \mathbf{A}_3 &= \mathbf{A} + \cos^2\theta\, \mathbf{Z} \end{aligned}\right\}, \quad \mathbf{A}_1 \mp i\mathbf{A}_2 = \mathbf{W}_\pm.$$

In general, one obtains for a left-handed isodoublet $\mathbf{Q} = \binom{\mathbf{U}_L}{\mathbf{D}_L}$ with hypercharge number y and two corresponding right-handed isosinglets $(\mathbf{U}_R, \mathbf{D}_R)$ with hypercharge numbers $y \pm \frac{1}{2}$ as the interaction with the vector particle fields,

$$\begin{aligned} &[y\mathbf{Q}\check{\sigma}\mathbf{Q}^\star + (y+\tfrac{1}{2})\mathbf{U}_R\sigma\mathbf{U}_R^\star + (y-\tfrac{1}{2})\mathbf{D}_R\sigma\mathbf{D}_R^\star]\mathbf{A}_0 + \mathbf{Q}\check{\sigma}\tfrac{\vec{\tau}}{2}\mathbf{Q}^\star\, \vec{\mathbf{A}} \\ &\quad = \left[(y+\tfrac{1}{2})\mathbf{\Psi}_U\gamma\overline{\mathbf{\Psi}}_U + (y-\tfrac{1}{2})\mathbf{\Psi}_D\gamma\overline{\mathbf{\Psi}}_D\right]\mathbf{A} \\ &\quad + \left[\mathbf{\Psi}_U\gamma\frac{1-4(y+\frac{1}{2})\sin^2\theta + i\gamma_5}{4}\overline{\mathbf{\Psi}}_U - \mathbf{\Psi}_D\gamma\frac{1+4(y-\frac{1}{2})\sin^2\theta + i\gamma_5}{4}\overline{\mathbf{\Psi}}_D\right]\mathbf{Z} \\ &\quad + \mathbf{U}_L\check{\sigma}\mathbf{D}_L^\star\mathbf{W}_- + \mathbf{D}_L\check{\sigma}\mathbf{U}_L^\star\mathbf{W}_+, \end{aligned}$$

where the parity combinations have been used in Dirac fields $\mathbf{\Psi}_{U,D}$, e.g., for $\mathbf{\Psi}_U$,

$$\begin{aligned} \mathbf{\Psi}_U\gamma\overline{\mathbf{\Psi}}_U &= \mathbf{U}_L\check{\sigma}\mathbf{U}_L^\star + \mathbf{U}_R\sigma\mathbf{U}_R^\star, \\ i\mathbf{\Psi}_U\gamma\gamma_5\overline{\mathbf{\Psi}}_U &= \mathbf{U}_L\check{\sigma}\mathbf{U}_L^\star - \mathbf{U}_R\sigma\mathbf{U}_R^\star. \end{aligned}$$

This leads for the leptons with $y = -\frac{1}{2}$ to

$$\begin{aligned} &(-\tfrac{1}{2}\mathbf{l}\check{\sigma}\mathbf{l}^\star - \mathbf{e}\sigma\mathbf{e}^\star)\mathbf{A}_0 + \mathbf{l}\check{\sigma}\tfrac{\vec{\tau}}{2}\mathbf{l}^\star\vec{\mathbf{A}} \\ &\quad = -\,\mathbf{\Psi}_e\gamma\overline{\mathbf{\Psi}}_e\mathbf{A} + \left(\tfrac{1}{2}\boldsymbol{\nu}_e\check{\sigma}\boldsymbol{\nu}_e^\star - \mathbf{\Psi}_e\gamma\frac{1-4\sin^2\theta + i\gamma_5}{4}\overline{\mathbf{\Psi}}_e\right)\mathbf{Z} \\ &\quad + \boldsymbol{\nu}_e\check{\sigma}\mathbf{e}_L^\star\mathbf{W}_- + \mathbf{e}_L\check{\sigma}\boldsymbol{\nu}_e^\star\mathbf{W}_+. \end{aligned}$$

A "sterile neutrino" remains "sterile." The quark fields with $y = \frac{1}{6}$ have the electroweak interactions

$$\left(\frac{1}{6}\mathbf{q}\check{\sigma}\mathbf{q}^\star + \frac{2}{3}\mathbf{u}\sigma\mathbf{u}^\star - \frac{1}{3}\mathbf{d}\sigma\mathbf{d}^\star\right)\mathbf{A}_0 + \mathbf{q}\check{\sigma}\frac{\vec{\tau}}{2}\mathbf{q}^\star\ \vec{\mathbf{A}} = \left(\frac{2}{3}\mathbf{\Psi}_u\gamma\overline{\mathbf{\Psi}}_u - \frac{1}{3}\mathbf{\Psi}_d\gamma\overline{\mathbf{\Psi}}_d\right)\mathbf{A}$$
$$+\left(\mathbf{\Psi}_u\gamma\frac{1-\frac{8}{3}\sin^2\theta+i\gamma_5}{4}\overline{\mathbf{\Psi}}_u - \mathbf{\Psi}_d\gamma\frac{1-\frac{4}{3}\sin^2\theta+i\gamma_5}{4}\overline{\mathbf{\Psi}}_d\right)\mathbf{Z}$$
$$+\ \mathbf{u}_L\check{\sigma}\mathbf{d}_L^\star\mathbf{W}_- + \mathbf{d}_L\check{\sigma}\mathbf{u}_L^\star\mathbf{W}_+.$$

The electroweak model contains many basically unknown parameters, especially the gauge field normalizations $g_{1,2}^2$ and the ground-state or electroweak mass unit M^2. The weak breakdown mass can be replaced by the experimentally determined Fermi constant for the four fermion interactions as the low-energy limit of the charged weak interaction, i.e., for the propagator:

$$\begin{aligned} q^2 \to 0: &\quad -\frac{g_2^2}{q^2-m_W^2} \to \frac{g_2^2}{m_W^2} = \frac{2}{M^2},\\ \text{experiment:} &\quad M \sim 169\ \frac{\text{GeV}}{c^2}. \end{aligned}$$

To determine the electroweak orthogonal triangle, one needs one constant in addition to the experimentally determined fine structure constant, e.g., the experimentally determined Weinberg angle:

$$\text{experiment:}\left\{\begin{array}{ll} \frac{g^2}{4\pi} = \alpha_S & \sim \frac{1}{137} \\ \sin^2\theta & \sim 0.23 \end{array}\right\} \Rightarrow \left(\frac{1}{g_1^2}, \frac{1}{g_2^2}\Big|\frac{1}{\gamma^2}, \frac{1}{g^2}\right) \sim (8.4, 2.5|1.9, 10.9),$$

from which the dual electroweak mass triangle can be computed:

$$\begin{aligned} (m_W^2, m_1^2|m_0^2, m_Z^2) &= (g_2^2, g_1^2|g^2, \gamma^2)\frac{M^2}{2} = \left(\frac{1}{\sin^2\theta}, \frac{1}{\cos^2\theta}\Big|1, \frac{4}{\sin^2 2\theta}\right)\frac{g^2M^2}{2},\\ (m_W, m_1|m_0, m_Z) &\sim (2.1, 1.2|1, 2.4)\ 37\ \frac{\text{GeV}}{c^2}. \end{aligned}$$

The weak boson masses are in good agreement with the experimental results:

$$m_W \sim 80\ \frac{\text{GeV}}{c^2},\quad m_Z \sim 91\ \frac{\text{GeV}}{c^2}.$$

6.6.2 Transmutation from Hyperisospin to Electromagnetic Symmetry

The electroweak symmetry "breakdown" (rearrangement) is a transmutation from a hyperisopin $\mathbf{U}(2)$-compatible framework for the interaction to a formulation with the remaining electromagnetic fixgroup $\mathbf{U}(1)_+$-symmetry for the particles (see Chapter 7):

$$\{t \in \mathbf{U}(2) \mid \langle\mathbf{\Phi}\rangle(x) = \begin{pmatrix} 0 \\ M \end{pmatrix} = t\begin{pmatrix} 0 \\ M \end{pmatrix}\} \cong \mathbf{U}(1)_+.$$

The $\mathbf{U}(2)/\mathbf{U}(1)_+$-isomorphic orbit of the Higgs field in the Hilbert space $\mathbb{C}^2$ provides translation-dependent Lie parameters for the fixgroup classes (symmetric space):

$$\begin{aligned} \mathbf{\Phi}(x) &= e^{\frac{i}{2}\gamma_\perp(x)} \frac{\mathbf{1}_2 - \tau^3}{2} R(x), \\ \mathbb{R}^4 \ni x \longmapsto \frac{i}{2}\gamma_\perp(x) &= i\frac{-\gamma_3(x)\mathbf{1}_2 + \vec{\gamma}(x)\vec{\tau}}{2} \in \log \mathbf{U}(2), \\ \mathbb{R}^4 \ni x \longmapsto V(x) &= v\left(\frac{\mathbf{\Phi}(x)}{R(x)}\right) = e^{\frac{i}{2}\gamma_\perp(x)} \in \mathbf{U}(2). \end{aligned}$$

The representations of the electromagnetic orientation manifold (Goldstone or ground-state manifold) on a vector space $W \cong \mathbb{C}^{1+2T}$ with $\mathbf{U}(2)$-representation,

$$\begin{aligned} (\mathbf{U}(2)/\mathbf{U}(1)_+)_{\text{repr}} &= \mathcal{G}^3 \longrightarrow \mathbf{U}(1+2T), \tfrac{\mathbf{\Phi}}{R} \longmapsto D(v(\tfrac{\mathbf{\Phi}}{R})), \quad R = |\mathbf{\Phi}|, \\ v(\tfrac{\mathbf{\Phi}}{R}) = \tfrac{1}{R}\begin{pmatrix} \mathbf{\Phi}_2^\star & \mathbf{\Phi}^1 \\ -\mathbf{\Phi}_1^\star & \mathbf{\Phi}^2 \end{pmatrix} &= e^{\frac{i}{2}\gamma_\perp} \in \mathbf{U}(2)/\mathbf{U}(1)_+, \end{aligned}$$

are products of the fundamental representation. A "left" hyperisospin $\mathbf{U}(2)$ action gives the representation with the $\mathbf{U}(2)$-transformed Higgs vector up to a "right" action with the electromagnetic fixgroup $\mathbf{U}(1)_+$:

$$\begin{aligned} u = e^{i\frac{\gamma_0 \mathbf{1}_2 + i\vec{\gamma}\vec{\tau}}{2}} \in \mathbf{U}(2) \Rightarrow \quad & u \circ v(\tfrac{\mathbf{\Phi}}{R}) = v(\tfrac{u.\mathbf{\Phi}}{R}) \circ t(u), \\ & \text{with } t(u) = e^{i2\gamma_0 \frac{\mathbf{1}_2 + \tau^3}{2}} \in \mathbf{U}(1)_+. \end{aligned}$$

The $\mathcal{G}^3$-representations are decomposable into transmutators from $\mathbf{U}(2)$-vectors (boldface) to $\mathbf{U}(1)_+$-active vectors with $\mathcal{G}^3$-frozen components (underlined) $\underline{\alpha} = 1, 2$:

$$\begin{aligned} &\underline{\mathbf{vec}}_{\mathbf{U}(2)} \ni W \overset{\mathbf{U}(1)_+}{\cong} \bigoplus_\iota W^\iota, \quad W^\iota \in \underline{\mathbf{vec}}_{\mathbf{U}(1)_+}, \\ &\left.\begin{aligned} W \ni \mathbf{E}^\alpha &= D(V)^\alpha_{\underline{\alpha}} E^{\underline{\alpha}} \in \bigoplus_\iota W^\iota, \\ W^T \ni \mathbf{E}^\star_\alpha &= E^\star_{\underline{\alpha}} D(V^\star)^{\underline{\alpha}}_\alpha \in \bigoplus_\iota W^{\iota T} \end{aligned}\right\} \text{ with } \mathbf{E}^\star_\alpha \mathbf{E}^\alpha = E^\star_{\underline{\alpha}} E^{\underline{\alpha}}, \end{aligned}$$

especially for the Higgs field, where the dilation degree of freedom R constitutes the frozen field:

$$\begin{aligned} (\epsilon^{\alpha\beta}\mathbf{\Phi}^\star_\beta, \mathbf{\Phi}^\alpha) &= (V^\alpha_1, V^\alpha_2) R \quad \cong V \circ \left(\tfrac{\mathbf{1}_2 + \tau^3}{2}, \tfrac{\mathbf{1}_2 - \tau^3}{2}\right) R, \\ (\epsilon_{\alpha\beta}\mathbf{\Phi}^\beta, \mathbf{\Phi}^\star_\alpha) &= R\ (V^{\star 1}_\alpha, V^{\star 2}_\alpha) \cong R\left(\tfrac{\mathbf{1}_2 + \tau^3}{2}, \tfrac{\mathbf{1}_2 - \tau^3}{2}\right) \circ V^\star. \end{aligned}$$

Hence, $\mathbf{U}(2)$-invariants have the same form in the frozen $\mathbf{U}(1)_+$-active fields as in the boldface unfrozen $\mathbf{U}(2)$-active ones. For example, the Yukawa coupling above with a $\mathbf{U}(2)$-doublet fermion field $\mathbf{Q}$, like the left-handed lepton fields, is written with the $\mathbf{U}(1)_+$-fields $Q^{1,2} = V^{\star 1,2}_\alpha \mathbf{Q}^\alpha = (U_L, D_L)$:

$$\begin{aligned} \mathbf{L}^{\text{Yuk}}(\mathbf{\Phi}) &= -g_D(\mathbf{D}_R\mathbf{\Phi}\mathbf{Q}^\star + \mathbf{Q}\mathbf{\Phi}^\star\mathbf{D}^\star_R) \ - g_U(\mathbf{U}_R\mathbf{\Phi}^\star\mathbf{Q}^\star + \mathbf{Q}\mathbf{\Phi}\mathbf{U}^\star_R) \\ &= -g_D R(\mathbf{D}_R D^\star_L + D_L\mathbf{D}^\star_R) \ - g_U R(\mathbf{U}_R U^\star_L + U_L\mathbf{U}^\star_R). \end{aligned}$$

The group element $V(x) = v(\frac{\mathbf{\Phi}(x)}{R(x)}) \in \mathbf{U}(2)/\mathbf{U}(1)_+$ defines a gauge transformation with a translation-parametrized Lie algebra element as representative of the fix-Lie algebra classes $\log \mathbf{U}(2)/\log \mathbf{U}(1)_+$:

$$\begin{aligned} \mathbb{R}^4 &\longrightarrow \log \mathbf{U}(2), \quad x \longmapsto l(V(x)),\\ l(V) &= (\partial V) \circ V^\star = i\frac{-(\partial\gamma_3)\mathbf{1}_2 + (\partial\gamma_a)[\delta_{ab}\frac{\sin\gamma}{\gamma} + \epsilon_{abc}\frac{\gamma_c}{\gamma}\frac{1-\cos\gamma}{\gamma} + \frac{\gamma_a\gamma_b}{\gamma^2}(1-\frac{\sin\gamma}{\gamma})]\tau^b}{2}. \end{aligned}$$

The Lie algebra-valued $\mathbf{U}(2)$-gauge fields, in the (2×2) Pauli representation,

$$\begin{aligned} &\mathbb{R}^4 \longrightarrow \log \mathbf{U}(2), && x \longmapsto \underline{\mathbf{A}}(x) = i\frac{\mathbf{A}_0(x)\mathbf{1}_2 + \vec{\mathbf{A}}(x)\vec{\tau}}{2},\\ &\mathbb{R}^4 \longmapsto \mathbf{U}(2), && x \longmapsto U(x),\\ &\mathbb{R}^4 \longrightarrow \log \mathbf{U}(2), && x \longmapsto l(U(x)), \quad l(U) = (\partial U) \circ U^\star, \end{aligned}$$

with the transformation behavior

$$\underline{\mathbf{A}} \longmapsto U \circ \underline{\mathbf{A}} \circ U^\star + l(U),$$

are stripped of the Higgs field-provided $\mathbf{U}(2)/\mathbf{U}(1)_+$-degrees of freedom:

$$\mathbb{R}^4 \longrightarrow \log \mathbf{U}(1)_+, \quad x \longmapsto \underline{A}(x), \quad \underline{A} = l(V^\star) + V^\star \circ \underline{\mathbf{A}} \circ V.$$

There remains only the electromagnetic $\mathbf{U}(1)_+$-gauge degree of freedom,

$$\underline{A} \longmapsto V^\star \circ U \circ \underline{\mathbf{A}} \circ U^\star \circ V + l(V^\star \circ U)$$

parametrized with spacetime translations,

$$\begin{aligned} \mathbb{R}^4 \ni x &\longmapsto V^\star(x) \circ U(x) \in \mathbf{U}(1)_+,\\ \mathbb{R}^4 \ni x &\longmapsto l(V^\star(x) \circ U(x)) \in \log \mathbf{U}(1)_+. \end{aligned}$$

The Higgs field derivative,

$$(\partial - \underline{\mathbf{A}})\mathbf{\Phi} = V(\partial - \underline{A})\frac{\mathbf{1}_2 - \tau^3}{2}R,$$

leads with the ground-state-characterizing mass $\langle R\rangle = M$ to the mass terms for the spin-1 particles (weak bosons) in the vector fields related to the three ground-state degrees of freedom $\mathbf{U}(2)/\mathbf{U}(1)_+$.

6.7 Lie Group Coset Bundles

A *principal bundle* is a Lie group bundle $H(\mathbb{M}) \in \underline{\mathbf{lgrp}}_{\mathbb{K}}(\mathbb{M})$, where the typical fiber $H \in \underline{\mathbf{lgrp}}_{\mathbb{K}}$ coincides with its structural group. The structural group has to act on itself either by left or by right translations. An example is the frame bundle of a manifold $\mathbb{M} \in \underline{\mathbf{dif}}_{\mathbb{R}}$ with $H = \mathbf{GL}(n, \mathbb{R})$.

A Lie group with a closed subgroup defines the associated *coset bundle*, a principal bundle with the subgroup as structural group:

$$G \supseteq H: \quad \underline{\mathbf{lgrp}}_{\mathbb{R}} \ni G \longmapsto H(\mathbb{M}) \in \underline{\mathbf{lgrp}}_{\mathbb{R}}(\mathbb{M}), \text{ with } \mathbb{M} = (G/H)_r \in \underline{\mathbf{dif}}_{\mathbb{R}}.$$

The base manifold is constituted by a representative for each coset (H-equivalence class):

$$\pi : G \longrightarrow G/H \longrightarrow \mathbb{M} = (G/H)_r \subseteq G, \quad g \longmapsto gH \longmapsto g_r = (gH)_r, \\ \text{with } g_r H = gH.$$

The local fiber is the fixgroup $G_{g_r} \cong H$ of the representative. As manifold, a coset bundle is isomorphic to the full group:

$$G \cong H(\mathbb{M}) = \bigcup_{g_r \in \mathbb{M}=(G/H)_r} (g_r, G_{g_r}), \quad G_{g_r} \cong H.$$

Examples are the spheres, hyperboloids, and Euclidean spaces with the rotation group $\mathbf{SO}(s)$ as typical fiber and the corresponding manifold isomorphies. They can be visualized for $s = 2$, e.g., the antipodally identified 2-sphere with all diameters,

$$\begin{aligned} \mathbf{SO}(1+s) &\cong \mathbf{SO}(s)(\Omega^s), \quad \text{with} \quad \Omega^s \cong \mathbf{SO}(1+s)/\mathbf{SO}(s), \\ \mathbf{SO}_0(1,s) &\cong \mathbf{SO}(s)(\mathcal{Y}^s), \quad \text{with} \quad \mathcal{Y}^s \cong \mathbf{SO}_0(1,s)/\mathbf{SO}(s), \\ \mathbf{SO}(s) \vec{\times} \mathbb{R}^s &\cong \mathbf{SO}(s)(\mathbb{R}^s), \quad \text{with} \quad \mathbb{R}^s \cong \mathbf{SO}(s) \vec{\times} \mathbb{R}^s/\mathbf{SO}(s). \end{aligned}$$

The coset bundle notation of a direct product group is $H_1 \times H_2 \cong H_1(H_2)$. It is a trivial bundle.

The left multiplication with a group element $k \in G$ on the base with the representatives gives the shifted representative up to the right multiplication with a subgroup element (*Wigner element*). It defines an isomorphism f of the group coset bundle:

$$\begin{array}{ccc} H(\mathbb{M}) & \xrightarrow{f} & H(\mathbb{M}) \\ \pi \downarrow & & \downarrow \pi \\ \mathbb{M} & \xrightarrow[L_k]{} & \mathbb{M} \end{array}, \qquad \begin{aligned} & L_k : \mathbb{M} \longrightarrow \mathbb{M}, \quad g_r \longmapsto (kg)_r \\ & \text{with} \begin{cases} kg_r = (kg)_r h(k, g_r) \in G, \\ (kg)_r = (kgH)_r \in \mathbb{M}, \\ h(k, g_r) \in H, \end{cases} \\ & f|_{G_{g_r}} : H \cong G_{g_r} \longrightarrow G_{(kg)_r} \cong H, \\ & \qquad h_r \longmapsto h_r h(k, g_r). \end{aligned}$$

$$(G/H)_r = \mathbb{M},$$

The Lie algebra bundle for a Lie group coset bundle,

$$\mathbb{M} = (G/H)_r : \quad \underline{\mathbf{lgrp}_{\mathbb{R}}}(\mathbb{M}) \ni H(\mathbb{M}) \longmapsto \log H(\mathbb{M}) \in \underline{\mathbf{lag}_{\mathbb{R}}}(\mathbb{M}),$$

comes with the subgroup-H gauge transformations in a connection.

6.8 Electroweak Spacetime

Reissner spacetime with metric, curvature and Ricci tensor (see Chapter 2):

$$\mathbf{g} = \left(1 - \frac{2\ell_m}{\rho} + \frac{\ell_z^2}{\rho^2}\right) dt^2 - \frac{d\rho^2}{1 - \frac{2\ell_m}{\rho} + \frac{\ell_z^2}{\rho^2}} - \rho^2 d\omega^2,$$

$$\mathcal{R}^{dabc}(t,\rho,\vec{\omega}) \cong \frac{1}{\rho^2}\left(\begin{array}{c|c|c|c} \frac{2\ell_m}{\rho}-\frac{3\ell_z^2}{\rho^2} & 0 & 0 & 0 \\ \hline 0 & -\left(\frac{\ell_m}{\rho}-\frac{\ell_z^2}{\rho^2}\right)\mathbf{1}_2 & 0 & 0 \\ \hline 0 & 0 & -\frac{2\ell_m}{\rho}+\frac{\ell_z^2}{\rho^2} & 0 \\ \hline 0 & 0 & 0 & \left(\frac{\ell_m}{\rho}-\frac{\ell_z^2}{\rho^2}\right)\mathbf{1}_2 \end{array}\right),$$

$$\mathcal{R}^{ab}_{\bullet}(t,\rho,\vec{\omega}) \cong \frac{\ell_z^2}{\rho^4}\left(\begin{array}{c|c|c} 1 & 0 & 0 \\ \hline 0 & -1 & 0 \\ \hline 0 & 0 & \mathbf{1}_2 \end{array}\right),$$

has a local $\mathbf{SO}(2)$-invariance with the decomposition of the Lorentz Lie algebra and tangent Minkowski spacetime:

$$\begin{aligned} \log \mathbf{SO}_0(1,3) \cong \mathbb{R}^6: \quad & 6 \overset{\mathbf{SO}(2)}{=} 1 \oplus 2 \oplus 1 \oplus 2, \\ \mathbb{R}^4: \quad & 4 \overset{\mathbf{SO}(2)}{=} 1 \oplus 1 \oplus 2. \end{aligned}$$

The local $\mathbf{SO}(2)$-invariance is reflected by the invariance of the energy-momentum tensor of the charged matter fields with respect to a local electromagnetic $\mathbf{SO}(2) \cong \mathbf{U}(1)$-transformation (see Chapter 5), e.g., for a scalar field,

$$\begin{aligned} \mathbf{T}^{ab}(\mathbf{\Phi}) &= (\partial^b + iz\mathbf{A}^b)\mathbf{\Phi}^*(\partial^a - iz\mathbf{A}^a)\mathbf{\Phi} \\ &\quad - \eta^{ab}[(\partial^c + iz\mathbf{A}^c)\mathbf{\Phi}^*(\partial_c - iz\mathbf{A}_c)\mathbf{\Phi} - m^2\mathbf{\Phi}^*\mathbf{\Phi}]. \end{aligned}$$

This suggests the *identification of the local invariance group $H_{\mathbf{g}}$ of the spacetime metrical tensor* $\mathbf{g}$ *with the internal (chargelike) operation group for spacetime fields.* If such an identification is taken as the guiding principle, there are some suggestive arguments to distinguish a spacetime manifold, which shows an integrative symbiosis of spacetimelike (horizontal) and chargelike (vertical) transformation groups.

An immediate argument against such an approach is the identification of an external subgroup, i.e., a Lorentz subgroup $H_{\mathbf{g}} \subset \mathbf{SO}_0(1,3) \sim \mathbf{SL}(2,\mathbb{C})$ with an internal operation group, i.e., a group action on the chargelike degrees of freedom as given by hypercharge $\mathbf{U}(1)$, isospin $\mathbf{SU}(2)$, and color $\mathbf{SU}(3)$ in the standard model of particles.

The situation is more subtle: There is a conceptual difference between a group and the action of the group on itself: The left–right multiplications of the group G define the bi-regular action with the group doubling $G \times G$ (see Chapter 8). Such a doubling will be used for the dichotomy of external and internal operations with the prominent example of the isomorphic but action-different group $\mathbf{SU}(2)$ for spin and isospin in $\mathbf{SU}(2) \times \mathbf{SU}(2)$. This doubling also occurs in the Cartan classification of the globally symmetric Riemannian manifolds, used there for the compact and noncompact type II manifolds (see Chapter 2).

If the local invariance group $H_{\mathbf{g}} \subseteq \mathbf{SO}_0(1,3)$ of a real four-dimensional spacetime with causal signature $(\mathbb{M}^{(1,3)}, \mathbf{g})$ is identified with internal gauge operations involving the electromagnetic $\mathbf{U}(1) \cong \mathbf{SO}(2) \subset H_{\mathbf{g}}$, it has to be brought in connection with the *real four-dimensional hyperisospin group* $\mathbf{U}(2)$ *of the electroweak standard model* with the centrally[1] correlated groups $\mathbf{U}(1)$ for hypercharge and $\mathbf{SU}(2)$ for isospin. The center classes of $\mathbf{U}(2)$ constitute a rotation group $\mathbf{SO}(3)$, a Lorentz subgroup. It is taken as the local invariance group of spacetime:

$$\begin{aligned}\mathbf{U}(2) &= \mathbf{U}(\mathbf{1}_2) \circ \mathbf{SU}(2) \cong \tfrac{\mathbf{U}(1)\times\mathbf{SU}(2)}{\mathbb{I}(2)},\\ \mathbf{U}(2)/\mathbf{U}(1) &\cong \mathbf{SU}(2)/\mathbb{I}(2) \cong \mathbf{SO}(3) = H_{\mathbf{g}} \subset \mathbf{SO}_0(1,3).\end{aligned}$$

A local invariance group $\mathbf{SO}(3) \sim \mathbf{SU}(2)$ characterizes position submanifolds $\mathbb{M}^{(0,3)} \subset \mathbb{M}^4$ with maximal hyperbolic, flat, and spherical motion groups, as used for Friedmann universes (see Chapter 1):

$$\mathbb{M}^{(0,3)} \cong \begin{cases} \mathcal{Y}^3 \cong & \mathbf{SO}_0(1,3)/\mathbf{SO}(3) \cong \mathbf{SL}(2,\mathbb{C})/\mathbf{SU}(2),\\ \mathbb{R}^3 \cong & \mathbf{SO}(3) \vec{\times} \mathbb{R}^3/\mathbf{SO}(3) \cong \mathbf{SU}(2) \vec{\times} \mathbb{R}^3/\mathbf{SU}(2),\\ \Omega^3 \cong & \mathbf{SO}(4)/\mathbf{SO}(3) \cong \frac{\mathbf{SU}(2)\times\mathbf{SU}(2)}{\mathbf{SU}(2)}.\end{cases}$$

The hyperbolic and spherical position manifolds constitute a type II noncompact–compact pair in the Cartan classification. Such pairs use for the compact II_c-partner the doubling $\mathbb{K} = \frac{K\times K}{\mathrm{diag}\,(K\times K)}$ of a group K, here of $K = \mathbf{SU}(2)$, and, in the doubling, the classes of the diagonal group, here $\Omega^3 \cong \mathbb{SU}(2)$, and, for the noncompact II_{nc}-partner $\frac{K\times K^*}{K}$ with the Weyl complexification in the Lie algebra $\log K \oplus \log K^* \cong (i\mathbb{R} \oplus \mathbb{R})^s$, here $\mathcal{Y}^3 \cong \mathbf{SL}(2,\mathbb{C})/\mathbf{SU}(2)$. The three-dimensional nonflat position manifolds are isomorphic to the type I orthogonal Riemannian spaces of subtype *BD I*.

A real four-dimensional operational manifold for spacetime as a transitive homogeneous space consisting of one orbit $G_{\mathbf{g}}/\mathbf{SO}(3)$ requires a real sevven-dimensional motion group $G_{\mathbf{g}} \supset \mathbf{SO}(3)$. An extension of the isotropy group $\mathbf{SU}(2)$ to $\mathbf{U}(2)$, whereof $\mathbf{SO}(3)$ describes the $\mathbf{U}(\mathbf{1}_2)$-classes, requires a real eight-dimensional supgroup $\mathbf{G} \supset \mathbf{U}(2)$:

$$\mathbb{M}^4 \cong G_{\mathbf{g}}/\mathbf{SO}(3) \cong \mathbf{G}/\mathbf{U}(2), \quad G_{\mathbf{g}} \cong \mathbf{G}/\mathbf{U}(1).$$

The nonflat position manifolds, hyperbolic $\mathcal{Y}^3$ and spherical Ω^3, can be extended with an additional hypercharge fixgroup $\mathbf{U}(1)$ as follows:

$$\mathbb{M}^{(1,3)} \cong \mathbf{GL}(2,\mathbb{C})/\mathbf{U}(2), \quad \mathbb{M}^{(0,4)} \cong \frac{\mathbf{U}(2)\times\mathbf{U}(2)}{\mathbf{U}(2)}.$$

Only noncompact *electroweak spacetime* $\mathbb{D}(2)$, i.e., the classes of the hyperisospin group in the extended Lorentz group, with a hyperbolic position submanifold has a causal structure:

$$\begin{aligned}\mathbb{D}(2) = \mathbb{M}^{(1,3)} &\cong \mathbf{D}(1) \times \mathcal{Y}^3,\\ \mathbb{U}(2) = \mathbb{M}^{(0,4)} &\cong \mathbf{U}(1) \times \Omega^3.\end{aligned}$$

[1]The $\mathbf{SU}(n)$-center is isomorphic to the multiplicative cyclotomic group $\mathbb{I}(n) = \{z \in \mathbb{C} \mid z^n = 1\}$ and to the additive group $\mathbb{Z}_n = \mathbb{Z}/n\mathbb{Z}$. $\mathbf{U}(\mathbf{1}_n) \subset \mathbf{GL}(n,\mathbb{C})$ is the phase group for a complex n-dimensional vector space.

The definition of electroweak spacetime $\mathbb{D}(2) \cong \mathbf{D}(1) \times \mathbf{SO}_0(1,3)/\mathbf{SO}(3)$ does not involve the real eight-dimensional color group $\mathbf{SU}(3)$ as a constitutive operation group.

Electroweak spacetime is the lowest-dimensional noncompact nonabelian case $n = 2$ of the noncompact–compact manifold pairs,

$$\begin{array}{lcll} \mathbb{D}(n) & = \mathbf{GL}(n,\mathbb{C})/\mathbf{U}(n) & \cong & \mathbf{D}(1) \times \mathbf{SL}(n,\mathbb{C})/\mathbf{SU}(n), \\ \mathbb{U}(n) & \frac{\mathbf{U}(n)\times\mathbf{U}(n)}{\mathbf{U}(n)} & \cong & \mathbf{U}(1) \times \frac{\mathbf{SU}(n)\times\mathbf{SU}(n)}{\mathbf{SU}(n)}. \end{array}$$

The abelian case $n = 1$ gives the groups $\mathbf{D}(1) = \mathbb{D}(1)$ and $\mathbf{U}(1) = \mathbb{U}(1)$. The motion groups of these manifolds can be considered as coset bundles with $\mathbb{D}(n)$ and $\mathbb{U}(n)$ as base manifolds and the local invariance group $\mathbf{U}(n)$ as the typical fiber:

$$\mathbf{GL}(n,\mathbb{C}) \cong \mathbf{U}(n)(\mathbb{D}(n)), \quad \mathbf{U}(n) \times \mathbf{U}(n) \cong \mathbf{U}(n)(\mathbb{U}(n)).$$

$\mathbb{D}(n)$ is constituted by the classes of the maximal compact group $\mathbf{U}(n)$ in the general linear group $\mathbf{GL}(n,\mathbb{C})$. It uses the polar decomposition of $\mathbf{GL}(n,\mathbb{C})$ into nonabelian phases $\mathbf{U}(n)$ and positive moduli $\mathbb{D}(n)$. The compact group has a Lie algebra $\log \mathbf{U}(n) \cong (i\mathbb{R})^{n^2}$ with n^2 imaginary parameters; $\mathbb{D}(n)$ is the positive cone in the C^*-algebra $\mathbf{AL}(\mathbb{C}^n)$ of the complex $(n \times n)$-matrices: The real n^2-dimensional vector subspace with the hermitian $(n \times n)$-matrices,

$$\mathbb{R}^{n^2} = \{x \in \mathbf{AL}(\mathbb{C}^n) \bigm| x = x^\star\} \in \underline{\mathbf{vec}}_{\mathbb{R}},$$

has a causal order by a positive spectrum:

$$\mathbf{AL}(\mathbb{C}^n) \ni x \succeq 0 \iff x = x^*, \ \mathbf{spec}\, x \geq 0 \iff x \in \mathbb{R}^{n^2}_+.$$

The hermitian matrices parametrize the translations of real n^2-dimensional Minkowski spacetime and are isomorphic to the tangent spaces of $\mathbb{D}(n)$, which itself is parametrizable by the spacetime translations of the open future cone $\mathbb{D}(n) \simeq \mathbb{R}^{n^2}_+$ (see Chapter 11).

Chapter 7

Relativities and Homogeneous Spaces

This chapter discusses the dichotomy and connection of external and internal operations in real four-dimensional electroweak spacetime $\mathbb{D}(2) = \mathbf{D}(1) \times \mathcal{Y}^3 = \mathbf{GL}(2,\mathbb{C})/\mathbf{U}(2)$ will be under the label *unitary relativity*. To see its general and specific structures, unitary relativity will be considered as one example in five relativities: "up-down" or *perpendicular relativity* as realized after discovering the earth's surface to be spherical; then rotation or space and time relativity, as used in what we call *special relativity*; then Lorentz group or flat Minkowski spacetime relativity or, also, with Wigner's definition, *particle relativity* as an important ingredient (local inertial systems) of general relativity; then *electromagnetic relativity* as used for the particle definition in the standard model of electroweak interactions [58]; and, finally, unitary relativity with "spacetimelike" and "chargelike" operations.

Relativity will be defined by operation group classes, e.g., in special relativity, the distinction of your rest system determines a decomposition of spacetime translations into time and position translations. Compatible with this decomposition is your position rotation group $\mathbf{SO}(3)$ as a subgroup of the orthochronous Lorentz group $\mathbf{SO}_0(1,3)$. There are as many decompositions of spacetime into time and position as there are rotation groups in a Lorentz group. The rotation group classes are parametrizable by the points of a one-shell three-dimensional hyperboloid $\mathcal{Y}^3 \cong \mathbf{SO}_0(1,3)/\mathbf{SO}(3)$ that give the momenta (velocities) for all your possible motions, i.e., by your mass hyperboloid $q_0^2 - \vec{q}^2 = m^2$. Another example: The perpendicularities of mankind, if earthbound, are characterized by the axial rotation groups in a rotation group and parametrizable by the two coordinates of the earth's surface $\Omega^2 \cong \mathbf{SO}(3)/\mathbf{SO}(2)$.

H. Saller, *Operational Spacetime: Interactions and Particles*,
Fundamental Theories of Physics 163, DOI 10.1007/978-1-4419-0898-8_8,

Now in general: The distinction of an "idolized" operation subgroup[1] H in a "general" operation group G picks one element in the G-symmetric space G/H, which stands for the relativity of the "idolized" group, called H-*relativity.* Relativity groups H, i.e., the local fixgroups in G/H, are candidates for gauge groups. An "idolization" [1] goes, negatively, with the "narrow-minded" assumption of an absolute point of view, e.g., absolute up-down, absolute time, absolute Minkowski spacetime (particle universality), or, positively, with the distinction of a smaller operation symmetry, enforced, e.g., by initial or boundary conditions. Important examples are degenerate ground states ("spontaneous symmetry breakdown"), where, by distinguishing one ground state from the symmetric space G/H as the -degeneracy manifold, an "interaction-symmetry" G is reduced to a "particle-symmetry" H, e.g., the ground states of superconductivity, superfluidity, ferromagnetism, and the electroweak standard model.

All of this gives the first four columns of the following table, which together with the last one will be discussed with their representations in more detail ahead

Relativity	"General" group G $(r_c + r, r)$	"Idolized" subgroup H(fiber)	Homogeneous space $\mathbb{M} \cong G/H$ (base manifold)	Relativity manifold parameters
Axial rotation (perpendicular) relativity	$\mathbf{SO}(3)$ $\sim \mathbf{SU}(2)$ $(1,0)$	$\mathbf{SO}(2)$	2-sphere $\Omega^2 \cong \mathbf{SO}(3)/\mathbf{SO}(2)$ $\cong \mathbf{SU}(2)/\mathbf{SO}(2)$	two spherical coordinates
Rotation (special) relativity	$\mathbf{SO}_0(1,3)$ $\sim \mathbf{SL}(2,\mathbb{C})$ $(2,1)$	$\mathbf{SO}(3)$ $\sim \mathbf{SU}(2)$	3-hyperboloid $\mathcal{Y}^3 \cong \mathbf{SO}_0(1,3)/\mathbf{SO}(3)$ $\cong \mathbf{SL}(2,\mathbb{C})/\mathbf{SU}(2)$	three momenta
Lorentz group (particle) relativity	$\mathbf{GL}(4,\mathbb{R})$ $(4,4)$	$\mathbf{O}(1,3)$	tetrad or metric manifold $\mathcal{M}^{10} \cong \mathbf{GL}(4,\mathbb{R})/\mathbf{O}(1,3)$ $\cong \mathbf{D}(1) \times \mathbf{SO}_0(3,3)/\mathbf{SO}_0(1,3)$	10 components for metrical tensor
Electro-magnetic relativity	$\mathbf{U}(2)$ $(2,0)$	$\mathbf{U}(1)_+$	Goldstone manifold $\mathcal{G}^3 \cong \mathbf{U}(2)/\mathbf{U}(1)_+$	three weak coordinates
Unitary (electroweak) relativity	$\mathbf{GL}(2,\mathbb{C})$ $(4,2)$	$\mathbf{U}(2)$	future 4-cone $\mathbb{D}(2) \cong \mathbf{GL}(2,\mathbb{C})/\mathbf{U}(2)$ $\cong \mathbf{D}(1) \times \mathcal{Y}^3$	four spacetime coordinates

Coset bundles $H(\mathbb{M})$ for five relativities

Somewhat in accordance with the historical development, the "general" operations of one relativity can constitute the "idolized" group of the next relativity as seen in the two chains ending in full general linear groups, a real one for spacetime concepts, from flat to spherical earth to special and general relativity, and a complex one for interactions, from electromagnetic to electroweak transformations and their spacetime (gauge) dependence:

$$\begin{array}{llll} \mathbf{SO}(2) & \subset \mathbf{SO}(3) & \subset \mathbf{SO}_0(1,3) & \subset \mathbf{GL}(4,\mathbb{R}), \\ \mathbf{U}(1)_+ & \subset \mathbf{U}(2) & \subset \mathbf{GL}(2,\mathbb{C}). & \end{array}$$

[1]In general, H is assumed to be no normal subgroup. Otherwise, it "disappears" in the quotient group $G_0 = G/H$.

There is a connection of the two chains on the level of the orthochronous Lorentz group $\mathbf{SO}_0(1,3)$ and its cover group $\mathbf{SL}(2,\mathbb{C})$ by

$$\mathbf{SO}_0(1,3) \cong \mathbf{GL}(2,\mathbb{C})/\mathbf{GL}(1,\mathbb{C}) \cong \mathbf{SL}(2,\mathbb{C})/\{\pm\mathbf{1}_2\}.$$

All groups in the five relativities considered are real Lie groups. All "general" groups have reductive Lie algebras[2]; for perpendicular and rotation relativity they are even simple. Perpendicular and electromagnetic relativity have a compact "general" group. With the exception of Lorentz group relativity, all "idolized" groups are compact subgroups. The second column contains the dimension $r_c + r$ of the maximal abelian subgroups, which is the rank of the group G generating Lie algebra $L = \log G$, and of the maximal noncompact abelian subgroups, i.e., the real rank r. The rank $r_c + r$ gives the number of independent invariants, rational or continuous, that characterize a G-representation. The real rank r is the maximal number of independent continuous invariants.

Unitary relativity $\mathbf{GL}(2,\mathbb{C})/\mathbf{U}(2)$, i.e., the complex linear relativization of the maximal compact subgroup with the internal "chargelike" hypercharge and isospin operations $\mathbf{U}(2)$, is parametrized by a noncompact real four-dimensional homogeneous space, called electroweak spacetime $\mathbb{D}(2)$. Unitary relativity is visible in the spacetime dependence of quantum fields, which represent the internal operations. The representations of unitary relativity $\mathbb{D}(2)$ are characterized by two continuous invariants, which, in appropriate units, can be taken as two masses. The $\mathbb{D}(2)$-representations determine the spacetime interactions with their normalization, especially the gauge interactions with their coupling constants, which are related to the ratio of the two invariants, and, for the $\mathbb{D}(2)$-tangent translations $\mathbb{R}^4$, the particles and their masses. The common language for interactions and elementary particles is the representation theory and harmonic analysis of unitary relativity (see Chapters 11 and 12).

There is a mathematical framework, almost tailored for relativities G/H: the theory of induced representations, pioneered by Frobenius [25], used for free particles by Wigner [62] and worked out for noncompact groups especially by Mackey [43]. There, a subgroup H-representation induces a full group G-representation leading to a $G \times H$-representation as subrepresentation of the two-sided regular $G \times G$-representation. Relativities G/H are acted on by the bi-regular subgroup $G \times H$. Such a dichotomic transformation property with a doubled group in $G \times G$ as group and "isogroup" is familiar, with respect to the Lorentz and the isospin group, $\mathbf{SU}(2) \times \mathbf{SU}(2)$ as spin and isospin, from the fields in the electroweak standard model. The theory is not easy to penetrate, especially for noncompact nonabelian groups. All the mathematical details are given in the textbooks by Helgason [36], Knapp [41], Folland [24] and, especially for distributions, by Treves [56].

[2]A Lie group of complex hermitian matrices is reductive. A finite-dimensional Lie algebra L with semisimple "square" $[L, L]$ is reductive. Then, $L = A \oplus S$ with abelian and semisimple Lie subalgebras.

In the following, only some motivating and qualititive mathematical remarks will be given with respect to this theory, which will be used in physical implementations. This chapter works with finite-dimensional relativity structures, which may not be so familiar in such a conceptual framework. After a parametrization of the relativity manifold G/H, its representations, called transmutators and closely related to n-beins, will be given, which mediate the transition from an idolized group H to the full group G. By products of the fundamental transmutators, all finite-dimensional relativity representations can be constructed as used, e.g., in the transition from the fields for the electroweak interactions to the asymptotic particles.

7.1 Parametrization of Relativity Manifolds

For real homogeneous relativity manifolds $\mathbb{M} \cong G/H$ with Lie groups, there are operation-induced parameters, e.g., the three momenta (velocities) for rotation (special) relativity or the three weak coordinates of electromagnetic relativity as used in the mass modes of the three weak bosons.

The action of a "general" group G on a set S, denoted by $\bullet$, decomposes S into disjoint orbits $G \bullet x$ for $x \in S$ that are isomorphic to subgroup classes $G \bullet x \cong G/H$, where the "idolized" group $H \cong G_x$ is the *fixgroup* (fixer, "little" group, isotropy group) of the G-action. The elements of homogeneous spaces $gH \in G/H$ are group subsets (cosets). The cosets have representatives $g_r = (gH)_r \in gH \in G/H$, denoted as $g_r \Subset G/H$, which can be characterized by what will be called *relativity parameters*, a real parametrization of the subgroup classes.

Relativity parameters can be obtained via orbit parametrizations. The real Lie groups considered are linear groups $H \subseteq G \subseteq \mathbf{GL}(V)$, acting on real or complex vector spaces V. The orbit $G \bullet x$ parametrizes the homogeneous space G/H by V-vectors and their components with respect to a basis:

$$\begin{aligned} &x \in V,\ H \cong G_x = \{g \in G \mid g \bullet x = x\} \\ &\Rightarrow G/H \cong G \bullet x \subseteq V. \end{aligned}$$

7.1.1 Goldstone Manifold for Weak Coordinates

The hypercharge-isospin group $\mathbf{U}(2)$ acts, in the defining representation, on a complex two-dimensional vector space:

$$\mathbf{U}(2) \ni u = e^{i\alpha_0} \begin{pmatrix} e^{i\alpha_3} \cos \frac{\theta}{2} & -e^{-i\varphi} \sin \frac{\theta}{2} \\ e^{i\varphi} \sin \frac{\theta}{2} & e^{-i\alpha_3} \cos \frac{\theta}{2} \end{pmatrix}.$$

Each nontrivial vector has a $\mathbf{U}(1)$-isomorphic fixgroup, e.g., e^2, which defines $\mathbf{U}(1)_+$ as an "idolized" electromagnetic subgroup:

$$\mathbb{C}^2 \cong V \ni e^2 = \begin{pmatrix} 0 \\ 1 \end{pmatrix} \Rightarrow \begin{pmatrix} e^{2i\alpha_0} & 0 \\ 0 & 1 \end{pmatrix} \in \mathbf{U}(2)_{e^2} = \mathbf{U}(1)_+.$$

The orbit points for the chosen "startvector," here $u \bullet e^2$, and for its $\mathbf{U}(2)$-orthonormal partner, here $(u \bullet e^2)_\perp$, give the two columns of the matrix parametrization $v \in \mathbf{U}(2)$ of the real three-dimensional *Goldstone manifold* $\mathcal{G}^3$:

$$\begin{aligned} \mathcal{G}^3 \cong \mathbf{U}(2)/\mathbf{U}(1)_+ &\cong \{((u \bullet e^2)_\perp, u \bullet e^2) = v \mid u \in \mathbf{U}(2)\}, \\ v &= \begin{pmatrix} e^{i(\alpha_3-\alpha_0)} \cos\frac{\theta}{2} & -e^{-i(\varphi-\alpha_0)} \sin\frac{\theta}{2} \\ e^{i(\varphi-\alpha_0)} \sin\frac{\theta}{2} & e^{-i(\alpha_3-\alpha_0)} \cos\frac{\theta}{2} \end{pmatrix}. \end{aligned}$$

In the standard model of electroweak interactions, the vector space $V \cong \mathbb{C}^2$ desribes the chargelike degrees of freedom of the Higgs field $\mathbf{\Phi} \in V$. The three weak parameters $(\alpha_3 - \alpha_0, \varphi - \alpha_0, \theta)$ parametrize electromagnetic relativity $\mathcal{G}^3$. As manifold, not as homogeneous space, $\mathcal{G}^3$ is isomorphic to Ω^3 and $\mathbf{SU}(2)$.

7.1.2 Orientation Manifolds of Metrical Tensors

With the exception of electromagnetic relativity in the last subsection, all relativity parameters will be given by the "general" group G-orbit of a metric that is invariant under the action of an "idolized" subgroup H. In this context, Weyl [59] called the homogeneous space G/H for H-relativity *orientation manifold of the metric* (bilinear or sesquilinear product), familiar from the orientations of the metrical ellipsoid of 3-position. The transformations involved constitute n-beins.

The invariance of a vector space metric $\mathbf{g}$ with respect to the action of a linear group H,

$$\mathbf{GL}(V) \supset H \ni h, \quad \mathbf{g}(x,y) \longmapsto \mathbf{g}(h \bullet x, h \bullet y) = \mathbf{g}(x,y) \text{ for all } x, y \in V,$$

gives the parametrization of the fixgroup classes G/H by the G-orbit of the metrical tensor $\mathbf{g}$:

$$\begin{aligned} H &= \{h \in G \subseteq \mathbf{GL}(V) \mid h \circ \mathbf{g} \circ h^* = \mathbf{g}\} \\ \Rightarrow G/H &\cong \{g \circ \mathbf{g} \circ g^* \mid g \in G\}. \end{aligned}$$

For $V \cong \mathbb{K}^n$, G/H is parametrized by $(n \times n)$-matrices.

7.1.3 Orientation Manifold of Metrical Hyperboloids

A real vector space $V \cong \mathbb{R}^n$ has a causality structure by embedding the cone of the positive numbers $\mathbb{R}_+ \longrightarrow V_+ \subset V$ into the "future" cone of the vector space $x \succeq 0 \iff x \in V_+$. A nontrivial "future" cone $V_+ \neq \{0\}$ can be defined by a bilinear symmetric form with "causal" signature $(t,s) = (1,s)$, invariant under the generalized Lorentz group $\mathbf{SO}_0(1,s)$. Such a causality structure for $V \cong \mathbb{R}^{1+s}$ is familiar for time $\mathbb{R}$ with total order and Minkowski spacetime $\mathbb{R}^4$ with the special relativistic partial order.

Any metrical tensor of $V \cong \mathbb{R}^4$ with causal signature $(1,3)$, e.g., an orthonormal Lorentz metrical tensor,

$$V \cong \mathbb{R}^4, \quad \eta = \begin{pmatrix} 1 & 0 \\ 0 & -\mathbf{1}_3 \end{pmatrix} \in V^T \vee V^T,$$

defines an "idolized" Lorentz group as its invariance group. Its $\mathbf{GL}(4,\mathbb{R})$-orbit leads to a parametrization of the orientation manifold for the metrical hyperboloid or for Lorentz group relativity with dimension $\binom{5}{2} = 10$:

$$\begin{aligned} \mathcal{M}^{10} \cong \mathbf{GL}(4,\mathbb{R})/\mathbf{O}(1,3) &\cong \{\mathbf{e} \circ \eta \circ \mathbf{e}^T = \mathbf{g} \mid \mathbf{e} \in \mathbf{GL}(4,\mathbb{R})\}, \\ \mathbf{g} &\cong \mathbf{g}^{il} = \mathbf{e}^i_a \eta^{ab} \mathbf{e}^l_b = \mathbf{g}^{li}, \text{ with } i, a = 0,1,2,3. \end{aligned}$$

In general, one has $\mathbf{GL}(1+s,\mathbb{R})/\mathbf{O}(1,s) \cong \mathcal{M}^{\binom{2+s}{2}}$ for $s = 0, 1, \ldots$.

7.1.4 2-Sphere for Perpendicular Relativity

With the local isomorphy of the rotation group to the spin group $\mathbf{SO}(3) \sim \mathbf{SU}(2)$, an "idolized" axial rotation subgroup $\mathbf{SO}(2) \subset \mathbf{SU}(2)$ is given by the invariance group of the hermitian and traceless Pauli matrix σ_3. Its $\mathbf{SU}(2)$-orbit leads to the 2-sphere parametrization of perpendicular relativity:

$$\begin{aligned} \sigma_3 = \begin{pmatrix} 1 & 0 \\ 0 & -1 \end{pmatrix}, \quad \Omega^2 \cong \mathbf{SO}(3)/\mathbf{SO}(2) &\cong \{u \circ \sigma_3 \circ u^\star = \tfrac{\vec{x}}{r} \mid u \in \mathbf{SU}(2)\}, \\ \tfrac{\vec{x}}{r} &\cong \tfrac{\vec{x}^\alpha_\beta}{r} = u^\alpha_A \sigma_3{}^A_B u^\star{}^B_\beta, \text{ with } \alpha, A = 1,2. \end{aligned}$$

The two angles (spherical coordinates) in the traceless hermitian matrix $\frac{\vec{x}}{r}$ can be parametrized by three position translations, with one condition for the determinant:

$$\begin{aligned} &\tfrac{\vec{x}}{r} = \tfrac{\vec{x}^\star}{r} = \tfrac{1}{r}\begin{pmatrix} x_3 & x_1 - ix_2 \\ x_1 + ix_2 & -x_3 \end{pmatrix} = \begin{pmatrix} \cos\theta & e^{-i\varphi}\sin\theta \\ e^{i\varphi}\sin\theta & -\cos\theta \end{pmatrix}, \\ &\text{with } \operatorname{tr}\tfrac{\vec{x}}{r} = 0 \text{ and } -\det\tfrac{\vec{x}}{r} = \tfrac{\vec{x}^2}{r^2} = 1. \end{aligned}$$

The restriction uses the rotation $\mathbf{SO}(3)$-invariant product $\vec{x}^2 = x_3^2 + x_1^2 + x_2^2$ in three dimensions.

7.1.5 Mass-Hyperboloid for Rotation Relativity

An "idolized" rotation group $\mathbf{SO}(3)$ in a Lorentz group $\mathbf{SO}_0(1,3)$ is characterized by a distinguished definite metric of a real three-dimensional vector space (position), e.g., $\mathbf{g} = \mathbf{1}_3$. Similarly, one can work with a sesquilinear scalar product δ of a complex two-dimensional space $V \cong \mathbb{C}^2$ invariant under the locally isomorphic spin group $\mathbf{SU}(2) \sim \mathbf{SO}(3)$ in the special linear group $\mathbf{SL}(2,\mathbb{C}) \sim \mathbf{SO}_0(1,3)$. The $\mathbf{SL}(2,\mathbb{C})$-orbit of the metric parametrizes rotation relativity by the points of an energy-momentum 3-hyperboloid:

$$\begin{aligned} \delta = \mathbf{1}_2, \quad \mathcal{Y}^3 \cong \mathbf{SO}_0(1,3)/\mathbf{SO}(3) &\cong \{s \circ \delta \circ s^\star = \tfrac{q}{m} \mid s \in \mathbf{SL}(2,\mathbb{C})\}, \\ \tfrac{q}{m} &\cong \tfrac{q^A_{\dot B}}{m} = s^A_\alpha \delta^\alpha_\beta s^\star{}^\beta_{\dot B}, \text{ with } A, \alpha = 1,2. \end{aligned}$$

The three real hyperbolic coordinates in the hermitian matrix $\frac{q}{m}$ can be chosen from four energy-momenta with one condition for the determinant:

$$\frac{q}{m} = \frac{q^\star}{m} = \frac{1}{m}\begin{pmatrix} q_0 + q_3 & q_1 - iq_2 \\ q_1 + iq_2 & q_0 - q_3 \end{pmatrix} = \begin{pmatrix} \cosh\psi + \cos\theta \sinh\psi & e^{-i\varphi}\sin\theta\sinh\psi \\ e^{i\varphi}\sin\theta\sinh\psi & \cosh\psi - \cos\theta\sinh\psi \end{pmatrix},$$
$$\text{with } \det\frac{q}{m} = \frac{q^2}{m^2} = 1.$$

The restriction of the four energy-momenta to the three momenta uses the $\mathbf{SO}_0(1,3)$-invariant bilinear form $q^2 = q_0^2 - \vec{q}^2$.

7.1.6 Spacetime Future Cone for Unitary Relativity

An "idolized" unitary group $\mathbf{U}(2)$, called a hyperisospin group, is a maximal compact subgroup of the general linear group $\mathbf{GL}(2,\mathbb{C})$, called the extended Lorentz group. It is given by the invariance group of a scalar product δ of a complex two-dimensional vector space. The $\mathbf{GL}(2,\mathbb{C})$-orbit defines four real parameters for unitary relativity, i.e., for the orientation manifold of the $\mathbf{U}(2)$-scalar product,

$$\begin{aligned}
\delta = \mathbf{1}_2,\ \ \mathbb{D}(2) \cong \mathbf{GL}(2,\mathbb{C})/\mathbf{U}(2) &\cong \{\chi\circ\delta\circ\chi^\star = x \mid \chi\in\mathbf{GL}(2,\mathbb{C})\},\\
x &\cong x^A_{\dot{B}} = \chi^A_\alpha \delta^\alpha_\beta \chi^{\star\,\beta}_{\dot{B}},\\
x &= x^\star = \begin{pmatrix} x_0 + x_3 & x_1 - ix_2 \\ x_1 + ix_2 & x_0 - x_3 \end{pmatrix},\ \text{with } A,\alpha = 1,2.
\end{aligned}$$

These four real orbit parameters characterize the strictly positive elements in the C^*-algebra of complex (2×2) matrices,

$$\begin{aligned}
x = \chi\circ\chi^\star \quad &\iff x = x^\star \text{ and } \mathbf{spec}\, x > 0\\
&\iff \det x = x^2 > 0 \text{ and } \operatorname{tr} x = 2x_0 > 0.
\end{aligned}$$

They describe the absolute modulus set in the polar decomposition of $\mathbf{GL}(2,\mathbb{C})$ into noncompact classes for the maximal compact group with the unitary phases,

$$\begin{aligned}
\mathbf{GL}(2,\mathbb{C}) &\ni \chi = |\chi|\circ u \in \mathbb{D}(2)\circ\mathbf{U}(2),\\
\mathbf{GL}(2,\mathbb{C})/\mathbf{U}(2) &\cong \mathbb{D}(2) \ni |\chi| = \sqrt{\chi\circ\chi^\star} = \sqrt{x}.
\end{aligned}$$

The positive matrices x are parametrizable by the points of the open future cone in flat Minkowski spacetime:

$$\mathbb{D}(2) \cong \mathbb{R}^4_+ = \{x\in\mathbb{R}^4 \mid x^2 > 0, x_0 > 0\}.$$

The cone manifold is embeddable into its own tangent space, the spacetime translations $\mathbb{R}^4 \supset \mathbb{D}(2)$. The translations inherit the action $x \longmapsto \chi\circ x\circ\chi^*$ of the dilation-extended orthochronous Lorentz group $\mathbf{GL}(2,\mathbb{C})/\mathbf{U}(1) \cong \mathbf{D}(1)\times \mathbf{SO}_0(1,3)$, which constitutes the homogeneous part of the extended Poincaré group $[\mathbf{D}(1)\times\mathbf{SO}_0(1,3)] \vec{\times} \mathbb{R}^4$.

7.2 Relativity Transitions by Transmutators

Elements of a relativity, i.e., of a homogeneous space G/H, are related to each other by the action of the full group G, e.g., different perpendicularities by rotations of the earth's surface or different nonrelativistic space–times by Lorentz transformations of a mass hyperboloid.

With real parameters for H-relativity G/H, the "general" group G is partly parametrized. Each coset can be given a defining representative $g_r \in gH \subseteq G$. Such representatives have a characteristic two-sided $G \times H$-transformation behavior in the group $G \times G$, called *relativity transition* or *transmutation* from the "general" group to the "idolized" group: A left multiplication of the representative with $k \in G$ hits the chosen representative $(kg)_r \in kgH$ up to a right multiplication with an H-element,

$$k \in G, \quad kg_r = k(gH)_r = (kgH)_r h(k, g_r) = (kg)_r h(k, g_r), \text{ with } h(k, g_r) \in H.$$

The left group action $k \in G$ is accompanied by a right action from the "idolized" subgroup $h(k, g_r) \in H$, which depends on both k and the representative g_r. It is called *Wigner element* and *Wigner subgroup-operation*, in generalization of the familiar Wigner rotation, which arises from a Lorentz transformation of a boost.

7.2.1 From Interactions to Particles

An example in which both electromagnetic relativity with the transition from hyperisospin to electromagnetic group $\mathbf{U}(2) \to \mathbf{U}(1)_+$ and rotation (special) relativity with the transition from Lorentz to spin group $\mathbf{SL}(2, \mathbb{C}) \to \mathbf{SU}(2)$ play a role is the transition from relativistic electroweak interaction fields to particles in the standard model,

$$\underset{\text{Lorentz}}{\mathbf{SL}(2, \mathbb{C})} \times \underset{\text{hypercharge-isospin}}{\mathbf{U}(2)} \longrightarrow \underset{\text{spin}}{\mathbf{SU}(2)} \times \underset{\text{electromagnetism}}{\mathbf{U}(1)_+}.$$

The lepton field in the minimal electroweak model connects, for each spacetime translation, the external Lorentz $\mathbf{SL}(2, \mathbb{C})$-degrees of freedom with the internal isospin $\mathbf{SU}(2)$-degrees of freedom and a hypercharge $\mathbf{U}(1)$-value $y = -\frac{1}{2}$:

$$\mathbb{R}^4 \ni x \longmapsto \mathbf{l}^A_\alpha(x), \text{ with } A, \alpha = 1, 2.$$

The transition from field to particles with respect to the internal degrees of freedom uses the ground-state degeneracy, implemented by the $\mathbf{U}(2)$-invariant condition $\langle \mathbf{\Phi}^\star \mathbf{\Phi} \rangle(x) = M^2 > 0$ of the Lorentz scalar Higgs field,

$$\mathbb{R}^4 \ni x \longmapsto \mathbf{\Phi}^\alpha(x).$$

The Higgs field is an isospin dublet with hypercharge $y = \frac{1}{2}$. It transmutes from "general" hyperisospin $\mathbf{U}(2)$-properties of the interaction fields to "idolized" electromagnetic $\mathbf{U}(1)_+$-properties of the particles, e.g., from the lepton field isodublet to the electron-positron field, an isosinglet with electromagnetic charge number $z = -1$:

$$\mathbf{U}(2) \longrightarrow \mathbf{U}(1)_+ : \ \mathbf{l}_\alpha^A(x) \longmapsto \mathbf{e}^A(x) = \frac{\mathbf{\Phi}^\alpha(x)}{|\mathbf{\Phi}|(x)} \mathbf{l}_\alpha^A(x) = \mathbf{l}_2^A(x) + \dots .$$

The "idolization" of an electromagnetic $\mathbf{U}(1)$ comes with the distinction of a ground state and the expansion of the Higgs transmutator (more ahead), e.g., $\frac{\mathbf{\Phi}^\alpha(x)}{|\mathbf{\Phi}|(x)} = \delta_2^\alpha + \dots$ for $e^2 = \binom{0}{1} \cong \delta_2^\alpha$ with $|\mathbf{\Phi}|(x) = \sqrt{\mathbf{\Phi}^\star\mathbf{\Phi}(x)}$.

With respect to external degrees of freedom, the transition from a left-handed Weyl field with Lorentz group $\mathbf{SL}(2,\mathbb{C})$-action to particles with mass $m^2 > 0$ and $\mathbf{SU}(2)$-spin requires a rest system. The related harmonic expansion (particle analysis) of the spacetime field with respect to eigenvectors involves the electron creation and positron annihilation operators $\mathrm{u}^a(\vec q)$ and $\mathrm{a}^{\star a}(\vec q)$, respectively, for spin directions $a = 1, 2$ and momentum $\vec q$ as translation eigenvalues:

$$\begin{aligned} \mathbf{SL}(2,\mathbb{C}) \longrightarrow \mathbf{SU}(2) : \ & \mathbf{e}^A(x) \longmapsto \mathrm{u}^a(\vec q),\ \mathrm{a}^{\star a}(\vec q) \\ \text{where } \ & \mathbf{e}^A(x) = {}^{\oplus}\!\!\int \frac{d^3q}{2q_0(2\pi)^3}\ s_a^A(\tfrac{q}{m})\ [e^{iqx}\mathrm{u}^a(\vec q) + e^{-iqx}\mathrm{a}^{\star a}(\vec q)], \\ \text{with } \ & q_0 = \sqrt{m^2 + \vec q^{\,2}}. \end{aligned}$$

The boost representation $s_a^A(\frac{q}{m})$, discussed ahead as a Weyl transmutator, connects the Lorentz group $\mathbf{SL}(2,\mathbb{C})$-action for fields with a rest system spin $\mathbf{SU}(2)$-action for massive particles.

Altogether in the field-particle transition, there are four types of transmutators $G \overset{\text{transmutator}}{\leftrightarrow} H$ involved with $G \times H$-transformations for four different group pairs $H \subset G$: the lepton field $\mathbf{l}$ with external–internal transformation behavior, the Lorentz scalar Higgs field $\mathbf{\Phi}$ as internal transmutator from interaction to particles, the boost representation s as corresponding external transmutator, and, finally, the creation and annihilation operators (u, a) with the external–internal properties of the particles (spin and charge)

Extended Lorentz group	$\mathbf{GL}(2,\mathbb{C})$	Hyperisospin group	$\mathbf{U}(2)$
Spin group	$\mathbf{SU}(2)$	Electromagnetic group	$\mathbf{U}(1)$

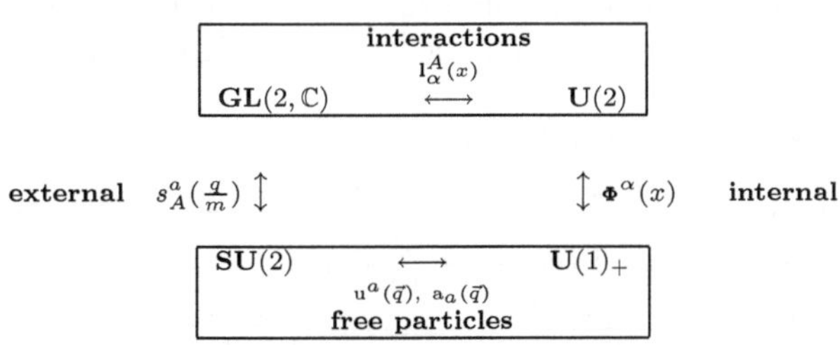

7.2.2 Pauli Transmutator

Perpendicular relativity is the base manifold of the coset bundle $\mathbf{SU}(2) \cong \mathbf{SO}(2)(\Omega^2)$ with the axial rotations as typical fiber. It is linearly represented by the fundamental *Pauli transmutator* from rotations to axial rotations,

$$\begin{aligned} \mathbb{R}^3 \supset \Omega^2 \ni \tfrac{\vec{x}}{r} &\longmapsto u(\tfrac{\vec{x}}{r}) \in \mathbf{SU}(2)/\mathbf{SO}(2), \\ u(\tfrac{\vec{x}}{r}) \circ \sigma_3 \circ u^\star(\tfrac{\vec{x}}{r}) &= \tfrac{\vec{x}}{r} = \tfrac{\sigma_a x_a}{r} \text{ with } r^2 = \vec{x}^2, \\ u(\tfrac{\vec{x}}{r}) &= e^{i\vec{\alpha}} = \mathbf{1}_2 \cos\alpha + i\tfrac{\vec{\alpha}}{\alpha}\sin\alpha \text{ with } \tan 2\alpha = \tan\theta = \tfrac{\sqrt{x_1^2+x_2^2}}{x_3} \\ &= \sqrt{\tfrac{x_3+r}{2r}}[\mathbf{1}_2 + i\tfrac{\vec{x}}{x_3+r}] = \tfrac{1}{\sqrt{2r(x_3+r)}}\begin{pmatrix} x_3+r & -x_1+ix_2 \\ x_1+ix_2 & x_3+r \end{pmatrix} \\ &= u(\varphi,\theta) = \begin{pmatrix} \cos\frac{\theta}{2} & -e^{-i\varphi}\sin\frac{\theta}{2} \\ e^{i\varphi}\sin\frac{\theta}{2} & \cos\frac{\theta}{2} \end{pmatrix}. \end{aligned}$$

A left action on the Pauli transmutator $u(\frac{\vec{x}}{r})$ with the spin group $\mathbf{SU}(2)$ gives the transmutator at the rotated point $O.\vec{x}$ on the 2-sphere up to a right action with the axial group $\mathbf{SO}(2)$ (Wigner axial rotation):

$$o \in \mathbf{SU}(2): \quad o \circ u(\tfrac{\vec{x}}{r}) = u(\tfrac{O.\vec{x}}{r}) \circ v(o, \tfrac{\vec{x}}{r})$$
$$\text{with} \quad \begin{cases} v(o,\frac{\vec{x}}{r}) \in \mathbf{SO}(2), \\ O.\vec{x} = o \circ \vec{x} \circ o^\star, \\ O_a^b = \frac{1}{2}\operatorname{tr}\sigma_a \circ o \circ \sigma_b \circ o^\star \in \mathbf{SO}(3). \end{cases}$$

The complicated explicit expression for the Wigner axial rotation can be computed from $v(o,\frac{\vec{x}}{r}) = u^\star(\frac{O.\vec{x}}{r}) \circ o \circ u(\frac{\vec{x}}{r})$.

7.2.3 Weyl Transmutators

Special relativity, e.g., the energy-momentum hyperboloid for a mass $m^2 > 0$, is the base manifold for the coset bundle $\mathbf{SL}(2,\mathbb{C}) \cong \mathbf{SU}(2)(\mathcal{Y}^3)$ with the spin rotations as typical fiber. The related Weyl representations of the boosts are a familiar example for a transmutator,

$$\begin{aligned} \mathbb{R}^4 \supset \mathcal{Y}^3 \ni \tfrac{q}{m} &\longmapsto s(\tfrac{q}{m}) \in \mathbf{SL}(2,\mathbb{C})/\mathbf{SU}(2), \\ s(\tfrac{q}{m}) \circ \mathbf{1}_2 \circ s^\star(\tfrac{q}{m}) &= \tfrac{q}{m} = \tfrac{\sigma^a q_a}{m} \text{ with } m^2 = q^2, \end{aligned}$$

and Weyl matrices, $\sigma^a = (\mathbf{1}_2, \vec{\sigma})$ (left) and $\check{\sigma}^a = (\mathbf{1}_2, -\vec{\sigma})$ (right). The explicit expressions involve the Lorentz dilations $e^{\beta\sigma_3} \in \mathbf{SO}_0(1,1)$ in addition to the Pauli transmutator for the two spherical degrees of freedom:

$$\begin{aligned} \tfrac{q}{m} &= u(\tfrac{\vec{q}}{|\vec{q}|}) \circ e^{\psi\sigma_3} \circ u^\star(\tfrac{\vec{q}}{|\vec{q}|}), \quad \psi = 2\beta, \\ e^{2\beta\sigma_3} &= \operatorname{diag}\tfrac{q}{m} = \tfrac{1}{m}\begin{pmatrix} q_0+|\vec{q}| & 0 \\ 0 & q_0-|\vec{q}| \end{pmatrix}, \quad \tanh 2\beta = \tfrac{|\vec{q}|}{q_0} = \tfrac{v}{c}, \\ s(\tfrac{q}{m}) &= u(\tfrac{\vec{q}}{|\vec{q}|}) \circ e^{\beta\sigma_3} = \mathbf{1}_2\cosh\beta + \tfrac{\vec{q}}{|\vec{q}|}\sinh\beta \\ &= \sqrt{\tfrac{q_0+m}{2m}}\left[\mathbf{1}_2 + \tfrac{\vec{q}}{q_0+m}\right] = \tfrac{1}{\sqrt{2m(q_0+m)}}\begin{pmatrix} q_0+q_3+m & q_1-iq_2 \\ q_1+iq_2 & q_0-q_3+m \end{pmatrix}. \end{aligned}$$

The left-handed *Weyl transmutator* $s(\frac{q}{m}) \in \mathbf{SL}(2,\mathbb{C})$ together with its right-handed partner $\hat{s} = s^{-1\star}$, i.e., $\hat{s}(\frac{q}{m}) = u(\frac{\vec{q}}{|\vec{q}|}) \circ e^{-\beta\sigma_3} \in \mathbf{SL}(2,\mathbb{C})$, are the two fundamental transmutators from Lorentz group to rotation subgroups. The restriction of the energy-momenta from four to three parameters by the on-shell hyperboloid $\mathcal{Y}^3$ condition $\frac{q^2}{m^2} = 1$ is expressed by the Dirac equation in energy-momentum space:

$$\left.\begin{array}{l} s(\frac{q}{m}) \circ \hat{s}^{-1}(\frac{q}{m}) = \frac{\sigma^a q_a}{m} \Rightarrow s(\frac{q}{m}) = \frac{\sigma^a q_a}{m} \circ \hat{s}(\frac{q}{m}) \\ \hat{s}(\frac{q}{m}) \circ s^{-1}(\frac{q}{m}) = \frac{\check{\sigma}^a q_a}{m} \Rightarrow \hat{s}(\frac{q}{m}) = \frac{\check{\sigma}^a q_a}{m} \circ s(\frac{q}{m}) \end{array}\right\} \Rightarrow (\gamma^a q_a - m)\mathbf{s}(\frac{q}{m}) = 0,$$

$$\text{with } \gamma^a = \begin{pmatrix} 0 & \sigma^a \\ \check{\sigma}^a & 0 \end{pmatrix}, \quad \mathbf{s}(\frac{q}{m}) = \begin{pmatrix} s(\frac{q}{m}) & 0 \\ 0 & \hat{s}(\frac{q}{m}) \end{pmatrix}.$$

The four columns of the (4×4) matrix $\mathbf{s}(\frac{q}{m})$ are familiar as solutions of the Dirac equation.

For the Pauli transmutator, the analogue to the Dirac equation is the condition $\vec{\sigma}\vec{x}u(\frac{\vec{x}}{r}) - u(\frac{\vec{x}}{r})\sigma_3 r = 0$, which restricts the three rotation parameters to two independent parameters for $\mathbf{SU}(2)/\mathbf{SO}(2)$.

A left action with the Lorentz group $\mathbf{SL}(2, \mathbb{C})$ gives the Weyl transmutator at the Lorentz-transformed energy-momenta $\Lambda.q$ on the hyperboloid $q^2 = m^2$, up to a right action by a Wigner spin $\mathbf{SU}(2)$-rotation:

$$\lambda \in \mathbf{SL}(2, \mathbb{C}): \quad \lambda \circ s(\frac{q}{m}) = s(\frac{\Lambda.q}{m}) \circ u(\frac{q}{m}, \lambda),$$
$$\text{with} \quad \begin{cases} u(\frac{q}{m}, \lambda) \in \mathbf{SU}(2), \\ \Lambda.q = \lambda \circ q \circ \lambda^\star, \\ \Lambda_a^b = \frac{1}{2} \operatorname{tr} \sigma_a \circ \lambda \circ \check{\sigma}^b \circ \lambda^\star \in \mathbf{SO}_0(1,3). \end{cases}$$

7.2.4 Higgs Transmutators

In the standard model of electroweak interactions, the three weak parameters for the Goldstone manifold of electromagnetic relativity as base manifold of the coset bundle $\mathbf{U}(2) \cong \mathbf{U}(1)(\mathcal{G}^3)$ are implemented by three chargelike degrees of freedom of the Higgs vector $\mathbf{\Phi}^\alpha \cong \begin{pmatrix} \mathbf{\Phi}^1 \\ \mathbf{\Phi}^2 \end{pmatrix} \in V \cong \mathbb{C}^2$ and its orthogonal $\tilde{\mathbf{\Phi}}^\alpha = \epsilon^{\alpha\beta}\mathbf{\Phi}^\star_\beta = \begin{pmatrix} \mathbf{\Phi}^\star_2 \\ -\mathbf{\Phi}^\star_1 \end{pmatrix}$:

$$\mathbb{C}^2 \supset \mathcal{G}^3 \ni \frac{\mathbf{\Phi}}{M} \longmapsto v(\frac{\mathbf{\Phi}}{M}) \in \mathbf{U}(2)/\mathbf{U}(1)_+,$$

$$v(\frac{\mathbf{\Phi}}{M}) = \begin{pmatrix} e^{i(\alpha_3-\alpha_0)}\cos\frac{\theta}{2} & -e^{-i(\varphi-\alpha_0)}\sin\frac{\theta}{2} \\ e^{i(\varphi-\alpha_0)}\sin\frac{\theta}{2} & e^{-i(\alpha_3-\alpha_0)}\cos\frac{\theta}{2} \end{pmatrix} = u(\varphi - \alpha_3, \theta) \circ e^{i(\alpha_3-\alpha_0)\sigma_3}$$
$$= \frac{1}{M}\begin{pmatrix} \mathbf{\Phi}^\star_2 & \mathbf{\Phi}^1 \\ -\mathbf{\Phi}^\star_1 & \mathbf{\Phi}^2 \end{pmatrix} \text{ with } \det v(\frac{\mathbf{\Phi}}{M}) = \frac{|\mathbf{\Phi}^1|^2 + |\mathbf{\Phi}^2|^2}{M^2} = 1.$$

The restriction from four to three real weak degrees of freedom uses the $\mathbf{U}(2)$-invariant scalar product $\langle\mathbf{\Phi}|\mathbf{\Phi}\rangle = |\mathbf{\Phi}|^2 = M^2$ of the Higgs vector space.

A left hypercharge-isospin action on the fundamental *Higgs transmutator* gives the transmutator at the $\mathbf{U}(2)$-transformed Higgs vector on the Goldstone manifold, accompanied by a Wigner electromagnetic $\mathbf{U}(1)_+$-transformation from the right:

$$u \in \mathbf{U}(2): \quad u \circ v(\frac{\mathbf{\Phi}}{M}) = v(\frac{u.\mathbf{\Phi}}{M}) \circ u_+, \text{ with } \begin{cases} u = e^{i\gamma_0}u_2 \in \mathbf{U}(1) \circ \mathbf{SU}(2), \\ u_+ = \begin{pmatrix} e^{i2\gamma_0} & 0 \\ 0 & 1 \end{pmatrix} \in \mathbf{U}(1)_+. \end{cases}$$

7.2.5 Real Tetrads (Vierbeins)

Lorentz group or particle relativity is the 10-parametric base manifold of the coset bundle $\mathbf{GL}(4,\mathbb{R}) \cong \mathbf{SO}_0(1,3)(\mathcal{M}^{10})$ with the Lorentz group as typical fiber. It is the $\mathbf{GL}(4,\mathbb{R})$-orbit of the orthonormal $\mathbf{O}(1,3)$-"idolized" Lorentz metric in a symmetric matrix η and diagonalizable to four principal axes with a transformation from a maximal compact subgroup $\mathbf{O}(4) \subset \mathbf{GL}(4,\mathbb{R})$ (six parameters):

$$\mathbf{g} = \mathbf{e}^T \circ \eta \circ \mathbf{e} = O_4^T \circ \operatorname{diag}\mathbf{g} \circ O_4, \text{ with } O_4 \in \mathbf{O}(4).$$

The diagonal part of the metrical hyperboloid, multiplied by the inverse metric η^{-1}, displays the remaining four dilation transformations from the maximal noncompact abelian subgroup:

$$\eta^{-1} \circ \operatorname{diag}\mathbf{g} \in \mathbf{D}(1) \times \mathbf{SO}_0(1,1)^3 \cong \mathbf{D}(1)^4 \subset \mathbf{GL}(4,\mathbb{R}).$$

The diagonal elements are four directional units, one for time and three for the metrical ellipsoid of the 3-position.

The operational decomposition of the metrical hyperboloid leads to the parametrization of the 10-dimensional tetrad $\mathbf{e}$ as a basis of real four-dimensional tangent spacetime $\mathbb{R}^4$ by four dilations and a six-dimensional rotation:

$$\mathcal{M}^{10} \ni \mathbf{g} \longmapsto \mathbf{e}(\mathbf{g}) \Subset \mathbf{GL}(4,\mathbb{R})/\mathbf{O}(1,3).$$

A general linear $\mathbf{GL}(4,\mathbb{R})$ left multiplication gives the tetrad for a transformed metrical tensor and a Wigner right transformation by the idolized Lorentz group $\mathbf{O}(1,3)$:

$$h \in \mathbf{GL}(4,\mathbb{R}): \quad h \circ \mathbf{e}(\mathbf{g}) = \mathbf{e}(h^T \circ \mathbf{g} \circ h) \circ \Lambda(h,\mathbf{g}), \text{ with } \Lambda(h,\mathbf{g}) \in \mathbf{O}(1,3).$$

7.2.6 Complex Dyads (Zweibeins)

Electroweak spacetime $\mathbb{D}(2)$ is the orientation manifold of $\mathbf{U}(2)$-scalar products (unitary relativity) as base manifold of the coset bundle $\mathbf{U}(2)(\mathbb{D}(2))$ with the hyperisospin group as the typical fiber:

$$x \in \mathbb{D}(2) \cong \mathbf{GL}(2,\mathbb{C})/\mathbf{U}(2) = \mathbf{D}(\mathbf{1}_2) \times \mathbf{SL}(2,\mathbb{C})/\mathbf{SU}(2) \cong \mathbf{D}(1) \times \mathcal{Y}^3.$$

It is transformed to an "idolized" diagonal scalar product by a Weyl transmutator $s(\frac{x}{\sqrt{x^2}}) \in \mathbf{SL}(2,\mathbb{C})$ for the three hyperbolic degrees of freedom and a dilation $\mathbf{D}(1) = \exp\mathbb{R} \cong \mathbb{R}$ for eigentime $e^{2\beta_0} = \sqrt{x^2}$:

$$\begin{aligned} x &= \begin{pmatrix} x_0 + x_3 & x_1 - ix_2 \\ x_1 + ix_2 & x_0 - x_3 \end{pmatrix} = s(\tfrac{x}{\sqrt{x^2}}) \circ e^{2\beta_0 \mathbf{1}_2} \circ s^\star(\tfrac{x}{\sqrt{x^2}}) = u(\tfrac{\vec{x}}{r}) \circ \operatorname{diag} x \circ u^\star(\tfrac{\vec{x}}{r}), \\ \operatorname{diag} x &= \begin{pmatrix} x_0 + r & 0 \\ 0 & x_0 - r \end{pmatrix} = e^{2(\beta_0 \mathbf{1}_2 + \beta\sigma_3)} \in \mathbf{D}(1) \times \mathbf{SO}_0(1,1), \\ &\text{with } e^{4\beta_0} = x^2, \ \tanh 2\beta = \tfrac{r}{x_0}. \end{aligned}$$

The diagonalization of the scalar products gives the fundamental transmutator from the extended Lorentz group $\mathbf{GL}(2,\mathbb{C})$ to the hyperisospin subgroup $\mathbf{U}(2)$. It contains a basis of the complex two-dimensional space and will be called, in analogy to a real tetrad or vierbein $\mathbf{e}(\mathbf{g}(x)) \Subset \mathbf{GL}(4,\mathbb{R})/\mathbf{SO}_0(1,3)$, a complex *dyad (zweibein)* $\chi(x) \Subset \mathbf{GL}(2,\mathbb{C})/\ \mathbf{U}(2)$. It is parametrized by the future cone spacetime points as orbit of the $\mathbf{U}(2)$-scalar product,

$$\mathbb{R}^4 \supset \mathbb{D}(2) \ni x \longmapsto \chi(x) \Subset \mathbf{GL}(2,\mathbb{C})/\mathbf{U}(2),$$

$$\chi(x) \circ \mathbf{1}_2 \circ \chi^\star(x) = x = s(\tfrac{x}{\sqrt{x^2}}) \circ e^{\beta_0 \mathbf{1}_2} = u(\tfrac{\vec{x}}{r}) \circ e^{\beta_0 \mathbf{1}_2 + \beta \sigma_3}.$$

The left action by the extended Lorentz group $\mathbf{GL}(2,\mathbb{C})$ as external transformation gives the dyad χ at a Lorentz-transformed and dilated spacetime point in the future cone, accompanied by an action from the right with an internal spacetime-dependent Wigner hyperisospin $\mathbf{U}(2)$-transformation:

$$g \in \mathbf{GL}(2,\mathbb{C}): \quad g \circ \chi(x) = \chi(e^{2\delta_0}\Lambda.x) \circ u(x,g),$$
$$\text{with} \quad \begin{cases} u(x,g) \in \mathbf{U}(2), \\ g = e^{\delta_0 + i\alpha_0}\lambda \in \mathbf{D}(1) \times \mathbf{U}(1) \circ \mathbf{SL}(2,\mathbb{C}), \\ g \circ x \circ g^\star = e^{2\delta_0}\lambda \circ x \circ \lambda^\star = e^{2\delta_0}\Lambda.x, \quad \Lambda \in \mathbf{SO}_0(1,3). \end{cases}$$

7.3 Linear Representations of Relativities

In the foregoing section, the classes $\mathbb{M} \cong G/H$ of the five relativities with linear groups $H \subset G$ were represented by linear transformations of the group G-defining vector spaces:

Relativity manifold fixgroup H	Fundamental transmutator $\mathbb{M} \longrightarrow G/H$
2-sphere $\vec{x}^2 = \ell^2$ axial rotations	Pauli transmutator $\Omega^2 \ni \frac{\vec{x}}{\ell} \longmapsto u_k^\alpha(\frac{\vec{x}}{\ell}) \Subset \mathbf{SU}(2)/\mathbf{SO}(2)$
Goldstone manifold $\Phi^\star\Phi = M^2$ electromagnetic group	Higgs transmutator $\mathcal{G}^3 \ni \frac{\Phi}{M} \longmapsto v_k^\alpha(\frac{\Phi}{M}) \Subset \mathbf{U}(2)/\mathbf{U}(1)_+$
3-hyperboloid $q^2 = m^2$ rotations	Weyl transmutators $\mathcal{Y}^3 \ni \frac{q}{m} \longmapsto s_\alpha^A(\frac{q}{m}) \Subset \mathbf{SL}(2,\mathbb{C})/\mathbf{SU}(2)$
Future 4-cone unitary group	dyad $\mathbb{D}(2) \ni x \longmapsto \chi_\alpha^A(x) \Subset \mathbf{GL}(2,\mathbb{C})/\mathbf{U}(2)$
Metric manifold Lorentz group	tetrad $\mathcal{M}^{10} \ni \mathbf{g} \longmapsto \mathbf{e}_a^j(\mathbf{g}) \Subset \mathbf{GL}(4,\mathbb{R})/\mathbf{O}(1,3)$

As for the G-representations, the products of these fundamental transmutators give the *finite-dimensional representations of the homogeneous spaces* G/H. For a compact group G, the finite-dimensional transmutators are complete for the harmonic analysis of the Hilbert spaces with the square-integrable functions $L^2(\mathbb{M}, V^T)$ of the orientation manifold of the relativity valued in a vector space V with H-action (more ahead).

7.3.1 Rectangular Transmutators

Representations of the "general" group G involve representations of the cosets G/H representatives,

$$\begin{aligned} G \ni g &\longmapsto D(g) \in \mathbf{GL}(V), \\ G/H \ni gH \ni g_r &\longmapsto D(g_r), \end{aligned}$$

e.g., for perpendicular relativity,

$$\begin{aligned} \mathbf{SU}(2) \ni u(\varphi,\theta,\chi) &= \begin{pmatrix} e^{i\frac{\chi+\varphi}{2}}\cos\frac{\theta}{2} & -e^{i\frac{\chi-\varphi}{2}}\sin\frac{\theta}{2} \\ e^{-i\frac{\chi-\varphi}{2}}\sin\frac{\theta}{2} & e^{-i\frac{\chi+\varphi}{2}}\cos\frac{\theta}{2} \end{pmatrix} \\ \longmapsto u(\varphi,\theta,\chi)_r = u(\varphi,\theta) &= \begin{pmatrix} \cos\frac{\theta}{2} & -e^{-i\varphi}\sin\frac{\theta}{2} \\ e^{i\varphi}\sin\frac{\theta}{2} & \cos\frac{\theta}{2} \end{pmatrix} \in \mathbf{SU}(2)/\mathbf{SO}(2) \cong \Omega^2. \end{aligned}$$

A G-representation by an $(n \times n)$ matrix, with an (8×8)-example for $\mathbf{SU}(3)$,

$$D(g) \in V \otimes V^T \cong \mathbb{C}^n \otimes \mathbb{C}^n \overset{\text{e.g.}}{\sim} \begin{pmatrix} \bullet&\bullet&\bullet&\bullet&\bullet&\bullet&\bullet&\bullet \\ \bullet&\bullet&\bullet&\bullet&\bullet&\bullet&\bullet&\bullet \\ \bullet&\bullet&\bullet&\bullet&\bullet&\bullet&\bullet&\bullet \\ \bullet&\bullet&\bullet&\bullet&\bullet&\bullet&\bullet&\bullet \\ \bullet&\bullet&\bullet&\bullet&\bullet&\bullet&\bullet&\bullet \\ \bullet&\bullet&\bullet&\bullet&\bullet&\bullet&\bullet&\bullet \\ \bullet&\bullet&\bullet&\bullet&\bullet&\bullet&\bullet&\bullet \\ \bullet&\bullet&\bullet&\bullet&\bullet&\bullet&\bullet&\bullet \end{pmatrix},$$

can be decomposed into H-representations, exemplified by the octet decomposition $8 = 2 \oplus 1 \oplus 3 \oplus 2$ into $\mathbf{SU}(2)$-representations,

$$V \overset{H}{\cong} \bigoplus_{\iota=1}^{k} V_\iota, \quad H \bullet V_\iota \subseteq V_\iota, \quad D(h) \overset{H}{\cong} \bigoplus_{\iota=1}^{k} d_\iota(h) \overset{\text{e.g.}}{\sim} \left(\begin{array}{cc|c|ccc|cc} \bullet&\bullet&0&0&0&0&0&0 \\ \bullet&\bullet&0&0&0&0&0&0 \\ \hline 0&0&\bullet&0&0&0&0&0 \\ \hline 0&0&0&\bullet&\bullet&\bullet&0&0 \\ 0&0&0&\bullet&\bullet&\bullet&0&0 \\ 0&0&0&\bullet&\bullet&\bullet&0&0 \\ \hline 0&0&0&0&0&0&\bullet&\bullet \\ 0&0&0&0&0&0&\bullet&\bullet \end{array}\right).$$

Thus, the G-representation matrices can be decomposed "lopsidedly" into rectangular $(n \times n_\iota)$ matrices, $n_\iota \leq n$,

$$D(g) = \bigoplus_{\iota=1}^{k} D_\iota(g) = \Big(D_1(g)\Big|D_2(g)\Big|\cdots\Big|D_k(g)\Big) \overset{\text{e.g.}}{\sim} \left(\begin{array}{cc|c|ccc|cc} \bullet&\bullet&\bullet&\bullet&\bullet&\bullet&\bullet&\bullet \\ \bullet&\bullet&\bullet&\bullet&\bullet&\bullet&\bullet&\bullet \\ \bullet&\bullet&\bullet&\bullet&\bullet&\bullet&\bullet&\bullet \\ \bullet&\bullet&\bullet&\bullet&\bullet&\bullet&\bullet&\bullet \\ \bullet&\bullet&\bullet&\bullet&\bullet&\bullet&\bullet&\bullet \\ \bullet&\bullet&\bullet&\bullet&\bullet&\bullet&\bullet&\bullet \\ \bullet&\bullet&\bullet&\bullet&\bullet&\bullet&\bullet&\bullet \\ \bullet&\bullet&\bullet&\bullet&\bullet&\bullet&\bullet&\bullet \end{array}\right).$$

The rectangular matrices have a left-right $G \times H$-action, e.g., the $\mathbf{SU}(3) \times \mathbf{SU}(2)$-action on octet-dublet, octet-singlet, octet-triplet, and octet-dublet. As mappings of the coset representatives $(G/H)_r$, they are called *transmutators*:

$$\begin{aligned} &(G/H)_r \ni g_r \longmapsto D_\iota(g_r) \in V \otimes V_\iota^T \cong \mathbb{C}^n \otimes \mathbb{C}^{n_\iota}, \\ &\text{with} \begin{cases} D_\iota(g_r h) &= D_\iota(g_r) \circ d_\iota(h), \ h \in H, \\ D_\iota(k g_r) &= D(k) \circ D_\iota(g_r) \\ &= D((kg)_r) \circ d_\iota(h(k, g_r)), \ k \in G. \end{cases} \end{aligned}$$

With bases of the G-vector space $|n;j\rangle \in V$ and the H-vector spaces $|n_\iota;a\rangle \in V_\iota$, one has in a Dirac notation with kets for vectors $|\ \rangle \in V$ and bras for linear forms $\langle\ | \in V_\iota^T$,

$$\begin{aligned} V \otimes V_\iota^T &\ni D_\iota(g_r) = |n;j\rangle D_\iota(g_r)^j_a \langle n_\iota;a|, \\ \text{e.g., } \mathbb{C}^8 \otimes \mathbb{C}^2 &\ni D_2(g_r) = |8;j\rangle D_2(g_r)^j_a \langle 2;a|, \text{ with } \begin{cases} j = 1,\ldots,8, \\ a = 1,2. \end{cases} \end{aligned}$$

The finite-dimensional transmutators are from $(n \times n_\iota)$-dimensional vector spaces $V \otimes V_\iota^T$ with representations of the bi-regular subgroup $G \times H \subseteq G \times G$. Those representations are unitary, i.e., Hilbert representations, only for the compact groups G, i.e., in the examples above, for perpendicular and electromagnetic relativity.

7.3.2 Representations of Perpendicular Relativity

For perpendicular relativity, all transmutators from rotations to axial rotations arise by the totally symmetric products, denoted by $\overset{2J}{\bigvee}$, of the fundamental Pauli transmutator $u(\frac{\vec{x}}{r}) \in \mathbf{SU}(2)$,

$$\begin{aligned} \mathbf{SU}(2)/\mathbf{SO}(2) \cong \Omega^2 &\longrightarrow \mathbf{SU}(1+2J), \\ \tfrac{\vec{x}}{r} &\longmapsto [J](\tfrac{\vec{x}}{r}) = \overset{2J}{\bigvee} u(\tfrac{\vec{x}}{r}),\ \vec{x}^2 = r^2. \end{aligned}$$

The irreducible spin $\mathbf{SU}(2)$-representations $[J]$ with dimensions $1+2J$ are decomposable into axial rotation $\mathbf{SO}(2)$-representations $(\pm n)$ with dimension 2 for $n \neq 0$ and two polarizations $\pm n$ (left- and right-circularly polarized):

$$\mathbf{irrep}\,\mathbf{SU}(2) \ni [J] \overset{\mathbf{SO}(2)}{\cong} \begin{cases} \bigoplus_{n=0,2,\ldots,}^{2J} (\pm n) & \text{for } J = 0,1,\ldots, \\ \bigoplus_{n=1,3,\ldots,}^{2J} (\pm n) & \text{for } J = \frac{1}{2},\frac{3}{2},\ldots, \end{cases}$$

e.g., for rotations acting on 3-position $\mathbb{R}^3$:

$$\begin{aligned} [1](\tfrac{\vec{x}}{r}) \cong O^s_r(\tfrac{\vec{x}}{r}) &= \tfrac{1}{2}\operatorname{tr} u(\tfrac{\vec{x}}{r}) \circ \sigma^s \circ u^\star(\tfrac{\vec{x}}{r}) \circ \sigma^r \\ &= \tfrac{1}{r}\left(\begin{array}{c|c} \delta^{\alpha\beta}r - \frac{x_\alpha x_\beta}{r+x_3} & x_\alpha \\ -x_\beta & x_3 \end{array} \right) \in \mathbf{SO}(3), \text{ with } \begin{cases} r = 1,2,3, \\ \alpha = 1,2, \end{cases} \end{aligned}$$

$$[1] \overset{\mathbf{SO}(2)}{\cong} (\pm 2) \oplus (0),$$

with the relations for the $\mathbf{SO}(3)$ and $\mathbf{SO}(2)$ metrical tensors:

$$O^r_{\alpha,3}(\tfrac{\vec{x}}{r})\delta_{rs}O^s_{\beta,3}(\tfrac{\vec{x}}{r}) = \left(\begin{array}{c|c} \delta_{\alpha\beta} & 0 \\ \hline 0 & 1 \end{array} \right),\quad O^r_\alpha(\tfrac{\vec{x}}{r})\delta^{\alpha\beta}O^s_\beta(\tfrac{\vec{x}}{r}) = \delta^{rs} - \tfrac{x_r x_s}{r^2}.$$

The symmetric square $u \vee u$ of the Pauli transmutator, in a Cartesian and a spherical basis,

$$O(\tfrac{\vec{x}}{r}) = \left(\begin{array}{c|c} \delta^{\alpha\beta} - \frac{x_\alpha x_\beta}{r(r+x_3)} & \frac{x_\alpha}{r} \\ -\frac{x_\beta}{r} & \frac{x_3}{r} \end{array} \right) \cong \left(\begin{array}{cc|c} e^{i\varphi}\cos^2\frac{\theta}{2} & -e^{i\varphi}\sin^2\frac{\theta}{2} & ie^{i\varphi}\frac{\sin\theta}{\sqrt{2}} \\ i\frac{\sin\theta}{\sqrt{2}} & i\frac{\sin\theta}{\sqrt{2}} & \cos\theta \\ -e^{-i\varphi}\sin^2\frac{\theta}{2} & e^{-i\varphi}\cos^2\frac{\theta}{2} & ie^{-i\varphi}\frac{\sin\theta}{\sqrt{2}} \end{array} \right),$$

displays in the third column $O_3^b(\frac{\vec{x}}{r})$ the spherical harmonics $\mathrm{Y}^1(\varphi,\theta) \sim \frac{\vec{x}}{r}$ as a basis for subspace $\mathbb{C}^3$ in the Hilbert space $L^2(\Omega^2)$ with the square-integrable functions on the 2-sphere. Its symmetric traceless products of power $J = 1, 2, \ldots$ give all *spherical harmonics* $\mathrm{Y}^J(\varphi,\theta) \sim (\frac{\vec{x}}{r})^J_{\text{traceless}}$, which arise as the $(1+2J)$-entries in one column of the $(1+2J)\times(1+2J)$-matrices for the representation $[J]$. The spherical harmonics are bases for the Hilbert spaces $\mathbb{C}^{1+2J} \subset L^2(\Omega^2)$ acted on by the irreducible $\mathbf{SO}(3)$-representations.

With respect to the bi-regular $\mathbf{SU}(2) \times \mathbf{SO}(2)$-transformation behavior, the four functions in the two columns of the (2×2)-Pauli transmutator,

$$u(\tfrac{\vec{x}}{r}) = \begin{pmatrix} \sqrt{\frac{x_3+r}{2r}} & -\frac{x_1-ix_2}{\sqrt{2r(x_3+r)}} \\ \frac{x_1+ix_2}{\sqrt{2r(x_3+r)}} & \sqrt{\frac{x_3+r}{2r}} \end{pmatrix} = \begin{pmatrix} \cos\frac{\theta}{2} & -e^{-i\varphi}\sin\frac{\theta}{2} \\ e^{i\varphi}\sin\frac{\theta}{2} & \cos\frac{\theta}{2} \end{pmatrix} \in \mathbb{C}^2 \otimes \mathbb{C}^2,$$

and the six functions in the first and second columns $O^b_{1,2}(\frac{\vec{x}}{r}) \cong O^b_{+,-}(\frac{\vec{x}}{r})$ above in a rectangular (3×2) matrix constitute bases for finite-dimensional Hilbert spaces $\mathbb{C}^2 \otimes \mathbb{C}^2$ and $\mathbb{C}^3 \otimes \mathbb{C}^2$ with $\mathbf{SU}(2)$-representations, acting on the left factors $\mathbb{C}^2$ and $\mathbb{C}^3$, and nontrivial $\mathbf{SO}(2)$-representations $\mathbf{SO}(2) \ni e^{i\alpha_3\sigma_3} \longmapsto e^{in\alpha_3\sigma_3}$, acting on the right factor $\mathbb{C}^2$. They span irreducible subspaces for the harmonic analysis of the Hilbert space $L^2(\Omega^2, \mathbb{C}^2)$ with the square-integrable mappings from the 2-sphere into a vector space with nontrivial $\mathbf{SO}(2)$-action.

In general, one has the *Peter–Weyl decompositions* [47] (see Chapter 8) into irreducible subspaces for $\mathbf{SU}(2) \times \mathbf{SO}(2)$-action:

$$V_n \cong \mathbb{C}^{2-\delta_{n0}} : \;\; L^2(\Omega^2, V_n) \overset{\text{dense}}{\cong} \bigoplus_{2J \geq n} \mathbb{C}^{1+2J} \otimes V_n.$$

The orthogonal sum goes over all $\mathbf{SU}(2)$-representation that contain the $\mathbf{SO}(2)$-representations on $V_n \cong \mathbb{C}, \mathbb{C}^2$ (Frobenius' reciprocity; more ahead). This generalizes the case for the spherical harmonics with $V_0 \cong \mathbb{C}$.

7.3.3 Representations of Electromagnetic Relativity

For the standard model of electroweak interactions, the Higgs-parametrized defining representation of the orientation manifold $\mathcal{G}^3$ of the electromagnetic group with hypercharge $y = \frac{1}{2}$ and isospin $T = \frac{1}{2}$ and its conjugate, i.e., the two Higgs transmutators,

$$\begin{aligned} [\tfrac{1}{2}||\tfrac{1}{2}](\tfrac{\mathbf{\Phi}}{M}) &= v(\tfrac{\mathbf{\Phi}}{M}) = \tfrac{1}{M}\left(\begin{array}{c|c} \mathbf{\Phi}_2^\star & \mathbf{\Phi}^1 \\ -\mathbf{\Phi}_1^\star & \mathbf{\Phi}^2 \end{array} \right) \in \mathbf{U}(2)/\mathbf{U}(1)_+, \\ [-\tfrac{1}{2}||\tfrac{1}{2}](\tfrac{\mathbf{\Phi}}{M}) &= v^\star(\tfrac{\mathbf{\Phi}}{M}), \end{aligned}$$

give, via their products, all irreducible representations of the Goldstone manifold:

$$\begin{aligned}
\mathbf{U}(2)/\mathbf{U}(1)_+ &\cong \mathcal{G}^3 \longrightarrow \mathbf{U}(1+2T), \quad \tfrac{\boldsymbol{\Phi}}{M} \longmapsto [\pm n + T||T](\tfrac{\boldsymbol{\Phi}}{M}),\\
\mathbf{irrep}\,\mathbf{U}(2) &\ni [\pm n + T||T] \cong [\pm 1||0]^n \otimes \bigvee^{2T} [\tfrac{1}{2}||\tfrac{1}{2}],\\
\mathbf{irrep}\,\mathbf{U}(1) &\ni [\pm 1||0] \cong [\pm\tfrac{1}{2}||\tfrac{1}{2}] \wedge [\pm\tfrac{1}{2}||\tfrac{1}{2}].
\end{aligned}$$

Because of the central correlation $\mathbf{SU}(2)\cap\mathbf{U}(\mathbf{1}_2) = \{\pm\mathbf{1}_2\}$ in $\mathbf{U}(2)$, the $\mathbf{U}(2)$-representations have the correlation of the hypercharge- and isospin-invariant $y = T \pm n$ with natural n, i.e., the two invariants (y, T) for the rank-2 $\mathbf{U}(2)$-transformations are either both integer or both half-integer, as visible in the colorless fields of the standard model.

The decomposition of a hyperisospin $\mathbf{U}(2)$-representation into irreducible representations of the electromagnetic group $\mathbf{U}(1)_+ \ni e^{i2\gamma_0} \longmapsto e^{zi2\gamma_0}$ is given with integer charge numbers $z \in \mathbb{Z}$:

$$\mathbf{U}(2) \ni [\pm n + T||T] \overset{\mathbf{U}(1)_+}{\cong} \bigoplus_{z=\pm n}^{\pm n+2T} [z],$$

$$\text{e.g.,} \quad \begin{cases} [\pm\tfrac{1}{2}||\tfrac{1}{2}] \cong [0] \oplus [\pm 1],\\ \quad [0||1] \cong [-1] \oplus [0] \oplus [1]. \end{cases}$$

v and $v^\star$ transmute from hyperisospin $\mathbf{U}(2)$-dublets to isospin $\mathbf{SU}(2)$-singlets with nontrivial charge numbers $z = \pm 1$ for electromagnetic $\mathbf{U}(1)_+$. The antisymmetric squares $v \wedge v$ and $v^\star \wedge v^\star$ of the fundamental Higgs transmutators have charge numbers $z = \pm 1$:

$$\begin{aligned}
[1||0](\tfrac{\boldsymbol{\Phi}}{M}) &= \tfrac{\tilde{\boldsymbol{\Phi}}^\star_\alpha \boldsymbol{\Phi}^\alpha}{M^2} \in \mathbf{U}(1), \text{ with } [1||0] \overset{\mathbf{U}(1)_+}{\cong} [1],\\
[-1||0](\tfrac{\boldsymbol{\Phi}}{M}) &= \tfrac{\boldsymbol{\Phi}^\star_\alpha \tilde{\boldsymbol{\Phi}}^\alpha}{M^2} \in \mathbf{U}(1), \text{ with } [-1||0] \overset{\mathbf{U}(1)_+}{\cong} [-1].
\end{aligned}$$

In the hypercharge trivial product $v\otimes v^\star$ of both fundamental Higgs transmutators, the columns transmute from a triplet to three isosinglets with charge $z \in \{-1, 0, 1\}$:

$$\begin{aligned}
[0||1](\tfrac{\boldsymbol{\Phi}}{M}) &= \tfrac{1}{2}\,\mathrm{tr}\,\tau^b \circ v(\tfrac{\boldsymbol{\Phi}}{M}) \circ \tau^a \circ v^\star(\tfrac{\boldsymbol{\Phi}}{M})\\
&= \left(\tfrac{\boldsymbol{\Phi}^\star \vec{\tau} \tilde{\boldsymbol{\Phi}} + \tilde{\boldsymbol{\Phi}}^\star \vec{\tau} \boldsymbol{\Phi}}{2M^2} \;\middle|\; \tfrac{\boldsymbol{\Phi}^\star \vec{\tau} \tilde{\boldsymbol{\Phi}} - \tilde{\boldsymbol{\Phi}}^\star \vec{\tau} \boldsymbol{\Phi}}{2iM^2} \;\middle|\; \tfrac{\tilde{\boldsymbol{\Phi}}^\star \vec{\tau} \tilde{\boldsymbol{\Phi}} - \boldsymbol{\Phi}^\star \vec{\tau} \boldsymbol{\Phi}}{2M^2} \right) \in \mathbf{SO}(3),\\
\text{with} \quad [0||1] &\overset{\mathbf{U}(1)_+}{\cong} [-1] \oplus [0] \oplus [1].
\end{aligned}$$

These three transmutators are used for the transition from the three isospin gauge fields in the electroweak standard model to the weak boson particles:

$$\begin{aligned}
\tau^a \mathbf{A}_a(x) &= \mathbf{A}(x) \longmapsto v^\star(\tfrac{\boldsymbol{\Phi}(x)}{M}) \circ \mathbf{A}(x) \circ v(\tfrac{\boldsymbol{\Phi}(x)}{M}) - v^\star(\tfrac{\boldsymbol{\Phi}(x)}{M}) \circ \partial v^\star(\tfrac{\boldsymbol{\Phi}(x)}{M})\\
&= (\mathbf{W}_-(x), \mathbf{W}_0(x), \mathbf{W}_+(x)) = (\mathbf{A}_-(x), \mathbf{A}_0(x), \mathbf{A}_+(x)) + \ldots\\
&\qquad\qquad = \delta^a_i + \ldots, \quad \text{with } [0||1](\tfrac{\boldsymbol{\Phi}(x)}{M})^a_i.
\end{aligned}$$

For the definition of particles, which requires the additional transmutation from Lorentz group to rotation group, the neutral field $\mathbf{W}_0$ is combined, in the Weinberg rotation, with the hypercharge gauge field.

Similar to perpendicular relativity, the Hilbert spaces of the square-integrable mappings of the compact Goldstone manifold $L^2(\mathcal{G}^3, V_z)$ into a Hilbert space $V_z \cong \mathbb{C}$ with electromagnetic action $\mathbf{U}(1)_+ \ni e^{i2\gamma_0} \longmapsto e^{zi2\gamma_0}$ have Peter–Weyl decompositions into finite-dimensional subspaces $\mathbb{C}^{1+2T} \otimes \mathbb{C}$ with irreducible representations of the bi-regular subgroup $\mathbf{U}(2) \times \mathbf{U}(1)$, where the isospin representations fulfill $2T \geq |z|$. The representation spaces are given by the columns in the products of the fundamental Higgs transmutators.

7.3.4 Representations of Rotation Relativity

For special relativity, all finite-dimensional representations of the 3-hyperboloid (boost representations), i.e., all finite-dimensional transmutators from Lorentz group to rotation group, can be built by the totally symmetric products of the two fundamental Weyl transmutators $s(\frac{q}{m}), \hat{s}(\frac{q}{m}) \in \mathbf{SL}(2, \mathbb{C})$:

$$\begin{aligned} \mathbf{SL}(2,\mathbb{C})/\mathbf{SU}(2) \cong \mathcal{Y}^3 &\longrightarrow \mathbf{SL}((1+2L)(1+2R), \mathbb{C}), \\ \tfrac{q}{m} &\longmapsto [L|R](\tfrac{q}{m}) = \bigvee^{2L} s(\tfrac{q}{m}) \otimes \bigvee^{2R} \hat{s}(\tfrac{q}{m}), \quad q^2 = m^2. \end{aligned}$$

The finite-dimensional irreducible Lorentz group representations $[L|R]$ with dimensions $(1+2L)(1+2R)$ can be decomposed into irreducible spin representations (see Chapter 5):

$$\mathbf{irrep}_{\text{fin}} \mathbf{SL}(2,\mathbb{C}) \ni [L|R] \overset{\mathbf{SU}(2)}{\cong} \bigoplus_{J=|L-R|}^{L+R} [J].$$

For example, the vector representation $s \otimes s^\star \cong \Lambda = [\frac{1}{2}|\frac{1}{2}]$ gives two irreducible transmutators from Lorentz group to rotation group, the first column for a spin-0 representation and the three remaining columns for a spin-1 representation:

$$\begin{aligned} [\tfrac{1}{2}|\tfrac{1}{2}](\tfrac{q}{m}) = \Lambda^a_b(\tfrac{q}{m}) &\cong \tfrac{1}{2} \operatorname{tr} s(\tfrac{q}{m}) \circ \sigma^a \circ s^\star(\tfrac{q}{m}) \circ \check{\sigma}_b \\ &= \tfrac{1}{m} \left(\begin{array}{c|c} q_0 & q_r \\ q_s & \delta_{rs} m + \frac{q_r q_s}{m+q_0} \end{array} \right) \in \mathbf{SO}_0(1,3), \text{ with } \begin{cases} r = 1,2,3, \\ a = 0,1,2,3, \end{cases} \\ [\tfrac{1}{2}|\tfrac{1}{2}] &\overset{\mathbf{SO}(3)}{\cong} [0] \oplus [1]. \end{aligned}$$

The four columns of the matrix $\Lambda^a_{0,r}(\frac{q}{m})$ relate to each other the metrical tensors of $\mathbf{SO}_0(1,3)$ and $\mathbf{SO}(3)$:

$$\Lambda^a_{0,r}(\tfrac{q}{m}) \eta_{ab} \Lambda^b_{0,s}(\tfrac{q}{m}) = \left(\begin{array}{c|c} 1 & 0 \\ \hline 0 & -\delta_{rs} \end{array} \right), \quad \Lambda^a_r(\tfrac{q}{m}) \delta^{rs} \Lambda^b_s(\tfrac{q}{m}) = -\eta^{ab} + \frac{q^a q^b}{m^2}.$$

The transmutator from Lorentz group to rotation group in the rectangular (4×3)-submatrix $\Lambda^a_r(\frac{q}{m}) \in \mathbb{R}^4 \otimes \mathbb{R}^3$ is used for a massive spin-1 particle in a Lorentz vector field, e.g., for the neutral weak boson and its Feynman propagator,

$$\begin{aligned}
\mathbf{Z}^a(x) &= {}^{\oplus}\!\!\int \frac{d^3q}{2q_0(2\pi)^3}\, \Lambda^a_r(\tfrac{q}{m})\, [e^{iqx}\mathrm{u}^r(\vec{q}) + e^{-iqx}\mathrm{u}^{\star r}(\vec{q})],\\
\langle 0|\{\mathbf{Z}^b(y), \mathbf{Z}^a(x)\} - \epsilon(x_0 - y_0)[\mathbf{Z}^b(y), \mathbf{Z}^a(x)]|0\rangle &= \frac{i}{\pi}\int \frac{d^4q}{(2\pi)^3}\, \frac{(-\eta^{ab} + \frac{q^a q^b}{m^2})}{q^2 + io - m^2} e^{iq(x-y)}.
\end{aligned}$$

In contrast to compact perpendicular relativity with the Hilbert space $L^2(\Omega^2)$ for the 2-sphere functions, the Hilbert space with the square-integrable functions on the 3-hyperboloid $L^2(\mathcal{Y}^3)$ has no finite-dimensional Hilbert subspaces with irreducible $\mathbf{SL}(2,\mathbb{C})$-representations. The monomials in the columns of the fundamental Weyl transmutators are not in $L^2(\mathcal{Y}^3)$. They give bases for finite-dimensional $\mathbf{SL}(2,\mathbb{C}) \times \mathbf{SU}(2)$-representations on $\mathbb{C}^{(1+2L)(1+2R)} \otimes \mathbb{C}^{1+2J}$, which are indefinite unitary for the noncompact Lorentz group $\mathbf{SL}(2,\mathbb{C})$. The spin $\mathbf{SU}(2)$-representation has to be contained in the $\mathbf{SL}(2,\mathbb{C})$-representation; i.e., $|L-R| \le J \le L+R$.

7.3.5 Representations of Unitary Relativity

All finite-dimensional representations of unitary relativity, i.e., of nonlinear spacetime $\mathbb{D}(2)$,

$$\begin{aligned}
\mathbb{D}(2) \cong \mathbf{GL}(2,\mathbb{C})/\mathbf{U}(2) &\cong \mathbf{D}(1) \times \mathbf{SO}_0(1,3)/\mathbf{SO}(3)\\
&\longrightarrow \mathbf{GL}((1+2L)(1+2R),\mathbb{C}),
\end{aligned}$$

use products of the two conjugated dyads, e.g.,

$$\begin{aligned}
\chi(x) &= u(\tfrac{\vec{x}}{r}) \circ e^{\beta_0 \mathbf{1}_2 + \beta\sigma_3} \in \mathbf{GL}(2,\mathbb{C})/\mathbf{U}(2),\\
\chi(x) \circ \mathbf{1}_2 \circ \chi^\star(x) &= u(\tfrac{\vec{x}}{r}) \circ e^{2(\beta_0 \mathbf{1}_2 + \beta\sigma_3)} \circ u^\star(\tfrac{\vec{x}}{r}) = x \in \mathbf{GL}(2,\mathbb{C}),\\
(\chi(x) \circ \mathbf{1}_2 \circ \chi^\star(x))^2 &= e^{4\beta_0} = x^2 \in \mathbf{D}(1).
\end{aligned}$$

The monomials in the dyads span finite-dimensional spaces with $\mathbf{GL}(2,\mathbb{C}) \times \mathbf{U}(2)$-representations, which are, because of the noncompact group $\mathbf{GL}(2,\mathbb{C})$, not Hilbert spaces. Hilbert spaces with faithful representations of nonlinear spacetime have to be infinite-dimensional. They will be considered in Chapter 11.

7.3.6 Representations of Lorentz Group Relativity

All finite-dimensional representations of the general linear group $\mathbf{GL}(4,\mathbb{R})$, the structure group of the tangent bundle of four-dimensional spacetimes,

$$\mathbf{GL}(4,\mathbb{R}) \cong \mathbf{D}(1) \times \{\pm \mathbf{1}_4\} \times \mathbf{SL}_0(\mathbb{R}^4), \quad \mathbf{SL}_0(\mathbb{R}^4) \sim \mathbf{SO}_0(3,3),$$

contain products of the fundamental representations of the 15-dimensional rank-3 special subgroup $\mathbf{SL}_0(\mathbb{R}^4)$, which, as a specialty of four spacetime dimensions and the isomorphy $A_3 \cong D_3$ in the Cartan classification, is locally isomorphic to the indefinite orthogonal group $\mathbf{SO}_0(3,3)$, the invariance group of the Killing form of the Lorentz group $\mathbf{SO}_0(1,3)$. It has the Iwazawa decomposition with maximal compact and maximal abelian noncompact group (see Chapter 8):

$$\mathbf{SO}_0(3,3) = [\mathbf{SO}(3) \times \mathbf{SO}(3)] \circ \mathbf{SO}_0(1,1)^3 \circ \exp \mathbb{R}^6.$$

The three fundamental representations of the cover group $\overline{\mathbf{SO}_0(3,3)}$ are the two four-dimensional spinor representations, which coincide with the defining $\mathbf{SL}_0(\mathbb{R}^4)$-representation on $V \cong \mathbb{R}^4$ and its dual, and a six-dimensional self-dual one:

$$\dim_{\mathbb{R}}[1,0,0] = 4, \quad \dim_{\mathbb{R}}[0,1,0] = \binom{4}{2} = 6, \quad \dim_{\mathbb{R}}[0,0,1] = \binom{4}{3} = 4.$$

The dimensions of the finite-dimensional irreducible representations are given by the Weyl formula:

$$\begin{aligned} \mathbf{irrep}_{\text{fin}}\overline{\mathbf{SO}_0(3,3)} &\ni [n_1,n_2,n_3] \cong \mathbb{N}^3, \\ \dim_{\mathbb{R}}[n_1,n_2,n_3] &= \tfrac{(n_1+1)(n_2+1)(n_3+1)(n_1+n_2+2)(n_3+n_2+2)(n_1+n_2+n_3+3)}{12}, \\ \text{dual reflection:} \quad &[n_1,n_2,n_3] \leftrightarrow [n_3,n_2,n_1]. \end{aligned}$$

Spaces with self-dual representations $[n,m,n]$ have an $\overline{\mathbf{SO}_0(3,3)}$-invariant symmetric bilinear form.

The finite-dimensional representations of Lorentz group relativity, parametrized by the metric manifold,

$$\mathbf{GL}(4,\mathbb{R})/\mathbf{O}(1,3) \cong \mathcal{M}^{10} \ni \mathbf{g} \longmapsto [n_1,n_2,n_3](\mathbf{g}) \in \mathbf{GL}(d_n,\mathbb{R}),$$

have decompositions with respect to an "idolized" Lorentz group. The three fundamental $\overline{\mathbf{SO}_0(3,3)}$-representations decompose into the two fundamental $\mathbf{SO}_0(1,3)$-representations, i.e., the four-dimensional Minkowski representation and the six-dimensional adjoint representation:

$$\begin{aligned} [1,0,0],[0,0,1] &\overset{\mathbf{SO}_0(1,3)}{\cong} [\tfrac{1}{2}|\tfrac{1}{2}], \\ [0,1,0] &\overset{\mathbf{SO}_0(1,3)}{\cong} [1|0] \oplus [0|1], \\ \text{adjoint } [1,0,1] &\overset{\mathbf{SO}_0(1,3)}{\cong} [1|0] \oplus [0|1] \oplus [1|1]. \end{aligned}$$

The Killing form of $\log \mathbf{SO}(3,3)$ has signature $(9,6)$.

The three fundamental $\mathbf{SL}(4,\mathbb{R})$-representations act as a left factor in $\mathbf{SL}(4,\mathbb{R}) \times \mathbf{SO}_0(1,3)$ on the tetrad and its antisymmetric products as fundamental transmutators from the general linear group to the Lorentz group:

$$\begin{aligned} [1,0,0] &\cong \mathbf{e}_a^j \Subset \mathbf{GL}(4,\mathbb{R})/\mathbf{O}(1,3), \\ \bigwedge^2[1,0,0] = [0,1,0] &\cong \epsilon_{jikl}\epsilon^{abcd}\mathbf{e}_a^j\mathbf{e}_b^i \in \mathbf{GL}(6,\mathbb{R}), \\ \bigwedge^3[1,0,0] = [0,0,1] &\cong \epsilon_{jikl}\epsilon^{abcd}\mathbf{e}_a^j\mathbf{e}_b^i\mathbf{e}_c^k \in \mathbf{GL}(4,\mathbb{R}), \\ \bigwedge^4[1,0,0] = [0,0,0] &\cong \det \mathbf{e} \in \mathbf{GL}(1,\mathbb{R}). \end{aligned}$$

The antisymmetric tetrad square $\mathbf{e} \wedge \mathbf{e}$ is a (6×6) transmutator acted on by the self-dual fundamental $\overline{\mathbf{SO}_0(3,3)}$-representation (from the left) and by the adjoint Lorentz group representation (from the right). The inverse tetrad contains the antisymmetric cube $\mathbf{e} \wedge \mathbf{e} \wedge \mathbf{e}$, divided by the determinant with power 4, which is an $\mathbf{SO}_0(3,3)$-scalar with nontrivial $\mathbf{D}(1)$-dilation properties.

The metrical tensor is the symmetrical tetrad square $\mathbf{e} \vee \mathbf{e}$:

$$\begin{array}{rcl} \bigvee^2 [1,0,0] = [2,0,0] & \cong & \mathbf{g}^{ji} = \eta^{ab}\mathbf{e}^j_a \mathbf{e}^i_b \in \mathbf{GL}(10,\mathbb{R}), \\ [2,0,0] & \overset{\mathbf{SO}_0(1,3)}{\cong} & [0|0] \oplus [1|1]. \end{array}$$

The curvature is like the quartic tetrad product $(\mathbf{e} \wedge \mathbf{e}) \vee (\mathbf{e} \wedge \mathbf{e})$ from a 20-dimensional representation:

$$\begin{array}{rcl} [0,1,0] \vee [0,1,0] & = & [0,0,0] \oplus [0,2,0], \\ \mathcal{R}^{ijkl} \cong [0,2,0] & \overset{\mathbf{SO}_0(1,3)}{\cong} & [0|0] \oplus [1|1] \oplus [2|0] \oplus [0|2]. \end{array}$$

Obviously, all those real finite-dimensional representation spaces of the noncompact bi-regular subgroup $\mathbf{GL}(4,\mathbb{R}) \times \mathbf{SO}_0(1,3)$ have no invariant Hilbert product. A harmonic analysis of, e.g., square-integrable functions $L^2(\mathcal{M}^{10})$ on the metric manifold $\mathcal{M}^{10} \cong \mathbf{GL}(4,\mathbb{R})/\mathbf{SO}_0(1,3)$ with infinite-dimensional Hilbert representations does not play a role in classical gravity.

Without rotation degrees of freedom in the abelian Lorentz group $\mathbf{SO}_0(1,1) \subset \mathbf{SO}_0(1,3)$, four-dimensional causal spacetimes $\mathbb{M}^{(1,3)}$ contain two-dimensional causal spacetimes $\mathbb{M}^{(1,1)}$ with a three-dimensional metrical manifold. The dyad $\mathbf{e}^{j=0,3}_{a=0,3}$ involves one compact rotation and two dilations for the metrical hyperbola:

$$\begin{array}{c} \mathcal{M}^3 \cong \mathbf{GL}(2,\mathbb{R})/\mathbf{O}(1,1),\ \mathbf{GL}(2,\mathbb{R}) = \mathbf{D}(1) \times \mathbf{SL}(2,\mathbb{R}), \\ \mathbf{SL}(2,\mathbb{R}) \sim \mathbf{SO}_0(1,2) \sim \mathbf{SU}(1,1). \end{array}$$

The group $\mathbf{SL}(2,\mathbb{R})$ has one fundamental two-dimensional spinor representation, e.g., the dyad, whose totally symmetric products of power $2J \in \mathbb{N}$ give all irreducible finite-dimensional representations with dimension $1 + 2J$ — no Hilbert representations. The decomposition of $\mathbf{SL}(2,\mathbb{R})$-representations into Lorentz group $\mathbf{SO}_0(1,1)$-representations is parallel to the $\mathbf{SO}(2)$-decomposition for $\mathbf{SU}(2)$-representations.

Classes of orthogonal groups constitute manifolds with q compact and p noncompact dimensions:

$$\begin{array}{c} \mathbb{M}^{(p,q)} = \mathbf{SO}_0(a,b)/\mathbf{SO}_0(c,d), \text{ with } \left\{ \begin{array}{l} p+q = \binom{a+b}{2} - \binom{c+d}{2} \geq 0, \\ p = ab - cd \geq 0, \end{array} \right. \\ \text{e.g., } \mathcal{M}^3 \cong \mathbf{GL}(2,\mathbb{R})/\mathbf{O}(1,1) \sim \mathbf{D}(1) \times \mathbf{SO}_0(1,2)/\mathbf{SO}_0(1,1) = \mathbf{D}(1) \times \mathcal{Y}^{(1,1)}, \\ \mathcal{M}^{10} \cong \mathbf{GL}(4,\mathbb{R})/\mathbf{O}(1,3) \sim \mathbf{D}(1) \times \mathbf{SO}_0(3,3)/\mathbf{SO}_0(1,3) = \mathbf{D}(1) \times \mathbb{M}^{(6,3)}. \end{array}$$

7.4 Relativity Representations by Induction

Finite-dimensional rectangular matrices, as discussed in the foregoing section, give all Hilbert representation spaces of homogeneous spaces G/H (H-relativity) only for compact groups $G \supseteq H$, e.g., for perpendicular and electromagnetic relativity. In general, the faithful $G \times H$-Hilbert representations of a locally compact relativity G/H, infinite-dimensional for noncompact G, can be induced by representations of an "idolized" subgroup H.

7.4.1 Induced Representations

Induced G-representations are $G \times H$-subrepresentations of the two-sided regular $G \times G$-representation. They are the extension of the left G-action $gH \stackrel{{}_kL}{\longmapsto} kgH$ on the right H-cosets in the form of linear transformations.

The vector spaces for G-representations, induced by subgroup H-representations, consist of *H-intertwiners* on the group $w : G \longrightarrow W$ with values in a Hilbert space with a unitary action of the "idolized" subgroup $d : H \longrightarrow \mathbf{U}(W)$. The G-action on the intertwiners is defined by left multiplication ${}_kL$. All of this is expressed in the followingcommutative diagram:

$$\begin{array}{ccc} G & \stackrel{{}_kL\times R_h}{\longrightarrow} & G \\ w\big\downarrow & & \big\downarrow {}_kw, \\ W & \underset{d(h)}{\longrightarrow} & W \end{array} \qquad \begin{array}{rl} g,k \in G,\ h \in H: & {}_kL \times R_h(g) = kgh^{-1}, \\ H\text{-intertwiner:} & w(gh^{-1}) = d(h).w(g), \\ G\text{-action:} & w \longmapsto {}_kw, \\ & {}_kw(g) = w(k^{-1}g). \end{array}$$

An H-intertwiner on the group $w \in W^{G/H}$ maps H-cosets of the group into H-orbits in the Hilbert space W. It is defined by its values on representatives $g_r \in (G/H)_r \subseteq G$. The G-action comes with the representative-dependent H-action of the related Wigner element $h(k, g_r) \in H$:

$$\begin{array}{ccc} (G/H)_r & \stackrel{{}_kL}{\longrightarrow} & (G/H)_r \\ w\big\downarrow & & \big\downarrow {}_kw, \\ W & \longrightarrow & W \end{array} \qquad \begin{array}{l} G \times W^{(G/H)_r} \longrightarrow W^{(G/H)_r}, \\ {}_{k^{-1}}w(g_r) = w(kg_r) = d(h^{-1}(k, g_r)).w((kg)_r), \\ \text{with } kg_r = (kg)_r h(k, g_r). \end{array}$$

In fiber bundle terms: The vector space bundle $H \bullet W(\mathbb{M})$ with base manifold $\mathbb{M} = (G/H)_r$ represents the coset bundle $H \bullet \mathbb{M}$. The "idolized" subgroup acts on the vector space $H \bullet W \in \underline{\mathbf{tvec}}_{\mathbb{C}}$; it is a "gauge group." The full group acts on the manifold base $G \bullet \mathbb{M} \in \underline{\mathbf{dif}}_{\mathbb{R}}$.

In general, the induced G-representation on the vector space $W^{(G/H)_r}$ is decomposable. Since the fixgroups for the left G-action on the right H-cosets are conjugates of H,

$$G_{gH} = \{k \in G \mid kgH = gH\} = gHg^{-1} \cong H,$$

each G-representation on $W^{(G/H)_r}$ and on its G-stable subspaces contains the inducing H-representation d.

With a G left-invariant measure $dg_r = dkg_r$ of the manifold $\mathbb{M} = (G/H)_r$, the intertwiners, in a bra vector notation $\langle w| : (G/H)_r \longrightarrow W$, have a direct integral expansion with the cosets as natural distributive basis $\langle g_r, a|$ and complex coefficients $w(g_r)_a \in \mathbb{C}$ (examples ahead):

$$\langle w| = {}^{\oplus}\!\!\int_{(G/H)_r} dg_r \; w(g_r)_a \langle g_r, a| \in W^{(G/H)_r} \text{ for } W \cong \mathbb{C}^n, a = 1, \ldots, n.$$

An *orthonormal distributive basis* is defined with a Dirac distribution $\delta(g_r, g'_r)$, supported by the relativity manifold G/H and normalized with respect to the invariant measure dg_r:

$$\begin{aligned} \langle g'_r, a'|g_r, a\rangle &= \delta_{aa'}\delta(g_r, g'_r), \\ \text{with } \langle w|g_r, a\rangle &= \int_{(G/H)_r} dg'_r \; \delta(g_r, g'_r) w(g'_r)_a = w(g_r)_a. \end{aligned}$$

The *G-invariant Hilbert product* integrates the Hilbert product of the value space W over the cosets:

$$W^{(G/H)_r} \times W^{(G/H)_r} \longrightarrow \mathbb{C}, \quad \| w \|^2 = \int_{(G/H)_r} dg_r \; \overline{w(g_r)_a} w(g_r)_a.$$

In the simplest case, the functions on the homogeneous space G/H for H-relativity are valued in the complex numbers as one-dimensional space $W = \mathbb{C}\langle 1|$ with trivial H-action $d_0(h) = 1$. They are expanded as a direct integral over the cosets with the corresponding function values:

$$\langle f| : (G/H)_r \longrightarrow \mathbb{C}, \quad \langle f| = {}^{\oplus}\!\!\int_{(G/H)_r} dg_r \; f(g_r)\langle g_r|.$$

They are matrix elements (coefficients) of G-representations D that contain a trivial H-representation $D \supseteq d_0$.

7.4.2 Transmutators as Induced Representations

The transmutators above, given by finite-dimensional rectangular matrices $V_D \otimes V_\iota^T \cong \mathbb{C}^n \otimes \mathbb{C}^{n_\iota}$, are acted on by $G \times H$-representations. A left G-action induces a right H-action by a Wigner element:

$$\begin{array}{ccc} (G/H)_r & \xrightarrow{\;{}_kL\;} & (G/H)_r \\ {\scriptstyle D_\iota}\downarrow & & \downarrow{\scriptstyle {}_kD_\iota}, \\ V_D \otimes V_\iota^T & \longrightarrow & V_D \otimes V_\iota^T \end{array} \qquad \begin{aligned} {}_{k^{-1}}D_\iota(g_r) &= D_\iota(kg_r) = D(k) \circ D_\iota(g_r) \\ &= D((kg)_r) \circ d_\iota(h(k, g_r)). \end{aligned}$$

The transmutator bundle $V_D \otimes V_\iota^T(\mathbb{M})$ has the G-space $\mathbb{M} \cong (G/H)_r$ as base manifold and, in the fiber, the linear transmutators with rectangular matrices from the G-space V_D to the H-space V_ι^T.

Transmutators can be used for a decomposition of any G-representation induced by a H-representation d_ι on a vector space with basis $\langle n_\iota; a| \in V_\iota^T$:

$$w_\iota : (G/H)_r \longmapsto V_\iota^T, \quad \langle w_\iota| = {}^{\oplus}\!\!\int_{(G/H)_r} dg_r \; w_\iota(g_r)_a \langle n_\iota; g_r, a|.$$

There occur all G-representations D, which contain the inducing H-representation d_ι. With a basis $|n_D; j\rangle \in V_D$, one obtains the harmonic D-components $\tilde{w}_\iota(D)_j$, which come with multiplicity n_D,

$$\begin{aligned}
\langle w_\iota^{\text{fin}}| &= \bigoplus_{D \supseteq d_\iota} n_D \tilde{w}_\iota(D)_j \; \langle D_\iota^j|, \text{ with} \begin{cases} \langle D_\iota^j| = {}^{\oplus}\!\!\int_{(G/H)_r} dg_r \; D_\iota(g_r)_a^j \langle n_\iota; g_r, a|, \\ \tilde{w}_\iota(D)_j = \langle w_\iota | n_D; j\rangle, \end{cases} \\
w_\iota^{\text{fin}}(g_r)_a &= \bigoplus_{D \supseteq d_\iota} n_D \tilde{w}_\iota(D)_j D(g_r)_a^j, \quad w_\iota^{\text{fin}}(kg_r)_a = \bigoplus_{D \supseteq d_\iota} n_D \tilde{w}_\iota(D)_j D(k)_k^j D(g_r)_a^k,
\end{aligned}$$

e.g., the harmonic analysis of functions with the harmonic D-components $\tilde{f}(D)_j$,

$$\langle f^{\text{fin}}| = \bigoplus_{D \supseteq d_0} n_D \tilde{f}(D)_j \langle D_0^j|, \text{ with} \begin{cases} \langle D_0^j| = {}^{\oplus}\!\!\int_{(G/H)_r} dg_r \; D(g_r)_0^j \langle g_r|, \\ \tilde{f}(D)_j = \langle f | D; j\rangle. \end{cases}$$

7.5 Hilbert Spaces of Compact Relativities

For a compact group $G \supseteq H$, the finite-dimensional rectangular transmutators are square-integrable on the manifolds $\mathbb{M} = (G/H)_r$. They are complete for the harmonic analysis of the group G and its homogeneous spaces G/H; i.e., they exhaust all square-integrable induced representations by orthogonal direct Peter–Weyl decompositions:

$$\text{compact } G: \begin{cases} \langle w_\iota| = \langle w_\iota^{\text{fin}}|, \\ L^2(\mathbb{M}, V_\iota^T) \overset{\text{dense}}{\cong} \bigoplus_{D \supseteq d_\iota} n_D \; V_D \otimes V_\iota^T, \\ \text{id}_{(V_\iota^T)^{G/H}} \cong \bigoplus_{D \supseteq d_\iota} n_D \; \text{id}_{V_D} = \bigoplus_{D \supseteq d_\iota} n_D \; |n_D; j\rangle\langle n_D; j|. \end{cases}$$

The representation structure of the full group G is determined by that of the inducing subgroup; for compact groups, there is *Frobenius' reciprocity theorem*: The number n_D of equivalent irreducible G-representations of type D on $L^2(G/K, V_\iota^T)$, induced by an irreducible subgroup K-representations d_ι on V_ι, is given by the number of d_ι-equivalent K-representations in D, i.e., $n_D = n_{d_\iota \subseteq D}$. Therefore, there are n_D^2 K-representations in G-representations of type D. For example, an $\mathbf{SU}(2)$-dublet representation induces on $L^2(\mathbf{SU}(3)/\mathbf{SU}(2), \mathbb{C}^2)$ no $\mathbf{SU}(3)$-singlet, one $\mathbf{SU}(3)$-triplet, one $\mathbf{SU}(3)$-antitriplet, two $\mathbf{SU}(3)$-octet representations, etc. Therefore, for $\mathbf{SU}(2)$-dublet representations: 1 in $\mathbf{SU}(3)$-triplets and 1 in antitriplets, 2^2 in octets, etc.

In the examples discussed earlier, all representation matrix elements of the compact groups $\mathbf{U}(2)$ and $\mathbf{SU}(2)$ are square-integrable with the finitely decomposable Hilbert spaces for electromagnetic relativity $L^2(\mathcal{G}^3, V_z)$ and perpendicular relativity $L^2(\Omega^2, V_{|z|})$: The complex functions for perpendicular relativity, i.e., $L^2(\Omega^2, V_{|z|}) = L^2(\Omega^2)$ for $z=0$, $V_0 \cong \mathbb{C}$, are the spherical harmonics[3] as products of the three matrix elements $\vec{\omega} \longmapsto \sqrt{\frac{4\pi}{3}}\mathrm{Y}^1_a(\vec{\omega}) \in \mathbb{C}$ in the middle column with trivial representations of $\mathbf{SO}(2) \ni e^{i\chi}$ and a triplet representation of $\mathbf{SO}(3)$:

$$\begin{pmatrix} e^{i(\chi+\varphi)}\cos^2\frac{\theta}{2} & ie^{i\varphi}\frac{\sin\theta}{\sqrt{2}} & -e^{-i(\chi-\varphi)}\sin^2\frac{\theta}{2} \\ ie^{i\chi}\frac{\sin\theta}{\sqrt{2}} & \cos\theta & ie^{-i\chi}\frac{\sin\theta}{\sqrt{2}} \\ -e^{i(\chi-\varphi)}\sin^2\frac{\theta}{2} & ie^{-i\varphi}\frac{\sin\theta}{\sqrt{2}} & e^{-i(\chi+\varphi)}\cos^2\frac{\theta}{2} \end{pmatrix} \in \mathbf{SO}(3),$$

$$\begin{aligned} \mathbf{SU}(2)/\mathbf{SO}(2) \cong \Omega^2 \ni \tfrac{\vec{x}}{r} = \vec{\omega} \longmapsto [J](\vec{\omega})^a_0 &= \sqrt{\tfrac{4\pi}{1+2J}}\mathrm{Y}^J_a(\vec{\omega}) \in \mathbb{C} \\ \text{for } J = 0,1,2,\dots &\text{ with } a = -J,\dots,J, \\ O \in \mathbf{SO}(3): \ [J](O)^a_b \mathrm{Y}^J_b(\vec{\omega}) &= \mathrm{Y}^J_a(O.\vec{\omega}). \end{aligned}$$

There is *Schur orthogonality* [52, 24, 41] with the *Plancherel distribution* (see Chapter 8) given by the dimension $1+2J$ of the irreducible representation space,

$$\int_{\Omega^2} \frac{d^2\omega}{4\pi}\, \overline{[J](\vec{\omega})^a_0}\, [J'](\vec{\omega})^{a'}_0 = \int_{\Omega^2} d^2\omega\, \frac{\overline{\mathrm{Y}^J_a(\vec{\omega})}}{\sqrt{1+2J}}\frac{\mathrm{Y}^{J'}_{a'}(\vec{\omega})}{\sqrt{1+2J'}} = \frac{1}{1+2J}\delta^{JJ'}\delta_{aa'}.$$

It involves the rotation-invariant normalizable Haar measure and the distributive basis of the 2-sphere,

$$\begin{aligned} \int_{\Omega^2} d^2\omega &= \int_0^{2\pi} d\varphi \int_{-1}^{1} d\cos\theta = 4\pi, \\ \langle\vec{\omega}'|\vec{\omega}\rangle &= \delta(\vec{\omega}-\vec{\omega}') = \delta(\varphi-\varphi')\tfrac{1}{\sin\theta}\delta(\theta-\theta'). \end{aligned}$$

The spherical harmonics exhaust the square-integrable 2-sphere functions:

$$L^2(\Omega^2) \ni \langle f| = {}^{\oplus}\!\!\int_{\Omega^2} d^2\omega f(\vec{\omega})\langle\vec{\omega}|, \quad f(\vec{\omega}) = \bigoplus_{J=0}^{\infty} \tilde{f}(J)_a \mathrm{Y}^J_a(\vec{\omega}).$$

The Haar measure can be rewritten with a 2-sphere-supported Dirac distribution:

$$\begin{aligned} \langle [J]^a_0| \sim {}^{\oplus}\!\!\int_{\Omega^2} d^2\omega\, \mathrm{Y}^J(\vec{\omega})\langle\vec{\omega}| &= {}^{\oplus}\!\!\int d^3x\, \delta(\vec{x}^2-1)(\vec{x})^J_{\text{traceless}}\langle\vec{x}|, \\ \text{with } \left(\tfrac{\vec{x}}{|\vec{x}|}\right)^J_{\text{traceless}} &= [J](\vec{\omega}). \end{aligned}$$

There are no finite-dimensional faithful Hilbert representations of noncompact Lie groups. For example, the three matrix elements in the middle

[3] In the Euler angle parametrization, both the middle column and the middle row define the $\mathbf{SO}(3)$-action on the 2-sphere $\Omega^2 \cong \mathbf{SO}(3)/\mathbf{SO}(2) \cong \mathbf{SO}(2)\backslash\mathbf{SO}(3)$. The central element $\theta \longmapsto \cos\theta$ parametrizes the double-coset space, the 1-sphere $\Omega^1 \cong \mathbf{SO}(2)\backslash\mathbf{SO}(3)/\mathbf{SO}(2) \cong \mathbf{SO}(2)$ and is a spherical Ω^2-function.

column with trivial representations of $\mathbf{SO}(2) \ni e^{i\chi}$, i.e., the noncompact hyperbolic partners of the spherical harmonics $\mathrm{Y}^1_a(\varphi, \theta)$ for $i\theta = \psi$,

$$\begin{pmatrix} e^{i(\chi+\varphi)} \cosh^2 \frac{\psi}{2} & e^{i\varphi} \frac{\sinh\psi}{\sqrt{2}} & -e^{-i(\chi-\varphi)} \sinh^2 \frac{\psi}{2} \\ e^{i\chi} \frac{\sinh\psi}{\sqrt{2}} & \cosh\psi & e^{-i\chi} \frac{\sinh\psi}{\sqrt{2}} \\ -e^{i(\chi-\varphi)} \sinh^2 \frac{\psi}{2} & e^{-i\varphi} \frac{\sinh\psi}{\sqrt{2}} & e^{-i(\chi+\varphi)} \cosh^2 \frac{\psi}{2} \end{pmatrix} \in \mathbf{SO}_0(1,2),$$

are complex functions on the 2-hyperboloid $\mathbf{SO}_0(1,2)/\mathbf{SO}(2) \cong \mathcal{Y}^2 \longrightarrow \mathbb{C}$ with a triplet representation of $\mathbf{SO}_0(1,2)$, not square-integrable.

7.6 Flat Spaces as Orthogonal Relativities

Representations of a Lie group can be induced from any subgroup. Sometimes there are distinguished subgroups, e.g., for the infinite-dimensional Hilbert representations of the affine groups $\mathbf{SO}_0(t,s) \vec{\times} \mathbb{R}^n$ (semidirect products), with translations or (energy-)momenta $x, q \in \mathbb{R}^n$, $x^2 = x_t^2 - x_s^2$, and the homogeneous groups — rotations for $ts = 0$ or the causality-compatible Lorentz groups for $t = 1$. They are used for flat spaces as *orthogonal relativities*:

$$\mathbf{SO}_0(t,s) \vec{\times} \mathbb{R}^n / \mathbf{SO}_0(t,s) \cong \mathbb{R}^n, \;\; n = t + s \geq 2.$$

$H \vec{\times} \mathbb{R}^n$-representations are inducible from representations of direct product subgroups $H_0 \times \mathbb{R}^n$ with the homogeneous fixgroups $H_0 \subseteq H$ for the different types of translations or (energy-)momenta, which, for the nontrivial cases $q \neq 0$, all have the real dimension $\binom{n-1}{2}$:

$$\mathbf{SO}_0(t,s)_q = \{g \in \mathbf{SO}_0(t,s) \mid g.q = q\} \cong \begin{cases} \mathbf{SO}_0(t,s) & \text{for } q = 0, \\ \mathbf{SO}_0(t-1,s) & \text{for } q^2 > 0;\ t \geq 1, \\ \mathbf{SO}_0(t,s-1) & \text{for } q^2 < 0;\ s \geq 1, \\ \mathbf{SO}_0(t-1,s-1) \vec{\times} \mathbb{R}^{n-2} & \\ \qquad \text{for } q^2 = 0,\ q \neq 0;\ t,s \geq 1. & \end{cases}$$

For example, the Hilbert representations of the Euclidean group $\mathbf{SO}(3) \vec{\times} \mathbb{R}^3$ for nonrelativistic scattering are inducible by the Hilbert representations of the direct product $\mathbf{SO}(2) \times \mathbb{R}^3$ with the axial rotation subgroups $\mathbf{SO}(2)$ around the momentum direction $\frac{\vec{q}}{|\vec{q}|} = \vec{\omega} \in \Omega^2 \cong \mathbf{SO}(3)/\mathbf{SO}(2)$. The translation representations $\mathbb{R}^3 \ni \vec{x} \longmapsto e^{iP\vec{\omega}\vec{x}} \in \mathbf{U}(1)$ with invariant $P > 0$ use the directions on the manifold Ω^2 of $\mathbf{SO}(2)$-subgroups in $\mathbf{SO}(3)$ (see Chapter 8).

7.6.1 Particle Analysis of Special Relativity

The Hilbert representations of the Poincaré group $\mathbf{SO}_0(1,3) \vec{\times} \mathbb{R}^4$ and its cover group $\mathbf{SL}(2,\mathbb{C}) \vec{\times} \mathbb{R}^4$ (see Chapter 5) for free relativistic particles are induced, for massive particles $m > 0$ (translation-invariant), from $\mathbf{SU}(2) \times \mathbb{R}^4$ with the spin subgroups $\mathbf{SU}(2)$. The translation representations $\mathbb{R}^4 \ni x \longmapsto e^{imyx} \in \mathbf{U}(1)$ use the timelike direction to the point $\frac{q}{m} = \mathrm{y} \in \mathcal{Y}^3 \cong$

$\mathbf{SO}_0(1,3)/\mathbf{SO}(3)$ on the energy-momentum hyperboloid $q^2 = m^2$, which parametrizes the $\mathbf{SU}(2)$-subgroups in $\mathbf{SL}(2,\mathbb{C})$. For massless particles, the induction starts from the fixgroup (axial rotations around the momentum) in the fixgroup $\mathbf{SO}(2) \subset \mathbf{SO}(2) \vec{\times} \mathbb{R}^2 \subseteq \mathbf{SO}_0(1,3) \vec{\times} \mathbb{R}^4$.

The massive particle analysis of free quantum fields with respect to the eigenvectors for spin rotations involves non-Hilbert representations of the Lorentz group via the rectangular transmutators acted on by $\mathbf{SL}(2,\mathbb{C}) \times \mathbf{SU}(2)$-representations. For example, the representation $\Lambda = [\frac{1}{2}|\frac{1}{2}]$ of the Lorentz group in the $\mathbf{SO}_0(1,3) \times \mathbf{SO}(3)$-representation on $\mathbb{C}^4 \otimes V_\iota^T$ for a special relativistic vector field and the tensor representation $\Lambda \wedge \Lambda = [1|0] \oplus [0|1]$ on $\mathbb{C}^6 \otimes V_\iota^T$ for its field strength,

$$\begin{aligned}
\mathbf{Z}^a(0) &= {}^{\oplus}\!\!\int \frac{d^3q}{2q_0(2\pi)^3} \qquad \Lambda_r^a(\tfrac{q}{m}) \qquad [\mathrm{u}^r(\vec{q}) + \mathrm{u}^{\star r}(\vec{q})],\\
i\mathbf{F}^{ab}(0) &= {}^{\oplus}\!\!\int \frac{d^3q}{2q_0(2\pi)^3}\, \Lambda_0^c(\tfrac{q}{m})\epsilon_{cd}^{ab}\Lambda_r^d(\tfrac{q}{m})[\mathrm{u}^r(\vec{q}) - \mathrm{u}^{\star r}(\vec{q})],\\
\text{with } q_0 &= \sqrt{m^2+\vec{q}^{\,2}},\ \ \epsilon_{cd}^{ab} = \delta_c^a\delta_d^b - \delta_d^a\delta_c^b,\ r=1,2,3,\ a=0,1,2,3,
\end{aligned}$$

are both induced by an adjoint $\mathbf{SO}(3)$-representation $[1]$ on $V_\iota \cong \mathbb{C}^3$ and its dual V_ι^T. These spin representations act on the creation and annihilation operators $\mathrm{u}^r(\vec{q}), \mathrm{u}^{\star r}(\vec{q})$ for a massive particle with momentum $\vec{q}$ and spin-1 directions $r = 1,2,3$. The action of the creation operators on the Fock ground state $|0\rangle$ gives dual distributive bases of the special relativistic manifold, i.e., of the energy-momentum hyperboloid $\mathcal{Y}^3 \cong \mathbf{SO}_0(1,3)/\mathbf{SO}(3)$ for mass $m^2 > 0$. The distributive orthogonality is given by the Fock expectation value $\langle 0|\ldots|0\rangle$:

$$\begin{aligned}
|1;\vec{q},r\rangle &= \mathrm{u}^r(\vec{q})|0\rangle \in V_\iota(\vec{q}),\quad \langle 1;\vec{q},r| = \langle 0|\mathrm{u}^{\star r}(\vec{q}) \in V_\iota^T(\vec{q}),\\
\langle 1;\vec{p},s|1;\vec{q},r\rangle &= \langle 0|\mathrm{u}^{\star s}(\vec{p})\mathrm{u}^r(\vec{q})|0\rangle = \delta^{rs}2q_0(2\pi)^3\delta(\vec{q}-\vec{p}).
\end{aligned}$$

The four- and six-dimensional Lorentz group representations do not act on Hilbert spaces. This can be seen at the transmutators from the Lorentz group to the rotation group, e.g., $\{\frac{q}{m} \longmapsto \Lambda_a^j(\frac{q}{m})\}$, which are not square-integrable $L^2(\mathcal{Y}^3)$ on the energy-momentum hyperboloid.

The Lorentz invariant nonnormalizable measure of the 3-hyperboloid in the momentum parametrization can be written as an integral with a $\mathcal{Y}^3$-supported Dirac distribution:

$$m^2 \int d^3\mathrm{y} = \int \frac{d^3q}{\sqrt{m^2+\vec{q}^{\,2}}} = \int d^4q\ \vartheta(q_0)\ 2\delta(q^2-m^2) = m^2|\mathcal{Y}^3| = \infty.$$

The finite-dimensional Lorentz group $\mathbf{SL}(2,\mathbb{C})$-representations that contain a trivial rotation group $\mathbf{SU}(2)$-representation are the Minkowski representations $[\frac{n}{2}|\frac{n}{2}]$ (see Chapter 5). They act on vector spaces $\mathbb{C}^{(1+n)(1+n)}$, $n = 0,1,\ldots$, with the spin-representation decomposition:

$$\mathbf{irrep}_{\mathrm{Mink}}\mathbf{SO}_0(1,3) \ni [\tfrac{n}{2}|\tfrac{n}{2}] \overset{\mathbf{SO}(3)}{\cong} \bigoplus_{J=0}^{n}[J],\ \text{e.g.,} \left\{\begin{array}{l} [0|0] \cong [0],\\ [\frac{1}{2}|\frac{1}{2}] \cong [0] \oplus [1],\\ [1|1] \cong [0] \oplus [1] \oplus [2]. \end{array}\right.$$

They are used for the expansion of the Minkowski representations in the complex functions on energy-momentum hyperboloids,

$$\begin{aligned}
\langle[\tfrac{n}{2}|\tfrac{n}{2}]_0^{j_1\ldots j_n}|: \quad & \mathcal{Y}^3 \longrightarrow \mathbb{C} \text{ for } n=0,1,2,\ldots,\\
\langle[\tfrac{n}{2}|\tfrac{n}{2}]_0^{j_1\ldots j_n}| &= {}^{\oplus}\!\!\int \frac{d^3q}{2q_0(2\pi)^3}\, [\tfrac{n}{2}|\tfrac{n}{2}]_0^{j_1\ldots j_n}(\vec{q})\langle\vec{q}|, \text{ with } q_0=\sqrt{m^2+\vec{q}^2}\\
&= {}^{\oplus}\!\!\int \frac{d^4q}{(2\pi)^3}\, \vartheta(q_0)\delta(q^2-m^2)[\tfrac{n}{2}|\tfrac{n}{2}]_0^{j_1\ldots j_n}(q)\langle q|,
\end{aligned}$$

and arise as contributions of the Feynman propagators for trivial translations, e.g., for a spin-0 particle in a scalar field $\mathbf{\Phi}$, a spin-$\frac{1}{2}$-particle in a Dirac field $\mathbf{\Psi}$, and a spin-1 particle in a vector field $\mathbf{Z}$:

$$\begin{array}{rr}
\mathbf{\Phi}(0): & \langle[0|0]| = {}^{\oplus}\!\!\int \frac{d^4q}{(2\pi)^3}\, \vartheta(q_0)\delta(q^2-m^2)\langle q|,\\
\mathbf{\Psi}(0): & \langle[0|0]| \oplus \gamma_a\langle[\tfrac{1}{2}|\tfrac{1}{2}]_0^a| = {}^{\oplus}\!\!\int \frac{d^4q}{(2\pi)^3}\, \vartheta(q_0)\delta(q^2-m^2)(\mathbf{1}_4+\frac{\gamma_a q^a}{m})\langle q|,\\
\mathbf{Z}^a(0): & \langle[1|1]_0^{ab}| = {}^{\oplus}\!\!\int \frac{d^4q}{(2\pi)^3}\, \vartheta(q_0)\delta(q^2-m^2)(-\eta^{ab}+\frac{q^aq^b}{m^2})\langle q|.
\end{array}$$

The spacetime translation-dependent fields, e.g., a massive vector field,

$$\mathbb{R}^4 \ni x \longmapsto \mathbf{Z}^a(x), \mathbf{F}^{ab}(x) = \epsilon^{ab}_{cd}\frac{\partial^c}{m}\mathbf{Z}^d(x),$$

involve $e^{iqx}\mathrm{u}^r(\vec{q})$ and $e^{-iqx}\mathrm{u}^{\star r}(\vec{q})$, which are the translation orbits $\mathbb{R}^4 \ni x \longmapsto e^{\pm iqx} \in \mathbf{U}(1)$ for a representation of the Poincaré group $\mathbf{SO}_0(1,3) \vec{\times} \mathbb{R}^4$. This leads to the spacetime translation representation coefficients with $\langle q|x\rangle = e^{iqx}$ and $\langle x|q\rangle = e^{-iqx}$ as the on-shell part of the Feynman propagator:

$$\begin{aligned}
\langle[1|1]_0^{ab}|x\rangle + \langle x|[1|1]_0^{ab}\rangle &= \int \frac{d^4q}{(2\pi)^3}\, \delta(q^2-m^2)(-\eta^{ab}+\frac{q^aq^b}{m^2})e^{iqx}\\
&= \langle 0|\{\mathbf{Z}^a(0),\mathbf{Z}^b(x)\}|0\rangle.
\end{aligned}$$

Chapter 8

Representation Coeffficients

Internal chargelike ("vertical") operations come from compact groups, e.g., hypercharge $\mathbf{U}(1)$, isospin $\mathbf{SU}(2)$, and color $\mathbf{SU}(3)$. In the electroweak and strong standard model, they are implemented as gauge transformations that, via gauge fields in covariant derivatives, accompany the translations (see Chapter 6). External spacetimelike ("horizontal") operations come from noncompact groups: For example, interaction-free vectors are acted on by Hilbert representations of flat space operation groups, e.g., of the Euclidean group $\mathbf{SO}(3) \vec{\times} \mathbb{R}^3$ in nonrelativistic quantum mechanical scattering theory, or of the Poincaré (cover) group $\mathbf{SL}(2,\mathbb{C}) \vec{\times} \mathbb{R}^4$ for elementary particles in relativistic quantum fields.

Each group determines its Hilbert spaces, finite- or infinite-dimensional, where its action can be represented by definite unitary automorphisms. A Hilbert space, e.g., the Fock space for translations and free particles, may not be appropriate for another group, e.g., for bound states or for the implementation of interactions.

Each representation of a *compact Lie group* K is equivalent to a representation $D : K \longrightarrow \mathbf{U}(V)$ in the unitary group of a Hilbert space V. It is decomposable into irreducible finite-dimensional ones, $V \cong \mathbb{C}^d$. The invariants and eigenvalues, such as quantum numbers for charge, isospin, color, etc., are rational and ultimately connected to *integer winding numbers* $z \in \mathbb{Z}$ from the dual group of $\mathbf{U}(1)$, which characterize the irreducible representations $\mathbf{U}(1) \ni e^{i\alpha} \longmapsto e^{iz\alpha} \in \mathbf{U}(1)$. A maximal abelian compact subgroup with direct factors $\mathbf{U}(1) \cong \mathbf{SO}(2)$ is called a *Cartan torus*, e.g., $\mathbf{SO}(2)^r \subset \mathbf{SU}(1+r)$ or, for hyperisospin, $\mathbf{U}(1)_+ \times \mathbf{U}(1)_- \subset \mathbf{U}(2)$. The representations (characters) of the Cartan tori are basic for the representations of a compact group.

The theory of the Hilbert representations of *noncompact Lie groups*, infinite-dimensional if faithful, is more complicated and difficult. A noncompact semisimple group has an Iwasawa decomposition $G = K \circ A \circ N$ into a maximal compact subgroup K, a maximal noncompact abelian subgroup A,

H. Saller, *Operational Spacetime: Interactions and Particles*, Fundamental Theories of Physics 163, DOI 10.1007/978-1-4419-0898-8_9,

and a subgroup N with nilpotent Lie algebra. The factor subgroups are unique up to isomorphy, e.g.,

$$\begin{aligned} 1 \le t \le s: \ \mathbf{SO}_0(t,s) &= [\mathbf{SO}(t) \times \mathbf{SO}(s)] \circ \mathbf{SO}_0(1,1)^t \circ \exp \mathbb{R}^{t(s-1)}, \\ \mathbf{SL}(1+r,\mathbb{R}) &= \mathbf{SO}(1+r) \circ \mathbf{SO}_0(1,1)^r \circ \exp \mathbb{R}^{\binom{1+r}{2}}, \\ \mathbf{SL}(1+r,\mathbb{C}) &= \mathbf{SU}(1+r) \circ \mathbf{SO}_0(1,1)^r \circ \exp \mathbb{R}^{(1+r)r}. \end{aligned}$$

The group A is isomorphic to a translation group $A \cong \mathbf{D}(1)^r \cong \mathbf{SO}_0(1,1)^r \cong \mathbb{R}^r$ and called a *Cartan plane.* Its dimension r is the real rank of the noncompact group G and gives the maximal number of representation characterizing invariants from a continuous spectrum, called *continuous invariants.* In physics, they can be used as basic units for the related operations.

Hilbert representations of a Lie group contain, as basic substructures, Hilbert representations of its Cartan tori and planes, which determine, respectively, the rational and continuous representation invariants and eigenvalues, e.g., one rational and one continuous invariant for a Cartan cylinder with axial rotations and dilations $\mathbf{SO}(2) \times \mathbf{SO}_0(1,1) \subset \mathbf{SO}_0(1,3)$ in the Lorentz group.

The cyclic Hilbert representations $\mathbf{cycrep}_+\mathbb{R}^r$ (definition ahead) of a Cartan plane with translations are characterized by essentially bounded *positive-type functions* (*scalar product or Hilbert metric-inducing functions*) $\mathbb{R}^r \ni x \longmapsto d(x)$, $d \in L^\infty(\mathbb{R}^r)_+$. Such positive-type functions are, in a sense, the noncompact analogue to the representation-characterizing integer winding numbers $\mathbb{Z}^r$ for Cartan tori. Scalar product–inducing functions are Fourier-transformed positive Radon measures (distributions, densities) $\mathcal{M}(\check{\mathbb{R}}^r)_+$ of the group dual $\check{\mathbb{R}}^r$, i.e., of the irreducible, but not faithful translation representations $\mathbb{R}^n \ni x \longrightarrow e^{iqx} \in \mathbf{U}(1)$ with a continuous eigenvalue iq for "energy-momentum" $q \in \mathbb{R}^n$.

8.1 Finite-dimensional Representations

In this section, all vector spaces are assumed to be finite-dimensional over the scalars $\mathbb{K} \in \{\mathbb{R}, \mathbb{C}\}$.

8.1.1 Metrics of Representation Spaces

If a vector space is acted on by a group G or a Lie algebra L,

$$\begin{aligned} D: \ & G \longrightarrow \mathbf{GL}(V), \quad & G \times V \longrightarrow V, \quad & g \bullet v = D(g)v, \\ \mathcal{D}: \ & L \longrightarrow V \otimes V^T, \quad & L \times V \longrightarrow V, \quad & l \bullet v = \mathcal{D}(l)v, \end{aligned}$$

the dual vector space V^T has the dual representations $G \ni g \longmapsto \check{D}(g) = D(g^{-1})^T$ and $L \ni l \longmapsto \check{\mathcal{D}}(l) = -\mathcal{D}(l)^T$. With a trivial action on the scalars, $g \bullet \alpha = \alpha \in \mathbb{K}$, $l \bullet \alpha = 0$, this defines the group action on the full tensor algebra $\bigotimes(V \oplus V^T)$ by $g \bullet (a_1 \otimes a_2) = (g \bullet a_1) \otimes (g \bullet a_2)$ and the Lie algebra action by Leibniz's rule, $l \bullet (a_1 \otimes a_2) = (l \bullet a_1) \otimes a_2 + a_1 \otimes (l \bullet a_2)$.

An (anti-)symmetric inner product of a vector space is *invariant* under the corresponding action for

$$\zeta(\ ,\ ) : V \times V \longrightarrow \mathbb{K}, \quad \begin{cases} \zeta(v, g \bullet w) &= \zeta(g^{-1} \bullet v, w), \quad g \in G, \\ \zeta(v, l \bullet w) &= -\zeta(l \bullet v, w), \quad\ l \in L. \end{cases}$$

A product representation on $V_1 \otimes V_2$ with invariant forms has $\zeta_1 \otimes \zeta_2$ with the product matrix elements $\zeta_1 \otimes \zeta_2(v_1 \otimes v_2, w_1 \otimes w_2) = \zeta_1(v_1, w_1)\zeta_2(v_2, w_2)$ as an invariant form.

Any invariant subspace has its invariant ζ-orthogonal subspace partner,

$$\begin{aligned} &W \subseteq V : \ G \bullet W \subseteq W,\ L \bullet W \subseteq W, \\ &W_\zeta^\perp = \{v \in V \mid \zeta(W, v) = \{0\}\} \Rightarrow G \bullet W_\zeta^\perp \subseteq W_\zeta^\perp,\ \ L \bullet W_\zeta^\perp \subseteq W_\zeta^\perp. \end{aligned}$$

Therefore, if an irreducible nontrivial representation has a nontrivial invariant form, the inner product must be nondegenerate:

$$V \text{ irreducible} \Rightarrow \zeta \text{ nondegenerate} \iff V_\zeta^\perp = \{0\}.$$

With Schur's lemma, an irreducible, complex finite-dimensional representation of a group or Lie algebra can have, up to scalar multiples, only one invariant linear or only one invariant conjugate linear dual isomorphism ζ, possibly both or none. With two $\zeta_{1,2}$, one has $\zeta_1 \circ \zeta_2^{-1} = \alpha\, \mathrm{id}_V$.

A nondegenerate inner product defines a (conjugate) linear dual isomorphism $\zeta : V \longrightarrow V^T$, $v \longmapsto \zeta(\ , v)$ to the dual space and an invariant product ζ^{-1} of V^T with $\zeta^{-1}(\omega, \theta) = \zeta(\zeta^{-1}.\omega, \zeta^{-1}.\theta)$. The representations are called *(conjugate) linear self-dual*:

$$\begin{array}{ccc} V & \xrightarrow{D(g),\ \mathcal{D}(l)} & V \\ \zeta \downarrow & & \downarrow \zeta \\ V^T & \xrightarrow[\check{D}(g),\ \check{\mathcal{D}}(l)]{} & V^T \end{array}, \quad \begin{aligned} D(g^{-1}) &= \zeta^{-1} \circ D(g)^T \circ \zeta = D(g)^*, \\ -\mathcal{D}(l) &= \zeta^{-1} \circ \mathcal{D}(l)^T \circ \zeta \ \ = \mathcal{D}(l)^*. \end{aligned}$$

With an inner product ζ of V and ζ^{-1} of its dual V^T, and the corresponding product of the scalars, linear $\alpha\beta$ or antilinear $\overline{\alpha}\beta$, there is an invariant inner product of all finite tensor powers $\bigotimes^k V \otimes \bigotimes^l V^T$.

The properties of nondegenerate inner products determine the groups,

$$\zeta(v, u) = \begin{cases} \zeta(u, v) \\ -\zeta(u, v) \\ \overline{\zeta(u, v)} \end{cases} \Rightarrow D[G] \subseteq \begin{cases} \mathbf{O}(V, \zeta), & \text{orthogonally self-dual}, \\ \mathbf{Sp}(V, \zeta), & \text{symplectically self-dual}, \\ \mathbf{U}(V, \zeta), & \text{unitarily self-dual}, \end{cases}$$

and the Lie algebras, e.g., $\mathcal{D}[L] \subseteq \log \mathbf{O}(V, \zeta)$. The diagonalization of orthogonal and unitary ζ displays its signature (p, q).

A *complex* representation of a *real* Lie group $G_\mathbb{R}$ or Lie algebra $L_\mathbb{R}$ has to come with a conjugation of the representation space, i.e., it has to be unitarily self-dual:

$$\begin{aligned} D[G_\mathbb{R}] &\subseteq \mathbf{U}(p, q), & &\text{with } D(g)^* = D(g^{-1}), \\ \mathcal{D}[L_\mathbb{R}] &\subseteq \log \mathbf{U}(p, q), & &\text{with } \mathcal{D}(l)^* = -\mathcal{D}(l). \end{aligned}$$

A real Lie algebra representation in compact $\mathbf{O}(n)$ or $\mathbf{U}(n)$ is *positive self-dual*, i.e., a *Hilbert representation*. Then the double trace is strictly positive and the associated inner product is strictly negative:

$$\begin{aligned} \operatorname{tr} \mathcal{D}(l)^\star \circ \mathcal{D}(l) &= -\operatorname{tr} \mathcal{D}(l) \circ \mathcal{D}(l) \geq 0, \\ \text{notation: } \zeta(v,w) &= \langle v|w\rangle \text{ with } \langle v|v\rangle = \| v \|^2 \geq 0 \iff v \neq 0. \end{aligned}$$

Some examples: The irreducible $\mathbf{SU}(n)$-representations have an invariant antilinear scalar product as the product of the group-defining scalar product $\zeta_n(v,w) = \langle v|w\rangle$ on $V \cong \mathbb{C}^n$. The $\mathbf{SU}(2)$- and $\mathbf{SL}(2,\mathbb{C})$-spinor representations on $V \cong \mathbb{C}^2$ have the invariant indefinite antisymmetric linear "metric" $\epsilon(e^A, e^B) = \begin{pmatrix} 0 & 1 \\ -1 & 0 \end{pmatrix}$, given by the volume elements. There is no invariant bilinear form for the (anti)quark representation of $\mathbf{SU}(3)$. The $\mathbf{SU}(3)$-octet representation has the Killing form $\kappa_8 \cong \zeta_3 \otimes \zeta_3^{-1}$ as invariant definite symmetric linear metric — more generally, $\kappa_{n^2-1} \cong \zeta_n \otimes \zeta_n^{-1}$ for $\mathbf{SU}(n)/\mathbb{I}(n)$ on $V \cong \mathbb{R}^{n^2-1}$. The Minkowski $\mathbf{SO}_0(1,3)$-representation has the invariant indefinite symmetric linear Lorentz "metric" $\eta \cong \epsilon \otimes \epsilon^{-1}$ as the product of the dual volume elements.

The natural isomorphisms (dual representations) for transposed endomorphism Lie algebras, $\mathbf{AL}(V) \cong \mathbf{AL}(V^T)$, $l \leftrightarrow -l^T$, and automorphism groups, $\mathbf{GL}(V) \cong \mathbf{GL}(V^T)$, $g \leftrightarrow g^{-1T}$, do not imply that these isomorphisms have to arise as products $\zeta \otimes \zeta^{-1}$ with a dual isomorphism $V \overset{\zeta}{\cong} V^T$.

8.1.2 Metrics of Lie Algebras

Now, the special case of linear metrics for a Lie algebra with adjoint action:

An invariant linear form $\omega \in L^T$ is trivial on its commutator $\omega([L,L]) = \{0\}$. Since $[L,L] = L$ for a semisimple Lie algebra, such an L has no nontrivial invariant linear form. Each representation of a semisimple Lie algebra is traceless, e.g., the adjoint one (structure constants):

$$\begin{aligned} &\mathcal{D} : L \longrightarrow \mathbf{AL}(V), \quad L \longrightarrow \mathbb{K}, \quad l \longmapsto \operatorname{tr} \mathcal{D}(l), \\ &\text{semisimple } L \Rightarrow \begin{cases} \operatorname{tr} \mathcal{D}(l) = 0, \\ \operatorname{tr} \operatorname{ad} l^a = \epsilon_b^{ab} = 0. \end{cases} \end{aligned}$$

Any representation of a Lie algebra has as *associate inner product* $\kappa_{\mathcal{D}}$ the symmetric "double trace," invariant under the adjoint action,

$$\kappa_{\mathcal{D}}(\ ,\ ) : L \times L \longrightarrow \mathbb{K}, \quad \kappa_{\mathcal{D}}(l,m) = \operatorname{tr} \mathcal{D}(l) \circ \mathcal{D}(m).$$

Associated to the adjoint Lie algebra representation is the *Killing form* $\kappa(l,m) = \operatorname{tr} \operatorname{ad} l \circ \operatorname{ad} m$. An associate inner product can be diagonalized with $\kappa_{\mathcal{D}}(l^a, l^b) = \pm\delta^{ab}, 0$.

Precisely for semisimple Lie algebras, the Killing form is nondegenerate. For a complex, simple Lie algebra, the Killing form is, up to a scalar factor, the unique invariant inner product.

An abelian Lie algebra has a trivial Killing form. However, the associated inner products $\kappa_{\mathcal{D}}$ are not neccessarily trivial for all its representations. Nonsemisimple Lie algebras, e.g., the abelian Lie algebra $\mathbb{K}$, can have representations with nondegenerate inner products $\kappa_{\mathcal{D}}$.

An invariant linear inner product γ of a complex vector space with a faithful irreducible representation of a simple Lie algebra defines an invariant inner product of the endomorphisms $\mathbf{AL}(V) \cong V \otimes V^T$:

$$\Gamma = \gamma \otimes \gamma^{-1} : \mathbf{AL}(V) \times \mathbf{AL}(V) \longrightarrow \mathbb{C}.$$

It has to coincide, up to a nontrivial factor, on the image of the Lie algebra $\mathcal{D}[L] \subseteq \mathbf{AL}(V)$ with the associated inner product $\kappa_{\mathcal{D}}$. In addition, γ has to be either *symmetric or antisymmetric*:

$$\begin{aligned} \kappa_{\mathcal{D}}(l,m) &= \operatorname{tr} \mathcal{D}(l) \circ \mathcal{D}(m) = e^{\alpha}\Gamma(l,m), \quad \alpha \in \mathbb{C}, \\ \Gamma &= \Gamma^T \iff \gamma = \gamma^T \text{ or } \gamma = -\gamma^T. \end{aligned}$$

A semisimple real *Lie algebra* is *compact* with strictly negative Killing form. Compact Lie algebras have bases with totally antisymmetric structure constants. Also, noncompact Lie algebras, e.g., $\mathbb{R}$, can have Hilbert representations. All complex representations of compact Lie algebras are semisimple and decomposable into irreducible Hilbert representations, all of which are finite-dimensional (*theorem of Weyl*).

8.2 Group Algebras and Representation Spaces

The equivalence classes of the irreducible complex Hilbert representations of a group $G \longrightarrow \mathbf{U}(V)$ constitute the *group dual* $\check{G} = \mathbf{irrep}_+ G$. Its determination, especially for noncompact nonabelian groups, may be a difficult task. Group representation matrix elements are group functions, called *representation coefficients*. For an orientation, it is useful to recapitulate in this section, in general, the relevant vector spaces and algebras for group functions [24].

8.2.1 Finite Groups

For a finite group, compact with the discrete topology, the complex linear combinations[1] of group elements $f = \bigoplus_{g \in G} g f(g) \in \mathbb{C}^G$ can be seen as group functions $g \longmapsto f(g)$. They constitute the group algebra $\mathbb{C}^G = \{G \longrightarrow \mathbb{C}\}$ with the group elements as natural basis $\mathbb{C}^G \cong \mathbb{C}^{\operatorname{card} G}$. The group product induces an algebra structure with *convolution product* $f_1 * f_2 \in \mathbb{C}^G$. The minimal two-sided ideals of the group algebra are generated by the conjugacy classes $G(k) = \{gkg^{-1} \mid g \in G\}$ of the group [40]. They characterize

[1] Sometimes, for distinction, the addition of linear independent vectors is written as $|v\rangle \oplus |w\rangle$.

the irreducible representations with dimension d_ι of the group G by finite unitary groups in $\mathbf{U}(d_\iota) \subseteq \mathbf{U}(\operatorname{card} G)$, and of the group algebra $\mathbb{C}^G$ by the corresponding full matrix ideals with dimensions d_ι^2:

$$\text{finite } G: \quad \mathbb{C}^G \cong \bigoplus_{\iota=1}^{n} \mathbb{C}^{d_\iota} \otimes \mathbb{C}^{d_\iota}, \quad \operatorname{card} G = \sum_{\iota=1}^{n} d_\iota^2.$$

A characteristic and familiar example is the Young tableaux–related irreducible representations of the permutation group $\mathbf{G}(n)$ with cardinality $n!$.

8.2.2 Algebras and Vector Spaces for Locally Compact Groups

More general: For a real finite-dimensional Lie group G, complex-valued mappings (functions, distributions, measures) of the group have to be considered [57,28]. Summation over group elements is expressed by invariant integration with *Haar measure*, a counting measure for finite groups.

A (direct) integral over group elements uses a left or right Haar measure $\int_G dg = \int dg$ — both unique up to a scalar and related to each other by the modular function in the dilation representation $G \ni g \longmapsto \Delta(g) \in \mathbf{D}(1)$. A non-unimodular group has to have a normal subgroup with dilation classes $G/N \cong \mathbf{D}(1)$.

The group functions contain all representation matrix elements. The mappings $f = {}^{\oplus}\!\!\int dg \; g f(g)$ constitute group representation spaces; they inherit the group multiplication in the *left-right regular (bi-regular) representation of* $G \times G$,

$$\begin{aligned} (G \times G) \bullet G &\longrightarrow G, \quad (g_1, g_2) \bullet g = g_1 g g_2^{-1}, \\ f &\longmapsto (g_1, g_2) \bullet f = {}_{g_1}f_{g_2} = {}^{\oplus}\!\!\int dg \; g f(g_1^{-1} g g_2). \end{aligned}$$

With a Haar measure basis, all group measures can be characterized by (generalized) functions $\mu(g)dg$. Distributions have, via duality to their functions, the dual action

$$\langle (g_1, g_2) \bullet \mu, f \rangle = \langle \mu, (g_1, g_2) \bullet f \rangle, \quad \int dg \; \mu(g_1 g g_2^{-1}) f(g) = \int dg \; \mu(g) f(g_1^{-1} g g_2).$$

As exemplified by the finite groups with the convolution algebra $\mathbb{C}^G$, also in the general case, all group representation spaces come from a convolution algebra with group functions.

Two products are important for the (generalized) group functions $\mu = {}^{\oplus}\!\!\int dg \; g\mu(g)$ of a locally compact group (unimodular $\Delta(g) = 1$ if necessary). The group functions inherit the group multiplication $g_1 g_2 = g \in G$ as *convolution product* (where defined),

$$\begin{aligned} \mu_1 * \mu_2 &= [{}^{\oplus}\!\!\int dg \; g\mu_1(g)] \; * \; [{}^{\oplus}\!\!\int dg \; g\mu_2(g)] &&= {}^{\oplus}\!\!\int dg \; g(\mu_1 * \mu_2)(g), \\ \mu_1 * \mu_2(g) &= \int dg_1 dg_2 \; \mu_1(g_1) \; \delta(g_1 g_2 g^{-1}) \mu_2(g_2) &&= \int dg_1 \; \mu_1(g_1) \mu_2(g_1^{-1} g). \end{aligned}$$

The associative convolution is abelian if, and only if, the group multiplication is abelian. From the complex numbers, the group functions inherit the abelian *pointwise multiplication* (where defined), important for product representations,

$$\mu_1 \cdot \mu_2(g) = \mu_1(g)\mu_2(g).$$

With respect to the convolution and pointwise product, the *Lebesgue Banach spaces* $L^p(G)$, $1 \le p \le \infty$, with the classes of Haar measure almost everywhere defined absolute p-integrable functions, i.e., $(\| f \|_p)^p = \int dg\, |f(g)|^p < \infty$, are related to each other as follows:

$$1 \le p, r, s \le \infty : \left\{ \begin{array}{ll} L^p(G) * L^r(G) \subseteq L^s(G), & \text{with } \frac{1}{p} + \frac{1}{r} - \frac{1}{s} = 1, \\ L^p(G) \cdot L^r(G) \subseteq L^s(G), & \text{with } \frac{1}{p} + \frac{1}{r} - \frac{1}{s} = 0. \end{array} \right.$$

They are left-right modules, respectively, for the absolute integrable group function classes $L^1(G)$, a convolution algebra, and for the essentially bounded group functions $L^\infty(G)$, i.e., $|f(g)| < \infty$ almost everywhere, a pointwise product algebra,

$$L^1(G) * L^1(G) \subseteq L^1(G), \quad L^\infty(G) \cdot L^\infty(G) = L^\infty(G).$$

They are left-right convolution modules even for the *Radon measures* $\mathcal{M}(G)$ of the group (definition ahead), a unital convolution Banach algebra,

$$\begin{array}{rl} \mathcal{M}(G) * \mathcal{M}(G) &= \mathcal{M}(G), \\ 1 \le p \le \infty : \quad \mathcal{M}(G) * L^p(G) * \mathcal{M}(G) &= L^p(G). \end{array}$$

The Radon measures, in the form of Radon distributions with Haar measure $\omega(g)dg$, embed the group by Dirac measures:

$$\begin{array}{rl} G \ni k \longmapsto \delta_k \in \mathcal{M}(G) \text{ with } \langle \delta_k, f \rangle &= \int \delta_k(g)dg\, f(g) = f(k), \\ \delta_k * \delta_l &= \delta_{kl}, \quad \delta_1 = \delta, \\ \delta_k(g) &= \delta(gk^{-1}). \end{array}$$

Dirac distributions and Haar measures are "inverse to each other," especially their normalizations:

$$\langle \delta_k, 1 \rangle = \int \delta_k(g)dg = 1 = \int \delta(g)dg.$$

The left-right action of a group on itself is embedded in the left-right convolution module property of the function spaces for the Radon group measures,

$$\begin{array}{rl} \mathcal{M}(G) * L^p(G) * \mathcal{M}(G) &\longrightarrow L^p(G), \quad f \longmapsto \mu_1 * f * \mu_2, \\ \mu_1 * f * \mu_2(g) &= \int dg_1 \int dg_2\, \mu_1(g_1) f(g_1^{-1} g g_2) \mu_2(g_2^{-1}), \\ \text{e.g., } \delta_k * f * \delta_l(g) &= f(k^{-1} g l^{-1}). \end{array}$$

The Radon measures can be defined as the dual $\mathcal{M}(G) = \mathcal{C}_c(G)'$ of the compactly supported continuous functions $\mathcal{C}_c(G)$; these functions are dense in all $L^p(G)$, $1 \le p < \infty$. $\mathcal{M}(G)$ contains the function algebra $L^1(G)$ as two-sided ideal. The involutive convolution algebra $\mathcal{C}_c(G)$ is a subspace of the bounded continuous functions $\mathcal{C}_b(G)$, which, in their turn, can be considered as a closed, in general, proper subspace of the essentially bounded functions,

$$\begin{array}{l} L^1(G) \subseteq \mathcal{M}(G) \supset G, \\ \mathcal{C}_c(G) \subseteq \mathcal{C}_b(G) \ \subseteq L^\infty(G). \end{array}$$

As suggested by the measure $d^n x$ in the integration with a distribution, every Radon distribution of an open real set $T \subseteq \mathbb{R}^n$ is a finite sum of derivatives up to order n of locally essentially bounded functions [56],

$$T \subseteq \mathbb{R}^n: \quad \mathcal{M}(T) \subseteq \left\{ \sum_{N=0}^{n} \alpha_N \partial^N L^\infty(T) \right\},$$

e.g., the Dirac distribution as derivation of the step and sign functions $L^\infty(\mathbb{R}) \ni \vartheta, \epsilon \notin \mathcal{M}(\mathbb{R})$ or the Yukawa potential,

$$\begin{array}{rlll} \mathbb{R} & \text{with } dx: & \frac{d}{dx}\vartheta(x) & = \frac{d}{dx}\frac{\epsilon(x)}{2} = \delta(x), \\ \frac{\mathbf{SO}(3)\ \vec{\times}\ \mathbb{R}^3}{\mathbf{SO}(3)} \cong \mathbb{R}^3 & \text{with } d^3x: & \frac{d}{dr^2}e^{-mr} & = -\frac{m}{2r}e^{-mr}, \\ \frac{\mathbf{SO}_0(1,3)\ \vec{\times}\ \mathbb{R}^4}{\mathbf{SO}_0(1,3)} \cong \mathbb{R}^4 & \text{with } d^4x: & (\frac{d}{dx^2})^N\vartheta(x^2) & = \delta^{(N-1)}(x^2), \quad N = 1, 2. \end{array}$$

All the (generalized) function vector spaces and algebras considered have an *involution*:

$$\mu \leftrightarrow \hat{\mu} \text{ with } \hat{\mu}(g) = \overline{\mu(g^{-1})} \text{ for unimodular } G.$$

With a group representation $D : G \longrightarrow \mathbf{GL}(V)$, there may be a representation of the three convolution group algebras $A(G) \in \{\mathcal{C}_c(G),\ L^1(G),\ \mathcal{M}(G)\}$ in the endomorphism algebra $\mathbf{AL}(V)$, for a function or a Radon distribution μ,

$$\begin{aligned} D : A(G) \longrightarrow \mathbf{AL}(V),\ \mu \longmapsto D(\mu) &= \int dg\ D(g)\mu(g), \\ D(\mu_1 * \mu_2) &= D(\mu_1) \circ D(\mu_2), \\ D(\delta_k) &= D(k). \end{aligned}$$

In the following, the unital convolution algebra $\mathcal{M}(G)$ with the Radon measures, and its ideal the Lebesgue functions $L^1(G)$, and the unital abelian pointwise product algebra $L^\infty(G)$ with the essentially bounded functions play the most important roles. With the two-sided convolutive action of $\mathcal{M}(G)$ and

$L^1(G)$ and the pointwise action of $L^\infty(G)$, these spaces will be used for the group G and the group dual $\check{G}$,

$*$	$\mathcal{M}(G)$	$L^1(G)$	$L^\infty(G)$
$\mathcal{M}(G)$	$\mathcal{M}(G)$	$L^1(G)$	$L^\infty(G)$
$L^1(G)$	$L^1(G)$	$L^1(G)$	$L^\infty(G)$
$L^\infty(G)$	$L^\infty(G)$	$L^\infty(G)$	–

$\cdot$	$L^\infty(G)$	$L^1(G)$	$\mathcal{M}(G)$
$L^\infty(G)$	$L^\infty(G)$	$L^1(G)$	$\mathcal{M}(G)$
$L^1(G)$	$L^1(G)$	–	–
$\mathcal{M}(G)$	$\mathcal{M}(G)$	–	–

Convolution product
$\mu_1 * \mu_2(g)$
from group product
$G \times G \longrightarrow G$

Pointwise product
$\mu_1 \cdot \mu_2(g)$
for product representations
$\mathbf{rep}\, G \times \mathbf{rep}\, G \longrightarrow \mathbf{rep}\, G$

The "divergences" of quantum field theory have their origin in undefined pointwise multiplications of Radon measures $\mathcal{M}(\mathbb{R}^4) \cdot \mathcal{M}(\mathbb{R}^4)$ of spacetime translations.

8.3 Schur Product of Group Functions

For a vector space with group functions, the value at the neutral element, if defined, is a linear form:

$$\mathcal{F}(G) \ni f \longmapsto f(1) = \int dg\ \delta(g) f(g) \in \mathbb{C}.$$

Two vector spaces are *put in duality by a bilinear form* $W \times V \ni (w,v) \longmapsto d(w,v) \in \mathbb{K}$. It defines a mapping to the linear forms $W \ni w \longmapsto w_d = d(w,\ ) \in V^T$ with $\langle w_d, v\rangle = d(w,v)$. For finite-dimension, all duality structures are described by the algebraic dual V^T and the dual product $V^T \times V \ni (\theta, v) \longmapsto \langle \theta, v\rangle \in \mathbb{K}$, for dual bases $\langle \check{e}^a, e_b\rangle = \delta^a_b$. That is more complicated for infinite-dimension and continuous linear forms from the topological dual $V' \subseteq V^T$.

For bilinear and sesquilinear forms of vector spaces with (generalized) group functions, the *Schur product* is used. It is the *convolution product at the neutral element*, if defined,

$$\begin{aligned} \mathcal{F}_1(G) \times \mathcal{F}_2(G) \ni (\mu_1, \mu_2) \longmapsto\quad & \mu_1 * \mu_2(1) = \int dg\ \mu_1(g^{-1})\mu_2(g), \\ & \hat{\mu}_1 * \mu_2(1) = \int dg\ \overline{\mu_1(g)}\mu_2(g). \end{aligned}$$

It does not have to be a complex number and may be a distribution, e.g., on the group dual $\check{G}$ with the Plancherel measure (more ahead).

8.3.1 Duality for Group Function Spaces

The duality for the Lebesgue spaces is given by the Schur product, valued in the essentially bounded functions, at the neutral group element,

$$\begin{aligned} L^p(G)' &= L^r(G), \text{ with } L^p(G) * L^r(G) \subseteq L^\infty(G) \\ &\text{for } \tfrac{1}{p} + \tfrac{1}{r} = 1,\ 1 < p, r < \infty, \\ L^p(G) \times L^r(G) &\longrightarrow \mathbb{C},\ \langle f_p, f_r\rangle = f_p * f_r(1) = \int dg\ f_p(g^{-1}) f_r(g). \end{aligned}$$

The space $L^2(G)$ with the square-integrable functions is self-dual.

The essentially bounded functions $L^\infty(G)$ constitute the dual space for the Lebesgue function algebra $L^1(G)$ and the Radon distributions for the compactly supported functions, $\mathcal{M}(G) = \mathcal{C}_c(G)'$. The Radon distributions are put into duality also with the essentially bounded functions by $\langle \mu, d\rangle = \mu * d(1)$:

$$\begin{aligned} L^1(G)' &= L^\infty(G), \quad \text{with } L^1(G) * L^\infty(G) \subseteq L^\infty(G),\\ L^1(G) \subseteq \mathcal{M}(G) &\subseteq L^\infty(G)', \quad \text{with } \mathcal{M}(G) * L^\infty(G) \subseteq L^\infty(G). \end{aligned}$$

8.3.2 Hilbert Metrics of Cyclic Representation Spaces

For a *cyclic* ("one orbit-based") *Hilbert representation* of a locally compact group G, the representation Hilbert space is the closure of the $\mathbb{C}$-span of the G-orbit $V = \overline{\mathbb{C}^{(G\bullet|c\rangle)}}$ of a vector $|c\rangle$, which is called a *cyclic vector.* With the fixgroup H of a cyclic vector, $G \bullet |c\rangle \cong G/H$, one has $V \cong \overline{\mathbb{C}^{(G/H)}}$. All vectors $|c'\rangle \in G \bullet |c\rangle$ of the orbit are cyclic. An invariant vector $G \bullet |c\rangle = |c\rangle$ gives a one-dimensional trivial representation space $\mathbb{C}|c\rangle \cong \mathbb{C}$. For example, the Fock state vector, cyclic for translations, is nondegenerate $\mathbb{R}^n \bullet |0\rangle = |0\rangle$.

Ground-state vectors in physics are cyclic; they determine cyclic G-representations. With a proper subgroup $H \neq G$ as fixgroup, e.g., $\mathbf{U}(1)_+ \subset \mathbf{U}(2)$ in the electroweak standard model (see Chapter 6) or $\{1\} \subset \mathbf{U}(1)$ in superconductivity and the chiral model of Nambu and Jona-Lasinio (see Chapter 9), the ground-state is *degenerate*; all vectors of the ground-state orbit, called *degeneracy manifold*, $|c'\rangle \in G \bullet |c\rangle \cong G/H$, are possible ground-state vectors.

The wave function conditions in quantum mechanics concerning their behavior for infinity and near by singular points are more generally considered in the framework of Lebesgue function spaces. All cyclic Hilbert spaces can be constructed from the Lebesgue function algebra $L^1(G)$ with its dual $L^\infty(G)$: A *positive-type function* [21, 24] or *metric- (scalar product) inducing function* is an essentially bounded function $d \in L^\infty(G)$, which defines a positive product of the convolution algebra $L^1(G)$:

$$d \in L^\infty(G)_+ \iff \begin{array}{l} \langle f|f\rangle_d = \int dg_1 dg_2\, \overline{f(g_1)} d(g_1^{-1} g_2) f(g_2) \geq 0 \\ \text{for all } f \in L^1(G). \end{array}$$

A positive-type function is locally almost everywhere a continuous bounded function,

$$L^\infty(G)_+ \overset{dg}{=} \mathcal{C}_b(G)_+.$$

Connected to each positive-type function $d \in L^\infty(G)_+$ is a cyclic Hilbert space: The induced scalar product is the value of the convolution $L^1(G) * L^\infty(G) * L^1(G) \subseteq L^\infty(G)$ at the neutral group element $1 \in G$:

$$L^1(G) \times L^1(G) \longrightarrow \mathbb{C}, \quad \langle f|f'\rangle_d = \langle \hat{f} * f'\rangle_d = \hat{f} * d * f'(1), \quad \text{with } \hat{f}(g) = \overline{f(g^{-1})}.$$

With this product, the algebra functions define a pre-Hilbert space and, by canonical Cauchy completion of the nontrivial norm classes $|f\rangle_d$, a Hilbert space $\overline{|L^1(G)\rangle}_d$. There exists a cyclic vector $|c\rangle \in \overline{|L^1(G)\rangle}_d$ whose positive-type function is the expectation value $d(g) = \langle c|g \bullet |c\rangle$.

All diagonal matrix elements of a Hilbert representation $D : G \longrightarrow \mathbf{U}(V)$ define continuous positive-type functions,

$$0 \neq |v\rangle \in V, \quad G \ni g \longmapsto d_v(g) = \langle v|D(g)|v\rangle, \quad d_v \in \mathcal{C}_b(G)_+,$$

e.g., $\mathbb{R} \ni x \longmapsto d^{iq}(x) = e^{iqx} \in \mathbf{U}(1)$. Compact group examples are the diagonal elements in the $\mathbf{SU}(2)$- and $\mathbf{SO}(3)$-matrices:

$$u = \begin{pmatrix} e^{i\frac{\varphi+\chi}{2}} \cos\frac{\theta}{2} & ie^{i\frac{\varphi-\chi}{2}} \sin\frac{\theta}{2} \\ ie^{-i\frac{\varphi-\chi}{2}} \sin\frac{\theta}{2} & e^{-i\frac{\varphi+\chi}{2}} \cos\frac{\theta}{2} \end{pmatrix} \in \mathbf{SU}(2),$$

$$u \vee u = \begin{pmatrix} e^{i(\varphi+\chi)} \cos^2\frac{\theta}{2} & ie^{i\varphi}\frac{\sin\theta}{\sqrt{2}} & -e^{i(\varphi-\chi)} \sin^2\frac{\theta}{2} \\ ie^{i\chi}\frac{\sin\theta}{\sqrt{2}} & \cos\theta & ie^{-i\chi}\frac{\sin\theta}{\sqrt{2}} \\ -e^{-i(\varphi-\chi)} \sin^2\frac{\theta}{2} & ie^{-i\varphi}\frac{\sin\theta}{\sqrt{2}} & e^{-i(\varphi+\chi)} \cos^2\frac{\theta}{2} \end{pmatrix} \in \mathbf{SO}(3).$$

The *representation normalization* of a positive-type functions at the neutral element, then called a *state*, is related to a quantum theoretical probability normalization,

$$G \ni 1 \longmapsto d_v(1) = \langle v|D(1)|v\rangle = \langle v|v\rangle.$$

According to Gel'fand and Raikov [27], there is *a surjection from the positive-type functions (representation metrics) to the equivalence classes of cyclic Hilbert representations*:

$$L^\infty(G)_+ \longrightarrow \mathbf{cycrep}_+ G \supseteq \check{G}.$$

In general, an essentially bounded function $d \in L^\infty(G)$ defines a bilinear form on the Radon measures and its subspaces, $\mathcal{M}(G) * L^\infty(G) * \mathcal{M}(G) = L^\infty(G)$:

$$\mathcal{M}(G) \times \mathcal{M}(G) \longrightarrow \mathbb{C}, \quad \omega * d * \omega'(1) = \int dg_1 dg_2\, \omega(g_1^{-1}) d(g_1^{-1} g_2) \omega'(g_2).$$

The essentially bounded functions $L^\infty(G)$ are ordered. The *cone* $L^\infty(G)_+$ with the positive-type functions is convex. A positive-type function is *reflection-symmetric*, i.e., *unitary*, and bounded by the neutral element value:

$$\begin{aligned} &d, d' \in L^\infty(G)_+,\ \alpha, \alpha' \geq 0 \Rightarrow \alpha d + \alpha' d' \in L^\infty(G)_+, \\ &\quad d = \hat{d} \iff \overline{d} = d^-, \text{ with } d^-(g) = d(g^{-1}),\ \overline{d}(g) = \overline{d(g)}, \\ &|d(g)| \leq d(1). \end{aligned}$$

The unit function $G \ni g \longmapsto d^1(g) = 1$ characterizes the trivial group representation. Functions of the inverse, i.e., conjugate functions, are for *dual (inverse-transposed) representations*:

$$\text{duality: } d \leftrightarrow d^- = \overline{d}, \text{ self-dual: } d = d^- = \overline{d}.$$

The pointwise product of two positive-type functions gives a positive-type function for the product representation (see Chapter 9). The extremal continuous states, i.e., there exist only trivial cone combinations with states, $d = \alpha d' + (1 - \alpha)d''$, $\alpha \geq 0$, with $\alpha \in \{0, 1\}$, are called *pure states.* They give the equivalence classes of the irreducible Hilbert representations. For an irreducible Hilbert representation, every nontrivial vector $|v\rangle \neq 0$ is cyclic.

Positive-type functions are the "continuous" extension of positive sesquilinear forms (scalar products) of finite-dimensional vector spaces with $\langle v|w\rangle = \zeta^{vw} = \overline{\zeta^{wv}} = \overline{v}^i \zeta_{ij} w^j$. Finite-dimensional scalar products can be unitarily diagonalized $\zeta = \xi^* \circ \operatorname{diag} \zeta \circ \xi$ with $\xi \in \mathbf{U}(n)$.

The familiar Hilbert spaces with square-integrable functions are included as follows: An absolute square of an L^2-function is a positive-type function:

$$L^2(G) * L^2(G) \subseteq L^\infty(G), \quad L^2(G) \ni \xi \longmapsto d = \hat{\xi} * \xi \in L^\infty(G)_+.$$

If, and only if, a positive-type function is the absolute square of a square-integrable one, the Hilbert space can be constructed with square-integrable group functions:

$$\langle f|f'\rangle_d = \hat{f} * \hat{\xi} * \xi * f'(1) = \langle \xi * f|\xi * f'\rangle_\delta \text{ with } \xi * L^1(G) \subseteq L^2(G).$$

In general, the mapping above $L^2(G) \longrightarrow L^\infty(G)_+$ is not surjective; i.e., not all cyclic representations d can be characterized by a square-integrable G-function ξ_d.

In analogy to positive-type functions, *a positive-type Radon distribution* defines a positive product of functions from the convolution algebra with compact support and the Hilbert space $\overline{|\mathcal{C}_c(G)\rangle_\omega}$ with function classes, e.g., for compact groups G,

$$\begin{aligned} \omega \in \mathcal{M}(G)_+ : \quad & \mathcal{C}_c(G) \times \mathcal{C}_c(G) \longrightarrow \mathbb{C}, \\ & \langle f|f'\rangle_\omega = \langle \hat{f} * f'\rangle_\omega = \hat{f} * \omega * f'(1) = \int dg_1 dg_2\, \overline{f(g_1)} \omega(g_1^{-1} g_2) f'(g_2). \end{aligned}$$

Positive-type functions yield only cyclic representations; positive-type measures yield more general representations. For example, the Dirac distribution at the unit δ leads to $L^2(G) = \overline{|\mathcal{C}_c(G)\rangle_\delta}$.

8.3.3 Induced Positive-Type Measures

The embedding of a positive-type Radon distribution of a closed subgroup of a locally compact group $H \subseteq G$ defines a positive-type Radon G-distribution [24]:

$$\begin{aligned} & \mathcal{M}(H)_+ \ni \omega_H \longmapsto \omega_G \in \mathcal{M}(G)_+, \\ & \text{with } \langle \omega_G, f\rangle = \langle \omega_H, f|_H\rangle = \int_H \omega_H(h) dh f(h) \text{ for } f \in \mathcal{C}_c(G). \end{aligned}$$

For non unimodular groups, the embedded measure has to be multiplied by $\sqrt{\frac{\Delta_G(h)}{\Delta_H(h)}}$ with the modular functions. If the positive-type distribution

ω_H characterizes the class $[d^H]$ of the regular Hilbert representation of the subgroup, then ω_G characterizes the class $[\mathrm{ind}_H^G d^H]$ of the induced Hilbert representation of G. The induced scalar product comes from

$$\begin{aligned} f, f' \in \mathcal{C}_c(G): \langle f|f'\rangle_{\omega_G} &= \langle \hat{f} * f'|_H\rangle_{\omega_H} = \int_G dg \int_H dh\, \overline{f(g)}\omega_H(h) f'(gh) \\ &= \int_{G/H} dgH \int_{H\times H} dh_1 dh_2\, \overline{f(gh_1)}\, \omega_H(h_1^{-1}h_2)\, f'(gh_2). \end{aligned}$$

Examples are given by the functions on a (semi)direct product group:

$$G = K \vec{\times} H: \ \langle f|f'\rangle_{\omega_G} = \int_K dk \int_{H\times H} dh_1 dh_2\, \overline{f(k,h_1)}\, \omega_H(h_1^{-1}h_2)\, f'(k,h_2),$$

in the simplest case for the abelian product group $\mathbb{R}^{k+s}$ and their Fourier transforms, e.g., for time and position translations with $(X, x) \to (x_0, \vec{x})$ and $(Q, q) = (q_0, \vec{q})$:

$$\begin{aligned} G{=}\mathbb{R}^k \oplus \mathbb{R}^s: \langle f|f'\rangle_{\omega_{k+s}} &= \int d^k X \int d^s x_1 d^s x_2\, \overline{f(X, x_1)}\, \omega_s(x_2 - x_1)\, f'(X, x_2) \\ &= \int \frac{d^k Q}{(2\pi)^t} \int \frac{d^s q}{(2\pi)^s}\, \overline{\tilde{f}(Q,q)}\, \tilde{\omega}_s(q)\, \tilde{f}'(Q,q). \end{aligned}$$

8.4 Harmonic Analysis of Representations

A set with group action is the disjoint union of its orbits, $\biguplus_\iota G/H_\iota$, each characterized by a fixgroup $G \bullet x_\iota \cong G/H_\iota$. The linear extension: A Hilbert representation space is a direct sum $\bigoplus_\iota \overline{\mathbb{C}^{(G/H_\iota)}}$ of cyclic ones. An irreducible representation is cyclic, but the converse is not rtrue:

$$\mathbf{rep}\, G \supset \mathbf{cycrep}\, G \supset \mathbf{irrep}\, G.$$

Any cyclic representation and, a fortiori, any irreducible Hilbert representation of G distinguish a subgroup $H \subseteq G$.

A cyclic representation of a locally compact group is a direct integral ${}^{\oplus}\!\int_{\check{G}} d\check{g}$ of irreducible ones. Corresponding to the Haar measure of a locally compact group, unique up to a constant factor, there is the positive *Plancherel measure* ${}^{\oplus}\!\int_{\check{G}} d\check{g} = {}^{\oplus}\!\int d\check{g}$ of the group dual $\check{G} = \mathbf{irrep}_+ G$ with the equivalence classes of the irreducible Hilbert representations, defined up to normalization and unique for a given Haar measure. For a unimodular group, the renormalization are inverse to each other: $\mathcal{M}(G)_+ \times \mathcal{M}(\check{G})_+ \ni (dg, d\check{g}) \to (e^\lambda dg, e^{-\lambda} d\check{g})$. The Plancherel measure is a measure of the group invariants, which, for a Lie group, are generated by multilinear Lie algebra forms. Topologies of $\check{G}$ are not discussed here.

Familiar examples are the translation groups $\mathbb{R}^n$ and Haar measure $d^n x$ with the energy-momenta $\check{\mathbb{R}}^n$ as dual groups and associated Plancherel measure $d^n \frac{q}{2\pi}$ with the Fourier transformation from translation functions $L^1(\mathbb{R}^n)$ to energy-momentum functions $\mathcal{C}_0(\check{\mathbb{R}}^n)$, vanishing at infinity (more ahead):

$$\tilde{f}(q) = \int d^n x\, e^{-iqx}\, f(x) \text{ and } f(x) = \int d^n \tfrac{q}{2\pi}\, e^{iqx}\, \tilde{f}(q).$$

In general, the unitary *Fourier transformation* or *harmonic analysis* associates functions of the group dual, called *harmonic components*, to appropriate group function spaces $\mathcal{F}(G)$:

$$\mathbf{F} : \mathcal{F}(G) \ni f \longmapsto \mathbf{F}f = \tilde{f} \in {}^{\oplus}\!\!\int d\check{g}\ V_{\check{g}} \otimes V'_{\check{g}}.$$

The unitary irreducible G-representation, $G \ni g \longmapsto D_{\check{g}}(g) \in \mathbf{U}(V_{\check{g}})$, $D_{\check{g}} \cong \check{g} \in \check{G}$, acts on the Hilbert space $V_{\check{g}}$. The Fourier transform $\tilde{f}(\check{g})$ gives the coefficients of this representation:

$$\check{G} \ni \check{g} \longmapsto \tilde{f}(\check{g}) = \int dg\ D_{\check{g}}(g^{-1})f(g) = \langle \check{g}|f\rangle = D_{\check{g}} * f(1).$$

The function f can be decomposed with its Fourier integral. Its normalization is the integral with a trace over the irreducible components:

$$G \ni g \longmapsto f(g) = \int d\check{g}\ \operatorname{tr} D_{\check{g}}(g)\tilde{f}(\check{g}), \quad f(1) = \int d\check{g}\ \operatorname{tr} \tilde{f}(\check{g}).$$

The use of representation equivalence classes makes all this rather troublesome for nonabelian groups.

The Fourier transform is a conjugation-compatible algebra morphism. It is injective, i.e., invertible on the image $\mathbf{F}(\mathcal{F}(G))$:

$$\begin{aligned} \mathbf{F}(f + f') &= \mathbf{F}f + \mathbf{F}f', & \mathbf{F}(\alpha f) &= \alpha\mathbf{F}f,\\ \mathbf{F}(f * f') &= \mathbf{F}f \cdot \mathbf{F}f, & \mathbf{F}(\overline{f}) &= \overline{\mathbf{F}f}. \end{aligned}$$

For compact groups, one has as Fourier transformable functions $\mathcal{F}(G) = L^1(G)$, for noncompact groups (second countable, unimodular, type I), a restriction to $\mathcal{F}(G) = L^1(G) \cap L^2(G)$.

The Plancherel measure of a group dual is defined by and allows an orthogonal direct integral decomposition of the both-sided regular $G \times G$-representation, and the left- and right-regular G-representation,

$$\begin{aligned} L^2(G) &\cong {}^{\oplus}\!\!\int d\check{g}\ V_{\check{g}} \otimes V'_{\check{g}}, \quad \mathrm{id}_{L^2(G)} \cong {}^{\oplus}\!\!\int d\check{g}\ D_{\check{g}} \otimes \overline{D}_{\check{g}},\\ f = {}^{\oplus}\!\!\int dg\ gf(g) &= \tilde{f} = {}^{\oplus}\!\!\int d\check{g}\ D_{\check{g}} \otimes \tilde{f}(\check{g}) \text{ (where defined)},\\ \text{right-, left-regular: } R &\cong {}^{\oplus}\!\!\int d\check{g}\ D_{\check{g}} \otimes \mathbf{1}, \quad L \cong {}^{\oplus}\!\!\int d\check{g}\ \mathbf{1} \otimes \overline{D}_{\check{g}}. \end{aligned}$$

The *Parseval formula* for the scalar product and the *Fourier inversion* are

$$\begin{aligned} f_{1,2} \in \mathcal{F}(G): \quad \int dg\ \overline{f_1(g)}f_2(g) &= \int d\check{g}\ \operatorname{tr} \overline{\tilde{f}_1(\check{g})}\tilde{f}_2(\check{g}) = \langle f_1|f_2\rangle,\\ f(g) &= \int d\check{g}\ \operatorname{tr} D_{\check{g}}(g)\tilde{f}(\check{g}). \end{aligned}$$

For noncompact groups, the inversion holds only for a subspace of functions.

A direct decomposition of a Hilbert representation D displays the *normalizations* (cardinal multiplicities for compact groups) of the irreducible components $D_{\check{g}}$ by a positive *spectral distribution* ρ_D *of* $\check{G}$:

$$D = {}^{\oplus}\!\!\int d\check{g}\ D_{\check{g}}\ \rho_D(\check{g}).$$

A positive-type function d for a cyclic representation is an integral of positive-type functions $d_{\check{g}}$ for its irreducible components, i.e., a matrix element for a cyclic vector $d_{\check{g}}(g) = \langle c|D_{\check{g}}(g)|c\rangle$, with a positive Plancherel distribution $\tilde{d}$ of the group dual,

$$d(g) = \int d\check{g}\ d_{\check{g}}(g)\ \tilde{d}(\check{g}).$$

For a compact group with normalized Haar measure, the Plancherel measure is the counting measure $\bigoplus_{\iota\in\check{G}} d_\iota$ with the dimensions $d_\iota = \dim_{\mathbb{C}} V_\iota$ of the irreducible representation spaces, i.e., the number of columns or lines in the full matrix algebra $\mathbb{C}^{d_\iota}\otimes\mathbb{C}^{d_\iota}$. The Plancherel measure for a noncompact group has continuous support. For noncompact nonabelian groups, determining the Plancherel measure is difficult. The support of the Plancherel measure is the reduced dual; it must not be the full group dual; if not, the group is nonamenable. There may be irreducible Hilbert representations, not in the support of the Plancherel measure.

The Schur product for cyclic representations integrates their positive-type functions over the group:

$$\begin{aligned} d_{1,2} \in L^\infty(G)_+ : \quad \{d_2|d_1\} &= \int dg\ \overline{d_2(g)}d_1(g) = \int d\check{g}\ \ \operatorname{tr}\overline{\tilde{d}_2(\check{g})}\tilde{d}_1(\check{g}) \\ &= \hat{d}_2 * d_1(1) \in \mathcal{M}(\check{G}). \end{aligned}$$

It is a *Plancherel distribution*, i.e., a distribution of the group dual with the Plancherel measure. The inverse Plancherel measure is the *Plancherel density.*

8.5 Schur Orthogonality for Compact Groups

The Lebesgue function spaces for a compact group K are all subspaces of the convolution algebra with the absolute integrable function classes,

$$L^\infty(K) \subseteq L^p(K) \subseteq L^q(K) \subseteq L^1(K) \ \text{ for } \infty \geq p \geq q \geq 1.$$

All representations can be formulated as acting on square-integrable functions $L^2(K)$. For example, the central element in the $\mathbf{SO}(3)$-matrix above and the corresponding elements for the irreducible $(1+2L)$-dimensional representations are metric-inducing functions, the Legendre polynomials $\mathrm{P}^L \in L^1(\mathbf{SO}(3))_+$:

$$\begin{aligned} \langle L;0|O(\chi,\varphi,\theta)|L;0\rangle &= \tfrac{1}{2^L L!}(-\tfrac{\partial}{\partial\cos\theta})^L \sin^{2L}\theta = \mathrm{P}^L(\cos\theta) \\ &= \sum_{k=0}^{L} \tfrac{(L+k)!}{(L-k)!}\tfrac{(-\sin^2\frac{\theta}{2})^k}{(k!)^2} = \begin{cases} 1, \\ 1-\sin^2\frac{\theta}{2} = \cos\theta, \\ 1-6\sin^2\frac{\theta}{2}+6\sin^4\frac{\theta}{2} \\ \quad = \frac{3}{2}(\cos^2\theta - \frac{1}{3}), \ldots . \end{cases} \end{aligned}$$

They define the scalar product for the Hilbert spaces $\overline{|L^1(\mathbf{SO}(3))\rangle_{\mathrm{P}^L}} \cong \mathbb{C}^{1+2L}$.

There is the *Peter–Weyl decomposition* [47] of the algebra with the square-integrable functions into the countably many finite-dimensional full matrix algebras as irreducible group algebra representations with the columns (or lines) the irreducible Hilbert spaces:

$$\text{compact } K: \quad L^2(K) \overset{\text{dense}}{=} \bigoplus_{\iota=1}^{\infty} \mathbb{C}^{d_\iota} \otimes \mathbb{C}^{d_\iota}, \quad \langle \xi_\iota | \xi_{\iota'} \rangle = \tfrac{1}{d_\iota} \delta_{\iota\iota'}.$$

It is a special case of Frobenius' reciprocity (see Chapter 7) for the K-representation induced by the trivial representation of the trivial subgroup,

$$L^2(K) = L^2(K/\{1\}, \mathbb{C}) \overset{\text{dense}}{\cong} \bigoplus_{\iota=1}^{\infty} d_\iota \mathbb{C}^{d_\iota}.$$

The algebras are *Schur-orthogonal* to each other with respect to the *Schur scalar product*, defined with normalized group measure,

$$\begin{aligned} D_\iota : K &\longrightarrow \mathbf{U}(V_\iota), \; V_\iota \cong \mathbb{C}^{d_\iota}, \\ \{D_{\iota'} | D_\iota\} &= \hat{D_{\iota'}} \circledast D_\iota(1) = \textstyle\int_K dk \; \overline{D_{\iota'}(k)} \otimes D_\iota(k) = \delta_{\iota\iota'} \tfrac{1}{d_\iota} \mathbf{1}_{d_\iota^2}. \end{aligned}$$

More explicit: Two representation matrix elements of inequivalent irreducible Hilbert representations are orthogonal, with $v_\iota, w_\iota \in V_\iota$:

$$\textstyle\int_K dk \; \overline{\langle v_{\iota'} | D_{\iota'}(k) | w_{\iota'} \rangle} \langle v_\iota | D_\iota(k) | w_\iota \rangle = \begin{cases} 0, & D_\iota \not\cong D_{\iota'}, \\ \frac{1}{d_\iota} \langle v_\iota | v_{\iota'} \rangle \langle w_{\iota'} | w_\iota \rangle, & D_\iota = D_{\iota'}. \end{cases}$$

The *Plancherel measure* of an irreducible representation D_ι of a compact group with normalized Haar measure is a counting measure, given by the number d_ι of equivalent irreducible representations $\mathbb{C}^{d_\iota}$ in one algebra $\mathbb{C}^{d_\iota} \otimes \mathbb{C}^{d_\iota}$, which coincides with its dimension, $\Pi^K(D_\iota) = d_\iota$. The Schur orthogonality involves the Plancherel density (inverse Plancherel measure).

The Schur orthogonality is conceptually different from a possible orthonormality of the vectors in the representation space, e.g., for a Hilbert basis $\{e^j \mid j = 1, \ldots, \iota\}$,

$$\langle e^i | e^j \rangle = \delta^{ij} \Rightarrow \textstyle\int_K dk \; \overline{D_{\iota'}(k)^{j'}_{i'}} \; D_\iota(k)^j_i = \delta_{\iota\iota'} \tfrac{1}{d_\iota} \delta_{ii'} \delta^{jj'}.$$

The Schur orthogonality is exemplified for $\mathbf{SU}(2) \sim \mathbf{SO}(3)$ with Plancherel density $d_J = 1 + 2J$ by the positive-type square-integrable functions arising as diagonal matrix elements of the 2- and three-dimensional representation:

$$\int d^3k = \int_{-2\pi}^{2\pi} \frac{d\chi}{4\pi} \int_0^{2\pi} \frac{d\varphi}{2\pi} \int_{-1}^{1} \frac{d\cos\theta}{2} = 1,$$

$$\begin{cases} \displaystyle\int d^3k \; |e^{i\frac{\chi+\varphi}{2}} \cos\frac{\theta}{2}|^2 = \tfrac{1}{2}, \\ \displaystyle\int d^3k \; |e^{i(\chi+\varphi)} \cos^2\frac{\theta}{2}|^2 = \tfrac{1}{3}, \\ \displaystyle\int d^3k \; e^{-i(\chi+\varphi)} \cos^2\frac{\theta}{2} \; e^{i\frac{\chi+\varphi}{2}} \cos\frac{\theta}{2} = 0. \end{cases}$$

Positive-type functions, normalized as states (representation normalization), do not have to be Schur-normalized. A familiar example is given by the Legendre polynomials P^L, which are normalized positive-type functions for irreducible representations $[L]$ of the rotations $\mathbf{SO}(3)$. They are the $(m, m') = (0,0)$-components in the rotation group representing matrices, $\mathrm{P}^L = [L]_0^0$. The Legendre polynomials are Schur-orthogonal for different angular momenta L and have as Schur norm the inverse of the $1+2L$ dimension of the representation space:

$$\begin{aligned} |L\}(\theta) &= \mathrm{P}^L(\cos\theta), \quad \mathrm{P}^L(1) = 1, \\ \{L'|L\} &= \int_{-1}^{1} \tfrac{d\cos\theta}{2}\ \mathrm{P}^{L'}(\cos\theta)\mathrm{P}^L(\cos\theta) = \delta^{LL'} \tfrac{1}{1+2L}. \end{aligned}$$

The transition to the spherical harmonics $\mathrm{Y}_m^L = |L;m\rangle$ involves the multiplication with the inverse square root of its Schur norm:

$$\begin{aligned} |L;0\rangle &= \mathrm{Y}_0^L = \sqrt{\tfrac{1+2L}{|\Omega^2|}}\mathrm{P}^L, \\ \langle L';m'|L;m\rangle &= \int d^2\omega\ \overline{\mathrm{Y}_{m'}^{L'}(\varphi,\theta)}\mathrm{Y}_m^L(\varphi,\theta) = \delta^{LL'}\delta_{mm'}. \end{aligned}$$

With the additional normalization with the 2-sphere area $|\Omega^2| = 4\pi$, equal for all L, the spherical harmonics $\{\mathrm{Y}_m^L \mid m = -L,\dots,L\}$ are an orthonormal basis for the irreducible $\mathbf{SO}(3)$-representation space with angular momentum L.

8.6 Translation Representations

In the following, the structures for locally compact groups, as given above, are specialized to the noncompact abelian translation groups $\mathbb{R}^n$, e.g., to a Cartan plane A in a semisimple Lie group $G = K \circ A \circ N$ or to translations in an affine group $G \vec{\times} \mathbb{R}^n$. The (energy-)momenta constitute the group dual with the equivalence classes of the irreducible Hilbert representations $x \longmapsto e^{iqx} = \langle x|q\rangle \in \mathbf{U}(1)$. The translation group has the distinction of being isomorphic to its group dual $\check{\mathbb{R}}^n$, an abelian group,

$$x \in \mathbb{R}^n \cong \check{\mathbb{R}}^n = (i\mathbb{R})^n \ni iq.$$

All faithful Hilbert representations of the translations $\mathbb{R}$ are infinite-dimensional.

8.6.1 Fourier Transformation

The harmonic components of a translation distribution, its *Fourier components*, are given by the Fourier transform:

$$\begin{aligned} \langle q|\mu\rangle = \tilde{\mu}(q) &= \int \tfrac{d^n x}{N}\ e^{-iqx}\mu(x) = \int \tfrac{d^n x}{N}\ \langle q|x\rangle\langle x|\mu\rangle, \\ \int \tfrac{N d^n q}{(2\pi)^n}\ |\tilde{\mu}(q)|^2 &= \int \tfrac{d^n x}{N}\ |\mu(x)|^2, \text{ where defined.} \end{aligned}$$

If the Haar measure of the translations $\mathbb{R}^n$ is normalized as $\frac{d^n x}{N}$, the corresponding Plancherel measure of the (energy-)momenta $\check{\mathbb{R}}^n$ is $\frac{N d^n q}{(2\pi)^n}$ with a free normalization factor $N > 0$, e.g., $N = 1$ or $N = (2\pi)^n$ or $N = (2\pi)^{\frac{n}{2}}$.

The Fourier transforms of the three relevant spaces $L^1(\mathbb{R}^n)$, $\mathcal{M}(\mathbb{R}^n)$, and $L^\infty(\mathbb{R}^n)$ are as follows: According to a theorem of Lebesgue, the Fourier transformation of the convolution algebra $L^1(\mathbb{R}^n)$ is an injective algebra morphism, with a dense range, but not surjective, into the continuous functions $\mathcal{C}_0(\check{\mathbb{R}}^n)$, vanishing at infinity. The Fourier transformation can be extended to the Radon measure algebra $\mathcal{M}(\mathbb{R}^n)$ with values in the bounded continuous functions $\mathcal{C}_b(\check{\mathbb{R}}^n)$. Positive Radon measures with spectral energy-momentum distributions and the continuous positive-type functions are bijective (*Bochner's theorem* [8]):

$$\text{Fourier:} \quad \begin{cases} \widetilde{L^p(\mathbb{R}^n)} \subseteq L^r(\check{\mathbb{R}}^n), \ \frac{1}{p} + \frac{1}{r} = 1, \ 1 \le p \le 2, \ \ \infty \ge r \ge 2, \\ \widetilde{L^1(\mathbb{R}^n)} = \dot{\mathcal{C}}_0(\check{\mathbb{R}}^n) \text{ dense in } \mathcal{C}_0(\check{\mathbb{R}}^n) \text{ (Lebesgue)}, \\ \widetilde{\mathcal{M}(\mathbb{R}^n)}_+ = \mathcal{C}_b(\check{\mathbb{R}}^n)_+ \overset{d^n q}{=} L^\infty(\check{\mathbb{R}}^n)_+ \quad \text{(Bochner)}, \\ \widetilde{\mathcal{M}(\mathbb{R}^n)} = \dot{\mathcal{C}}_b(\check{\mathbb{R}}^n) = \text{complex span of } \mathcal{C}_b(\check{\mathbb{R}}^n)_+. \end{cases}$$

The convolution and pointwise product are Fourier-compatible (with convenient measure normalization $N = 1$),

$$\mu(x) = \int \frac{d^n q}{(2\pi)^n} \, \tilde{\mu}(q) e^{iqx}, \quad \tilde{\mu}(q) = \int d^n x \, \mu(x) e^{-iqx}, \quad \delta(x) = \int \frac{d^n q}{(2\pi)^n} e^{iqx},$$

$$\tilde{\tilde{\mu}} = \mu, \quad \widetilde{\mu_1 \cdot \mu_2} = \tilde{\mu}_1 * \tilde{\mu}_2, \quad \begin{cases} \mu_1 \cdot \mu_2(x) = \int \frac{d^n q}{(2\pi)^n} \, \tilde{\mu}_1 * \tilde{\mu}_2(q) e^{iqx}, \\ \tilde{\mu}_1 * \tilde{\mu}_2(q) = \int \frac{d^n p}{(2\pi)^n} \, \tilde{\mu}_1(p) \tilde{\mu}_2(q - p), \end{cases}$$

for the representation relevant spaces are exchanged in the spaces with the harmonic components:

$G = \mathbb{R}^n$

$*$	$\mathcal{M}(\mathbb{R}^n)$	$L^1(\mathbb{R}^n)$	$\dot{\mathcal{C}}_b(\mathbb{R}^n)$
$\mathcal{M}(\mathbb{R}^n)$	$\mathcal{M}(\mathbb{R}^n)$	$L^1(\mathbb{R}^n)$	$\dot{\mathcal{C}}_b(\mathbb{R}^n)$
$L^1(\mathbb{R}^n)$	$L^1(\mathbb{R}^n)$	$L^1(\mathbb{R}^n)$	$\dot{\mathcal{C}}_b(\mathbb{R}^n)$
$\dot{\mathcal{C}}_b(\mathbb{R}^n)$	$\dot{\mathcal{C}}_b(\mathbb{R}^n)$	$\dot{\mathcal{C}}_b(\mathbb{R}^n)$	–

$\mu_1 * \mu_2(x)$

$\cdot$	$\dot{\mathcal{C}}_b(\mathbb{R}^n)$	$L^1(\mathbb{R}^n)$	$\mathcal{M}(\mathbb{R}^n)$
$\dot{\mathcal{C}}_b(\mathbb{R}^n)$	$\dot{\mathcal{C}}_b(\mathbb{R}^n)$	$L^1(\mathbb{R}^n)$	$\mathcal{M}(\mathbb{R}^n)$
$L^1(\mathbb{R}^n)$	$L^1(\mathbb{R}^n)$	–	–
$\mathcal{M}(\mathbb{R}^n)$	$\mathcal{M}(\mathbb{R}^n)$	–	–

$\mu_1 \cdot \mu_2(x)$

$\updownarrow$ **F** Fourier $\updownarrow$

$\check{G} = \check{\mathbb{R}}^n$

$\cdot$	$\dot{\mathcal{C}}_b(\check{\mathbb{R}}^n)$	$\dot{\mathcal{C}}_0(\check{\mathbb{R}}^n)$	$\mathcal{M}(\check{\mathbb{R}}^n)$
$\dot{\mathcal{C}}_b(\check{\mathbb{R}}^n)$	$\dot{\mathcal{C}}_b(\check{\mathbb{R}}^n)$	$\dot{\mathcal{C}}_0(\check{\mathbb{R}}^n)$	$\mathcal{M}(\check{\mathbb{R}}^n)$
$\dot{\mathcal{C}}_0(\check{\mathbb{R}}^n)$	$\dot{\mathcal{C}}_0(\check{\mathbb{R}}^n)$	$\dot{\mathcal{C}}_0(\check{\mathbb{R}}^n)$	$\mathcal{M}(\check{\mathbb{R}}^n)$
$\mathcal{M}(\check{\mathbb{R}}^n)$	$\mathcal{M}(\check{\mathbb{R}}^n)$	$\mathcal{M}(\check{\mathbb{R}}^n)$	–

$\tilde{\mu}_1 \cdot \tilde{\mu}_2(q)$
from group product

$*$	$\mathcal{M}(\check{\mathbb{R}}^n)$	$\dot{\mathcal{C}}_0(\check{\mathbb{R}}^n)$	$\dot{\mathcal{C}}_b(\check{\mathbb{R}}^n)$
$\mathcal{M}(\check{\mathbb{R}}^n)$	$\mathcal{M}(\check{\mathbb{R}}^n)$	$\dot{\mathcal{C}}_0(\check{\mathbb{R}}^n)$	$\dot{\mathcal{C}}_b(\check{\mathbb{R}}^n)$
$\dot{\mathcal{C}}_0(\check{\mathbb{R}}^n)$	$\dot{\mathcal{C}}_0(\check{\mathbb{R}}^n)$	–	–
$\dot{\mathcal{C}}_b(\check{\mathbb{R}}^n)$	$\dot{\mathcal{C}}_b(\check{\mathbb{R}}^n)$	–	–

$\tilde{\mu}_1 * \tilde{\mu}_2(q)$
for product representations

The Fourier transformation defines algebra isomorphisms:

$$\mathcal{M}(\mathbb{R}^n) \overset{\mathbf{F}}{\cong} \dot{\mathcal{C}}_b(\check{\mathbb{R}}^n) \overset{d^n q}{=} L^\infty(\check{\mathbb{R}}^n), \quad L^1(\mathbb{R}^n) \overset{\mathbf{F}}{\cong} \dot{\mathcal{C}}_0(\check{\mathbb{R}}^n).$$

Dual and sesquilinear products use the convolution product at the trivial translation or at the trivial (energy-)momentum as integrals over all translations or over all (energy-)momenta. For a sesquilinear form, the conjugated (generalized) function is used:

$$\begin{aligned}
\langle \mu_1, \mu_2 \rangle &= \int d^n x\ \mu_1(-x)\mu_2(x) &&= \mu_1 * \mu_2(0) \\
&= \int \tfrac{d^n q}{(2\pi)^n}\ \tilde{\mu}_1(q)\tilde{\mu}_2(q) &&= \tilde{\mu}_1^- \underset{(2\pi)^n}{*} \tilde{\mu}_2(0), \\
\langle \mu_1 | \mu_2 \rangle &= \langle \overline{\mu}_1, \mu_2 \rangle, \text{ with } \hat{\mu}(x) &&= \overline{\mu^-(x)} = \overline{\mu(-x)} = \int \tfrac{d^n q}{(2\pi)^n}\ \overline{\tilde{\mu}(q)} e^{iqx}.
\end{aligned}$$

8.6.2 Cyclic Translation Representations

The continuous translation positive-type functions, isomorphic to the positive (energy-)momentum Radon measures, characterize all cyclic translation representations in the bijection

$$\mathcal{C}_b(\mathbb{R}^n)_+ \cong \mathcal{M}(\check{\mathbb{R}}^n)_+ \leftrightarrow \textbf{cycrep}_+\mathbb{R}^n.$$

The Hilbert product induced by a function positive-type function,

$$\mathcal{C}_b(\mathbb{R}^n)_+ \ni d \leftrightarrow \tilde{d} \in \mathcal{M}(\check{\mathbb{R}}^n)_+, \quad d(x) = \int \tfrac{d^n q}{(2\pi)^n}\ \tilde{d}(q)\ e^{iqx},$$

can be transformed into an integration of the pointwise product of the harmonic components with a representation characteristic positive (energy-)momentum measure $\tilde{d}(q)\frac{d^n q}{(2\pi)^n}$, which involves a positive Radon distribution as spectral measure,

$$\begin{aligned}
L^1(\mathbb{R}^n) \underset{d}{*} L^1(\mathbb{R}^n) &\longrightarrow \mathbb{C} \overset{\mathbf{F}}{\leftrightarrow} \dot{\mathcal{C}}_0(\check{\mathbb{R}}^n) \underset{\tilde{d}}{\cdot} \dot{\mathcal{C}}_0(\check{\mathbb{R}}^n) \longrightarrow \mathbb{C}, \\
\langle f | f' \rangle_d &= \int d^n x_1 d^n x_2\ \overline{f(x_1)} d(x_2 - x_1) f'(x_2) = \hat{f} * d * f'(0) \\
&= \int \tfrac{d^n q}{(2\pi)^n}\ \overline{\tilde{f}(q)}\ \tilde{d}(q)\ \tilde{f}'(q) = \overline{\tilde{f}} \cdot \tilde{d} \cdot \tilde{f}'(\check{\mathbb{R}}^n).
\end{aligned}$$

A finite product gives a finite Radon measure $\overline{\tilde{f}} \cdot \tilde{d} \cdot \tilde{f}'$ of the (energy-)momenta.

The product can be extended to the Radon distributions and their Fourier-transformed bounded (energy-)momentum functions

$$\begin{aligned}
\mathcal{M}(\mathbb{R}^n) \underset{d}{*} \mathcal{M}(\mathbb{R}^n) &\longrightarrow \mathbb{C} \overset{\mathbf{F}}{\leftrightarrow} \dot{\mathcal{C}}_b(\check{\mathbb{R}}^n) \underset{\tilde{d}}{\cdot} \dot{\mathcal{C}}_b(\check{\mathbb{R}}^n) \longrightarrow \mathbb{C}, \\
\langle \omega | \omega' \rangle_d &= \int \tfrac{d^n q}{(2\pi)^n}\ \overline{\tilde{\omega}(q)}\ \tilde{d}(q)\ \tilde{\omega}'(q).
\end{aligned}$$

The transition to a Hilbert space with square-integrable $\mathbb{R}^n$-functions requires the positive-type function to be an absolute square of a square-integrable function:

$$\begin{aligned}
d = \hat{\xi} * \xi \iff \tilde{d} = |\tilde{\xi}|^2 \text{ with } L^2(\mathbb{R}^n) \ni \xi \leftrightarrow \tilde{\xi} \in L^2(\check{\mathbb{R}}^n), \\
\xi * L^1(\mathbb{R}^n) \subseteq L^2(\mathbb{R}^n) \leftrightarrow \tilde{\xi} \cdot \dot{\mathcal{C}}_0(\check{\mathbb{R}}^n) \subseteq L^2(\check{\mathbb{R}}^n).
\end{aligned}$$

The Hilbert space $H = \overline{|L^1(\mathbb{R}^n)\rangle}_d$ with the cyclic translation representation has a *distributive basis*, labeled by the eigenvalues (energy-momenta),

$$\{|q\rangle_d = |d;q\rangle \mid q \in \mathbb{R}^n\}: \left\{ \begin{array}{rl} {}^{\oplus}\!\int \frac{d^nq}{(2\pi)^n} |q\rangle_d \langle q|_d &= \mathrm{id}_H, \\ H \ni |f\rangle_d = |d;\tilde{f}\rangle &= {}^{\oplus}\!\int \frac{d^nq}{(2\pi)^n} \tilde{f}(q)|q\rangle_d, \\ \text{with } \tilde{f}(q) &= \langle q|f\rangle_d, \\ \langle f'|f\rangle_d &= \langle d;\tilde{f}'|d;\tilde{f}\rangle, \\ \langle q'|q\rangle_d = \langle d;q'|d;q\rangle &= \tilde{d}(q)(2\pi)^n \delta(q-q'), \\ D(x)|q\rangle_d &= e^{iqx}|q\rangle_d. \end{array} \right.$$

If normalizable, the integration of a distributive basis over the eigenvalues gives a cyclic vector:

$$|1\rangle_d = |d;1\rangle = {}^{\oplus}\!\int \frac{d^nq}{(2\pi)^n} |q\rangle_d: \quad d(x) = \langle 1|D(x)|1\rangle_d \begin{array}{l} = \int \frac{d^nq}{(2\pi)^n} \frac{d^nq'}{(2\pi)^n} e^{iqx} \langle q'|q\rangle_d \\ = \int \frac{d^nq}{(2\pi)^n} \tilde{d}(q)\, e^{iqx}. \end{array}$$

The Schur product of two translation representations, characterized by positive-type functions, integrates over the translations

$$\begin{array}{rlcl} \{d_2|d_1\} &= \int d^nx\, \overline{d_2(x)} d_1(x) &=& \int \frac{d^nq}{(2\pi)^n} \tilde{d}_1(q)\tilde{d}_2(q) \\ &= \quad \hat{d}_2 * d_1(0) &=& \tilde{d}_1^- \underset{(2\pi)^n}{*} \tilde{d}_2(0). \end{array}$$

It is a Plancherel distribution of the (energy-)momenta.

8.6.3 Spherical and Hyperbolic Positive-Type Functions

The positive Radon measures of the dual group $iq \in \check{\mathbb{R}}$ with the momenta (or energies) are characterized by their support. The extremal states for the irreducible translation representations in $\mathbf{U}(1)$ acting on one-dimensional Hilbert spaces are the Dirac measures $\delta_P \in \mathcal{M}(\check{\mathbb{R}})_+$ supported by the real momentum P for the invariant imaginary eigenvalue $iP \in i\mathbb{R}$:

$$\mathbb{R} \ni x \longmapsto d^{iP}(x) = e^{iPx} = \int dq\, \delta(q-P)e^{iqx} = \oint \frac{dq}{2i\pi} \frac{1}{q-P} e^{iqx}.$$

The Dirac distribution restricts the induced scalar product to one component:

$$\begin{array}{rl} f,f' \in L^1(\mathbb{R}): \quad \langle f|f'\rangle_d &= \int dx_1 dx_2\, \overline{f(x_1)} e^{iP(x_2-x_1)} f'(x_2) \\ &= \int dq\, \overline{\tilde{f}(q)}\, \delta(q-P)\, \tilde{f}'(q) = \overline{\tilde{f}(P)}\, \tilde{f}'(P), \end{array}$$

$$\overline{|L^1(\mathbb{R})\rangle}_{d^{iP}} \cong \mathbb{C}, \text{ with basis } \{|P\rangle\} \left\{ \begin{array}{rl} \langle P|P\rangle &= 1, \\ D(x)|P\rangle &= e^{iPx}|P\rangle, \\ \text{cyclic vector: } |P;1\rangle &= |P\rangle. \end{array} \right.$$

The Schur product of the irreducible Hilbert representations gives the Plancherel density (inverse Plancherel measure):

$$\text{Schur product: } \{P'|P\} = \int dx\, e^{-iP'x} e^{iPx} = \delta(\tfrac{P-P'}{2\pi}).$$

The self-dual representations on two-dimensional Hilbert spaces are cyclic with the reflection $E \leftrightarrow -E$ invariant positive Radon distributions, supported by two reflected eigenvalues $\pm iE \in i\mathbb{R}$ and positive invariant E^2, as used for time translations in the harmonic oscillator,

$$\begin{aligned}
\mathbb{R} \ni t \longmapsto d^{E^2}(t) &= \int dq\ |q|\delta(q^2 - E^2)e^{iqt} = \cos Et,\\
|q|\delta(q^2 - E^2) &= \frac{\delta(q-E)+\delta(q+E)}{2},\\
\overline{|L^1(\mathbb{R})\rangle}_{d^{E^2}} \cong \mathbb{C}^2,\ \text{with basis } \{|E\rangle, |-E\rangle\} &\left\{\begin{aligned}
\langle \pm E|\pm E\rangle &= 1,\\
\langle E|-E\rangle &= 0,\\
D(t)|\pm E\rangle &= e^{\pm iEt}|\pm E\rangle,\\
\text{cyclic vector: } |E^2;1\rangle &= \frac{|E\rangle \oplus |-E\rangle}{\sqrt{2}},\\
\langle E^2;1|E^2;1\rangle &= 1,
\end{aligned}\right.\\
\text{Schur product: } \{E'^2|E^2\} &= \int dt\ \cos E't \cos Et = \tfrac{1}{2}[\delta(\tfrac{E-E'}{2\pi}) + \delta(\tfrac{E+E'}{2\pi})]\\
&= \delta(\tfrac{|E|-|E'|}{\pi}).
\end{aligned}$$

Irreducible faithful translation representations $x \longmapsto e^{Qx}$ with real eigenvalue $Q \in \mathbb{R}$ are not unitary. From the residual form of the *positive-type functions for the circle* $\mathbf{SO}(2)$ and its finite two-dimensional representations,

$$\mathbb{R} \ni x \longmapsto d^{P^2}(x) = \cos Px = \oint \frac{dq}{i\pi}\frac{q}{q^2-P^2}e^{iqx},$$

one obtains by the transition $P \to iQ$ the *positive-type functions for the hyperbola* $\mathbf{SO}_0(1,1) \cong \mathbb{R}$ and its cyclic Hilbert representations with negative invariant $-Q^2$:

$$\begin{aligned}
\mathbb{R} \ni x \longmapsto d^{-Q^2}(x) &= \int \frac{dq}{\pi}\frac{|Q|}{q^2+Q^2}e^{iqx} = e^{-|Qx|},\\
\frac{1}{\pi}\frac{|Q|}{q^2+Q^2} &= \frac{1}{2i\pi}\left(\frac{1}{q-i|Q|} - \frac{1}{q+i|Q|}\right).
\end{aligned}$$

They characterize a scalar product for an infinite-dimensional Hilbert space:

$$\begin{aligned}
f, f' \in L^1(\mathbb{R}):\quad \langle f|f'\rangle_d &= \int dx_1 dx_2\ \overline{f(x_1)}e^{-|Q(x_2-x_1)|}f'(x_2)\\
&= \int \frac{dq}{\pi}\ \overline{\tilde{f}(q)}\ \frac{|Q|}{q^2+Q^2}\ \tilde{f}'(q),\\
\overline{|L^1(\mathbb{R})\rangle}_{d^{-Q^2}} \cong L^2(\mathbb{R}) &\text{ with distributive basis } \{|-Q^2;q\rangle \bigm| q \in \mathbb{R}\},\\
&\left\{\begin{aligned}
\langle -Q^2;q'|-Q^2;q\rangle &= \frac{2|Q|}{q^2+Q^2}\delta(\tfrac{q-q'}{2\pi}),\\
D(x)|-Q^2;q\rangle &= e^{iqx}|-Q^2;q\rangle,\\
\text{cyclic vector: } |-Q^2;1\rangle &= {}^{\oplus}\!\!\int \frac{dq}{2\pi}|-Q^2;q\rangle,\\
\langle -Q^2;1|-Q^2;1\rangle &= 1.
\end{aligned}\right.
\end{aligned}$$

In contrast to the spherical positive-type functions, the hyperbolic ones for the noncompact group are not orthogonal for different invariants:

$$\text{Schur product: } \left\{\begin{aligned}
\{-Q'^2|-Q^2\} &= \int dx\ e^{-|Q'|x}e^{-|Q|x} = \frac{2}{|Q|+|Q'|},\\
\{-Q^2|-Q^2\} &= \frac{1}{|Q|}.
\end{aligned}\right.$$

The singularities of the function in the Radon measure (imaginary momenta $\pm iQ$) are not eigenvalues for unitary translation representations. The positive-type function is an absolute square of L^2-functions $\tilde{d}^{-Q^2} = |\tilde{\xi}|^2 \in L^2(\breve{\mathbb{R}})$.

8.6.4 Breit–Wigner Functions

Time translations together with their reflection $t \longmapsto -t$ constitute a non-abelian semidirect group:

$$\begin{aligned} \mathbb{I}(2) \vec{\times} \mathbb{R} \ni (\epsilon, t) &\longmapsto \left(\begin{array}{c|c} \epsilon & t \\ \hline 0 & 1 \end{array}\right) \in \mathbf{GL}(2,\mathbb{R}), \\ \epsilon \in \mathbb{I}(2) = \{1,-1\},\ & (\epsilon_1, t_1) \circ (\epsilon_2, t_2) = (\epsilon_1\epsilon_2, t_1 + \epsilon_1 t_2). \end{aligned}$$

The reflection group $\mathbb{I}(2)$ is discrete. Therefore, and with Wigner, *time reflection* can be represented in quantum theories by an *antilinear* transformation. For the complex energies $q + i\Gamma \in \mathbb{C}$ as eigenvalues for all irreducible time translation representations (not only unitary), the reflection group is implemented by the identity and the complex conjugation:

$$\mathbb{I}(2) = \{1,-1\} \longrightarrow \{\mathrm{id}_{\mathbb{C}}, *\}: \quad \left\{ \begin{array}{c} \mathbb{I}(2) \times \mathbb{C} \longrightarrow \mathbb{C}, \\ -1 \bullet (q + i\Gamma) = (q + i\Gamma)^* = q - i\Gamma. \end{array}\right.$$

A complex energy has as reflection orbit $\{E + i\Gamma, E - i\Gamma\}$ with fixgroup $\mathbb{I}(2)$ for real energies $\Gamma = 0$ and trivial fixgroup $\{1\} \subset \mathbb{I}(2)$ for nontrivial width $\Gamma \neq 0$.

The characteristic functions for the *future and past* have an energy pole in the *upper and lower complex energy plane*, respectively,

$$\begin{aligned} \vartheta(\pm t) &= \pm \int \frac{dq}{2i\pi} \frac{1}{q \mp io} e^{iqt} = \left\{ \begin{array}{ll} 1, & \pm t > 0, \\ 0, & \pm t < 0, \end{array}\right. \\ \vartheta(t) + \vartheta(-t) &= 1, \quad \vartheta(t) - \vartheta(-t) = \epsilon(t) = \tfrac{t}{|t|}. \end{aligned}$$

Summing the time representations on a nontrivial $\mathbb{I}(2)$-orbit gives the *Breit–Wigner states*, which are used for unstable particles. They are both the sum and the convolution product of its future and past contributions:

$$\begin{aligned} \Gamma > 0: \quad \mathbb{R} \ni t \longmapsto \int \frac{dq}{2i\pi} \left(\frac{1}{q-E-i\Gamma} - \frac{1}{q-E+i\Gamma} \right) e^{iqt} &= \int \frac{dq}{\pi} \frac{\Gamma}{(q-E)^2+\Gamma^2} e^{iqt} \\ &= \vartheta(t) e^{i(E+i\Gamma)t} + \vartheta(-t) e^{i(E-i\Gamma)t} \\ &= [\vartheta(t)\sqrt{2\Gamma} e^{i(E+i\Gamma)t}] * [\vartheta(-t)\sqrt{2\Gamma} e^{i(E-i\Gamma)t}] = e^{iEt-\Gamma|t|}. \end{aligned}$$

For a nontrivial width, the energy-distribution supporting pair $E \pm i\Gamma$ with the representation characterizing two reals (E, Γ) are eigenvalues of irreducible time translations, however not of unitary ones.

The Breit–Wigner states are Fourier-transformed positive energy densities. The trivial width limit gives the Dirac distribution for stable states,

$$\begin{aligned} \Gamma \geq 0: \ |E, \Gamma\}(t) &= e^{iEt-\Gamma|t|} = \int dq\ \delta_\Gamma(q - E) e^{iqt}, \\ \delta_\Gamma(q - E) &= \frac{1}{\pi} \frac{\Gamma}{(q-E)^2+\Gamma^2} = \delta(q - E) * \frac{2\Gamma}{q^2+\Gamma^2}, \\ \delta_0(q - E) &= \delta(q - E). \end{aligned}$$

Breit–Wigner states are products of spherical and hyperbolic positive-type functions. In the language of residues in the complex energy plane, time translation representations without reflection are obtained in the distributive limit of vanishing width $0 < \Gamma = o \to 0$, i.e., in the Breit–Wigner approximation of the Dirac distribution,

$$\begin{aligned}\Gamma = 0: \quad \mathbb{R} \ni t \longmapsto \oint \tfrac{dq}{2i\pi} \tfrac{1}{q-E} e^{iqt} &= \int dq\, \delta(q-E) e^{iqt} = e^{iEt},\\ \tfrac{1}{2i\pi}\left(\tfrac{1}{q-E-io} - \tfrac{1}{q-E+io}\right) &= \delta(q-E).\end{aligned}$$

The Breit–Wigner energy densities characterize the product of infinite-dimensional Hilbert spaces:

$$\mathbb{R} \ni t \longmapsto e^{iEt-\Gamma|t|}: \begin{cases} \text{distributive Hilbert space basis: } \{|E,\Gamma;q\rangle \bigm| q \in \check{\mathbb{R}}\},\\ \text{translation action: } D(t)|E,\Gamma;q\rangle = e^{iqt}|E,\Gamma;q\rangle,\\ \text{time reflection: } \alpha|E,\Gamma;q\rangle \leftrightarrow \langle E,\Gamma;q|\overline{\alpha}, \quad \alpha \in \mathbb{C},\\ \text{Hilbert product: } \langle E,\Gamma;q'|E,\Gamma;q\rangle = \frac{2\Gamma}{(q-E)^2+\Gamma^2}\delta(\frac{q-q'}{2\pi}).\end{cases}$$

The time reflection $-1 \ni \mathbb{I}(2)$ acts antilinearly with the exchange of bra $\langle \ldots |$ for *in-going or past* and ket $| \ldots \rangle$ for *out-going or future.*

The proper Hilbert vectors use energy functions, square-integrable with the Breit–Wigner functions,

$$|E,\Gamma;\tilde{f}\rangle = \int \tfrac{dq}{2\pi} \tilde{f}(q)|E,\Gamma;q\rangle, \quad \langle E,\Gamma;\tilde{f}'|E,\Gamma;\tilde{f}\rangle = \int \tfrac{dq}{2\pi}\, \overline{\tilde{f}'(q)}\, \tfrac{2\Gamma}{(q-E)^2+\Gamma^2}\, \tilde{f}(q).$$

The sum over the distributive basis, i.e., with $\tilde{f}(q) = 1$, gives a cyclic vector:

$$|E,\Gamma;1\rangle = \int \tfrac{dq}{2\pi}|E,\Gamma;q\rangle, \quad \langle E,\Gamma;1|t \bullet |E,\Gamma;1\rangle = e^{iEt-\Gamma|t|}.$$

Two Breit–Wigner states are not Schur-orthogonal for a nontrivial width sum $\Gamma + \Gamma' > 0$. The Schur norm is the lifetime,

$$\text{Schur product:} \begin{cases} \{E',\Gamma'|E,\Gamma\} &= \int dt\; e^{i(E-E')t-(\Gamma+\Gamma')|t|}\\ &= \frac{2(\Gamma+\Gamma')}{(E-E')^2+(\Gamma+\Gamma')^2} = 2\pi\delta_{\Gamma+\Gamma'}(E-E'),\\ \{E,\Gamma|E,\Gamma\} &= \frac{1}{\Gamma}.\end{cases}$$

8.6.5 Gaussian Functions

With the Fourier transformation–effected isomorphy of the square-integrable functions of translations and its dual momentum group,

$$L^2(\mathbb{R}) \ni f \leftrightarrow \tilde{f} \in L^2(\check{\mathbb{R}}),$$

there exists a *Fourier self-dual representation,* where the scalar product–inducing function $d_1 \in L^2(\mathbb{R})$ and its position-momentum expectations coincide, in appropriate normalization, with *their Fourier transforms*. This determines the *Gaussian positive-type function*:

$$\left.\begin{aligned} d_1 &= \tilde{d}_1 : \\ d_1(x) &= \int \frac{dq}{\sqrt{2\pi}}\, d_1(q)\, e^{iqx}, \\ xd_1(x) &= \int \frac{dq}{\sqrt{2\pi}}\, (-iq) d_1(q)\, e^{iqx} \\ \Rightarrow \frac{d}{dq} d_1(q) &= -q d_1(q), \\ d_1(0) &= 1 \end{aligned}\right\} \Rightarrow \begin{aligned} d_1(q) &= e^{-\frac{q^2}{2}}, \\ \text{or } e^{-\pi x^2} &= \int dq\; e^{-\pi q^2}\, e^{2i\pi q x}. \end{aligned}$$

The Gaussian positive-type function is used for the position representation coefficients (Schrödinger functions) of the harmonic oscillator,

$$\mathbb{R} \ni x \longmapsto d_1(x) = e^{-\frac{x^2}{2}}.$$

Its Schur normalization is

$$\{d_1 | d_1\} = \int dx\; e^{-x^2} = \sqrt{\pi}.$$

The cyclic Hilbert representations for Fourier self-dual representations of Euclidean translations $\mathbb{R}^s \cong \mathbf{SO}(s) \vec{\times} \mathbb{R}^s/\mathbf{SO}(s)$ use products of the basic Gaussian positive-type function:

$$\mathbb{R}^s \ni \vec{x} \longmapsto d_s(\vec{x}) = e^{-\frac{\vec{x}^2}{2}} = d_1(x_1) \cdots d_1(x_s) = \int \frac{d^s q}{\sqrt{(2\pi)^s}}\, d_s(\vec{q})\, e^{i\vec{q}\vec{x}}.$$

The additional rotation operations $\mathbf{SO}(s)$ are Fourier self-dually represented with the homogeneous position translation polynomials,

$$\begin{aligned} s \geq 2 : \mathbb{R}^s \ni \vec{x} \longmapsto \quad & \vec{x}\, e^{-\frac{r^2}{2}} = \int \frac{d^s q}{(2\pi)^{\frac{s}{2}}}\, (-i\vec{q})\, e^{-\frac{\vec{q}^2}{2}}\, e^{i\vec{q}\vec{x}}, \\ \vec{x} \longmapsto \quad & P^k(\vec{x})\, e^{-\frac{r^2}{2}} = \int \frac{d^s q}{(2\pi)^{\frac{s}{2}}}\, P^k(-i\vec{q})\, e^{-\frac{\vec{q}^2}{2}}\, e^{i\vec{q}\vec{x}} \in V \cong \bigvee^k \mathbb{C}^s, \\ \text{with} \quad & P^k(\vec{x}) = x^{a_1} \cdots x^{a_k} \in \bigvee^k \mathbb{R}^s. \end{aligned}$$

The irreducible $\mathbf{SU}(s)$-representations $[k, 0, \ldots, 0]$ ($s - 2$ zeros) act on the totally symmetric products $P^k(\vec{x})$ of the defining representation space $[1, 0, \ldots, 0]$ with complex dimension $\binom{k+s-1}{k}$.

The $\mathbf{SU}(s)$-representations are decomposable into $\mathbf{SO}(s)$-representations (see Chapter 4 for $s = 3$). Their representation coefficients contain the *harmonic* $\mathbf{SO}(s)$*-polynomials*, which are generated by vectors of the defining $\mathbf{SO}(s)$-representation on $\mathbb{R}^s$ (next section). By complexification, one has a representation on a complex vector space $\mathbb{C}^s$ where the Cartan subgroup operations are diagonalizable as familiar from the rotations $\mathbf{SO}(3)$ with the $\mathbf{SO}(2)$ eigenvectors ($x^1 \pm ix^2 = re^{\pm i\varphi} \sin\theta, x^3 = r\cos\theta$).

8.7 Harmonic Representations of Orthogonal Groups

Product representations of the orthogonal group $\mathbf{SO}_0(t,s)$ act on the totally symmetric tensor products $\bigvee^L V$ of a vector space, e.g., with (energy-)momenta $q \in V \cong \mathbb{R}^n$, $n = t + s = 2, 3, \dots$, and its complexification $\mathbb{C} \otimes V \cong \mathbb{C}^n$:

$$\dim_{\mathbb{R}} \bigvee^L V = \binom{n+L-1}{L} = \frac{\Gamma(n+L)}{\Gamma(1+L)\Gamma(n)}.$$

$\bigvee^L V$ and its complexification $\mathbb{C} \otimes \bigvee^L V$ are irreducible, e.g., for the action of $\mathbf{SL}(n,\mathbb{R})$ and $\mathbf{SU}(n)$, respectively, and in general, with irreducible representation $[L, 0, \dots, 0]$ ($n-1$ entries), for the Lie algebra A_{n-1} and its real forms.

In general, $\bigvee^L V$ is decomposable for the action of orthogonal groups. The irreducible subspaces are acted on by the *harmonic* $\mathbf{SO}_0(t,s)$*-representations*, denoted by $(\mathbf{SO}_0(t,s))^L$. They have the *harmonic polynomials* (totally symmetric, homogeneous, and "traceless") as bases $(q)^L$,

$$(q)^L = \{(q)^L_{a_1 \dots a_L} \mid a_k = 1, \dots, n\},\quad L = 0, 1, \dots, \qquad \begin{cases} (q)^0 &= 1, \\ (q)^1 &= q = \{q_a \mid a = 1, \dots, n\}, \\ (q)^2 &= (q \vee q) = \{q_a q_b - \frac{\eta_{ab}}{n} q^2\}, \\ (q)^3 &= \{q_a q_b q_c - \frac{\eta_{ab} q_c + \eta_{ac} q_b + \eta_{bc} q_a}{n} q^2\}, \\ \dots . \end{cases}$$

Since "traceless" harmonic polynomials are translation-invariant, i.e., they have a trivial action of the Laplacian,

$$\partial^2 = \eta_{ab} \partial^a \partial^b, \quad \partial^2 (q)^L = \{0\}.$$

Because of the quadratic metrical invariant $q^2 = \eta^{ab} q_a q_b$, the decomposition of the totally symmetric vector space powers into orthogonally irreducible subspaces has to distinguish between even and odd powers L — for the dimensions, independent of the signature:

$$t + s = n: \quad \binom{n+L-1}{L} = \begin{cases} \sum\limits_{k=0,2,\dots,L} \dim_{\mathbb{R}} (\mathbf{SO}_0(t,s))^k, & L = 0, 2, \dots, \\ \sum\limits_{k=1,3,\dots,L} \dim_{\mathbb{R}} (\mathbf{SO}_0(t,s))^k, & L = 1, 3, \dots . \end{cases}$$

This gives the dimensions of the harmonic representations,

$$\dim_{\mathbb{R}} (\mathbf{SO}_0(t,s))^L = \binom{n+L-1}{L} - \binom{n+L-3}{L-2} = \binom{n-2+L}{n-2} \frac{n-2+2L}{n-2+L},$$

$t+s=n$	2	3	4	5
$\dim_{\mathbb{R}} (\mathbf{SO}_0(t,s))^L$	2	$1+2L$	$(1+L)^2$	$\frac{(1+L)(2+L)(3+2L)}{6}$

For example, the rotation group $\mathbf{SO}(3)$ leads to irreducible $(1+2L)$-dimensional representations $(\mathbf{SO}(3))^L = [L]$, the harmonic Lorentz group polynomials to irreducible $(1+L)^2$-dimensional Minkowski representations $(\mathbf{SO}_0(1,3))^L = [\frac{L}{2}|\frac{L}{2}]$, and its compact partner to the irreducible $(1+L)^2$-dimensional Kepler representations $(\mathbf{SO}(4))^L = (\frac{L}{2},\frac{L}{2})$.

Each polynomial is a linear combination of powers of the metrical invariant $(q^2)^L \in I(V)$ with harmonic polynomials $h_L \in H(V)$:

$$\begin{aligned}
\mathbb{R}[q_1,\ldots,q_n] &= \bigvee V = \bigoplus_{L\geq 0} \bigvee^{L} V = H(V)I(V),\\
P(q) &= \sum_{L=0}^{k} h_L(q)(q^2)^L, \text{e.g., } 3q_aq_b = 3(q_aq_b - \tfrac{\eta_{ab}}{n}q^2) + 3\tfrac{\eta_{ab}}{n}q^2.
\end{aligned}$$

For compact groups, the $\mathbf{SO}(s)$-representations are Hilbert. There, the *spherical harmonics* are defined by $(\frac{\vec{q}}{\sqrt{\vec{q}^2}})^L = (\omega_{s-1}^{\to})^L$.

The vector space V to be considered may be a Lie algebra $x \in L \cong \mathbb{R}^n$ or its linear forms $q \in L^T$, with, respectively, the adjoint and coadjoint action of the corresponding Lie group $\operatorname{Ad} G \subseteq \mathbf{GL}(n,\mathbb{R})$, $\operatorname{Ad} G \cong G/\operatorname{centr} G$. For a semisimple Lie algebra, the adjoint group is a subgroup of the Killing form invariance group $\mathbf{SO}_0(t,s)$. The harmonic polynomial $\mathbf{SO}_0(t,s)$-representations are decomposable into $\operatorname{Ad} G$-representations.

The *harmonic* $\mathbf{O}(t,s)$*-units* (*projectors*), related to the corresponding harmonic polynomials, arise by their harmonic derivation, i.e., derivations and polynomials are dual to each other,

$$(\mathbf{1}_n)^L = \frac{(\frac{\partial}{\partial q})^L \otimes (q)^L}{\Gamma(1+L)} : \quad \begin{cases} (\mathbf{1}_n)^0 &= 1,\\ (\mathbf{1}_n)^1 &= \mathbf{1}_n \cong \delta^a_d,\\ (\mathbf{1}_n)^2 &\cong \frac{\delta^a_d\delta^b_e + \delta^a_e\delta^b_d}{2} - \frac{\eta^{ab}\eta_{de}}{n},\\ \ldots . \end{cases}$$

8.8 Hilbert Metrics for Flat Manifolds

The representations of flat manifolds with translations describe interaction-free structures, asymptotic scattering for Euclidean position, and free particles for Minkowski spacetime.

Cyclic Hilbert representations of translations arise in irreducible Hilbert representations of affine groups $\mathbf{SO}_0(t,s) \vec{\times} \mathbb{R}^n$. The induced representations collect, via direct integrals over the orbits of a compact translation fix-group in the homogeneous group, e.g., over spheres $\Omega^s \cong \mathbf{SO}(1+s)/\mathbf{SO}(s)$ or hyperboloids $\mathcal{Y}^s \cong \mathbf{SO}_0(1,s)/\mathbf{SO}(s)$, the translation representations with Dirac distributions of the (energy-)momentum invariants. This will be given for Euclidean groups $\mathbf{SO}(1+s) \vec{\times} \mathbb{R}^{1+s}$, isomorphic to Galilei groups as contractions of Lorentz groups $\mathbf{SO}_0(1,1+s)$, and for Poincaré groups $\mathbf{SO}_0(1,s) \vec{\times} \mathbb{R}^{1+s}$.

8.8.1 Euclidean Position for Nonrelativistic Scattering

The position translation representations in the free scattering structures have a momentum $P^2 > 0$ as translation invariant for the Euclidean group $\mathbf{SO}(3) \vec{\times} \mathbb{R}^3$. The spherical Bessel functions are positive-type functions for an irreducible representation with $P = |P|$, normalized by $j_0(0) = 1$:

$$\mathbb{R}^3 \ni \vec{x} \longmapsto d^3(P\vec{x}) = j_0(Pr) = \frac{\sin Pr}{Pr} = \int \frac{d^3q}{2\pi P} \, \delta(\vec{q}^2 - P^2)e^{i\vec{q}\vec{x}}, d^3 \in \mathcal{C}_b(\mathbb{R}^3)_+$$

The $L^2(\Omega^2)$-isomorphic Hilbert spaces for scattering states are infinite-dimensional.

More general, the irreducible infinite-dimensional Hilbert representations of the Euclidean groups $\mathbf{SO}(1+s) \vec{\times} \mathbb{R}^{1+s}$, $s \geq 0$, are, for a nontrivial translation invariant and for $s \geq 1$, inducible with translation fixgroup $\mathbf{SO}(s)$. The scalar representation coefficients for the Euclidean spaces $\mathbf{SO}(1+s) \vec{\times} \mathbb{R}^{1+s}/\mathbf{SO}(1+s) \cong \mathbb{R}^{1+s}$,

$$P > 0: \quad \mathbb{R}^{1+s} \ni \vec{x} \longmapsto D^{1+s}(P^2r^2) \begin{aligned}[t] &= \int d^{1+s}q \; \delta(\vec{q}^2 - P^2)e^{-i\vec{q}\vec{x}} \\ &= \frac{P^{s-1}}{2} \int d^s\omega \; e^{-iP\vec{\omega}_s\vec{x}} \\ &= \frac{P^{s-1}}{2} \int d^s\omega \; \cos P\vec{\omega}_s\vec{x}, \end{aligned}$$

use the Fourier-transformed measure of the momentum direction sphere $\vec{\omega}_s = \frac{\vec{q}}{|\vec{q}|} \in \Omega^s \cong \mathbf{SO}(1+s)/\mathbf{SO}(s)$. The scalar representation coefficient can be normalized as a positive-type function for a cyclic translation representation where the momentum sphere has the invariant as the intrinsic unit:

$$\begin{aligned} \text{state: } d^{1+s}(P\vec{x}) &= \int \frac{2d^{1+s}q}{|\Omega^s|P^{s-1}} \, \delta(\vec{q}^2 - P^2)e^{-i\vec{q}\vec{x}} \\ &= \int \frac{d^s\omega}{|\Omega^s|} \, \cos P\vec{\omega}_s\vec{x}, \quad d^{1+s}(0) = 1, \end{aligned}$$

$$L^2(\Omega^s) \text{ with distributive basis } \{|P^2; \vec{\omega}_s\rangle \,\big|\, \vec{\omega}_s \in \Omega^s\},$$

$$\begin{cases} \langle P^2; \vec{\omega}_s{}' | P^2; \vec{\omega}_s\rangle &= |\Omega^s|\delta(\vec{\omega}_s - \vec{\omega}_s{}'), \\ D(\vec{x})|P^2; \vec{\omega}_s\rangle &= e^{iP\vec{\omega}_s\vec{x}}|P^2; \vec{\omega}_s\rangle, \\ \text{cyclic vector: } |P^2; 1\rangle &= {}^{\oplus}\!\!\int \frac{d^s\omega}{|\Omega^s|}|P^2; \vec{\omega}_s\rangle, \\ \langle P^2; 1|P^2; 1\rangle &= 1. \end{cases}$$

For $s \geq 2$, the scalar representation coefficients arise by a 2-sphere spread via derivations $-\frac{d}{d\frac{r^2}{4\pi}}$,

$$D^{1+s}(r^2) = -\frac{d}{d\frac{r^2}{4\pi}} D^{s-1}(r^2) = \frac{\pi \mathcal{J}_{\frac{s-1}{2}}(r)}{(\frac{r}{2\pi})^{\frac{s-1}{2}}},$$

and embed the self-dual $\mathbb{R}$-representation matrix element:

$$\mathbb{R} \ni x \longmapsto D^1(r^2) = \int dq \; \delta(q^2 - 1)e^{-iqx} = \cos r.$$

For odd-dimensional spaces with $\mathbf{SO}(1 + 2R)$, they involve half-integer-index (spherical) Bessel functions, whereas integer-index Bessel functions are used for even-dimensional spaces with $\mathbf{SO}(2R)$, both with rank R:

$$D^{1+s}(r^2) = \begin{cases} 2^{2R-1}\frac{\pi j_{R-1}(r)}{(\frac{r}{2\pi})^{R-1}} = \left(-\frac{d}{d\frac{r^2}{4\pi}}\right)^R \cos r, \\ \qquad \text{for } 1+s = 1+2R = 1,3,\dots, \\ \frac{\pi \mathcal{J}_{R-1}(r)}{(\frac{r}{2\pi})^{R-1}} = \left(-\frac{d}{d\frac{r^2}{4\pi}}\right)^{R-1} \pi\mathcal{J}_0(r), \\ \qquad \text{for } 1+s = 2R = 2,4,\dots. \end{cases}$$

The integrals sum over the embedded $\mathbb{R}$-representation coefficients:

$$\begin{aligned} D^2(r^2) &= \int_0^\pi d\chi \ \cos(r\cos\chi) = \pi\mathcal{J}_0(r), \\ D^3(r^2) &= \pi\int_{-1}^1 d\zeta \ \cos(r\zeta) = 2\pi j_0(r) = 2\pi\frac{\sin r}{r} = -\frac{d}{d\frac{r^2}{4\pi}}\cos r. \end{aligned}$$

The 2-sphere spread from one to three dimensions is illustrated by the scalar integral formula with the spherical Bessel function j_0:

$$\int d^3q \ f(\vec{q}^2)e^{i\vec{q}\vec{x}} = -\frac{2\pi}{r}\frac{\partial}{\partial r}\int dq \ f(q^2) \ e^{iqr}|_{r=|\vec{x}|} = 4\pi\int_0^\infty q^2dq \ f(q^2) \ j_0(qr).$$

$\mathbf{SO}(1+s)$-nontrivial degrees of freedom $\mathbb{R}^{1+s} \cong \mathbb{R}_+ \times \Omega^s$ use derivations,

$$(\vec{q})^L \sim (i\frac{\partial}{\partial\vec{x}})^L, \quad L = 0,1,\dots, \ \frac{\partial}{\partial\vec{x}} = \frac{\vec{x}}{r}\frac{\partial}{\partial r} = 2\vec{x}\frac{\partial}{\partial r^2}.$$

The related coefficients,

$$\mathbb{R}^s \ni \vec{x} \longmapsto |P^2, L\}(\vec{x}) = \int \frac{2d^{1+s}q}{|\Omega^s|P^{s-1}} \ (\tfrac{\vec{q}}{P})^L\delta(\vec{q}^2 - P^2)e^{i\vec{q}\vec{x}},$$

contain products for the rotations $\mathbf{SO}(1+s)$ (harmonic polynomials) and translations $\mathbb{R}^{1+s}$. There is Schur orthogonality for different translation and rotation invariants. The Schur product involves as Plancherel distribution the inverse of the translation measure $\frac{P^s dP}{(2\pi)^{1+s}}|\Omega^s|$:

$$\begin{aligned} \{P'^2, L'|P^2, L\} &= (\tfrac{2\pi}{P})^s\delta(\tfrac{P-P'}{\pi})\int \frac{4d^{1+s}q}{|\Omega^s|^2} \ (\vec{q})^L \otimes (\vec{q})^{L'}\delta(\vec{q}^2 - 1) \\ &= \frac{2}{|\Omega^s|}(\tfrac{2\pi}{P})^s\delta(\tfrac{P-P'}{\pi})\delta^{LL'}\frac{\Gamma(1+L)}{2^L}\frac{\Gamma(\frac{1+s}{2})}{\Gamma(\frac{1+s}{2}+L)}(\mathbf{1}_{1+s})^L. \end{aligned}$$

and the multiplicity factors for the harmonic polynomials and units (see Chapter 10):

$$\int \frac{2d^{1+s}q}{|\Omega^s|} \ (\vec{q})^L \otimes (\vec{q})^{L'}\delta(\vec{q}^2 - 1) = \delta^{LL'}\frac{\Gamma(1+L)}{2^L}\frac{\Gamma(\frac{1+s}{2})}{\Gamma(\frac{1+s}{2}+L)}(\mathbf{1}_{1+s})^L.$$

The harmonic analysis for representations of the Euclidean group is familiar from the nonrelativistic scattering $\mathbf{SO}(3) \ \vec{\times} \ \mathbb{R}^3$ with the angular momentum decomposition of a plane wave with $z = r\cos\theta$ and translation-invariant momentum P, involving Legendre polynomials, i.e., spherical harmonics, and spherical Bessel functions j_L:

$$e^{iPz} = \sum_{L=0}^\infty (1+2L)i^L \mathrm{P}^L(\cos\theta)j_L(Pr), \quad \begin{cases} j_L(r) = \int_{-1}^1 \frac{d\zeta}{2i^L} \ \mathrm{P}^L(\zeta)e^{iPr\zeta} \\ \qquad = r^L(-\frac{1}{r}\frac{d}{dr})^L\frac{\sin r}{r}, \\ \mathrm{P}^L(\cos\theta) = \sqrt{\frac{4\pi}{1+2L}}\mathrm{Y}_0^L(0,\theta). \end{cases}$$

8.8.2 Minkowski Spacetime for Relativistic Particles

The cyclic spacetime translation representations for free particles with mass $m^2 \geq 0$ as translation invariant of the Poincaré group $\mathbf{SO}_0(1,3) \vec{\times} \mathbb{R}^4$ with the positive on-shell distribution of the Feynman propagator are similar,

$$\mathbb{R}^4 \ni x \longmapsto d^{(1,3)}(mx) = \int \frac{d^4q}{\mathcal{V}^3 m^2} \, \delta(q^2 - m^2) e^{iqx},$$

normalized with a constant $\mathcal{V}^3 > 0$, e.g., $\mathcal{V}^3 m^2 = (2\pi)^3$. The $L^2(\mathcal{Y}^3)$-isomorphic Hilbert spaces for free particles are infinite-dimensional.

The irreducible Hilbert representations of the Poincaré group $\mathbf{SO}_0(1,s) \vec{\times} \mathbb{R}^{1+s}$, $s \geq 1$, for positive translation invariant, e.g., massive particle representations for $m > 0$, are inducible with translation fixgroup $\mathbf{SO}(s)$. They come as the Fourier-transformed measure of the energy-momentum hyperboloid with the directions $\frac{q}{|q|} = \pm \mathrm{y}_s \in \mathcal{Y}^s_\pm \cong \mathbf{SO}_0(1,s)/\mathbf{SO}(s)$ as eigenvalues. The scalar representation coefficients for the Minkowski translations read, with $|x| = \sqrt{|x^2|}$,

$$\begin{aligned} m^2 > 0: \ \mathbb{R}^{1+s} \ni x \longmapsto D^{(1,s)}(m^2x^2) &= \int d^{1+s}q \, \delta(q^2 - m^2) e^{iqx} \\ &= m^{s-1} \int d^s\mathrm{y} \, \cos m\mathrm{y}_s x, \\ D^{(1,s)}(x^2) = \frac{d}{d\frac{x^2}{4\pi}} D^{(1,s-2)}(x^2) &= \frac{-\vartheta(x^2)\pi\mathcal{N}_{-\frac{s-1}{2}}(|x|) + \vartheta(-x^2)2\mathcal{K}_{\frac{s-1}{2}}(|x|)}{|\frac{x}{2\pi}|^{\frac{s-1}{2}}}. \end{aligned}$$

The hyperbolic invariant m is used as intrinsic unit in the representation coefficients:

$$\begin{aligned} &d^{(1,s)}(mx) = \int \frac{d^{1+s}q}{\mathcal{V}^s m^{s-1}} \, \delta(q^2 - m^2) e^{iqx} = \int \frac{d^s\mathrm{y}}{\mathcal{V}^s} \, \cos m\mathrm{y}_s x. \\ &L^2(\mathcal{Y}^s) \text{ with distributive basis } \{|m^2;\mathrm{y}_s\rangle \mid \mathrm{y}_s \in \mathcal{Y}^s\} \\ &\qquad \begin{cases} \langle m^2;\mathrm{y}'_s|m^2;\mathrm{y}_s\rangle &= \mathcal{V}^s\delta(\mathrm{y}_s - \mathrm{y}'_s), \\ D(x)|m^2;\mathrm{y}_s\rangle &= e^{im\mathrm{y}_s x}|m^2;\mathrm{y}_s\rangle. \end{cases} \end{aligned}$$

Since the hyperboloid volume is infinite, the positive-type function is not in $L^\infty(\mathbb{R}^{1+s})$:

$$d^{(1,s)}(0) = \int \frac{d^{1+s}q}{\mathcal{V}^s} \, \delta(q^2 - 1) = \int \frac{d^s\mathrm{y}}{\mathcal{V}^s} = \frac{|\mathcal{Y}^s|}{\mathcal{V}^s}.$$

A formal analogy to the Euclidean case above with the sphere volume is obtained with an "infinite normalization" $\mathcal{V}^s = |\mathcal{Y}^s| = \infty$.

In particle physics, the momentum parametrization (geodesic polar coordinates; see Chapter 2) with the manifold isomorphy $\mathcal{Y}^s \cong \mathbb{R}^s$ is more familiar than the hyperbolic parametrization with the rapidity $\psi = \operatorname{arsinh} \frac{|\vec{q}|}{m}$:

$$\begin{aligned} &\mathrm{y}_s = \frac{1}{m}\begin{pmatrix} \sqrt{m^2 + \vec{q}^2} \\ \vec{q} \end{pmatrix} = \begin{pmatrix} \cosh\psi \\ \sinh\psi \, \vec{\omega}_{s-1} \end{pmatrix}, \quad \begin{cases} d\mathrm{y}_s^2 &= \frac{d|\vec{q}|^2}{m^2+\vec{q}^2} + \frac{\vec{q}^2}{m^2} d\omega_{s-1}^2, \\ d^s\mathrm{y} &= \frac{d^s\vec{q}}{m^s\sqrt{1+\frac{\vec{q}^2}{m^2}}}, \end{cases} \\ &\{|m^2;\vec{q}\rangle \mid \vec{q} \in \mathbb{R}^s\} \text{ with } \langle m^2;\vec{q}'|m^2;\vec{q}\rangle = m^s\sqrt{1 + \tfrac{\vec{q}^2}{m^2}}\mathcal{V}^s\delta(\vec{q} - \vec{q}'). \end{aligned}$$

The "cyclic vector" e.g., for a free scalar field (see Chapter 5),

$$\begin{aligned} m^2 {}^{\oplus}\!\!\int \frac{d^s y}{2}\ (|m^2; y_s\rangle \oplus \langle m^2; y_s|) &= {}^{\oplus}\!\!\int \frac{d^3 q}{2q_0(2\pi)^3}\ (|m^2; \vec{q}\rangle \oplus \langle m^2; \vec{q}\,|) \\ &= \mathbf{\Phi}(0)(|0\rangle \oplus \langle 0|), \end{aligned}$$

is not normalizable and, therefore, not in the Hilbert space. There exist cyclic vectors as limits of normalizable functions on the momentum hyperboloid $\mathcal{Y}^s$.

The representation coefficients embed time and 1-position representations

$$\begin{aligned} \mathbb{R} \ni t &\longmapsto \int dq\ \delta(q^2 - 1) e^{iqt} &= \cos t, \\ \mathbb{R} \ni z &\longmapsto \int \frac{dq}{\pi}\ \frac{1}{q^2+1} e^{-iqz} &= e^{-|z|}, \end{aligned}$$

for $s \geq 2$ as 2-sphere spreads. For odd dimension and $\mathbf{SO}_0(1, 2R)$, they involve half-integer-index functions, hyperbolic Macdonald functions, and spherical Bessel functions:

$$\begin{aligned} D^{(1,2)}(x^2) &= 2\pi \frac{-\vartheta(x^2) \sin|x| + \vartheta(-x^2) e^{-|x|}}{|x|}, \\ D^{(1,2R)}(x^2) &= \left(\frac{\partial}{\partial \frac{x^2}{4\pi}}\right)^R [\vartheta(x^2) \cos|x| + \vartheta(-x^2) e^{-|x|}] \\ &= \frac{\vartheta(x^2)(-1)^R \pi \mathcal{J}_{R-\frac{1}{2}}(|x|) + \vartheta(-x^2) 2\mathcal{K}_{R-\frac{1}{2}}(|x|)}{|\frac{x}{2\pi}|^{R-\frac{1}{2}}} \text{ for} \\ & \quad 1 + s = 1 + 2R = 3, 5, \ldots. \end{aligned}$$

For even spacetime dimension and $\mathbf{SO}_0(1, 2R-1)$ they start with the rank-$R = 1$ Poincaré group by integrating $\mathbb{R}$-representation coefficients on a hyperbola, leading to timelike oscillations and a spacelike hyperbolic fall-off:

$$\begin{aligned} D^{(1,1)}(x^2) &= \int d\psi\ [\vartheta(x^2) \cos(|x| \cosh\psi) + \vartheta(-x^2) e^{-|x| \cosh\psi}] \\ &= -\vartheta(x^2) \pi \mathcal{N}_0(|x|) + \vartheta(-x^2) 2\mathcal{K}_0(|x|). \end{aligned}$$

A 2-sphere spread gives the Hilbert representation of the rank-2 Poincaré group with Minkowski translations and, in general, the integer-index functions:

$$\begin{aligned} D^{(1,2R-1)}(x^2) &= \left(\frac{\partial}{\partial \frac{x^2}{4\pi}}\right)^{R-1} [-\vartheta(x^2) \pi \mathcal{N}_0(|x|) + \vartheta(-x^2) 2\mathcal{K}_0(|x|)] \\ &= \frac{\vartheta(x^2)(-1)^R \pi \mathcal{N}_{R-1}(|x|) + \vartheta(-x^2) 2\mathcal{K}_{R-1}(|x|)}{|\frac{x}{2\pi}|^{R-1}} \text{ for} \\ & \quad 1 + s = 2R = 2, 4, \ldots. \end{aligned}$$

$\mathbf{SO}_0(1, s)$-nontrivial coefficients use derivations

$$(q)^L \sim (-i\frac{\partial}{\partial x})^L, \quad L = 0, 1, \ldots, \quad \frac{\partial}{\partial x} = 2x \frac{\partial}{\partial x^2}$$

and the products for the two factors — Lorentz group with harmonic poynomials and translation group,

$$\mathbb{R}^{1+s} \ni x \longmapsto |m^2, L\}(x) = \int \frac{d^{1+s} q}{\mathcal{V}^s m^{s-1}}\ (\frac{q}{m})^L \delta(q^2 - m^2) e^{iqx}.$$

For different translation and Lorentz group invariants, the representation coefficients are Schur-orthogonal with a Plancherel distribution containing the inverse of the measure $\frac{m^s dm}{(2\pi)^{1+s}}$ for the translation invariants,

$$\{m'^2, L'|m^2, L\} = \frac{|\mathcal{Y}^s|}{(\mathcal{V}^s)^2}(\tfrac{2\pi}{m})^s \delta(\tfrac{m-m'}{\pi})\delta^{LL'}\frac{\Gamma(1+L)}{2^L}\frac{\Gamma(\frac{1+s}{2})}{\Gamma(\frac{1+s}{2}+L)}(\mathbf{1}_{1+s})^L,$$

and the multiplicity factors for the harmonic polynomials and units,

$$\int d^{1+s}q\ (q)^L \otimes (q)^{L'}\delta(q^2-1) = |\mathcal{Y}^s|\delta^{LL'}\frac{\Gamma(1+L)}{2^L}\frac{\Gamma(\frac{1+s}{2})}{\Gamma(\frac{1+s}{2}+L)}(\mathbf{1}_{1+s})^L.$$

The on-shell contributions of Feynman propagators for particle fields contain, in general, linear combinations of harmonic $\mathbf{SO}(1,3)$ energy-momentum polynomials (see Chapter 5).

8.9 Parabolic Subgroups

For a noncompact semisimple group G with known representations of its maximal compact group K, there remains the representation of the compact group classes G/K. G-representations are inducible from those of parabolic subgroups.

With an Iwasawa decomposition $G = K \circ A \circ N = \check{N} \circ A \circ K$, the K-classes can be parametrized by the triagonal group $A \circ N \cong G/K$ with maximal noncompact abelian A and nilpotent Lie algebra $\log N$ or by $\check{N} \circ A$ with the negative transposed $\log \check{N}$. A *minimal parabolic subgroup* $P = [K_0 \times A] \circ N$ extends the Cartan plane A by its centralizer $Z_G(A) = K_0 \subseteq K$ in the compact group. One can visualize, in appropriate bases, parabolic subgroups as block-triagonal,

$$\begin{aligned}
G &= K \circ A \circ N = \check{N} \circ [K_0 \times A] \circ N, \quad \log \check{N} = -(\log N)^T,\\
[K_0 \times A] \circ N &= \begin{pmatrix} \times & \times & n & n & n & n\\ \times & \times & n & n & n & n\\ 0 & 0 & \times & \times & \times & n\\ 0 & 0 & \times & \times & \times & n\\ 0 & 0 & \times & \times & \times & n\\ 0 & 0 & 0 & 0 & 0 & \times \end{pmatrix} \supseteq \begin{pmatrix} \times & n & n & n & n & n\\ 0 & \times & n & n & n & n\\ 0 & 0 & \times & n & n & n\\ 0 & 0 & 0 & \times & n & n\\ 0 & 0 & 0 & 0 & \times & n\\ 0 & 0 & 0 & 0 & 0 & \times \end{pmatrix}\\
&= A \circ N,\\
\dim_{\mathbb{R}} G &= \dim_{\mathbb{R}} K_0 + \dim_{\mathbb{R}} A + 2\dim_R N, \quad \dim_{\mathbb{R}} K\\
&= \dim_{\mathbb{R}} K_0 + \dim_R N,
\end{aligned}$$

with $K_0 \times A$ block-diagonal (entries $\times$) and N with diagonal $\mathbf{1}$ and entries n strictly above and to the right of the blocks, with the examples

$$\begin{aligned}
\mathbf{SO}_0(1,s) &\supset [\mathbf{SO}(s-1) \times \mathbf{SO}_0(1,1)] \circ \exp \mathbb{R}^{s-1}, \ s = 2,3,\dots,\\
\mathbf{SL}(2,\mathbb{C}) &\supset [\mathbf{SO}(2) \times \mathbf{SO}_0(1,1)] \circ \exp \mathbb{R}^2 \ni \left(\begin{array}{c|c} e^{i\alpha_3+\beta_3} & n_1 + in_2\\ \hline 0 & e^{-(i\alpha_3+\beta_3)} \end{array}\right),\\
\mathbf{SL}_0(1+r,\mathbb{R}) &\supset \mathbf{SO}_0(1,1)^r \circ \exp \mathbb{R}^{\binom{1+r}{2}},\\
\mathbf{SL}(1+r,\mathbb{C}) &\supset [\mathbf{SO}(2) \times \mathbf{SO}_0(1,1)]^r \circ \exp \mathbb{R}^{r(1+r)}.
\end{aligned}$$

A *parabolic subgroup* of G is a closed subgroup $\mathcal{P} \supseteq P$ containing a minimal parabolic subgroup. It has the *Langlands decomposition*

$$\mathcal{P} = [\mathcal{K}_0 \times \mathcal{A}] \circ \mathcal{N}, \quad \log \mathcal{P} = \log \mathcal{K}_0 \oplus \log \mathcal{A} \oplus \log \mathcal{N} \text{ (as vector space)},$$

with the properties for the mutually orthogonal Lie subalgebras:

$$\begin{array}{rl} \text{noncompact abelian:} & \log \mathcal{A}, \ \text{nilpotent: } \log \mathcal{N}, \\ \log \mathcal{A}\text{-centralizer in } \log G: & \log \mathcal{K}_0 \oplus \log \mathcal{A} = \{l \in \log G \mid [l, \log \mathcal{A}] = \{0\}\}, \\ \mathcal{N}\text{-normalizing:} & [\log \mathcal{K}_0 \oplus \log \mathcal{A}, \log \mathcal{N}] \subseteq \log \mathcal{N}. \end{array}$$

The Lie algebra can be decomposed with upper and lower triagonal matrices into eigenspaces for the adjoint action of a commuting family of hermitian operators involving $\log \mathcal{A}$ with the roots $R_+ \cup (-R_+)$. The adjoint action decomposes the nilpotent $\log \mathcal{N}$ into an orthogonal sum of common eigenspaces of $\log \mathcal{A}$:

$$\begin{aligned} \log G &= \log \check{\mathcal{N}} \oplus \log \mathcal{K}_0 \oplus \log \mathcal{A} \oplus \log \mathcal{N}, \\ \log \mathcal{N} &= \bigoplus_{\omega \in R_+} \log \mathcal{N}_\omega, \quad \log \check{\mathcal{N}} = \bigoplus_{\omega \in R_+} \log \mathcal{N}_{-\omega}, \\ \log \mathcal{N}_\omega &= \{n \in \log \mathcal{N} \mid [a, n] = \omega(a) n \text{ for all } a \in \log \mathcal{A}\}. \end{aligned}$$

The nilpotent Lie algebra defines the *nilpotent root sum* ρ_N, which involves the multiplicities n_ω of the positive roots (eigenvalues collection) of $\log \mathcal{A}$ on $\log \mathcal{N}$:

$$\rho_N = \tfrac{1}{2} \sum_{\omega \in R_+} n_\omega \omega \in (\log \mathcal{A})^T.$$

The induction of representations of affine groups $H \vec{\times} \mathbb{R}^n$ (flat spaces) by representations of direct product subgroups $H_0 \times \mathbb{R}^n$ (see Chapter 7) is a good introduction to the parabolic induction for representations of a semisimple noncompact Lie group G (also for a reductive Lie group with suitable interpretations). In comparing $K \circ A \circ N$ with affine groups $H \vec{\times} \mathbb{R}^n$, the subgroups $\mathcal{K}_0 \times \mathcal{A}$ are the analogues to the direct product subgroups $H_0 \times \mathbb{R}^n$ with translation fixgroups H_0. The Hilbert representations of the parabolic subgroups with trivial $\mathcal{N}$-representation,

$$\mathcal{P} = [\mathcal{K}_0 \times \mathcal{A}] \circ \mathcal{N} \ni u e^a n \longmapsto e^{iqa} D(u) \ni \mathbf{U}(W),$$

induce Hilbert representations of the full group G acting by left translations on the $\mathcal{P}$-intertwiners,

$$\{w : G \longrightarrow W \mid w(k u e^a n) = e^{-(\rho_N + iq)a} D(u^{-1}).w(k)\}.$$

They involve a modular function $e^a \longmapsto e^{\rho_N a}$ for the non-unimodular measure of the parabolic subgroup.

8.9.1 Discrete and Continuous Invariants

For a semisimple Lie group, the irreducible *discrete series Hilbert representations* nontrivially and discretely support the Plancherel measure. They correspond to representations (characters) of a Cartan torus (determined up to conjugacy) and have square-integrable coefficients — therefore also called *square-integrable representations.* There are countably many equivalence classes of them. For a compact group, there are only discrete series representations, and there is only one Cartan torus class. For noncompact groups, there are finitely many classes of Cartan tori, e.g., $\mathbf{SO}(2)$ for $\mathbf{SL}(2,\mathbb{R})$, sometimes none, e.g., for $\mathbf{SL}(2,\mathbb{C})$. Coefficients of square-integrable representations D_ι display Schur orthogonality,

$$\begin{aligned} \{D_{\iota'}|D_\iota\} = \hat{D}_{\iota'} \circledast D_\iota(1) &= \textstyle\int_G dg\ \overline{D_{\iota'}(g)} \otimes D_\iota(g) = \delta_{\iota\iota'} \frac{1}{c_\iota} \mathbf{1}_{d_\iota^2}, \\ \textstyle\int_G dg\ \langle v_\iota|D_\iota(g)|w_\iota\rangle \overline{\langle v_{\iota'}|D_{\iota'}(g)|w_{\iota'}\rangle} &= \begin{cases} 0, & D_\iota \not\cong D_{\iota'} \\ \frac{1}{c_\iota}\langle v_\iota|v_{\iota'}\rangle\langle w_{\iota'}|w_\iota\rangle, & D_\iota = D_{\iota'}. \end{cases} \end{aligned}$$

A noncompact group has a *formal finite degree* c_ι for the representation D_ι, which, for a compact group, is the quotient of its dimension d_ι and the group volume, both finite,

$$c_\iota = \frac{d_\iota}{\int_G dg} \text{ for a compact group } G.$$

For a noncompact group, both volume and dimension are infinite.

Noncompact groups have Hilbert representations that continuously and nontrivially support the Plancherel measure or have trivial Plancherel measure (*nonamenable representations*). The continuously contributing ones in the reduced group dual can be induced from (conjugacy classes of) parabolic subgroups $[\mathcal{K}_0 \times \mathcal{A}] \circ \mathcal{N} \subset G$.

8.10 Eigenfunctions of Homogeneous Spaces (Spherical Functions)

Spherical functions, defined in the following, are positive-type, i.e., scalar product–inducing functions for a special class of irreducible G-representations (definition ahead). The representation normalized Legendre polynomials in the spherical harmonics $\sqrt{\frac{4\pi}{1+2L}}\mathrm{Y}_0^L(\theta,\varphi) = \mathrm{P}^L(\cos\theta)$ as the name-giving example are given ahead.

For a compact subgroup $K \subseteq G$ of a connected Lie group, the *spherical functions* [36] are defined as nontrivial *normalized K-invariant joint eigenfunctions* (continuous functions of homogenous spaces G/K with compact group classes),

$$\Phi : G/K \longrightarrow \mathbb{C}, \begin{cases} \Phi(K) = 1, & \text{representation-normalized,} \\ \Phi(kgK) = \Phi(gK), & K\text{-left invariance,} \\ \mathtt{D}\Phi = I(\mathtt{D})\Phi, & \text{eigenfunction for each } \mathtt{D} \in \mathtt{D}(G/K), \end{cases}$$

or with the class projection $d^I = \Phi \circ \pi_K$:

$$d^I : G \longrightarrow \mathbb{C}, \quad \begin{cases} d^I(1) = 1, & \text{representation-normalized,} \\ d^I(kgk') = d^I(g), & K\text{-bi-invariance,} \\ \mathtt{D}d^I = I(\mathtt{D})d^I, & \text{eigenfunction for each } \mathtt{D} \in \mathtt{D}_K(G). \end{cases}$$

In this definition, $\mathtt{D}(G)$ denotes the *left-invariant differential operators* on G, $\mathtt{D}_K(G)$ the subspace of those that are also right-K-invariant, and $\mathtt{D}(G/K)$ the algebra of the left-G-invariant differential operators on the homogeneous space G/K (see Chapter 10). For a semisimple Lie algebra, the invariants constitute the center of the enveloping algebra.

The *joint eigenfunctions* are from the complex vector space

$$V^I = \{f \in \mathcal{C}^\infty(G/K) \,\big|\, \mathtt{D}f = I(\mathtt{D})f \text{ for each } \mathtt{D} \in \mathtt{D}(G/K)\}$$

with characterizing representation invariants $I(\mathtt{D}) \in \mathbb{C}$. Each space of joint eigenfunctions contains exactly one spherical function d^I. The *harmonic functions* (harmonic polynomials for a compact group G) $f \in V^0$ with trivial invariants, i.e., $\mathtt{D}f = 0$, contain the spherical function $d^1 = 1$. A joint eigenfunction is characterized by the product property with its spherical function,

$$f, d^I \in V^I: \quad f(g_1K)d^I(g_2K) = \int_K dk\ f(g_1kg_2K).$$

With their class K-property and with the integral notation for the projection,

$$d^I : G \xrightarrow{\pi_K} G/K \xrightarrow{\Phi} \mathbb{C}, \quad g \longmapsto gK \longmapsto {}^{\oplus}\!\!\int_K dk\ d^I(gk) = \Phi(gK),$$

spherical functions can be written as G-representation coefficients integrated over the compact group $\int_K dk = 1$, as exemplified earlier.

Spherical functions are defined equivalently by their G/K*-coset-representation* property, with normalized Haar measure $\int_K dk = 1$,

$$d^I \in \mathcal{C}(G),\ d^I \neq 0: \quad g_{1,2} \in G: \quad d^I(g_1)d^I(g_2) = \int_K dk\ d^I(g_1kg_2),$$

which reflects the representation property $D(g_1K) \circ D(g_2K) = D(g_1Kg_2K)$.

Now the connection to positive-type functions: A Hilbert representation $D : G \longrightarrow \mathbf{U}(V)$ is called K*-spherical* if there exists a vector with compact fixgroup K. A spherical function on G/K defines a scalar product for an *irreducible spherical representation*, and vice versa: The diagonal elements of a normalized vector with fixgroup K (therefrom the bi-invariance) in an irreducible spherical Hilbert representation constitute a positive-type spherical function:

$$\begin{aligned} &d_v(g) = \langle v|D(g)|v\rangle, \text{ with } G_v = K, \ \ \langle v|v\rangle = 1, \\ &\text{bijection: spherical } L^\infty(G)_+ \cong \mathbf{irrep}_+ G \text{ spherical.} \end{aligned}$$

8.10.1 Simple Examples

The simplest spherical functions are the unitary exponentials for the one-dimensional abelian Lie groups with trivial compact subgroup $K = \{1\}$:

$$\mathbf{U}(1),\quad z \in \mathbb{Z},\quad d^z(0) = 1,$$

$d^z(\alpha) = e^{i\alpha z}$
$\frac{d}{di\alpha} e^{i\alpha z} = z e^{i\alpha z}$
$\int_{-\pi}^{\pi} \frac{d\alpha}{2\pi}\, e^{iz\alpha} e^{-iz'\alpha} = \delta_{zz'}$
$\sum_{z\in\mathbb{Z}} e^{iz\alpha} e^{-iz\alpha'} = \delta(\frac{\alpha-\alpha'}{2\pi})$

$$\mathbb{R} \cong \mathbf{D}(1),\quad iq \in i\mathbb{R},\quad d^{iq}(0) = 1,$$

$d^{iq}(x) = e^{iqx}$
$\frac{d}{dx} e^{iqx} = iq e^{iqx}$
$\int dx\, e^{iqx} e^{-iq'x} = \delta(\frac{q-q'}{2\pi})$
$\int \frac{dq}{2\pi}\, e^{iqx} e^{-iqx'} = \delta(x - x')$

Here and in the following examples, the left-invariant differential operator stands in the second line. In the third and fourth lines, the Schur orthogonality and the completeness are given, involving the Haar and Plancherel measures.

Rank-1 nonabelian examples with compact subgroup $K = \mathbf{SO}(2)$ are the Legendre polynomials (spherical harmonics, analysis on Ω^2) with the Casimir invariant $L(1+L)$ as spherical functions for compact symmetric spaces

$$\begin{aligned} \Omega^2 &\cong \mathbf{SO}(3)/\mathbf{SO}(2),\\ \mathrm{P}^L(\cos\theta) &= \sum_{k=0}^{L} \frac{(L+k)!}{(L-k)!} \frac{(-z)^k}{(k!)^2},\\ L = 0, 1, \ldots,\quad z &= \sin^2 \frac{\theta}{2},\\ \mathrm{P}^L(1) &= 1, \end{aligned}$$

$\mathrm{P}^L(\cos\theta) = \int_{-\pi}^{\pi} \frac{d\varphi}{2\pi} (\cos\theta + i\cos\varphi \sin\theta)^L$
$\xi = \cos\theta,\quad \frac{d}{d\xi}(1-\xi^2)\frac{d\mathrm{P}^L}{d\xi} = -L(1+L)\mathrm{P}^L$
$\int_{-1}^{1} \frac{d\xi}{2}\, \mathrm{P}^L(\xi)\, \mathrm{P}^{L'}(\xi) = \frac{1}{1+2L}\delta_{LL'}$
$\sum_{L=0}^{\infty} (1+2L)\mathrm{P}^L(\xi)\, \mathrm{P}^L(\xi') = \delta(\frac{\xi-\xi'}{2})$

and the *Legendre functions* with continuous invariants as bounded spherical functions for noncompact symmetric spaces as used for the harmonic analysis (Fock–Mehler transformation) of the hyperboloid $\mathcal{Y}^2$ (more later):

$$\begin{aligned} \mathcal{Y}^2 &\cong \mathbf{SO}_0(1,2)/\mathbf{SO}(2),\\ \mathrm{P}^{iQ-\frac{1}{2}}(\cosh\psi) &= \sum_{k=0}^{\infty} \frac{\Gamma(iQ+\frac{1}{2}+k)}{\Gamma(iQ+\frac{1}{2}-k)} \frac{(-z)^k}{(k!)^2},\\ Q \in \mathbb{R},\quad z &= -\sinh^2 \frac{\psi}{2},\\ \mathrm{P}^{iQ-\frac{1}{2}}(1) &= 1,\\ \Pi^2(Q^2) = \left|\frac{\Gamma(iQ+\frac{1}{2})}{\Gamma(iQ)}\right|^2 &= Q \tanh \pi Q, \end{aligned}$$

$\mathrm{P}^{iQ-\frac{1}{2}}(\cosh\psi) = \int_{-\pi}^{\pi} \frac{d\varphi}{2\pi} (\cosh\psi + \cos\varphi \sinh\psi)^{iQ-\frac{1}{2}}$
$\zeta = \cosh\psi,\quad \frac{d}{d\zeta}(\zeta^2-1)\frac{d\mathrm{P}^{iQ-\frac{1}{2}}}{d\zeta} = -(Q^2+\frac{1}{4})\mathrm{P}^{iQ-\frac{1}{2}}$
$\int_1^{\infty} d\zeta\, \mathrm{P}^{iQ-\frac{1}{2}}(\zeta)\, \mathrm{P}^{-iQ'-\frac{1}{2}}(\zeta) = \frac{1}{\Pi^2(Q^2)}\delta(Q-Q')$
$\int_0^{\infty} \Pi^2(Q^2) dQ\, \mathrm{P}^{iQ-\frac{1}{2}}(\zeta)\, \mathrm{P}^{-iQ-\frac{1}{2}}(\zeta') = \delta(\zeta-\zeta')$

$$\left.\begin{aligned} f(\zeta) &= \int_0^{\infty} \Pi^2(Q^2) dQ\, \mathrm{P}^{iQ-\frac{1}{2}}(\zeta)\tilde{f}(Q),\\ \tilde{f}(Q) &= \int_1^{\infty} d\zeta\, \mathrm{P}^{-iQ-\frac{1}{2}}(\zeta) f(\zeta), \end{aligned}\right\} \quad \int_1^{\infty} d\zeta\, |f(\zeta)|^2 = \int_0^{\infty} \Pi^2(Q^2) dQ\, |\tilde{f}(Q)|^2.$$

The spherical functions for the analysis of the Euclidean plane are the Bessel functions with one momentum invariant (considered already, more ahead):

$$\begin{aligned} \mathbb{R}^2 &\cong \mathbf{SO}(2) \,\vec{\times}\, \mathbb{R}^2/\mathbf{SO}(2),\\ \mathcal{J}_0(|P|r) &= \sum_{k=0}^{\infty} \frac{(Pz)^{2k}}{(k!)^2},\\ P^2 \geq 0,\quad z^2 &= -\frac{r^2}{4},\\ \mathcal{J}_0(0) &= 1, \end{aligned}$$

$\mathcal{J}_0(\lvert P\vec{x}\rvert) = \int_{-\pi}^{\pi} \frac{d\varphi}{2\pi}\, e^{i\lvert P\vec{x}\rvert\cos\varphi} = \int \frac{d^2q}{\pi} \delta(\vec{q}^2 - P^2) e^{i\vec{q}\vec{x}}$
$\vec{\partial}^2 \mathcal{J}_0(\lvert P\vec{x}\rvert) = -P^2 \mathcal{J}_0(\lvert P\vec{x}\rvert)$
$\int d^2x\, \mathcal{J}_0(\lvert P\vec{x}\rvert)\mathcal{J}_0(\lvert P'\vec{x}\rvert) = \frac{1}{\lvert P\rvert}\delta(\frac{\lvert P\rvert - \lvert P'\rvert}{2\pi})$
$\int_0^{\infty} dP^2\, \mathcal{J}_0(\lvert P\rvert r)\mathcal{J}_0(\lvert P\rvert r') = \frac{1}{r}\delta(r-r')$

8.10.2 Eigenfunctions of Compact Groups

All irreducible representations of a *compact group* U are spherical, i.e., characterized by a spherical function for a compact fixgroup K. The spherical functions on a compact homogeneous space U/K are precisely the character class functions of the irreducible spherical U-representations,

$$U \ni u \longmapsto d^I(u) = \operatorname{tr} D(Ku^{-1}) = \operatorname{tr} \int_K dk\ D(ku^{-1}).$$

They are positive-type functions with the bijection

$$\text{spherical } L^\infty(U)_+ \cong \mathbf{irrep}\, U.$$

8.10.3 Eigenfunctions of Euclidean Groups

The spherical functions on an *affine group with compact homogeneous group* K and translations $V \cong \mathbb{R}^{1+s}$, considered earlier, are characterized by the K-invariant momentum $P \in V^T$,

$$K \vec{\times} \mathbb{R}^{1+s}/K \cong \mathbb{R}^{1+s} \ni \vec{x} \longmapsto d^{iP}(\vec{x}) = e^{i\langle P, K\bullet\vec{x}\rangle} = \int_K dk\ e^{i\langle P, k\bullet\vec{x}\rangle},$$

e.g., on the Euclidean groups with real rank 1:

$$\mathbb{R}^{1+s} \cong \mathbf{SO}(1+s) \times \mathbb{R}^{1+s}/\mathbf{SO}(1+s),$$
$$d^{P^2,1+s}(r) = \sum_{k=0}^{\infty} \frac{\Gamma(\frac{1+s}{2})}{\Gamma(\frac{1+s}{2}+k)} \frac{(Pz)^{2k}}{k!},$$
$$P^2 \geq 0,\ z^2 = -\frac{r^2}{4},$$
$$d^{P^2,1+s}(0) = 1,$$

$d^{P^2,1+s}(\lvert\vec{x}\rvert) = \int \frac{d^s\omega}{\lvert\Omega^s\rvert}\ e^{iP\vec{\omega}_s\vec{x}}$ $= \int \frac{2d^{1+s}q}{\lvert\Omega^s\rvert\lvert P\rvert^{s-1}}\ \delta(\vec{q}^2 - P^2)e^{i\vec{q}\vec{x}} = \Gamma(\frac{1+s}{2}) \frac{\mathcal{J}_{\frac{s-1}{2}}(\lvert P\vec{x}\rvert)}{(\frac{\lvert P\vec{x}\rvert}{2})^{\frac{s-1}{2}}}$
$\vec{\partial}^2 d^{P^2,1+s} = -P^2 d^{P^2,1+s}$
$\int d^{1+s}x\ d^{P^2,1+s}(\lvert\vec{x}\rvert) d^{P'^2,1+s}(\lvert\vec{x}\rvert) = \frac{1}{\lvert\Omega^s\rvert}(\frac{2\pi}{\lvert P\rvert})^s \delta(\frac{\lvert P\rvert - \lvert P'\rvert}{2\pi})$
$\int_0^\infty \lvert\Omega^s\rvert(\frac{P}{2\pi})^{1+s} dP\ d^{P^2,1+s}(r) d^{P^2,1+s}(r') = \frac{1}{\lvert\Omega^s\rvert r^s}\delta(r - r')$

8.10.4 Eigenfunctions of Noncompact Groups

For a *noncompact semisimple Lie group* G with Iwasawa decomposition

$$G = N \circ A \circ K \ni g = e^{n(g)} e^{a(g)} k(g) = \begin{pmatrix} 1 & 0 \\ n(g) & 1 \end{pmatrix} \circ \begin{pmatrix} e^{\beta(g)} & 0 \\ 0 & e^{-\beta(g)} \end{pmatrix} \circ k(g),$$

the Hilbert representations, which contribute continuously to the Plancherel measure, can be characterized, according to Harish–Chandra, by the spherical functions of the noncompact homogeneous space $G/K \cong N \circ A$. They have $r = \dim_{\mathbb{R}} A$ complex invariants Q from the complexified weight space,

$$\{d^{iQ+\rho_N} \in \mathcal{C}(G/K) \mid Q \in \mathbb{C} \otimes (\log A)^T = (\log A_{\mathbb{C}})^T \cong \mathbb{C}^r\},$$
$$G \ni g = e^{n(g)} e^{a(g)} k(g) \longmapsto d^{iQ+\rho_N}(g) = e^{\langle iQ+\rho_N, a(gK)\rangle}$$
$$= \int_K dk\ e^{\langle iQ+\rho_N, a(gk)\rangle}.$$

The dilation factor $e^{\langle \rho_N, a(gK)\rangle}$ contains the nilpotent root sum ρ_N.

An example is the bounded Legendre functions on the hyperboloids $\mathcal{Y}^s$, $s \geq 2$, with real rank 1:

$$\mathbf{SO}_0(1,s)/\mathbf{SO}(s) \cong \mathcal{Y}^s \ni \mathrm{y}_s = \begin{pmatrix} \cosh\psi \\ \sinh\psi\ \cos\varphi \\ \sinh\psi\ \sin\varphi\ \vec{\omega}_{s-2} \end{pmatrix} \text{ with } \vec{\omega}_{s-2} \in \Omega^{s-2},$$

$$\mathrm{y}_s \longmapsto e^{\langle iQ-\frac{s-1}{2},\ \mathbf{SO}(2)\bullet\psi\rangle} = \mathrm{P}^{iQ-\frac{s-1}{2}}(\cosh\psi) = \int_{-\pi}^{\pi} \frac{d\varphi}{2\pi}(\cosh\psi + \cos\varphi\sinh\psi)^{iQ-\frac{s-1}{2}}.$$

$s-1$ is the dimension of the nilpotent Lie algebra $\mathbb{R}^{s-1}$ and, therefore, the multiplicity n_ω of the one positive root $\omega \in (\log\mathbf{SO}_0(1,1))^T$. The compact group integration goes over $\mathbf{SO}(2)$.

In the general orthogonality and L^2-completeness,

$$\begin{aligned} \mathrm{P}^\lambda(\cosh\psi) &= \sum_{k=0}^{\infty} \frac{\Gamma(\lambda+k+1)}{\Gamma(\lambda-k+1)} \frac{(-z)^k}{(k!)^2}, \\ \lambda \in \mathbb{C}, \quad z &= -\sinh^2\frac{\psi}{2}, \\ \mathrm{P}^\lambda(1) &= 1, \\ \text{for } \mathcal{Y}^s &\cong \mathbf{SO}_0(1,s)/\mathbf{SO}(s), \\ \lambda &= iQ - \frac{s-1}{2},\ Q \in \mathbb{R}, \end{aligned}$$

$\mathrm{P}^\lambda(\cosh\psi) = \int_{-\pi}^{\pi} \frac{d\varphi}{2\pi}(\cosh\psi + \cos\varphi\sinh\psi)^\lambda$
$\zeta = \cosh\psi,\quad \frac{d}{d\zeta}(\zeta^2-1)\frac{d\mathrm{P}^\lambda}{d\zeta} = \lambda(1+\lambda)\mathrm{P}^\lambda,$
$\int_1^\infty d\zeta\ \mathrm{P}^{iQ-\frac{s-1}{2}}(\zeta)\ \mathrm{P}^{-iQ'-\frac{s-1}{2}}(\zeta) = \frac{1}{\Pi^s(Q^2)}\delta(Q-Q')$
$\int_0^\infty \Pi^s(Q^2)dQ\ \mathrm{P}^{iQ-\frac{s-1}{2}}(\zeta)\ \mathrm{P}^{-iQ-\frac{s-1}{2}}(\zeta') = \delta(\zeta-\zeta')$

the Plancherel measure for the harmonic analysis of hyperboloid functions [55] has no discrete series contributions. It has to distinguish between even and odd dimensions:

$$\Pi^s(Q^2) = \left|\frac{\Gamma(iQ+\frac{s-1}{2})}{\Gamma(iQ)}\right|^2 = \begin{cases} \Gamma(R-\frac{1}{2})^2 \times Q^2\frac{\tanh\pi Q}{\pi Q}\prod_{k=1}^{R-1}\left(1+\frac{4Q^2}{(2k-1)^2}\right), \\ \qquad s = 2R = 2,4,6,\ldots, \\ \Gamma(R-1)^2 \times Q^2\prod_{k=1}^{R-2}\left(1+\frac{Q^2}{k^2}\right), \\ \qquad s = 2R-1 = 3,5,7,\ldots. \end{cases}$$

For the minimal nonabelian odd-dimensional case $\mathcal{Y}^3$ with the proper Lorentz group and Plancherel measure $\int_0^\infty dQ\ Q^2$, the principal series $\mathbf{SO}_0(1,3)$- representations are used.

Spherical functions are equal if, and only if, their Cartan invariants are connected by the Weyl group; the r invariants are characterized by Weyl group orbits:

$$\begin{aligned} \mathrm{S} \in \mathrm{Weyl}(G) &: d^{iQ_1+\rho_N} = d^{iQ_2+\rho_N} \iff Q_1 = \mathrm{S}.Q_2, \\ \mathrm{Weyl}(G) &.Q \subset (\log A_{\mathbb{C}})^T. \end{aligned}$$

Spherical functions of noncompact spaces do not have to be bounded. The Legendre functions above for real invariant $Q \in \mathbb{R}$ are bounded, $\mathrm{P}^{iQ-\frac{s-1}{2}} \in \mathcal{C}_b(\mathcal{Y}^s)$. This illustrates the general property: A spherical function is bounded if, and only if, the invariants are in the tube $i(\log A)^T + C(\rho_N)$, where $C(\rho_N)$ is the real convex hull of the Weyl orbit of the root sum ρ_N.

Spherical functions of noncompact spaces G/K have a tangent space parametrization: The translations $\log G/K \cong \log N \circ A = V \cong \mathbb{R}^s$ are acted on by the adjoint G-representation, which defines the semidirect tangent group,

$$G_{\text{tan}} = K \vec{\times} \log G/K \ni (k, \vec{x}), \text{ with } k \bullet \vec{x} = k \circ \vec{x} \circ k^{-1}.$$

Examples are the Euclidean groups $\mathbf{SO}(s) \vec{\times} \mathbb{R}^s$ for the hyperboloids, which are manifold-isomorphic to vector spaces $\mathcal{Y}^s \cong \mathbf{SO}_0(1,s)/\mathbf{SO}(s) \cong \mathbb{R}^s$. With the Killing form $\langle \vec{q}|\vec{x}\rangle$, restricted to the tangent space $\log A \circ N = V$, there exists [36], for each spherical function of G/K, a unique K-invariant function in a *tangent parametrization* with r real invariants $m \in \log A$, using translations and momenta $\vec{x}, \vec{q} \in V$,

$$\begin{aligned}
&\{d^{iQ+\rho_N} \mid Q \in (\log A_{\mathbb{C}})^T \cong \mathbb{C}^r\} \cong \{d^{im,J} \mid m \in \log N \circ A \cong \mathbb{R}^s\},\\
&G \ni g = e^n e^a k = e^{\vec{x}} k \longmapsto d^{iQ+\rho_N}(g) = e^{\langle iQ+\rho_N, a(gK)\rangle} = \int_K dk\ e^{\langle iQ+\rho_N, a(gk)\rangle}\\
&\qquad = d^{im,J}(\vec{x}) = \frac{e^{i\langle \vec{q}|K\bullet\vec{x}\rangle}}{\sqrt{J(\vec{x})}} = \frac{\int_K dk\ e^{i\langle \vec{q}|k\bullet\vec{x}\rangle}}{\sqrt{J(\vec{x})}}.
\end{aligned}$$

The local normalization of the exponent involves the ratio $J(\vec{x})$ of the volume elements in V and G/K.

8.11 Hilbert Metrics for Hyperboloids and Spheres

In contrast to the representations of flat spaces above (scattering waves, free particles), Hilbert representations of hyperboloids and spheres with nonabelian degrees of freedom use higher-order momentum poles. This will be exemplified first by the nonrelativistic hydrogen atom bound waves, which represent the noncompact nonabelian group $\mathbf{SO}_0(1,3)$ and start with momentum dipoles.

8.11.1 Hyperbolic Position in the Hydrogen Atom

The nonrelativistic dynamics $\mathbf{H} = \frac{\vec{\mathbf{p}}^2}{2} - \frac{1}{\mathbf{r}}$ with the Coulomb–Kepler potential has a rotation and a Lenz–Runge "perihelion" invariance (see Chapter 4). The measure of the unit 3-sphere as the manifold of the orientations of the rotation group $\mathbf{SO}(3)$ in the invariance group $\mathbf{SO}(4)$ for bound waves has a momentum parametrization by a dipole (see Chapter 2):

$$\frac{1}{\sqrt{\vec{q}^2+1}}\begin{pmatrix}1\\ i\vec{q}\end{pmatrix} \in \Omega^3 \subset \mathbb{R}^4 \Rightarrow |\Omega^3| = \int d^3\omega = \int \frac{2d^3q}{(\vec{q}^2+1)^2} = 2\pi^2.$$

A sphere radius $\frac{1}{|Q|}$ with curvature Q^2 is implemented by $\int \frac{2d^3q}{(\vec{q}^2+Q^2)^2} = \frac{2\pi^2}{|Q|}$.

The Fourier-transformed Ω^3-measure, gives the hydrogen ground-state function as a scalar representation coefficient of 3-position space. It is a positive-type hyperbolic function:

$$\mathbf{SO}_0(1,3)/\mathbf{SO}(3) \cong \mathcal{Y}^3 \cong \mathbb{R}^3 \ni \vec{x} \longmapsto \int \frac{d^3q}{\pi^2} \frac{|Q|}{(\vec{q}^2+Q^2)^2} e^{-i\vec{q}\vec{x}} = e^{-|Q|r}.$$

In the bound waves, position as a noncompact hyperboloid is represented in $L^2(\mathcal{Y}^3)$ in the form of Fourier-transformed Ω^3-measures with a continuous invariant Q^2 for the imaginary "momenta" $\vec{q}^2 = -Q^2$ on a 2-sphere Ω^2 and a rational rotation invariant $2J \in \mathbb{N}$. Hyperbolic position is isomorphic as manifold, not as symmetric space, to the translations $\mathcal{Y}^3 \cong \mathbb{R}^3$.

The Kepler bound waves are coefficients of infinite-dimensional cyclic representations of the Lorentz group $\mathbf{SO}_0(1,3)$. With the Cartan subgroups $\mathbf{SO}(2) \times \mathbf{SO}_0(1,1)$, the irreducible representations are characterized by one integer and one continuous invariant. In the language of induced representations, the bound waves of the hydrogen atom are rotation $\mathbf{SO}(3)$-intertwiners on the group $\mathbf{SO}_0(1,3)$ ($\mathcal{Y}^3$-functions) with values in Hilbert spaces with $\mathbf{SO}(3)$-representations in $(1+2J)^2$-dimensional $\mathbf{SO}(4)$-representations.

The rotation dependence $\vec{x}$ is effected by momentum derivation $i\frac{\partial}{\partial \vec{q}}$ of the Ω^3-measure:

$$\vec{x}e^{-r} = \int \frac{d^3q}{\pi^2} \frac{4i\vec{q}}{(1+\vec{q}^2)^3} e^{-i\vec{q}\vec{x}} \text{ with } \frac{4\vec{q}}{(1+\vec{q}^2)^3} = -\frac{\partial}{\partial \vec{q}} \frac{1}{(1+\vec{q}^2)^2}.$$

The 3-vector factor $\frac{2\vec{q}}{1+\vec{q}^2} = \frac{\vec{q}}{|\vec{q}|} \sin\chi$ is uniquely supplemented to a parametrization of the unit 3-sphere by a normalized 4-vector:

$$\begin{pmatrix} \cos\chi \\ \frac{\vec{q}}{|\vec{q}|} i \sin\chi \end{pmatrix} = \frac{1}{1+\vec{q}^2} \begin{pmatrix} 1-\vec{q}^2 \\ 2i\vec{q} \end{pmatrix} = \begin{pmatrix} p_0 \\ i\vec{p} \end{pmatrix} = p \in \Omega^3 \subset \mathbb{R}^4, \;\; p_0^2 + \vec{p}^2 = 1.$$

The unit 4-vector $\mathrm{Y}^{(\frac{1}{2},\frac{1}{2})}(p) \sim \begin{pmatrix} p_0 \\ i\vec{p} \end{pmatrix} \in \Omega^3$ is the analogue to the unit 3-vector $\mathrm{Y}^1(\frac{\vec{q}}{|\vec{q}|}) \sim \frac{\vec{q}}{|\vec{q}|} \in \Omega^2$ used for the buildup of the 2-sphere harmonics $\mathrm{Y}^L(\frac{\vec{q}}{|\vec{q}|}) \sim (\frac{\vec{q}}{|\vec{q}|})^L$. Analoguously, the Ω^3-harmonics are the totally symmetric traceless products $\mathrm{Y}^{(J,J)}(p) \sim (p)^{2J}$, e.g., the nine independent components in the (4×4)-matrix and the decomposition $9 \overset{\mathbf{SO}(3)}{=} 1 \oplus 3 \oplus 5$:

$$\begin{aligned} \mathrm{Y}^{(1,1)}(p) \sim (p)^2_{jk} &= p_j p_k - \frac{\delta_{jk}}{4} \cong \left(\begin{array}{c|c} \frac{3p_0^2-\vec{p}^2}{4} & ip_0 p_a \\ \hline ip_0 p_b & p_a p_b - \frac{\delta_{ab}}{4} \end{array} \right), \\ \text{with } p_a p_b - \frac{\delta_{ab}}{4} &= p_a p_b - \frac{\delta_{ab}}{3}\vec{p}^2 - \frac{\delta_{ab}}{3} \frac{3p_0^2-\vec{p}^2}{4} \text{ for } p^2 = 1. \end{aligned}$$

The Kepler bound waves in $(1+2J)^2$-multiplets for $\mathbf{SO}(4)$ come with momentum poles of order $2+2J$:

$$\frac{1}{Q} = 1+2J : \mathcal{Y}^3 \ni \vec{x} \longmapsto \int \frac{d^3q}{\pi^2} \frac{1}{(1+\vec{q}^2)^2} (p)^{2J} e^{-i\vec{q}Q\vec{x}} \text{ with } \begin{cases} p = \frac{1}{1+\vec{q}^2} \begin{pmatrix} 1-\vec{q}^2 \\ 2i\vec{q} \end{pmatrix}, \\ E = -\frac{Q^2}{2}. \end{cases}$$

The radius of the 3-sphere is the $\mathbf{SU}(2)$-multiplicity $\frac{1}{Q} = 1 + 2J$ with curvature $Q^2 = \frac{1}{(1+2J)^2}$, or, in the hydrogen atom $(\frac{1}{(1+2J)\ell_R})^2$ with the Bohr length $\ell_R = \frac{\hbar}{m_e c \alpha_S^2} \sim 6.8 \times 10^{-8}$ m as unit.

The Fourier transformations with the 3-sphere measure

$$\mu(\vec{x})e^{-r} = \int \frac{1}{(1+\vec{q}^{\,2})^2} \frac{d^3q}{\pi^2} \tilde{\mu}(\vec{q}) e^{-i\vec{q}\vec{x}} :$$

$\mu(\vec{x})$	$\tilde{\mu}(\vec{q})$
1	1
r	$\frac{3-\vec{q}^{\,2}}{1+\vec{q}^{\,2}}$
$\frac{\vec{x}}{2}$	$\frac{2i\vec{q}}{1+\vec{q}^{\,2}}$
$\frac{r^2}{3}$	$\frac{4(1-\vec{q}^{\,2})}{(1+\vec{q}^{\,2})^2}$
$r\vec{x}$	$\frac{2i\vec{q}\,(5-\vec{q}^{\,2})}{(1+\vec{q}^{\,2})^2}$
$\vec{x} \otimes \vec{x} - \mathbf{1}_3 \frac{r^2}{3}$	$-\frac{6(\vec{q}\otimes\vec{q}-\mathbf{1}_3 \frac{\vec{q}^{\,2}}{3})}{(1+\vec{q}^{\,2})^2}$

are used for the ground-state with a dipole,

$$\frac{1}{Q} = 1: \quad e^{-Qr} = \int \frac{d^3q}{\pi^2} \frac{Q}{(Q^2+\vec{q}^{\,2})^2} e^{-i\vec{q}\vec{x}} = \int \frac{d^3q}{\pi^2} \frac{1}{(1+\vec{q}^{\,2})^2} e^{-i\vec{q}Q\vec{x}},$$

the bound-state quartet with $\mathrm{Y}^{(\frac{1}{2},\frac{1}{2})}(p)$ and tripole, leading to Laguerre polynomials,

$$\begin{aligned} \frac{1}{Q} = 2: \quad \int \frac{d^3q}{\pi^2} \frac{1}{(1+\vec{q}^{\,2})^2} \begin{pmatrix} p_0 \\ i\vec{p} \end{pmatrix} e^{-i\vec{q}Q\vec{x}} &= \int \frac{d^3q}{\pi^2} \frac{1}{(1+\vec{q}^{\,2})^3} \begin{pmatrix} 1 - \vec{q}^{\,2} \\ 2i\vec{q} \end{pmatrix} e^{-i\vec{q}Q\vec{x}} \\ &= \begin{pmatrix} \frac{Qr-1}{2} \\ \\ \frac{Q\vec{x}}{2} \end{pmatrix} e^{-Qr} = \begin{pmatrix} -\frac{1}{4}\mathrm{L}_1^1(2Qr) \\ \frac{Q\vec{x}}{2}\,\mathrm{L}_2^0(2Qr) \end{pmatrix} e^{-Qr}, \end{aligned}$$

and the bound-state nonet with $\mathrm{Y}^{(1,1)}(p)$ and quadrupole:

$$\begin{aligned} \frac{1}{Q} = 3: \int \frac{d^3q}{\pi^2} \frac{1}{(1+\vec{q}^{\,2})^2} \begin{pmatrix} 3p_0^2 - \vec{p}^{\,2} \\ ip_0\vec{p} \\ 3\vec{p}\otimes\vec{p} - \mathbf{1}_3\vec{p}^{\,2} \end{pmatrix} e^{-i\vec{q}Q\vec{x}} &= \int \frac{d^3q}{\pi^2} \frac{4}{(1+\vec{q}^{\,2})^4} \begin{pmatrix} 3(\frac{1-\vec{q}^{\,2}}{2})^2 - \vec{q}^{\,2} \\ i\vec{q}\,\frac{1-\vec{q}^{\,2}}{2} \\ 3\vec{q}\otimes\vec{q} - \mathbf{1}_3\vec{q}^{\,2} \end{pmatrix} e^{-i\vec{q}Q\vec{x}} \\ = \begin{pmatrix} 1 - 2Qr + \frac{2Q^2r^2}{3} \\ \frac{Qr-2}{3}\frac{Q\vec{x}}{2} \\ \frac{Q^2}{2}(\mathbf{1}_3\frac{r^2}{3} - \vec{x}\otimes\vec{x}) \end{pmatrix} e^{-Qr} &= \begin{pmatrix} \frac{1}{3}\mathrm{L}_1^2(2Qr) \\ -\frac{Q\vec{x}}{2}\,\frac{1}{6}\mathrm{L}_3^1(2Qr) \\ \frac{Q^2}{2}(\mathbf{1}_3\frac{r^2}{3} - \vec{x}\otimes\vec{x})\mathrm{L}_5^0(2Qr) \end{pmatrix} e^{-Qr}. \end{aligned}$$

The Schur product for the wave functions involves the harmonic $\mathbf{SO}(3)$-momentum polynomials with the corresponding multipoles at the invariants:

$$\begin{aligned} \{-Q_{L'}^2, L' | -Q_L^2, L\}_3 &= \int \frac{d^3q}{\pi^2} \frac{(2\vec{q})^L}{(\vec{q}^{\,2}+Q_L^2)^{2+L}} \otimes \frac{(2\vec{q})^{L'}}{(\vec{q}^{\,2}+Q_{L'}^2)^{2+L'}} \\ &= \delta^{LL'} \frac{\Gamma(1+L)\Gamma(\frac{3}{2})}{\Gamma(\frac{3}{2}+L)} \int \frac{d^3q}{\pi^2} \frac{(2\vec{q}^{\,2})^L}{(\vec{q}^{\,2}+Q_L^2)^{4+2L}} \, (\mathbf{1}_3)^L \\ &= \delta^{LL'} \frac{1}{1+L} \frac{1}{2^{3+L}|Q_L|^{5+2L}} \, (\mathbf{1}_3)^L. \end{aligned}$$

8.11.2 Representations of Hyperboloids and Spheres

Distributions of s-dimensional momenta $\vec{q} \in \mathbb{R}^s$ with the action of the rotation group $\mathbf{SO}(s)$ are used for representations [55, 54] of the hyperboloids $\mathcal{Y}^s$ and spheres Ω^s. Flat spaces and hyberboloids are isomorphic as manifolds:

$$\begin{array}{c} \mathbb{R}^s \cong \mathcal{Y}^s \text{ with } \mathbb{R}^{1+s} \ni \begin{pmatrix} \sqrt{\vec{q}^2+1} \\ \vec{q} \end{pmatrix} = \begin{pmatrix} \cosh\psi \\ \sinh\psi\ \omega_{s-1} \end{pmatrix} \in \mathcal{Y}^s \\ \frac{d^s q}{\sqrt{\vec{q}^2+1}} = \frac{q^{s-1}dq}{\sqrt{q^2+1}}\, d^{s-1}\omega = d^s y = (\sinh\psi)^{s-1} d\psi\ d^{s-1}\omega. \end{array}$$

The residual representations of nonabelian noncompact hyperboloids and compact spheres with $s \geq 2$ have to embed the nontrivial representations of the abelian groups with continuous and integer dual invariants, respectively:

$$\begin{array}{l} \mathbf{SO}_0(1,1) \cong \mathcal{Y}^1 \ni x \longmapsto \int \frac{dq}{\pi} \frac{|Q|}{q^2+Q^2} e^{-iqx} = e^{-|Qx|}, \\ \mathbf{SO}(2) \cong \Omega^1 \ni e^{ix} \longmapsto \begin{cases} \int \frac{dq}{i\pi} \frac{P}{q^2-io-P^2} e^{-iqx} = e^{iP|x|}, \\ \int dq\ |q| \delta(P^2-\vec{q}^2) e^{-iqx} = \cos Px, \\ P = 0,1,2,\ldots. \end{cases} \end{array}$$

The invariant poles $\{\pm iP\}$ and $\{\pm Q\}$ on the discrete sphere $\Omega^0 = \{\pm 1\}$ are embedded, for the nonabelian case, in singularity spheres $\Omega^{s-1} \cong \mathbf{SO}(s)/\mathbf{SO}(s-1)$, whose action groups arise in the Iwasawa decomposition,

$$\begin{array}{r} \mathbf{SO}_0(1,s) = \mathbf{SO}(s) \circ \mathbf{SO}_0(1,1) \circ \exp \mathbb{R}^{s-1}, \\ \text{parabolic subgroup: } [\mathbf{SO}(s-1) \times \mathbf{SO}_0(1,1)] \circ \exp \mathbb{R}^{s-1}. \end{array}$$

The Lorentz groups for even dimension have a unique Cartan subgroup type,

$$\mathbf{SO}_0(1,2R-1) \supseteq \mathbf{SO}(2)^{R-1} \times \mathbf{SO}_0(1,1),\ s = 2R-1 = 1,3,5,\ldots,$$

in contrast to odd dimensions, e.g., $\mathbf{SO}_0(1,1)$ and $\mathbf{SO}(2)$ for $\mathbf{SO}_0(1,2)$. The real rank 1 of the orthogonal groups $\mathbf{SO}_0(1,2R-1)$ gives the real (noncompact) rank 1 for the odd-dimensional hyperboloids, i.e., one continuous noncompact invariant. The noncompact–compact pairs $(\mathcal{Y}^{2R-1}, \Omega^{2R-1})$ with odd-dimensional hyperboloids and spheres will be considered as a generalization of the minimal and characteristic nonabelian case $(\mathcal{Y}^3, \Omega^3)$ with nontrivial rotations as used for the nonrelativistic hydrogen atom above.

The coefficients of residual representations of hyperboloids $\mathcal{Y}^{2R-1}$ use the Fourier-transformed measure of the momentum sphere Ω^{2R-1} with singularity sphere $\{\vec{q} \in \mathbb{R}^{2R-1} \,|\, \vec{q}^2 = -Q^2 < 0\} \cong \Omega^{2R-2}$ with continuous noncompact invariant Q^2 and for imaginary "momenta" as eigenvalues. $\mathbf{SO}(2R)$-multiplets arise via the sphere parametrization $\frac{1}{\vec{q}^2+Q^2}\begin{pmatrix} Q^2-\vec{q}^2 \\ 2i|Q|\vec{q} \end{pmatrix} \in \Omega^{2R-1} \subset \mathbb{R}^{2R}$,

$$\begin{array}{l} \text{for } \mathcal{Y}^{2R-1},\ R = 1,2,\ldots,\ \frac{2}{|\Omega^{2R-1}|} = \frac{\Gamma(R)}{\pi^R}, \\ \vec{x} \longmapsto \begin{cases} \int \frac{2d^{2R-1}q}{|\Omega^{2R-1}|} \frac{|Q|}{(\vec{q}^2+Q^2)^R} e^{-i\vec{q}\vec{x}} = e^{-|Q|r}, \\ \int \frac{2d^{2R-1}q}{|\Omega^{2R-1}|} \frac{R}{(\vec{q}^2+Q^2)^{R+1}} \begin{pmatrix} Q^2-\vec{q}^2 \\ 2i|Q|\vec{q} \end{pmatrix} e^{-i\vec{q}\vec{x}} = \begin{pmatrix} 1-R+|Q|r \\ \vec{x} \end{pmatrix} e^{-|Q|r}. \end{cases} \end{array}$$

The order of the singularity is related to the rank R of the acting group $\mathbf{SO}_0(1,2R-1)$, i.e., the dimension of the Cartan subgroups $\mathbf{SO}(2)^{R-1}\times\mathbf{SO}_0(1,1)$:

$$\frac{d^{2R-1}q}{(\vec{q}^{\,2}+Q^2)^R}=\left(\frac{\vec{q}^{\,2}}{\vec{q}^{\,2}+Q^2}\right)^R\frac{d|\vec{q}|}{|\vec{q}|}\,d^{2(R-1)}\omega.$$

Each state $\{\vec{x}\longmapsto e^{-|Q|r}\}\in L^\infty(\mathbf{SO}_0(1,2R-1))_+$ with invariant $Q^2>0$ characterizes an infinite-dimensional Hilbert space with a faithful cyclic representation of $\mathbf{SO}_0(1,2R-1)$ as familiar for $R=2$ from the principal series representations of the Lorentz group $\mathbf{SO}_0(1,3)$. The positive-type function defines the Hilbert product:

$$\begin{aligned}\text{distributive basis: }&\{|-Q^2;\vec{q}\rangle\mid\vec{q}\in\mathbb{R}^{2R-1}\},\\ \text{scalar product distribution: }&\langle -Q^2;\vec{q}\,'|-Q^2;\vec{q}\rangle=\frac{|Q|}{(\vec{q}^{\,2}+Q^2)^R}\frac{|\Omega^{2R-1}|}{2}\delta(\vec{q}-\vec{q}\,'),\\ \text{Hilbert vectors: }&|-Q^2;f\rangle=\oplus\!\!\int\frac{2d^{2R-1}q}{|\Omega^{2R-1}|}\,f(\vec{q})|-Q^2;\vec{q}\rangle,\\ &\langle -Q^2;f'|-Q^2;f\rangle=\int\frac{2d^{2R-1}q}{|\Omega^{2R-1}|}\,\overline{f'(\vec{q})}\frac{|Q|}{(\vec{q}^{\,2}+Q^2)^R}f(\vec{q}).\end{aligned}$$

There is a representation of each abelian noncompact subgroup in the Cartan decomposition $\mathcal{Y}^{2R-1}\sim\mathbf{SO}_0(1,1)\times\Omega^{2R-2}$ with the action on a distributive basis and hence on the Hilbert vectors:

$$\begin{aligned}\mathbf{SO}_0(1,1)\text{-representations for all }\vec{\omega}\in\Omega^{2R-2}:\ e^{-\vec{\omega}\vec{x}}&\longmapsto e^{-i|\vec{q}|\vec{\omega}\vec{x}}\\ &=e^{-i\vec{q}\vec{x}}\in\mathbf{U}(1),\\ \text{action of all }\mathbf{SO}_0(1,1):\ |-Q^2;\vec{q}\rangle&\longmapsto e^{-i\vec{q}\vec{x}}|-Q^2;\vec{q}\rangle,\\ \text{cyclic vector: }|-Q^2;1\rangle&=\oplus\!\!\int\frac{2d^{2R-1}q}{|\Omega^{2R-1}|}\,|-Q^2;\vec{q}\rangle\\ \text{with }\int\frac{4d^{2R-1}q\,d^{2R-1}q'}{|\Omega^{2R-1}|^2}\,\langle -Q^2;\vec{q}\,'|e^{-i\vec{q}\vec{x}}|-Q^2;\vec{q}\rangle&=e^{-|Q|r}.\end{aligned}$$

The scalar product of $L^1(\mathcal{Y}^{2R-1})$ involves the positive-type function, e.g., for three-dimensional position $R=2$ with intrinsic unit $Q^2=1$,

$$\begin{aligned}\langle -1;f'|-1;f\rangle=\int\frac{d^3q}{\pi^2}\,\overline{f'(\vec{q})}\,\frac{1}{(\vec{q}^{\,2}+1)^2}\,f(\vec{q})&=\int d^3x_1d^3x_2\,\overline{\tilde{f}'(\vec{x}_2)}\\ &\quad e^{-|\vec{x}_1-\vec{x}_2|}\tilde{f}(\vec{x}_1),\\ \text{with }f(\vec{q})&=\int d^3x\,\tilde{f}(\vec{x})e^{i\vec{q}\vec{x}}.\end{aligned}$$

It can be put in the form of square-integrability $L^2(\mathcal{Y}^3)$ by absorption of the square-integrable square root of the positive-type function:

$$\begin{aligned}L^1(\mathcal{Y}^3)\longrightarrow L^2(\mathcal{Y}^3),\quad \tilde{f}(\vec{x})&\longmapsto\tilde{\psi}(\vec{x})=\tilde{\xi}*\tilde{f}(\vec{x}),\\ f(\vec{q})&\longmapsto\psi(\vec{q})=\frac{\sqrt{8\pi}}{\vec{q}^{\,2}+1}f(\vec{q}),\\ \text{with }\int\frac{d^3q}{(2\pi)^3}\frac{8\pi}{(\vec{q}^{\,2}+1)^2}e^{-i\vec{q}\vec{x}}&=e^{-r}=\tilde{d}(\vec{x})=\tilde{\xi}*\tilde{\xi}(\vec{x}),\\ \int\frac{d^3q}{(2\pi)^3}\frac{\sqrt{8\pi}}{\vec{q}^{\,2}+1}\,e^{-i\vec{q}\vec{x}}&=\frac{e^{-r}}{\sqrt{2\pi}r}=\tilde{\xi}(\vec{x})\\ \Rightarrow\langle -1;f'|-1;f\rangle=\int\frac{d^3q}{\pi^2}\,\overline{f'(\vec{q})}\,\frac{1}{(\vec{q}^{\,2}+1)^2}\,f(\vec{q})&=\int\frac{d^3q}{(2\pi)^3}\,\overline{\psi'(\vec{q})}\,\psi(\vec{q})\\ &=\int d^3x\,\overline{\tilde{\psi}'(\vec{x})}\,\tilde{\psi}(\vec{x}).\end{aligned}$$

Therefore, all infinite-dimensional Hilbert spaces for different continuous invariants $Q^2 > 0$ can be transformed to subspaces of one Hilbert space $L^1(\mathcal{Y}^{2R-1}) \longrightarrow L^2(\mathcal{Y}^{2R-1}) \cong L^2(\mathbb{R}^{2R-1})$.

Hyperboloid representations are characterized by an integer invariant $L \in \mathbb{N}$ for the harmonic $\mathbf{O}(2R-1)$-momentum polynomials and a continuous invariant $Q > 0$ related to $\mathbf{SO}_0(1,1) \subseteq \mathbf{SO}_0(1,2R-1)$:

$$|-Q^2, L\}_{2R-1}(\vec{x}) = \int \frac{2d^{2R-1}q}{|\Omega^{2R-1}|} \frac{(2\vec{q})^L}{(\vec{q}^2+Q^2)^{R+L}} e^{-i\vec{q}\vec{x}}.$$

The Schur product displays orthogonality for different rotation invariants (see Chapters 9 and 10):

$$\begin{aligned}\{-Q'^2, L'|-Q^2, L\}_{2R-1} &= \int \frac{2d^{2R-1}q}{|\Omega^{2R-1}|} \frac{(2\vec{q})^L}{(\vec{q}^2+Q^2)^{R+L}} \otimes \frac{(2\vec{q})^{L'}}{(\vec{q}^2+Q'^2)^{R+L'}} \\ &= \delta^{LL'} \frac{\Gamma(1+L)\Gamma(R-\frac{1}{2})}{\Gamma(R-\frac{1}{2}+L)} \\ &\quad \int \frac{2d^{2R-1}q}{|\Omega^{2R-1}|} \frac{(2q^2)^L}{(\vec{q}^2+Q^2)^{R+L}(\vec{q}^2+Q'^2)^{R+L}} (\mathbf{1}_{2R-1})^L \\ &= \delta^{LL'} \frac{\Gamma(1+L)\Gamma(R)}{\Gamma(R+L)} \frac{2^L}{QQ'(Q+Q')^{2R+2L-1}} (\mathbf{1}_{2R-1})^L.\end{aligned}$$

States with equal rotation invariants $L = L'$ and different continuous invariants $Q \neq Q'$ are not orthogonal. The orthogonality of the $\mathcal{Y}^3$-representation coefficients with different invariants in the hydrogen atom is a consequence of the different rotation invariants. The Schur normalization of the hydrogen wave functions is used for the probability interpretation.

The positive-type functions for $L = 0$ are representation-normalized with a factor $|Q|$:

$$\int \frac{2d^{2R-1}q}{|\Omega^{2R-1}|} \frac{|Q|}{(\vec{q}^2+Q^2)^R} = 1.$$

The representation normalization of the $L \neq 0$ functions will be discussed in Chapter 10.

The corresponding matrix elements of representations of odd-dimensional spheres are obtained by a real–imaginary transition. They involve multipole Feynman distributions (derived Dirac distributions) with supporting singularity sphere Ω^{2R-2} for real momenta with, for appropriate normalization, integer invariants $\vec{q}^2 = L^2$, $L = 0, 1, 2, \ldots$,

$$\text{for } \Omega^{2R-1}, R = 1, 2, \ldots,$$
$$\vec{x} \longmapsto \begin{cases} \int \frac{2d^{2R-1}q}{i|\Omega^{2R-1}|} \frac{L}{(\vec{q}^2 - io - L^2)^R} e^{-i\vec{q}\vec{x}} = e^{iLr}, \\ \int \frac{d^{2R-1}q}{\pi^{R-1}} L\delta^{(R-1)}(L^2 - \vec{q}^2) e^{-i\vec{q}\vec{x}} = \cos Lr. \end{cases}$$

In contrast to hyperboloids $\mathcal{Y}^s \cong \mathbb{R}^s$, spheres are compact. Therefore, the Fourier-transformed positive-type functions on the spheres for $\mathbf{SO}(1+s)$-representations do not have to give positive Radon measures of the momenta.

The irreducible $\mathbf{SO}(1+s)$-representation spaces used for the sphere Ω^s are finite-dimensional. A basis is given by the spherical harmonics $(\vec{\omega_s})^L$, $L=0,1,\dots$, with $r \sim \theta$:

$$\vec{\omega_0} = 1,\quad \vec{\omega_s} = \begin{pmatrix} \cos\theta \\ \sin\theta\ \vec{\omega_{s-1}} \end{pmatrix} \in \Omega^s \subset \mathbb{R}^{1+s},\quad s = 1, 2\dots.$$

In the nontrivial case, $L \neq 0$, $\Omega^1 = \mathbf{SO}(2)$ is acted on by the two-dimensional $\mathbf{SO}(2)$-representations $\begin{pmatrix} \cos L\theta & -\sin L\theta \\ \sin L\theta & \cos L\theta \end{pmatrix}$, Ω^2 by the $(1+2L)$-dimensional harmonic $\mathbf{SO}(3)$-representations $[L]$, and Ω^3 by the $(1+L)^2$-dimensional harmonic $\mathbf{SO}(4)$-representations $(\frac{L}{2}, \frac{L}{2})$. Different invariants L characterize Schur-orthogonal subspaces of the infinite-dimensional Hilbert space $L^2(\Omega^s)$.

Mathematical Tools

8.12 Spherical, Hyperbolic, Feynman, and Causal Distributions

The Lebesgue measure $\frac{d^n q}{(2\pi)^n}$ is the Plancherel measure for the irreducible translation representations $\mathbb{R}^n \ni x \longmapsto e^{iqx} \in \mathbf{U}(1)$ and Haar measure $d^n x$.

The circle has different parametrizations, e.g.,

$$\Omega^1 \ni \begin{pmatrix} q_0 \\ iq \end{pmatrix},\ \text{for semicircle:} \begin{pmatrix} \cos\theta \\ i\sin\theta \end{pmatrix}_{-\frac{\pi}{2}}^{\frac{\pi}{2}} = \begin{pmatrix} \sqrt{1-q^2} \\ iq \end{pmatrix}_{-1}^{1} = \frac{1}{\sqrt{1+p^2}} \begin{pmatrix} 1 \\ ip \end{pmatrix}_{-\infty}^{\infty}.$$

Therefore, the compact classes of orthogonal groups $\mathbf{SO}(1+s)/\mathbf{SO}(s) \cong \Omega^s$ (unit sphere) have volume

$$\begin{aligned} |\Omega^s| &= \int d^s\omega = \int d^{1+s}q\ 2\delta(q_0^2 + \vec{q}^2 - 1) = \int_{\vec{q}^2 \le 1} \frac{d^s q}{\sqrt{1-\vec{q}^2}} \\ &= \int_0^\pi (\sin\theta)^{s-1} d\theta \int d^{s-1}\omega = \int \frac{2d^s p}{(1+\vec{p}^2)^{\frac{1+s}{2}}} \end{aligned}$$

$$|\Omega^s| = \frac{2\pi^{\frac{1+s}{2}}}{\Gamma(\frac{1+s}{2})}: \begin{cases} |\Omega^{2R}| = \frac{(4\pi)^R \Gamma(R)}{\Gamma(2R)} = 2, 4\pi, \frac{8\pi^2}{3}, \dots, \\ |\Omega^{2R-1}| = \frac{2\pi^R}{\Gamma(R)} = 2\pi, 2\pi^2, \dots, \end{cases} \quad \frac{|\Omega^{s-2}|}{|\Omega^s|} = \frac{s-1}{2\pi},$$

$$\text{polar decomposition:}\quad q = |q|\vec{\omega}_s \text{ with } |q|^2 = q_0^2 + \vec{q}^2,\ \vec{\omega}_s \in \Omega^s,$$
$$\int d^{1+s}q = \int_0^\infty |q|^s d|q| \int d^s\omega.$$

The Gamma function has the following properties, where defined,

$$z \in \mathbb{C}: \begin{cases} \Gamma(z) = \int_0^\infty dx\ e^{-x} x^{z-1} = \overline{\Gamma(\overline{z})}, \\ \Gamma(1+z) = z\Gamma(z),\ \Gamma(1+N) = N!,\ N = 0, 1, \dots, \\ \Gamma(z)\Gamma(1-z) = \frac{\pi}{\sin \pi z},\quad \frac{\Gamma(\frac{1}{2}+z)}{\Gamma(\frac{1}{2})} = 2^{1-2z}\frac{\Gamma(2z)}{\Gamma(z)}. \end{cases}$$

The spherical–hyperbolic, i.e., compact–noncompact partner transition,

for $\mathbf{SO}(1+s)/\mathbf{SO}(s) \cong \Omega^s : (i\theta, i\vec{q}, i\vec{p}) \leftrightarrow (\psi, \vec{q}, \vec{p})$ for $\mathcal{Y}^s \cong \mathbf{SO}_0(1,s)/\mathbf{SO}(s)$, gives the parametrizations and volume elements of the unit hyperboloids with noncompact classes of orthogonal groups,

$$\begin{aligned}
&\mathcal{Y}^1 \ni \begin{pmatrix} \vartheta(q_0)q_0 \\ q \end{pmatrix}, \quad \begin{pmatrix} \cosh\psi \\ \sinh\psi \end{pmatrix}_0^\infty = \begin{pmatrix} \sqrt{1+q^2} \\ q \end{pmatrix}_0^\infty = \frac{1}{\sqrt{1-p^2}}\begin{pmatrix} 1 \\ p \end{pmatrix}_0^1,\\
&\int d^s\mathrm{y} = \int \vartheta(q_0)\, d^{1+s}q\ 2\delta(q_0^2 - \vec{q}^{\,2} - 1) = \int \frac{d^sq}{\sqrt{1+\vec{q}^{\,2}}}\\
&\qquad = \int_0^\infty (\sinh\psi)^{s-1} d\psi \int d^{s-1}\omega\\
&\qquad = \int_{\vec{p}^{\,2}\le 1} \frac{2d^sp}{(1-\vec{p}^{\,2})^{\frac{1+s}{2}}},\\
&\text{“polar” decomposition:}\quad q = |q|\mathrm{y}_s \text{ with } |q|^2 = q_0^2 - \vec{q}^{\,2},\ \mathrm{y}_s \in \mathcal{Y}^s,\\
&\qquad \int \vartheta(q_0)\vartheta(q^2) d^{1+s}q = \int_0^\infty |q|^s d|q| \int d^s\mathrm{y}.
\end{aligned}$$

The measures of the momentumlike hyperboloids with the noncompact classes of noncompact groups $\mathbf{SO}_0(1,s)/\mathbf{SO}_0(1,s-1) \cong \mathcal{Y}^{(1,s-1)}$ are

$$\int d^s\mathrm{s} = \int d^{1+s}q\ 2\delta(q_0^2 - \vec{q}^{\,2} + 1) = 2\int_{\vec{q}^{\,2}\ge 1} \frac{d^sq}{\sqrt{\vec{q}^{\,2}-1}} = \int_{-\infty}^{\infty} (\cosh\psi)^{s-1} d\psi \int d^{s-1}\omega,$$

$$\begin{aligned}
&\text{“polar” decomposition:}\quad q = |q|\mathrm{s}_s \text{ with } |q|^2 = -q_0^2 + \vec{q}^{\,2},\ \mathrm{s}_s \in \mathcal{Y}^{(1,s-1)},\\
&\qquad \int \vartheta(-q^2) d^{1+s}q = \int_0^\infty |q|^s d|q| \int d^s\mathrm{s}.
\end{aligned}$$

Distributions with real poles have real–imaginary decompositions with the principal value a_{P}:

$$\begin{aligned}
a \in \mathbb{R},\ \nu \in \mathbb{R},\ \nu \ne -1, -2, \cdots: &\quad \frac{1}{(a-io)^\nu} = \frac{1}{|a|^\nu}[\vartheta(a) + \vartheta(-a)e^{i\nu\pi}],\\
N = 0, 1, 2, \cdots: &\quad \frac{\Gamma(1+N)}{(a-io)^{1+N}} = \frac{\Gamma(1+N)}{a_{\mathrm{P}}^{1+N}} + i\pi\delta^{(N)}(-a).
\end{aligned}$$

The Dirac “on-shell” and the principal value (with q_{P}^2) “off-shell” distributions are imaginary and real part of the *(anti-)Feynman distributions*:

$$\begin{aligned}
\log(q^2 \mp io - \mu^2) &= \log|q^2 - \mu^2| \mp i\pi\vartheta(\mu^2 - q^2)\\
\frac{\Gamma(1+N)}{(q^2 \mp io - \mu^2)^{1+N}} &= -(-\frac{\partial}{\partial q^2})^{1+N} \log(q^2 \mp io - \mu^2)\\
&= (-\frac{\partial}{\partial q^2})^N \frac{1}{q^2 \mp io - \mu^2}\\
&= \frac{\Gamma(1+N)}{(q_{\mathrm{P}}^2 - \mu^2)^{1+N}} \pm i\pi\delta^{(N)}(\mu^2 - q^2)\\
\text{for } \mu^2 \in \mathbb{R} &\quad \text{and } N = 0, 1, \ldots.
\end{aligned}$$

Feynman distributions are possible for any signature $\mathbf{O}(t,s)$ with positive or negative invariant μ^2.

Characteristic for and compatible only with the orthochronous Lorentz group $\mathbf{SO}_0(1,s)$ are the *advanced (future) and retarded (past) causal energy-momentum distributions* with positive invariant m^2 only. They are distinguished by their energy q_0 behavior:

$$\begin{aligned}
\log((q \mp io)^2 - m^2) &= \log|q^2 - m^2| \mp i\pi\epsilon(q_0)\vartheta(m^2 - q^2)\\
\frac{\Gamma(1+N)}{((q \mp io)^2 - m^2)^{1+N}} &= -(-\frac{\partial}{\partial q^2})^{1+N} \log((q \mp io)^2 - m^2)\\
&= (-\frac{\partial}{\partial q^2})^N \frac{1}{(q \mp io)^2 - m^2}\\
&= \frac{\Gamma(1+N)}{(q_{\mathrm{P}}^2 - m^2)^{1+N}} \pm i\pi\epsilon(q_0)\delta^{(N)}(m^2 - q^2)\\
\text{for } m^2 \ge 0 &\quad \text{and } (q \mp io)^2 = (q_0 \mp io)^2 - \vec{q}^{\,2}.
\end{aligned}$$

The principal value integration is off-shell:

$$\int \frac{d^{1+s}q}{\pi} \frac{\Gamma(1+N)}{(q_{\rm P}^2-\mu^2)^{1+N}} e^{iqx} = i\epsilon(x_0) \int d^{1+s}q\ \epsilon(q_0)\delta^{(N)}(\mu^2 - q^2)e^{iqx}.$$

For the advanced and retarded integrations, one obtains the Fourier transforms:

$$\begin{aligned}\int \frac{d^{1+s}q}{\pi} \frac{\Gamma(1+N)}{[(q\mp io)^2-m^2]^{1+N}} e^{iqx} &= \int d^{1+s}q\ [\tfrac{1}{\pi} \frac{\Gamma(1+N)}{(q_{\rm P}^2-m^2)^{1+N}}\\ &\pm i\epsilon(q_0)\delta^{(N)}(m^2 - q^2)]e^{iqx}\\ &= 2\vartheta(\pm x_0) \int \frac{d^{1+s}q}{\pi} \frac{\Gamma(1+N)}{(q_{\rm P}^2-m^2)^{1+N}} e^{iqx}\\ &= \pm 2i\vartheta(\pm x_0) \int d^{1+s}q\ \epsilon(q_0)\delta^{(N)}(m^2 - q^2)e^{iqx}.\end{aligned}$$

8.13 Residual Distributions

The causal structure of the reals and its unitary representations occur in the Fourier-transformed causal measures,

$$m, \nu \in \mathbb{R}: \qquad \int \frac{dq}{2i\pi} \frac{\Gamma(1-\nu)}{(q-io-m)^{1-\nu}} e^{iqx} = \vartheta(x) \frac{e^{imx}}{(ix)^\nu}.$$

Here and in the following, the integrals hold wherever the Γ-functions are defined.

8.13.1 Macdonald, Bessel, and Neumann Functions

The scalar distributions for the definite orthogonal groups in general dimension with real and imaginary singularities on spheres Ω^s with $\vec{q}^{\,2} = \pm 1$ define the *Macdonald functions* $\mathcal{K}_\nu$ and, for imaginary argument, the corresponding *Hankel (Bessel with Neumann) functions* $\mathcal{H}_\nu^{1,2} = \mathcal{J}_\nu \pm i\mathcal{N}_\nu$:

$$\begin{array}{c}\mathbf{O}(1+s),\\ s = 0,1,2,\ldots,\\ r = \sqrt{\vec{x}^{\,2}},\quad \nu \in \mathbb{R},\end{array} \left\{\begin{aligned} \int \frac{d^{1+s}q}{\pi^{\frac{1+s}{2}}} \frac{\Gamma(\frac{1+s}{2}-\nu)}{(\vec{q}^{\,2})^{\frac{1+s}{2}-\nu}} e^{i\vec{q}\vec{x}} &= \frac{\Gamma(\nu)}{(\frac{r}{2})^{2\nu}},\\ \int \frac{d^{1+s}q}{\pi^{\frac{1+s}{2}}} \frac{\Gamma(\frac{1+s}{2}-\nu)}{(\vec{q}^{\,2}+1)^{\frac{1+s}{2}-\nu}} e^{i\vec{q}\vec{x}} &= \frac{2\mathcal{K}_\nu(r)}{(\frac{r}{2})^\nu},\\ \int \frac{d^{1+s}q}{i\pi^{\frac{1+s}{2}}} \frac{\Gamma(\frac{1+s}{2}-\nu)}{(\vec{q}^{\,2}-io-1)^{\frac{1+s}{2}-\nu}} e^{i\vec{q}\vec{x}} &= -i\frac{2\mathcal{K}_\nu(-ir)}{(\frac{r}{2})^\nu} = \frac{\pi[\mathcal{J}_\nu + i\mathcal{N}_\nu](r)}{(\frac{r}{2})^\nu}.\end{aligned}\right.$$

The angle integration is different for odd and even dimensions with $d\theta$ and $d\cos\theta = d\zeta$:

$$\begin{aligned} s \geq 1: \int \frac{d^{1+s}q}{|\Omega^{s-1}|}\ \mu(\vec{q}^{\,2})e^{i\vec{q}\vec{x}} &= \int_0^\infty q^s dq\ \mu(q^2) \int_0^\pi (\sin\theta)^{s-1} d\theta\ e^{iqr\cos\theta}\\ &= \left\{\begin{array}{lll} \int_0^\infty q^{2R-1}dq\ \mu(q^2) & \int_0^\pi (1-\cos^2\theta)^{R-1}d\theta\ e^{iqr\cos\theta}, & s = 2R-1,\\ -\int_0^\infty q^{2R}dq\ \mu(q^2) & \int_{-1}^1 (1-\zeta^2)^{R-1}d\zeta\ e^{iqr\zeta}, & s = 2R.\end{array}\right.\end{aligned}$$

The integrals can be obtained by 2-sphere spread from the values for $R = 1$.

All (half-)integer-index functions can be obtained by derivation ("2-sphere spread") $\frac{d}{dr^2} = \frac{1}{2r}\frac{d}{dr}$:

$$\mathbb{R}_+ \ni r \longmapsto \frac{(2\mathcal{K}_\nu,\ \pi\mathcal{J}_\nu,\ \pi\mathcal{N}_\nu)(r)}{r^\nu} = \begin{cases} \left(-2\frac{d}{dr^2}\right)^N \sqrt{2\pi}(e^{-r},\ \cos r,\ \sin r), \\ \qquad \nu+\frac{1}{2} = N = 0,1,2,\ldots, \\ \left(-2\frac{d}{dr^2}\right)^N \Big(2\mathcal{K}_0(r),\ \pi\mathcal{J}_0(r),\ \pi\mathcal{N}_0(r)\Big), \\ \qquad \nu = N = 0,1,2,\ldots, \end{cases}$$

$$\begin{aligned} \frac{(2\mathcal{K}_{\nu+1},\ \pi\mathcal{J}_{\nu+1},\ \pi\mathcal{N}_{\nu+1})(r)}{(\frac{r}{2})^{\nu+1}} &= -4\frac{d}{dr^2}\frac{(2\mathcal{K}_\nu,\ \pi\mathcal{J}_\nu,\ \pi\mathcal{N}_\nu)(r)}{(\frac{r}{2})^\nu}, \\ (\mathcal{K}_{-N},\mathcal{J}_{-N},\mathcal{N}_{-N})(r) &= (\mathcal{K}_N,(-1)^N\mathcal{J}_N,(-1)^N\mathcal{N}_N)(r), \\ (\mathcal{K}_{-N-\frac{1}{2}},\mathcal{J}_{-N-\frac{1}{2}},\mathcal{N}_{-N-\frac{1}{2}})(r) &= (\mathcal{K}_{N+\frac{1}{2}},(-1)^{N+1}\mathcal{N}_{N+\frac{1}{2}},(-1)^N\mathcal{J}_{N+\frac{1}{2}})(r). \end{aligned}$$

The half-integer index functions for $\nu > 0$ start from $\nu = -\frac{1}{2}$ with $\mathbb{R}$-representations, whereas the integer-index functions start from $\nu = 0$, which involves a finite integration of $\mathbb{R}$-representations:

$$\text{for } \nu = -\tfrac{1}{2}:\ \cos r \begin{array}{l} \nearrow \\ \searrow \end{array} \begin{array}{rll} \frac{\sin r}{r} &= \int_0^1 d\zeta\ \cos\zeta r = -2\frac{d}{dr^2}\cos r & \text{for } \nu = \frac{1}{2}, \\ \mathcal{J}_0(r) &= \int_0^\pi \frac{d\theta}{\pi}\ \cos(r\cos\theta) & \text{for } \nu = 0. \end{array}$$

For $\nu \geq 0$, only the Bessel functions are regular at $r = 0$:

$$\begin{aligned} 2\mathcal{K}_0(r) &= \sum_{k=0}^\infty \frac{(\frac{r^2}{4})^k}{(k!)^2}[2\chi_k - \log\tfrac{r^2}{4}] \\ &= \mathcal{H}_0^1(ir) = i\pi[\mathcal{J}_0(ir) + i\mathcal{N}_0(ir)], \end{aligned} \quad \begin{cases} -\pi\mathcal{N}_0(r) = \sum_{k=0}^\infty \frac{(-\frac{r^2}{4})^k}{(k!)^2}[2\chi_k - \log\frac{r^2}{4}], \\ \mathcal{J}_0(r) = \sum_{k=0}^\infty \frac{(-\frac{r^2}{4})^k}{(k!)^2}, \end{cases}$$

$$\begin{aligned} \text{Euler's constant } \chi_0 &= \Gamma'(1) = \lim_{k\to\infty}[\log k - (1 + \tfrac{1}{2} + \ldots + \tfrac{1}{k})] \\ &= -0.5772\ldots, \\ \chi_k &= \Gamma'(1) + 1 + \tfrac{1}{2} + \ldots + \tfrac{1}{k},\quad k = 1,2,\ldots. \end{aligned}$$

For integer ν, the index N characterizes the small distance behavior,

$$\mathcal{J}_N(r) = r^N(-2\tfrac{d}{dr^2})^N\mathcal{J}_0(r) = (\tfrac{r}{2})^N\sum_{k=0}^\infty \frac{(-\frac{r^2}{4})^k}{k!(N+k)!}.$$

The half-integer-index functions start from the exponentials. The noncompact and compact self-dual representations of the reals come with imaginary and real poles in the complex plane:

$$\mathbf{O}(1):\quad \int \frac{dq}{\pi}\ \frac{1}{q^2 - io \pm 1}e^{iqx} = \begin{cases} e^{-|x|}, & \text{poles at } q = \pm i \quad \text{(hyperbolic)}, \\ ie^{i|x|}, & \text{poles at } q = \pm 1 \quad \text{(spherical)}. \end{cases}$$

They involve the positive-type function for the basic self-dual spherical representation $\mathbb{R} \ni x \longmapsto \cos x$ and the basic self-dual hyperbolic one $\mathbb{R} \ni x \longmapsto e^{-|x|}$.

The integer-index functions begin with two-dimensional momentum integrals, which integrate over the $\mathbb{R}$-representation coefficients:

$$\begin{array}{c} \mathbf{O}(2): \\ r=\sqrt{\vec{x}^{\,2}} \end{array} \left\{ \begin{array}{rll} \int \frac{d^2q}{\pi} \frac{1}{\vec{q}^{\,2}+1} e^{i\vec{q}\vec{x}} &= 2\mathcal{K}_0(r) &= \int d\psi \; e^{-r\cosh\psi}, \\ \int \frac{d^2q}{i\pi} \frac{1}{\vec{q}^{\,2}-io-1} e^{i\vec{q}\vec{x}} &= \pi[\mathcal{J}_0 + i\mathcal{N}_0](r) &= -i\int d\psi \; e^{ir\cosh\psi}, \\ \int d^2q \; \delta(\vec{q}^{\,2}-1) e^{i\vec{q}\vec{x}} &= \pi\mathcal{J}_0(r) &= \int d\psi \; \sin(r\cosh\psi) \\ & &= \int_0^\pi d\theta \; \cos(r\cos\theta), \\ \int \frac{d^2q}{\pi} \frac{1}{\vec{q}^{\,2}_{\mathrm{P}}-1} e^{i\vec{q}\vec{x}} &= -\pi\mathcal{N}_0(r) &= \int d\psi \; \cos(r\cosh\psi), \end{array} \right.$$

Those functions are combined for the indefinite case with two-dimensional energy-momentum integrals:

$$\begin{array}{c} \mathbf{O}(1,1): \\ x^2 = x_0^2 - x_3^2, \\ |x| = \sqrt{|x^2|}, \end{array} \left\{ \begin{array}{rl} \int \frac{d^2q}{i\pi} \frac{1}{-q^2-io+1} e^{iqx} &= 2\mathcal{K}_0(\sqrt{-x^2+io}) \\ &= \sum_{k=0}^{\infty} \frac{(-\frac{x^2}{4})^k}{(k!)^2} [2\chi_k - \log \frac{-x^2+io}{4}] \\ &= \vartheta(-x^2) 2\mathcal{K}_0(|x|) - \vartheta(x^2) \\ &\quad \pi[\mathcal{N}_0 + i\mathcal{J}_0](|x|), \\ \int d^2q \; \delta(q^2-1) e^{iqx} &= \sum_{k=0}^{\infty} \frac{(-\frac{x^2}{4})^k}{(k!)^2} [2\chi_k - \log \frac{|x^2|}{4}] \\ &= \vartheta(-x^2) 2\mathcal{K}_0(|x|) - \vartheta(x^2)\pi\mathcal{N}_0(|x|), \\ \int \frac{d^2q}{\pi} \frac{1}{-q_{\mathrm{P}}^2+1} e^{iqx} &= -i\epsilon(x_0) \int d^2q \; \epsilon(q_0)\delta(q^2-1) e^{iqx} \\ &= \vartheta(x^2)\pi \sum_{k=0}^{\infty} \frac{(-\frac{x^2}{4})^k}{(k!)^2} = \vartheta(x^2)\pi\mathcal{J}_0(|x|), \\ \int \frac{d^2q}{i\pi} \frac{1}{q^2-io+1} e^{iqx} &= 2\mathcal{K}_0(\sqrt{x^2+io}) \\ &= \sum_{k=0}^{\infty} \frac{(\frac{x^2}{4})^k}{(k!)^2} [2\chi_k - \log \frac{x^2+io}{4}] \\ &= \vartheta(x^2) 2\mathcal{K}_0(|x|) \\ &\quad -\vartheta(-x^2)\pi[\mathcal{N}_0 + i\mathcal{J}_0](|x|). \end{array} \right.$$

By analytic continuation and orthogonally invariant embedding, one obtains for indefinite orthogonal groups

$$\begin{array}{c} \mathbf{O}(t,s): \\ t \geq 1, \; s \geq 1, \\ n = t+s, \\ n = 2,3,\ldots, \\ x^2 = \vec{x}^{\,2}_t - \vec{x}_s^2, \\ |x| = \sqrt{|x^2|}, \end{array} \left\{ \begin{array}{rl} \int \frac{d^nq}{i^t\pi^{\frac{n}{2}}} \frac{\Gamma(\frac{n}{2}-\nu)}{(-q^2-io)^{\frac{n}{2}-\nu}} e^{iqx} &= \frac{\Gamma(\nu)}{(\frac{-x^2+io}{4})^\nu}, \\ \int \frac{d^nq}{i^t\pi^{\frac{n}{2}}} \frac{\Gamma(\frac{n}{2}-\nu)}{(-q^2-io+1)^{\frac{n}{2}-\nu}} e^{iqx} & \\ &= \frac{\vartheta(-x^2)2\mathcal{K}_\nu(|x|) - \vartheta(x^2)\pi[\mathcal{N}_{-\nu} + i\mathcal{J}_{-\nu}](|x|)}{|\frac{x}{2}|^\nu} \\ &\quad -\delta_\nu^N i\pi \sum_{k=1}^{N} \frac{1}{(N-k)!} \delta^{(k-1)}(-\frac{x^2}{4}). \end{array} \right.$$

Equivalent formulas, also for $\mathbf{O}(t,s)$, are obtained by the exchange $(t,s,q^2,x^2) \leftrightarrow (s,t,-q^2-x^2)$. For integers $N=1,2,\dots$, there arise, via the phase of the logarithm, $x^2=0$ supported Dirac distributions

$$\begin{aligned}\log(-x^2-io) &= \log|x^2| - i\pi\vartheta(x^2),\\ \left(-\frac{\partial}{\partial\frac{x^2}{4}}\right)^k\vartheta(x^2) &= \delta^{(k-1)}(-\tfrac{x^2}{4}),\ k=1,2,\dots.\end{aligned}$$

The residual normalizations for positive and negative invariants a are

$$\begin{array}{l}\mathbf{O}(t,s):\\ n=1,2,\dots,\\ a\in\mathbb{R},\end{array}\left\{\begin{aligned}\int\frac{d^nq}{i^t\pi^{\frac{n}{2}}}\frac{\Gamma(\frac{n}{2}-\nu)}{(-q^2-io+a)^{\frac{n}{2}-\nu}} &= \int\frac{d^nq}{i^s\pi^{\frac{n}{2}}}\frac{\Gamma(\frac{n}{2}-\nu)}{(q^2-io+a)^{\frac{n}{2}-\nu}}\\ &= \frac{\Gamma(-\nu)}{(a-io)^{-\nu}},\\ \int\frac{2d^nq}{i^t|\Omega^{n-1}|}\frac{1}{(-q^2-io+a)^{\frac{n}{2}-\nu}} &= \int\frac{2d^nq}{i^s|\Omega^{n-1}|}\frac{1}{(q^2-io+a)^{\frac{n}{2}-\nu}}\\ &= \frac{\Gamma(\frac{n}{2})\Gamma(-\nu)}{\Gamma(\frac{n}{2}-\nu)}\frac{1}{(a-io)^{-\nu}}.\end{aligned}\right.$$

Hyperbolically invariant distributions are used for $(1,s)$-spacetime with the general Lorentz groups,

$$\mathbf{O}(1,s):\left\{\begin{aligned}\int\frac{d^{1+s}q}{i\pi^{\frac{1+s}{2}}}\frac{\Gamma(\frac{1+s}{2}-\nu)}{(-q^2-io)^{\frac{1+s}{2}-\nu}}e^{iqx} &= \frac{\Gamma(\nu)}{(\frac{-x^2+io}{4})^{\nu}},\\ \int\frac{d^{1+s}q}{i\pi^{\frac{1+s}{2}}}\frac{\Gamma(\frac{1+s}{2}-\nu)}{(-q^2-io+1)^{\frac{1+s}{2}-\nu}}e^{iqx} &= \frac{\vartheta(-x^2)2\mathcal{K}_\nu(|x|)-\vartheta(x^2)\pi[\mathcal{N}_{-\nu}+i\mathcal{J}_{-\nu}](|x|)}{|\frac{x}{2}|^\nu}\\ &\quad -\delta_\nu^N\, i\pi\sum_{k=1}^N\frac{1}{(N-k)!}\delta^{(k-1)}(-\tfrac{x^2}{4}),\\ \int\frac{d^{1+s}q}{i^s\pi^{\frac{1+s}{2}}}\frac{\Gamma(\frac{1+s}{2}-\nu)}{(q^2-io+1)^{\frac{1+s}{2}-\nu}}e^{iqx} &= \frac{\vartheta(x^2)2\mathcal{K}_\nu(|x|)-\vartheta(-x^2)\pi[\mathcal{N}_{-\nu}+i\mathcal{J}_{-\nu}](|x|)}{|\frac{x}{2}|^\nu}\\ &\quad -\delta_\nu^N\, i\pi\sum_{k=1}^N\frac{1}{(N-k)!}\delta^{(k-1)}(\tfrac{x^2}{4}).\end{aligned}\right.$$

With respect to hyperbolic differential equations with $\mathbf{SO}_0(1,s)$, Huygens' principle with spherical $\mathbf{SO}(s)$-boundary conditions holds for odd dimensions $1+s$, but not, however, for even spacetime dimensions [36].

8.13.2 Some Special Cases

Now special cases: For $\nu=-\frac{1}{2}$, there are *no singularities*:

$$\mathbf{O}(1+s):\left\{\begin{aligned}\int\frac{2d^{1+s}q}{|\Omega^{1+s}|}\frac{1}{(\vec q^{\,2})^{\frac{2+s}{2}}}e^{i\vec q\vec x} &= -r,\\ \int\frac{2d^{1+s}q}{|\Omega^{1+s}|}\frac{1}{(\vec q^{\,2}+1)^{\frac{2+s}{2}}}e^{i\vec q\vec x} &= e^{-r},\\ \int\frac{2d^{1+s}q}{i|\Omega^{1+s}|}\frac{1}{(\vec q^{\,2}-io-1)^{\frac{2+s}{2}}}e^{i\vec q\vec x} &= e^{ir},\end{aligned}\right.$$

$$\mathbf{O}(1,s): \begin{cases} \int \frac{2d^{1+s}q}{i|\Omega^{1+s}|} \frac{1}{(-q^2-io)^{\frac{2+s}{2}}} e^{iqx} = & -|x|[\vartheta(-x^2)+i\vartheta(x^2)], \\ \int \frac{2d^{1+s}q}{i|\Omega^{1+s}|} \frac{1}{(-q^2-io+1)^{\frac{2+s}{2}}} e^{iqx} = & \vartheta(-x^2)e^{-|x|}+\vartheta(x^2)e^{-i|x|}, \\ \int \frac{2d^{1+s}q}{i^s|\Omega^{1+s}|} \frac{1}{(q^2-io+1)^{\frac{2+s}{2}}} e^{iqx} = & \vartheta(x^2)e^{-|x|}+\vartheta(-x^2)e^{-i|x|}. \end{cases}$$

For $\nu = 0$, there is a *logarithmic singularity* in $\mathcal{K}_0$ and $\mathcal{N}_0$:

$$\mathbf{O}(1+s): \begin{cases} \int \frac{2d^{1+s}q}{|\Omega^s|} \frac{1}{(\vec{q}^2+1)^{\frac{1+s}{2}}} e^{i\vec{q}\vec{x}} = 2\mathcal{K}_0(r), \\ \int \frac{2d^{1+s}q}{i|\Omega^s|} \frac{1}{(\vec{q}^2-io-1)^{\frac{1+s}{2}}} e^{i\vec{q}\vec{x}} = \pi[\mathcal{J}_0+i\mathcal{N}_0](r), \end{cases}$$

$$\mathbf{O}(1,s): \begin{cases} \int \frac{2d^{1+s}q}{i|\Omega^s|} \frac{1}{(-q^2-io+1)^{\frac{1+s}{2}}} e^{iqx} \\ \qquad = \vartheta(-x^2)2\mathcal{K}_0(|x|) - \vartheta(x^2)\pi[\mathcal{N}_0+i\mathcal{J}_0](|x|), \\ \int \frac{2d^{1+s}q}{i^s|\Omega^s|} \frac{1}{(q^2-io+1)^{\frac{1+s}{2}}} e^{iqx} \\ \qquad = \vartheta(x^2)2\mathcal{K}_0(|x|) - \vartheta(-x^2)\pi[\mathcal{N}_0+i\mathcal{J}_0](|x|). \end{cases}$$

The Fourier-transformed Dirac part of the *simple poles* is used for representations of the affine groups $\mathbf{SO}(1+s) \vec{\times} \mathbb{R}^{1+s}$ and $\mathbf{SO}_0(1,s) \vec{\times} \mathbb{R}^{1+s}$:

$$\mathbf{O}(1+s): \begin{cases} \int \frac{d^{1+s}q}{\pi^{\frac{1+s}{2}}} \frac{1}{\vec{q}^2} e^{i\vec{q}\vec{x}} = \frac{\Gamma(\frac{s-1}{2})}{(\frac{r}{2})^{s-1}}, \\ \int \frac{d^{1+s}q}{\pi^{\frac{1+s}{2}}} \frac{1}{\vec{q}^2+1} e^{i\vec{q}\vec{x}} = \frac{2\mathcal{K}_{\frac{s-1}{2}}(r)}{(\frac{r}{2})^{\frac{s-1}{2}}}, \\ \int \frac{d^{1+s}q}{i\pi^{\frac{1+s}{2}}} \frac{1}{\vec{q}^2-io-1} e^{i\vec{q}\vec{x}} = \frac{\pi[\mathcal{J}_{\frac{s-1}{2}}+i\mathcal{N}_{\frac{s-1}{2}}](r)}{(\frac{r}{2})^{\frac{s-1}{2}}}, \end{cases}$$

$$\mathbf{O}(1,s): \begin{cases} \int \frac{d^{1+s}q}{i\pi^{\frac{1+s}{2}}} \frac{1}{-q^2-io} e^{iqx} = \frac{\Gamma(\frac{s-1}{2})}{(\frac{-x^2+io}{4})^{\frac{s-1}{2}}}, \\ \int \frac{d^{1+s}q}{i\pi^{\frac{1+s}{2}}} \frac{1}{-q^2-io+1} e^{iqx} = \frac{\vartheta(-x^2)2\mathcal{K}_{\frac{s-1}{2}}(|x|)-\vartheta(x^2)\pi[\mathcal{N}_{-\frac{s-1}{2}}+i\mathcal{J}_{-\frac{s-1}{2}}](|x|)}{|\frac{x}{2}|^{\frac{s-1}{2}}} \\ \qquad\qquad -\delta^{2R}_{1+s}\, i\pi \sum_{k=1}^{R-1} \frac{1}{(R-1-k)!}\delta^{(k-1)}(-\frac{x^2}{4}), \\ \int \frac{d^{1+s}q}{i^s\pi^{\frac{1+s}{2}}} \frac{1}{q^2-io+1} e^{iqx} = \frac{\vartheta(x^2)2\mathcal{K}_{\frac{s-1}{2}}(|x|)-\vartheta(-x^2)\pi[\mathcal{N}_{-\frac{s-1}{2}}+i\mathcal{J}_{-\frac{s-1}{2}}](|x|)}{|\frac{x}{2}|^{\frac{s-1}{2}}} \\ \qquad\qquad -\delta^{2R}_{1+s}\, i\pi \sum_{k=1}^{R-1} \frac{1}{(R-1-k)!}\delta^{(k-1)}(\frac{x^2}{4}). \end{cases}$$

The lightcone-supported Dirac distributions arise for even-dimensional spacetime with nonflat position, i.e., for $(1,s) = (1,3), (1,5), \ldots$.

The one-dimensional pole integrals are spread to odd dimensions starting with $1+s=3$ and with a singularity at $|x|=0$:

$$\mathbf{O}(3): \begin{cases} \int \frac{d^3q}{\pi^2} \, \frac{1}{\vec{q}^{\,2}} e^{i\vec{q}\vec{x}} = \frac{2}{r}, \\ \int \frac{d^3q}{\pi^2} \frac{1}{\vec{q}^{\,2}+1} e^{i\vec{q}\vec{x}} = -\frac{\partial}{\partial \frac{r^2}{4}} e^{-r} = 2\frac{e^{-r}}{r}, \\ \int \frac{d^3q}{i\pi^2} \frac{1}{\vec{q}^{\,2}-io-1} e^{i\vec{q}\vec{x}} = \frac{\partial}{\partial \frac{r^2}{4}} e^{ir} = 2\frac{e^{ir}}{ir}, \end{cases}$$

$$\mathbf{O}(1,2): \begin{cases} \int \frac{d^3q}{i\pi^2} \, \frac{1}{-q^2-io} e^{iqx} = 2\frac{\vartheta(-x^2)-i\vartheta(x^2)}{|x|}, \\ \int \frac{d^3q}{i\pi^2} \frac{1}{-q^2-io+1} e^{iqx} = 2\frac{\vartheta(-x^2)e^{-|x|}-\vartheta(x^2)ie^{-i|x|}}{|x|}, \\ -\int \frac{d^3q}{\pi^2} \frac{1}{q^2-io+1} e^{iqx} = 2\frac{\vartheta(x^2)e^{-|x|}-\vartheta(-x^2)ie^{-i|x|}}{|x|}. \end{cases}$$

The *dipoles in three dimensions* are without singularity:

$$\mathbf{O}(3): \begin{cases} \int \frac{d^3q}{\pi^2} \frac{1}{(\vec{q}^{\,2})^2} e^{i\vec{q}\vec{x}} = -r, \\ \int \frac{d^3q}{\pi^2} \frac{1}{(\vec{q}^{\,2}+1)^2} e^{i\vec{q}\vec{x}} = e^{-r}, \\ \int \frac{d^3q}{i\pi^2} \frac{1}{(\vec{q}^{\,2}-io-1)^2} e^{i\vec{q}\vec{x}} = e^{ir}, \end{cases}$$

$$\mathbf{O}(1,2): \begin{cases} \int \frac{d^3q}{i\pi^2} \, \frac{1}{(-q^2-io)^2} e^{iqx} = |x|[-\vartheta(-x^2)+i\vartheta(x^2)], \\ \int \frac{d^3q}{i\pi^2} \frac{1}{(-q^2-io+1)^2} e^{iqx} = \vartheta(-x^2)e^{-|x|}+\vartheta(x^2)e^{-i|x|}, \\ -\int \frac{d^3q}{\pi^2} \frac{1}{(q^2-io+1)^2} e^{iqx} = \vartheta(x^2)e^{-|x|}+\vartheta(-x^2)e^{-i|x|}. \end{cases}$$

The two-dimensional integrals are spread to even dimensions, starting with $1+s=4$:

$$\mathbf{O}(4): \begin{cases} \int \frac{d^4q}{\pi^2} \, \frac{1}{\vec{q}^{\,2}} e^{i\vec{q}\vec{x}} = \frac{4}{r^2}, \\ \int \frac{d^4q}{\pi^2} \frac{1}{\vec{q}^{\,2}+1} e^{i\vec{q}\vec{x}} = -\frac{\partial}{\partial \frac{r^2}{4}} 2\mathcal{K}_0(r) = \frac{2\mathcal{K}_1(r)}{\frac{r}{2}}, \\ \int \frac{d^4q}{i\pi^2} \frac{1}{\vec{q}^{\,2}-io-1} e^{i\vec{q}\vec{x}} = -\frac{\partial}{\partial \frac{r^2}{4}} \pi[\mathcal{J}_0+i\mathcal{N}_0](r) = \frac{\pi[\mathcal{J}_1+i\mathcal{N}_1](r)}{\frac{r}{2}}, \end{cases}$$

$$\mathbf{O}(1,3): \begin{cases} \int \frac{d^4q}{i\pi^2} \, \frac{1}{-q^2-io} e^{iqx} = \frac{4}{-x^2+io}, \\ \int \frac{d^4q}{i\pi^2} \frac{1}{-q^2-io+1} e^{iqx} = \frac{\vartheta(-x^2)2\mathcal{K}_1(|x|)+\vartheta(x^2)\pi[\mathcal{N}_1+i\mathcal{J}_1](|x|)}{\frac{|x|}{2}} - i\pi\delta(\frac{x^2}{4}), \\ -\int \frac{d^4q}{i\pi^2} \frac{1}{q^2-io+1} e^{iqx} = \frac{\vartheta(x^2)2\mathcal{K}_1(|x|)+\vartheta(-x^2)\pi[\mathcal{N}_1+i\mathcal{J}_1](|x|)}{\frac{|x|}{2}} - i\pi\delta(\frac{x^2}{4}). \end{cases}$$

Dipoles for four spacetime dimensions lead to maximally logarithmic singularities.

8.14 Hypergeometric Functions

A hypergeometric function,

$$\alpha,\gamma,z\in\mathbb{C}:\quad\begin{cases}{}_0\mathrm{F}_1(\gamma;z) &= \sum_{k=0}^{\infty}\frac{\Gamma(\gamma)}{\Gamma(\gamma+k)}\frac{z^k}{k!},\\ {}_1\mathrm{F}_1(\alpha;\gamma;z) &= \sum_{k=0}^{\infty}\frac{\Gamma(\alpha+k)}{\Gamma(\alpha)}\frac{\Gamma(\gamma)}{\Gamma(\gamma+k)}\frac{z^k}{k!},\\ {}_2\mathrm{F}_1(\alpha_1,\alpha_2;\gamma;z) &= \sum_{k=0}^{\infty}\frac{\Gamma(\alpha_1+k)}{\Gamma(\alpha_1)}\frac{\Gamma(\alpha_2+k)}{\Gamma(\alpha_2)}\frac{\Gamma(\gamma)}{\Gamma(\gamma+k)}\frac{z^k}{k!},\end{cases}$$

is in the case of at least one nonpositive integer α_j, e.g., $\alpha_1\in\{-1,-2,\dots\}$, a polynomial of degree $|\alpha_1|$. Functions of type ${}_1F_1$ were used for quantum mechanical wave functions (see Chapter 4).

The spherical Bessel functions for the Euclidean spaces $\mathbb{R}^s$ are hypergeometric functions of type ${}_0F_1$:

$$\begin{aligned}\Gamma(\tfrac{s}{2})\frac{\mathcal{J}_{\frac{s-2}{2}}(|P|r)}{\left(\frac{|P|r}{2}\right)^{\frac{s-2}{2}}} &= \int\frac{d\omega^{s-1}}{|\Omega^{s-1}|}\,e^{i|P|\vec{\omega}_{s-1}\vec{x}} = {}_0\mathrm{F}_1(\tfrac{s}{2};z)\\ &= \sum_{k=0}^{\infty}\frac{\Gamma(\frac{s}{2})}{\Gamma(\frac{s}{2}+k)}\frac{z^k}{k!},\quad z=-\frac{P^2r^2}{4}.\end{aligned}$$

The Legendre functions for the hyperboloids $\mathcal{Y}^s$ are hypergeometric functions of type ${}_2F_1$ with $\lambda=iQ-\frac{s-1}{2}$:

$$\begin{aligned}\mathrm{P}^\lambda(\cosh\psi) &= \int_{-\pi}^{\pi}\frac{d\varphi}{2\pi}(\cosh\psi+\cos\varphi\sinh\psi)^\lambda\\ &= {}_2\mathrm{F}_1(1+\lambda,-\lambda;1;z) = \sum_{k=0}^{\infty}\frac{\Gamma(\lambda+k+1)}{\Gamma(\lambda-k+1)}\frac{(-z)^k}{(k!)^2},\quad z=-\sinh^2\frac{\psi}{2}.\end{aligned}$$

The Legendre polynomials for the 2-sphere Ω^2 are the compact partners with the transition $\lambda\to L=0,1,\dots$ and $\psi\to i\theta$:

$$\begin{aligned}\mathrm{P}^L(\cos\theta) &= \int_{-\pi}^{\pi}\frac{d\varphi}{2\pi}(\cos\theta+i\cos\varphi\sin\theta)^L\\ &= {}_2\mathrm{F}_1(1+L,-L;1;z) = \sum_{k=0}^{L}\frac{(L+k)!}{(L-k)!}\frac{(-z)^k}{(k!)^2},\quad z=\sin^2\frac{\theta}{2}.\end{aligned}$$

Precisely for the minimal nonabelian case $s=2$, the Legendre $\mathcal{Y}^2$-functions are real:

$$\begin{aligned}\lambda=iQ-\tfrac{s-1}{2},\ Q\in\mathbb{R}:\lambda(1+\lambda)\in\mathbb{R} &\iff s=2\Rightarrow-\lambda(1+\lambda)=Q^2+\tfrac{1}{4},\\ \mathbb{R}\ni{}_2F_1(\tfrac{1}{2}+iQ,\tfrac{1}{2}-iQ;1;z) &= \mathrm{P}^{iQ-\frac{1}{2}}(\cosh\psi)=\mathrm{P}^{-iQ-\frac{1}{2}}(\cosh\psi)\\ &= \sum_{k=0}^{\infty}\alpha_k(Q)\frac{z^k}{(k!)^2},\quad z=-\sinh^2\frac{\psi}{2},\\ \text{with }\alpha_k(Q)=(-1)^k\frac{\Gamma(iQ+\frac{1}{2}+k)}{\Gamma(iQ+\frac{1}{2}-k)} &= \prod_{n=1}^{k}[Q^2+\tfrac{(2n-1)^2}{4}].\end{aligned}$$

In the Inönü-Wigner contraction $\mathbf{SO}_0(1,1) \to \mathbb{R}$ ("flattening"),

$$\begin{aligned}
\mathbf{SO}_0(1,1) \ni \Lambda(\psi) &= \begin{pmatrix} \cosh\psi & Q\sinh\psi \\ \frac{1}{Q}\sinh\psi & \cosh\psi \end{pmatrix}, \\
\lim_{Q\to\infty}\Lambda(\psi) &= \begin{pmatrix} 1 & Px \\ 0 & 1 \end{pmatrix} \in e^{\mathbb{R}} \cong \mathbb{R} \text{ with } \psi = \tfrac{Px}{Q}, \\
&\text{similarly } \lim_{Q\to\infty}\mathbf{SO}_0(1,s) = \mathbf{SO}(s) \vec{\times} \mathbb{R}^s, \text{ with } \psi = \tfrac{Pr}{Q},
\end{aligned}$$

all hyperbolic Legendre functions lead to Bessel functions for trivial index:

$$\begin{aligned}
&\lim_{Q\to\infty}\tfrac{1}{(-Q^2)^k}\tfrac{\Gamma(iQ+\rho+k)}{\Gamma(iQ+\rho-k)} = 1, \quad \lim_{Q\to\infty} Q^2\sinh^2\tfrac{\psi}{2} = \tfrac{P^2r^2}{4} \text{ for } \psi = \tfrac{Pr}{Q}, \\
&\lim_{Q\to\infty}\mathrm{P}^{iQ+\rho}(\cosh\psi) = {}_0F_1(1;Pr) = \mathcal{J}_0(Pr) \\
&\qquad = \int_{-\pi}^{\pi}\tfrac{d\varphi}{2\pi}\, e^{iPr\cos\varphi} = \sum_{k=0}^{\infty}\tfrac{(Pr)^k}{(k!)^2}.
\end{aligned}$$

There is the connection of Macdonald (Bessel, Neumann) functions for $\nu \in \mathbb{R}$ with exponentially multiplied hypergeometric functions,

$$\begin{aligned}
\mathbb{C} \ni z \longmapsto 2\mathcal{K}_\nu(z) &= \sqrt{4\pi}(2z)^\nu\ {}_1G_1(\tfrac{1}{2}+\nu;1+2\nu;2z)e^{-z} \\
&= \Big[\ \Gamma(-\nu)(\tfrac{z}{2})^\nu\ {}_1F_1(\tfrac{1}{2}+\nu;1+2\nu;2z) \\
&\qquad + \Gamma(\nu)\ (\tfrac{z}{2})^{-\nu}\ {}_1F_1(\tfrac{1}{2}-\nu;1-2\nu;2z)\Big]e^{-z} \\
&= \tfrac{\sqrt{\pi}}{\sin\pi\nu}\left[-\sum_{k=0}^{\infty}\tfrac{\Gamma(\frac{1}{2}+\nu+k)}{\Gamma(1+2\nu+k)}\tfrac{(2z)^{k+\nu}}{k!} + \sum_{k=0}^{\infty}\tfrac{\Gamma(\frac{1}{2}-\nu+k)}{\Gamma(1-2\nu+k)}\tfrac{(2z)^{k-\nu}}{k!}\right]e^{-z},
\end{aligned}$$

where $\Gamma(\nu)\Gamma(1-\nu) = \frac{\pi}{\sin\pi\nu}$ and

$$\begin{aligned}
{}_1G_1(\alpha;\gamma;z) &= \tfrac{\Gamma(1-\gamma)}{\Gamma(\alpha-\gamma+1)}\ {}_1F_1(\alpha;\gamma;z) + \tfrac{\Gamma(\gamma-1)}{\Gamma(\alpha)}\ z^{1-\gamma}\ {}_1F_1(1+\alpha-\gamma;2-\gamma;z) \\
&= \tfrac{1}{\Gamma(\alpha)}\int_0^\infty dQ\ e^{-Qz}Q^{\alpha-1}(1+Q)^{\gamma-\alpha-1} \text{ if } \mathrm{Re}\,\alpha > 0.
\end{aligned}$$

The functions ${}_1F_1(\alpha;\gamma;z) = {}_1F_1(\gamma-\alpha;\gamma;z)e^z$ and ${}_1G_1(\alpha;\gamma;z)$ are linearly independent solutions of the *confluent hypergeometric equation*

$$[zd_z^2 + (\gamma - z)d_z - \alpha]f(z) = 0.$$

Chapter 9

Convolutions and Product Representations

The Feynman integrals in special relativistic quantum field theories involve convolutions of energy-momentum distributions. The on-shell parts for translation representations give product representation coefficients of the Poincaré group, i.e., energy-momentum distributions for free states (multiparticle measures, discussed ahead). The off-shell interaction contributions ("virtual particles") are not convolutable; this is the origin of the "divergence" problem in quantum field theories with interactions. With respect to Poincaré group representations, the convolution of Feynman propagators embedding the pointwise product of interactions, e.g. $(\frac{e^{-mr}}{r})^2$, makes no sense.

The pointwise product algebra of the essentially bounded complex functions of a real Lie group, $L^\infty(G) = L^\infty(G) \cdot L^\infty(G)$, characterizes its representations. The cone of positive-type functions $d = \hat{d} \in L^\infty(G)_+$ induces the scalar products for cyclic Hilbert representations (see Chapter 8). Its conjugation property $d \leftrightarrow d^- = \overline{d}$ connects dual representations. The pointwise product of two positive-type functions is a positive-type function for the product representation:

$$d_1 \cdot d_2(g) = \langle c_1|D_1(g)|c_1\rangle\langle c_2|D_2(g)|c_2\rangle = \langle c_1, c_2|D_1 \otimes D_2(g)|c_1, c_2\rangle.$$

With the trivial representation and its constant positive type function $d^1 = 1$ as unit for the pointwise product, the cyclic Hilbert representation classes $L^\infty(G)_+$ constitute a monoid.

The energies for time translations $\mathbb{R}$ and the momenta for position translations $\mathbb{R}^3$ are, as eigenvalues, the characters $q \in \check{\mathbb{R}}^n$ (representation classes, dual group) of the additive group $\mathbb{R}^n$. The Radon (energy-)momentum measures are a convolution algebra, which reflects the pointwise multiplication

H. Saller, *Operational Spacetime: Interactions and Particles*,

Fundamental Theories of Physics 163, DOI 10.1007/978-1-4419-0898-8_10,

property of the essentially bounded function classes:

$$\begin{aligned} L^\infty(\mathbb{R}^n) \cdot L^\infty(\mathbb{R}^n) &= L^\infty(\mathbb{R}^n), \quad \mathcal{M}(\check{\mathbb{R}}^n) * \mathcal{M}(\check{\mathbb{R}}^n) = \mathcal{M}(\check{\mathbb{R}}^n), \\ d_1 \cdot d_2(x) &= \int \frac{d^n q}{(2\pi)^n} \, \tilde{d}_1 * \tilde{d}_2(q) e^{iqx}. \end{aligned}$$

Product representations of $\mathbb{R}^n$ come with the product of representation coefficients, i.e., with the convolution $*$ of (energy-)momentum distributions:

$$* \sim \delta(q_1 + q_2 - q).$$

The convolution adds (energy-)momenta of singularity manifolds, which support imaginary and real eigenvalues for compact and noncompact representation invariants.

After exemplifying a composite structure with spacetime product representations by a Nambu–Goldstone field in the case of a chirally degenerate ground-state, this chapter considers the (energy-)momentum convolution structure of time, position, and spacetime representations.

9.1 Composite Nambu–Goldstone Bosons

Electrodynamics and gravity, in the flat spacetime approach, are implemented by massless fields. Also, a *degenerate ground-state* ("spontaneous symmetry breakdown") comes with long-range interactions, characterized, qualitatively, by the "broken" symmetries of a Lie group G, i.e., by the degeneracy-effecting operations in the classes G/H with respect to the remaining "unbroken" symmetry for a distinguished subgroup $H \subseteq G$, and, quantitatively, by a dilation scale, a "breakdown mass." The long-range interactions come in representations of the "unbroken" local group H.

As illustrated, e.g., by superfluid helium III, a dynamics, invariant under a group G, can have different ground-states (phases) with degeneracy orbits G/H_ι for different invariance groups $H_\iota \subseteq G$. This is in some analogy to different boundary conditions for a classical dynamics or different cosmological models (see Chapter 1).

A "dynamical breakdown" establishes the degeneracy-implementing massless fields as product representations of spacetime. In the model of Nambu and Jona-Lasinio, the massless chiral Goldstone boson for a $\mathbf{U}(1)$-degeneracy ("breakdown") is a Lorentz pseudoscalar bound state of self-interacting massive Dirac fermions.

9.1.1 Chirality

The relative phase of the $\mathbf{SL}(\mathbb{C}^2)$-irreducible left- and right-handed Weyl spinors in a Dirac spinor is affected by the *chiral transformations*,

$$\begin{aligned} \mathbf{U}(1): \quad \boldsymbol{\Psi} &\longmapsto e^{\gamma_5 \frac{\alpha}{2}} \boldsymbol{\Psi}, \quad \overline{\boldsymbol{\Psi}} \longmapsto \overline{\boldsymbol{\Psi}} e^{\gamma_5 \frac{\alpha}{2}}, \\ \boldsymbol{\Psi} &= \begin{pmatrix} \frac{\mathbf{1}_4 + i\gamma_5}{2} \\ \frac{\mathbf{1}_4 - i\gamma_5}{2} \end{pmatrix} \boldsymbol{\Psi} = \begin{pmatrix} \mathbf{l} \\ \mathbf{r} \end{pmatrix} \longmapsto \begin{pmatrix} e^{i\frac{\alpha}{2}} \mathbf{l} \\ e^{-i\frac{\alpha}{2}} \mathbf{r} \end{pmatrix}. \end{aligned}$$

The $16 = 1+4+6+4+1$ elements of the Clifford algebra $\{\gamma_a, \gamma_b\} = 2\eta_{ab}\mathbf{1}_4$ for Minkowski spacetime, represented by bilinear Dirac spinors, involve five different couplings:

$$\begin{array}{l} \mathbf{1}_4,\ \gamma_a,\ \gamma_{ab} = -\frac{1}{4}[\gamma_a, \gamma_b],\ \gamma_5\gamma_a,\ \gamma_5 = \frac{\epsilon^{abcd}}{4!}\gamma_a\gamma_b\gamma_c\gamma_d, \\ \text{with } \{\gamma_a, \gamma_5\} = 0, \quad \gamma_5^2 = -\mathbf{1}_4, \end{array}$$

with vector γ_a and axial vector $\gamma_5\gamma_a$ as chiral invariants, and two chiral $\mathbf{SO}(2)$-dublets, given by scalar and pseudoscalar $(\mathbf{1}_4, \gamma_5)$, and by the tensor via $(\gamma_{ab}, \gamma_5\gamma_{ab} = \frac{1}{2}\epsilon_{abcd}\gamma^{cd})$, e.g.,

$$\mathbf{U}(1): \begin{pmatrix} \overline{\mathbf{\Psi}}\mathbf{1}_4\mathbf{\Psi} \\ \overline{\mathbf{\Psi}}\gamma_5\mathbf{\Psi} \end{pmatrix} \longmapsto \begin{pmatrix} \cos\alpha & \sin\alpha \\ -\sin\alpha & \cos\alpha \end{pmatrix} \begin{pmatrix} \overline{\mathbf{\Psi}}\mathbf{1}_4\mathbf{\Psi} \\ \overline{\mathbf{\Psi}}\gamma_5\mathbf{\Psi} \end{pmatrix}, \quad \overline{\mathbf{\Psi}}\frac{\mathbf{1}_4 \pm i\gamma_5}{2}\mathbf{\Psi} = \begin{pmatrix} \mathbf{r}^*\mathbf{l} \\ \mathbf{l}^*\mathbf{r} \end{pmatrix} \longmapsto \begin{pmatrix} e^{i\alpha}\,\mathbf{r}^*\mathbf{l} \\ e^{-i\alpha}\mathbf{l}^*\mathbf{r} \end{pmatrix}.$$

The five quartic Lorentz scalar couplings,

$$\begin{pmatrix} \mathbf{1}_4 \otimes \mathbf{1}_4 \\ \gamma_a \otimes \gamma^a \\ \gamma_{ab} \otimes \gamma^{ab} \\ \gamma_5\gamma_a \otimes \gamma_5\gamma^a \\ \gamma_5 \otimes \gamma_5 \end{pmatrix} = \begin{pmatrix} s \\ v \\ t \\ a \\ p \end{pmatrix},$$

have, by the involutive Fierz recoupling[1] (via index exchange $MN \leftrightarrow NM$), two Fierz-symmetric and three Fierz-antisymmetric linear combinations as eigenvectors:

$$\begin{pmatrix} s \\ v \\ t \\ a \\ p \end{pmatrix}_{KL}^{MN} \overset{\text{Fierz}}{=} F \begin{pmatrix} s \\ v \\ t \\ a \\ p \end{pmatrix}_{KL}^{NM},$$

$$F = \left(\begin{array}{cc|c|cc} \frac{1}{4} & \frac{1}{4} & -\frac{1}{8} & \frac{1}{4} & -\frac{1}{4} \\ 1 & -\frac{1}{2} & 0 & \frac{1}{2} & 1 \\ \hline -3 & 0 & -\frac{1}{2} & 0 & 3 \\ \hline 1 & \frac{1}{2} & 0 & -\frac{1}{2} & 1 \\ -\frac{1}{4} & \frac{1}{4} & \frac{1}{8} & \frac{1}{4} & \frac{1}{4} \end{array}\right), \quad F^2 = \mathbf{1}_5,$$

$$\Rightarrow \left(\begin{array}{c} s - p - \frac{t}{6} \\ s - p + \frac{t}{2} \\ s + p + \frac{v+a}{2} \\ \hline s + p - \frac{v+a}{2} \\ v - a \end{array}\right) \overset{\text{Fierz}}{\leftrightarrow} \left(\begin{array}{c} s - p - \frac{t}{6} \\ -(s - p + \frac{t}{2}) \\ s + p + \frac{v+a}{2} \\ \hline -(s + p - \frac{v+a}{2}) \\ -(v - a) \end{array}\right).$$

With the chiral invariants,

$$\mathbf{U}(1): \quad (v, a, s+p) \longmapsto (v, a, s+p),$$

one can combine the chiral invariant 4-fermion couplings. The two Fierz antisymmetric ones $\{s + p - \frac{v+a}{2}, v - a\}$ are nontrivial for Fermi Dirac fields. The chiral invariant coupling with a scalar term $s + p - \frac{v+a}{2}$ is used in the nonlinear interaction of the model of Nambu and Jona-Lasinio with the equation of motion:

$$i\gamma\partial\mathbf{\Psi} = \frac{1}{2}\Gamma\mathbf{\Psi}(\overline{\mathbf{\Psi}}\Gamma\mathbf{\Psi}), \text{ with } \Gamma \otimes \Gamma = \frac{1}{4}(\mathbf{1}_4 \otimes \mathbf{1}_4 + \gamma_5 \otimes \gamma_5 - \frac{\gamma_a \otimes \gamma^a + \gamma_5\gamma_a \otimes \gamma_5\gamma^a}{2}).$$

[1] The numbers in the 5×5-matrix are Lorentz group recoupling coefficients analogous to $9j$-symbols for the rotation group.

This quartic interaction is distinguished as the square of the radial part $\mathbf{R}^2 = \mathbf{\Phi}\mathbf{\Phi}^*$ in the scalar and pseudoscalar Dirac field product, which yield the analogue of the $\mathbf{U}(1) \times \mathbf{D}(1)$-factorization of a basic chiral Higgs field $\mathbf{\Phi} = \mathbf{r}^*\mathbf{l} = e^{i\boldsymbol{\alpha}}\mathbf{R}$:

$$\overline{\mathbf{\Psi}}\frac{\mathbf{1}_4 \pm i\gamma_5}{2}\mathbf{\Psi} = e^{\pm i\boldsymbol{\alpha}}\mathbf{R} \Rightarrow \begin{cases} \mathbf{R}^2 &= (\overline{\mathbf{\Psi}}\frac{\mathbf{1}_4 + i\gamma_5}{2}\mathbf{\Psi})(\overline{\mathbf{\Psi}}\frac{\mathbf{1}_4 - i\gamma_5}{2}\mathbf{\Psi}) = (\mathbf{r}^*\mathbf{l})(\mathbf{l}^*\mathbf{r}) \\ &= \frac{1}{2}(\overline{\mathbf{\Psi}}\Gamma\mathbf{\Psi})(\overline{\mathbf{\Psi}}\Gamma\mathbf{\Psi}), \\ e^{\pm i\boldsymbol{\alpha}} &= \frac{\overline{\mathbf{\Psi}}\frac{\mathbf{1}_4 \pm i\gamma_5}{2}\mathbf{\Psi}}{\mathbf{R}}. \end{cases}$$

It arises in the corresponding classical Lagrangian:

$$\mathbf{L}(\mathbf{\Psi}) = i\overline{\mathbf{\Psi}}\gamma\partial\mathbf{\Psi} - \frac{1}{4}(\overline{\mathbf{\Psi}}\Gamma\mathbf{\Psi})(\overline{\mathbf{\Psi}}\Gamma\mathbf{\Psi}).$$

9.1.2 Chiral Degeneracy

An interaction-free Dirac field has a Fock ground-state vector $|0\rangle$ for its Feynman propagator:

$$\langle 0|\mathbf{\Psi}(y)\overline{\mathbf{\Psi}}(x)|0\rangle_{\text{Feynman}} = \rho(m^2)\frac{i}{\pi}\int \frac{d^4q}{(2\pi)^3}\frac{\gamma q + m}{q^2 + io - m^2}e^{iq(x-y)} = d^m(x-y).$$

An interaction determines the normalization of the Poincaré group representation: A coupling constant $g_0\Gamma \otimes \Gamma$ can be absorbed into the normalization $\rho(m^2) \longmapsto g_0\rho(m^2)$.

A nontrivial mass term $m \neq 0$ characterizes a chirally $\mathbf{U}(1)$-degenerate ground-state. The local group (unbroken subgroup, fixgroup of the ground-state orbit) is trivial, $G/H = \mathbf{U}(1)/\{1\}$. Particles are chiral singlets. The nonlinear field equation yields, in a first-order approximation, a self-consistency condition ("gap equation") for the mass m as the chiral breakdown parameter:

$$\begin{aligned} i\gamma\partial\mathbf{\Psi} &= -\Gamma\mathbf{\Psi}\,\mathrm{tr}\,\Gamma d^m(0) + \tfrac{1}{2} : \Gamma\mathbf{\Psi}(\overline{\mathbf{\Psi}}\Gamma\mathbf{\Psi}) := -m\mathbf{\Psi} + \ldots \\ &\Rightarrow m = \tfrac{1}{4}\,\mathrm{tr}\,\mathbf{1}_4 d^m(0) = m\frac{i}{\pi}\int \frac{d^4q}{(2\pi)^3}\frac{\rho(m^2)}{q^2 + io - m^2}. \end{aligned}$$

Here, the distribution d^m of the spacetime translations, used for a free particle, is taken for the trivial translation $x = 0$. With the volume of the mass hyperboloid $\int d^4q\ \delta(q^2 - 1) = |\mathcal{Y}^3|$, it is "divergent"; $d^m(0)$ does not make sense. A perturbative approach with free fields is inappropriate for a bound-state problem. The model of Nambu and Jona-Lasinio is nonrenormalizable.

The Fock ground-state vector $|0\rangle$ for free fields in flat spacetime has to be replaced by a ground-state vector $|\mathcal{C}\rangle$ from a *chiral ground-state manifold* $\mathbf{U}(1)$ for fields with an interaction. Such a ground-state will be implemented in the form of a regularization, e.g., by a dipole at mass M^2 in the regularized Feynman propagator:

$$\begin{aligned} \langle \mathcal{C}|\mathbf{\Psi}(y)\overline{\mathbf{\Psi}}(x)|\mathcal{C}\rangle_{\text{Feynman}} &= d^{M^2,m}(x-y) \\ &= \rho(m^2)\frac{i}{\pi}\int \frac{d^4q}{(2\pi)^3}\frac{(m^2 - M^2)^2}{(q^2 + io - M^2)^2}\frac{\gamma q + m}{q^2 + io - m^2}e^{iq(x-y)}. \end{aligned}$$

Other regularizations are possible.

With respect to the spacetime translation representations, the dipole regularization involves a ghost pair (Witt pair) with an indefinite $\mathbf{U}(1,1)$-metric (see Chapter 4). It cannot be interpreted as a particle:

$$\begin{aligned}\frac{(m^2-M^2)^2}{(q^2-M^2)^2(q^2-m^2)} &= \frac{1}{q^2-m^2} - \frac{1}{q^2-M^2} - \frac{m^2-M^2}{(q^2-M^2)^2} \\ &\to \frac{1}{q^2-m^2} \text{ for } M^2 \to \infty.\end{aligned}$$

The modification of the free-field propagator for flat spacetime by the regularization, here by the dipole with the "flattening" (contraction) $d^{M^2,m} \to d^m$ for $M^2 \to \infty$, will be related to the transition to representations of curved spacetime. This connects a nontrivial curvature and interaction (see Chapter 11).

The dipole-regularized Feynman propagator can be used as a function of the spacetime translations $\mathbb{R}^4 \ni x \longmapsto d^{M^2,m}(x)$, defined for the trivial translation in the modified consistency equation,

$$\begin{aligned}m\mathbf{1}_4 &= \langle\mathcal{C}|\mathbf{\Psi}(x)\overline{\mathbf{\Psi}}(x)|\mathcal{C}\rangle_{\text{Feynman}} = d^{M^2,m}(0) \\ &= m\mathbf{1}_4\frac{i}{\pi}\int \frac{d^4q}{(2\pi)^3}\frac{(m^2-M^2)^2}{(q^2+io-M^2)^2}\frac{\rho(m^2)}{q^2+io-m^2} \\ &= m\mathbf{1}_4\frac{\rho(m^2)}{8\pi^2}\left(M^2-m^2-m^2\log\frac{M^2}{m^2}\right).\end{aligned}$$

For a chiral degeneracy, the consistency equation is a representation normalization. It determines the ratio of the chiral breakdown mass and the regularization mass in terms of the normalization factor $\rho(m^2)$:

$$m \neq 0: \quad \frac{1}{m}d^{M^2,m}(0) = \mathbf{1}_4 \Rightarrow \frac{M^2}{m^2} - 1 - \log\frac{M^2}{m^2} = \frac{8\pi^2}{m^2\rho(m^2)}.$$

The normalization factor is trivial for infinite regularization mass.

9.1.3 Massless Chiral Boson

The *Green's distribution* of the free Dirac equation with mass,

$$(i\gamma\partial + m)\kappa^m(x) = \delta(x) \Rightarrow \kappa^m(x) = -\int \frac{d^4q}{(2\pi)^4}\frac{\gamma q+m}{q^2+io-m^2}e^{iqx},$$

is not a function. *Only for a free theory (flat spacetime) can the Feynman propagator be identified, up to a constant factor, with the Green's distribution* $d^m = -2i\rho(m^2)\kappa^m$. In the regularization used, it is related to the Feynman propagator of the interacting Dirac field by a convolution with a dipole regulator:

$$\begin{aligned}d^{M^2,m}(x) &= -d^{M^2} * \kappa^m(x) = -\int d^4y\, d^{M^2}(x-y)\kappa^m(y), \\ d^{M^2}(x) &= \rho(m^2)\frac{i}{\pi}\int \frac{d^4q}{(2\pi)^3}\frac{(m^2-M^2)^2}{(q^2+io-M^2)^2}e^{iqx}.\end{aligned}$$

The Green's distribution inverts the free-field differential operator:

$$\begin{aligned}(i\gamma\partial + m)\mathbf{\Psi} &= \tfrac{1}{2} : \mathbf{\Gamma\Psi}(\overline{\mathbf{\Psi}}\mathbf{\Gamma\Psi}) :, \\ \mathbf{\Psi}(x) &= \tfrac{1}{2}\int d^4y\, \kappa^m(x-y) : \mathbf{\Gamma\Psi}(\overline{\mathbf{\Psi}}\mathbf{\Gamma\Psi})(y) : .\end{aligned}$$

The double-dot prescription $:\cdots:$ is defined by the subtraction of the mass term as the leading bilinear ground-state contribution. The bilinear local products of the Dirac field for bosonic matrix elements with a state vector $|B\rangle$,

$$\Gamma(x) = \langle\mathcal{C}|\overline{\mathbf{\Psi}}\Gamma\mathbf{\Psi}(x)|B\rangle \quad \text{for } \Gamma \in \{\mathbf{1}_4, \gamma_a, \gamma_5\gamma_a, \gamma_5\},$$

have the equations of motion

$$\Gamma_1(x) = \tfrac{1}{2}\int d^4y\ \langle\mathcal{C}|\overline{\mathbf{\Psi}}(x)\Gamma_1\kappa^m(x-y) : \Gamma\mathbf{\Psi}(\overline{\mathbf{\Psi}}\Gamma\mathbf{\Psi})(y) : |B\rangle.$$

They can be linearized with the Feynman propagator of the interacting Dirac field to yield first-order eigenvalue equations,

$$\Gamma_1(x) = \kappa_{\Gamma_1}^{\Gamma_2} * \Gamma_2(x) = \int d^4y\ \kappa_{\Gamma_1}^{\Gamma_2}(x-y)\ \Gamma_2(y),$$

with the *tangent kernel matrix* κ (see 10):

$$\kappa_{\Gamma_1}^{\Gamma_2}(x) = -\operatorname{tr}(\Gamma_1 \otimes \Gamma_2) \circ (\kappa^m \otimes d^{M^2,-m})(x) = -\operatorname{tr}\Gamma_1\kappa^m(x)\Gamma_2 d^{M^2,-m}(x).$$

The corresponding convolution product for the energy-momenta (translation eigenvalues) distributions gives the distributions for eigenvalues of the product representations:

$$\begin{aligned}
\tilde{\Gamma}_1(q) &= \tilde{\kappa}_{\Gamma_1}^{\Gamma_2}(q)\tilde{\Gamma}_2(q), \quad [\delta_{\Gamma_1}^{\Gamma_2} - \tilde{\kappa}_{\Gamma_1}^{\Gamma_2}(q)]\tilde{\Gamma}_2(q) = 0,\\
\text{with } \tilde{\kappa}_{\Gamma_1}^{\Gamma_2}(q) &= -\operatorname{tr}(\Gamma_1 \otimes \Gamma_2) \circ (\tilde{\kappa}^m \circledast \tilde{d}^{M^2,-m})(q)\\
&= \rho(m^2)\tfrac{i}{\pi}\operatorname{tr}\int \tfrac{d^4p}{(2\pi)^3}\Gamma_1 \tfrac{\gamma(p-q)+m}{(p-q)^2+io-m^2}\Gamma_2 \tfrac{(m^2-M^2)^2}{(p^2+io-M^2)^2}\tfrac{\gamma p+m}{p^2+io-m^2}.
\end{aligned}$$

The convolution $\circledast$ contains a tensor product for Clifford algebra elements.

The product $\kappa^m(x) \otimes d^m(-x) = -2i\rho(m^2)\kappa^m(x) \otimes \kappa^{-m}(x)$ of the Green's distribution and the free-field propagator is not defined because of the distributional off-shell contributions (more ahead). With the regularization for the quantization of the interacting Dirac field, the product is defined.

The scalar product and pseudoscalar product $(\overline{\mathbf{\Psi}}\mathbf{1}_4\mathbf{\Psi}, \overline{\mathbf{\Psi}}\gamma_5\mathbf{\Psi})$ constitute a chiral dublet. The degenerate ground-state is characterized by the particle mass,

$$\langle\mathcal{C}|\overline{\mathbf{\Psi}}\tfrac{\mathbf{1}_4+i\gamma_5}{2}\mathbf{\Psi}|\mathcal{C}\rangle\langle\mathcal{C}|\overline{\mathbf{\Psi}}\tfrac{\mathbf{1}_4-i\gamma_5}{2}\mathbf{\Psi}|\mathcal{C}\rangle = 4m^2;$$

i.e., the degeneracy manifold is a circle $\Omega^1 = \mathbf{SO}(2) \cong \mathbf{U}(1)$ with radius $2m$. A scalar ground-state, defined by

$$\begin{aligned}
\langle\mathcal{C}|\overline{\mathbf{\Psi}}\mathbf{1}_4\mathbf{\Psi}|\mathcal{C}\rangle &= 4m,\\
\langle\mathcal{C}|\overline{\mathbf{\Psi}}\gamma_5\mathbf{\Psi}|\mathcal{C}\rangle &= 0,
\end{aligned} \qquad \langle\mathcal{C}|\mathbf{r}^*\mathbf{l}|\mathcal{C}\rangle = \langle\mathcal{C}|\mathbf{l}^*\mathbf{r}|\mathcal{C}\rangle = 2m,$$

strips the particles of the chiral degree of freedom. In a first-order approximation, the pseudoscalar is the infinitesimal chiral field, i.e., the Goldstone degree of freedom,

$$\begin{aligned}
e^{i\boldsymbol{\alpha}} &= 1 + i\boldsymbol{\alpha} + \cdots = \frac{\overline{\mathbf{\Psi}}(\mathbf{1}_4+i\gamma_5)\mathbf{\Psi}}{2\mathbf{R}} = 1 + i\frac{\overline{\mathbf{\Psi}}\gamma_5\mathbf{\Psi}}{4m} + \ldots, \text{ with}\\
\Rightarrow \quad \boldsymbol{\alpha} &= \frac{\overline{\mathbf{\Psi}}\gamma_5\mathbf{\Psi}}{4m} + \ldots,
\end{aligned}$$

for the degeneracy transformation of the ground-state vector $|\mathcal{C}\rangle$. The *equation of motion for the composite Nambu–Goldstone field* as the corresponding matrix element with the "pion" $|\pi\rangle$ is given as follows:

$$\begin{aligned}
\gamma_5(x) &= \langle\mathcal{C}|\overline{\mathbf{\Psi}}\gamma_5\mathbf{\Psi}(x)|\pi\rangle,\\
\tilde\gamma_5(q) &= \tilde\kappa^{\gamma_5}_{\gamma_5}(q^2)\tilde\gamma_5(q),\quad [1-\tilde\kappa^{\gamma_5}_{\gamma_5}(q^2)]\tilde\gamma_5(q)=0,\\
\text{with } \tilde\kappa^{\gamma_5}_{\gamma_5}(q^2) &= \frac{\rho(m^2)}{4}\frac{i}{\pi}\,\mathrm{tr}\int\frac{d^4p}{(2\pi)^3}\gamma_5\frac{\gamma(p-q)+m}{(p-q)^2+io-m^2}\gamma_5\frac{(m^2-M^2)^2}{(p^2+io-M^2)^2}\frac{\gamma p+m}{p^2+io-m^2}.
\end{aligned}$$

With $\gamma_5(\gamma p+m)\gamma_5=\gamma p-m$, it has a mass zero $q^2=0$ solution $\tilde\kappa^{\gamma_5}_{\gamma_5}(0)=1$ if compared with the consistency equation for the chiral breakdown, $m\neq 0$:

$$\frac{i}{\pi}\int\frac{d^4q}{(2\pi)^3}\frac{(m^2-M^2)^2}{(q^2+io-M^2)^2}\frac{\rho(m^2)}{q^2+io-m^2}=\left\{\begin{aligned}
\frac{1}{4m}\,\mathrm{tr}\int\frac{d^4q}{(2\pi)^4}\,\tilde d^{M^2,m}(q) &= 1,\\
\frac{1}{4}\,\mathrm{tr}\int\frac{d^4q}{(2\pi)^4}\,\tilde\kappa^{-m}(q)\tilde d^{M^2,m}(q) &= \tilde\kappa^{\gamma_5}_{\gamma_5}(0).
\end{aligned}\right.$$

The consistency equation for the fermion mass as the degeneracy parameter and the eigenvalue equation for the massless field coincide.

The expansion at the mass zero solution determines the normalization $\rho(0)$ of the corresponding spacetime translation representation as the residue at the pole:

$$\frac{1}{1-\tilde\kappa^{\gamma_5}_{\gamma_5}(q^2)}=\frac{1}{q^2}\frac{q^2}{1-\tilde\kappa^{\gamma_5}_{\gamma_5}(q^2)}=\frac{\rho(0)}{q^2}+\dots \text{ with } \frac{1}{\rho(0)}=-\frac{\partial\tilde\kappa^{\gamma_5}_{\gamma_5}}{\partial q^2}(0).$$

The characteristic structures of the chiral model do not depend on a "perturbative" expansion. They are given by the relation between the normalization, in the "gap" equation, of the spacetime representation by the regularized Feynman propagator and the ensuing massless solution via the singularity of the resolvent $\frac{1}{1-\tilde\kappa^{\gamma_5}_{\gamma_5}(q^2)}$ with the normalized kernel $\tilde\kappa^{\gamma_5}_{\gamma_5}$. This will be discussed in more detail in Chapter 10.

9.2 Convolutions for Abelian Groups

Product representations of translations $\mathbb{R}^n$ with sum and difference of the energy-momentum invariants arise by the pointwise product of positive-type functions $L^\infty(\mathbb{R}^n)_+ \overset{d^nx}{=} \mathcal{C}_b(\mathbb{R}^n)_+$ or the convolution of positive energy-momentum Radon measures $\mathcal{M}(\check{\mathbb{R}}^n)_+$.

The simplest case is given for one-dimensional translations, e.g., for time translations $t\in\mathbb{R}$ with an addition of the energy invariants in the irreducible and self-dual representations:

$$\begin{aligned}
\mathcal{C}_b(\mathbb{R})_+\cdot\mathcal{C}_b(\mathbb{R})_+=\mathcal{C}_b(\mathbb{R})_+ &\left\{\begin{aligned}
e^{im_1t}\cdot e^{im_2t} &= e^{im_+t},\\
2\cos m_1t\cdot 2\cos m_2t &= 2\cos m_+t+2\cos m_-t
\end{aligned}\right.\\
\text{with } m_\pm &= m_1\pm m_2,\\
\mathcal{M}(\check{\mathbb{R}})_+ * \mathcal{M}(\check{\mathbb{R}})_+=\mathcal{M}(\check{\mathbb{R}})_+ &\left\{\begin{aligned}
&\delta(q-m_1)*\delta(q-m_2)=\delta(q-m_+),\\
&2|q|\delta(q^2-m_1^2)*2|q|\delta(q^2-m_2^2)\\
&\quad=2|q|\delta(q^2-m_+^2)+2|q|\delta(q^2-m_-^2).
\end{aligned}\right.
\end{aligned}$$

9.2.1 Convolutions with Linear Invariants

The convolution product for the two causal function algebras, conjugate and orthogonal to each other, and the Dirac convolution algebra is summarized with the residually normalized representation functions and the integration contours:

$$\vartheta(\pm t)e^{imt} = \pm\int \frac{dq}{2i\pi}\,\frac{1}{q\mp io-m}e^{iqt},$$

Causal time $\mathbf{D}(1)$ and energies $\mathbb{R}$
$(\overset{1}{*}, q) = (\pm\frac{*}{2i\pi}, q\mp io)$ causal, orthogonal
$\frac{1}{q-m_1} \overset{1}{*} \frac{1}{q-m_2} = \frac{1}{q-m_+}$
$\delta(q-m_1)*\delta(q-m_2)=\delta(q-m_+)$

The *normalization factor for the convolution product (residual normalization)* is the 1-sphere measure as used in the residue:

$$\oint \frac{dq}{2i\pi} = \text{res}, \quad \frac{*}{2\pi} \cong \frac{1}{|\Omega^1|}\delta(q_1+q_2-q).$$

There is the more general convolution

$$\frac{\Gamma(1+\nu_1)}{(q-m_1)^{1+\nu_1}} \overset{1}{*} \frac{\Gamma(1+\nu_2)}{(q-m_2)^{1+\nu_2}} = \frac{\Gamma(1+\nu_1+\nu_2)}{(q-m_+)^{1+\nu_1+\nu_2}},$$

which generalizes the integer-power derivatives $(\frac{\partial}{\partial m})^N$ for nontrivial nildimensions $N=1,2,\dots$ to real powers $\nu\in\mathbb{R}$ wherever the Γ-functions are defined.

9.2.2 Convolutions with Self-Dual Invariants

The causal distributions with compact dual invariants

$$\pm\frac{1}{i\pi}\frac{q}{(q\mp io)^2-m^2} = |m|\delta(q^2-m^2)\pm\frac{1}{i\pi}\frac{q}{q_{\mathrm{P}}^2-m^2} = \pm\frac{1}{2i\pi}\left(\frac{1}{q\mp io-|m|}+\frac{1}{q\mp io+|m|}\right)$$

keep the property of constituting orthogonal convolution algebras, conjugate to each other:

$$\vartheta(\pm t)2\cos mt = \pm\int\frac{dq}{i\pi}\frac{q}{(q\mp io)^2-m^2}e^{iqt},$$

Causal time $\mathbf{D}(1)$ and energies $\mathbb{R}$
$(\overset{1}{*}, q^2) = (\pm\frac{*}{i\pi}, (q\mp io)^2)$ causal, orthogonal
$\frac{q}{q^2-m_1^2} \overset{1}{*} \frac{q}{q^2-m_2^2} = \frac{q}{q^2-m_+^2}+\frac{q}{q^2-m_-^2}$

Since the Feynman energy distributions combine advanced and retarded distributions,

$$\pm\frac{1}{i\pi}\frac{|m|}{q^2\mp io-m^2} = |m|\delta(q^2-m^2)\pm\frac{1}{i\pi}\frac{|m|}{q_{\mathrm{P}}^2-m^2} = \pm\frac{1}{2i\pi}\left(\frac{1}{q\mp io-|m|}-\frac{1}{q\pm io+|m|}\right),$$

they constitute convolution algebras, conjugate to each other, however not orthogonal, $e^{+i|m_1t|}\cdot e^{-i|m_2t|}\neq 0$:

$$e^{\pm i|mt|} = \pm\int\frac{dq}{i\pi}\frac{|m|}{q^2\mp io-m^2}e^{iqt},$$

Bicone time $\mathbb{R}_+ \uplus \mathbb{R}_-$ and energies $\mathbb{R}$
$(\overset{1}{*}, q^2) = (\pm\frac{*}{i\pi}, q^2\mp io)$ Feynman, not orthogonal
$\frac{\lvert m_1\rvert}{q^2-m_1^2} \overset{1}{*} \frac{\lvert m_2\rvert}{q^2-m_2^2} = \frac{\lvert m_+\rvert}{q^2-m_+^2}$

The faithful Hilbert representations of $\mathcal{Y}^1 \cong \mathbf{SO}_0(1,1) \cong \mathbb{R}$ (one-dimensional abelian position) with Fourier-transformed Ω^1-measures and noncompact dual invariants constitute a convolution algebra:

$$e^{-|mz|} = \int \frac{dq}{\pi} \frac{|m|}{q^2+m^2} e^{-iqz},$$

Position $\mathcal{Y}^1$ and "momenta" $\mathbb{R}$
$\vert\Omega^1\vert = 2\pi, \quad \overset{1}{*} = \frac{*}{\pi}$
$\frac{\vert m_1\vert}{q^2+m_1^2} \overset{1}{*} \frac{\vert m_2\vert}{q^2+m_2^2} = \frac{\vert m_+\vert}{q^2+m_+^2}$

9.3 Convolutions for Position Representations

Representations of Euclidean, spherical, and hyperbolic spaces are characterized by *singularity spheres* with real momenta (imaginary eigenvalues) for free and scattering structures, and imaginary "momenta" (real eigenvalues) for bound structures. The convolution of the related "momentum" functions reflect pointwise multiplications of Macdonald and Hankel (Bessel with Neumann) functions (see Chapter 8):

$$\mathbf{O}(s): \begin{cases} \int \frac{d^sq}{i\pi^{\frac{s}{2}}} \frac{\Gamma(\frac{s}{2}-\nu)}{(\vec q^2-io-1)^{\frac{s}{2}-\nu}} e^{i\vec q\vec x} = \frac{\pi[\mathcal{J}_\nu+i\mathcal{N}_\nu](r)}{(\frac{r}{2})^\nu}, & \text{spherical,}\\ \int \frac{d^sq}{\pi^{\frac{s}{2}}} \delta(\vec q^2-1) e^{i\vec q\vec x} = \frac{\mathcal{J}_{\frac{s-2}{2}}(r)}{(\frac{r}{2})^{\frac{s-2}{2}}}, & \text{Euclidean,}\\ \int \frac{d^sq}{\pi^{\frac{s}{2}}} \frac{\Gamma(\frac{s}{2}-\nu)}{(\vec q^2+1)^{\frac{s}{2}-\nu}} e^{i\vec q\vec x} = \frac{2\mathcal{K}_\nu(r)}{(\frac{r}{2})^\nu}, & \text{hyperbolic.} \end{cases}$$

9.3.1 Convolutions for Euclidean Spaces

Interaction-free product structures convolute Dirac distributions for cyclic translation representations. In contrast to the convolution of Dirac distributions for self-dual invariants with basic spherically self-dual two-dimensional representations,

$$\begin{aligned} \text{abelian } \mathbb{R}: \quad & 2|q|\delta(q^2-P_1^2) * 2|q|\delta(q^2-P_2^2)\\ & = 2|q|\delta(q^2-P_-^2) + 2|q|\delta(q^2-P_+^2),\\ \text{with} \quad & P_\pm = |P_1| \pm |P_2|, \end{aligned}$$

the convolution of Dirac distributions for the infinite-dimensional representations of the Euclidean groups, $s \geq 1$, with the sphere radii as momentum invariants $\vec q^2 = P^2 > 0$ leads to position translation representations with the momentum sphere radii between the invariants, $P_-^2 \leq \vec q^2 \leq P_+^2$,

$$\begin{aligned} \mathbf{SO}(1+s) \,\vec\times\, \mathbb{R}^{1+s}: \quad & \delta(\vec q^2-P_1^2) \frac{*}{|\Omega^{s-1}|} \delta(\vec q^2-P_2^2)\\ s = 1, 2, \ldots, \quad & = \frac{[-\Delta(\vec q^2)]^{\frac{s-2}{2}}}{(2|\vec q|)^{s-1}} \vartheta(P_+^2-\vec q^2)\vartheta(\vec q^2-P_-^2). \end{aligned}$$

The convolution product is normalized with the $(s-1)$-sphere. There arises a momentum-dependent normalization factor that contains the characteristic *two-particle convolution function*:

$$\Delta(\vec{q}^2) = \Delta(\vec{q}^2, P_1^2, P_2^2) = (\vec{q}^2 - P_+^2)(\vec{q}^2 - P_-^2).$$

It is symmetric in the three invariants involved:

$$\Delta(a,b,c) = a^2 + b^2 + c^2 - 2(ab + ac + bc) = (a + b - c)^2 - 4ab.$$

The minimal cases $s = 1, 2$ are characteristic for even- and odd-dimensional positions:

Scattering in two dimensions involves the pointwise product of Bessel functions $\mathcal{J}_0$,

$$\begin{aligned}
\mathbf{SO}(2) \vec{\times} \mathbb{R}^2 : \quad & \delta(\vec{q}^2 - P_1^2) \tfrac{*}{2} \delta(\vec{q}^2 - P_2^2) \\
& = \tfrac{2}{\sqrt{(P_+^2 - \vec{q}^2)(\vec{q}^2 - P_-^2)}} \vartheta(P_+^2 - \vec{q}^2)\vartheta(\vec{q}^2 - P_-^2), \\
\mathcal{J}_0(r) & = \int \tfrac{d^2q}{\pi}\, \delta(\vec{q}^2 - 1) e^{i\vec{q}\vec{x}}, \quad \int \tfrac{d^2x}{4\pi} \mathcal{J}_0(r) e^{i\vec{q}\vec{x}} = \delta(\vec{q}^2 - 1), \\
\mathcal{J}_0(|P_1|r)\mathcal{J}_0(|P_2|r) & = \int_0^\pi \tfrac{d\theta}{\pi}\, \mathcal{J}_0(|P|r) \text{ with } P^2 = P_1^2 + P_2^2 - 2P_1P_2\cos\theta \\
& = \tfrac{1}{\pi} \int_{P_-^2}^{P_+^2} \tfrac{dP^2}{\sqrt{(P_+^2 - P^2)(P^2 - P_-^2)}} \mathcal{J}_0(|P|r) \quad \text{for } P_1P_2 \neq 0.
\end{aligned}$$

One obtains for the $\mathbf{SO}(3) \vec{\times} \mathbb{R}^3$ scattering representations in three dimensions:

$$\begin{aligned}
\mathcal{C}_b(\mathbb{R}^3)_+ \cdot \mathcal{C}_b(\mathbb{R}^3)_+ = \mathcal{C}_b(\mathbb{R}^3)_+ : & \tfrac{\sin P_1 r}{r} \cdot \tfrac{\sin P_2 r}{r} = \tfrac{\cos P_- r - \cos P_+ r}{2r^2}, \\
\mathcal{M}(\check{\mathbb{R}}^3)_+ * \mathcal{M}(\check{\mathbb{R}}^3)_+ = \mathcal{M}(\check{\mathbb{R}}^3)_+ : & \delta(\vec{q}^2 - P_1^2) \tfrac{*}{2\pi} \delta(\vec{q}^2 - P_2^2) \\
& = \tfrac{2}{|\vec{q}|} \vartheta(P_+^2 - \vec{q}^2)\vartheta(\vec{q}^2 - P_-^2).
\end{aligned}$$

The square of a representation is a normalized positive-type function:

$$s = 3 : \quad \left(\tfrac{\sin Pr}{Pr}\right)^2 = \tfrac{1 - \cos 2Pr}{2(Pr)^2}, \quad \delta(\vec{q}^2 - P^2) \tfrac{*}{2\pi} \delta(\vec{q}^2 - P^2) = \tfrac{2}{|\vec{q}|} \vartheta(4P^2 - \vec{q}^2).$$

9.3.2 Convolutions for Odd-Dimensional Hyperboloids

Cyclic representations of the even-dimensional Lorentz group $\mathbf{SO}_0(1, 2R-1)$, $2R = 2, 4, \ldots$, for the hyperboloid $\mathcal{Y}^{2R-1} \cong \mathbf{SO}_0(1, 2R-1)/\mathbf{SO}(2R-1)$ with real rank 1 and noncompact invariant $\vec{q}^2 = -m^2 < 0$ are characterized by continuous positive-type functions $\mathcal{C}_b(\mathcal{Y}^{2R-1})_+ \overset{d^{2R-1}x}{\cong} L^\infty(\mathcal{Y}^{2R-1})_+$:

$$\mathcal{Y}^{2R-1} \cong \mathbb{R}^{2R-1} \ni \vec{x} \longmapsto \int \tfrac{2d^{2R-1}q}{|\Omega^{2R-1}|} \tfrac{|m|}{(\vec{q}^2 + m^2)^R} e^{i\vec{q}\vec{x}} = e^{-|m|r}.$$

Nontrivial properties for the maximal compact group, the rotations $\mathbf{SO}(2R-1)$, $R \geq 2$, arise by derivations $\frac{\partial}{\partial \vec{x}} \sim i\vec{q}$. These Lorentz group representations start from the characteristic hyperbolic exponentials for the maximal noncompact abelian subgroup with imaginary singularities $q = \pm im$:

$$\mathbf{SO}_0(1,1) \cong \mathbb{R} \ni x \longmapsto \int \tfrac{dq}{\pi} \tfrac{|m|}{q^2 + m^2} e^{-iqx} = e^{-|mx|}.$$

The representations are faithful and cyclic, but not irreducible. They are square-integrable and not Schur-orthogonal for different invariants $m_1^2 \neq m_2^2$.

Product representations $e^{-|m_1|r} \cdot e^{-|m_2|r} = e^{-|m_+|r}$ convolute the positive momentum measures. The measure of the associated compact unit sphere Ω^{2R-1} is used for the residual normalization (more on the normalization ahead). The representations of three-dimensional hyperbolic position $\mathcal{Y}^3$ use the Fourier-transformed Ω^3-measure, familiar from the nonrelativistic hydrogen Schrödinger functions (see Chapter 8). The radii of the "momentum" spheres as invariants are added up in the convolution

$$e^{-|m|r} = \int \frac{d^3q}{\pi^2} \frac{|m|}{(\vec{q}^{\,2}+m^2)^2} e^{i\vec{q}\vec{x}},$$

Position $\mathcal{Y}^3 \cong \mathbf{SO}_0(1,3)/\mathbf{SO}(3)$ and "momenta" $\mathbb{R}^3$ with $\mathbf{SO}(3)$
$\lvert\Omega^3\rvert = 2\pi^2, \quad \overset{3}{*} = \frac{*}{\pi^2}$
$\frac{\lvert m_1\rvert}{(\vec{q}^{\,2}+m_1^2)^2} \overset{3}{*} \frac{\lvert m_2\rvert}{(\vec{q}^{\,2}+m_2^2)^2} = \frac{\lvert m_+\rvert}{(\vec{q}^{\,2}+m_+^2)^2}$

In general, the representations of odd-dimensional hyperboloids $\mathcal{Y}^{2R-1}$ come with Fourier-transformed Ω^{2R-1}-measures and imaginary singularity sphere Ω^{2R-2} for the "momentum" eigenvalues. The sphere measures can be obtained by invariant momentum derivatives:

$$\left(-\frac{\partial}{\partial \vec{q}^{\,2}}\right)^{R-1} \frac{|m|}{\vec{q}^{\,2}+m^2} = \Gamma(R) \frac{|m|}{(\vec{q}^{\,2}+m^2)^R}, \quad R = 1, 2, \ldots.$$

Product representations arise by the convolution with the sphere volume as residual normalization:

$$e^{-|m|r} = \int \frac{2d^{2R-1}q}{|\Omega^{2R-1}|} \frac{|m|}{(\vec{q}^{\,2}+m^2)^R} e^{i\vec{q}\vec{x}},$$

Position $\mathcal{Y}^{2R-1} \cong \mathbf{SO}_0(1,2R-1)/\mathbf{SO}(2R-1),\ 2R-1 = 1,3,\ldots,$ and "momenta" $\mathbb{R}^{2R-1}$ with $\mathbf{SO}(2R-1)$
$\lvert\Omega^{2R-1}\rvert = \frac{2\pi^R}{\Gamma(R)}, \quad \overset{2R-1}{*} = \frac{*\,2}{\lvert\Omega^{2R-1}\rvert}$
$(\frac{\partial}{\partial\vec{q}})^{L_1} \frac{\lvert m_1\rvert}{(\vec{q}^{\,2}+m_1^2)^R} \overset{2R-1}{*} (\frac{\partial}{\partial\vec{q}})^{L_2} \frac{\lvert m_2\rvert}{(\vec{q}^{\,2}+m_2^2)^R} = (\frac{\partial}{\partial\vec{q}})^{L_1+L_2} \frac{\lvert m_+\rvert}{(\vec{q}^{\,2}+m_+^2)^R}$ for $L = 0, 1, \ldots$.

Via momentum derivatives $\frac{\partial}{\partial \vec{q}}$, the convolutions may involve tensor products for $\mathbf{SO}(2R-1)$-representations.

9.3.3 Convolutions for Odd-Dimensional Spheres

The representations of odd-dimensional spheres use a singularity sphere Ω^{2R-2} with real momentum eigenvalues in the convolutions

$$e^{\pm i|m|r} = \pm \int \frac{2d^{2R-1}q}{i|\Omega^{2R-1}|} \frac{|m|}{(\vec{q}^{\,2} \mp io - m^2)^R} e^{i\vec{q}\vec{x}},$$

Sphere $\Omega^{2R-1} \cong \mathbf{SO}(2R)/\mathbf{SO}(2R-1),\ 2R-1 = 1, 3, \dots,$ and momenta $\mathbb{R}^{2R-1}$ with $\mathbf{SO}(2R-1)$
$\lvert\Omega^{2R-1}\rvert = \frac{2\pi^R}{\Gamma(R)},\quad (\overset{2R-1}{*}, \vec{q}^{\,2}) = (\pm\frac{*\,2}{i\lvert\Omega^{2R-1}\rvert}, \vec{q}^{\,2} \mp io)$ not orthogonal
$(\frac{\partial}{\partial\vec{q}})^{L_1} \frac{\lvert m_1\rvert}{(\vec{q}^{\,2}-m_1^2)^R} \overset{2R-1}{*} (\frac{\partial}{\partial\vec{q}})^{L_2} \frac{\lvert m_2\rvert}{(\vec{q}^{\,2}-m_2^2)^R} = (\frac{\partial}{\partial\vec{q}})^{L_1+L_2} \frac{\lvert m_+\rvert}{(\vec{q}^{\,2}-m_+^2)^R}$ for $L = 0, 1, \dots.$

9.4 Residual Normalization

The abelian convolutions $* \cong \delta(q_1+q_2-q)$ of the energy-momentum measures above with the $\mathbb{R}^n$-Lebesgue measures $d^n q$ as basis and the invariants as pole singularities in additional factors, e.g., $\frac{dq}{\pi}\frac{|m|}{q^2+m^2}$, are "rationalized" with respect to the product representations in such a way that the spherical degrees do not show up (no π's). This *representation normalization* (see Chapter 8) results from spheres and related rotation groups in the definition of *higher-dimensional residues.*

Since the convolution with a Dirac distribution amounts to a residue (where defined),

$$f(q) = \int dp\ \delta(p-q)f(p) = \oint \frac{dp}{2i\pi}\frac{f(p)}{p-q},$$

the convolution normalization for the time representation coefficients is given by the normalization of the residue of the real pole $q = m$,

$$\mathcal{D}^1:\ \ \frac{*}{2i\pi}\ \text{from}\ \int \frac{dq}{2i\pi}\frac{1}{q-io-m}e^{iqx_0} = \vartheta(x_0)e^{imx_0};$$

2π is the length of the unit circle $\Omega^1 \cong \mathbf{U}(1)$, the compact representation image of $\mathbf{D}(1)$. It normalizes the energy Plancherel measure $\frac{dq}{2\pi}$ for the time translation Haar measure dx_0.

The convolution normalization for $\mathcal{Y}^1 \cong \mathbf{SO}_0(1,1)$ is determined by the residual normalization in the faithful cyclic representation coefficient whose self-duality causes the factor 2 for the two imaginary poles $q = \pm i|m|$:

$$\mathcal{Y}^1:\ \ \frac{*}{\pi}\ \text{from}\ \int \frac{dq}{\pi}\frac{|m|}{q^2+m^2}e^{-iqx} = e^{-|mx|}.$$

For nontrivial invariant $m^2 > 0$, one can replace

$$\frac{dq}{\pi}\frac{|m|}{q^2+m^2} = \frac{dq}{\pi}\frac{|q|}{q^2+m^2} = \frac{dq^2}{2\pi}\frac{1}{q^2+m^2}.$$

In general for odd-dimensional hyperboloids, the *self-dual residual normalization* of the rotation scalar positive-type functions uses *half the measures* of the corresponding spheres,

$$\mathcal{Y}^{2R-1}:\ \ \overset{2R-1}{*} = \frac{*\,2}{|\Omega^{2R-1}|}\ \text{from}\ \int \frac{2d^{2R-1}q}{|\Omega^{2R-1}|}\frac{|m|}{(\vec{q}^{\,2}+m^2)^R}e^{i\vec{q}\vec{x}} = e^{-|m|r},$$

and, analogously, the residual normalization for odd-dimensional spheres:

$$\Omega^{2R-1}: \quad \overset{2R-1}{*} = \pm\frac{*\,2}{i|\Omega^{2R-1}|} \text{ from } \pm\int \frac{2d^{2R-1}q}{i|\Omega^{2R-1}|}\,\frac{|m|}{(\vec{q}^{\,2}\mp io-m^2)^R}e^{i\vec{q}\vec{x}} = e^{\pm i|m|r}.$$

The momentum eigenvalues lie on a sphere $\{\vec{q}\in\check{\mathbb{R}}^{2R-1} \mid \vec{q}^{\,2} = \mp m^2\} \cong \Omega^{2R-2}$.

For the nonabelian case $R\ge 2$, the normalization $\frac{2}{|\Omega^{2R-1}|} = \frac{\Gamma(R)}{\pi^R}$ for the curved spaces differs from the "flat" normalization $(\frac{2}{|\Omega^1|})^{2R-1} = \frac{1}{\pi^{2R-1}}$ for self-dually represented translations $\mathbb{R}^{2R-1} \longrightarrow \mathbf{SO}(2)^{2R-1}$.

9.5 Convolution of Feynman Measures

The convolution of (energy-)momentum distributions adds the space(-time) translation eigenvalues to the eigenvalue $q=\sum\limits_{j=1}^{k} q_j$ of the product representation:

$$\mu_1 * \cdots * \mu_k(q) = \int d^{1+s}q_1\cdots d^{1+s}q_k\;\delta\Big(\sum_{j=1}^{k} q_j - q\Big)\prod_{j=1}^{k}\mu_j(q_j).$$

In generalizing the formula $\frac{\Gamma(\nu_1)\Gamma(\nu_2)}{\Gamma(\nu_1+\nu_2)} = \int_0^1 d\zeta\;\zeta^{\nu_1-1}(1-\zeta)^{\nu_2-1}$ for the beta function, a convolution with denominator singularities can be performed by joining first the denominators,

$$\begin{aligned}
\Gamma(\nu_1)\Gamma(\nu_2) &= \int_0^1 d\zeta_1\int_0^1 d\zeta_2\;\delta(\zeta_1+\zeta_2-1)\zeta_1^{\nu_1-1}\zeta_2^{\nu_2-1}\Gamma(\nu_1+\nu_2),\\
\frac{\Gamma(\nu_1)\cdots\Gamma(\nu_k)}{R_1^{\nu_1}\cdots R_k^{\nu_k}} &= \int_0^1 d\zeta_1\cdots\int_0^1 d\zeta_k\;\delta(\zeta_1+\cdots+\zeta_k-1)\frac{\zeta_1^{\nu_1-1}\cdots\zeta_k^{\nu_k-1}\Gamma(\nu_1+\cdots+\nu_k)}{(\zeta_1R_1+\cdots+\zeta_kR_k)^{\nu_1+\cdots+\nu_k}},\\
&\nu_j\in\mathbb{R},\quad \nu_j\neq 0,-1,-2,\dots,
\end{aligned}$$

e.g., for multipole distributions,

$$\begin{aligned}
\frac{2q}{-q^2+m^2}\frac{1}{(-q^2+M^2)^\nu} &= \frac{\partial}{\partial q}\int_0^1 d\zeta\;\frac{\zeta^{\nu-1}}{[-q^2+\zeta M^2+(1-\zeta)m^2]^\nu}\\
&= \frac{\partial}{\partial q}\int_{M^2}^{m^2}\frac{d\kappa^2}{m^2-M^2}\left(\frac{m^2-\kappa^2}{m^2-M^2}\right)^{\nu-1}\frac{1}{(-q^2+\kappa^2)^\nu}.
\end{aligned}$$

Here and in the following, the convolutions exist only for pole orders where the involved Γ-functions are defined. Elsewhere "divergences" arise.

The product of two (energy-) momentum distributions,

$$\frac{\Gamma(\nu_1)}{(-q_1^2+m_1^2)^{\nu_1}}\times\frac{\Gamma(\nu_2)}{(-q_2^2+m_2^2)^{\nu_2}} = \int_0^1 d\zeta\;\frac{\zeta^{\nu_1-1}(1-\zeta)^{\nu_2-1}\Gamma(\nu_1+\nu_2)}{[\zeta(-q_1^2+m_1^2)+(1-\zeta)(-q_2^2+m_2^2)]^{\nu_1+\nu_2}},$$

can be written with the center of mass (energy-)momentum q and relative (energy-)momentum p:

$$\zeta q_1^2+(1-\zeta)q_2^2 = (p+\tfrac{2\zeta-1}{2}q)^2+\zeta(1-\zeta)q^2 \text{ with } \begin{cases} q_1+q_2 = q,\\ q_1-q_2 = 2p,\\ d^nq_1d^nq_2 = d^nq\;d^np.\end{cases}$$

The convoluted distributions must all be of the same type, e.g., either all Feynman $q^2 + io$ or all anti-Feynman $q^2 - io$. For two *equal-type Feynman distributions* $\frac{1}{-q^2 \mp io + m^2}$ with positive or negative invariants $m^2 \in \mathbb{R}$, the product inherits this Feynman type for both the center-of-mass and the relative (energy-)momentum distributions:

$$q_1^2 + io, \text{ with } q_2^2 + io \Rightarrow \zeta q_1^2 + (1-\zeta)q_2^2 + io \Rightarrow q^2 + io,\ p^2 + io.$$

This is in contrast to the causal integration prescriptions $(q \pm io)^2 = (q_0 \pm io)^2 - \vec{q}^{\,2} = q^2 \pm ioq_0$ for $\mathbf{SO}_0(1,s)$-spacetimes, where there is no definite integration prescription for the relative energy-momenta integration,

$$\begin{aligned} \zeta(q_1 + io)^2 + (1-\zeta)(q_2 + io)^2 = (p + \tfrac{2\zeta - 1}{2}q)^2 + io\epsilon(2\zeta - 1)p_0 \\ + \zeta(1-\zeta)q^2 + ioq_0, \\ \text{with } \epsilon(2\zeta - 1) \in \{\pm 1\} \text{ for } \zeta \in [0,1]. \end{aligned}$$

The convolution product for causal measures will be defined by using the Feynman integration for the relative energy-momenta, i.e., $p^2 \pm io$. The center-of-mass prescription inherits the prescription of the factors $(q \pm io)^2$.

The convolution product of two Feynman distributions, normalized by $\frac{1}{i^t \pi^{\frac{n}{2}}}$,

$$\begin{aligned} &\frac{\Gamma(\frac{n}{2}+\nu_1)}{(-q^2 - io + m_1^2)^{\frac{n}{2}+\nu_1}} \frac{*}{i^t \pi^{\frac{n}{2}}} \frac{\Gamma(\frac{n}{2}+\nu_2)}{(-q^2 - io + m_2^2)^{\frac{n}{2}+\nu_2}} \\ &= \int_0^1 d\zeta \int \frac{d^n p}{i^t \pi^{\frac{n}{2}}} \frac{\zeta^{\frac{n}{2}+\nu_1-1}(1-\zeta)^{\frac{n}{2}+\nu_2-1}\Gamma(n+\nu_1+\nu_2)}{[-p^2 - io - \zeta(1-\zeta)q^2 + \zeta m_1^2 + (1-\zeta)m_2^2]^{n+\nu_1+\nu_2}}, \end{aligned}$$

is the *residue with respect to the relative energy-momentum* $p = \frac{q_1 - q_2}{2}$ *dependence*,

$\mathbf{O}(t, n-t)$ with $n = 1, 2, \ldots;\ t = 0, 1, \ldots,\ \lvert\Omega^{n-1}\rvert = \frac{2\pi^{\frac{n}{2}}}{\Gamma(\frac{n}{2})}$ for $m^2 \in \mathbb{R}$
$\int \frac{d^n q}{i^t \pi^{\frac{n}{2}}} \frac{\Gamma(\frac{n}{2}+\nu)}{(-q^2 - io + m^2)^{\frac{n}{2}+\nu}} = \frac{\Gamma(\nu)}{(m^2 - io)^\nu}$
$(\frac{\partial}{\partial q})^{L_1} \frac{\Gamma(\frac{n}{2}+\nu_1)}{(-q^2 - io + m_1^2)^{\frac{n}{2}+\nu_1}} \frac{*}{i^t \pi^{\frac{n}{2}}} (\frac{\partial}{\partial q})^{L_2} \frac{\Gamma(\frac{n}{2}+\nu_2)}{(-q^2 - io + m_2^2)^{\frac{n}{2}+\nu_2}}$ $= (\frac{\partial}{\partial q})^{L_1+L_2} \int_0^1 d\zeta \frac{\zeta^{\frac{n}{2}+\nu_1-1}(1-\zeta)^{\frac{n}{2}+\nu_2-1}\Gamma(\frac{n}{2}+\nu_1+\nu_2)}{[-\zeta(1-\zeta)q^2 - io + \zeta m_1^2 + (1-\zeta)m_2^2]^{\frac{n}{2}+\nu_1+\nu_2}}$

Nontrivial $\mathbf{O}(t, n-t)$-properties are effected by the convolution-compatible (energy-)momentum derivatives,

$$\begin{aligned} \frac{\partial}{\partial q} = 2q\frac{\partial}{\partial q^2},\ \frac{\partial}{\partial q} \otimes q &= \mathbf{1}_n + q \otimes q\ 2\frac{\partial}{\partial q^2}, \\ \frac{\partial}{\partial q} \otimes \frac{\partial}{\partial q} &= (\mathbf{1}_n + q \otimes q\ 2\frac{\partial}{\partial q^2})2\frac{\partial}{\partial q^2}, \ldots, \end{aligned}$$

e.g., in the form of harmonic polynomials $(\frac{\partial}{\partial q})^L$, $L = 1, 2, \ldots$. The derivatives, acting on multipoles, raise the pole order,

$$\frac{\partial}{\partial q} \frac{\Gamma(R)}{(-q^2 + m^2)^R} = \frac{2q\ \Gamma(1+R)}{(-q^2 + m^2)^{1+R}}.$$

By derivations with respect to the (energy-)momentum invariants,

$$(\frac{\partial}{\partial q^2})^k(-\frac{\partial}{\partial m_1^2})^{k_1}(-\frac{\partial}{\partial m_2^2})^{k_2}\frac{\Gamma(\nu)}{[-\zeta(1-\zeta)q^2+\zeta m_1^2+(1-\zeta)m_2^2]^\nu}$$
$$=\frac{\Gamma(\nu+k+k_1+k_2)\zeta^{k+k_1}(1-\zeta)^{k+k_2}}{[-\zeta(1-\zeta)q^2+\zeta m_1^2+(1-\zeta)m_2^2]^{\nu+k+k_1+k_2}},$$

the integrals can be reduced for (half-)integer powers to a (cubic root) quadratic polynomial in the denominator:

$$\frac{n}{2}+\nu_1+\nu_2=\begin{cases} N \Rightarrow \frac{1}{P(\zeta)}, \\ \frac{1}{2}+N \Rightarrow \frac{1}{P(\zeta)^{\frac{3}{2}}}=\frac{4}{(-q^2-io+m_-^2)(-q^2-io+m_+^2)}\frac{d^2\sqrt{P(\zeta)}}{d\zeta^2}, \end{cases}$$
with $N=1,2,\dots$ and $P(\zeta)=-\zeta(1-\zeta)q^2-io+\zeta m_1^2+(1-\zeta)m_2^2$.

The integration is completely different for integer and half-integer powers: For half-integer powers, the ζ-integration compensates the m_-^2-pole from the discriminant:

$$\int_0^1\frac{d\zeta}{P(\zeta)^{\frac{3}{2}}}=\frac{4}{(-q^2-io+m_-^2)(-q^2-io+m_+^2)}\frac{d\sqrt{P(\zeta)}}{d\zeta}\Big|_0^1=\frac{4|m_+|}{m_+^2-m_-^2}\frac{1}{-q^2-io+m_+^2},$$
$$\int_0^1 d\zeta\,\frac{\zeta^{N-1}(1-\zeta)^{N-1}\Gamma(N+\frac{1}{2})}{\Gamma(\frac{3}{2})[-\zeta(1-\zeta)q^2-io+\zeta m_1^2+(1-\zeta)m_2^2]^{N+\frac{1}{2}}}=\frac{4|m_+|}{m_+^2-m_-^2}\frac{\Gamma(N)}{(-q^2-io+m_+^2)^N}.$$

This has been used above for odd-dimensional spaces like time $\mathbb{R}$ and positions $\mathcal{Y}^3,\mathbb{R}^3$, and Ω^3. The integral $\int_0^1\frac{d\zeta}{-\zeta(1-\zeta)q^2-io+\zeta m_1^2+(1-\zeta)m_2^2}$ for integer powers, characteristic for even-dimensional spaces, has more structure. It is discussed next.

9.6 Convolutions for Even-Dimensional Spacetimes

The convolution product for even-dimensional spacetimes $\mathbf{SO}_0(1,2R-1)$ will be given for Feynman measures that hold for real m^2. The convolutions of Cartan energy-momentum pole distributions,

$\mathbf{SO}_0(1,1)$-spacetime, $\lvert\Omega^1\rvert=2\pi$
$(\overset{2}{*},q^2)=(\pm\frac{*}{i\pi},q^2\pm io)$
$\frac{1}{-q^2+m_1^2}\overset{2}{*}\frac{1}{-q^2+m_2^2}=\int_0^1 d\zeta\frac{1}{-\zeta(1-\zeta)q^2+\zeta m_1^2+(1-\zeta)m_2^2}$

involves the definition of the convolution product of the causal (advanced and retarded) measures, restricted to $m^2\geq 0$, from the pointwise product of Bessel functions $\vartheta(x)\mathcal{J}_N(|mx|)$ by analytic continuation of the convolution product of the Feynman measures as given above, i.e., with Feynman integration for the relative energy-momenta.

In contrast to the factors for time and position, both odd-dimensional with real rank 1, the convolutions for minimal two-dimensional spacetime produces *no pole singularities* (zero-dimensional) for product representations.

The residual products of even-dimensional spaces display pole distributions only before the *finite ζ-integration over an invariant singularity line* (one-dimensional). They can be written with spectral functions, e.g., for one vanishing mass,

$$\begin{array}{rl} \int_0^1 d\zeta \, \frac{1}{-\zeta q^2+m^2} &= \int_{m^2}^\infty \frac{d\kappa^2}{\kappa^2} \, \frac{1}{-q^2+\kappa^2}, \\ \int_0^1 d\zeta \, \frac{1-\zeta}{-\zeta q^2+m^2} &= \int_{m^2}^\infty \frac{d\kappa^2}{\kappa^2} \, \frac{\kappa^2-m^2}{\kappa^2} \frac{1}{-q^2+\kappa^2}. \end{array}$$

After ζ-integration logarithms arise. The logarithm is typical for a finite integration [4], e.g., for a function holomorphic on the integration curve (where defined),

$$\int_\beta^\alpha dz \; f(z) = \sum \text{res}[f(z) \log \tfrac{z-\alpha}{z-\beta}], \left\{ \begin{array}{rl} \int_\beta^\infty dz \; f(z) &= -\sum \text{res}[f(z)\log(z-\beta)], \\ \int_{-\infty}^\infty dz \; f(z) &= 2i\pi \sum \text{res} f(z), \end{array} \right.$$

with the sum of all residues in the closed complex plane, cut along the integration curve, e.g.,

$$\begin{array}{rl} \int_0^1 \frac{d\zeta}{-\zeta q^2+m^2} &= \sum \text{res} \left[\frac{1}{-\zeta q^2+m^2} \log \frac{\zeta-1}{\zeta} \right] = \frac{\log(1-\frac{q^2}{m^2})}{-q^2}, \\ \int_0^1 d\zeta \, \frac{1-\zeta}{-\zeta q^2+m^2} &= \sum \text{res} \left[\frac{1-\zeta}{-\zeta q^2+m^2} \log \frac{\zeta-1}{\zeta} \right] = \frac{(1-\frac{m^2}{q^2}) \log(1-\frac{q^2}{m^2})-1}{-q^2}. \end{array}$$

In the second case, there is a nontrivial residue at the holomorphic point $\zeta = \infty$.

The corresponding structures for Minkowski spacetime as a minimal case with nontrivial rotation degrees of freedom are as follows:

$\mathbf{SO}_0(1,3)$-spacetime, $\lvert\Omega^3\rvert = 2\pi^2$
$(\overset{4}{*}, q^2) = (\pm \frac{*}{i\pi^2}, q^2 \pm io)$
$\frac{\partial}{\partial q^2} \frac{1}{-q^2+m_1^2} \overset{4}{*} \frac{\partial}{\partial q^2} \frac{1}{-q^2+m_2^2} = \frac{\partial}{\partial q^2} \int_0^1 d\zeta \frac{1}{-\zeta(1-\zeta)q^2+\zeta m_1^2+(1-\zeta)m_2^2}$
$= \frac{1}{(-q^2+m_1^2)^2} \overset{4}{*} \frac{1}{(-q^2+m_2^2)^2} = \int_0^1 d\zeta \frac{\zeta(1-\zeta)}{[-\zeta(1-\zeta)q^2+\zeta m_1^2+(1-\zeta)m_2^2]^2}$

In general, one obtains the even-dimensional spacetime distributions of energy-momenta by relativistically compatible 2-sphere spread. Measures for higher-dimensional spacetimes are obtained by derivation of order $R-1$, the rank of the maximal compact group $\mathbf{SO}(2R-1)$,

$$R = 1, 2, \cdots : \quad \frac{1}{\Gamma(R)} \left(\frac{\partial}{\partial q^2}\right)^{R-1} \frac{1}{-q^2+m^2} = \frac{1}{(-q^2+m^2)^R},$$

with the convolution

$\mathbf{SO}_0(1, 2R-1)$-spacetime, $2R = 2, 4, \ldots,$ $\lvert\Omega^{2R-1}\rvert = \frac{2\pi^R}{\Gamma(R)}$
$(\overset{2R}{*}, q^2) = (\pm \frac{*2}{i\lvert\Omega^{2R-1}\rvert}, q^2 \pm io)$
$(\frac{\partial}{\partial q})^{L_1} \frac{1}{(-q^2+m_1^2)^R} \overset{2R}{*} (\frac{\partial}{\partial q})^{L_2} \frac{1}{(-q^2+m_2^2)^R}$
$= (\frac{\partial}{\partial q})^{L_1+L_2} \frac{1}{\Gamma(R)} (\frac{\partial}{\partial q^2})^{R-1} \int_0^1 d\zeta \frac{1}{-\zeta(1-\zeta)q^2+\zeta m_1^2+(1-\zeta)m_2^2}$

In contrast to the residual normalization of odd-dimensional eigenvalue spaces (energy-momenta), the residual normalization for even-dimensional spacetime (energy-momenta) uses the volume of the odd-dimensional unit sphere Ω^{2R-1}, the compact partner of the embedded position hyperboloid $\mathcal{Y}^{2R-1}$.

9.7 Feynman Propagators

A Feynman propagator for flat spacetime, e.g., for a free massive scalar field (see Chapter 5),

$$\langle 0|\{\mathbf{\Phi}(y),\mathbf{\Phi}(x)\}|0\rangle - \epsilon(x_0-y_0)\langle 0|[\mathbf{\Phi}(y),\mathbf{\Phi}(x)]|0\rangle = \int \frac{d^4q}{i\pi(2\pi)^3}\,\frac{\rho(m^2)}{-q^2-io+m^2}e^{iq(x-y)},$$

is the sum of a representation coefficient of the Poincaré group $\mathbf{SO}_0(1,3)\;\vec{\times}\,\mathbb{R}^4$ for particles, supported by all translations $\mathbb{R}^4$, and the causally ordered quantization, supported by the causal bicone $\mathbb{R}^4_+\cup\mathbb{R}^4_-$.

The Fourier-transformed energy-momentum distributions in Feynman propagators give Macdonald $\mathcal{K}$ and Neumann $\mathcal{N}$ functions for the on-shell contribution (real) and, in addition to Bessel functions $\mathcal{J}$ in the off-shell contribution (imaginary), lightcone-supported Dirac distributions δ for even-dimensional spacetime with nonabelian position rotations, i.e., for $1+s=2R=4,6,\ldots$. The Dirac distributions arise by derivations of the characteristic causal function $\vartheta(x^2)$:

$$\begin{aligned}\int \frac{d^{2R}q}{i\pi}\,\frac{1}{-q^2-io+1}e^{iqx} &= \left(\frac{\partial}{\partial\frac{x^2}{4\pi}}\right)^{R-1}\left[\vartheta(-x^2)2\mathcal{K}_0(|x|)-\vartheta(x^2)\pi(\mathcal{N}_0+i\mathcal{J}_0)(|x|)\right]\\ &= \left(\frac{\partial}{\partial\frac{x^2}{4\pi}}\right)^{R-1}\int d\psi\,\left[\vartheta(-x^2)e^{-|x|\cosh\psi}+\vartheta(x^2)e^{-i|x|\cosh\psi}\right].\end{aligned}$$

The projection of a free-field Feynman propagator to time and position by position and time integration, respectively, displays a translation representation coefficient $\cos mx_0$ only for the on-shell part $\delta(q^2-m^2)$, whereas the principal value P off-shell part $\frac{1}{i\pi}\frac{1}{-q_{\mathrm{P}}^2+m^2}$ with causal support in spacetime is position-projected to the exponential potential $e^{-|m|r}$ for Cartan spacetime $R=1$ and to the Yukawa potential $\frac{e^{-|m|r}}{r}$ for Minkowski spacetime $R=2$:

$$\begin{aligned}\begin{pmatrix}\int\frac{d^{2R-1}x}{(2\pi)^{2R-1}}\\ \int\frac{dx_0}{2\pi}\end{pmatrix}\int\frac{d^{2R}q}{i\pi}\,\frac{1}{-q^2-io+1}e^{iqx} &= \begin{pmatrix}\int\frac{dq_0}{i\pi}\,\frac{1}{-q_0^2-io+1}e^{iq_0x_0}\\ \int\frac{d^{2R-1}q}{i\pi}\,\frac{1}{\vec{q}^{\,2}+1}e^{-i\vec{q}\vec{x}}\end{pmatrix}\\ &= \begin{pmatrix}\cos x_0\\ 0\end{pmatrix} - i\begin{pmatrix}\epsilon(x_0)\sin x_0\\ \left(-\frac{\partial}{\partial\frac{r^2}{4\pi}}\right)^{R-1}e^{-r}\end{pmatrix}.\end{aligned}$$

The off-shell contributions ("virtual particles") are not coefficients of Poincaré group representations,

$$\begin{aligned}
\int \frac{d^{2R}q}{\pi}\ \frac{1}{-q_{\rm P}^2+1}e^{iqx} &= -i\epsilon(x_0)\int d^{2R}q\ \epsilon(q_0)\delta(q^2-1)e^{iqx}\\
&= \left(\frac{\partial}{\partial\frac{x^2}{4\pi}}\right)^{R-1}\vartheta(x^2)\pi\mathcal{J}_0(|x|),\\
(\partial^2+1)\int \frac{d^{2R}q}{\pi}\ \frac{1}{-q_{\rm P}^2+1}e^{iqx} &= \frac{1}{\pi}\delta(\tfrac{x}{2\pi}).
\end{aligned}$$

They are the relevant part of the Green's distributions associated with free-field equations.

9.7.1 Convolutions of Feynman Propagators

Feynman integrals involve convolutions of energy-momentum distributions for pointwise products of spacetime distributions. There arise undefined ("divergent") products of generalized functions. In general, they do not make sense since the tempered distributions $\mathcal{S}'(\mathbb{R}^n)$ do not constitute a convolution algebra.

In a schematic notation for Feynman propagators with the real on-shell contribution, supported by all translations $\mathbb{R}^{1+s}$, and the imaginary off-shell contribution, causally supported and, for even-dimensions $2R = 4, 6, \ldots$, with Dirac distributions,

$$\int \frac{d^{2R}q}{i\pi}\ \frac{1}{-q^2-io+m^2}e^{iqx} = \vartheta_s\mathcal{K}+\vartheta_t\mathcal{N}+i\vartheta_t(\mathcal{J}+\delta),\quad \vartheta_s=\vartheta(-x^2),\quad \vartheta_t=\vartheta(x^2),$$

the convolution gives as real and imaginary contributions:

$$\begin{aligned}
[\vartheta_s\mathcal{K}+\vartheta_t\mathcal{N}+i\vartheta_t(\mathcal{J}+\delta)] &\quad\cdot\quad [\vartheta_s\mathcal{K}+\vartheta_t\mathcal{N}+i\vartheta_t(\mathcal{J}+\delta)]\\
\sim \frac{1}{i\pi}\frac{1}{-q^2-io+m^2} &\quad *\quad \frac{1}{i\pi}\frac{1}{-q^2-io+m^2}\\
= \left[\delta(q^2-m^2)+\frac{1}{i\pi}\frac{1}{-q_{\rm P}^2+m^2}\right] &\quad *\quad \left[\delta(q^2-m^2)+\frac{1}{i\pi}\frac{1}{-q_{\rm P}^2+m^2}\right]
\end{aligned}$$

$$\begin{aligned}
= \left[\delta(q^2-m^2)*\delta(q^2-m^2)+\frac{1}{q_{\rm P}^2-m^2}*\frac{1}{q_{\rm P}^2-m^2}\right] &\quad+\quad \delta(q^2-m^2)*\frac{1}{i\pi}\frac{1}{-q_{\rm P}^2+m^2}\\
\sim [\vartheta_s\mathcal{K}^2+\vartheta_t(\mathcal{N}^2+\mathcal{J}^2+\mathcal{J}\cdot\delta+\delta^2)] &\quad+\quad i\vartheta_t(\mathcal{N}\cdot\mathcal{J}+\mathcal{N}\cdot\delta).
\end{aligned}$$

The "divergent" parts δ^2 with the pointwise product of $\mathcal{S}'(\mathbb{R}^4)$-distributions, as familiar from a perturbation expansion quantum electrodynamics, do not make sense:

$$\begin{aligned}
&\left[\delta(q^2-m_1^2)+\frac{1}{i\pi}\frac{1}{-q_{\rm P}^2+m_1^2}\right] \quad * \quad \left[\delta(q^2-m_2^2)+\frac{1}{i\pi}\frac{1}{-q_{\rm P}^2+m_2^2}\right]\\
\overset{\mathbb{R}^4}{\sim}\ &\left[\frac{1}{x_{\rm P}^2}+i\delta(x^2)+\cdots\right] \quad\cdot\quad \left[\frac{1}{x_{\rm P}^2}+i\delta(x^2)+\cdots\right].
\end{aligned}$$

The "divergent" parts $\delta(q^2-m_1^2) * \frac{1}{-q_{\rm P}^2+m_2^2}$ and $\frac{1}{q_{\rm P}^2-m_1^2} * \frac{1}{-q_{\rm P}^2+m_2^2}$ contain the causally supported off-shell contribution from the energy-momentum principal value $\frac{1}{-q_{\rm P}^2+m^2} \overset{\mathbb{R}^4}{\sim} \delta(x^2)+\cdots$. They are not coefficients of translation product representations like the meaningful on-shell convolution $\delta(q^2-m_1^2) * \delta(q^2-m_2^2)$.

9.7.2 Convolutions for Free Particles

The on- and off-shell contributions are convoluted as follows:

$$\mathbf{O}(1,s),\ m^2 \geq 0 : (F_{m^2}, F^\star_{m^2}) = \pm\frac{1}{i\pi}\frac{1}{-q^2 \mp io + m^2} = \delta_{m^2} \pm i\mathrm{P}_{m^2},$$
$$\begin{pmatrix} F_{m_1^2} * F_{m_2^2} \\ F^\star_{m_1^2} * F^\star_{m_2^2} \end{pmatrix} = \delta_{1*2} \pm i\mathrm{P}_{1*2},$$

with the Dirac and principal value contribution

$$\begin{aligned} \delta_{1*2} &= \delta_{m_1^2} * \delta_{m_2^2} - \mathrm{P}_{m_1^2} * \mathrm{P}_{m_2^2}, \\ \mathrm{P}_{1*2} &= \delta_{m_1^2} * \mathrm{P}_{m_2^2} + \mathrm{P}_{m_1^2} * \delta_{m_2^2}. \end{aligned}$$

The Fourier-transformed principal part of a Feynman distribution for an orthochronous group $\mathbf{SO}_0(1,s)$ can be written as an order function times an on-shell part:

$$F_{m^2} = \frac{1}{i\pi}\frac{1}{-q^2 - io + m^2} = \delta_{m^2} + i\mathrm{P}_{m^2} : \quad \begin{cases} \delta_{m^2} &= \frac{\delta_{|m|} + \delta_{-|m|}}{2}, \\ \mathrm{P}_{m^2} &\sim i\epsilon(x_0)\epsilon_{m^2}, \\ \epsilon_{m^2} &= \frac{\delta_{|m|} - \delta_{-|m|}}{2}. \end{cases}$$

In the principal value convolution contribution of two Feynman propagators for spacetime $\mathbb{R}^{1+s}$ the order function drops out for $\epsilon(x_0)^2 = 1$:

$$\begin{aligned} \pm\frac{1}{i\pi}\frac{1}{-q^2 \mp io + m_1^2} * \pm\frac{1}{i\pi}\frac{1}{-q^2 \mp io + m_2^2} &= 2\left[\vartheta(+q_0)\delta(q^2 - m_1^2) * \vartheta(+q_0)\delta(q^2 - m_2^2)\right. \\ &\quad \left. + \vartheta(-q_0)\delta(q^2 - m_1^2) * \vartheta(-q_0)\delta(q^2 - m_2^2)\right] \\ &\quad \pm\frac{1}{i\pi}\left[\delta(q^2 - m_1^2) * \frac{1}{-q_{\mathrm{P}}^2 + m_2^2} + \frac{1}{-q_{\mathrm{P}}^2 + m_1^2} * \delta(q^2 - m_2^2)\right]. \end{aligned}$$

The principal value square is also an on-shell convolution only. The convolution of translation representation coefficients from the real part of the propagator (free particles) gives corresponding coefficients for product representations (product of free particles):

$$\begin{aligned} \delta_{1*2} &= \delta_{m_1^2} * \delta_{m_2^2} - \mathrm{P}_{m_1^2} * \mathrm{P}_{m_2^2} \\ &= \delta_{m_1^2} * \delta_{m_2^2} + \epsilon_{m_1^2} * \epsilon_{m_2^2} \quad = \frac{\delta_{|m_1|} * \delta_{|m_2|} + \delta_{-|m_1|} * \delta_{-|m_2|}}{2}. \end{aligned}$$

The set with all $(1+s)$-dimensional "filled-up" forward (backward) energy-momentum hyperboloids is an additive cone. Therefore, the distributions supported by positive and negative energy-momentum are convolution algebras, however, not orthogonal to each other:

$$\begin{aligned} &\{q \succeq |m_1|\} + \{q \succeq |m_2|\} = \{q \succeq |m_+|\}, \\ &\text{with } \delta_{\pm|m|} \sim 2|m|\vartheta(\pm q_0)\delta(q^2 - m^2) \in \mathcal{D}'(\mathbb{R}^{1+s}_\pm) \in \underline{\mathbf{aag}}_{\mathbb{C}} \text{ (convolution product)}. \end{aligned}$$

The convolution for abelian time with self-dual invariants $m_{1,2}^2 > 0$,

$$\begin{aligned} \text{abelian } \mathbb{R} : \quad &\vartheta(\pm q_0)|q_0|\delta(q_0^2 - m_1^2) * \vartheta(\pm q_0)|q_0|\delta(q_0^2 - m_2^2) \\ &= \vartheta(\pm q_0)|q_0|\delta(q_0^2 - m_+^2), \end{aligned}$$

is embedded into the convolution of nonabelian hyperboloids for product representations of the translations (the real part δ_{1*2} for simple pole Feynman propagators). With the hyberboloid "radii" as energy-momentum invariants $q^2 = m^2 \ge 0$,

$$\begin{aligned}\mathbf{SO}_0(1,s)\ \vec{\times}\,\mathbb{R}^{1+s}:\ &\vartheta(\pm q_0)\delta(q^2-m_1^2)\ \underset{|\Omega^{s-1}|}{*}\ \vartheta(\pm q_0)\delta(q^2-m_2^2)\\ &= \frac{\Delta(q^2)^{\frac{s-2}{2}}}{(2|q|)^{s-1}}\vartheta(\pm q_0)\vartheta(q^2-m_+^2),\quad s=1,2,\ldots,\end{aligned}$$

they involve the *two-particle threshold factor*:

$$\Delta(q^2) = \Delta(q^2, m_1^2, m_2^2) = (q^2 - m_+^2)(q^2 - m_-^2).$$

For nontrivial position, the convolution (phase space integral) of s-dimensional on-shell hyperboloids (particle measures) does not lead to s-dimensional on-shell hyperboloids $\delta(q^2 - m_+^2)$. It leads to translation representations with energy-momenta over the free particle threshold at the mass sum $q^2 \ge m_+^2$, i.e., $\vec{q}^{\,2} = q_0^2 - m_+^2 \ge 0$, with $m_\pm = |m_1| \pm |m_2|$. Here, the energy is enough to produce two free particles with masses $m_{1,2}$ and momentum $(\vec{q_1} + \vec{q_2})^2 \ge 0$.

The minimal cases $s = 1, 2$ are characteristic for even- and odd-dimensional spacetime:

$$\begin{aligned}\mathbf{SO}_0(1,1)\ \vec{\times}\,\mathbb{R}^2:\ &\vartheta(\pm q_0)\delta(q^2-m_1^2)\underset{2}{*}\vartheta(\pm q_0)\delta(q^2-m_2^2)\\ &= \frac{1}{\sqrt{(q^2-m_+^2)(q^2-m_-^2)}}\vartheta(\pm q_0)\vartheta(q^2-m_+^2),\\ \mathbf{SO}_0(1,2)\ \vec{\times}\,\mathbb{R}^3:\ &\vartheta(\pm q_0)\delta(q^2-m_1^2)\ \underset{2\pi}{*}\ \vartheta(\pm q_0)\delta(q^2-m_2^2)\\ &= \frac{1}{2|q|}\vartheta(\pm q_0)\vartheta(q^2-m_+^2).\end{aligned}$$

The Poincaré group $\mathbf{SO}_0(1,3)\ \vec{\times}\ \mathbb{R}^4$ is the minimal case with nonabelian rotations:

$$\begin{aligned}\mathcal{M}(\check{\mathbb{R}}^4)_+ * \mathcal{M}(\check{\mathbb{R}}^4)_+ = \mathcal{M}(\check{\mathbb{R}}^4)_+:\ &\vartheta(q_0)\delta(q^2-m_1^2)\ \underset{4\pi}{*}\ \vartheta(q_0)\delta(q^2-m_2^2)\\ &= \frac{\sqrt{(q^2-m_+^2)(q^2-m_-^2)}}{q^2}\ \vartheta(q_0)\vartheta(q^2-m_+^2).\end{aligned}$$

Such convolutions arise, e.g., as the nondivergent on-shell contribution in the quantum electrodynamical vacuum polarization by electron-positron pairs.

9.7.3 Off-Shell Convolution Contributions

Energy-momentum convolutions combine the points on the hyperbolic-spherical singularity surfaces for particle-interaction structures, determined by the invariants. The characteristic new feature is the on-shell–off-shell convolution, i.e., of compact with noncompact invariants. The convolution contribution in the mixed terms is not for product representations of the spacetime translations,

$$\mathrm{P}_{1*2} = \delta_{m_1^2} * \mathrm{P}_{m_2^2} + \mathrm{P}_{m_1^2} * \delta_{m_2^2}.$$

The divergences in Minkowski space arise from the mixed terms (mathematically meaningless)

$$\delta(q^2 - m_1^2) * \frac{1}{-q_{\rm P}^2+m_2^2} \overset{\mathbb{R}^4}{\sim} \frac{1}{x_{\rm P}^2}\cdot\delta(x^2)+\cdots.$$

Only for trivial position does the principal value part also add the invariant poles:

$$s=0: \quad {\rm P}_{1*2} \sim i\epsilon(t)\frac{\delta_{|m_1|}*\delta_{|m_2|}-\delta_{-|m_1|}*\delta_{-|m_2|}}{2} \sim {\rm P}_{m_+^2}.$$

The characteristic effect of a convolution of noncompact with compact invariant comes in the principal value part for nontrivial position degrees of freedom:

$$\begin{array}{ccccc} \delta(q^2-m^2) & \sim & \vartheta(q^2-m^2), & & \\ \frac{1}{-q_{\rm P}^2+m^2} & \sim & \vartheta(q^2-m^2) & + & \vartheta(-q^2+m^2) \\ & & \cup & & \cup \\ & & \text{compact (free)} & + & \text{noncompact.} \\ & & e^{imt} & & e^{-|mz|} \end{array}$$

The denominator polynomial in the above convolution square has two energy-momentum-dependent zeros,

$$\begin{aligned} P(\zeta) &= -\zeta(1-\zeta)q^2+\zeta m_1^2+(1-\zeta)m_2^2 = q^2[\zeta-\zeta_1(q^2)][\zeta-\zeta_2(q^2)], \\ \zeta_{1,2}(q^2) &= \frac{q^2-m_+m_-\pm\sqrt{\Delta(q^2)}}{2q^2} \text{ with } m_\pm = |m_1|\pm|m_2|, \end{aligned}$$

which are either both real or complex conjugate to each other according to the sign of the discriminant $\Delta(q^2)$ (two-particle threshold factor):

$$\Delta(q^2) = (q^2-m_+^2)(q^2-m_-^2): \left\{ \begin{array}{c} \vartheta(\Delta(q^2)) = \vartheta(q^2-m_+^2)+\vartheta(m_-^2-q^2), \\ \vartheta(-\Delta(q^2)) = \vartheta(m_+^2-q^2)\vartheta(q^2-m_-^2). \end{array} \right.$$

Furthermore, real zeros, in the case of $\Delta(q^2)\geq 0$, are in the integration interval $\zeta\in[0,1]$ only for energy-momenta over the threshold $\vartheta(q^2-m_+^2)$,

$$\begin{aligned} \zeta_{1,2}(m_+^2) &= \frac{|m_2|}{|m_1|+|m_2|} \in [0,1], \\ \zeta_{1,2}(m_-^2) &= \frac{-|m_2|}{|m_1|-|m_2|} \notin [0,1], \end{aligned}$$

and graphically:

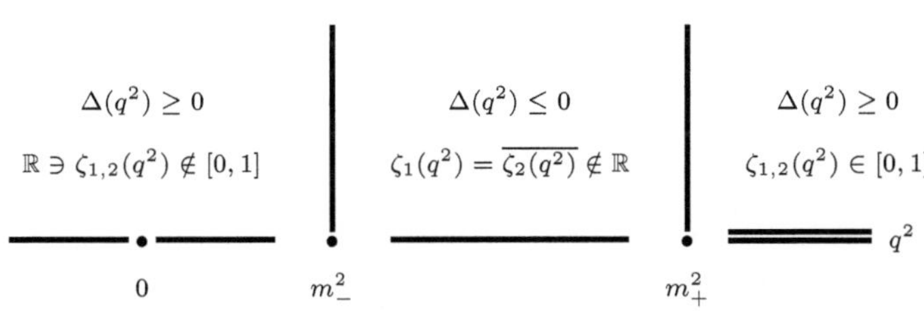

Therefore, the convolution of two energy-momentum Feynman pole distributions contains the following as relevant contributions, valid for general signature $\mathbf{O}(t,s)$:

$$\begin{aligned}
\int_0^1 \frac{d\zeta}{P(\zeta)} &= \int_0^1 d\zeta \frac{1}{-\zeta(1-\zeta)q^2 - io + \zeta m_1^2 + (1-\zeta)m_2^2}\\
&= \int_0^1 d\zeta \left[\frac{1}{-\zeta(1-\zeta)q_{\rm P}^2 + \zeta m_1^2 + (1-\zeta)m_2^2} + i\pi\delta\Big(-\zeta(1-\zeta)q^2 + \zeta m_1^2 + (1-\zeta)m_2^2\Big)\right]\\
&= \frac{2}{\sqrt{|\Delta(q^2)|}} \left[\begin{array}{rl} \vartheta(m_-^2 - q^2) & \log\left|\frac{\Sigma(q^2) - 2\sqrt{\Delta(q^2)}}{m_+^2 - m_-^2}\right| \\ + \vartheta(q^2 - m_-^2)\vartheta(m_+^2 - q^2) & \arctan\frac{2\sqrt{-\Delta(q^2)}}{\Sigma(q^2)} \\ + i\pi\, \vartheta(q^2 - m_+^2) & \end{array}\right],
\end{aligned}$$

$$\text{with}\quad \Delta(q^2) = (q^2 - m_+^2)(q^2 - m_-^2),\quad \Sigma(q^2) = (q^2 - m_+^2) + (q^2 - m_-^2).$$

The convolution product depends on the two variables $\{q^2 - m_+^2, q^2 - m_-^2\}$.

The part with $\vartheta(q^2 - m_+^2)$ (here imaginary) is the on-shell convolution, adding positive and negative energy contributions (above). The convolution of compact with noncompact invariants, original for indefinite signature spacetime, (here real part) shows up for energy-momenta under the threshold $\vartheta(m_+^2 - q^2)$ and for $\vartheta(m_-^2 - q^2)$ containing the momentumlike contribution $\vartheta(-q^2)$. This is illustrated by two equal masses:

$$\int_0^1 d\zeta \frac{1}{-\zeta(1-\zeta)q^2 - io + m^2} = \frac{2}{\sqrt{|q^2(q^2 - 4m^2)|}} \left[\begin{array}{ll} \vartheta(-q^2) & \log\left|\frac{2m^2 - q^2 + \sqrt{q^2(q^2 - 4m^2)}}{2m^2}\right| \\ + \vartheta(q^2)\vartheta(4m^2 - q^2) & \arctan\frac{\sqrt{q^2(4m^2 - q^2)}}{q^2 - 2m^2} \\ + i\pi\, \vartheta(q^2 - 4m^2) & \end{array}\right].$$

Chapter 10

Interactions and Kernels

It is physically obvious that interactions and bound-states have a close connection. Both are described, in distinction to free objects, by nonabelian operations of nonflat position. In contrast to position functions, which are representation coefficients of position groups, position interactions are formalized by linear mappings of position functions and are, in general, position distributions. The relation between bound-states and interactions will be seen in parallel to the relation between Lie groups and Lie algebras.

Examples for the Lie algebra–interaction relationship are given by Hamiltonians in mechanics, $\mathbf{H} = \frac{\vec{\mathbf{p}}^2}{2m} + V(\vec{\mathbf{x}})$, where a potential for an interaction contributes to the time translation Lie algebra $\mathbf{H} \sim \frac{d}{dt}$. For example, the convolutive action of the Yukawa and Kepler potential on the ground-state function $\vec{x} \longmapsto e^{-mr}$, $m > 0$, of the nonrelativistic hydrogen atom, a positive-type $L^\infty(\mathcal{Y}^3)$-function, representing hyperbolic 3-position $\mathcal{Y}^3 \cong \mathbb{R}^3$, leads to $L^\infty(\mathcal{Y}^3)$-functions:

$$\begin{aligned} -\frac{e^{-mr}}{r} \frac{*}{4\pi} e^{-Mr} &= \frac{e^{-Mr}}{M^2-m^2} + \frac{2M}{(M^2-m^2)^2} \frac{e^{-Mr}-e^{-mr}}{r}, \\ -\frac{1}{r} \frac{*}{4\pi} e^{-Mr} &= \frac{e^{-Mr}}{M^2} + \frac{2}{M^2} \frac{e^{-Mr}-1}{Mr}. \end{aligned}$$

Also, gauge interactions represent Lie algebras by charges, i.e., position-integrated currents $L \ni l \longmapsto iQ = i \int d^3x \, \mathbf{J}_0(\vec{x}) \in W \otimes W^T$, which occur in gauge vertices $\mathbf{A}^a \mathbf{J}_a$ with gauge fields (see Chapter 6). The gauge field–induced interactions show up in the off-shell contribution of their propagators, e.g., for electrodynamics $\int dx_0 \int \frac{d^4q}{\pi(2\pi)^2} \frac{\alpha_S}{-q^2-io} e^{iqx} = \frac{\alpha_S}{r}$ (see Chapter 5).

The transition from bound-states (functions) to interactions (distributions) is illustrated for three-dimensional position, $L^\infty(\mathcal{Y}^3) \longrightarrow \mathcal{M}(\mathcal{Y}^3)$, where derivations of the ground-state wave function of the nonrelativistic hydrogen atom lead to the Yukawa potential and the Yukawa force, which are Radon measures of the position hyperboloid,

$$\frac{\partial}{\partial \vec{x}} \frac{\partial}{\partial r^2} e^{-mr} = -m \frac{\partial}{\partial \vec{x}} \frac{e^{-mr}}{2r} = m \frac{\vec{x}}{r} \frac{1+mr}{2r^2} e^{-mr}.$$

H. Saller, *Operational Spacetime: Interactions and Particles*,
Fundamental Theories of Physics 163, DOI 10.1007/978-1-4419-0898-8_11,

As will be discussed ahead, there is an interpretation in terms of group representations for the fact that the pointwise product of the Kepler potential $\frac{1}{r}$ with a bound-state wave function e^{-mr} for representation-characterizing invariant m^2 gives a Yukawa potential $\frac{1}{r}\cdot e^{-mr}$. The Kepler potential is the inverse Laplacian, $\frac{1}{r} = -\frac{4\pi}{\vec{\partial}^2}$, and the Yukawa potential and the bound-state function are inverse Laplacians of order 1 and 2 with invariant m^2:

$$\begin{aligned} (-\vec{\partial}^2 + m^2)\frac{e^{-mr}}{r} &= 4\pi\delta(\vec{x}), \\ (-\vec{\partial}^2 + m^2)^2 e^{-mr} &= 2m\ 4\pi\delta(\vec{x}). \end{aligned}$$

Inverted differential operators (derivations) define kernels. The transition from bound-states to interactions is expressible by the action of a Green's kernel on Lie group representation coefficients.

In a residual representation, the (energy-)momentum distributions for bound state functions and interaction distributions have singularities at the invariants, in the example above where the derivative of the bound-state function is the Yukawa potential:

$$\begin{aligned} (1, \tfrac{2m}{r})e^{-mr} &= \int \frac{d^3q}{2\pi^2}\frac{2m}{(\vec{q}^2+m^2)^{2-N}}e^{-i\vec{q}\vec{x}} \text{ for } N = (0,1), \\ \text{with } (-\vec{\partial}^2 + m^2)e^{-mr} &= \tfrac{2m}{r}e^{-mr}. \end{aligned}$$

The structures of interactions as implemented by tangent kernels, considered in this chapter mainly for time $\mathbf{D}(1) \cong \mathbb{R}$ and position $\mathcal{Y}^3$, both with real rank 1, and exemplified by the nonrelativistic hydrogen atom and the model of Nambu and Jona-Lasinio (see Chapter 9), will be used for the more complicated spacetime operations $\mathbb{D}(2)$ with real rank 2 in Chapter 10.

The dual of a real finite-dimensional Lie algebra $L = \log G$, i.e., its linear forms L^T, is more easily accessible than the group dual $\check{G} = \mathbf{irrep}\, G$, which, in general, is not a group, but, e.g., a cone or a direct sum of cones. $\mathbb{C} \otimes L^T$ contains all eigenvalues of the Lie algebra action. The invariants that characterize the group dual are given by multilinear forms of the eigenvalues in the complexified tensor algebra $\mathbb{C} \otimes \bigotimes L^T$.

A Lie group and Lie algebra act on themselves and coadjointly on the dual Lie algebra, with the actions denoted by $\bullet$ in the schema:

$\bullet$	G	L	L^T
G	G	L	L^T
L	L	L	L^T
L^T	L^T	L^T	$-$

$*$	$\mathcal{M}(G)$	$L^1(G)$	$L^\infty(G)$
$\mathcal{M}(G)$	$\mathcal{M}(G)$	$L^1(G)$	$L^\infty(G)$
$L^1(G)$	$L^1(G)$	$L^1(G)$	$L^\infty(G)$
$L^\infty(G)$	$L^\infty(G)$	$L^\infty(G)$	$-$

Both a Lie group and the enveloping algebra of its Lie algebra can be realized in one algebra: A matrix group $G \subseteq \mathbf{GL}(V)$ for a finite-dimensional vector space $V \cong \mathbb{C}^n$ has a minimal endomorphism subalgebra $A(G, L) \subseteq \mathbf{AL}(V)$, which embeds both the algebra $\mathbb{C}^{(G)}$, generated by the group, and the algebra $\mathbb{C}^{(L)}$, generated by its Lie algebra $L = \log G \subseteq \mathbf{AL}(V)$. $\mathbb{C}^{(L)}$ is the endomorphism image of the enveloping algebra $\mathbf{E}(L) \longrightarrow \mathbf{AL}(V)$. The algebra $A(G, L) \supseteq \mathbb{C}^{(G)} \cup \mathbb{C}^{(L)}$ is generalized by the group measure algebra $\mathcal{M}(G) \supseteq L^1(G)$. The convolution algebra $\mathcal{M}(G)$ embeds the group G by Dirac measures. It is in duality with $L^\infty(G)$. Tangent (Lie algebra) kernels of a Lie group G are Radon measures $\mathcal{M}(G)$.

10.1 Invariant Differential Operators

Differential operators [36] acting on functions of a Lie group and its symmetric spaces $\mathbb{M} \cong G/H$ are related to its Lie algebra $\log G \cong \mathbb{R}^n$ with the left-invariant vector fields, i.e., to the derivations $v \in \mathbf{T}(\mathbb{M})$ from its tangent spaces (see Chapter 2), built, e.g., by $(\partial^a)_{a=1}^n$ in holonomic bases.

The enveloping algebra $\mathbf{E}(L)$ with the associative products of the Lie algebra elements, modulo the identification of the Lie bracket with commutator, is mapped into the composition algebra, built by derivatives, e.g., $\alpha_{kj}\partial^k \otimes \partial^j \longmapsto \alpha_{kj}\partial^k\partial^j$. The invariant differential operators are determined by the center of the enveloping algebra.

For simple Lie algebras, the invariant Laplace–Beltrami operator (see Chapter 2) corresponds to the invariant Casimir operator $\kappa^{-1} = \kappa_{ab} l^a \otimes l^b$ as inverse Killing form $\kappa(l, m) = \operatorname{tr} \operatorname{ad} l \circ \operatorname{ad} m$, represented on vector spaces with manifold functions. It defines eigenfunctions or eigendistributions with eigenvalues and invariants.

The Laplace–Beltrami operator of a Riemannian manifold $(\mathbb{M}, \mathbf{g})$, $\mathbf{g} = \mathbf{g}^{kj} dx_k \otimes dx_j$, $\mathbf{g}^{-1} = \mathbf{g}_{kj}\partial^k \otimes \partial^j$, is invariant under a diffeomorphism if, and only if, it is an element of the motion (global invariance) group $\varphi \in G_{\mathbf{g}}$, i.e., an isometry, $\mathbf{g}(\varphi.x, \varphi.y) = \mathbf{g}(x, y)$,

$$\begin{aligned}
\partial_{\mathbf{g}}^2 &: \mathcal{C}(\mathbb{M}^{(t,s)}) \longrightarrow \mathcal{C}(\mathbb{M}^{(t,s)}), \\
\partial_{\mathbf{g}}^2 &= \frac{1}{\sqrt{|\mathbf{g}|}}\partial^k \sqrt{|\mathbf{g}|}\ \ \mathbf{g}_{kj}\partial^j = \mathbf{g}_{kj}(\partial^k\partial^j - \Gamma_i^{kj}\partial^i) \\
&= \partial^k \mathbf{g}_{kj}\partial^j + \mathbf{g}_{kj}(\partial^k \log\sqrt{|\mathbf{g}|})\partial^j.
\end{aligned}$$

As the inverse metric $\mathbf{g}^{-1}$, it has a degree-2 contribution, and for $|\det \mathbf{g}| = |\mathbf{g}| \neq$ constant an additional contribution for the dilation group in $\mathbf{GL}(\mathbb{R}^n) = \mathbf{D}(1) \times \mathbf{SL}(\mathbb{R}^n)$, which contains the derivative of the dilation degree of freedom of the metric. The $\mathbf{D}(1)$-factor also arises in the invariant measure $\sqrt{|\mathbf{g}|}d^{t+s}x$. Locally, there exist stationary coordinates, where the Laplace–Beltrami operator can be brought to the simple flat space form $\partial_{\mathbf{g}}^2|_{x^0} = \eta_{ab}\partial^a\partial^b$.

For a scalar spacetime field, the gravitative interaction is described by $\partial_{\mathbf{g}}^2\mathbf{\Phi} = -m^2\mathbf{\Phi}$ (see Chapter 5). The negative mass $-m^2$ is the invariant of the Laplace–Beltrami operator, also for nonflat spacetime.

For translations, the algebra $\mathtt{D}(\mathbb{R}^n)$ of the invariant differential operators consists of the operators $\mathbb{C}[(\partial^a)_{a=1}^n]$ with constant coefficients, generated by the derivatives. The joint eigenfunctions are the exponentials; the eigenspace representations are one-dimensional and irreducible.

The joint eigenfunctions on the cosets G/K of a Lie group for a compact subgroup $\{\mathtt{D}f = I(\mathtt{D})f \text{ for each } \mathtt{D} \in \mathtt{D}(G/K)\}$ are characterized by spherical functions (see Chapter 8). The algebra of the invariant differential operators $\mathtt{D}(G/K)$ on cosets of a maximal compact subgroup K is a commutative polynomial ring with algebraically independent generators $\mathbb{C}[(\mathtt{D}^A)_{A=1}^r]$ whose degrees d_A are canonically determined. r is the real rank of G/K.

For manifolds with constant curvature (maximal symmetry) $\mathcal{R} = k\mathbf{g} \wedge \mathbf{g}$, the polynomials $\mathbb{C}[\partial_{\mathbf{g}}^2]$ of the Laplace–Beltrami operator constitute the invariant differential operators, i.e., for the general hyperboloids $\mathcal{Y}^{(t,s)} \cong \mathbf{SO}_0(t,1+s)/\mathbf{SO}_0(t,s)$ with $k=1$ and $\mathcal{Y}^{(s,t)} \cong \mathbf{SO}_0(1+t,s)/\mathbf{SO}_0(t,s)$ with $k=-1$, and for the flat manifolds $\mathbf{SO}_0(t,s) \vec{\times} \mathbb{R}^n/\mathbf{SO}_0(t,s)$ with $k=0$ and Laplace–Beltrami operator $\vec{\partial}_t^2 - \vec{\partial}_s^2$.

The Laplace–Beltrami operator for spheres, Euclidean spaces, and time-like hyperboloids with dimension $s \geq 1$ can be decomposed into a radial part and a spherical part,

$$\begin{pmatrix} \Omega^s \\ \mathbb{R}^s \\ \mathcal{Y}^s \end{pmatrix} \sim \begin{pmatrix} \mathbf{SO}(2) \\ \mathbb{R} \\ \mathbf{SO}_0(1,1) \end{pmatrix} \times \Omega^{s-1},$$

as familiar from the flat 3-position $\vec{\mathbf{p}}^2 = \mathbf{p}_r^2 + \frac{\vec{\mathbf{L}}^2}{r^2} \cong -\vec{\partial}^2$ with radial and angular momentum invariant $\vec{\mathbf{L}}^2 \cong -\frac{\partial^2}{\partial \omega_2^2}$ on Ω^2 and $\frac{\partial^2}{\partial \omega_1^2} = \frac{\partial^2}{\partial \varphi^2} \cong -\mathbf{L}_3^2$ on Ω^1 (see Chapter 4). The metrics and the operators in the Euler parametrization display the general polar decomposition,

$$\mathbf{g} = \begin{pmatrix} d\omega_s^2 \\ d\vec{x}_s^2 \\ d\mathrm{y}_s^2 \end{pmatrix} = \begin{pmatrix} d\theta^2 \\ dr^2 \\ d\psi^2 \end{pmatrix} + \begin{pmatrix} \sin^2\theta \\ r^2 \\ \sinh^2\psi \end{pmatrix} d\omega_{s-1}^2,$$

$$\partial_{\mathbf{g}}^2 = \begin{pmatrix} \frac{\partial^2}{\partial \omega_s^2} \\ \vec{\partial}_s^2 \\ \frac{\partial^2}{\partial \mathrm{y}_s^2} \end{pmatrix} = \begin{pmatrix} \frac{\partial^2}{\partial \theta^2} \\ \frac{\partial^2}{\partial r^2} \\ \frac{\partial^2}{\partial \psi^2} \end{pmatrix} + \begin{pmatrix} \frac{1}{\sin^2\theta} \\ \frac{1}{r^2} \\ \frac{1}{\sinh^2\psi} \end{pmatrix} \frac{\partial^2}{\partial \omega_{s-1}^2} + (s-1)\begin{pmatrix} \cot\theta \frac{\partial}{\partial\theta} \\ \frac{1}{r}\frac{\partial}{\partial r} \\ \coth\psi \frac{\partial}{\partial\psi} \end{pmatrix},$$

where the $(s-1)$-proportional degree-1 contribution contains the logarithm of the dilation factor in the measure:

$$(s-1)\begin{pmatrix} \cot\theta \\ \frac{1}{r} \\ \coth\psi \end{pmatrix} = \begin{pmatrix} \frac{d}{d\theta}\log\sin^{s-1}\theta \\ \frac{d}{dr}\log r^{s-1} \\ \frac{d}{d\psi}\log\sinh^{s-1}\psi \end{pmatrix}, \quad \sqrt{|\mathbf{g}|}\, d^s x = \begin{pmatrix} d^s\omega \\ d^s x \\ d^s \mathrm{y} \end{pmatrix} = \begin{pmatrix} \sin^{s-1}\theta d\theta \\ r^{s-1} dr \\ \sinh^{s-1}\psi d\psi \end{pmatrix} d^{s-1}\omega.$$

They are explicitly in two and three dimensions:

$$\Omega^{2,3}: \left\{ \begin{array}{ll} \frac{\partial^2}{\partial \omega_2^2} & = \frac{1}{\sin^2\theta}[(\sin\theta \frac{\partial}{\partial\theta})^2 + \frac{\partial^2}{\partial\varphi^2}], \\ \frac{\partial^2}{\partial \omega_3^2} & = \frac{1}{\sin\theta}(\frac{\partial^2}{\partial\theta^2} - 1)\sin\theta + \frac{1}{\sin^2\theta}\frac{\partial^2}{\partial\omega_2^2}, \end{array} \right.$$

$$\mathbb{R}^{2,3}: \left\{ \begin{array}{ll} \vec{\partial}_2^2 & = \frac{1}{r^2}[(r\frac{\partial}{\partial r})^2 + \frac{\partial^2}{\partial\varphi^2}], \\ \vec{\partial}_3^2 & = \frac{1}{r}\frac{\partial^2}{\partial r^2} r + \frac{1}{r^2}\frac{\partial^2}{\partial\omega_2^2}, \end{array} \right.$$

$$\mathcal{Y}^{2,3}: \left\{ \begin{array}{ll} \frac{\partial^2}{\partial \mathrm{y}_2^2} & = \frac{1}{\sinh^2\psi}[(\sinh\psi \frac{\partial}{\partial\psi})^2 + \frac{\partial^2}{\partial\varphi^2}], \\ \frac{\partial^2}{\partial \mathrm{y}_3^2} & = \frac{1}{\sinh\psi}(\frac{\partial^2}{d\psi^2} + 1)\sinh\psi + \frac{1}{\sinh^2\psi}\frac{\partial^2}{\partial\omega_2^2}. \end{array} \right.$$

For a three-dimensional flat space, the spherical-hyperbolic analogue formulation is $\frac{1}{r}(\frac{\partial^2}{\partial r^2} - k)r = \frac{1}{r}\frac{\partial^2}{\partial r^2}r$ with $k = 0$.

In geodesic polar and Cartesian coordinates, the metric, measure, and Laplace–Beltrami operator look like the following:

$$
\begin{aligned}
\text{for } k = \begin{pmatrix} 1 \\ 0 \\ -1 \end{pmatrix}: \quad & \begin{pmatrix} \sin\theta \\ r \\ \sinh\psi \end{pmatrix} = \rho = \frac{r}{1+\frac{k}{4}r^2}, \\
\text{metric:} \quad & \begin{pmatrix} d\omega_s^2 \\ d\vec{x}_s^2 \\ d\mathrm{y}_s^2 \end{pmatrix} = \frac{d\rho^2}{1-k\rho^2} + \rho^2 d\omega_{s-1}^2 = \frac{dr^2 + r^2 d\omega_{s-1}^2}{(1+k\frac{r^2}{4})^2}, \\
\text{measure:} \quad & \begin{pmatrix} d^s\omega \\ d^s x \\ d^s \mathrm{y} \end{pmatrix} = \frac{\rho^{s-1} d\rho}{\sqrt{1-k\rho^2}}\, d^{s-1}\omega = \frac{r^{s-1} dr}{(1+k\frac{r^2}{4})^{s-1}}\, d^{s-1}\omega, \\
\text{Laplacian:} \quad & \begin{pmatrix} \frac{\partial^2}{\partial\omega_s^2} \\ \vec{\partial}_s^2 \\ \frac{\partial^2}{\partial \mathrm{y}_s^2} \end{pmatrix} = (1-k\rho^2)\frac{\partial^2}{\partial\rho^2} + \frac{1}{\rho^2}\frac{\partial^2}{\partial\omega_{s-1}^2} + \frac{d}{d\rho}\log[\rho^{s-1}e^{-\frac{ks}{2}\rho^2}]\frac{\partial}{\partial\rho} \\
& = (1+\frac{k}{4}r^2)^2(\frac{\partial^2}{\partial r^2} + \frac{1}{r^2}\frac{\partial^2}{\partial\omega_{s-1}^2}) \\
& \quad + \frac{d}{dr}\log[r^{s-1}e^{\frac{k}{4}r^2(1-k\frac{s-3}{16}r^2)}]\frac{\partial}{\partial r}.
\end{aligned}
$$

10.2 Kernels

Kernels are defined as linear mappings of vector spaces that contain (generalized) functions on measure spaces:

$$\kappa : \mathcal{F}_1 \longrightarrow \mathcal{F}_2, \quad f \longmapsto \kappa(f), \quad \kappa(f)(y) = \int dx\ \kappa(y,x)f(x).$$

They generalize the structure of mappings and tensor products of finite-dimensional vector spaces $\{\kappa : V_1 \longrightarrow V_2 \mid \text{linear}\} \cong V_2 \otimes V_1^T$ with $\kappa = \kappa_a^j e^a \otimes \check{e}_j$ for bases.

A kernel $\kappa : \mathcal{F}_1 \longrightarrow \mathcal{F}_2$ is *semiregular* if its action leads to functions $\mathcal{F}_2$. A *regular* kernel has to be semiregular for both κ and its transposed,

$$\kappa^T : \mathcal{F}_2' \longrightarrow \mathcal{F}_1', \quad \omega \longmapsto \kappa^T(\omega), \quad \kappa^T(\omega)(x) = \int dy\ \omega(y)\kappa(y,x),$$

e.g., a semiregular (anti)symmetric kernel $\kappa : \mathcal{F} \longrightarrow \mathcal{F}'$, $\kappa(x,y) = \pm\kappa(y,x)$ for $(x,y) \in \mathbb{R}^n \times \mathbb{R}^n$ or a semiregular diagonal kernel $\kappa(y-x)$.

A kernel is *regularized* if $(y,x) \longmapsto \kappa(y,x)$ is a function. Then it is regular. The inverse is not true, as seen at the Dirac distribution $\delta(y-x)$.

10.2.1 Kernels for Bilinear Forms

With dual finite-dimensional vector spaces, a linear mapping $\kappa : V \longrightarrow V^T$ gives a bilinear V-form $\kappa = \kappa^{ij}\check{e}_i \otimes \check{e}_j \in (V \otimes V)^T \cong V^T \otimes V^T$. As a generalization, the kernel theorems of L. Schwartz for nuclear vector spaces [56] establish the isomorphy of, on the one hand, linear continuous mappings

$\kappa : \mathcal{F} \longrightarrow \mathcal{F}'$ of function spaces on a real set into a dual space with distributions and, on the other hand, the same distribution type $\mathcal{F}'$ on the product set, which can be "factorized" in the form of a completized tensor product:

$$\begin{aligned}
&\{\kappa : \mathcal{F}(T^n) \longrightarrow \mathcal{F}(T^m)' \mid \text{linear, continuous}\} \\
&\qquad\qquad \cong \mathcal{F}(T^m \times T^n)' \cong \mathcal{F}(T^m)'\overline{\otimes}\mathcal{F}(T^n)', \\
&f_n \longmapsto \kappa(f_n), \quad \kappa(f_n)(y) = \textstyle\int_{T^n} d^n x \; \kappa(y,x) f_n(x) \text{ for } y \in T^m,
\end{aligned}$$

$$\text{e.g., for } \mathcal{F} = \mathcal{S},\ \mathcal{C}^\infty,\ \mathcal{C}_c^\infty \text{ with } \mathcal{F}' = \mathcal{S}',\ \mathcal{E}',\ \mathcal{D}' \text{ and open } T^{n,m} \subseteq \mathbb{R}^{n,m}.$$

The transposed kernel is given by

$$\begin{aligned}
&\kappa^T : \mathcal{F}(T^m) \longrightarrow \mathcal{F}(T^n)', \\
&f_m \longmapsto \kappa^T(f_m), \quad \kappa^T(f_m)(x) = \textstyle\int_{T^m} d^m y \; f_m(y)\kappa(y,x) \text{ for } x \in T^n.
\end{aligned}$$

In a basis notation, one can write $\kappa = \kappa(y,x)\ d^m y\ d^n x \ \in \mathcal{F}(T^m \times T^n)'$. Kernels with $\kappa(x,y) = \pm\kappa(y,x)$ for $T^n = T^m$ are *(anti)symmetric.*

Kernels $\mathcal{F} \stackrel{\kappa}{\longrightarrow} \mathcal{F}'$ for dual spaces define a bilinear form:

$$\begin{aligned}
f_1 \otimes f_2 \in \mathcal{F} \otimes \mathcal{F} : \langle \kappa, f_1 \otimes f_2 \rangle &= \langle f_1, \kappa(f_2) \rangle = \langle \kappa^T(f_1), f_2 \rangle \\
&= \textstyle\int dydx\ f_1(y)\kappa(y,x) f_2(x) \in \mathbb{C}.
\end{aligned}$$

10.2.2 Group Kernels

The group product–induced convolution product defines *group kernels* as linear mappings of group functions or distributions:

$$\begin{aligned}
\kappa \in \mathcal{F}(G) : \quad \mathcal{F}_1(G) \stackrel{\kappa}{\longrightarrow} \mathcal{F}_2(G),\ \mu \longmapsto \kappa * \mu,\ \mu * \kappa, \\
\text{with } \kappa * \mu(g) = \textstyle\int dg_1\ \kappa(g g_1^{-1})\mu(g_1).
\end{aligned}$$

The left–right action of the Radon distributions $\mathcal{M}(G)$ on group functions and on itself $\mathcal{F}(G) \in \{L^p(G), \mathcal{M}(G)\}$ involves the left and right multiplications with the group elements $G \ni k \longmapsto \delta_k \in \mathcal{M}(G)$:

$$\begin{aligned}
\mathcal{M}(G) * \mathcal{F}(G) * \mathcal{M}(G) \longrightarrow \mathcal{F}(G), \quad \text{e.g.,} \quad &\mu \longmapsto \delta_k * \mu * \delta_l, \\
&\delta_k * \mu * \delta_l(g) = \mu(k^{-1} g l^{-1}).
\end{aligned}$$

Group kernels, depending on the quotient of group elements, i.e., on the diagonal group, $G \times G \ni (g_1, g_2) \longmapsto \kappa(g_2 g_1^{-1})$ (see Chapter 2), are called *diagonal.* For translations, they depend on the difference $\mathbb{R}^n \times \mathbb{R}^n \ni (y,x) \longmapsto \kappa(y-x)$, e.g., the symmetric Dirac distribution $\delta(y-x)$ or the Kepler potential $\frac{\gamma_0}{|\vec{y}-\vec{x}|}$.

For a manifold with cosets $G/K \cong \mathbb{M} \subseteq \mathbb{R}^n$, one has as convolution products and as pointwise products for the Fourier transforms with the crossover

Fourier correspondence $(\dot{\mathcal{C}}_b(\mathbb{R}^n), \mathcal{M}(\mathbb{R}^n)) \overset{\mathbf{F}}{\leftrightarrow} (\mathcal{M}(\check{\mathbb{R}}^n), \dot{\mathcal{C}}_b(\check{\mathbb{R}}^n))$ of bounded functions and Radon distributions (see Chapter 8):

$*$	$\mathcal{M}(\mathbb{R}^n)$	$L^1(\mathbb{R}^n)$	$L^\infty(\mathbb{R}^n)$
$\mathcal{M}(\mathbb{R}^n)$	$\mathcal{M}(\mathbb{R}^n)$	$L^1(\mathbb{R}^n)$	$L^\infty(\mathbb{R}^n)$
$L^1(\mathbb{R}^n)$	$L^1(\mathbb{R}^n)$	$L^1(\mathbb{R}^n)$	$L^\infty(\mathbb{R}^n)$
$L^\infty(\mathbb{R}^n)$	$L^\infty(\mathbb{R}^n)$	$L^\infty(\mathbb{R}^n)$	–

$\overset{\mathbf{F}}{\leftrightarrow}$

$\cdot$	$\dot{\mathcal{C}}_b(\check{\mathbb{R}}^n)$	$\dot{\mathcal{C}}_0(\check{\mathbb{R}}^n)$	$\mathcal{M}(\check{\mathbb{R}}^n)$
$\dot{\mathcal{C}}_b(\check{\mathbb{R}}^n)$	$\dot{\mathcal{C}}_b(\check{\mathbb{R}}^n)$	$\dot{\mathcal{C}}_0(\check{\mathbb{R}}^n)$	$\mathcal{M}(\check{\mathbb{R}}^n)$
$\dot{\mathcal{C}}_0(\check{\mathbb{R}}^n)$	$\dot{\mathcal{C}}_0(\check{\mathbb{R}}^n)$	$\dot{\mathcal{C}}_0(\check{\mathbb{R}}^n)$	$\mathcal{M}(\check{\mathbb{R}}^n)$
$\mathcal{M}(\check{\mathbb{R}}^n)$	$\mathcal{M}(\check{\mathbb{R}}^n)$	$\mathcal{M}(\check{\mathbb{R}}^n)$	–

$$\widetilde{\mu_1 * \mu_2} = \tilde{\mu}_1 \cdot \tilde{\mu}_2, \quad \mathcal{M}(\mathbb{R}^n) \overset{\mathbf{F}}{\cong} \dot{\mathcal{C}}_b(\check{\mathbb{R}}^n) \overset{d^n q}{\cong} L^\infty(\check{\mathbb{R}}^n).$$

This defines group kernels (distributions and functions):

$$\begin{array}{lll} \kappa \in \mathcal{M}(\mathbb{M}): & \mathcal{M}(\mathbb{M}) \xrightarrow{\kappa} \mathcal{M}(\mathbb{M}), & L^p(\mathbb{M}) \xrightarrow{\kappa} L^p(\mathbb{M}), \quad 1 \le p \le \infty, \\ f \in L^1(\mathbb{M}): & \mathcal{M}(\mathbb{M}) \xrightarrow{f} L^1(\mathbb{M}), & L^p(\mathbb{M}) \xrightarrow{f} L^p(\mathbb{M}), \\ d \in L^\infty(\mathbb{M}): & \mathcal{M}(\mathbb{M}) \xrightarrow{d} L^\infty(\mathbb{M}), & L^1(\mathbb{M}) \xrightarrow{d} L^\infty(\mathbb{M}). \end{array}$$

Convolution algebras $A(\mathbb{M}) \subseteq \mathcal{M}(\mathbb{M}), L^1(\mathbb{M})$ with Radon distributions or absolute integrable functions give algebras with diagonal kernels $A(\mathbb{M} \times \mathbb{M})$.

10.3 Green's Kernels

The *inverse (dual) of derivations* $L^p(\mathbb{M}) \ni f \longmapsto \mathtt{D}f \in L^p(\mathbb{M})$ for functions on a manifold $\mathbb{M}$ with invariant measure $d^{\mathbb{M}}x$ and associated Dirac distribution $\delta_{\mathbb{M}}$ define *Green's kernels* $\kappa = \mathtt{D}^{-1}$, assumed as diagonal Radon distributions:

$$\mathtt{D} \longmapsto \mathtt{D}^{-1} = \kappa \in \mathcal{M}(\mathbb{M}), \quad \left\{ \begin{array}{l} x, y \in \mathbb{M}: \ \mathtt{D}_y \kappa(y,x) = \delta_{\mathbb{M}}(y,x) \\ \text{with } \int d^{\mathbb{M}}x \ \delta_{\mathbb{M}}(y,x) f(x) = f(y), \end{array} \right.$$

$$\mathcal{M}(\mathbb{M}) * L^p(\mathbb{M}) \subseteq L^p(\mathbb{M}), \quad \mathtt{D}f(x) = g(x) \Rightarrow \left\{ \begin{array}{ll} f(y) & = \kappa * g(y) \\ & = \int d^n x \ \kappa(y-x) g(x), \\ \tilde{f}(q) & = \tilde{\kappa} \cdot \tilde{g}(q). \end{array} \right.$$

Such kernels are familiar from differential equations of motion, as used for the composite massless Nambu–Goldstone fields (see Chapter 9). Green's kernels are determined up to a solution of the homogeneous equation $\mathtt{D}\kappa_0 = 0$, which may be related to a trivial eigenvalue or a trivial invariant, i.e., for a harmonic function, e.g., the harmonic polynomials of the groups $\mathbf{O}(t,s)$. Invariant differential operators $\mathtt{D}(\mathbb{M})$ lead to invariant kernels $\mathtt{D}^{-1}(\mathbb{M})$.

Kernels generalize the *dual structure* of finite-dimensional endomorphisms:

$$\begin{array}{rl} f & = f^b_a e^a \in \check{e}^b \in V \otimes V^T \cong \mathbf{AL}(V), \quad V \cong \mathbb{K}^n, \\ \mathtt{D}, \kappa \in \mathbf{AL}(V): & \kappa = \mathtt{D}^{-1} \iff \langle \mathtt{D}, \kappa \rangle = \operatorname{tr} \mathtt{D} \circ \kappa = \mathtt{D}^b_a \kappa^a_b = 1. \end{array}$$

The trace-inverse (dual) endomorphism is determined up to $\kappa_0 \in \mathbf{AL}(V)$ with $\operatorname{tr} \mathtt{D} \circ \kappa_0 = 0$ (more ahead).

The composition of two differential operators, e.g., for the product in the enveloping algebra of a Lie algebra, leads to the convolution product of the corresponding Green's kernels, which, for a translation parametrization, is the pointwise product of the Fourier transforms, if defined,

$$
\begin{aligned}
&(\mathtt{D}_1\mathtt{D}_2)^{-1} = \mathtt{D}_2^{-1} * \mathtt{D}_1^{-1} = \kappa_2 * \kappa_1 : \\
&\begin{array}{r}\mathtt{D}_j\kappa_j(x) = \delta(x),\\ \mathtt{D}_1\mathtt{D}_2\kappa_{12}(x) = \delta(x)\end{array} \Rightarrow \mathtt{D}_2\kappa_{12}(x) = \kappa_1(x) \Rightarrow \left\{\begin{array}{l}\kappa_{12}(x) = \kappa_2 * \kappa_1(x),\\ \tilde\kappa_{12}(q) = \tilde\kappa_2 \cdot \tilde\kappa_1(q).\end{array}\right.
\end{aligned}
$$

In correspondence to the composition algebra $\mathtt{D}(\mathbb{M})$ with the invariant differential operators, generated by fundamental ones, there are fundamental invariant kernels that generate by convolution or, for the Fourier transforms, by pointwise product the algebra $\mathtt{D}^{-1}(\mathbb{M})$ with the invariant kernels.

10.3.1 Linear Kernels for Spacetime

Linear kernels are defined as inverse derivatives of degree 1 $\frac{1}{\partial}$. They are Fourier-transformed simple (energy-)momentum poles at the trivial invariant, to be taken from the following expressions (see Chapter 8):

$$
\mathbb{R}: \quad \int \frac{dq}{2i\pi}\frac{\Gamma(1-\nu)}{(q-io)^{1-\nu}} e^{iqx} = \vartheta(x)\frac{1}{(ix)^\nu}, \quad \nu \in \mathbb{R},
$$

$$
\mathbf{SO}(s): \quad \int \frac{d^s q}{\pi^{\frac{s}{2}}}\begin{pmatrix}1\\ i\vec q\end{pmatrix}\frac{\Gamma(\frac{s}{2}-\nu)}{(\vec q^{\,2})^{\frac{s}{2}-\nu}} e^{-i\vec q\vec x} = \begin{pmatrix}\frac{\Gamma(\nu)}{(\frac{r}{2})^{2\nu}}\\ \frac{\vec x}{2}\frac{\Gamma(1+\nu)}{(\frac{r}{2})^{2(1+\nu)}}\end{pmatrix}, \quad s = 1, 2, \ldots,
$$

$$
\mathbf{SO}_0(1,s): \quad \int \frac{d^{1+s}q}{i\pi^{\frac{1+s}{2}}}\begin{pmatrix}1\\ iq\end{pmatrix}\frac{\Gamma(\frac{1+s}{2}-\nu)}{(-q^2-io)^{\frac{1+s}{2}-\nu}} e^{iqx} = \begin{pmatrix}\frac{\Gamma(\nu)}{\left(\frac{-x^2+io}{4}\right)^\nu}\\ \frac{x}{2}\frac{\Gamma(1+\nu)}{\left(\frac{-x^2+io}{4}\right)^{1+\nu}}\end{pmatrix}, \quad s=0, 1, \ldots.
$$

The inverse of the invariant derivative $\frac{d}{dx} \in \mathtt{D}(\mathbb{R})$ is half the sign function for *principal value* integration and the characteristic function for the future and past for *advanced and retarded* integration:

$$
\mathbb{R}: \quad \left\{\begin{array}{l}
\frac{d}{dx}\begin{pmatrix}\frac{\epsilon(x)}{2}\\ \pm\vartheta(\pm x)\end{pmatrix} = \delta(x),\\
\begin{pmatrix}(\frac{d}{dx})^{-1}\\ (\frac{d}{dx})_\pm^{-1}\end{pmatrix} = \begin{pmatrix}\frac{\epsilon(x)}{2}\\ \pm\vartheta(\pm x)\end{pmatrix} = \int \frac{dq}{2i\pi}\begin{pmatrix}\frac{1}{q_{\mathrm{P}}}\\ \frac{1}{q\mp io}\end{pmatrix} e^{iqx}, \quad \frac{1}{q_{\mathrm{P}}} = \frac{q}{q^2+o^2}.
\end{array}\right.
$$

The difference between the $\mathbb{R}$-kernels is constant: $2\vartheta(\pm x) \mp \epsilon(x) = 1$.

The inverse derivatives of Euclidean spaces are not invariant for $s \geq 1$:

$$
\mathbf{SO}(s): \quad \left\{\begin{array}{rl}
\frac{\partial}{\partial \vec x_s}\frac{1}{|\Omega^{s-1}|}\frac{\vec x}{r^s} &= \delta(\vec x) \text{ with } |\Omega^{s-1}| = \frac{2\pi^{\frac{s}{2}}}{\Gamma(\frac{s}{2})},\\
(\frac{\partial}{\partial \vec x_s})^{-1} = \frac{1}{|\Omega^{s-1}|}\frac{\vec x}{r^s} &= \int \frac{d^s q}{(2\pi)^s}\frac{i\vec q}{\vec q^{\,2}} e^{-i\vec q\vec x}.
\end{array}\right.
$$

An example is the Kepler force $\frac{1}{4\pi}\frac{\vec x}{r^3}$ in $\mathbb{R}^3$-position.

The Feynman measures of Minkowski spacetimes, not invariant for $s \geq 2$,

$$\mathbf{SO}_0(1,s): \quad \int \frac{d^{1+s}q}{(2\pi)^{1+s}} \frac{q}{-q^2-io} e^{iqx} = \frac{1}{|\Omega^s|} \frac{x}{(-x^2+io)^{\frac{1+s}{2}}},$$

give as linear kernels with principal value integration and odd spacetime dimensions $1+2R = 1, 3, \ldots$:

$$\begin{aligned}
\mathbf{SO}_0(1,2R): \quad & \int \frac{d^{1+2R}q}{(2\pi)^{1+2R}} \frac{q}{-q^2-io} e^{iqx} = \frac{1}{|\Omega^{2R}|} \frac{x}{|x|^{1+2R}} [\vartheta(-x^2) + i(-1)^R \vartheta(x^2)],\\
(\tfrac{\partial}{\partial x_{1+2R}})^{-1} = & \int \frac{d^{1+2R}q}{i(2\pi)^{1+2R}} \frac{q}{q_{\mathrm{P}}^2} e^{iqx} = \frac{(-1)^R}{|\Omega^{2R}|} \vartheta(x^2) \frac{x}{|x|^{1+2R}}.
\end{aligned}$$

They start with half the sign distribution $(\frac{\partial}{\partial x_1})^{-1} = \frac{x}{2|x|} = \frac{\epsilon(x)}{2}$. The kernels for even spacetime dimensions $2R = 2, 4, \ldots$ involve a Dirac distribution on the lightcone:

$$\begin{aligned}
\mathbf{SO}_0(1,2R-1): \quad & \int \frac{d^{2R}q}{(2\pi)^{2R}} \frac{q}{-q^2-io} e^{iqx} = \frac{x}{|\Omega^{2R-1}|} \left[-\frac{i\pi}{\Gamma(R)} \delta^{(R-1)}(x^2) + \frac{1}{(-x_{\mathrm{P}}^2)^R} \right],\\
(\tfrac{\partial}{\partial x_{2R}})^{-1} = & \int \frac{d^{2R}q}{i(2\pi)^{2R}} \frac{q}{q_{\mathrm{P}}^2} e^{iqx} = \frac{x}{2\pi^{R-1}} \delta^{(R-1)}(x^2).
\end{aligned}$$

They start with $(\frac{\partial}{\partial x_2})^{-1} = \frac{x}{2}\delta(x^2)$.

The linear kernels with support by the future and past are

$$\begin{aligned}
(\tfrac{\partial}{\partial x_{1+2R}})^{-1}_{\pm} &= \int \frac{d^{1+2R}q}{i(2\pi)^{1+2R}} \frac{q}{(q\mp io)^2} e^{iqx}\\
&= 2\vartheta(\pm x_0)(\tfrac{\partial}{\partial x_{1+2R}})^{-1} \quad = 2\vartheta(\pm x_0) \frac{(-1)^R}{|\Omega^{2R}|} \vartheta(x^2) \frac{x}{|x|^{1+2R}},\\
(\tfrac{\partial}{\partial x_{2R}})^{-1}_{\pm} &= \int \frac{d^{2R}q}{i(2\pi)^{2R}} \frac{q}{(q\mp io)^2} e^{iqx}\\
&= 2\vartheta(\pm x_0)(\tfrac{\partial}{\partial x_{2R}})^{-1} \quad = 2\vartheta(\pm x_0) \frac{x}{2\pi^{R-1}} \delta^{(R-1)}(x^2).
\end{aligned}$$

The difference between the kernels with Feynman and causal integration prescription is an on-shell contribution, i.e., a solution of the homogeneous equation $(\partial^2 + m^2)^{1+N} f_0 = 0$,

$$\int \frac{d^{1+s}q}{2i\pi} \left[\frac{\Gamma(1+N)}{(q^2-io-m^2)^{1+N}} + \frac{\Gamma(1+N)}{[(q-io)^2-m^2]^{1+N}} \right] e^{iqx} = \int d^{1+s}q \; \vartheta(q_0)\delta^{(N)}(m^2-q^2) e^{iqx}$$

10.3.2 Laplace–Beltrami Kernels

The inverse $\frac{1}{\partial_{\mathbf{g}}^2} \in \mathtt{D}^{-1}(\mathbb{M})$ of the Laplace–Beltrami operator of a Riemannian manifold $(\mathbb{M}, \mathbf{g})$ is a kernel, invariant under the global group $G_{\mathbf{g}}$: For dimension $s \geq 2$, the spheres, Euclidean spaces, and timelike hyperboloids with real rank 1 have a parametrization with geodesic polar coordinates: There the radial function $R \longmapsto \rho(R)$ of the invariant measure is used in the characterizing radial-spherical decomposition of the Laplace–Beltrami operator:

$$\begin{aligned}
\text{for } k &= \begin{pmatrix}1\\0\\-1\end{pmatrix} \sim \begin{pmatrix}\Omega^s\\ \mathbb{R}^s\\ \mathcal{Y}^s\end{pmatrix}:\\
\partial^2_{k,s} &= \frac{\partial^2}{\partial R^2} + \frac{d\log\rho^{s-1}}{dR}\frac{\partial}{\partial R} + \frac{1}{\rho^2}\frac{\partial^2}{\partial\omega^2_{s-1}} = \frac{1}{\rho^{s-1}}\frac{\partial}{\partial R}\rho^{s-1}\frac{\partial}{\partial R} + \frac{1}{\rho^2}\frac{\partial^2}{\partial\omega^2_{s-1}},\\
\sqrt{|\mathbf{g}|}d^s x &= \rho^{s-1}dRd^{s-1}\omega,\\
\text{with } R &= \begin{pmatrix}\theta\\ r\\ \psi\end{pmatrix},\quad \rho(R) = \begin{pmatrix}\sin\theta\\ r\\ \sinh\psi\end{pmatrix},\\
\text{e.g. } s = 2,3:\quad \partial^2_{k,2} &= \tfrac{1}{\rho^2}[(\rho\tfrac{\partial}{\partial R})^2 + \tfrac{\partial^2}{\partial\varphi^2}],\quad \partial^2_{k,3} = \tfrac{1}{\rho}(\tfrac{\partial^2}{\partial R^2} - k)\rho + \tfrac{1}{\rho^2}\tfrac{\partial^2}{\partial\omega^2_2}.
\end{aligned}$$

The *Laplace–Beltrami kernel* is given for the noncompact Euclidean spaces and hyperboloids with the Riemannian metric $d(x,y) = \int_x^y \sqrt{\mathbf{g}}$ as follows:

$$\begin{aligned}
k = 0,-1:\quad \partial^2_{k,s}\frac{1}{\partial^2_{k,s}}(x,y) &= \delta^s_k(x,y),\\
\frac{1}{\partial^2_{k,s}}(x,y) &= \frac{1}{|\Omega^{s-1}|}\int_1^{d(x,y)}\frac{dR}{\rho^{s-1}(R)}.
\end{aligned}$$

The explicit expressions for the inverse of the Laplacian $\vec{\partial}^2_s$ for Euclidean positions $\mathbb{R}^s$ with the radial function $\rho(R) = r$ involve the Kepler potential $\frac{1}{r}$ in $\mathbb{R}^3$-position:

$$\frac{|\Omega^{s-1}|}{\vec{\partial}^2_s}(\vec{x},\vec{y}) = \int_1^{|\vec{x}-\vec{y}|}\frac{dr}{r^{s-1}} \Rightarrow \begin{cases}
-\frac{|\Omega^{s-1}|}{\vec{\partial}^2_s}(\vec{x}) = \frac{1}{s-2}(\frac{1}{r^{s-2}} - 1)\\
\qquad = |\Omega^{s-1}|\int\frac{d^sq}{(2\pi)^s}\frac{1}{\vec{q}^2}e^{-i\vec{q}\vec{x}} - \frac{1}{2-s},\\
-\frac{2\pi}{\vec{\partial}^2_2}(\vec{x}) = \log\frac{1}{r},\\
-\frac{4\pi}{\vec{\partial}^2_3}(\vec{x}) = \frac{1}{r} - 1.
\end{cases}$$

The homogeneous term gives the two-dimensional Laplace kernel in the limit $s \to 2$. For the hyperboloids $\mathcal{Y}^s$ one uses as integrals over the radial function $\rho(R) = \sinh\psi$:

$$\text{for } \frac{|\Omega^{s-1}|}{\partial^2_{-1,s}}:\quad \begin{cases}
s = 3,4,\cdots: & -\int\frac{d\psi}{\sinh^{s-1}\psi} = \frac{1}{s-2}\left[\frac{\cosh\psi}{\sinh^{s-2}\psi}\right.\\
& \qquad\left.+(s-3)\int\frac{d\psi}{\sinh^{s-3}\psi}\right],\\
s = 2: & -\int\frac{d\psi}{\sinh\psi} = \log\frac{\cosh\psi}{\sinh\psi},\\
s = 3: & -\int\frac{d\psi}{\sinh^2\psi} = \frac{\cosh\psi}{\sinh\psi}.
\end{cases}$$

The Laplace kernels for flat spaces with general signature can be taken from

$$\mathbf{SO}_0(t,s)\,\vec{\times}\,\mathbb{R}^n:\quad \begin{cases}
\int\frac{d^nq}{i^t\pi^{\frac{n}{2}}}\frac{\Gamma(\frac{n}{2}-\nu)}{(-q^2-io)^{\frac{n}{2}-\nu}}e^{iqx} = \frac{\Gamma(\nu)}{(\frac{-x^2+io}{4})^{\nu}},\\
\text{e.g. } \frac{1}{\partial_t^2-\vec{\partial}_s^2}(x) = \int\frac{d^nq}{(2\pi)^n}\frac{1}{-q^2-io}e^{iqx}\\
\qquad = \frac{i^t}{4\pi^{\frac{n}{2}}}\frac{\Gamma(\frac{n}{2}-1)}{(-x^2+io)^{\frac{n}{2}-1}}.
\end{cases}$$

For noncompact groups, the distributions with $\frac{1}{q^2}$ have to be defined by a complex integration prescription, e.g., by Feynman $q^2 \pm io$.

10.4 Tangent Kernels and Interactions

For a manifold with cosets $G/K \cong \mathbb{M} \subseteq \mathbb{R}^n$, a linear transformation of the bounded functions $L^\infty(\mathbb{R}^n) \overset{d^n x}{\cong} \dot{\mathcal{C}}_b(\mathbb{R}^n)$ by the pointwise product with a Green's kernel $\mathtt{D}^{-1} = \kappa^0$,

$$\mathcal{M}(\mathbb{M}) \ni \kappa^0 : L^\infty(\mathbb{M}) \longrightarrow \mathcal{M}(\mathbb{M}),\ d \longmapsto \kappa^0 \cdot d = \kappa^d,$$

gives *tangent kernels* $\kappa \in \mathcal{M}(\mathbb{M} \times \mathbb{M})$ that act on group functions,

$$\mathcal{M}(\mathbb{M}) \ni \kappa^d : \quad L^p(\mathbb{M}) \longrightarrow L^p(\mathbb{M}), \quad f \longmapsto \kappa^d * f.$$

For physics, the Radon distributions $\kappa^0 \cdot L^\infty(\mathbb{M})$ are called $\mathbb{M}$-*interactions.*

Of special interest is the orbit $L^\infty(G)_+ \ni d \overset{\kappa^0}{\longmapsto} \kappa^d \in \mathcal{M}(\mathbb{M})$ of the cone with the positive-type functions for the cyclic group representations: In analogy to the transition from a Lie group representation to the associated Lie algebra representation, a Green's kernel associates a tangent kernel (an interaction) to each cyclic group representation. A Green's kernel $\kappa^0 = \kappa^0 \cdot 1$ as the image of the unit function $d = 1$ is associated with the trivial group representation.

The Fourier transforms $\tilde{\kappa}^0 * \tilde{d}$ of tangent kernels $\kappa^0 \cdot d$ are functions of the translation eigenvalues ((energy-)momenta),

$$L^\infty(\mathbb{R}^n) \ni d \quad \longmapsto \quad \kappa^d = \kappa^0 \cdot d \in \mathcal{M}(\mathbb{R}^n)$$

$\cdot$	$\mathcal{M}(\mathbb{R}^n)$	$L^\infty(\mathbb{R}^n)$
$\mathcal{M}(\mathbb{R}^n)$	$-$	$\mathcal{M}(\mathbb{R}^n)$
$L^\infty(\mathbb{R}^n)$	$\mathcal{M}(\mathbb{R}^n)$	$L^\infty(\mathbb{R}^n)$

$$\mathbf{F} \updownarrow \qquad\qquad \updownarrow \mathbf{F}$$

$$\mathcal{M}(\check{\mathbb{R}}^n) \ni \tilde{d} \quad \longmapsto \quad \tilde{\kappa}^d = \tilde{\kappa}^0 * \tilde{d} \in L^\infty(\check{\mathbb{R}}^n)$$

$*$	$L^\infty(\check{\mathbb{R}}^n)$	$\mathcal{M}(\check{\mathbb{R}}^n)$
$L^\infty(\check{\mathbb{R}}^n)$	$-$	$L^\infty(\check{\mathbb{R}}^n)$
$\mathcal{M}(\check{\mathbb{R}}^n)$	$L^\infty(\check{\mathbb{R}}^n)$	$\mathcal{M}(\check{\mathbb{R}}^n)$

10.4.1 Position Interactions

The abelian group $\mathbf{D}(1) \cong \mathbb{R}$, used for causal *time*, is isomorphic to its Lie algebra. The Green's kernel, inverting the invariant derivative $\frac{d}{dx_0}$, acts as identity; the abelian causal group, more generally. the abelian translations $\mathbb{R}^n$, induces only trivial interactions,

$$\begin{aligned} \frac{d}{dx_0}\vartheta(x_0) &= \delta(x_0), \\ \vartheta(x_0)e^{imx_0} \overset{\vartheta(x_0)}{\longmapsto} \vartheta(x_0)e^{imx_0} &= \int \frac{dq}{2i\pi}\, \frac{1}{q-io-m} e^{iqx_0} = \left(\frac{d}{dx_0} - im\right)^{-1}, \\ \tilde{\kappa}^0(q) = \frac{1}{q}, \quad \tilde{\kappa}^m(q) &= \frac{1}{q-m}. \end{aligned}$$

$$\boxed{\begin{array}{c} \frac{1}{q} \overset{1}{*} \mathcal{D}^1 = \log \mathcal{D}^1 \\ (\overset{1}{*}, q) = (\frac{*}{2i\pi}, q - io) \\ \hline \frac{1}{q} \quad \overset{1}{*} \quad \frac{1}{q-m} = \frac{1}{q-m} \end{array}}$$

Time kernels

Nontrivial interactions are given by tangent kernels of *nonabelian (curved) positions*: For odd-dimensional hyperboloids $\mathcal{Y}^{2R-1}$ with nonabelian degrees of freedom $R \geq 2$, the invariant kernels invert the powers of the Laplacian:

$$N = 1, \ldots, R-1: \quad \frac{(4\pi)^R \Gamma(R-N)}{2\pi(-\vec{\partial}^2)^{R-N}} = \int \frac{d^{2R-1}q}{\pi^R} \frac{\Gamma(R-N)}{(\vec{q}^2)^{R-N}} e^{-i\vec{q}\vec{x}} = \frac{\Gamma(N-\frac{1}{2})}{\sqrt{\pi}} \left(\frac{2}{r}\right)^{2N-1}.$$

The action on the positive-type functions $\vec{x} \longmapsto e^{-|m|r}$ gives Yukawa-like potentials for $N = 1, \ldots, R-1$ as inverse Laplacian with invariant m^2:

$$\begin{aligned} \int \frac{d^{2R-1}q}{\pi^R} \frac{\Gamma(R-N)}{(\vec{q}^2+m^2)^{R-N}} e^{-i\vec{q}\vec{x}} &= \frac{(4\pi)^R \Gamma(R-N)}{2\pi(-\vec{\partial}^2+m^2)^{R-N}} = \left(-\frac{d}{d\frac{r^2}{4}}\right)^N \frac{e^{-|m|r}}{|m|} \\ &= \left(\frac{1}{|m|}, \frac{2}{r}, \frac{4(1+|m|r)}{r^3}, \ldots\right) e^{-|m|r}, \\ \text{for } N &= 0, 1, 2, \ldots, R-1. \end{aligned}$$

$N = 0$ characterizes positive-type functions, i.e., $\mathbf{SO}_0(1, 2R-1)$-representation coefficients. The kernel for $N = 1$ with integration $\frac{d^{2R-1}q}{(\vec{q}^2)^{R-1}} = d|\vec{q}|\, d^{2(R-1)}\omega$ is the Kepler potential $\frac{1}{(-\vec{\partial}^2)^{R-1}} \sim \frac{1}{r}$. The Yukawa potentials as position kernels constitute the position interactions:

$$\tilde{\kappa}^0(\vec{q}^2) = \frac{1}{(\vec{q}^2)^{R-N}}, \quad \tilde{\kappa}^{-m^2}(\vec{q}^2) = \frac{1}{(\vec{q}^2+m^2)^{R-N}}, \; N = 1, \ldots, R-1.$$

$$\boxed{\begin{array}{c} \frac{1}{(\vec{q}^2)^{R-N}} \overset{2R-1}{*} \mathcal{Y}^{2R-1} = \log \mathcal{Y}^{2R-1} \\ \overset{2R-1}{*} = \frac{*\,2}{|\Omega^{2R-1}|}, \quad 2R-1 = 3, 5, \ldots \\ \hline \frac{1}{(\vec{q}^2)^{R-N}} \quad \overset{2R-1}{*} \quad \frac{|m|}{(\vec{q}^2+m^2)^R} = \frac{1}{(\vec{q}^2+m^2)^{R-N}} \end{array}}$$

Position kernels (interactions)

10.5 Duality and Normalization

Bi- or sesquilinear forms and dual or scalar products of (generalized) functions, e.g., essentially bounded functions $L^\infty(\mathbb{R}^n)$, put in duality with Radon distributions $\mathcal{M}(\mathbb{R}^n)$,

$$\mathcal{M}(\mathbb{R}^n) * L^\infty(\mathbb{R}^n) \subseteq L^\infty(\mathbb{R}^n), \quad \langle \kappa, d\rangle = \kappa * d(0) = \tilde{\kappa}^- \tfrac{*}{(2\pi)^n} \tilde{d}(0),$$

are given by the Schur product, i.e., by convolution products at the neutral element (see Chapter 8). Two elements that are not orthogonal, $\langle \kappa, d\rangle \neq 0$, can be *normalized as a dual pair* $\langle \kappa, d\rangle = 1$.

10.5.1 Schur Products of Feynman Measures

Schur products of $\mathbf{O}(t,s)$-scalar Feynman measures, $t+s=n$, with real invariants $m^2 \in \mathbb{R}$, e.g., for one factor with a trivial invariant, are (see Chapter 9):

$$\begin{aligned}
&\int \frac{d^nq}{i^t\pi^{\frac{n}{2}}} \frac{\Gamma(\frac{n}{2}+\nu_1)}{(-q^2-io+m_1^2)^{\frac{n}{2}+\nu_1}} \frac{\Gamma(\frac{n}{2}+\nu_2)}{(-q^2-io+m_2^2)^{\frac{n}{2}+\nu_2}}\\
&= \int_0^1 d\zeta \, \frac{\zeta^{\frac{n}{2}+\nu_1-1}(1-\zeta)^{\frac{n}{2}+\nu_2-1}\Gamma(\frac{n}{2}+\nu_1+\nu_2)}{[\zeta m_1^2+(1-\zeta)m_2^2-io]^{\frac{n}{2}+\nu_1+\nu_2}}\\
&= \int_{m_2^2}^{m_1^2} \frac{d\kappa^2}{m_1^2-m_2^2} \frac{(\kappa^2-m_2^2)^{\frac{n}{2}+\nu_1-1}(m_1^2-\kappa^2)^{\frac{n}{2}+\nu_2-1}\Gamma(\frac{n}{2}+\nu_1+\nu_2)}{(m_1^2-m_2^2)^{n+\nu_1+\nu_2-2}(\kappa^2-io)^{\frac{n}{2}+\nu_1+\nu_2}},\\
&\int \frac{d^nq}{i^t\pi^{\frac{n}{2}}} \frac{\Gamma(\frac{n}{2}+\nu_1)}{(-q^2-io+m^2)^{\frac{n}{2}+\nu_1}} \frac{1}{(-q^2-io)^{\nu_2}} = \frac{\Gamma(\frac{n}{2}-\nu_2)}{\Gamma(\frac{n}{2})} \frac{\Gamma(\nu_1+\nu_2)}{(m^2-io)^{\nu_1+\nu_2}}.
\end{aligned}$$

They can be written with a measure over an invariant line:

$$\int_0^1 d\zeta = \int_{m_2^2}^{m_1^2} \frac{d\kappa^2}{m_1^2-m_2^2} \text{ with } \kappa^2 = \zeta m_1^2 + (1-\zeta)m_2^2 \in [m_2^2, m_1^2].$$

Schur products with orthogonal harmonic $\mathbf{O}(t,s)$-polynomials $(q)^L$ (see Chapter 8) are proportional to the harmonic units (projectors) $(\mathbf{1}_n)^L$. They involve characteristic *multiplicity factors*,

$$\int d^nq \; \mu(q^2)(q)^L \otimes (q)^{L'} = \delta^{LL'} \frac{\Gamma(1+L)}{2^L} \frac{\Gamma(\frac{n}{2})}{\Gamma(\frac{n}{2}+L)} \int d^nq \; \mu(q^2)(q^2)^L \; (\mathbf{1}_n)^L,$$

with the simplest examples for $L = 1, 2$:

$$\begin{aligned}
&\int d^nq \; \mu(q^2) q^a \; q_b = \frac{1}{n} \quad \int d^nq \; \mu(q^2) q^2 \quad \delta^a_b,\\
&\int d^nq \; \mu(q^2)(q^aq^c - \frac{\eta^{ac}}{n}q^2)(q_bq_d - \frac{\eta_{bd}}{n}q^2)\\
&= \frac{2}{n(n+2)} \int d^nq \; \mu(q^2)(q^2)^2 (\frac{\delta^a_b\delta^c_d+\delta^a_d\delta^c_b}{2} - \frac{\eta^{ac}\eta_{bd}}{n}).
\end{aligned}$$

The distributions μ are characterized by invariants of the $\mathbf{O}(t,s)$-representations acting on $(q)^L$. There remain scalar integrals, for example:

$$\begin{aligned}
&\int \frac{d^nq}{i^t\pi^{\frac{n}{2}}} \frac{\Gamma(\frac{n}{2}+\nu_1)\,(-q^2-io)^\lambda\; \Gamma(\frac{n}{2}+\nu_2)}{(-q^2-io+m_1^2)^{\frac{n}{2}+\nu_1}(-q^2-io+m_2^2)^{\frac{n}{2}+\nu_2}}\\
&= \frac{\Gamma(\frac{n}{2}+\lambda)}{\Gamma(\frac{n}{2})} \int_0^1 d\zeta \; \frac{\zeta^{\frac{n}{2}+\nu_1-1}(1-\zeta)^{\frac{n}{2}+\nu_2-1}\Gamma(\frac{n}{2}+\nu_1+\nu_2-\lambda)}{[\zeta m_1^2+(1-\zeta)m_2^2-io]^{\frac{n}{2}+\nu_1+\nu_2-\lambda}},\\
&\int \frac{d^nq}{i^t\pi^{\frac{n}{2}}} \frac{\Gamma(\frac{n}{2}+\nu+\lambda)(-q^2-io)^\lambda}{(-q^2-io+m^2)^{\frac{n}{2}+\nu+\lambda}} = \frac{\Gamma(\frac{n}{2}+\lambda)}{\Gamma(\frac{n}{2})} \frac{\Gamma(\nu)}{(m^2-io)^\nu}.
\end{aligned}$$

10.5.2 Representation Normalizations for Spheres and Hyperboloids

The definite orthogonal groups $\mathbf{O}(0,s)$ use $-q^2 = \vec{q}^{\,2}$ for the Schur products:

$$\begin{aligned}
\mathbf{O}(s): &\int \frac{d^s q}{\pi^{\frac{s}{2}}} \frac{\Gamma(\frac{s}{2}+\nu+L)\ (\vec{q})^L \otimes (\vec{q})^{L'}}{(\vec{q}^{\,2}-io+m^2)^{\frac{s}{2}+\nu+L}} &&= \delta^{LL'} \frac{\Gamma(1+L)}{2^L} \frac{\Gamma(\nu)}{(m^2-io)^\nu} (\mathbf{1}_s)^L,\\
\mathbf{O}(2R-1): &\int \frac{2d^{2R-1}q}{|\Omega^{2R-1}|} \frac{(\vec{q})^L \otimes (\vec{q})^{L'}}{(\vec{q}^{\,2}-io+m^2)^{R+L}} &&= \delta^{LL'} \frac{\Gamma(1+L)}{2^L} \frac{\Gamma(R)}{\Gamma(R+L)} \frac{1}{(m^2-io)^{\frac{1}{2}}} (\mathbf{1}_{2R-1})^L,\\
\mathbf{O}(2R): &\int \frac{2d^{2R}q}{|\Omega^{2R-1}|} \frac{(\vec{q})^L \otimes (\vec{q})^{L'}}{(\vec{q}^{\,2}-io+m^2)^{1+R+L}} &&= \delta^{LL'} \frac{\Gamma(1+L)}{2^L} \frac{\Gamma(R)}{\Gamma(R+L)} \frac{1}{m^2-io} (\mathbf{1}_{2R})^L.
\end{aligned}$$

The products for the spheres $\vec{\omega}_{s-1} \in \Omega^{s-1}$ and the spherical harmonics $(\vec{\omega}_{s-1})^L$ are

$$\begin{aligned}
\int \frac{d^s q}{\pi^{\frac{s}{2}-1}} \delta^{(N+L)}(m^2-\vec{q}^{\,2})(\vec{q})^L \otimes (\vec{q})^{L'} &= \delta^{LL'} \frac{\Gamma(1+L)}{2^L} \frac{\Gamma(1-\frac{s}{2}+N)}{i(-m^2)^{1-\frac{s}{2}+N}} (\mathbf{1}_s)^L\\
&= \delta^{LL'} \frac{\Gamma(1+L)}{2^L} \frac{\pi}{\Gamma(\frac{s}{2}-N)(m^2)^{1-\frac{s}{2}+N}} (\mathbf{1}_s)^L,\\
\int \frac{2d^s q}{|\Omega^{s-1}|} \delta^{(N+L)}(m^2-\vec{q}^{\,2})(\vec{q})^L \otimes (\vec{q})^{L'} &= \delta^{LL'} \frac{\Gamma(1+L)}{2^L} \frac{\Gamma(\frac{s}{2})}{\Gamma(\frac{s}{2}-N)} (m^2)^{\frac{s}{2}-1-N} (\mathbf{1}_s)^L,\\
N+L=0: \int \frac{d\omega^{s-1}}{|\Omega^{s-1}|} (\vec{\omega}_{s-1})^L \otimes (\vec{\omega}_{s-1})^{L'} &= \delta^{LL'} \frac{\Gamma(1+L)}{2^L} \frac{\Gamma(\frac{s}{2})}{\Gamma(\frac{s}{2}+L)} (\mathbf{1}_s)^L.
\end{aligned}$$

The energy invariants for the bound waves of the nonrelativistic hydrogen atom are given by momentum multipole singularities $2E = \vec{q}^{\,2} = -Q^2$ in the Fourier-transformed wave functions with the harmonic $\mathbf{SO}(3)$-polynomials $(\vec{q})^L$:

$$\begin{aligned}
&\mathcal{Y}^3 \cong \mathbf{SO}_0(1,3)/\mathbf{SO}(3) \ni \vec{x} \longmapsto |-Q_L^2, L\}_3(\vec{x}) \sim \int \frac{d^3 q}{\pi^2} \frac{(\vec{q})^L}{(\vec{q}^{\,2}+Q_L^2)^{2+L}} e^{i\vec{q}\vec{x}},\\
&\frac{1}{Q_L^2} = (1+L)^2, \quad L = 2J = 0, 1, \ldots .
\end{aligned}$$

In a *noncompact–compact reciprocity*, the continuous curvature invariants Q^2 of the $\mathbf{SO}_0(1,3)$-representations for rank-1 position $\mathcal{Y}^3 \supset \mathbf{SO}_0(1,1)$ are determined by the dimensions $(1+L)^2$ of $\mathbf{SO}(4)$-representation spaces. As will be shown in the following, this relation between continuous and integer invariants reflects a *representation normalization* of the hyperboloid functions.

The harmonic representation coefficients of odd-dimensional hyperboloids $\mathcal{Y}^{2R-1} \cong \mathbf{SO}_0(1,2R-1)/\mathbf{SO}(2R-1)$, $R = 1, 2, \ldots$,

$$\mathcal{Y}^{2R-1} \ni \vec{x} \longmapsto |-Q_L^2, L\}_{2R-1}(\vec{x}) = \int \frac{2d^{2R-1}q}{|\Omega^{2R-1}|} \frac{\rho_{2R-1,L}(2\vec{q})^L}{(\vec{q}^{\,2}+Q_L^2)^{R+L}} e^{i\vec{q}\vec{x}},$$

with $\mathbf{SO}_0(1,1)$-representation invariants Q_L^2 and normalization factors $\rho_{2R-1,L}$, have as dual products with harmonic $\mathbf{O}(2R-1)$-polynomials $(\vec{q})^L$ and harmonic projectors $(\mathbf{1}_{2R-1})^L$:

$$\begin{aligned}
\{L'|-Q_L^2, L\}_{2R-1} &= \rho_{2R-1,L} \int \frac{2d^{2R-1}q}{|\Omega^{2R-1}|} \frac{(2\vec{q})^L}{(\vec{q}^{\,2}+Q_L^2)^{R+L}} \otimes (\vec{q})^{L'}\\
&= \delta^{LL'} \rho_{2R-1,L} \frac{\Gamma(1+L)\Gamma(R)}{\Gamma(R+L)} \frac{1}{|Q_L|} (\mathbf{1}_{2R-1})^L.
\end{aligned}$$

The normalization conditions for the dual products, starting with the scalar case $|-Q_0^2,0\}_{2R-1}(0)=1$, give equations for the invariants:

$$\{L|-Q_L^2,L\}_{2R-1}=(\mathbf{1}_{2R-1})^L\Rightarrow \rho_{2R-1,L}\frac{\Gamma(1+L)\Gamma(R)}{\Gamma(R+L)}\frac{1}{|Q_L|}=1.$$

The representation normalizations differ from the wave function normalizations by the Schur products $\{-Q_{L'}^2,L'|-Q_L^2,L\}_{2R-1}$ (see Chapter 8).

For the abelian case, $R=1$, without rotation degrees of freedom, all invariants are required to coincide:

$$\mathcal{Y}^1=\mathbf{SO}_0(1,1)\cong\mathbb{R}:\quad \rho_{1,L}=|Q_L|=|Q|\text{ for all }L.$$

If *self-dual residual normalizations*, where the powers 2^L match the denominator powers $(\vec q^{\,2}+Q_L^2)^L$, are used for all hyperboloids, the representation invariants Q_L^2 for $\mathbf{SO}_0(1,2R-1)$ are determined by rotation $\mathbf{SO}(2R-1)$-invariants via dimensions of symmetric vector space products:

$$\begin{aligned}\frac{|Q|}{|Q_L|}&=\frac{\Gamma(R+L)}{\Gamma(1+L)\Gamma(R)}=\left\{\begin{array}{rl}1,&R=1,\\1+L,&R=2,\\\frac{(2+L)(1+L)}{2},&R=3,\\\dots,&\end{array}\right.\\ \binom{R+L-1}{L}&=\frac{\Gamma(R+L)}{\Gamma(1+L)\Gamma(R)}=\dim_{\mathbb{K}}\bigvee^L\mathbb{K}^R,\\ L=0:\quad |Q_0|&=|Q|\text{ for all }R.\end{aligned}$$

The representation-normalized coefficients are

$$\begin{aligned}\vec x\longmapsto|-Q_L^2,L\}_{2R-1}(\vec x\,)&=\int\frac{2d^{2R-1}q}{|\Omega^{2R-1}|}\frac{(2\vec q\,)^L|Q_0|}{(\vec q^{\,2}+Q_L^2)^{R+L}}e^{i\vec q\vec x},\\ \frac{(2\vec q\,)^L|Q_0|}{(\vec q^{\,2}+Q_L^2)^{R+L}}\overset{2R-1}{*}(\vec q\,)^L\Big|_{\vec q=0}&=(\mathbf{1}_{2R-1})^L,\ L=0,1,2,\dots.\end{aligned}$$

The invariants for the nonrelativistic hydrogen atom (Q_L^2,L) show an equipartition $Q_0^2=(1+L)^2Q_L^2$, similar to a flux quantization, of the basic invariant for the scalar $L=0$ case to the dimensions of the symmetric products of Pauli spinor spaces $\mathbb{C}^2$:

$$R=2:\quad \frac{|Q_0|}{|Q_L|}=1+L=\dim_{\mathbb{C}}\bigvee^L\mathbb{C}^2.$$

The basic invariant Q_0^2 is not determined. It can be normalized to 1 or used as the energy (-momentum) unit, e.g., in the hydrogen atom given by the Rydberg energy $\frac{1}{2}M_Rc^2\sim 14$ eV with $M_R=m_e(\alpha_S)^2$ (see Chapter 4).

10.6 Kernel Resolvents and Eigenvalues

The Fourier transform of the convolutive action of an interaction, given by a tangent kernel κ from the convolution algebra with the Radon group measures, on a vector space with group representations coefficients, given by

Lebesgue functions f on the group, parametrized by translations $x \in \mathbb{R}^n \supseteq \mathbb{M} \cong G/K$,

$$\mathcal{M}(\mathbb{R}^n) \ni \kappa : \quad L^p(\mathbb{R}^n) \longrightarrow L^p(\mathbb{R}^n), \quad f \longmapsto \kappa * f,$$

is a pointwise action of an (energy-)momentum function on the corresponding Fourier-transformed functions, if defined, e.g., between the convolution and pointwise product algebra as Fourier partners $L^1(\mathbb{R}^n) \ni f \stackrel{\mathbf{F}}{\longmapsto} \tilde{f} \in \dot{\mathcal{C}}_0(\breve{\mathbb{R}}^n)$. With the eigenvalue equations,

$$\kappa * f = f, \quad \tilde{\kappa} \cdot \tilde{f} = \tilde{f},$$

the invariant eigenvalues, linear or quadratic, are given by the singularities of the *kernel resolvent*:

$$\text{linear:} \quad \begin{cases} \text{kernel resolvent:} & q \longmapsto \text{Res}\, \tilde{\kappa}(q) = \frac{1}{1-\tilde{\kappa}(q)}, \\ \text{invariants:} & \textbf{spec}\, \tilde{\kappa} = \{I \mid \tilde{\kappa}(I) = 1\}, \end{cases}$$

$$\text{quadratic:} \quad \begin{cases} \text{kernel resolvent:} & q^2 \longmapsto \text{Res}\, \tilde{\kappa}(q^2) = \frac{1}{1-\tilde{\kappa}(q^2)}, \\ \text{invariants:} & \textbf{spec}\, \tilde{\kappa} = \{I^2 \mid \tilde{\kappa}(I^2) = 1\}. \end{cases}$$

The transition from kernel to resolvent, connected with a geometric series, is parallel to the exponential transition from Lie algebra to group representation,

$$\begin{array}{ccc} \tilde{l} & \longmapsto & l \\ \downarrow & & \downarrow \\ \frac{1}{1-\tilde{l}} = \sum_{k\geq 0} \tilde{l}^k & \longmapsto & e^l = \sum_{k\geq 0} \frac{l^k}{k!}, \end{array}$$

as illustrated for the abelian group $\mathbf{D}(1)$,

$$\begin{array}{ccccc} \frac{m}{q} & \longmapsto & imt & = \oint \frac{dq}{2i\pi} \frac{m}{q^2} e^{iqt} & \\ \downarrow & & & \downarrow & \text{with } \oint \frac{dq}{2i\pi} \frac{1}{q} \left(\frac{m}{q}, \frac{1}{1-\frac{m}{q}} \right) e^{iqt}. \\ \frac{1}{1-\frac{m}{q}} & \longmapsto & e^{imt} & = \oint \frac{dq}{2i\pi} \frac{1}{q-m} e^{iqt} & \end{array}$$

A decomposition with respect to translation representations is given by the principal parts of the resolvent at the invariants with the *residues as normalizations*,

$$\begin{array}{lllll} \text{linear:} & \text{Res}_I \tilde{\kappa}(q) & = \frac{\rho(I)}{q-I} & \text{with } \frac{1}{\rho(I)} & = -\frac{\partial \tilde{\kappa}}{\partial q}(I), \\ \text{quadratic:} & \text{Res}_{I^2} \tilde{\kappa}(q^2) & = \frac{\rho(I^2)}{q^2-I^2} & \text{with } \frac{1}{\rho(I^2)} & = -\frac{\partial \tilde{\kappa}}{\partial q^2}(I^2). \end{array}$$

10.6.1 Spacetime Normalization in the Nambu–Jona-Lasinio Model

The composite massless chiral Goldstone modes in the model of Nambu–Jona-Lasinio, as described in Chapter 9, arise in a product involving a Feynman propagator of interacting fields $d^{M^2,m}(x-y) = \langle \mathcal{C}|\boldsymbol{\Psi}(y)\overline{\boldsymbol{\Psi}}(x)|\mathcal{C}\rangle_{\text{Feynman}}$ with a cyclic ground-state vector $|\mathcal{C}\rangle$ from the degeneracy manifold $\mathbf{U}(1)$. It is an $L^\infty(\mathbb{R}^4)$-function of the spacetime translations,

$$\begin{aligned}\mathbb{R}^4 \ni x \longmapsto d^{M^2,m}(x) = \;& \rho(m^2)\tfrac{i}{\pi}\int \tfrac{d^4q}{(2\pi)^3}\tfrac{(m^2-M^2)^2}{(q^2+io-M^2)^2}\tfrac{\gamma q+m}{q^2+io-m^2}e^{iqx}\\ & \text{(dipole regularization)},\end{aligned}$$

and comes as convolution $d^{M^2,m} = -d^{M^2} * \kappa^m$ of a dipole with the Green's kernel κ^m of the equation of motion with the basic Dirac field for a free spin-$\frac{1}{2}$ massive particle. The Green's kernel, i.e., the free-field Feynman propagator, is a spacetime distribution $\mathcal{M}(\mathbb{R}^4)$:

$$(i\gamma\partial + m)\kappa^m(x) = \delta(x) \Rightarrow \kappa^m(x) = -\int \tfrac{d^4q}{(2\pi)^4}\tfrac{\gamma q+m}{q^2+io-m^2}e^{iqx}.$$

The scalar part of the propagator of the interacting Dirac field is Schur-normalized as the spacetime representation coefficient by the mass consistency equation for $m \neq 0$:

$$\begin{aligned}1 = \tfrac{1}{4m}\,\mathrm{tr}\, d^{M^2,m}(0) &= \tfrac{i}{\pi}\int \tfrac{d^4q}{(2\pi)^3}\tfrac{(m^2-M^2)^2}{(q^2+io-M^2)^2}\tfrac{\rho(m^2)}{q^2+io-m^2}\\ &= \tfrac{m^2\rho(m^2)}{8\pi^2}\left(\tfrac{M^2}{m^2} - 1 - \log\tfrac{M^2}{m^2}\right).\end{aligned}$$

The lowest-order equations for the translation invariants of the spacetime product representations (bound-states) are given for the functions of the spacetime translation $\mathbb{R}^4 \ni x \longmapsto \Gamma(x) = \langle \mathcal{C}|\overline{\boldsymbol{\Psi}}\Gamma\boldsymbol{\Psi}(x)|B\rangle$ (see Chapter 9):

$$\text{for } \Gamma_{1,2} \in \{\mathbf{1}_4, \gamma_a, \gamma_5\gamma_a, \gamma_5\}: \quad \begin{cases}\Gamma_1(x) &= \kappa_{\Gamma_1}^{\Gamma_2} * \Gamma_2(x),\\ \tilde{\Gamma}_1(q) &= \tilde{\kappa}_{\Gamma_1}^{\Gamma_2}\cdot\tilde{\Gamma}_2(q).\end{cases}$$

The eigenvalue equations involve a convolution with the spacetime kernel matrix $L^p(\mathbb{R}^4) \xrightarrow{\kappa *} L^p(\mathbb{R}^4)$ or a multiplication with its energy-momentum function matrix $\tilde{\kappa}$:

$$\begin{aligned}\kappa_{\Gamma_1}^{\Gamma_2}(x) &= -\,\mathrm{tr}\,\Gamma_1\kappa^m(x)\Gamma_2 d^{M^2,-m}(x),\\ \tilde{\kappa}_{\Gamma_1}^{\Gamma_2}(q) &= \tfrac{\rho(m^2)}{16\pi^2}\,\mathrm{tr}\,(\Gamma_1\otimes\Gamma_2)\circ\left(\tfrac{\gamma q-m}{q^2-m^2}\overset{4}{\circledast}\tfrac{(m^2-M^2)^2}{(q^2-M^2)^2}\tfrac{\gamma q+m}{q^2-m^2}\right).\end{aligned}$$

The *translation invariants* are given by the singularities of the resolvent, i.e., the solutions of the characteristic equation of the eigenvalue-dependent (4×4)-matrix $\tilde{\kappa}_{\Gamma_1}^{\Gamma_2}$:

$$\tfrac{1}{1-\tilde{\kappa}} \Rightarrow \left\{I^2 \mid \det\left[\delta_{\Gamma_1}^{\Gamma_2} - \tilde{\kappa}_{\Gamma_1}^{\Gamma_2}(I)\right] = 0\right\}.$$

The $\mathbb{R}^4$-function for the composite chiral pseudoscalar field has as equation of motion in a first-order approximation,

$$\text{for } \langle\mathcal{C}|\overline{\mathbf{\Psi}}\gamma_5\mathbf{\Psi}(x)|\pi\rangle = \gamma_5(x): \quad \left\{ \begin{array}{ll} \gamma_5 &= \kappa_{\gamma_5}^{\gamma_5} * \gamma_5, \\ \tilde{\gamma}_5 &= \tilde{\kappa}_{\gamma_5}^{\gamma_5} \cdot \tilde{\gamma}_5. \end{array} \right.$$

It involves the action of the chiral spacetime interaction (kernel) with an energy-momentum function for the translation invariants:

$$d^{M^2,m} \overset{\kappa^m}{\longmapsto} \kappa_{\gamma_5}^{\gamma_5}, \text{ with } \kappa_{\gamma_5}^{\gamma_5}(x) = -\tfrac{1}{4}\,\mathrm{tr}\,\gamma_5\kappa^m(x)\gamma_5 d^{M^2,-m}(x) = \int \tfrac{d^4q}{(2\pi)^4}\tilde{\kappa}_{\gamma_5}^{\gamma_5}(q^2)e^{iqx}.$$

It can be simplified with the chiral properties,

$$\gamma_5(\gamma p + m)\gamma_5 = \gamma p - m, \quad \gamma_5\kappa^m\gamma_5 = \kappa^{-m}.$$

The eigenvalue equation $[\tilde{\kappa}_{\gamma_5}^{\gamma_5}(q^2) - 1]\tilde{\gamma}_5(q) = 0$ has a solution $\tilde{\kappa}_{\gamma_5}^{\gamma_5}(0) = 1$ for a Poincaré group representation with mass $q^2 = 0$, if compared with the consistency equation for the chiral breakdown parameter $m \neq 0$,

$$-\int d^4x\ \kappa^{-m}(-x)d^{M^2,m}(x) = -\int \tfrac{d^4q}{(2\pi)^4}\ \tilde{\kappa}^{-m}(q)\tilde{d}^{M^2,m}(q) = \tfrac{1}{m}d^{M^2,m}(0) = \mathbf{1}_4.$$

This equation is interpretable as the Schur normalization of the regularized Feynman propagator as the spacetime representation coefficient with the Green's kernel of the free equation of motion in the form of a dual pair:

$$\tilde{\kappa}_{\gamma_5}^{\gamma_5}(0) = -\kappa^{-m} * d^{M^2,m}(0) = -\tilde{\kappa}^{-m} \tfrac{*}{(2\pi)^4}\ \tilde{d}^{M^2,m}(0) = \mathbf{1}_4.$$

Chapter 11

Electroweak Spacetime

Unitary relativity in two complex dimensions is parametrized by electroweak spacetime $\mathbb{D}(2) \cong \mathbf{GL}(\mathbb{C}^2)/\mathbf{U}(2)$ as the noncompact real four-dimensional base manifold of a coset bundle $\mathbf{U}(2)(\mathbb{D}(2))$ for the global group $\mathbf{GL}(\mathbb{C}^2)$ with the typical fiber $\mathbf{U}(2) = \mathbf{U}(\mathbf{1}_2) \circ \mathbf{SU}(2)$ as local group, i.e., the real four-dimensional internal (chargelike) hyperisospin operations (see Chapter 6). In the classical manifold interpretation, electroweak spacetime $\mathbf{D}(1) \times \mathcal{Y}^3$ is a flat Friedmann universe with a hyperbolically curved position. Its representations are characterized by two continuous invariants for causal time $\mathbf{D}(1) \cong \mathbb{R}_+$ and hyperbolic position $\mathcal{Y}^3 \cong \mathbb{R}^3$. The Hilbert representations, as used for quantum theory, are infinite-dimensional for a nontrivial action of the external operations $\mathbf{D}(1) \times \mathbf{SO}_0(1,3)$. They will be related to spacetime particles and interactions in Chapter 12.

11.1 Operational Spacetime

Metrical tensors contain representation coefficients of the motion (global) group of a Riemannian manifold. For example, the metrical position coefficients in Friedmann universes, parametrizing classes of the local group $\mathbf{SO}(3)$ (see Chapters 1 and 2),

$$\mathbf{g} = d\tau^2 - R^2(\tau)d\sigma_k^2, \quad d\sigma_k^2 = \frac{d\vec{x}^2}{(1+k\frac{\vec{x}^2}{4})^2} = \begin{pmatrix} d\theta^2 \\ dr^2 \\ d\psi^2 \end{pmatrix} + \begin{pmatrix} \sin^2\theta \\ r^2 \\ \sinh^2\psi \end{pmatrix} d\omega_2^2,$$

$$\text{with} \left\{ \begin{array}{rl} \mathbf{SO}(4)/\mathbf{SO}(3) \cong \Omega^3, & k = 1, \\ \mathbf{SO}(3) \vec{\times} \mathbb{R}^3/\mathbf{SO}(3) \cong \mathbb{R}^3, & k = 0, \\ \mathbf{SO}_0(1,3)/\mathbf{SO}(3) \cong \mathcal{Y}^3, & k = -1, \end{array} \right.$$

are from a real 10-dimensional Hilbert representation $(\frac{1}{2}, \frac{1}{2}) \vee (\frac{1}{2}, \frac{1}{2})$ of the compact motion group $\mathbf{SO}(4)$ for the spherical case Ω^3 with an irreducible one-dimensional contribution $(0,0)$ and a nine-dimensional one $(1,1)$ and from a trivial representation of the Euclidean group $\mathbf{SO}(3) \vec{\times} \mathbb{R}^3$ for flat

H. Saller, *Operational Spacetime: Interactions and Particles*,
Fundamental Theories of Physics 163, DOI 10.1007/978-1-4419-0898-8_12,

position $\mathbb{R}^3$. For the noncompact hyperboloid $\mathcal{Y}^3$, the metrical coefficients come from the real 10-dimensional Lorentz group $\mathbf{SO}_0(1,3)$-representation,

$$[\tfrac{1}{2}|\tfrac{1}{2}] \vee [\tfrac{1}{2}|\tfrac{1}{2}] = [0|0] \oplus [1|1] \overset{\mathbf{SO}(3)}{\cong} 2\times[0] \oplus [1] \oplus [2] \overset{\mathbf{SO}(2)}{\cong} 4\times(0) \oplus 2\times(\pm 1) \oplus (\pm 2),$$

which is not a Hilbert representation.

The irreducible Hilbert representations of the motion group $\mathbf{SO}(1+s)$, used for spheres $\Omega^s \ni \vec{\omega}_s$, act on the spherical harmonics $(\vec{\omega}_s)^L$ with natural winding number invariants $L \in \mathbb{N}$ (see Chapter 8), e.g., for Einstein's static universe (see Chapter 1), the $\mathbf{SO}(4)$-representations $(\frac{L}{2}, \frac{L}{2})$,

$$\mathbb{R}\times\Omega^3 \ni (\tau, \vec{\omega}_3) \longmapsto (1, (\vec{\omega}_3)^L) \in \mathbb{C}^{(1+L)^2} \subset L^2(\Omega^3).$$

Positive-type functions for cyclic Hilbert representations of the noncompact positions $\mathbb{R}^3$ and $\mathcal{Y}^3$, isomorphic as manifolds, are given by Fourier-transformed momentum Dirac distributions (nonrelativistic scattering) and dipoles (nonrelativistic hydrogen), in general for odd-dimensional maximally symmetric positions (see Chapter 8):

$$\begin{aligned}\mathbb{R}^{2R-1}: \quad \vec{x} &\longmapsto \int \frac{2d^{2R-1}q}{|\Omega^{2R-2}|}\frac{1}{|P|^{2R-3}}\,\delta(\vec{q}^{\,2}-P^2)e^{-i\vec{q}\vec{x}} = (-2\tfrac{d}{dP^2r^2})^{R-1}\cos Pr\\ &\qquad = \cos Pr,\ \frac{\sin Pr}{Pr}, \dots,\\ \mathcal{Y}^{2R-1}: \quad \vec{x} &\longmapsto \int \frac{2d^{2R-1}q}{|\Omega^{2R-1}|}\frac{|Q|}{(\vec{q}^{\,2}+Q^2)^R}e^{-i\vec{q}\vec{x}} = e^{-|Q|r},\end{aligned}$$

$$\mathbb{R}^{2R-1} \cong \mathcal{Y}^{2R-1} \text{ from } \mathbb{R}^{2R} \ni \begin{pmatrix}\sqrt{\vec{x}^2+1}\\ \vec{x}\end{pmatrix} = \begin{pmatrix}\cosh\psi\\ \sinh\psi\ \omega_{2\vec{R}-2}\end{pmatrix} \in \mathcal{Y}^{2R-1}.$$

The continuous momentum translation invariants $P^2 \in \mathbb{R}_+$ are for flat Euclidean positions $\mathbb{R}^{2R-1}$, and the continuous curvature invariants $Q^2 \in \mathbb{R}_+$ are for hyperbolic positions $\mathcal{Y}^{2R-1}$.

In Hilbert representations of spacetime, the positive-type functions for position have to be considered together with Hilbert representations of time $\tau \longmapsto R(\tau)$, built by the irreducible ones $\tau \longmapsto e^{im\tau}$.

Electroweak spacetime is constituted by the classes of the hyperisospin operations $\mathbf{U}(2) = \mathbf{U}(\mathbf{1}_2)\circ\mathbf{SU}(2)$ in the extended Lorentz group operations $\mathbf{GL}(\mathbb{C}^2) = \mathbf{D}(\mathbf{1}_2)\times\mathbf{U}(\mathbf{1}_2)\circ\mathbf{SL}(\mathbb{C}^2)$ with the dilation group $\mathbf{D}(1) = \exp\mathbb{R}$, the phase group $\mathbf{U}(1) = \exp i\mathbb{R}$, and the Lorentz (cover) group $\mathbf{SL}(\mathbb{C}^2)$:

$$\begin{aligned}\mathbb{D}(2) &\cong \mathbf{GL}(\mathbb{C}^2)/\mathbf{U}(2) \cong \mathbf{D}(1)\times\mathcal{Y}^3,\\ \mathcal{Y}^3 &\cong \mathbf{SL}(\mathbb{C}^2)/\mathbf{SU}(2) \cong \mathbf{SO}_0(1,3)/\mathbf{SO}(3).\end{aligned}$$

It is the real four-dimensional member $\mathbb{D}(2) = \mathcal{D}^4$ in two series: the causal *unitary series* with the real n^2-dimensional unitary classes of general linear groups,

$$\mathbb{D}(n) \cong \mathbf{GL}(\mathbb{C}^n)/\mathbf{U}(n) \cong \mathbf{D}(1)\times\mathbf{SL}(\mathbb{C}^n)/\mathbf{SU}(n),$$

and the causal *orthogonal series* with real $(1+s)$-dimensional manifolds, embedding hyperboloids,

$$\mathcal{D}^{1+s} \cong \mathbf{D}(1) \times \mathcal{Y}^s, \quad \mathcal{Y}^s \cong \mathbf{SO}_0(1,s)/\mathbf{SO}(s).$$

One direct factor is the totally ordered abelian dilation group, called *causal group* or *(eigen)time*,

$$n=1, \ s=0: \quad \mathbb{D}(1) = \mathcal{D}^1 = \mathbf{D}(1) \cong e^{\mathbb{R}} \cong \mathbb{R}.$$

The position factors are globally symmetric spaces (see Chapter 2) of non-compact type $K \times K^*/K$ in the case of $\mathbf{SL}(\mathbb{C}^n)/\mathbf{SU}(n)$ and of type $BD\ I$ for $\mathbf{SO}_0(1,s)/\mathbf{SO}(s)$.

The two series meet exactly for $(n,s) = (1,0)$ and for $(n,s) = (2,3)$. Abelian unitary relativity in one dimension, $\mathbf{D}(1) \cong \mathbf{GL}(\mathbb{C})/\mathbf{U}(1)$, can be related to quantum mechanics as a theory of time $\mathbf{D}(1)$ realizations in Hilbert spaces with probability amplitudes. Correspondingly, nonabelian unitary relativity in two dimensions, $\mathbb{D}(2) \cong \mathbf{GL}(\mathbb{C}^2)/\mathbf{U}(2)$, will be connected with quantum field theory as a theory of spacetime realizations.

The Iwasawa factorizations $G = K \circ A \circ N$ of the involved special linear groups into maximal compact group, maximal abelian noncompact group and a subgroup with a nilpotent Lie algebra, and the minimal parabolic subgroups $G \supseteq (K_0 \times A) \circ N$ are

$$\begin{aligned}
\mathbf{SL}(\mathbb{C}^n) &= \mathbf{SU}(n) \circ \mathbf{SO}_0(1,1)^{n-1} \circ \exp \mathbb{R}^{n(n-1)} \\
&\supseteq [\mathbf{SO}(2) \times \mathbf{SO}_0(1,1)]^{n-1} \circ \exp \mathbb{R}^{n(n-1)}, \\
\mathbf{SO}_0(1,s) &= \mathbf{SO}(s) \circ \mathbf{SO}_0(1,1) \circ \exp \mathbb{R}^{s-1} \\
&\supseteq [\mathbf{SO}(s-1) \times \mathbf{SO}_0(1,1)] \circ \exp \mathbb{R}^{s-1}, \\
\mathbf{SL}(\mathbb{C}^2) \sim \mathbf{SO}_0(1,3) &= \mathbf{SO}(3) \circ \mathbf{SO}_0(1,1) \circ \exp \mathbb{R}^2 \\
&\supseteq [\mathbf{SO}(2) \times \mathbf{SO}_0(1,1)] \circ \exp \mathbb{R}^2.
\end{aligned}$$

Therefore, the manifolds $\mathbb{D}(n)$ for $n \geq 1$ and $\mathcal{D}^{1+s}$ for $s \geq 1$ have, respectively, real rank n and 2 as the dimensions of a maximal abelian noncompact subgroup $\mathbf{D}(1) \times \mathbf{SO}_0(1,1)^{n-1}$ and $\mathbf{D}(1) \times \mathbf{SO}_0(1,1)$. The real rank gives the maximal number of the representation-characterizing invariants from a continuous spectrum.

11.2 Representations of Electroweak Spacetime

In the following, electroweak spacetime will be considered as a member of the orthogonal series with the real rank-2 *causal spacetimes* $\mathcal{D}^{1+s} \cong \mathbf{D}(1) \times \mathcal{Y}^s$, $s \geq 1$. A parametrization of $\mathcal{D}^{1+s}$ is possible by the open future cone of Minkowski spacetime with Lorentz group action $\mathbf{SO}_0(1,s) \vec{\times} \mathbb{R}^{1+s}$:

$$\begin{aligned}
\mathcal{D}^{1+s} &= \left\{ x = \begin{pmatrix} x_0 \\ \vec{x} \end{pmatrix} = e^{\psi_0} \begin{pmatrix} \cosh \psi \\ \sinh \psi \ \vec{\omega}_{s-1} \end{pmatrix} \bigm| \vec{\omega}_{s-1} \in \Omega^{s-1}, \ \psi_0, \psi \in \mathbb{R} \right\} \\
&= \mathbb{R}_+^{1+s} = \{ x \in \mathbb{R}^{1+s} \bigm| x^2 > 0, \ x_0 > 0 \}.
\end{aligned}$$

The *future cone translation parametrization* of causal spacetime is a finite-dimensional and nonunitary representation, e.g., the Weyl representation of electroweak spacetime,

$$\begin{aligned}
\mathcal{D}^4 \ni x &= \begin{pmatrix} x_0+x_3 & x_1-ix_2 \\ x_1+ix_2 & x_0-x_3 \end{pmatrix} = u(\tfrac{\vec{x}}{r}) \circ \begin{pmatrix} x_0+r & 0 \\ 0 & x_0-r \end{pmatrix} \circ u^\star(\tfrac{\vec{x}}{r}), \quad x_0 \pm r > 0,\\
\mathbf{D}(1)\times\mathbf{SO}_0(1,1) &\ni \begin{pmatrix} x_0+r & 0 \\ 0 & x_0-r \end{pmatrix} = \begin{pmatrix} e^{\psi_0+\psi} & 0 \\ 0 & e^{\psi_0-\psi} \end{pmatrix}, \quad e^{\psi} = \sqrt{\tfrac{x_0+r}{x_0-r}},\\
\Omega^2 \ni \tfrac{\vec{x}}{r} &\longmapsto u(\tfrac{\vec{x}}{r}) = \begin{pmatrix} \cos\frac{\theta}{2} & -e^{-i\varphi}\sin\frac{\theta}{2} \\ e^{i\varphi}\sin\frac{\theta}{2} & \cos\frac{\theta}{2} \end{pmatrix}\\
&= \frac{1}{\sqrt{2r(x_3+r)}} \begin{pmatrix} x_3+r & -x_1+ix_2 \\ x_1+ix_2 & x_3+r \end{pmatrix} \in \mathbf{SU}(2).
\end{aligned}$$

The dilations $\mathbf{D}(1) \cong \mathbb{R}_+$ constitute, in a relativistic framework, the group for strictly positive "eigentime":

$$e^{\psi_0} = \tau = |x| = \vartheta(x)\sqrt{x^2} \in \mathbf{D}(1), \text{ with } \vartheta(x) = \vartheta(x_0)\vartheta(x^2).$$

The hyperboloids $\mathcal{Y}^s \cong \mathbb{R}^s$ with $e^{\psi_0}|\sinh\psi| = r$ are the position submanifolds.

In contrast to the $\mathbf{D}(1)\times\mathbf{SO}_0(1,1)$-related group parameters (ψ_0,ψ), the translation parametrization (x_0, r) is appropriate for a transition $\mathbb{R}^4_+ \subset \mathbb{R}^4$ to tangent space structures with particles.

$\mathcal{D}^{1+s}$ is acted on by $\mathbf{D}(1)\times\mathbf{SO}_0(1,s)$. For the future cone, foliated by position hyperboloids $\mathcal{Y}^s$, the action of the causal group $\mathbf{D}(1)$ may be called *hyperbolic hopping*, from position hyperboloid to position hyperboloid,

$$\mathbf{D}(1): \quad \begin{pmatrix} x_0 \\ x_3 \end{pmatrix} = e^{\psi_0}\begin{pmatrix} \cosh\psi \\ \sinh\psi \end{pmatrix} \longmapsto e^{\lambda_0}\begin{pmatrix} x_0 \\ x_3 \end{pmatrix} = e^{\lambda_0+\psi_0}\begin{pmatrix} \cosh\psi \\ \sinh\psi \end{pmatrix},$$

and the action of the dilative Lorentz subgroup $\mathbf{SO}_0(1,1) \subseteq \mathbf{SO}_0(1,s)$ *hyperbolic stretching*, inside each position hyperboloid,

$$\mathbf{SO}_0(1,1): \quad \begin{pmatrix} x_0 \\ x_3 \end{pmatrix} = e^{\psi_0}\begin{pmatrix} \cosh\psi \\ \sinh\psi \end{pmatrix} \longmapsto \begin{pmatrix} \cosh\lambda & \sinh\lambda \\ \sinh\lambda & \cosh\lambda \end{pmatrix}\begin{pmatrix} x_0 \\ x_3 \end{pmatrix} = e^{\psi_0}\begin{pmatrix} \cosh(\lambda+\psi) \\ \sinh(\lambda+\psi) \end{pmatrix}.$$

The rotations $\mathbf{SO}(s-1)$ act on each position hyperboloid.

As classical manifold, $\mathcal{D}^{1+s} = \mathbf{D}(1)\times\mathcal{Y}^s$ has a Robertson–Walker metric for a hyperbolic Friedmann universe (see Chapter 1):

$$\mathbf{g} = T^2(\psi_0)(d\psi_0^2 - d\mathrm{y}_s^2) = d\tau^2 - R^2(\tau)d\mathrm{y}_s^2.$$

Spacetime $\mathcal{D}^4 \cong \mathbb{R}^4_+$ has trivial curvature:

$$\begin{aligned}
\mathcal{R}^{dabc} &\cong -\frac{1}{T^2}\left(\begin{array}{c|c} [(\frac{T'}{T})^2 - \frac{T''}{T}]\mathbf{1}_3 & 0 \\ \hline 0 & [(\frac{T'}{T})^2 - 1]\mathbf{1}_3 \end{array}\right)\\
&\cong -\frac{1}{R^2}\left(\begin{array}{c|c} \ddot{R}R\mathbf{1}_3 & 0 \\ \hline 0 & (1-\dot{R}^2)\mathbf{1}_3 \end{array}\right) = 0 \text{ for } T(\psi_0) = e^{\psi_0},\ R(\tau) = \tau.
\end{aligned}$$

The linear expansion factor (position "radius"), $R(\tau) = \tau \in \mathbf{D}(1)$, cannot be unitarily analyzed with invariant masses $\tau \longmapsto e^{im\tau}$. Also, the metrical coefficients of the hyperboloid do not belong to Hilbert representations.

In the following, $\mathcal{D}^4$ is not considered as a classical manifold, e.g., as a cosmological model, but as a parametrization of the spacetime operations.

The complex functions of the future cone,

$$\mathbf{GL}(\mathbb{C}^2)/\mathbf{U}(2) \cong \mathbb{R}^4_+ \ni \vartheta(x)x \longmapsto \vartheta(x)f(x) \in \mathbb{C},$$

are the coefficients of unitary relativity. They arise in all representations of $\mathbf{GL}(\mathbb{C}^2)$ containing trivial hyperisospin $\mathbf{U}(2)$-representations.

The hyperisospin $\mathbf{U}(2)$-induced representations of the extended Lorentz group $\mathbf{GL}(\mathbb{C}^2)$ are subrepresentations of the two-sided regular representation of $\mathbf{GL}(\mathbb{C}^2)$. They act on $\mathbf{U}(2)$-intertwiners, $w : \mathbf{GL}(\mathbb{C}^2) \longrightarrow W$ with $w(gu^{-1}) = d(u)w(g)$, i.e., on mappings of the causal cone $\mathbb{R}^4_+ \longrightarrow W$ that connect spacetime points, parametrizing unitary classes, with hyperisospin orbits in a Hilbert space W with $\mathbf{U}(2)$-representation $d(u) \in \mathbf{U}(W)$.

The action group $\mathbf{SL}(\mathbb{C}^2) \times \mathbf{U}(2)$ with external and internal transformations as subgroup of the both-sided regular $\mathbf{GL}(\mathbb{C}^2) \times \mathbf{GL}(\mathbb{C}^2)$ action is realized in the electroweak standard model (see Chapter 6): The representations are faithful for $\mathbf{SL}(\mathbb{C}^2) \times \mathbf{U}(2)$ on the left-handed isodoublet lepton field, for $\mathbf{SO}_0(1,3) \times \mathbf{SO}(3)$ on the isospin gauge vector field, for $\mathbf{SL}(\mathbb{C}^2) \times \mathbf{U}(1)$ on the right-handed isosinglet lepton field, and for $\mathbf{SO}_0(1,3) \times \{1\}$ on the hypercharge gauge vector field. With the notable exception of the Higgs field, all isospin $\mathbf{SU}(2)$-representations of the standard model fields are isomorphic to subrepresentations of their Lorentz group $\mathbf{SL}(\mathbb{C}^2)$-representations.

11.2.1 Harmonic Analysis of the Causal Cartan Plane

Causal spacetime $\mathcal{D}^{1+s} \cong \mathbf{D}(1) \times \mathcal{Y}^s$ has real rank 2 as dimension of a Cartan plane $\mathcal{D}^2 = \mathbf{D}(1) \times \mathbf{SO}_0(1,1)$, with causal operations and Lorentz dilations as maximal noncompact abelian group.

The characteristic function of the causal Cartan line $\mathbf{D}(1)$ and plane $\mathcal{D}^2$ can be unitarily analyzed (Fourier-transformed) with an advanced energy-momentum measure with a pole at a trivial invariant $q_0 = 0$ and $q^2 = 0$, respectively,

$$\begin{aligned}
\text{for } \mathbf{D}(1) \text{ and } \mathbf{SO}_0(1,1): \quad \vartheta(x_0) &= \int \tfrac{dq_0}{2i\pi} \tfrac{1}{q_0 - io} e^{iq_0x_0}, \\
\text{for } \mathbf{D}(1) \times \mathbf{SO}_0(1,1): \quad \vartheta(x) &= \int \tfrac{d^2q}{2\pi^2} \tfrac{1}{-(q-io)^2} e^{iqx}, \\
\text{with } \quad (q-io)^2 &= (q_0 - io)^2 - q_3^2.
\end{aligned}$$

The Fourier transform of an energy-momentum function, holomorphic in the lower (upper) complex energy q_0 plane, i.e., with $q_0 \mp io$, is valued in the future (past) cone, i.e., supported by causal line and plane with $\vartheta(\pm x_0)$.

In a Lorentz group action-compatible translation parametrization, the harmonic analysis of the Cartan plane uses $\mathbf{SO}_0(1,1)$-invariants $q^2 = m^2$ as energy-momentum singularities:

$$\mathbf{D}(1) \text{ and } \mathbf{SO}_0(1,1): \quad \left\{ \begin{array}{ll} \int \frac{dq_0}{2i\pi} \frac{1}{q_0 - io - m} e^{iq_0 x_0} & = \vartheta(x_0) e^{imx_0}, \\ \int \frac{dq_0}{2i\pi} \frac{q_0}{(q_0 - io)^2 - m^2} e^{iq_0 x_0} & = \vartheta(x_0) \cos mx_0, \end{array} \right.$$

$$\mathbf{D}(1) \times \mathbf{SO}_0(1,1): \quad \int \frac{d^2 q}{2\pi^2} \frac{1}{-(q-io)^2 + m^2} e^{iqx} = \vartheta(x) \mathcal{J}_0(|mx|).$$

For more than one dimension, $t + s \geq 2$, the (anti-)Feynman energy-momentum measures with $\frac{1}{-q^2 \pm io + m^2}$, possible for any $\mathbf{SO}_0(t,s)$-metric with any real invariant $m^2 \in \mathbb{R}$, cannot be combined by causal (advanced and retarded) measures with $\frac{1}{-(q_0 \mp io)^2 + \vec{q}^{\,2} + m^2}$ that require an $\mathbf{SO}_0(1,s)$-metric and a positive invariant $m^2 \geq 0$.

The Bessel functions with half-integer index, starting with

$$\cos mx = \tfrac{1}{2}(e^{imx} + e^{-imx}) = \int dq \; |q| \delta(q^2 - m^2) e^{iqx} = \sum_{k=0}^{\infty} \frac{(-m^2 x^2)^k}{(2k)!},$$

and an invariant m^2, normalized with $|\Omega^0| = 2$ for the two irreducible $\mathbf{D}(1)$-representations on a 0-sphere $\{\pm im\} \cong \Omega^0$, are for odd-dimensional spaces with (energy-)momenta as eigenvalues, the corresponding functions with integer index for even-dimensional spaces. They start with $\mathcal{J}_0$ and an invariant circle $\{im \cos\theta\} \cong \Omega^1$, normalized with $|\Omega^1| = 2\pi$ (see Chapter 8). The integer-index Bessel functions integrate over $\mathbb{R}$-representation coefficients,

$$\begin{aligned} \mathcal{J}_0(mx) &= \tfrac{1}{2\pi} \int_{-\pi}^{\pi} d\theta \; \cos(mx \cos\theta) = \tfrac{1}{\pi} \int d\psi \; \sin(|mx| \cosh\psi) \\ &= \tfrac{1}{\pi} \int_{-m}^{m} \frac{dq}{\sqrt{m^2 - q^2}} \; \cos qx = \sum_{k=0}^{\infty} \frac{(-\frac{m^2 x^2}{4})^k}{(k!)^2}, \end{aligned}$$

with, for causal Cartan $(1,1)$-spacetime, the projections on oscillating representation coefficients of the causal operation $\mathbf{D}(1)$ (time) and on exponentially decreasing representation coefficients of the Lorentz dilations $\mathbf{SO}_0(1,1)$ (position):

$$\mathcal{D}^2 \longrightarrow \left\{ \begin{array}{rll} \mathbf{D}(1): & \int dx_3 \; \vartheta(x) \mathcal{J}_0(|mx|) & = \vartheta(x_0) \frac{\sin mx_0}{m}, \\ \mathbf{SO}_0(1,1): & \int dx_0 \; \vartheta(x) \mathcal{J}_0(|mx|) & = \frac{e^{-|mx_3|}}{|m|}. \end{array} \right.$$

The invariants for the $\mathbf{D}(1)$-oscillations and the $\mathbf{SO}_0(1,1)$-fall-off must not to coincide. Two continuous invariants for the rank-2 causal plane are implemented, in a residual representation, by two poles in the complex energy-momentum plane for the $L^2(\mathcal{D}^2)$-functions:

$$\mathbf{D}(1) \times \mathbf{SO}_0(1,1) \ni \vartheta(x) x \longmapsto \int \frac{d^2 q}{2\pi^2} \frac{1}{[-(q-io)^2 + M^2][(q-io)^2 - m^2]} e^{iqx}.$$

The two poles can be taken as the endpoints of a *finite* $\mathbf{SO}_0(1,1)$*-invariant singularity line* $q^2 = \kappa^2 \in [M^2, m^2]$, characteristic for even-dimensional spaces,

$$\begin{aligned}\int \frac{d^2q}{2\pi^2} \frac{1}{[-(q-io)^2+M^2][(q-io)^2-m^2]}e^{iqx} &= -\int_0^1 d\zeta \int \frac{d^2q}{2\pi^2} \frac{1}{[-(q-io)^2+\zeta M^2+(1-\zeta)m^2]^2}e^{iqx}\\ &= -\int_{M^2}^{m^2} \frac{d\kappa^2}{m^2-M^2} \int \frac{d^2q}{2\pi^2} \frac{1}{[-(q-io)^2+\kappa^2]^2}e^{iqx}\\ &= \vartheta(x)\int_{M^2}^{m^2} \frac{d\kappa^2}{m^2-M^2} \frac{\partial}{\partial \kappa^2}\mathcal{J}_0(|\kappa x|)\\ &= \vartheta(x)\frac{\mathcal{J}_0(|mx|)-\mathcal{J}_0(|Mx|)}{m^2-M^2}.\end{aligned}$$

Ahead, the pointwise product of the two energy-momentum poles, i.e., the convolution product of the two spacetime functions,

$$\int \frac{d^2q}{2\pi^2} \frac{1}{[-(q-io)^2+M^2][(q-io)^2-m^2]}e^{iqx} = -\vartheta(x)\mathcal{J}_0(|mx|) \stackrel{*}{2} \vartheta(x)\mathcal{J}_0(|Mx|),$$

will be related to the product structure of the represented group, for eigentime $\mathbf{D}(1)$ and for 1-position $\mathbf{SO}_0(1,1) \cong \mathcal{Y}^1$.

In the non-Lorentz-compatible direct product form, the residual representations of the group for Cartan spacetime give the product of two functions, for example,

$$\begin{aligned}\mathbf{D}(1) \times \mathbf{SO}_0(1,1) \ni (\vartheta(x_0)x_0, x_3) \longmapsto \int \frac{dq_0dq_3}{(2i\pi)^2} \frac{2iq_0}{[q_3^2+M^2][(q_0-io)^2-m^2]}\\ e^{i(q_0x_0-q_3x_3)} = \vartheta(x_0)\cos mx_0 \frac{e^{-|Mx_3|}}{|M|}.\end{aligned}$$

11.2.2 Harmonic Analysis of Even-Dimensional Causal Spacetimes

A causal Cartan plane is a maximal noncompact abelian group in the product of the causal group (eigentime) and a Lorentz group for nontrivial position dimension,

$$\mathbf{D}(1) \times \mathbf{SO}_0(1,1) \subseteq \mathbf{D}(1) \times \mathbf{SO}_0(1,s), \quad s = 1,2,\ldots.$$

As familiar from Cartan's B- and D-series, the orthogonal groups $\mathbf{SO}_0(1,s)$ come in two basically different types, those for odd dimensions and those for even dimensions.

The lowest-dimensional nonabelian Lorentz groups for odd and even dimensions are locally isomorphic, respectively, to the lowest-dimensional nontrivial special real and special complex groups:

$$\mathbf{SO}_0(1,2) \sim \mathbf{SL}(\mathbb{R}^2), \quad \mathbf{SO}_0(1,3) \sim \mathbf{SL}(\mathbb{C}^2).$$

For even spacetime dimensions $1+s = 2R$, the noncompact Cartan subgroups $\mathbf{SO}(2)^{R-1} \times \mathbf{SO}_0(1,1)$ nontrivially fill the whole diagonal (diagonalized in the complex), whereas for the more complicated odd-dimensional case $1+s=2R-1$ with two different types of Cartan subgroups, either noncompact $\mathbf{SO}(2)^{R-2}\times\mathbf{SO}_0(1,1)$ or compact $\mathbf{SO}(2)^{R-1}$ (tori), there remains one trivial diagonal unit, starting with

$$\mathbf{SO}_0(1,1)\ni\begin{pmatrix}e^{\psi}&0\\0&e^{-\psi}\end{pmatrix},\quad \mathbf{SO}_0(1,3)\ni\begin{pmatrix}e^{\psi}&0&0&0\\0&e^{i\varphi}&0&0\\0&0&e^{-i\varphi}&0\\0&0&0&e^{-\psi}\end{pmatrix},$$
$$\mathbf{SO}_0(1,2)\ni\begin{pmatrix}e^{\psi}&0&0\\0&1&0\\0&0&e^{-\psi}\end{pmatrix},\ \begin{pmatrix}1&0&0\\0&e^{i\varphi}&0\\0&0&e^{-i\varphi}\end{pmatrix}.$$

The Cartan subgroup structure is revelant for the existence and nonexistence of discrete series representations in the case of $\mathbf{SO}_0(1,2R)$ and $\mathbf{SO}_0(1,2R-1)$, respectively (see Chapter 8).

The universal cover group of an odd-dimensional group $\mathbf{SO}_0(t,s)$, $t+s=2R-1=3,5,\dots$, has one fundamental spinor representation with dimension 2^{R-1}, e.g., the $\mathbf{SU}(2)$-Pauli spinors for the rotations $\mathbf{SO}(3)$, whereas for even-dimensional groups $t+s=2R=4,6,\dots$, there are two fundamental spinor representations with dimension 2^{R-1}, e.g., the left- and right-handed $\mathbf{SL}(\mathbb{C}^2)$-Weyl spinors for the Lorentz group $\mathbf{SO}_0(1,3)$.

The characteristic future functions in the translation cone parametrizations have the harmonic analysis with an advanced energy-momentum measure $(q-io)^2=(q_0-io)^2-\vec{q}^{\,2}$:

$$\mathbf{SO}_0(1,s):\ \begin{cases}\vartheta(x)\begin{pmatrix}\frac{x}{|x|}\\|x|\end{pmatrix}=\int\frac{d^{2R-1}q}{|\Omega^{2R-1}|}\frac{1}{[-(q-io)^2]^R}\begin{pmatrix}iq\\1\end{pmatrix}e^{iqx},\\ \qquad 1+s=2R-1=1,3,\dots,\\ \vartheta(x)\begin{pmatrix}1\\x\end{pmatrix}=\int\frac{d^{2R}q}{\pi|\Omega^{2R-1}|}\frac{1}{[-(q-io)^2]^R}\begin{pmatrix}1\\\frac{2iqR}{-(q-io)^2}\end{pmatrix}e^{iqx},\\ \qquad 1+s=2R=2,4,\dots.\end{cases}$$

The spherical degrees of freedom show up in the order of the pole and in the normalization of the integration $d^{1+s}q$ by the measures of the n-dimensional unit spheres $|\Omega^n|=\frac{2\pi^{\frac{1+n}{2}}}{\Gamma(\frac{1+n}{2})}$ with the full-dimensional $|\Omega^{1+s}|=|\Omega^{2R-1}|$ for odd dimensions and the product $|\Omega^1|\frac{|\Omega^{2R-1}|}{2}$ for even dimensions.

The analogue to the one-dimensional dilation-invariant residual energy- or momentum measure $\frac{dq}{i|\Omega^1|q}$, normalized with the circle $|\Omega^1|=2\pi$, is, for even dimensions, the dilation and $\mathbf{SO}_0(1,2R-1)$-invariant doubled spherically normalized energy-momentum measure $\frac{2d^{2R}q}{i|\Omega^1||\Omega^{2R-1}|(q^2)^R}$, starting with $\frac{d^2q}{2i\pi^2q^2}$ for the Cartan plane. The hyperbolic $\mathbf{SO}_0(1,2R-1)$-invariant measures, not dilation-invariant for $m^2\neq 0$, contain the derivative $(\frac{\partial}{\partial q^2})^{R-1}$ of the Cartan plane measures, "compensating" the corresponding factor $(q^2)^{R-1}$ in the measure $d^{2R}q$:

$$\frac{d^{2R}q\ \Gamma(R)}{\pi^R(-q^2+m^2)^R}:\ \begin{cases}\frac{\Gamma(R)}{(-q^2+m^2)^R}=(\frac{\partial}{\partial q^2})^{R-1}\frac{1}{-q^2+m^2},\\ 2\vartheta(q)d^{2R}q=d^{2R-1}\mathrm{y}\ (q^2)^{R-1}dq^2.\end{cases}$$

The invariant measures of unitary relativity,

$$\mathbb{D}(n) = \mathbf{GL}(\mathbb{C}^n)/\mathbf{U}(n): \quad \frac{d^{n^2}q}{(\det q)^n}, \text{ with } q = q^* \in \mathbf{GL}(\mathbb{C}^n),$$

start with $\frac{dq}{q}$ for real $q \in \mathbf{D}(1)$ and the Lorentz invariant $\frac{d^4q}{(\det q)^2} = \frac{d^4q}{(q^2)^2}$ for energy-momenta $q \in \mathbb{D}(2)$.

The causal spacetime coefficients with one invariant are

$$\begin{aligned}
\mathbf{SO}_0(1,2R-2): &\int \frac{d^{2R-1}q}{|\Omega^{2R-1}|} \frac{1}{[-(q-io)^2+m^2]^R} \begin{pmatrix} iq \\ 1 \end{pmatrix} e^{iqx} \\
&= \vartheta(x) \begin{pmatrix} \frac{x}{|x|} \cos|mx| \\ \frac{\sin m|x|}{m} \end{pmatrix}, \\
\mathbf{SO}_0(1,2R-1): &\int \frac{d^{2R}q}{\pi|\Omega^{2R-1}|} \frac{1}{[-(q-io)^2+m^2]^R} \begin{pmatrix} 1 \\ \frac{2iqR}{-(q-io)^2+m^2} \end{pmatrix} e^{iqx} \\
&= \vartheta(x) \begin{pmatrix} 1 \\ x \end{pmatrix} \mathcal{J}_0(|mx|).
\end{aligned}$$

The translation representation $x\vartheta(x) \in \mathcal{D}^{2R}$ has an order-$(1+R)$ pole with trivial invariant $m^2 = 0$.

The Fourier transform of the advanced integration is, up to the order function, the corresponding principal value off-shell integration:

$$\begin{aligned}
\int \frac{d^{1+s}q}{\pi} \frac{\Gamma(1+N)}{[-(q-io)^2+m^2]^{1+N}} e^{iqx} &= 2\vartheta(x_0) \int \frac{d^{1+s}q}{\pi} \frac{\Gamma(1+N)}{(-q_{\mathrm{P}}^2+m^2)^{1+N}} e^{iqx}, \\
\int \frac{d^{1+s}q}{\pi} \frac{\Gamma(1+N)}{(-q_{\mathrm{P}}^2+m^2)^{1+N}} e^{iqx} &= i\epsilon(x_0) \int d^{1+s}q\, \epsilon(q_0) \delta^{(N)}(q^2-m^2) e^{iqx}.
\end{aligned}$$

Up to the causal order function $\vartheta(x)$, the $\mathbf{SO}_0(1,2R-1)$-representation coefficients for noncompact spacetime with hyperbolic position $\mathcal{Y}^{2R-1}$ coincide with the $\mathbf{SO}(2R)$-coefficients for its compact spherical partner Ω^{2R-1}. The coefficients for the spheres are given by the Dirac contribution:

$$\begin{aligned}
\mathbf{SO}(2R): \quad &\int \frac{2d^{2R}q}{i\pi|\Omega^{2R-1}|} \frac{1}{(\vec{q}^2-io-m^2)^R} e^{i\vec{q}\vec{x}} = [\mathcal{J}_0 + i\mathcal{N}_0](|m\vec{x}|), \\
&\int \frac{d^{2R}q}{\pi^R} \delta^{(R-1)}(\vec{q}^2-m^2) e^{i\vec{q}\vec{x}} = \mathcal{J}_0(|m\vec{x}|).
\end{aligned}$$

Even-dimensional spacetime $\mathcal{D}^{2R}$ with two continuous invariants (real rank $r = 2$) is represented as the Fourier-transformed product of two energy-momentum distributions, one with a simple pole and the other one with a pole of order $R = 1, 2, \ldots$, in the Lorentz scalar $L^2(\mathcal{D}^{2R})$-functions:

$$\begin{aligned}
\mathcal{D}^{2R} \ni \vartheta(x)x \longmapsto &\int \frac{d^{2R}q}{\pi|\Omega^{2R-1}|} \frac{1}{[-(q-io)^2+M^2]^R[(q-io)^2-m^2]} e^{iqx} \\
&= -\int_0^1 d\zeta\, \zeta^{R-1} \int \frac{d^{2R}q}{\pi|\Omega^{2R-1}|} \frac{R}{[-(q-io)^2+\zeta M^2+(1-\zeta)m^2]^{R+1}} e^{iqx} \\
&= -\int_{M^2}^{m^2} d^R\kappa^2 \int \frac{d^{2R}q}{\pi|\Omega^{2R-1}|} \frac{R}{[-(q-io)^2+\kappa^2]^{R+1}} e^{iqx} \\
&= \vartheta(x) \int_{M^2}^{m^2} d^R\kappa^2 \frac{\partial}{\partial\kappa^2} \mathcal{J}_0(|\kappa x|).
\end{aligned}$$

The spacetime coefficients contain the characteristic rank-dependent integration with a *measure over a line with* $\mathbf{SO}_0(1,2R-1)$*-invariants*:

$$\begin{aligned}
&\int_0^1 d\zeta\, \zeta^{R-1} = \int_{M^2}^{m^2} d^R\kappa^2 \text{ with } \kappa^2 = \zeta M^2 + (1-\zeta)m^2, \\
&\text{where } d^R\kappa^2 = \frac{d\kappa^2}{m^2-M^2} \left(\frac{m^2-\kappa^2}{m^2-M^2} \right)^{R-1}, \quad R = 1, 2, \ldots.
\end{aligned}$$

Lorentz group $\mathbf{SO}_0(1,2R-1)$-properties, and nontrivial properties with respect to the maximal compact group $\mathbf{SO}(2R-1)$ for $R \geq 2$, are obtained by derivations $\frac{\partial}{\partial x} = 2x\frac{\partial}{\partial x^2} \sim iq$ and harmonic $\mathbf{SO}_0(1,2R-1)$-polynomials (see Chapter 12).

The simple pole embeds the representation of the abelian time operations $\mathbf{D}(1) \cong \mathbb{R}$ with rank 1. The representation of hyperbolic position $\mathcal{Y}^{2R-1} \cong \mathbb{R}^{2R-1}$ with the rank-R Lorentz group $\mathbf{SO}_0(1,2R-1)$ as motion group is embedded by the order R pole. With the $r=2$ (real rank) continuous invariants and the $r_c = R-1$ (imaginary rank) discrete invariants of the orthogonal group $\mathbf{SO}(2R-1)$ with a Cartan torus $\mathbf{SO}(2)^{R-1}$, the acting group $\mathbf{D}(1) \times \mathbf{SO}_0(1,2R-1)$ has rank $r + r_c = R+1$.

Obviously, the representations of spacetime $\mathcal{D}^{2R}$ differ from those of corresponding flat spacetime $\mathbf{SO}_0(1,2R-1) \vec{\times} \mathbb{R}^{2R}/\mathbf{SO}_0(1,2R-1) \cong \mathbb{R}^{2R}$, used with $R=2$ for particles. They are not Feynman propagators. Higher-order poles, here at $q^2 = M^2$ for nonflat position with $R \geq 2$, cannot be connected with particle propagators like $\frac{1}{q^2+io-m^2}$; M^2 cannot be used as particle mass, i.e., as invariant for tangent translations $\mathbb{R}^{2R}$.

Unitary relativity $\mathcal{D}^4 = \mathbf{D}(1) \times \mathcal{Y}^3$ with three space dimensions $s=3$ as the minimal nonabelian case has imaginary rank 1 for the $\mathbf{SO}(3)$-rotation degrees of freedom. The $\mathcal{D}^4$-representation coefficients with two continuous invariants (real rank 2) involve a position-representing dipole [33], e.g., with Lorentz group vector properties,

$$\begin{aligned}\mathbf{GL}(\mathbb{C}^2)/\mathbf{U}(2) &\cong \mathbb{R}^4_+ \ni \vartheta(x)x \longmapsto \int \frac{d^4q}{2i\pi^3} \frac{q}{[-(q-io)^2+M^2]^2[(q-io)^2-m^2]} e^{iqx} \\ &= \frac{\partial}{\partial x} \int_{M^2}^{m^2} d^2\kappa^2 \int \frac{d^4q}{2\pi^3} \frac{2}{[-(q-io)^2+\kappa^2]^3} e^{iqx} \\ &= -\vartheta(x)\frac{\partial}{\partial x} \int_{M^2}^{m^2} d^2\kappa^2 \frac{\partial}{\partial \kappa^2} \mathcal{J}_0(|\kappa x|) \\ &= \vartheta(x)\frac{x}{m^2-M^2} \frac{\partial}{\partial \frac{x^2}{4}} \left[\mathcal{J}_0(|Mx|) - \frac{\partial}{\partial \frac{x^2}{4}} \frac{\mathcal{J}_0(|Mx|)-\mathcal{J}_0(|mx|)}{M^2-m^2} \right].\end{aligned}$$

The normalization of harmonic spacetime representations with tensor properties $(q)^L$ will be discussed below and in the next chapter.

11.3 Time and Position Subrepresentations

A group representation represents all subgroups. The projections of the representations of unitary relativity $\mathbf{GL}(\mathbb{C}^2)/\mathbf{U}(2)$, i.e., of causal spacetime $\mathcal{D}^4 = \mathbf{D}(1) \times \mathcal{Y}^3$ with the action of $\mathbf{D}(1) \times \mathbf{SO}_0(1,3)$, to those of the factors causal group and Lorentz group lead, respectively, to representations of free particles and of interactions: Free particles are related to representations of the causal group $\mathbf{D}(1) \cong \mathbb{R}$, Lorentz compatibly embedded in representations of causal spacetime $\mathcal{D}^4$, whereas interactions are related to embedded representations of hyperbolic position $\mathcal{Y}^3$, the isospin $\mathbf{SU}(2)$-classes of the Lorentz group $\mathbf{SL}(\mathbb{C}^2)$.

11.3.1 Projection to Subgroup Representations

For a closed Lie subgroup $H \subseteq G$ and Haar measures, the (generalized) G-functions can be projected [36] to (generalized) G/H-functions by integration over the subgroup (where defined):

$$\begin{aligned} \mu(g) &\longmapsto \mu(gH) = \int_H dh\ \mu(gh), \\ \text{e.g., } \delta_k(g) &\longmapsto \delta_k(gH) = \int_H dh\ \delta(k^{-1}gh). \end{aligned}$$

Under certain conditions, related to unimodularity [36], there is the integral decomposition with respect to the subgroup with suitably normalized invariant measures:

$$\int_G dg\ \mu(g) = \int_{G/H} dgH\ \mu(gH) = \int_{G/H} dgH \int_H dh\ \mu(gh).$$

Examples are representations of translations $\mathbb{R}^n$, which are projected to representations of subgroups $\mathbb{R}^{n-k}$ by integrations over the subgroup $\mathbb{R}^k$:

$$\begin{aligned} \mathbb{R} &\longrightarrow \{0\}: & e^{iqx} &\longmapsto \int \frac{dx}{2\pi}\ e^{iqx} = \delta(q) \cong 1, \\ \mathbb{R}^n &\longrightarrow \mathbb{R}^{n-k}: & e^{ipy+iqx} &\longmapsto \int \frac{d^kx}{(2\pi)^k}\ e^{ipy+iqx} = [\delta(q)]^k e^{ipy} \cong e^{ipy}. \end{aligned}$$

Other examples are (semi)direct product groups $G = K\ \vec{\times}\ H$ with normal subgroups H and subgroups $K \cong G/H$: Representations of Euclidean groups can be projected to lower-dimensional ones, e.g., for three-dimensional position translations with the chain of positive-type functions for the subgroups $j_0 \longmapsto \mathcal{J}_0 \longmapsto \cos$,

$$\begin{aligned} \mathbf{SO}(3)\ \vec{\times}\ \mathbb{R}^3 \longrightarrow \mathbf{SO}(2)\ \vec{\times}\ \mathbb{R}^2 : &\int \frac{dx_3}{2\pi}\ j_0(Pr) \\ &= \int \frac{dx_3}{2\pi} \int \frac{d^3q}{2\pi P}\ \delta(\vec{q}^{\,2} - P^2) e^{i\vec{q}\vec{x}} \\ &= \int \frac{d^2q}{2\pi P}\ \delta(\vec{q}^{\,2} - P^2) e^{i\vec{q}\vec{x}} = \frac{\mathcal{J}_0(|P\vec{x}|)}{2P}, \\ \longrightarrow \mathbb{R} : &\int \frac{dx_2}{2\pi}\ \mathcal{J}_0(|P\vec{x}|) = \frac{\cos Px_1}{|P|}. \end{aligned}$$

Particle representations of the Poincaré group have nontrivial projections for time translations and trivial ones for the Euclidean group with momentum $\vec{q} = 0$:

$$\begin{aligned} \mathbf{SO}_0(1,3)\ \vec{\times}\ \mathbb{R}^4 &\longrightarrow \mathbb{R} : & \int \frac{d^3x}{(2\pi)^3} \int d^4q\ \delta(q^2 - m^2) e^{iqx} &= \frac{\cos mx_0}{|m|}, \\ \mathbf{SO}_0(1,3)\ \vec{\times}\ \mathbb{R}^4 &\longrightarrow \mathbf{SO}(3)\ \vec{\times}\ \mathbb{R}^3 : & \int dx_0 \int d^4q\ \delta(q^2 - m^2) e^{iqx} &= 0. \end{aligned}$$

The decomposition with respect to time representation shows the positive-type functions (spherical Bessel function) of irreducible representations of the Euclidean group for nontrivial momenta $\vec{q}^{\,2} = q_0^2 - m^2 > 0$:

$$\begin{aligned} \mathbf{SO}_0(1,3)\ \vec{\times}\ \mathbb{R}^4 &\supset [\mathbf{SO}(3)\ \vec{\times}\ \mathbb{R}^3] \times \mathbb{R}, \\ \int \frac{d^4q}{2\pi}\ \delta(q^2 - m^2) e^{iqx} &= \int_m^\infty dq_0\ \frac{\sin\sqrt{q_0^2 - m^2}\ r}{r}\ \cos q_0 x_0. \end{aligned}$$

11.3.2 Projection to and Embedding of Particles and Interactions

The representation coefficients of the acting product group for even-dimensional causal spacetime $\mathcal{D}^{2R}$, $R = 1, 2, \dots$,

$$\mathbf{D}(1) \times \mathbf{SO}_0(1, 2R-1) \quad \text{for} \left\{ \begin{array}{ccll} \mathbf{D}(1) & \times & \mathcal{Y}^{2R-1} & = \mathcal{D}^{2R} \\ \cong \mathbb{R}_+ & \times & \mathbb{R}^{2R-1} & = \mathbb{R}_+^{2R}, \\ \mathcal{M}(\mathbb{R}_+^{2R}) & * & L^2(\mathbb{R}_+^{2R}) & = L^2(\mathbb{R}_+^{2R}), \end{array} \right.$$

are the convolution products of a causally embedded coefficient d_{2R} for the Lorentz group $\mathbf{SO}_0(1, 2R-1)$ with a bounded function $|\mathcal{J}_0(r)| \leq \mathcal{J}_0(0) = 1$ and a Lorentz compatibly embedded Radon measure κ^1 for the causal group (eigentime) $\mathbf{D}(1)$:

$$\begin{aligned} \begin{pmatrix} d_{2R}(x) \\ \kappa^1(x) \end{pmatrix} &\sim \int \frac{d^{2R}q}{\pi|\Omega^{2R-1}|} \begin{pmatrix} \frac{1}{[-(q-io)^2+m^2]^R} \\ \frac{iq}{-(q-io)^2+m^2} \end{pmatrix} e^{iqx} \\ &= \begin{pmatrix} 1 \\ \frac{x}{2}\Gamma(R)\left(\frac{\partial}{\partial \frac{x^2}{4}}\right)^R \end{pmatrix} \vartheta(x)\mathcal{J}_0(|mx|), \\ d_{2R} * \kappa^1(x) &\sim \int \frac{d^{2R}q}{\pi|\Omega^{2R-1}|} \frac{iq}{[-(q-io)^2+M^2]^R[-(q-io)^2+m^2]} e^{iqx}. \end{aligned}$$

The Radon distributions for $N = 0, 1, \dots$ have the explicit form

$$\begin{aligned} &\left(\frac{\partial}{\partial \frac{x^2}{4}}\right)^{1+N} \vartheta(x^2)\mathcal{J}_0(|mx|) \\ &= \sum_{k=-N}^{0} \frac{(m^2)^k}{k!} \delta^{(N+k)}\left(-\frac{x^2}{4}\right) + \vartheta(x^2)\left(\frac{\partial}{\partial \frac{x^2}{4}}\right)^{1+N} \mathcal{J}_0(|mx|). \end{aligned}$$

The factors in the non-Lorentz-compatible direct product representations of Cartan spacetime $\mathbf{D}(1) \times \mathbf{SO}_0(1,1) \ni (\vartheta(x_0)x_0, x_3) \longmapsto \vartheta(x_0) \cos mx_0 \; e^{-|Mx_3|}$ are obtained from the Lorentz-compatible (generalized) functions of causal spacetime by projection to (generalized) functions of time and of hyperbolic position:

$$\mathcal{D}^{2R} \ni \vartheta(x)x \longmapsto \vartheta(x)\mu(x) \Rightarrow \left\{ \begin{array}{llll} \mathbf{D}(1) \ni \vartheta(x_0)x_0 & \longmapsto \int d^{2R-1}x & \vartheta(x)\mu(x), \\ \mathcal{Y}^{2R-1} \ni \; \vec{x} & \longmapsto \int \; dx_0 & \vartheta(x)\mu(x). \end{array} \right.$$

The time projections by integration over position, i.e., for trivial momenta $\int d^{2R-1}\frac{x}{2\pi} e^{-i\vec{q}\vec{x}} = \delta(\vec{q})$,

$$\begin{aligned} &\text{for } \mathbf{D}(1): \; \int \frac{\pi|\Omega^{2R-1}|d^{2R-1}x}{(2\pi)^{2R}} \begin{pmatrix} 1 \\ \frac{x}{2}\Gamma(R)\left(\frac{\partial}{\partial \frac{x^2}{4}}\right)^R \end{pmatrix} \vartheta(x)\mathcal{J}_0(|mx|) \\ &= \int \frac{dq_0}{2\pi} \begin{pmatrix} \frac{1}{[-(q_0-io)^2+m^2]^R} \\ \frac{iq_0}{-(q_0-io)^2+m^2} \end{pmatrix} e^{iq_0x_0} = \vartheta(x_0) \begin{pmatrix} \frac{1}{\Gamma(R)}\left(-\frac{\partial}{\partial m^2}\right)^{R-1} \frac{\sin mx_0}{m} \\ \cos mx_0 \end{pmatrix}, \end{aligned}$$

contain in the lower component a representation coefficient $\vartheta \cos$ for time:

$$\begin{aligned} \mathbf{D}(1) \ni \vartheta(x_0)x_0 \longmapsto \quad & \vartheta(x_0)\cos mx_0 &&= \int \frac{dq_0}{2i\pi} \frac{q_0}{(q_0-io)^2-m^2} e^{iq_0x_0}, \\ \hookrightarrow \quad & \frac{x}{2}\Gamma(R)\left(\frac{\partial}{\partial \frac{x^2}{4}}\right)^R \vartheta(x)\mathcal{J}_0(|mx|) && \\ & &&= \int \frac{d^{2R}q}{\pi|\Omega^{2R-1}|} \frac{iq}{-(q-io)^2+m^2} e^{iqx}. \end{aligned}$$

The real invariants $q_0 = \pm m$ can be used for free particles as translation invariant $q^2 = m^2$ in a Poincaré group representation. For $R \geq 2$, the upper component displays matrix elements of indefinite unitary faithful translation representations [9] (half-integer-index spherical Bessel functions [57]):

$$\begin{aligned}\left(-\frac{\partial}{\partial m^2}\right)^{R-1} \frac{\sin m x_0}{m} &= \left(-\frac{\partial}{\partial m^2}\right)^{R-2} \frac{\sin m x_0 - m x_0 \cos m x_0}{2m^3} \\ &= x_0^{2R-1}\left(-\frac{\partial}{\partial t^2}\right)^{R-1} \frac{\sin t}{t},\ t = m x_0.\end{aligned}$$

The projections on the position hyperboloid with the $\mathbf{SO}_0(1, 2R-1)$-coefficients by integration over time, i.e., for trivial energy $\int \frac{dx_0}{2\pi} e^{i q_0 x_0} = \delta(q_0)$,

$$\begin{aligned}\text{for } \mathcal{Y}^{2R-1}: \int dx_0 &\begin{pmatrix} 1 \\ \frac{x}{2}\Gamma(R)\left(\frac{\partial}{\partial \frac{x^2}{4}}\right)^R \end{pmatrix} \vartheta(x)\mathcal{J}_0(|mx|) \\ &= \int \frac{2d^{2R-1}q}{|\Omega^{2R-1}|} \begin{pmatrix} \frac{1}{(\vec{q}^{\,2}+m^2)^R} \\ \frac{i\vec{q}}{\vec{q}^{\,2}+m^2} \end{pmatrix} e^{-i\vec{q}\vec{x}} \\ &= \begin{pmatrix} \frac{e^{-|m|r}}{|m|} \\ \Gamma(R)\left(\frac{4m^2}{r^2}\right)^{R-1} |m| \left(-\frac{\partial}{\partial m^2}\right)^{R-1} \frac{\vec{x}}{r}\frac{e^{-|m|r}}{|m|} \end{pmatrix},\end{aligned}$$

give, in the upper component, an exponential potential as a positive-type function $\exp \in L^\infty(\mathcal{Y}^{2R-1})_+$ for hyperbolic position with imaginary invariants $|\vec{q}| = \pm im$. The inverse invariant of the $\mathbf{SO}_0(1,1)$-representations is the characteristic interaction range $\frac{1}{|m|}$:

$$\begin{aligned}\mathcal{Y}^{2R-1} \ni \vec{x} \longmapsto \quad & e^{-|m|r} &&= \int \frac{2d^{2R-1}q}{|\Omega^{2R-1}|} \frac{|m|}{(\vec{q}^{\,2}+m^2)^R} e^{-i\vec{q}\vec{x}}, \\ \hookrightarrow \quad & \vartheta(x)\mathcal{J}_0(|mx|) &&= \int \frac{d^{2R}q}{\pi|\Omega^{2R-1}|} \frac{1}{[-(q-io)^2+m^2]^R} e^{iqx}.\end{aligned}$$

The lower component involves generalized Yukawa forces (half-integer-index hyperbolic Macdonald functions [57]), i.e., exponential forces $\epsilon(x)e^{-|mx|}$ for Cartan spacetime $R = 1$ and Yukawa forces proper $\frac{\vec{x}}{r}\frac{1+|m|r}{2r^2}e^{-|m|r}$ for Minkowski spacetime $R = 2$:

$$\begin{aligned}\frac{\vec{x}}{(r^2)^{R-1}}\left(-\frac{\partial}{\partial m^2}\right)^{R-1} \frac{e^{-|m|r}}{|m|r} &= \frac{\vec{x}}{(r^2)^{R-2}}\left(-\frac{\partial}{\partial m^2}\right)^{R-2} \frac{1+|m|r}{2|m|^3 r^3} e^{-|m|r} \\ &= \vec{x}\left(-\frac{\partial}{\partial \rho^2}\right)^{R-1} \frac{e^{-\rho}}{\rho},\quad \rho = |m|r.\end{aligned}$$

11.4 Hilbert Spaces for Causal Spacetimes

The Hilbert space $L^2(\mathcal{D}^{2R})$ with square-integrable functions of causal spacetime $\vartheta(x)x \in \mathcal{D}^{2R} = \mathbf{D}(1) \times \mathcal{Y}^{2R-1}$ extends the Hilbert space $L^2(\mathcal{Y}^{2R-1})$ for hyperbolic position, used for position $\vec{x} \in \mathbb{R}^3 \cong \mathcal{Y}^3$ in the case of the bound states of nonrelativistic hydrogen atom (see Chapter 8).

The Hilbert spaces for the Poincaré group and free particles, built with the Fock state for the translations, e.g., $L^2(\mathcal{Y}^3) \times \mathbb{C}^{1+2J}$ for a massive spin-J

particle with the square-integrable functions of the momentum hyperboloid (see Chapters 5 and 8), are inappropriate for electroweak spacetime. This is in analogy to the different Hilbert spaces for free nonrelativistic scattering $L^2(\Omega^2)\times\mathbb{C}^2$ and the Hilbert space $L^2(\mathcal{Y}^3)$ for the nonrelativistic bound states of the periodic table.

11.4.1 Hardy Spaces for the Future and Past

The projection from all time translations $\mathbb{R}$ to future and past $\mathbb{R}_\pm$ goes with the transition from Dirac measure to the advanced and retarded measures, $\pm\frac{1}{i\pi}\frac{1}{q\mp io} = \delta(q) \pm \frac{1}{i\pi}\frac{1}{q_{\mathrm{P}}}$, as seen in the Fourier-transformed step and sign functions:

$$\vartheta(\pm t) = \frac{1\pm\epsilon(t)}{2} = \vartheta_\pm(t) = \pm\int \frac{dq}{2i\pi}\frac{1}{q\mp io}e^{iqt}, \left\{ \begin{array}{rl} \vartheta_\pm(t) & \overset{\mathbf{F}}{\leftrightarrow} \frac{\mp i}{q\mp io} = \tilde\vartheta_\pm(q), \\ \epsilon(t) = \frac{t}{|t|} & \overset{\mathbf{F}}{\leftrightarrow} -\frac{2i}{q_{\mathrm{P}}}, \\ 1 & \overset{\mathbf{F}}{\leftrightarrow} 2\pi\delta(q). \end{array}\right.$$

The Lebesgue spaces for the future and past can be obtained from the Lebesgue spaces for all time translations by projections with $\vartheta_\pm = \vartheta_\pm \cdot \vartheta_\pm$. Their Fourier transforms constitute the *Hardy spaces* $H^r_\pm(\check{\mathbb{R}})$ for energy functions $\tilde f_\pm$. The Hardy functions arise from the Lebesgue functions by a convolution with the Fourier-transformed step functions (Radon measures),

$$\frac{1}{p}+\frac{1}{r}=1,\ 1\le p\le 2,\ \ \infty\ge r\ge 2,$$

$$\begin{array}{ccc} L^p(\mathbb{R}) & \overset{\mathbf{F}}{\longrightarrow} & L^r(\check{\mathbb{R}}) \\ \vartheta_\pm\cdot \big\downarrow & & \big\downarrow \tilde\vartheta_\pm *, \\ L^p(\mathbb{R}_\pm) & \underset{\mathbf{F}}{\longrightarrow} & H^r_\pm(\check{\mathbb{R}}) \end{array} \qquad \begin{aligned} f(t) &= \int \frac{dq}{2\pi}\tilde f(q)e^{iqt} = f_+(t)+f_-(t), \\ f_\pm(t) &= \vartheta(\pm t)f(t) = \int \frac{dq}{2\pi}\tilde f_\pm(q\mp io)e^{iqt}, \\ \tilde f_\pm(q\mp io) &= \tilde\vartheta_\pm * \tilde f(q\mp io) \\ &= \pm\int \frac{dp}{2i\pi}\frac{1}{q\mp io-p}\tilde f(p). \end{aligned}$$

The *Hardy Hilbert spaces* $H^2_\pm(\check{\mathbb{R}})$ are Fourier-isomorphic to the square-integrable functions of the positive and negative lines (future and past),

$$\begin{gathered} L^2(\mathbb{R}) \supset L^2(\mathbb{R}_\pm) \cong H^2_\pm(\check{\mathbb{R}}) \subset L^2(\check{\mathbb{R}}), \\ L^2(\mathbb{R}) \cong L^2(\mathbb{R}_+) \oplus L^2(\mathbb{R}_-) \cong H^2_+(\check{\mathbb{R}}) \oplus H^2_-(\check{\mathbb{R}}) \cong L^2(\check{\mathbb{R}}), \end{gathered}$$

with the scalar products

$$\begin{aligned} \langle f|f'\rangle_\pm &= \int dt\ \overline{f_\pm(t)}f'_\pm(t) &&= \int \frac{dq}{2\pi}\ \overline{\tilde f_\pm(q\mp io)}\tilde f'_\pm(q\mp io) \\ &= \int dt\ \vartheta(\pm t)\overline{f(t)}f'(t) &&= \pm\int \frac{dq\,dq'}{(2\pi)^2}\ \overline{\tilde f(q)}\ \frac{1}{i}\frac{1}{q-q'\mp io}\ \tilde f'(q'). \end{aligned}$$

11.4.2 Hardy Spaces for Spacetime Cones

For Minkowski translations $\mathbf{SO}_0(1,s)\ \vec{\times}\ \mathbb{R}^{1+s}$, $s\ge 1$, the sum of the Hardy spaces of advanced and retarded energy-momentum functions is a proper subspace of the full Lebesgue space $L^p(\check{\mathbb{R}}^{1+s})$. The sum of the characteristic functions $\vartheta_\pm$ for the future and past is the characteristic function ϑ_c for the causal bicone (subindex c), equal to 1 for all translations $\mathbb{R}^{1+s}$ only for total order $s=0$:

$$\begin{aligned} s=1,2,\cdots: \quad & \mathbb{R}_c^{1+s} = \{x\in\mathbb{R}^{1+s} \mid x^2>0\} = \mathbb{R}_+^{1+s}\cup\mathbb{R}_-^{1+s} \neq \mathbb{R}^{1+s},\\ & \vartheta_\pm(x) = \vartheta(\pm x_0)\vartheta(x^2),\quad \vartheta_c(x)=\vartheta(x^2),\quad \epsilon_c(x)=\epsilon(x_0)\vartheta(x^2),\\ & \vartheta_\mp = \frac{\vartheta_c\pm\epsilon_c}{2},\quad \vartheta_{\pm,c}\cdot\vartheta_{\pm,c}=\vartheta_{\pm,c},\quad \epsilon_c\cdot\epsilon_c=\vartheta_c,\quad \epsilon_c\cdot\vartheta_c=\epsilon_c. \end{aligned}$$

The causal distributions for even-dimensional spacetimes have the Fourier transforms, with $(q-io)^2=(q_0-io)^2-\vec{q}^2$,

$$\mathbf{SO}_0(1,2R-1):\ \begin{cases} \vartheta_\pm(x) = \int \frac{d^{2R}q}{\pi|\Omega^{2R-1}|}\frac{1}{[-(q\mp io)^2]^R}e^{iqx},\\ \vartheta_c(x) = \int \frac{d^{2R}q}{\pi|\Omega^{2R-1}|}\frac{2}{(-q_P^2)^R}e^{iqx},\\ \epsilon_c(x) = \int \frac{d^{2R}q}{\pi^R} i\epsilon(q_0)\delta^{(R-1)}(q^2)e^{iqx}, \end{cases}$$

with the general relation between advanced (retarded) and principal value measures:

$$\begin{aligned} \int d^{1+s}q\ \frac{\Gamma(1+N)}{[(q\mp io)^2]^{1+N}}e^{iqx} &= \int d^{1+s}q\ \left[\frac{\Gamma(1+N)}{(q_P^2)^{1+N}} \pm i\pi\epsilon(q_0)\delta^{(N)}(-q^2)\right]e^{iqx}\\ &= 2\vartheta(\pm x_0)\int d^{1+s}q\ \frac{\Gamma(1+N)}{(q_P^2)^{1+N}}e^{iqx}. \end{aligned}$$

One obtains the *Hardy spaces* $H^r_\pm(\check{\mathbb{R}}^{2R})$ for energy-momentum functions of even-dimensional future and past cones:

$$\begin{array}{ccc} L^p(\mathbb{R}^{2R}) & \xrightarrow{\mathbf{F}} & L^r(\check{\mathbb{R}}^{2R})\\ \vartheta_\pm\cdot\ \downarrow & & \downarrow\ \tilde{\vartheta}_\pm *\\ L^p(\mathbb{R}^{2R}_\pm) & \xrightarrow[\mathbf{F}]{} & H^r_\pm(\check{\mathbb{R}}^{2R}) \end{array},\qquad \begin{aligned} f_\pm(x) &= \vartheta_\pm(x)f(x)\\ &= \int \frac{d^{2R}q}{(2\pi)^{2R}}\tilde{f}_\pm(q\mp io)e^{iqx},\\ \tilde{f}_\pm(q\mp io) &= \tilde{\vartheta}_\pm * \tilde{f}(q\mp io)\\ &= \int \frac{d^{2R}p}{\pi|\Omega^{2R-1}|}\frac{1}{[-(q\mp io-p)^2]^R}\tilde{f}(p). \end{aligned}$$

The *Hardy Hilbert spaces* $L^2(\mathcal{D}^{2R}) \cong L^2(\mathbb{R}^{2R}_\pm) \cong H^2_\pm(\check{\mathbb{R}}^{2R})$ have the following scalar products:

$$\begin{aligned} \langle f|f'\rangle_\pm = \int d^{2R}x\ \overline{f_\pm(x)}f'_\pm(x) \quad &= \int \frac{d^{2R}q}{(2\pi)^{2R}}\ \overline{\tilde{f}_\pm(q\mp io)}\tilde{f}'_\pm(q\mp io)\\ = \int d^{2R}x\ \vartheta_\pm(x)\overline{f(x)}f'(x) \quad &= \int \frac{d^{2R}q\ d^{2R}q'}{(2\pi)^{2R}\pi|\Omega^{2R-1}|}\ \overline{\tilde{f}(q)}\frac{1}{[-(q-q'\mp io)^2]^R}\tilde{f}'(q') \end{aligned}$$

Functions with different Lorentz group $\mathbf{SO}_0(1,2R-1)$-properties via harmonic polynomials $(q)^L$ are orthogonal to each other (see Chapter 12),

$$\int d^{2R}q\ \mu(q^2)(q)^L\otimes(q)^{L'} = \delta^{LL'}\frac{\Gamma(1+L)}{2^L}\frac{\Gamma(R)}{\Gamma(R+L)}\int d^{2R}q\ \mu(q^2)(q^2)^L\ (\mathbf{1}_{2R})^L.$$

The remaining convolutions for scalar Feynman measures at the neutral element are as follows (see Chapter 10):

			$(\overset{2R}{*}, q^2) = (\pm \frac{* \, 2}{i\vert\Omega^{2R-1}\vert}, q^2 \pm io),\ 2R = 2, 4, \dots$
$\frac{(-q^2)^\lambda}{(-q^2+m_1^2)^{R+\nu_1}}$	$\overset{2R}{*}$	$\frac{1}{(-q^2+m_2^2)^{R+\nu_2}}\Big\vert_{q=0}$	$= \frac{\Gamma(R+\lambda)\Gamma(R+\nu_1+\nu_2-\lambda)}{\Gamma(R+\nu_1)\Gamma(R+\nu_2)} \int_0^1 d\zeta \frac{\zeta^{R+\nu_1-1}(1-\zeta)^{R+\nu_2-1}}{[\zeta m_1^2+(1-\zeta)m_2^2]^{R+\nu_1+\nu_2-\lambda}}$
$\frac{(-q^2)^\lambda}{(-q^2+m^2)^{R+\nu_1}}$	$\overset{2R}{*}$	$\frac{1}{(-q^2+m^2)^{R+\nu_2}}\Big\vert_{q=0}$	$= \frac{\Gamma(R+\lambda)\Gamma(R+\nu_1+\nu_2-\lambda)}{\Gamma(R+\nu_1+\nu_2)} \frac{1}{(m^2)^{R+\nu_1+\nu_2-\lambda}}$
$(-q^2)^\lambda$	$\overset{2R}{*}$	$\frac{1}{(-q^2+m^2)^{R+\nu+\lambda}}\Big\vert_{q=0}$	$= \frac{\Gamma(R+\lambda)\Gamma(\nu)}{\Gamma(R+\nu+\lambda)} \frac{1}{(m^2)^\nu}$

11.5 Spacetime Interactions (Kernels)

Causal spacetime $\mathcal{D}^{2R} = \mathbf{D}(1)\times\mathcal{Y}^{2R-1}$ has the Lorentz compatibly embedded kernels of time and hyperbolic position.

The inverse spacetime derivatives involve lightcone-supported Dirac distributions:

$$\begin{aligned}
(4\pi)^R\Gamma(R-N)\ \frac{1}{(\partial^2)^{R-N}} &= \int \frac{d^{2R}q}{\pi^R} \frac{\Gamma(R-N)}{[-(q-io)^2]^{R-N}} e^{iqx} \\
&= \vartheta(x_0) 2\pi \times \begin{cases} \vartheta(x^2), & N=0, \\ \delta^{(N-1)}(-\frac{x^2}{4}), & N=1,\dots,R-1, \end{cases} \\
(4\pi)^R\Gamma(R+1-N)\ \frac{\partial}{(\partial^2)^{R+1-N}} &= \int \frac{d^{2R}q}{\pi^R} \frac{\Gamma(R+1-N)\ iq}{[-(q-io)^2]^{R+1-N}} e^{iqx} \\
&= \vartheta(x_0) \pi x \times \begin{cases} \vartheta(x^2), & N=0, \\ \delta^{(N-1)}(-\frac{x^2}{4}), & N=1,\dots,R. \end{cases}
\end{aligned}$$

The time and position projections of the spacetime interactions (kernels) lead back to the embedded kernels, e.g., to the characteristic causal function $\vartheta(x_0)$ for time and, for the $R = 1$ position, to the sign function $\epsilon(x) = \frac{x}{r}$, and, for $R \geq 2$, to the Kepler potential $\frac{1}{r}$:

$$\begin{aligned}
\frac{1}{\partial} \sim \frac{1}{q} = \frac{q}{q^2} : \qquad & \begin{pmatrix} \int \frac{|\Omega^{2R-1}|\ d^{2R-1}x}{(2\pi)^{2R}} \\ \int \frac{dx_0}{2\pi} \end{pmatrix} \\
& \int \frac{d^{2R}q}{i|\Omega^{2R-1}|}\ \frac{q}{(q-io)^2} e^{iqx} = \begin{pmatrix} \vartheta(x_0) \\ \frac{\vec{x}}{r} \frac{\Gamma(2R-1)}{2(r^2)^{R-1}} \end{pmatrix},
\end{aligned}$$

$$\begin{aligned}
\frac{1}{(\partial^2)^{R-1}} \sim \frac{1}{(q^2)^{R-1}} : \ & \begin{pmatrix} \int \frac{|\Omega^{2R-1}|\ d^{2R-1}x}{(2\pi)^{2R}} \\ \int \frac{dx_0}{2\pi} \end{pmatrix} \\
& \int \frac{d^{2R}q}{|\Omega^{2R-1}|}\ \frac{1}{[-(q-io)^2]^{R-1}} e^{iqx} = \begin{pmatrix} \vartheta(x_0) \frac{x_0^{2R-3}}{\Gamma(2R-2)} \\ \frac{1}{r} \end{pmatrix}.
\end{aligned}$$

For $R \geq 2$, the position projection of the time kernel gives a Kepler-like force $\frac{\vec{x}}{r^{2R-1}}$.

The spacetime interactions (kernels) in general are convolution products of spacetime Green's kernels with spacetime representation coefficients. They are computed by the following convolutions, wherever defined for $\nu \in \mathbb{R}$ (see Chapter 9):

$$(\overset{2R}{*}, q^2) = (\pm \frac{* \, 2}{i|\Omega^{2R-1}|}, q^2 \pm io), \ 2R = 2, 4, \ldots$$

$$\left(\frac{\partial}{\partial q}\right)^{L_1} \frac{\Gamma(\nu_1)}{(-q^2+m_1^2)^{\nu_1}} \overset{2R}{*} \left(\frac{\partial}{\partial q}\right)^{L_2} \frac{\Gamma(R+\nu_2)}{(-q^2+m_2^2)^{R+\nu_2}} = \Gamma(R)\left(\frac{\partial}{\partial q}\right)^{L_1+L_2} \int_0^1 d\zeta \ \frac{\zeta^{\nu_1-1}(1-\zeta)^{R+\nu_2-1}\Gamma(\nu_1+\nu_2)}{(-\zeta(1-\zeta)q^2+\zeta m_1^2+(1-\zeta)m_2^2)^{\nu_1+\nu_2}}$$

$$\frac{\Gamma(L_1+\nu_1)(q)^{L_1}}{(-q^2)^{L_1+\nu_1}} \overset{2R}{*} \left(\frac{\partial}{\partial q}\right)^{L_2} \frac{\Gamma(R+\nu_2)}{(-q^2+\kappa^2)^{R+\nu_2}} = \Gamma(R)\left(\frac{\partial}{\partial q}\right)^{L_2} \otimes (q)^{L_1} \int_0^1 d\zeta \ \frac{\zeta^{L_1+\nu_1-1}(1-\zeta)^{R-\nu_1-1}\Gamma(L_1+\nu_1+\nu_2)}{(-\zeta q^2+\kappa^2)^{L_1+\nu_1+\nu_2}}$$

$(q)^{L_1}$ can be any homogeneous energy-momentum polynomial. Nontrivial Lorentz group properties arise by derivations:

$$\frac{\partial}{\partial q} = 2q\frac{\partial}{\partial q^2}, \quad \frac{\partial}{\partial q} \otimes q = \mathbf{1}_{2R} + 2q \otimes q\frac{\partial}{\partial q^2}, \ldots$$

11.6 Normalization of Electroweak Spacetime

The representations of time $\mathbf{D}(1)$ and position $\mathcal{Y}^{2R-1} \supseteq \mathbf{SO}_0(1,1)$, both with real rank 1, come as Lorentz-compatible products in representations of real rank-2 spacetime $\mathcal{D}^{2R} = \mathbf{D}(1) \times \mathcal{Y}^{2R-1}$. The *representation normalization* of spacetime coefficients determines the ratio of the continuous invariants (masses) for time and position. The spacetime normalization is in analogy to the mass consistency ("gap") equation in the chiral model, i.e., to the representation normalization of the dipole-regularized propagator of the massive Dirac field. It extends the position normalization in the nonrelativistic hydrogen atom.

11.6.1 Central Correlation of Hypercharge and Isospin

An example for a relation of two discrete invariants, characterizing the representations of a rank-2 compact group product, is given by hyperisospin. The winding numbers of the $\mathbf{U}(2)$-embedded representations of hypercharge $\mathbf{U}(1)$ and isospin $\mathbf{SU}(2)$ are centrally correlated (see Chapters 6 and 7). The nontrivial intersection $\mathbb{I}(2) \cong \mathbf{U}(\mathbf{1}_2) \cap \mathbf{SU}(2)$ is the center of $\mathbf{SU}(2)$:

$$e^{i\alpha_0 \mathbf{1}_2} \circ e^{i\vec{\alpha}\vec{\tau}} \in \mathbf{U}(2) = \mathbf{U}(\mathbf{1}_2) \circ \mathbf{SU}(2) \cong \frac{\mathbf{U}(1)\times\mathbf{SU}(2)}{\mathbb{I}(2)}.$$

This can be seen at the two parameter pairs $(\alpha_0, \alpha_3) = (\pi, 0)$ and $(\alpha_0, \alpha_3) = (0, \pi)$, which yield the same group element $-\mathbf{1}_2$ in the two-ality group $\mathbb{I}(2) \cong \{\pm\mathbf{1}_2\}$. Any $\mathbf{U}(2)$-representation, denoted by hypercharge and isospin invariants $[y|T]$, has to be compatible with this central correlation. A direct product Cartan subgroup of $\mathbf{U}(2)$ is not given by $\mathbf{U}(\mathbf{1}_2) \circ \mathbf{U}(1)_3 \ni e^{i\alpha_0 \mathbf{1}_2 + i\alpha_3\tau_3}$, but by a product of projector subgroups $\mathbf{U}(1)_+ \times \mathbf{U}(1)_- \ni e^{i\alpha_+ \frac{\mathbf{1}_2+\tau_3}{2} + i\alpha_- \frac{\mathbf{1}_2-\tau_3}{2}}$.

If the defining representation and its dual $e^{-i\alpha_0\mathbf{1}_2+i\vec{\alpha}\vec{\tau}} \in \mathbf{U}(2)$ are denoted by $[\pm\frac{1}{2}|\frac{1}{2}]$, all irreducible $\mathbf{U}(2)$-representations display a correlation of hypercharge and isospin invariant as given by

$$\mathbf{irrep}\,\mathbf{U}(2) = \{[y|T] = [\pm n + T|T] \bigm| n, 2T = 0, 1, 2, \dots\}.$$

The representation-characterizing invariants y for hypercharge and T for isospin are either both integer or both half-integer; i.e., $y + T \in \mathbb{Z}$.

As a consequence in the electroweak standard model, the representations of the electromagnetic subgroup $\mathbf{U}(1)_+$ are characterized by integer winding (charge) numbers $y + T_3 = \pm n + T + T_3 \in \mathbb{Z}$.

All groups $\mathbf{U}(n) = \mathbf{U}(\mathbf{1}_n)\circ\mathbf{SU}(n)$ have a central correlation with a cyclotomic group $\mathbf{U}(\mathbf{1}_n)\cap\mathbf{SU}(n) \cong \mathbb{I}(n)$ (n-ality group), with similar consequences for the $\mathbf{U}(n)$-representations.

11.6.2 The Mass Ratio of Spacetime

The embedded representations of time $\mathbf{D}(1)$ come with Lorentz $\mathbf{SO}_0(1, 2R-1)$-properties, the fundamental one given by a vector:

$$\mathcal{D}^{2R} = \mathbf{D}(1) \times \mathcal{Y}^{2R-1} \ni \vartheta(x)x \longmapsto \int \frac{d^{2R}q}{i\pi|\Omega^{2R-1}|} \frac{2q}{[-(q-io)^2+M^2]^R[(q-io)^2-m^2]} e^{iqx}.$$

The doubled spherical normalization $\frac{2}{|\Omega^1||\Omega^{2R-1}|}$ contains the residual $\mathcal{Y}^{2R-1}$-normalization by the corresponding sphere (see Chapter 9).

The *mass ratio* $\mu^2 = \frac{M^2}{m^2}$ characterizes the representation of even-dimensional causal spacetime $\mathcal{D}^{2R}$ with real rank 2. It relates the two invariants for the Lorentz-compatibly embedded representations of the causal group (eigentime) $\mathbf{D}(1)$ with $q_0^2 = m^2$ (causal dilation-invariant) and of hyperbolic position $\mathcal{Y}^{2R-1}$ with $\vec{q}^{\,2} = -M^2$ (Lorentz dilation- or curvature-invariant).

The dual product with the vector kernel $\frac{1}{q}$ is given by the Schur product, i.e., by the convolution at trivial energy-momenta (see Chapter 9) with Feynman integration,

$$\frac{2q}{(-q^2+M^2)^R(q^2-m^2)} \overset{2R}{\circledast} \frac{1}{q}\Big|_{q=0} \text{ with } \overset{2R}{*} = \left(\frac{2}{i|\Omega^{2R-1}|}, q^2 + io\right).$$

It involves a tensor product $\overset{2R}{\circledast} = \otimes \overset{2R}{*}$ of spaces with a Lorentz group representation:

$$\begin{aligned}
\frac{2q}{(-q^2+\mu^2)^R(q^2-1)} \overset{2R}{\circledast} \frac{1}{q} &= \int_{\mu^2}^1 d^R\kappa^2 \, \frac{(-qR)}{(-q^2+\kappa^2)^{R+1}} \overset{2R}{\circledast} \frac{1}{q} \\
&= -\frac{\partial}{\partial q} \int_{\mu^2}^1 d^R\kappa^2 \, \frac{1}{(-q^2+\kappa^2)^R} \overset{2R}{\circledast} \frac{1}{q} \\
&= \frac{\partial}{\partial q} \otimes q \int_{\mu^2}^1 d^R\kappa^2 \int_0^1 d\xi \frac{(1-\xi)^{R-1}}{-\xi q^2+\kappa^2}.
\end{aligned}$$

The representation normalization condition comes with the vector unit $\frac{\partial}{\partial q} \otimes q = \mathbf{1}_{2R} \cong \delta_c^a$,

$$\mathbf{1}_{2R} = \frac{2q}{(-q^2+\mu^2)^R(q^2-1)} \overset{2R}{\circledast} \frac{1}{q}\Big|_{q=0} = -\frac{1}{R} \log_R \frac{M^2}{m^2} \, \mathbf{1}_{2R},$$

and the *"R-tail" of the logarithm*, which integrates the $\mathbf{SO}_0(1,2R-1)$-invariant inverse mass square $\frac{1}{\kappa^2}$ over the line $[M^2, m^2] \sim [\mu^2, 1]$:

$$\begin{aligned}
\log_R \mu^2 &= -\int \frac{2d^{2R}q}{i|\Omega^{2R-1}|(-q^2-io+\mu^2)^R(q^2+io-1)}\\
&= -\int_{\mu^2}^1 \frac{d^R\kappa^2}{\kappa^2} = -\frac{1}{1-\mu^2}\int_{\mu^2}^1 \frac{d\kappa^2}{\kappa^2}\left(\frac{1-\kappa^2}{1-\mu^2}\right)^{R-1} = -\int_0^1 \frac{d\zeta\, \zeta^{R-1}}{\zeta\mu^2+1-\zeta}\\
&= \frac{\log \mu^2}{(1-\mu^2)^R} + \sum_{k=1}^{R-1} \frac{(1-\mu^2)^{k-R}}{k} = -\sum_{k=R}^{\infty} \frac{(1-\mu^2)^{k-R}}{k},\\
\log_1 \mu^2 &= \frac{\log \mu^2}{1-\mu^2}, \quad \frac{1}{1-\mu^2}\log_{R+1}\mu^2 = \log_R \mu^2 + \frac{1}{R},\\
\log_R 1 &= -\frac{1}{R}, \quad \log_R \mu^2 = \log \mu^2 + \sum_{k=1}^{R-1}\frac{1}{k} + \ldots \quad \text{for } \mu^2 \ll 1.
\end{aligned}$$

For the nonabelian case, the residue $\int_{\mu^2}^1 \frac{d\kappa^2}{\kappa^2} = \int_{\log \mu^2}^0 d\log\kappa^2$ comes with the characteristic factor $(1-\kappa^2)^{R-1}$ from a Cartan torus $\mathbf{SO}(2)^{R-1} \subset \mathbf{SO}(2R-1)$.

"Tail" functions for exponents and logarithms are typical for representations of real rank-1 hyperboloids. For example, the Plancherel measure $d\kappa\, \Pi^{2R-1}(\kappa^2)$ of the irreducible $\mathbf{SO}_0(1,2R-1)$-representations, characterized by a continuous positive invariant κ^2, for the harmonic analysis [55] of functions $L^2(\mathcal{Y}^{2R-1})$ on nonabelian odd-dimensional position hyperboloids contains the "R-tail" of the hyperbolic function $\frac{\sinh \pi\kappa}{\pi\kappa}$ (see Chapter 8):

$$\mathcal{Y}^{2R-1}: \quad \left\{\begin{array}{ccccccc}
\Pi^{2R-1}(\kappa^2) &=& \left|\frac{\Gamma(R-1+i\kappa)}{\Gamma(i\kappa)}\right|^2 &=& \Gamma(R-1)^2\,\kappa^2 \prod_{k=1}^{R-2}\left(1+\frac{\kappa^2}{k^2}\right), && \\
\frac{1}{\Gamma(1-i\kappa)\Gamma(1+i\kappa)} &=& \frac{1}{\kappa^2|\Gamma(i\kappa)|^2} &=& \frac{\sinh \pi\kappa}{\pi\kappa} &=& \prod_{k=1}^{\infty}\left(1+\frac{\kappa^2}{k^2}\right).
\end{array}\right.$$

Also, the normalization equation in the chiral model of Nambu and Jona-Lasinio, where the representation of four-dimensional spacetime by the propagator of the interacting Dirac field is used with the dipole regularization (see Chapter 10), contains the Lorentz invariant factors $\frac{1}{(q^2-M^2)^2}\cdot\frac{m^2}{q^2-m^2}$, multiplied by a Fourier-transformed derivative $\gamma q + m$ and its inverse $\gamma_5 \frac{1}{\gamma q - m}\gamma_5 = \frac{1}{\gamma q + m}$. Both normalizations involve the 2-tail of the logarithm,

	Spacetime representation	Kernel	Representation normalization
Chiral $\mathbf{U}(1)$	$\int \frac{d^4q}{i\pi^2}\, d_{m^2}^{M^2}(q)(\gamma q + m)e^{iqx}$	$\frac{1}{\gamma q - m}$	$\int \frac{d^4q}{i\pi^2}\, d_{m^2}^{M^2}(q)\left[(\gamma q+m)\gamma_5\frac{1}{\gamma q - m}\gamma_5\right]$ $= \mathbf{1}_4 \int \frac{d^4q}{i\pi^2}\, d_{m^2}^{M^2}(q)$ $- \frac{\rho(m^2)(m^2-M^2)^2}{8\pi^2 m^2}\log_2\frac{M^2}{m^2} = 1$
$\mathbf{D}(1)\times\mathcal{Y}^3$	$\int \frac{d^4q}{i\pi^2}\, d_{m^2}^{M^2}(q)\; q\; e^{iqx}$	$\frac{1}{q}$	$\int \frac{d^4q}{i\pi^2}\, d_{m^2}^{M^2}(q)\;\left[q \otimes \frac{1}{q}\right]$ $= \frac{\mathbf{1}_4}{4}\int \frac{d^4q}{i\pi^2}\, d_{m^2}^{M^2}(q)$ $- \frac{1}{2}\log_2\frac{M^2}{m^2} = 1$

$$d_{m^2}^{M^2}(q) = \frac{1}{(q^2-M^2)^2}\cdot\frac{m^2}{q^2-m^2}, \qquad \int \frac{d^4q}{i\pi^2 m^4}\, d_{m^2}^{M^2}(q) = -\log_2\frac{M^2}{m^2}.$$

Usually, the invariant mass M^2 in the chiral model is considered to be a cutoff mass $\frac{M^2}{m^2} \to \infty$ for the regularization of the interacting Dirac field. It is related, by the gap equation, to the translation invariant m^2 (particle mass) and the normalization factor $\rho(m^2)$ of the basic representation of the spacetime translations $\mathbb{R}^4$. The mass M^2 in the electroweak spacetime model is not introduced as an ad hoc regularization mass — it is used as curvature for the representation of nonflat hyperbolic position, i.e., as invariant for the Lorentz dilations $\mathbf{SO}_0(1,1) \cong \mathcal{Y}^1 \subset \mathcal{Y}^3$.

In the normalization condition, the logarithm of the mass ratio goes with the rank, $\log \mu^2 \sim -R - \sum_{k=1}^{R-1} \frac{1}{k}$,

$$-R = \log_R \mu^2 = \begin{cases} \frac{\log \mu^2}{1-\mu^2} & \Rightarrow \mu^2 = 1, \quad R = 1, \\ \frac{\log \mu^2 + 1 - \mu^2}{(1-\mu^2)^2} & \Rightarrow \mu^2 \sim e^{-3}, \quad R = 2, \ldots. \end{cases}$$

11.6.3 Internal Multiplicities

Spacetime $\mathcal{D}^4 \cong \mathbf{GL}(\mathbb{C}^2)/\mathbf{U}(2)$ parametrizes hyperisospin classes. A spacetime coefficient for an induced $\mathbf{GL}(\mathbb{C}^2)$-representation can represent nontrivial internal $\mathbf{U}(2)$-properties; i.e., the matrix elements can come with representations (matrices) $\mathbf{I} \cong \mathbf{I}_A^B$ for spaces with a nontrivial action of the internal group,

$$\mathcal{D}^{2R} \ni \vartheta(x)x \longmapsto \int \frac{d^{2R}q}{i\pi|\Omega^{2R-1}|} \frac{2q \otimes \mathbf{I}}{[-(q-io)^2+\mu^2]^R[(q-io)^2-1]} e^{iqx}.$$

$q \otimes \mathbf{I}$ is a matrix for an $\mathbf{SL}(\mathbb{C}^2) \times \mathbf{SU}(2)$-representation. For $\mathcal{D}^4$ and a four-dimensional vector representation $[\frac{1}{2}|\frac{1}{2}]$ of the Lorentz group $\mathbf{SL}(\mathbb{C}^2)$, $q \cong q^a(\sigma_a)_A^{\dot{B}}$, the hyperisospin representation has to be isomorphic to a subrepresentation; i.e., it has to involve a singlet or triplet $\mathbf{SU}(2)$-representation $[\frac{1}{2}] \otimes [\frac{1}{2}] = [0] \oplus [1]$.

The corresponding kernel has to come with the dual representation $\check{\mathbf{I}}$, e.g., $\frac{1}{\mathbf{I} \otimes q} = \frac{q}{q^2} \otimes \check{\mathbf{I}}$. A convolution for a dual product with

$$\frac{1}{q \otimes \mathbf{I}} \circledast q \otimes \mathbf{I}$$

rearranges the energy-momenta of the factors to the sum (center-of-mass) energy-momenta $q_1 + q_2$ by integrating over the relative ones $q_1 - q_2$. Correspondingly, the tensor product of the internal factors (matrices) also has to be rearranged by Fierz recoupling F into the trace part and the traceless part, e.g., for isospin scalar couplings with $\mathbf{SU}(2)$-Pauli matrices $\vec{\tau}_A{}^B$, $A, B = 1, 2$,

$$\begin{array}{rll} \check{\mathbf{I}} \otimes \mathbf{I} & \stackrel{F}{=} \Gamma_j \otimes \Gamma_j, & \check{\mathbf{I}}_A^B \mathbf{I}_C^D = (\Gamma_j)_A^D (\Gamma_j)_C^B, \\ \text{e.g., } \mathbf{1}_2 \otimes \mathbf{1}_2 & \stackrel{F}{=} \frac{1}{2}(\mathbf{1}_2 \otimes \mathbf{1}_2 + \vec{\tau} \otimes \vec{\tau}), & \delta_A^B \delta_C^D = \frac{1}{2}(\delta_A^D \delta_C^B + \vec{\tau}_A{}^D \vec{\tau}_C{}^B). \end{array}$$

In general, with identity $\mathbf{1}_d \cong \delta_A^B$, $A, B = 1, \ldots, d$, and normalized traceless generalized Pauli matrices, there arises the internal multiplicity factor $\frac{1}{d}$ in the Fierz rearrangement:

$$\left.\begin{array}{rl} (\tau^a)_{a=1}^{d^2-1}, & \\ \text{with } \operatorname{tr}\tau^a & = 0 \\ \text{and } \operatorname{tr}\tau^a \circ \tau^b & = 2\delta^{ab} \end{array}\right\} \Rightarrow \mathbf{1}_d \otimes \mathbf{1}_d \stackrel{F}{=} \tfrac{1}{d}\mathbf{1}_d \otimes \mathbf{1}_d + \tfrac{1}{2}\tau^a \otimes \tau^a.$$

The internal multiplicity factors contribute to the equations for the mass ratios in the spacetime representation normalizations. For the Schur product, only the contribution with the unit matrix in the Fierz recoupling has to be taken; here $\frac{1}{d}\mathbf{1}_d \otimes \mathbf{1}_d = \frac{1}{d}\mathbf{1}_{d^2}$, etc. One obtains with internal identities $\mathbf{1}_d$ the normalization conditions for the vectoral spacetime representations:

$$\begin{aligned} \frac{2q\otimes\mathbf{1}_d}{(-q^2+\mu^2)^R(q^2-1)} \overset{2R}{\circledast} \frac{q\otimes\mathbf{1}_d}{q^2}\Bigg|_{q=0} &= \mathbf{1}_{2R} \otimes \mathbf{1}_{d^2} \\ \iff -\tfrac{1}{Rd}\log_R \tfrac{M^2}{m^2} &= 1. \end{aligned}$$

The arising of multiplicities in normalizations — here the external and internal dimensions in the product Rd — is familiar, not only from the Plancherel measure of compact groups (see Chapter 8), but also, e.g., from the vacuum polarization $\langle 0|\mathbf{J}^a(x)\mathbf{J}^b(y)|0\rangle$ in perturbative quantum electrodynamics (or more general, in gauge theories, abelian and nonabelian) with the number of the charged fields $\gamma \longrightarrow e^+e^-,\ \mu^+\mu^-, \ldots$ and the electromagnetic current $\mathbf{J} = \sum_{\iota=1}^d \mathbf{J}_\iota$. In the latter case, there arise also renormalization logarithms (see Chapter 5).

Chapter 12

Masses and Coupling Constants

The nonrelativistic hydrogen atom and the atomic spectrum characterize cyclic Hilbert representations for the analysis of hyperbolic position $\mathcal{Y}^3$, i.e., of the homogeneous space $\mathbf{SO}_0(1,3)/\mathbf{SO}(3)$ for rotation relativity. The particle spectrum is proposed to arise in an analysis of electroweak spacetime $\mathcal{D}^4$ as the homogeneous space for unitary relativity $\mathbf{D}(1) \times \mathcal{Y}^3 \cong \mathbf{GL}(\mathbb{C}^2)/\mathbf{U}(2)$, i.e., for position $\mathcal{Y}^3$ with additional causal (dilation) operations $\mathbf{D}(1)$.

The particle spectrum $(m^2, J, z) \in \mathbb{R}_+ \times \frac{\mathbb{N}}{2} \times \mathbb{Z}$ for flat spacetime is characterized by the continuous mass as invariant for translations $\mathbb{R}^4$ in the $\mathbf{D}(1) \times \mathbf{U}(1)$-extended Poincaré group $\mathbf{GL}(\mathbb{C}^2) \vec{\times} \mathbb{R}^4$, by (half-)integer spin or polarization for rotations $\mathbf{SU}(2)$ or $\mathbf{SO}(2)$ as translation fixgroup in the Lorentz cover group $\mathbf{SL}(\mathbb{C}^2)$, and by an integer charge number for electromagnetic windings $\mathbf{U}(1)$. The basic interactions are implemented by massless fields with characteristic coupling constants.

The eigentime $\mathbf{D}(1) \cong \mathbb{R}$ invariants for the representations of electroweak spacetime $\mathcal{D}^4 \cong \mathbf{D}(1) \times \mathcal{Y}^3 \cong \mathbb{R}^4_+$ are proposed to determine the mass of relativistic particles. Since the causal spacetime group $\mathbf{GL}(\mathbb{C}^2)$ has real rank 2, i.e., two characterizing continuous invariants $\{m^2, M^2\}$, the translation invariants are related to both the embedded causal group $\mathbf{D}(1)$ and Lorentz group $\mathbf{SO}_0(1,3)$-representations of 3-position $\mathcal{Y}^3$.

A classically oriented remark of Born (1962) with respect to a connection of curvature and particle masses:

> The masses of elementary particles, nuclei and electrons represent enormous concentrations of energy in very small regions of space. Hence one should presume that they will produce considerable local curvatures of space and corresponding gravitational fields. Can these fields explain the cohesive forces, which keep the particles together ... ?

H. Saller, *Operational Spacetime: Interactions and Particles*,
Fundamental Theories of Physics 163, DOI 10.1007/978-1-4419-0898-8_13,

Invariants for compact groups like electric charges, spin or polarization are taken from a rational spectrum; they are "quantum numbers" in the original sense. The particle masses and their ratios seem to come from a continuous spectrum, not from a rational one, i.e., in a group representation interpretation, as invariants of a noncompact group. If that is true, there has to be a structure that picks discrete values from a continuum, i.e., a structure for the "quantization" of continuous invariants.

Such a "discrete picking" arises for *products of one basic representation*, in the simplest case for the harmonic oscillator where the energy $E_1 \in \mathbb{R}$ of a 1-quantum state vector as continuous invariant for the abelian time translations $\mathbf{D}(1) \cong \mathbb{R} \longrightarrow \mathbf{U}(1)$ comes in integer multiples $E_k = kE_1$, $k \in \mathbb{N}$, for the product representations $|E_k\rangle = \overset{k}{\bigvee}|E_1\rangle$ (see Chapter 4).

A nonabelian example is the nonrelativistic hydrogen atom with its Hilbert representations of hyperbolic position $\mathcal{Y}^3 \cong \mathbf{SO}_0(1,3)/\mathbf{SO}(3)$. Here (Chapters 4 and 8), the bound-states $(\frac{L}{2}, \frac{L}{2})$ (Kepler or harmonic $\mathbf{SO}(4)$-representations) are the "leading" irreducible representations in the totally symmetric products $\overset{L}{\bigvee}(\frac{1}{2}, \frac{1}{2})$ of the fundamental $\mathbf{SO}(4)$-representation, the quartet $(\frac{1}{2}, \frac{1}{2})$ with $\mathbf{SO}(3)$-singlet and -triplet. The continuous negative energies (or imaginary "momenta" $\vec{q}^{\,2} = -Q^2 = 2E < 0$) as time translation invariants are, simultaneously, also invariants for the position group $\mathbf{SO}_0(1,1) \subset \mathbf{SO}_0(1,3)$. They are not equidistanced, but "quantized" in an equipartition ("flux quantization") of a basic energy $E_0 = E_L(1+L)^2$ to the dimensions of $\mathbf{SO}(4)$-product representation spaces.

As we will show for electroweak spacetime $\mathcal{D}^4 = \mathbf{D}(1) \times \mathcal{Y}^3$, the ratios of the particle masses as $\mathbf{D}(1)$-related invariants to one fixed $\mathbf{SO}_0(1,1)$-invariant as position curvature are "quantized" in a corresponding noncompact–compact reciprocity by multiplicities for product representations of external and internal spin-isospin $\mathbf{SU}(2) \times \mathbf{SU}(2)$-operations.

For a cyclic representation of a group G, the coset operations G/H, acting on a cyclic vector with fixgroup H, may be the origin of what, in a more phenomenological language, is called the degeneracy manifold of a ground-state with the Nambu–Goldstone degrees of freedom.

Gravity involves the dynamics of *dilations*, which occur in the Iwasawa decomposition $\mathbf{GL}(\mathbb{R}^4) = \mathbf{SO}(4) \circ \mathbf{D}(1)^4 \circ \exp \mathbb{R}^6$. In the tetrad, the dilations come as Lorentz vectors $(\mathbf{e}^a_j) \in \mathbf{GL}(\mathbb{R}^4)/\mathbf{O}(1,3)$. For example, the tetrad for the metric of Reissner spacetime (see Chapters 1 and 3) is built in the form of an $\mathbf{SO}_0(1,1)$-dilation by the Newton potential for a mass point and an electromagnetic contribution for a point charge:

$$\begin{aligned} \vec{\partial}^2\left(-\tfrac{\ell_m}{r} + \tfrac{\ell_z^2}{2r^2}\right) &= 4\pi\ell_m\delta(\vec{x}\,) + \left(\vec{\partial}\tfrac{\ell_z}{r}\right)^2 \\ e^{2\sigma_3\lambda_3} = \begin{pmatrix} \mathbf{g}^{tt} & 0 \\ 0 & -\mathbf{g}^{rr} \end{pmatrix}&, \quad \text{with } e^{2\lambda_3(r)} = 1 - \tfrac{2\ell_m}{r} + \tfrac{\ell_z^2}{r^2} \\ & \quad \text{for } 1 - \tfrac{2\ell_m}{r} + \tfrac{\ell_z^2}{r^2} > 0. \end{aligned}$$

Masses induce massless interactions: The phenomenon of nontrivial masses itself is interpretable as a ground-state degeneracy under dilations $\mathbf{D}(1)$ ("breakdown" of dilation symmetry). For the rearrangement of the spacetime dilation properties, the corresponding dilation degree of freedom is implemented via multiplets of the external–internal $\mathbf{SO}_0(1,3) \times \mathbf{SO}(3)$ for hyperisospin relativity $\mathbf{GL}(\mathbb{C}^2)/\mathbf{U}(2)$ as massless Nambu–Goldstone modes in, respectively, vector and tensor Lorentz group representations. Similarly to chiral Nambu–Goldstone modes, which are not chiral gauge fields, dilation Nambu–Goldstone modes [39] are not dilation gauge fields. A massless gauge field goes with the "unbroken" symmetries, a massless Nambu–Goldstone field with the "broken" ones. The *dilation rearranging massless scalar fields* are the polarization trivial nonparticle Coulomb and Newton degrees of freedom in the electroweak vector and gravitational tensor interactions. They are supplemented to full-fledged Lorentz group multiplets by gauge and polarized particle degrees of freedom. The representation normalizations determine the residues of the massless interactions, i.e., their coupling constants.

12.1 Harmonic Coefficients of Spacetime

The representation matrix elements of electroweak spacetime are convolution products (group products) of representation coefficients of eigentime and hyperbolic position, given by pointwise products in energy-momentum space. For nontrivial Lorentz group $\mathbf{SO}_0(1, 2R-1)$-properties, they come with harmonic energy-momentum polynomials $(q)^L$ (see Chapter 8):

$$\begin{aligned} \mathbf{D}(1) \times \mathcal{Y}^{2R-1} \ni \vartheta(x)x \longmapsto & |m_L^2, -M_L^2, L\}_{2R}(x) \\ &= \int \frac{2d^{2R}q}{i|\Omega^1||\Omega^{2R-1}|} \frac{2^{L-n(L)}(q)^L}{[-(q-io)^2+M_L^2]^{R+n(L)}[(q-io)^2-m_L^2]} e^{iqx}, \\ & L = 0,1,2,3,\ldots . \end{aligned}$$

The coefficients are from Lorentz compatibly embedded representations of the causal group $\mathbf{D}(1)$ with invariant m_L^2 and of $\mathbf{SO}_0(1, 2R-1)$ with invariant M_L^2 as the position $\mathcal{Y}^{2R-1}$-curvature. For the proper Lorentz group $\mathbf{SO}_0(1,3)$, the harmonic polynomials $(q)^L$ are acted on by the $(1+L)^2$-dimensional representations $[J|J]$, $J = \frac{L}{2} = 0, \frac{1}{2}, 1, \ldots$. The natural number L is a characterizing invariant for the maximal compact rotation group $\mathbf{SO}(2R-1)$, relevant for the nonabelian case $R \geq 2$.

The residual normalization of the harmonic representation coefficients with the powers $n(L)$ has been determined as follows: The additional multipole order $\frac{1}{(-q^2+M_L^2)^{n(L)}}$ for hyperbolic position has to distinguish between even and odd powers $(q)^L$, i.e., between integer and half-integer spin $J = \frac{L}{2}$,

$$n(L) = 0,1,2,\ldots = \left\{ \begin{array}{l} \frac{L}{2}, \\ \frac{L-1}{2}, \end{array} \right. \qquad L - 2n(L) = \left\{ \begin{array}{l} 0, \\ 1; \end{array} \right.$$

therefore, with mass ratios $\mu_L^2 = \frac{M_L^2}{m_L^2}$,

$$\frac{2^{L-n(L)}(q)^L}{(-q^2+\mu_L^2)^{R+n(L)}(q^2-1)} = \begin{cases} \frac{2^{\frac{L}{2}}(q)^L}{(-q^2+\mu_L^2)^{R+\frac{L}{2}}(q^2-1)}, & L=0,2,\dots, \\ \frac{2^{\frac{L+1}{2}}(q)^L}{(-q^2+\mu_L^2)^{R+\frac{L-1}{2}}(q^2-1)}, & L=1,3,\dots. \end{cases}$$

Similar to the nonrelativistic hydrogen atom with the normalized coefficients for the harmonic $\mathbf{SO}(4)$-representations $(\frac{L}{2},\frac{L}{2})$ (see Chapter 10),

$$\mathcal{Y}^{2R-1} \ni \vec{x} \longmapsto |-Q_L^2, L\}_{2R-1}(\vec{x}) = \int \frac{2d^{2R-1}q}{|\Omega^{2R-1}|} \frac{1}{(\vec{q}^2+Q_L^2)^R} \left(\frac{2}{\vec{q}^2+Q_L^2}\right)^L (\vec{q})^L e^{i\vec{q}\vec{x}}, \\ L=0,1,2,\dots,$$

the additional order of the hyperbolic pole and the power of 2 in the numerator coincide, $(\frac{2}{-q^2+\mu_L^2})^{n(L)}$ (self-dual residual normalization). However, for relativistic spacetime $\mathbf{D}(1)\times\mathcal{Y}^{2R-1}$, they come with the additional $\mathbf{D}(1)$-factors $(\frac{1}{q^2-1},\frac{2q}{q^2-1})$ in steps of two: order R-pole for $J=0,\frac{1}{2}$, order $R+1$-pole for $J=1,\frac{3}{2}$, etc.:

$$\frac{2^{\frac{L}{2}}(q)^L}{(-q^2+\mu_L^2)^{R+\frac{L}{2}}(q^2-1)} \sim \frac{1}{(-q^2+\mu_L^2)^R}\left(\frac{2}{-q^2+\mu_L^2}\right)^{\frac{L}{2}} (q)^L \ \frac{1}{q^2-1}, \quad L=0,2,\dots,$$

$$\frac{2^{\frac{L+1}{2}}(q)^L}{(-q^2+\mu_L^2)^{R+\frac{L-1}{2}}(q^2-1)} \sim \frac{1}{(-q^2+\mu_L^2)^R}\left(\frac{2}{-q^2+\mu_L^2}\right)^{\frac{L-1}{2}} (q)^{L-1}\frac{2q}{q^2-1}, \quad L=1,3,\dots.$$

It is useful to bring both poles together: The harmonic polynomials $(q)^L$ arise as leading terms by harmonic derivations $(\frac{\partial}{\partial q})^L$,

$$\begin{aligned} \frac{(q)^L\Gamma(k)}{(-q^2+\mu^2)^k(q^2-1)} &= -\int_{\mu^2}^1 d^k\kappa^2 \ \frac{(q)^L\Gamma(k+1)}{(-q^2+\kappa^2)^{k+1}} \\ &= -\frac{1}{2^L}(\frac{\partial}{\partial q})^L \int_{\mu^2}^1 d^k\kappa^2 \frac{\Gamma(k+1-L)}{(-q^2+\kappa^2)^{k+1-L}} + \dots, \end{aligned}$$

with the line-supported measures for the $\mathbf{SO}_0(1,2R-1)$-invariants (see Chapter 11):

$$d^k\kappa^2 = \frac{d\kappa^2}{1-\mu^2}\left(\frac{1-\kappa^2}{1-\mu^2}\right)^{k-1}, \quad k=1,2,\dots$$

This is used for the harmonic spacetime representation coefficients:

$$\frac{2^{\frac{L}{2}}(q)^L}{(-q^2+\mu_L^2)^{R+\frac{L}{2}}(q^2-1)} = -\frac{\Gamma(R-\frac{L-2}{2})}{2^{\frac{L}{2}}\Gamma(R+\frac{L}{2})}\left(\frac{\partial}{\partial q}\right)^L \int_{\mu_L^2}^1 d^{R+\frac{L}{2}}\kappa^2 \ \frac{1}{(-q^2+\kappa^2)^{R-\frac{L-2}{2}}} + \dots, \\ \text{for } L=0,2,\dots,$$

$$\frac{2^{\frac{L+1}{2}}(q)^L}{(-q^2+\mu_L^2)^{R+\frac{L-1}{2}}(q^2-1)} = -\frac{\Gamma(R-\frac{L-1}{2})}{2^{\frac{L-1}{2}}\Gamma(R+\frac{L-1}{2})}\left(\frac{\partial}{\partial q}\right)^L \int_{\mu_L^2}^1 d^{R+\frac{L-1}{2}}\kappa^2 \frac{1}{(-q^2+\kappa^2)^{R-\frac{L-1}{2}}} + \dots, \\ \text{for } L=1,3,\dots,$$

explicit for the fundamental vector $L = 1$ and for $L = 0, 2, 3$,

$$\begin{array}{lll}
L=0: & \frac{1}{(-q^2+\mu_0^2)^R(q^2-1)} = & -R \int_{\mu_0^2}^1 d^R\kappa^2 \, \frac{1}{(-q^2+\kappa^2)^{R+1}}, \\
L=1: & \frac{2q}{(-q^2+\mu_1^2)^R(q^2-1)} = & -\frac{\partial}{\partial q} \int_{\mu_1^2}^1 d^R\kappa^2 \, \frac{1}{(-q^2+\kappa^2)^R}, \\
L=2: & \frac{2(q)^2}{(-q^2+\mu_2^2)^{R+1}(q^2-1)} = & -\frac{1}{2R}\left(\frac{\partial}{\partial q}\right)^2 \int_{\mu_2^2}^1 d^{R+1}\kappa^2 \, \frac{1}{(-q^2+\kappa^2)^R} + \dots, \\
L=3: & \frac{4(q)^3}{(-q^2+\mu_3^2)^{R+1}(q^2-1)} = & -\frac{1}{2(R-1)R}\left(\frac{\partial}{\partial q}\right)^3 \int_{\mu_3^2}^1 d^{R+1}\kappa^2 \, \frac{1}{(-q^2+\kappa^2)^{R-1}} + \dots.
\end{array}$$

12.2 Translation Invariants as Particle Masses

In contrast to the linear energy spacing for time $\mathbf{D}(1)$ (harmonic oscillator),

$$\begin{array}{l}
\mathbf{D}(1) \cong \mathbb{R} \ni t \longmapsto e^{imt} = \oint \frac{dq}{2i\pi} \frac{1}{q-m} e^{iqt} \longmapsto (e^{imt})^k \in \mathbf{U}(1), \\
\{q = E_k = km \mid k = 0, 1, 2, \dots\},
\end{array}$$

there is no such simple regularity for the masses of spacetime particles as invariants for spacetime $\mathbf{D}(2)$. The equidistant energy pole structure of the product representations for the causal group is a peculiarity of abelian operations, exemplified by the Fock space for free particles.

For the causal group $\mathbf{D}(1)$, the convolution powers of the energy distribution for the defining representation with intrinsic unit m,

$$\tilde{d}(q) = \frac{1}{q-1}, \quad \tilde{d}^k = \underbrace{\tilde{d} \overset{1}{*} \tilde{d} \overset{1}{*} \cdots \overset{1}{*} \tilde{d}}_{k \text{ times}} = \frac{1}{q-k},$$

are acted on by the fundamental kernel:

$$\tilde{\kappa} \overset{1}{*} \tilde{\mathcal{D}}^1 = \log \tilde{\mathcal{D}}^1 : \quad \omega^k(q) = \frac{1}{q}\left(\overset{1}{*} \frac{1}{q-1}\right)^{k-1} = \frac{1}{q} \overset{1}{*} \frac{1}{q-(k-1)} = \frac{1}{q-(k-1)}.$$

The eigenvalue of a $\mathbf{D}(1)$-product representation $\tilde{d}^k(q) = \frac{1}{q-k}$ is given by the singularity $q = k$ with the eigenvalue equation $\frac{1}{\tilde{d}^k(q)} = q - k = 0$. With tangent kernels and $\log \tilde{\mathcal{D}}^1$-functions $\frac{1}{q-(k-1)}$, the eigenvalue is not given by the singularity $q = k - 1$ of ω^k but by the condition that, there, the resolvent $\frac{1}{1-\omega^k(q)}$ is singular (see Chapter 10):

$$\omega^k(q) = 1 \Rightarrow q = k = 1, 2, \dots.$$

In contrast to the product representations of the group with the invariants $\{q \mid \frac{1}{\tilde{d}^k(q)} = 0\}$, the normalized products with the kernels, leading to the invariant resolvent singularities $\{q \mid \omega^k(q) = 1\}$, can be generalized to the nonabelian case.

Analogously to the Schur products for the normalization of the position representations coefficients with the harmonic rotation group $\mathbf{SO}(2R-1)$-polynomials (see Chapter 10),

$$\{L| - Q_L^2, L\}_{2R-1} = (\vec{q})^L \overset{2R-1}{*} \frac{2^L(\vec{q})^L}{(\vec{q}^2+Q_L^2)^{R+L}}\Big|_{\vec{q}=0} = (\mathbf{1}_{2R-1})^L, \;\; L = 0, 1, 2, \dots,$$

the energy-momentum distributions for the spacetime representations with harmonic Lorentz group $\mathbf{SO}_0(1, 2R-1)$-polynomials $(q)^L$ are convoluted with the corresponding kernels, different for even and odd degree L, to yield the $\mathbf{D}(1)$-kernels of the spacetime tangent module:

$$\begin{aligned}
\tilde{\kappa} \overset{2R}{*} \tilde{\mathcal{D}}^{2R} &\subseteq \log \tilde{\mathcal{D}}^{2R} : \\
\omega_{2R}^L(q) &= \frac{(q)^L}{(q^2)^{L-n(L)}} \overset{2R}{*} \frac{2^{L-n(L)}(q)^L}{(-q^2+\mu_L^2)^{R+n(L)}(q^2-1)} \\
&= \begin{cases} \dfrac{(q)^L}{(q^2)^{\frac{L}{2}}} \overset{2R}{*} \dfrac{2^{\frac{L}{2}}(q)^L}{(-q^2+\mu_L^2)^{R+\frac{L}{2}}(q^2-1)}, & L = 0, 2, \dots, \\ \dfrac{(q)^L}{(q^2)^{\frac{L+1}{2}}} \overset{2R}{*} \dfrac{2^{\frac{L+1}{2}}(q)^L}{(-q^2+\mu_L^2)^{R+\frac{L-1}{2}}(q^2-1)}, & L = 1, 3, \dots. \end{cases}
\end{aligned}$$

In all cases, the energy-momentum powers in the convolution products of kernels and representation coefficients are those of the scalar case $L = 0$:

$$\begin{aligned}
\mathcal{Y}^{2R-1} &: & (\vec{q})^L \overset{2R-1}{*} \frac{2^L(\vec{q})^L}{(\vec{q}^2+Q_L^2)^{R+L}} &\sim 1 \overset{2R-1}{*} \frac{1}{(\vec{q}^2+Q_L^2)^R}, \\
\mathcal{D}^{2R} &: & \frac{(q)^L}{(q^2)^{L-n(L)}} \overset{2R}{*} \frac{2^{L-n(L)}(q)^L}{(-q^2+\mu_L^2)^{R+n(L)}(q^2-1)} &\sim 1 \overset{2R}{*} \frac{1}{(-q^2+\mu_L^2)^R(q^2-1)}.
\end{aligned}$$

The fundamental vector case $L = 1$ was considered in the last chapter:

$$\begin{array}{cc}
1 \overset{2R}{*} \dfrac{1}{(-q^2+\mu_0^2)^R(q^2-1)}, & \dfrac{(q)^2}{q^2} \overset{2R}{*} \dfrac{2(q)^2}{(-q^2+\mu_2^2)^{R+1}(q^2-1)}, \\
L = 0 \text{ (scalar)}, & L = 2 \text{ (tensor)}, \\[2ex]
\dfrac{q}{q^2} \overset{2R}{*} \dfrac{2q}{(-q^2+\mu_1^2)^R(q^2-1)}, & \dfrac{(q)^3}{(q^2)^2} \overset{2R}{*} \dfrac{4(q)^3}{(-q^2+\mu_3^2)^{R+1}(q^2-1)}, \\
L = 1 \text{ (vector)}, & L = 3.
\end{array}$$

The representation normalizations give $\mathbf{D}(1)$-eigenvalue equations. The $\mathbf{D}(1)$-invariants μ_ι^2 are given by those energy-momenta where the resolvent is singular,

$$\omega_{2R}^L(q) = \sum_\iota \mathcal{P}_\iota^L(q)\omega_{2R}^\iota(q^2), \;\; \{\mu_\iota^2 \mid \omega_{2R}^\iota(\mu_\iota^2) = 1\},$$

with a decomposition on the right-hand side in irreducible representations ι, which are $\mathbf{SO}_0(1, 2R-1)$-representations $(q)^{2J}$, $J = 0, \dots, L$, for the massless case $\mu_\iota^2 = 0$, and embedded $\mathbf{SO}(2R-1)$-representations for nontrivial masses.

12.3 Correlation of Spin and Masses

The normalization of the harmonic representations of spacetime $\mathcal{D}^{2R}$ relates, in a *noncompact–compact reciprocity*, the characterizing continuous and discrete invariants of the acting group $\mathbf{D}(1) \times \mathbf{SO}_0(1, 2R-1)$ to each other.

The convolution products of the harmonic spacetime representation coefficients with the $\mathbf{D}(1)$-kernels contain as leading terms:

$$\omega_{2R}^L(q, \kappa^2) = \begin{cases} \frac{(q)^L}{(q^2)^{\frac{L}{2}}} \overset{2R}{*} \frac{1}{\Gamma(R+\frac{L}{2})} (\frac{\partial}{\partial q})^L \frac{\Gamma(R-\frac{L-2}{2})}{(-q^2+\kappa^2)^{R-\frac{L-2}{2}}}, & L = 0, 2, \ldots, \\ \frac{(q)^L}{(q^2)^{\frac{L+1}{2}}} \overset{2R}{*} \frac{1}{\Gamma(R+\frac{L-1}{2})} (\frac{\partial}{\partial q})^L \frac{\Gamma(R-\frac{L-1}{2})}{(-q^2+\kappa^2)^{R-\frac{L-1}{2}}}, & L = 1, 3, \ldots, \end{cases}$$

with tensor coefficients and scalar integrals:

$$\begin{aligned} \omega_{2R}^L(q, \kappa^2) &= \frac{(\frac{\partial}{\partial q})^L \otimes (q)^L}{\Gamma(L+1)} \omega_{2R}^L(q^2, \kappa^2) \\ \omega_{2R}^L(q^2, \kappa^2) &= \begin{cases} \frac{\Gamma(L+1)\Gamma(R)}{\Gamma(\frac{L}{2})\Gamma(R+\frac{L}{2})} \int_0^1 d\zeta \, \frac{\zeta^{\frac{L-2}{2}} (1-\zeta)^{R+\frac{L-2}{2}}}{-\zeta q^2+\kappa^2}, & L = 0, 2, \ldots, \\ \frac{\Gamma(L+1)\Gamma(R)}{\Gamma(\frac{L+1}{2})\Gamma(R+\frac{L-1}{2})} \int_0^1 d\zeta \, \frac{\zeta^{\frac{L-1}{2}} (1-\zeta)^{R+\frac{L-3}{2}}}{-\zeta q^2+\kappa^2}, & L = 1, 3, \ldots. \end{cases} \end{aligned}$$

For the Schur products, i.e., at $q = 0$, the harmonic projectors (units) $(\mathbf{1}_{2R})^L$ for the Lorentz group $\mathbf{SO}_0(1, 2R-1)$ survive (see Chapter 8):

$$L = 0, 1, 2, 3, \cdots : \quad \begin{cases} \frac{(\frac{\partial}{\partial q})^L \otimes (q)^L}{\Gamma(L+1)} = (\mathbf{1}_{2R})^L, \\ \omega_{2R}^L(0, \kappa^2) = \frac{\Gamma(L+1)\Gamma(R)}{\Gamma(R+L)} \frac{1}{\kappa^2}. \end{cases}$$

In the normalization conditions of the harmonic spacetime representations,

$$(\mathbf{1}_{2R})^L = \{L|m_L^2, -M_L^2, L\}_{2R} = \begin{cases} \frac{(q)^L}{(q^2)^{\frac{L}{2}}} \overset{2R}{*} \frac{2^{\frac{L}{2}} (q)^L}{(-q^2+\mu_L^2)^{R+\frac{L}{2}} (q^2-1)} \Big|_{q=0}, \\ L = 0, 2, \ldots, \\ \frac{(q)^L}{(q^2)^{\frac{L+1}{2}}} \overset{2R}{*} \frac{2^{\frac{L+1}{2}} (q)^L}{(-q^2+\mu_L^2)^{R+\frac{L-1}{2}} (q^2-1)} \Big|_{q=0}, \\ L = 1, 3, \ldots, \end{cases}$$

the scalar contribution $\omega_{2R}^L(0, \kappa^2)$ with the Lorentz invariant inverse mass square $\frac{1}{\kappa^2}$ is integrated over the line, leading to logarithmic "tails" (see Chapter 11):

$$k = 1, 2, \cdots : \quad -\log_k \mu^2 = \int_{\mu^2}^1 \frac{d^k \kappa^2}{\kappa^2} = \frac{1}{1-\mu^2} \int_{\mu^2}^1 \frac{d\kappa^2}{\kappa^2} \left(\frac{1-\kappa^2}{1-\mu^2}\right)^{k-1}.$$

The resulting conditions for the mass ratios μ_L^2,

$$1 = -\frac{\Gamma(L+1)\Gamma(R)}{\Gamma(R+L)} \frac{1}{2^{\frac{L}{2}}} \int_{\mu_L^2}^1 \frac{d^{R+\frac{L}{2}}\kappa^2}{\kappa^2} = -\frac{\Gamma(L+1)\Gamma(R)}{\Gamma(R+L)} \frac{1}{2^{\frac{L}{2}}} \log_{R+\frac{L}{2}} \mu_L^2, \quad L = 0, 2, \ldots,$$

$$1 = -\frac{\Gamma(L+1)\Gamma(R)}{\Gamma(R+L)} \frac{1}{2^{\frac{L-1}{2}}} \int_{\mu_L^2}^1 \frac{d^{R+\frac{L-1}{2}}\kappa^2}{\kappa^2} = -\frac{\Gamma(L+1)\Gamma(R)}{\Gamma(R+L)} \frac{1}{2^{\frac{L-1}{2}}} \log_{R+\frac{L-1}{2}} \mu_L^2, \quad L = 1, 3, \ldots,$$

contain the multiplicity factor with the dimension of the totally symmetric Lth power of a vector space $\mathbb{K}^R$:

$$\int d^{2R}q\ \mu(q^2) \frac{(q)^{L'} \otimes (q)^L}{(q^2)^L} = \delta^{LL'} \frac{\Gamma(L+1)\Gamma(R)}{\Gamma(R+L)} \frac{1}{2^L} (\mathbf{1}_{2R})^L \int d^{2R}q\ \mu(q^2),$$

$$\dim_{\mathbb{K}} \bigvee^{L} \mathbb{K}^R = \binom{R+L-1}{L} = \frac{\Gamma(R+L)}{\Gamma(L+1)\Gamma(R)}.$$

The complete normalization conditions have to take into account additional internal multiplicity factors $d_{\text{int}}(L)$ (more ahead):

$$-\log_{R+n(L)} \mu_L^2 = 2^{n(L)} \binom{R+L-1}{L} d_{\text{int}}(L),$$

with, respectively, $n(L) = \frac{L}{2},\ \frac{L-1}{2} \in \mathbb{N}$.

12.4 Massless Interactions

Nontrivial masses rearrange ("break") the dilation properties ("invariance") $(x, q) \longmapsto (e^{\psi_0}x, e^{-\psi_0}q)$. The dilation rearrangement ("breakdown") for spacetime $\mathcal{D}^{2R}$ involves both the mass m^2, characterizing the embedded causal $\mathbf{D}(1)$-representations, and the mass M^2, characterizing the embedded Lorentz group $\mathbf{SO}_0(1, 2R-1)$-representations. The representation coefficients come with the energy-momentum measure

$$\left\{ \frac{d^{2R}q}{(-q^2+M^2)^R(q^2-m^2)} \mid m^2, M^2 > 0 \right\}.$$

On the *dilation degeneracy hyperbola* $\mathcal{Y}^1 = \mathbf{SO}_0(1, 1)$ for the two invariant masses with product $e^{2\lambda_0}$, a fixed ratio μ^2 determines one point:

$$\begin{pmatrix} m^2 & 0 \\ 0 & M^2 \end{pmatrix} = e^{\lambda_0} \begin{pmatrix} e^{\lambda} & 0 \\ 0 & e^{-\lambda} \end{pmatrix} \iff \begin{cases} e^{-2\lambda} = \frac{M^2}{m^2} = \mu^2, \\ e^{2\lambda_0} = M^2 m^2. \end{cases}$$

As a consequence of the degeneracy, one has to expect massless scalar modes for the dilation degree of freedom, which, for four-dimensional spacetime, are implemented by nonparticle degrees of freedom (Coulomb and Newton potential) in the vectorial electroweak and tensorial gravitational long-range interactions (see Chapter 5).

This is in analogy to the chiral phase degeneracy circle $\Omega^1 = \mathbf{SO}(2)$ with radius $2m$ and scalar propagator $\{\frac{d^4q}{(-q^2+M^2)^2(q^2-m^2)} \mid m^2, M^2 > 0\}$ where the chiral degree of freedom is realized by the pseudoscalar massless Nambu–Goldstone field (see Chapter 9).

12.4.1 Massless Vector Modes

The simplest nontrivial equation for $\mathbf{D}(1)$-eigenvalues contains the product of the vectorially embedded time representations, the cosines $\frac{2q}{q^2-m^2} \sim \cos mx_0$ (time projection, see Chapter 11) and, for trivial mass, the kernel $\frac{1}{q} \sim 1$, with the embedded hyperboloid representation, an exponential potential $\frac{1}{(-q^2+M^2)^R} \sim e^{-Mr}$ (position projection),

$$\omega^1_{2R}(q) \ = \frac{2q}{(-q^2+\mu_1^2)^R(q^2-1)} \overset{2R}{\circledast} \frac{1}{q} = \frac{\partial}{\partial q} \otimes q \int_{\mu_1^2}^1 d^R\kappa^2 \int_0^1 d\xi \ \frac{(1-\xi)^{R-1}}{-\xi q^2+\kappa^2}.$$

For $q^2 \neq 0$, it is decomposable with two projectors for the "spin" $\mathbf{SO}(2R-1)$-scalar and -vector, respectively (see Chapter 5),

$$\mathbf{1}_{2R} = \mathcal{S} + \mathcal{V}: \quad \begin{cases} \mathcal{V} = \mathbf{1}_{2R} - \frac{q\otimes q}{q^2}, \\ \mathcal{S} = \frac{q\otimes q}{q^2}, \end{cases}$$
$$\frac{\partial}{\partial q} \otimes q = \mathbf{1}_{2R} + 2q \otimes q \frac{\partial}{\partial q^2} = \mathcal{V} + \mathcal{S}\Big(1 + 2q^2 \frac{\partial}{\partial q^2}\Big),$$

for the q^2-dependent kernels,

$$\omega^1_{2R}(q) = \mathcal{V}\omega^{\mathcal{V}}_{2R}(q^2) + \mathcal{S}\omega^{\mathcal{S}}_{2R}(q^2): \quad \begin{cases} \omega^{\mathcal{V}}_{2R}(q^2) = \int_{\mu_1^2}^1 d^R\kappa^2 \int_0^1 d\xi \ \frac{(1-\xi)^{R-1}}{-\xi q^2+\kappa^2}, \\ \omega^{\mathcal{S}}_{2R}(q^2) \ = \Big(1 + 2q^2 \frac{\partial}{\partial q^2}\Big)\omega^{\mathcal{V}}_{2R}(q^2). \end{cases}$$

The resolvent with the two related eigenvalue equations is given as follows:

$$q^2 \neq 0: \quad \frac{\mathbf{1}_{2R}}{\mathbf{1}_{2R}-\omega^1_{2R}(q)} = \frac{\mathcal{S}}{1-\omega^{\mathcal{S}}_{2R}(q^2)} + \frac{\mathcal{V}}{1-\omega^{\mathcal{V}}_{2R}(q^2)}, \quad \begin{cases} \omega^{\mathcal{S}}_{2R}(q^2) = 1, \\ \omega^{\mathcal{V}}_{2R}(q^2) = 1. \end{cases}$$

The condition for the masslessness of the chiral $\mathbf{U}(1)$ Nambu–Goldstone mode coincides with the consistency condition ("gap equation") for the Dirac fermion mass as chiral breakdown parameter, which is a representation normalization for the regularized propagator (see Chapter 10). Similarily, the normalization condition above $\omega^1_{2R}(q) \overset{q=0}{=} \mathbf{1}_{2R}$ for the vector spacetime representation, leading to the determination of the mass ratio μ_1^2, can be read as an eigenvalue equation with a solution at $q^2 = 0$ for a Nambu–Goldstone dilation mode, i.e., for a Poincaré group representation with mass zero,

$$q^2 = 0 \Rightarrow \omega^{\mathcal{S}}_{2R}(0) = \omega^{\mathcal{V}}_{2R}(0) = 1.$$

The massless solution has the resolvent

$$\frac{\mathbf{1}_{2R}}{\mathbf{1}_{2R}-\omega_{2R}^1(q)} = \frac{\rho_{2R}^1(0)}{q^2}\mathbf{1}_{2R} + \dots.$$

The residue is the inverse of the negative derivative at the singularity. It can be simplified with the mass ratio condition $\log_R \mu_1^2 = -R$:

$$\begin{aligned} -\frac{1}{\rho_{2R}^1(0)} = \frac{\partial\omega_{2R}^{\mathcal{V}}}{\partial q^2}(0) &= \int_{\mu_1^2}^1 \frac{d^R\kappa^2}{(\kappa^2)^2}\int_0^1 d\xi\ \xi(1-\xi)^{R-1}\\ &= \frac{1}{R(R+1)}\left[\frac{1}{\mu_1^2} + (R-1)\log_R \mu_1^2\right]\\ &= \frac{1-R(R-1)\mu_1^2}{R(R+1)\mu_1^2}. \end{aligned}$$

Here, $\frac{1}{(\kappa^2)^2}$ has been integrated over the invariant line,

$$k = 1, 2, \dots:\quad \int_{\mu^2}^1 \frac{d^k\kappa^2}{(\kappa^2)^l} = \int_0^1 \frac{d\zeta\ \zeta^{k-1}}{(\zeta\mu^2+1-\zeta)^l} = \begin{cases} \frac{1}{k}, & l=0,\\ -\log_k \mu^2, & l=1,\\ \frac{1}{\mu^2} + (k-1)\log_k \mu^2, & l=2. \end{cases}$$

For small mass ratios, the residue of the massless $\mathbf{D}(1)$-representation is, up to a multiplicity factor, the ratio $\mu_1^2 = \frac{M_1^2}{m_1^2}$ of $\mathbf{SO}_0(1, 2R-1)$ to $\mathbf{D}(1)$-invariant:

$$\text{for } \mu_1^2 \ll 1:\quad -\rho_{2R}^1(0) \sim R(R+1)\mu_1^2.$$

There is an additional internal multiplicity factor $d_{\text{int}}(1)$ (more ahead).

12.4.2 Massless Tensor Modes

Generalizing the structures of the massless vector mode $L = 1$ in the foregoing subsection, the normalizations of the massless tensor modes use the derivatives of the kernels with the leading terms:

$$\frac{\partial\omega_{2R}^L}{\partial q^2}(q^2,\kappa^2) = \begin{cases} \frac{\Gamma(L+1)\Gamma(R)}{\Gamma(\frac{L}{2})\Gamma(R+\frac{L}{2})} \int_0^1 d\zeta\ \frac{\zeta^{\frac{L}{2}}(1-\zeta)^{R+\frac{L-2}{2}}}{(-\zeta q^2+\kappa^2)^2}, & L=2,4,\dots,\\ \frac{\Gamma(L+1)\Gamma(R)}{\Gamma(\frac{L+1}{2})\Gamma(R+\frac{L-1}{2})} \int_0^1 d\zeta\ \frac{\zeta^{\frac{L+1}{2}}(1-\zeta)^{R+\frac{L-3}{2}}}{(-\zeta q^2+\kappa^2)^2}, & L=1,3,\dots. \end{cases}$$

The scalar $L = 0$ convolution has no nontrivial q^2-dependence. For $L \geq 1$, the values at $q^2 = 0$,

$$\frac{\partial\omega_{2R}^L}{\partial q^2}(0,\kappa^2) = \frac{\Gamma(L+1)\Gamma(R)}{\Gamma(R+L+1)}\ \frac{1}{(\kappa^2)^2} \times \begin{cases} \frac{L}{2}, & L=2,4,\dots,\\ \frac{L+1}{2}, & L=1,3,\dots, \end{cases}$$

have to be integrated over the invariant line,

$$\frac{\partial\omega_{2R}^L}{\partial q^2}(0) = \frac{\Gamma(L+1)\Gamma(R)}{\Gamma(R+L+1)} \times \begin{cases} \frac{L}{2^{\frac{L+2}{2}}} \int_{\mu_L^2}^1 \frac{d^{R+\frac{L}{2}}\kappa^2}{(\kappa^2)^2}, & L=2,4,\dots,\\ \frac{L+1}{2^{\frac{L+1}{2}}} \int_{\mu_L^2}^1 \frac{d^{R+\frac{L-1}{2}}\kappa^2}{(\kappa^2)^2}, & L=1,3,\dots. \end{cases}$$

For small mass ratios μ_L^2 and internal multiplicities $d_{\rm int}(L)$, the residues $\rho_{2R}^L(0)$, i.e., the coupling constants of the massless dilation modes, are

$$\text{for } \mu_L^2 \ll 1: \quad -\rho_{2R}^L(0) \sim d_{\rm int}(L)\frac{\Gamma(R+L+1)}{\Gamma(L+1)\Gamma(R)} \times \begin{cases} \frac{2^{\frac{L+2}{2}}}{L}\,\mu_L^2, & L=2,4,\dots, \\ \frac{2^{\frac{L+1}{2}}}{L+1}\,\mu_L^2, & L=1,3,\dots. \end{cases}$$

12.5 Spacetime Masses and Normalizations

For electroweak spacetime $\mathcal{D}^4 = \mathbb{D}(2) \cong \mathbf{GL}(\mathbb{C}^2)/\mathbf{U}(2)$ with real dimension 4, real rank 2, and tangent Minkowski translations $\mathbb{R}^{1+3}$, the normalizations of the harmonic representations relate the ratios $\mu_L^2 = \frac{M_L^2}{m_L^2}$ of the continuous invariants for the Cartan plane $\mathbf{D}(1) \times \mathbf{SO}_0(1,1)$ and the dimensions of the spaces with the product representations of external–internal spin-isospin.

The product representations of position $\mathcal{Y}^3 \cong \mathbf{SO}_0(1,3)/\mathbf{SO}(3)$ with real dimension 3, real rank 1, and tangent translations $\mathbb{R}^3$ show an equipartition $Q_0^2 = Q_L^2(1+L)^2$ of the energies as $\mathbf{SO}_0(1,1) \subset \mathbf{SO}_0(1,3)$-boost invariants to the dimensions of $\mathbf{SO}(4)$-representation spaces (see Chapter 10). The noncompact–compact reciprocity of the nonrelativistic hydrogen atom,

$$\begin{array}{ll} \mathbf{SO}_0(1,3)\text{-invariants from} & \\ \text{position } \mathcal{Y}^3\text{-normalization:} & \frac{Q_0^2}{Q_L^2} = \frac{1}{\kappa_L^2} = (1+L)^2, \quad L=2J=0,1,\dots, \end{array}$$

is extended by integration with the measure of the Lorentz group invariants on a line $\int_{M^2}^{m^2} \sim \int_{\mu^2}^1$ to a "logarithmic" noncompact–compact reciprocity:

$$\begin{array}{ll} & \int_{\mu^2}^1 \frac{d^k\kappa^2}{\kappa^2} = -\log_k \mu^2, \\ \text{ratio of} & \\ \mathbf{D}(1)\times\mathbf{SO}_0(1,3)\text{-invariants} & -\log_{2+n(L)} \frac{M_L^2}{m_L^2} = 2^{n(L)}(1+L)d_{\rm int}(L), \\ \text{from spacetime} & \\ \mathbb{D}(2)\text{-normalization:} & n(L) = \begin{cases} \frac{L}{2}, & L=2J=0,2,\dots, \\ \frac{L-1}{2}, & L=2J=1,3,\dots. \end{cases} \end{array}$$

The Lorentz group properties of the spacetime $\mathbb{D}(2)$-representations $[J|J]$ lead to the external factors $1+2J$ as dimension of the maximal spin representation space. The internal multiplicities $d_{\rm int}(L)$ are determined by bi-regular $\mathbf{SL}(\mathbb{C}^2)\times\mathbf{SU}(2)$-representations: In the case of the Lorentz vector $L=1$ as the product of two $\mathbf{SL}(\mathbb{C}^2)$-Weyl spinors, the internal multiplicity is taken for the product of two $\mathbf{SU}(2)$-isospinors $[\frac{1}{2}]$, with Fierz recoupling $\stackrel{F}{=}$:

$$L=2J=1, \quad [\tfrac{1}{2}|\tfrac{1}{2}], \quad \begin{cases} [\frac{1}{2}]\otimes[\frac{1}{2}] = [0] \oplus [1], \\ \mathbf{1}_2 \otimes \mathbf{1}_2 \stackrel{F}{=} \frac{1}{d_{\rm int}(1)}\mathbf{1}_4 + \dots \\ \Rightarrow d_{\rm int}(1) = 2, \end{cases}$$

and in the case of the Lorentz tensor $L = 2$ as product of four $\mathbf{SL}(\mathbb{C}^2)$-Weyl spinors for the product of four $\mathbf{SU}(2)$-isospinors:

$$\begin{array}{c} L = 2J = 2, \\ [\frac{1}{2}|\frac{1}{2}] \vee [\frac{1}{2}|\frac{1}{2}] \\ \cong [1|1] \oplus [0|0], \end{array} \left\{ \begin{array}{rl} [\frac{1}{2}] \otimes [\frac{1}{2}] \otimes [\frac{1}{2}] \otimes [\frac{1}{2}] & = ([0] \oplus [1]) \otimes ([0] \oplus [1]) \\ & = 2 \times [0] \oplus 3 \times [1] \oplus [2], \\ \mathbf{1}_2 \otimes \mathbf{1}_2 \otimes \mathbf{1}_2 \otimes \mathbf{1}_2 & \stackrel{F}{=} (\frac{1}{2}\mathbf{1}_4 + \dots) \otimes (\frac{1}{2}\mathbf{1}_4 + \dots) \\ & \stackrel{F}{=} \frac{1}{d_{\text{int}}(2)}\mathbf{1}_{16} + \dots \\ \Rightarrow d_{\text{int}}(2) & = 2^4 = 16. \end{array} \right.$$

In general, the Fierz reordering of the product of 2^L units to the unit $\mathbf{1}_d$ leads to the following internal multiplicities:

$$L = 2J = 0, 1, 2, \dots : \underbrace{\mathbf{1}_2 \otimes \dots \otimes \mathbf{1}_2}_{2^L \text{ times}} \stackrel{F}{=} \frac{1}{d_{\text{int}}(L)}\mathbf{1}_d + \dots,$$

$$\text{with} \left\{ \begin{array}{rll} d & = 2^{(2^L)} & = 2, 4, 16, \dots, \\ d_{\text{int}}(L) & = 2^{\frac{L}{2} \cdot 2^L} & = 1, 2, 16, \dots \end{array} \right.$$

It is not obvious how to connect the invariants (m_L^2, M_L^2, L) of the representations of electroweak spacetime $\mathbb{D}(2) = \mathbf{D}(1) \times \mathcal{Y}^3$, on the one hand, with, on the other hand, the experimental masses and rotation invariants (spin, polarization) $J = \frac{L}{2}$ of four-dimensional tangent spacetime $\mathbb{R}^{1+3}$. In the following proposal for an orientation, it will be assumed that causal invariants m_L^2 for embedded $\mathbf{D}(1)$ representations with spin L are considered for one basic hyperbolic mass, i.e., for one fixed position $\mathcal{Y}^3$-curvature $M_L^2 = M^2$. This is in contrast to the position $\mathcal{Y}^3$-representations (Q_L^2, L) for the Kepler potential in the nonrelativistic hydrogen atom, where multipoles $\frac{(2\vec{q})^L}{(\vec{q}^2 + Q_L^2)^{2+L}}$ for L-dependent singularities are used.

The mass M^2 in the higher-order poles (m_L^2, M^2, L) for spacetime $\mathcal{D}^4$ cannot be related to a flat spacetime $\mathbb{R}^4$-particle; it may be connected with the invariant used as quark mass:

$$\mu_L^2 = \frac{M^2}{m_L^2}, \text{ with } M^2 = M_{\text{quark}}^2 \Rightarrow \frac{2^{L-n(L)}(q)^L \otimes \mathbf{1}_{d_{\text{int}}(L)}}{(-q^2 + M^2)^{2+n(L)}(q^2 - m_L^2)}.$$

Quarks and gluons are introduced for the parametrization of the strong interactions. With Wigner's particle definition, confined quarks and gluons are not particles. The mass of confined quarks and gluons is not a translation invariant.

The nonparticle degrees of freedom in the multipoles for electroweak spacetime $\mathbb{D}(2)$ *represent three-dimensional hyperbolic position* $\mathcal{Y}^3$. If, by abuse of language, they are also called "quarks," the quark mass is the position curvature. They have no additional internal $\mathbf{SU}(3)$-color symmetry. The theory could be extended by introducing, ad hoc, additional color degrees of freedom. Possibly, however, there exists an effective flat spacetime linearization of the degrees of freedom in nonparticle higher-order poles by simple poles, e.g., for dipoles with $\frac{q}{(q^2)^2} \sim \frac{1}{q^3}$ by the convolution product with three identical factors,

$$\int \frac{id^4q}{(2\pi)^4} \frac{1}{q^3} e^{iqx} = \frac{1}{\partial^3} = \frac{1}{\partial} * \frac{1}{\partial} * \frac{1}{\partial}.$$

To endow the simple pole parametrization with a property that enforces the poles into the original tripole product, additional internal ("color") degrees of freedom can be introduced in the "quark propagator" $\frac{\mathbf{1}_3}{\partial} = \int \frac{d^4q}{i(2\pi)^4} \frac{\mathbf{1}_3}{q} e^{iqx}$, which are confined for particles in the totally antisymmetric tensor product with convolution $\frac{\mathbf{1}_3}{\partial} \overset{*}{\wedge} \frac{\mathbf{1}_3}{\partial} \overset{*}{\wedge} \frac{\mathbf{1}_3}{\partial} = \frac{1}{\partial^3}$. In this case, confined color $\mathbf{SU}(3) \subset \mathbf{U}(3)$ is motivated by the higher-order poles for nonabelian curved three-dimensional position and stems from the antisymmetric cubic root of $\mathbf{U}(1)$. Color $\mathbf{SU}(3)$ is introduced as a unitary continuous generalization of the cubic roots and the cyclotomic group $\mathbb{I}(3)$.

In general: $\mathbf{U}(s)$ can be considered to be the sth antisymmetric root of $\mathbf{U}(1)$. $\mathbf{U}(1)$ embeds the cyclotomic group $\mathbb{I}(s)$ with the sth roots of 1, i.e., the $\mathbf{U}(s)$-center,

$$\bigwedge^{s} \mathbf{U}(s) \cong \mathbf{U}(1) \supset \mathbb{I}(s) = \{z \in \mathbb{C} \,\big|\, z^s = 1\} \cong \operatorname{centr} \mathbf{U}(s).$$

Position $\mathcal{Y}^s$, $s = 2R - 1 = 3, 5, \ldots$, would come with $\mathbf{SU}(s)$-quark multiplets — triplets, quintets, septets, etc., with $\mathbf{SU}(s)$-singlets in the totally antisymmetric convolution product $\overset{s}{\bigwedge}{}^{*} \frac{\mathbf{1}_s}{\partial} = \frac{1}{\partial^s}$. The connection between position dimension s and unitary group $\mathbf{U}(s)$ reminds us of the "color" degrees of freedom of an isotropic s-dimensional harmonic quantum oscillator (see Chapter 4).

For the vector representation of electroweak spacetime with multiplicity factor $2 \times 2 = 4$, the $\mathbf{D}(1)$ simple pole singularity at m_1^2 may be related to a Poincaré group representation for a lepton mass with spin-isospin $J = T = \frac{1}{2}$, i.e., to a *lepton spinor-isospinor field* $\mathbf{l}$:

$$L = 2J = 1, \quad \frac{2q \otimes \mathbf{1}_2}{(-q^2+M^2)^2(q^2-m_1^2)} : \quad \begin{cases} -\log_2 \mu_1^2 = 2d_{\text{int}}(1) = 4, \\ \mu_1 = \frac{M}{m_1} = \frac{M_{\text{quark}}}{m_{\text{lepton}}} \sim e^{-2.5} \sim \frac{1}{12}. \end{cases}$$

The corresponding nonparticle $\mathcal{Y}^3$ dipole degrees of freedom may be called *spinor-isospinor quark* $\mathbf{q}$.

For the tensor representation, there arises a multiplicity factor $2 \times 3 \times 2^4 = 96$ for the logarithmic tail, which, via the exponent, gives a huge mass ratio. Therefore, the $\mathbf{D}(1)$ simple pole singularity at m_2^2 may be related to Poincaré group representations for the Planck mass with spin-isospin $J = T = \{0, 1\}$ in a *vector-isovector field* $\mathbf{P}$ (for Planck):

$$L = 2J = 2, \quad \frac{2(q \otimes q - \frac{\mathbf{1}_4}{4} q^2) \otimes (\mathbf{1}_2 \otimes \mathbf{1}_2)}{(-q^2+M^2)^3(q^2-m_2^2)} : \begin{cases} -\log_3 \mu_2^2 = 6d_{\text{int}}(2) = 96, \\ \mu_2 = \frac{M}{m_2} = \frac{M_{\text{quark}}}{m_{\text{Planck}}} \sim e^{-49} \sim \frac{1}{1.9 \times 10^{21}}. \end{cases}$$

The corresponding nonparticle $\mathcal{Y}^3$ tripole degrees of freedom may be called *vector-isovector quark* $\mathbf{G}$.

For electroweak spacetime $\mathbb{D}(2)$, the residues (normalizations) of the massless solutions are

$$-\rho_4^L(0) \sim \begin{cases} 2^{\frac{L+2}{2}} \frac{(1+L)(2+L)}{L} d_{\text{int}}(L)\mu_L^2, & L = 2, 4, \ldots, \\ 2^{\frac{L+1}{2}} \; (2+L) \; d_{\text{int}}(L)\mu_L^2, & L = 1, 3, \ldots. \end{cases}$$

The massless vector field with four external Lorentz group degrees of freedom $[\frac{1}{2}|\frac{1}{2}]$ (see Chapter 5) and four internal ones (isospin singlet and triplet),

$$L = 1: \quad \frac{\mathbf{1}_4 \otimes \mathbf{1}_4}{\mathbf{1}_4 \otimes \mathbf{1}_4 - \omega_4^1(q)} = -\rho_4^1(0) \frac{\mathbf{1}_4 \otimes \mathbf{1}_4}{q^2} + \ldots,$$
$$-\rho_4^1(0) \sim 6d_{\text{int}}(1)\mu_1^2 = 12\mu_1^2 \sim \tfrac{1}{12},$$

can be related to the electroweak vector-isovector gauge fields $\mathbf{A}$ with the residue $-\rho_4^1(0)$ as gauge coupling constants. The experimental values in the electroweak $\mathbf{U}(2)$-standard model are (see Chapter 6)

$$(g_1^2, g_2^2 | g^2, \gamma^2) \sim \left(\tfrac{1}{8.4}, \tfrac{1}{2.5} \middle| \tfrac{1}{10.9}, \tfrac{1}{1.9}\right).$$

The electroweak standard model linearizes the $L = 2J = 1$ sector of electroweak spacetime with the basic equipment of flat spacetime fields:

Field	Symbol	$\mathbf{SL}(\mathbb{C}^2)$ $[L\|R]$	$\mathbf{SU}(2)$ $[T]$	Mass m^2
Spinor lepton	$\mathbf{l}$	$[\frac{1}{2}\|0]$	$[\frac{1}{2}]$	m_1^2
Spinor quark	$\mathbf{q}$	$[\frac{1}{2}\|0]$	$[\frac{1}{2}]$	M^2
Vector gauge	$\mathbf{A}$	$[\frac{1}{2}\|\frac{1}{2}]$	$[0] \oplus [1]$	0

$$\frac{M}{m_1} \sim \tfrac{1}{12}, \qquad -\rho_4^1(0) \sim \tfrac{1}{12}.$$

Electroweak linearization for vector $L = 2J = 1$

In a sense, the dichotomy of leptons and quarks reflects the embedded causal group for time and the embedded hyperbolic position. The quark-parametrized strong interactions describe the transition from a nonparticle multipole representation of curved position $\mathcal{Y}^3$ to simple pole particle structures, representing flat position $\mathbb{R}^3$.

Correspondingly, the massless tensor-isotensor field with 10 external degrees of freedom for Lorentz group multiplets $[\frac{1}{2}|\frac{1}{2}] \vee [\frac{1}{2}|\frac{1}{2}] = [0|0] \oplus [1|1]$ and 16 internal ones (two isospin singlets, three triplets, one quintet) may be related to the gravitational interaction $\mathbf{E}$ in the flat spacetime approach (see Chapter 5):

$$L = 2: \quad \frac{\mathbf{1}_{10} \otimes \mathbf{1}_{16}}{\mathbf{1}_{10} \otimes \mathbf{1}_{16} - \omega_4^2(q)} = -\rho_4^2(0) \frac{\mathbf{1}_{10} \otimes \mathbf{1}_{16}}{q^2} + \ldots,$$
$$-\rho_4^2(0) \sim 24d_{\text{int}}(2)\mu_2^2 = 384\mu_2^2 \sim \tfrac{1}{10^{40}}.$$

For the $L = 2J = 2$ sector of electroweak spacetime, a gravitoweak linearization can be constructed, in analogy to the $L = 2J = 1$ sector, with the basic equipment of flat spacetime fields:

Field	Symbol	$\mathbf{SL}(\mathbb{C}^2)$ $[L\|R]$	$\mathbf{SU}(2)$ $[T]$	Mass m^2	
Vector particle	$\mathbf{P}$	$[\frac{1}{2}\|\frac{1}{2}]$	$[0] \oplus [1]$	m_2^2	$\frac{M}{m_2} \sim \frac{1}{1.9\times 10^{21}}$,
Vector quark	$\mathbf{G}$	$[\frac{1}{2}\|\frac{1}{2}]$	$[0] \oplus [1]$	M^2	$-\rho_4^2(0) \sim \frac{1}{10^{40}}$.
Tensor gauge	$\mathbf{E}$	$[0\|0] \oplus [1\|1]$	$2 \times [0] \oplus 3 \times [1] \oplus [2]$	0	

Gravitoweak linearization for tensor $L = 2J = 2$

Analogously to the embedding of the electromagnetic interactions in an electroweak isoquartet with three short-range weak interactions, the gravitational interactions are embedded in a gravitoweak iso-16-plet, which, with Goldstone degeneracy $\mathcal{G}^3 \cong \mathbf{U}(2)/\mathbf{U}(1)_+$ of the ground-state, involves 15 short-range interactions in addition to the familiar massless gravity fields.

Bibliography

[1] F. Bacon, *Novum Organon* (1620), P. Urban and J. Gibson (tr. eds.), Open Court Publishing Company, Peru, IL, 1994.

[2] V. Bargmann, Irreducible representations of the Lorentz group, *Ann. Math.*, 48 (1947), 568–640.

[3] C. Becchi, A. Rouet, R. Stora, *Ann. Phys.* 98 (1976), 287.

[4] H. Behnke, F. Sommer, *Theorie der analytischen Funktionen einer komplexen Veränderlichen* (1962), Springer-Verlag, Berlin.

[5] G. Birkhoff, J.von Neumann, *Ann. Math.* 37 (1936), 823 (*The Logic of Quantum Mechanics*).

[6] J.D. Bjorken, S. Drell, *Relativistische Quantenmechanik* (1966), BI Hochschultaschenbücher, Mannheim.

[7] J.D. Bjorken, S. Drell, *Relativistische Quantenfelder* (1966), BI Hochschultaschenbücher, Mannheim.

[8] S. Bochner, Monotone Funktionen, Stieltjessche Integrale und harmonische Analyse, *Math. Ann.*, 108 (1933), 378–410.

[9] H. Boerner, *Darstellungen von Gruppen* (1955), Springer-Verlag, Berlin.

[10] N.N. Bogliubov, D.V. Shirkov, *Introduction to the Theory of Quantized Fields* (1959), Interscience, New York.

[11] M. Born, *Einstein's Theory of Relativity* (1962), Dover Publications, New York.

[12] N. Bourbaki, *Algebra I*, Chapters 1–3 (1989), Springer-Verlag, Berlin.

[13] N. Bourbaki, *Algebra II*, Chapters 4–7 (1990), Springer-Verlag, Berlin.

[14] N. Bourbaki, *Algèbre*, Chapitre 9 (Formes sesquilineaires et formes quadratiques) (1959), Hermann, Paris .

[15] N. Bourbaki, *Lie Groups and Lie Algebras*, Chapters 1–3 (1989), Springer-Verlag, Berlin.

[16] N. Bourbaki, *Groupes et Algèbres de Lie*, Chapitres 4–8 (1968–1975), Hermann, Paris.

[17] N. Bourbaki, *Intégration*, Chapitres 7–8 (1963), Hermann, Paris.

[18] N. Bourbaki, *Topological Vector Spaces*, Chapter 1–5 (1987), Springer-Verlag, Berlin.

[19] N. Bourbaki, *Théories Spectrales* (1967), Hermann, Paris.

[20] H. van Dam, M. Veltman, Massive and Massless Yang–Mills and Gravitational Fields, *Nuclear Physics* B22 (1970), 397–411.

[21] J. Dixmier, *C^*-Algebras* (1969, Engl. Trans. 1977), North-Holland Publishing Company, Amsterdam.

[22] A. Einstein, *Ideas and Opinions* (1954), Three River Press, New York (1995).

[23] V. Fock, Zur Theorie des Wasserstoffatoms, *Zeits. Phys.* 98 (1935), 145–154.

[24] G.B. Folland, *A Course in Abstract Harmonic Analysis* (1995), CRC Press, Boca Raton, FL.

[25] F.G. Frobenius, *Gesammelte Abhandlungen* (1968), Springer-Verlag, Berlin.

[26] W. Fulton, J. Harris, *Representation Theory* (1991), Springer.

[27] I.M. Gelfand and D.A. Raikov, Irreducible unitary representations of locally bicompact groups, *Mat. Sb.* 13 (55) (1942), 301–316.

[28] I.M. Gel'fand, G.E. Shilov, *Generalized Functions I (Properties and Operations)* (1958, Engl. trans. 1963), Academic Press, New York.

[29] I.M. Gel'fand, G.E. Shilov, *Generalized Functions II (Spaces of Fundamental and Generalized Functions)* (1958, Engl. trans. 1968), Academic Press, New York.

[30] I.M. Gel'fand, M.I. Graev, N.Ya. Vilenkin, *Generalized Functions V (Integral Geometry and Representation Theory)* (1962, Engl. trans. 1966), Academic Press, New York.

[31] I.M. Gelfand, M.A. Neumark, *Unitäre Darstellungen der klassischen Gruppen* (1950, German Trans. 1957), Akademie Verlag, Berlin.

[32] R. Gilmore, *Lie Groups, Lie Algebras and Some of Their Applications* (1974), John Wiley & Sons, New York.

[33] W. Heisenberg, *Einführung in die einheitliche Feldtheorie der Elementarteilchen* (1967), Hirzel, Stuttgart.

[34] W. Heisenberg, *Gesammelte Werke Abteilung C: Allgemeinverständliche Schriften, I–V*, (1984–1989), Piper, München.

[35] S. Helgason, *Differential Geometry, Lie Groups and Symmetric Spaces* (1978), Academic Press, New York.

[36] S. Helgason, *Groups and Geometric Analysis* (1984), Academic Press, New York.

[37] N.J. Hicks, *Notes on Differential Geometry*, (1965) D. van Nostrand Company, Princeton, NJ.

[38] E. Inönü, E. P. Wigner, On the contraction of groups and their representations, *Proc. NAS* 39 (1952), 510–525.

[39] C.J. Isham, A. Salam, J. Strathdee, Nonlinear realizations of space-time symmetries. Scalar and tensor gravity, *Ann. Phys.* 62 (1971), 98–119.

[40] A.A. Kirillov, *Elements of the Theory of Representations* (1976), Springer-Verlag, Berlin.

[41] A. Knapp, *Representation Theory of Semisimple Groups* (1986), Princeton University Press, Princeton, NJ.

[42] T. Kugo, I. Ojima, *Progr. Theor. Phys.* 60 (1978), 1869.

[43] G.W. Mackey, On induced representations of groups (1951), *Am. J. Math.* 73, 576–592.

[44] A. Messiah, *Quantum Mechanics I* (1965), North-Holland, Amsterdam.

[45] Y. Nambu, G. Jona-Lasinio, *Phys. Rev.* 122 (1961), 345–358; 124 (1961), 246–254.

[46] M.A. Neumark, *Lineare Darstellungen der Lorentzgruppe* (1958, German Trans. 1963), VEB Deutscher Verlag der Wissenschaften, Berlin.

[47] F. Peter, H. Weyl, Die Vollständigkeit der primitiven Darstellungen einer geschlossenen kontinuierlichen Gruppe, *Math. Ann.* 97 (1927), 737–755.

[48] H. Reichenbach, *The Philosophy of Space and Time* (1927), Dover, New York.

[49] C.E. Rickart, *General Theory of Banach Algebras* (1960), D. van Nostrand, Princeton, NJ.

[50] H. Saller, *Operational Quantum Theory I — Nonrelativistic Structures* (2006), Springer, New York.

[51] H. Saller, *Operational Quantum Theory II — Relativistic Structures* (2006), Springer, New York.

[52] I. Schur, Neue Begründung der Gruppencharaktere, *Sitzungsber. Preuss. Akad. 1905*, 406.

[53] R.U. Sexl, H.K. Urbantke, *Gravitation und Kosmologie* (1983), 2. Auflage, Bibliographisches Institut, Mannheim.

[54] T.O. Sherman, Fourier analysis on the sphere, *Trans. AMS* 209 (1975), 1–31

[55] R.S. Strichartz, Harmonic analysis on hyperboloids, *J. Funct. Anal.* 12 (1973), 341–383.

[56] F. Treves, *Topological Vector Spaces, Distributions and Kernels* (1967), Academic Press, New York.

[57] N.Ja. Vilenkin, A.U. Klimyk, *Representations of Lie Groups and Special Functions* (1991), Kluwer Academic Publishers, Dordrecht.

[58] S. Weinberg, A model of leptons, *Phys. Rev. Lett.* 18 (1967), 507.

[59] H. Weyl, *Raum, Zeit, Materie* (1923), Wissenschaftliche Buchgesellschaft, Darmstadt.

[60] H. Weyl, *The Theory of Groups and Quantum Mechanics* (1931), Dover, New York.

[61] H. Weyl, *The Classical Groups* (1946), Princeton University Press, Princeton, NJ.

[62] E. P. Wigner, On unitary representations of the inhomogeneous Lorentz group, *Ann. Math.* 40 (1939), 149–204.

[63] C. N. Yang, R. Mills, *Phys. Rev.* 96 (1954), 191.

Index

"Prokrustes transformations", 7

absolute derivative, 43
adjoint representation, 45
angular momentum, 83
annihilation operator, 97, 121

Becchi–Rouet–Stora
  charge, 132
Becchi–Rouet-Stora
  transformations, 130
Bessel functions, 230, 246
  spherical, 227
bi-regular, 206
  group action, 169
Bianchi identity, 42
Born, M., 3, 317
Bose–Fermi twin structure, 132
Breit–Wigner
  functions, 222
  states, 222
bundle
  associated, 148, 149
  connection, 152
  coset, 174
  Lie algebra, 168
  Lie group coset, 167
  nonrelativistic, 150
  principal, 167
  section, 149
  special relativistic, 150
  structural group, 150
  structural Lie algebra, 153
  trivialization, 149
  vector space, 148

canonically complexified, 75
Cartan
  plane, 202
  spacetime, 7
  torus, 201
Casimir operator, 279
causal
  Cartan plane, 299
  ordering, 120
  orthogonal spacetime, 297
  spacetime, 304
  support, 120
  unitary spacetime, 297
center of mass, 84
central correlation, 160
centrifugal potential, 83
chargelike, 1, 13, 147, 201
chiral
  chirality, 257
  degeneracy, 258
  Goldstone modes, 293
  transformations, 257
classical physics, 82
Clifford algebra, 257
color, 97, 328
compact Lie group, 201
compact–noncompact
  duality, 74
  reciprocity, 290, 323, 327
composite gauge fields, 156
confluent hypergeometric equation, 253
conformal vector field, 53
conjugacy classes, 205
connection
  adjoint, 49

affine, 40
internal, 156
Killing, 50
Riemannian, 40, 51
conserved quantities, 55
contraction
Inönü–Wigner, 25
convolution, 255
abelian groups, 261
Euclidean spaces, 263
Feynman measures, 267
Feynman propagators, 272
free particles, 273
hyperboloids, 264
linear invariants, 262
position representations, 263
product, 205, 206
selfdual invariants, 262
spacetimes, 269
spheres, 265
coordinates
Cartesian, 59, 281
Eddington–Finkelstein, 22
Eulerian, 58
geodesic polar, 58, 281
Kruskal, 22
stationary, 54
weak, 176
Copernicus, N., 5
coset space, 8
cosmological constant, 18, 26
Coulomb
degree of freedom, 126
interaction, 11
potential, 84, 128
covariant
derivation, 40
derivative, 114, 153
covariantly constant, 41
creation operator, 97, 121
current, 156
electromagnetic, 114
curvature, 17, 154
constant, 57
covariantly constant, 73
scalar, 54
tensor, 41
cyclic
Hilbert space, 8
vector, 210

de Sitter universes, 28
decomposition
Iwasawa, 202
Langlands, 232
definite metric, 106
degeneracy manifold, 210
degenerate ground-state, 210
Descartes, R., 3
diagonal group, 44
differential operators, 279
dilation
degeneracy, 324
group, 7
in gravity, 19
rearrangement, 319
dipole, 109, 251
Dirac
distribution, 120
equation, 183
field, 121
measure, 207
distinguished frames, 156
distribution
advanced, 244
causal, 244
Dirac, 207
Feynman, 244
hyperbolic, 244
Radon, 207
residual, 246
retarded, 244
spherical, 244
distributive basis, 195, 220
divergence, 53
divergencies, 141, 255
double trace, 204
dual derivation, 283
dual normalization, 288

duality of vector spaces, 209
dyad, 184

eigenvalues, 291
eigenvector notation, 9
Einstein tensor, 18, 34
Einstein, A., 1, 2, 111
electrodynamics, 86, 113
  current, 87
  field strength, 87, 113
  vector potential, 113
electroweak
  and strong gauge interactions, 158
  coupling constants, 163
  mass unit, 165
  spacetime, 170
  triangle, 163, 164
energy-momentum tensor, 88, 116, 117
enveloping algebra, 45
epistemology, 4, 93
equations of motion, 81
equivalence principle, 19
Erlangen program, 1
Euclidean space, 58
external, 1, 13, 147, 201
external derivative, 32

Fadeev–Popov
  fields, 133
  ghosts, 130
Faraday, M., 4
Feynman
  measures, 289
  propagator, 108, 120, 271
fiber
  bundle, 148
  bundle base, 148
  bundle morphism, 149
  bundle trivial, 148
  local, 148
  typical, 148
field, 4
  Hilbert metric, 121
  quantization, 121
  quantum, 111
field strengths, 154
Fierz recoupling, 257
fine structure constant, 6, 84
fixgroup, 38
flat
  position, 9
  spaces, 198
Fock
  space, 106
  state, 97, 107
Fock–Iwanenkov
  coefficients, 52
  connection, 115
Fock–Mehler transformation, 235
Fourier
  components, 217
  inversion, 214
  transformation, 214, 217
free particles, 119
Frobenius' reciprocity, 188, 196
future cone, 297

Galilei group, 9
gauge
  -fixing constant, 135
  coupling constants, 164, 330
  degree of freedom, 126
  field, 153, 159
  field normalization, 158
  fixing constant, 108
  interactions, 157
  transformation, 43, 153
  vertices, 157
Gell–Mann matrices, 98
geodesics, 81, 87
ghost pair, 109, 129
global group, 37
Goldstone manifold, 166, 176
gradient, 52
Grassmann algebra, 31
gravity
  Einstein, 17
  of fields, 115

Green's
  distribution, 259, 283
  kernel, 283
ground state
  degenerate, 161
  electroweak, 161
  manifold, 166
ground-state
  degenerate, 256
  vector, 210
group algebra, 205
group doubling, 169
group dual, 205

Haar measure, 206
Hankel functions, 246
Hardy spaces, 308
harmonic
  analysis, 213
  components, 214
  expansion of a field, 121
  polynomials, 95, 226
  representations, 104
  representations orthogonal groups, 225
  units (projectors), 226
harmonic oscillator, 8, 96
  Bose, 99
  Fermi, 99
Heisenberg
  group, 94
  Lie algebra, 93
Heisenberg, W., 10, 12
Helgason, S., 1
helicity, 9
Higgs field, 159, 162, 177
homogeneous space, 8
horizontal, 148
hydrogen atom, 10, 100, 238
hyperbolic
  hopping, 298
  stretching, 298
hyperboloid, 58, 244
  $\mathcal{Y}$, 7
hypergeometric functions, 252
hyperisospin correlation, 311

Inönü–Wigner contraction, 25
Inönü-Wigner contraction, 253
indefinite
  metric, 109
  unitary, 108
induced representations, 175, 194
information catalogue, 93
inner product
  invariant, 203
interactions, 277, 287
internal, 1, 13, 147, 201
  multiplicities, 314, 327
intertwiner, 194
invariants, 56
involution of functions, 208
isometry group, 37
isotropy group, 38
Iwasawa
  decomposition, 202, 231
  factorization, 297

Kant, I., 10
Kepler
  dynamics position representation, 101
  dynamics quantum algebraic, 102
  potential, 84, 238, 278
  representations, 104
Kepler, J., 5
kernel, 260, 277
  Green's, 283
  group, 282
  Laplace–Beltrami, 285
  linear, 284
  regular, 281
  regularized, 281
  resolvents, 291
  semiregular, 281
  tangent, 287
Killing form, 204, 279
Klein space, 1

Klein, F., 1
Knapp, A.W., 2

Laguerre polynomials, 96
Langlands decomposition, 232
Laplace
–Beltrami operator, 53, 279, 285
-Beltrami operator, 115
kernel, 85
Lebesgue Banach spaces, 207
left–right
multiplication, 169
left-invariant differential operator, 234
left-right
translations, 44
Legendre
functions, 235, 252
polynomials, 228, 252
Lenz-Runge
classes, 13
vector, 85
leptons, 159
Lie algebra
compact, 205
densities, 156
orthogonal symmetric, 73
Lie derivation, 31
Lie–Jacobi isomorphisms, 47
lifetime, 223
light ray bending, 91
linearized
connection coefficients, 118
Einstein equations, 118
gravitative interaction, 119
little group, 112
local group, 38, 114
London, F., 13
Lorentz covariant derivative, 115
Lorentz force, 87
Lorentz group, 6
extended, 296
of a Lie group, 46
representations, 123
Lorentz invariant measure, 121

Macdonald functions, 230, 246
manifold
affine locally symmetric, 73
differential, 29
Einstein, 57
globally symmetric, 76
maximal global symmetry, 57
parallelizable, 31
Riemannian, 33
rotation symmetric, 60
tangent structures, 30
with dimension 1, 62
with dimension 2, 62
with dimension 3, 64
with dimension 4, 67
mass points, 3
mass-hyperboloid, 178
massless
fields, 126
interactions, 324
tensor modes, 326
vector modes, 325
Maurer–Cartan formula, 32
maximally symmetric
position, 24, 86
spacetime, 27
Maxwell equations, 113
Maxwell, J.C., 4, 5
measure
Haar, 206
Plancherel, 213
metric
Kerr–Newman, 23
operational, 55
Reissner, 23
Robertson–Walker, 24, 70
Schwarzschild, 23
signature, 35
tangent space, 33, 55
metric-inducing function, 202, 210
metrical reflection, 34
metrical structures, 6
motion group, 37
multipoles, 238

n-bein, 30, 176
Nambu
–Goldstone field, 261
–Goldstone modes, 319
–Jona Lasinio model, 256, 293
Neumann functions, 246
Newton
constant, 6
potential, 84, 128
Newton, I., 3
noncompact Lie group, 202
nondecomposable representation, 108
nonparticle degrees of freedom, 328
nonrelativistic scattering, 9, 227
normalization
hyperboloids, 290
spacetime, 293
spheres, 290

off-shell, 120
off-shell convolution, 274
on-shell, 120
operation
chargelike, 1, 147
external, 1, 147
internal, 1, 147
spacetimelike, 1, 147
orientation manifold, 177
orthogonal symmetric, 73

parabolic subgroup, 231
Parseval formula, 214
particle, 106
definition, 9
masses, 321
spectrum, 317
particle analysis, 181, 198
Dirac field, 121
scalar field, 121
spin-1 field, 124
spin-2 field, 124
vector field, 134
Pauli exclusion principle, 99
Pauli-Villars regularization, 142
perihelion
invariant, 85
shift, 90
periodic table, 13
perturbative corrections, 141
Peter–Weyl decomposition, 188, 216
photons, 11
Plancherel density, 215, 216
Plancherel distribution, 197, 213
Euclidean groups, 228
Harish–Chandra, 236
Poincaré groups, 231
Planck
area, 18
constant, 94
mass, 6, 84
Planck, M., 6
Plato, 10
Poincaré group, 9, 86
of a Lie group, 46
representations, 136
Wigner classification, 138
pointlike particle, 93, 112
pointwise product, 207
Poisson–Lie-bracket, 83
position
distributions, 277
flat, 9, 24
hyperbolic, 13, 24, 240, 328
interactions, 287
kernels, 288
maximally symmetric, 24, 86
measure, 94
paraboloid, 21
spherical, 24, 26
positive-type functions, 202, 210
for hyperboloids, 221
for spheres, 221
Gaussian, 224
principal value, 120
probability amplitudes, 4
product representations, 255
pure gauge, 155
pure state, 212

quantization
of fields, 121
second, 112
quantum algebra
Bose, 106
Fermi, 106
quantum gauge fields, 130
quantum theory, 93
quark, 11, 14, 159
masses, 11
propagator, 329

Radon distribution, 207
positive-type, 212
relativistic particles, 229
relativity
electromagnetic, 173, 177, 188
electroweak, 173
linear representations, 185
Lorentz group, 191
orthogonal, 198
particle, 173
perpendicular, 173, 178, 187
rotation, 178, 190
special, 173
transitions, 180
unitary, 173, 179, 191
renormalizable, 141
renormalization-regularization, 141
representation
bi-regular, 206
coefficients, 112, 201, 205
cyclic, 210
cyclic translation, 219
discrete series, 137, 233
dual, 204, 212
Euclidean groups, 227
finite-dimensional, 202
Hilbert, 204
induced, 175
left-right regular, 206
metric, 202
nonamenable, 233
normalization, 211, 290
Poincaré groups, 229
principal series, 137
residual, 291
self-dual, 203
spacetime, 112
spherical, 234
square-integrable, 233
supplementary series, 137
translations, 217
residual normalization, 262, 266
resolvents, 291
Ricci tensor, 18, 42, 54

Schrödinger
equation, 95
wave function, 93
Schur
lemma, 203
orthogonality, 197, 216
product, 209, 216, 289
self-dual, 203
singularity
hyperboloids, 263
line, 301
naked, 92
spheres, 263
Sommerfeld, A., 6
spacetime
(anti-)de Sitter, 28
Cartan, 7, 299
Einstein, 26
electroweak, 14, 170, 175, 295
Friedmann, 24, 295
future cone, 179
global group, 295
harmonic analysis, 301
harmonic coefficients, 319
interaction projection, 306
interactions, 310
Kerr–Newman, 20
Kruskal, 20
local group, 295
mass ratio, 312
masses, 327
maximally symmetric, 27
motion group, 82

normalization, 311, 327
particle projection, 306
Reissner, 20
Schwarzschild, 20, 88
spacetime quantization, 3
spacetimelike, 1, 13, 147, 201
special functions, 8, 244
spectral distribution, 215
sphere, 58, 244
Ω, 7
spherical functions, 233
Euclidean groups, 236
spherical harmonics, 95, 188, 226
spin, 9
spin-mass correlation, 323
spinor-isospinor
lepton field, 329
quark field, 329
spontaneous symmetry breakdown, 161
standard model, 158
minimal, 159
stationary coordinates, 54
structural group for frames, 43
symmetric space, 8
symmetry rearrangement, 162, 319

tangent
Euclidean group, 34
groups, 226
kernel, 260
Poincaré group, 34
tensor algebra, 31
tetrad, 19, 184
as Goldstone field, 127
theorem
Bochner, 218
Gel'fand and Raikov, 211
kernel (Schwartz), 281
Lebesgue, 218
Weyl, 205
time reflection, 222
time translations, 81
torsion tensor, 41
translation
group, 7
invariants, 293, 321
transmutator, 176, 180, 195
Higgs, 180, 183, 189
Pauli, 181
rectangular, 186
Weyl, 181, 182
two-particle
convolution, 264
threshold factor, 274

unimodular group, 206

vacuum polarization, 142
vector potential, 87
vector-isovector
quark field, 329
field, 329
vertical, 148
vierbein, 184
virtual particle, 11
virtual particles, 120
volume element, 33

wave functions, 93
weak
boson masses, 165
coordinates, 176
weight diagrams, 103
Weinberg
angle, 163
rotation, 163
Weyl, H., 13
Wigner
axial rotation, 182
element, 168
group element, 180
rotation, 180
spin rotation, 183
Wigner, E., 9
winding number, 201
Witt pair, 109, 129

Yukawa potential, 120, 278

zweibein, 184

MIX
Papier aus verantwortungsvollen Quellen
Paper from responsible sources
FSC® C105338

If you have any concerns about our products,
you can contact us on
ProductSafety@springernature.com

In case Publisher is established outside the EU,
the EU authorized representative is:
Springer Nature Customer Service Center GmbH
Europaplatz 3, 69115 Heidelberg, Germany

Printed by Libri Plureos GmbH
in Hamburg, Germany